# THE THEORY OF UNIFORM ALGEBRAS

# THE THEORY OF UNIFORM ALGEBRAS

**Edgar Lee Stout**
*The University of Washington*

**Bogden & Quigley, Inc.**
**Publishers**

**Tarrytown-on-Hudson, New York / Belmont, California**

**1971**

*Text design by Science Bookcrafters, Inc.*

*Library of Congress Catalog Card No.: 73-164949*
*Standard Book No.: 0-8005-0018-0*

*Printed in the United States of America*

1 2 3 4 5 6 7 8 9 10—75 74 73 72 71

*to Janet*

# PREFACE

In this book I have set forth a portion of the theory of uniform algebras. Uniform algebras are commutative Banach algebras, and as this fact is crucial for much of the theory, I have devoted the first chapter to an exposition of those parts of the theory of commutative Banach algebras which are relevant to my purposes. Although it is true that at various points in the study of uniform algebras, *ad hoc* arguments can be devised which enable one to circumvent the general theory, it seems to me preferable to proceed in a reasonably systematic way, to develop at the beginning the pertinent part of the general theory of Banach algebras.

There is a vague feeling that either a uniform algebra is $\mathscr{C}(X)$ or else it consists of complex analytic functions on a space which admits some sort of complex structure. By now it is known that one cannot insist on the literal truth of this proposition, but nevertheless the point of view embodied in it has led to many of the most interesting developments in the subject. On the one hand, there is the search for analytic structure in spectra (maximal ideal spaces) as represented, for example, by the study of Gleason parts; on the other hand, there is a rather extensive collection of results that provide criteria for specific algebras to be $\mathscr{C}(X)$. These two problems, the problem of recognizing $\mathscr{C}(X)$ and the problem of discovering analytic structure in spectra, permeate most of the theory presented in the second and subsequent chapters.

As the text emphasizes, the interplay between analytic function theory and the theory of uniform algebras is fascinating and leads to deep results and interesting problems in both subjects. Many of the fundamental results in the general theory of commutative Banach algebras, for example the Shilov idempotent theorem, are simple applications of function theory. In the other direction, the study of algebras of functions leads to interesting problems in function theory. This is a well-known facet of the relation between abstract and classical methods: The abstract point of view leads one to ask questions which are posed initially in abstract language but whose answers can be obtained only by solving hard, concrete problems. It is also true, of course, that abstract methods can be brought to bear on classically conceived problems. At its best, this approach yields dazzling results, for example Gelfand's proof of Wiener's theorem on the reciprocal of an absolutely convergent Fourier series.

The prerequisites for reading most of this book are modest and consist of what can be described as the standard beginning curriculum in analysis. Thus, someone who has assimilated the elements of set theoretic topology, measure and integration theory, function theory, and functional analysis will be able to read most of the book. However, at certain points in the development, more specialized results are required. These results are summarized in five appendixes, appendixes that contain virtually no proofs because most of the results are available in standard sources.

This book was started during the academic year 1967–1968 while I held an Office of Naval Research Associateship, and some of the later work was supported by National Science Foundation grants at Yale and the University of Washington.

I am deeply indebted to my colleague, Stuart Jay Sidney, who read the manuscript with the closest attention. His efforts have rescued me from many slips and have led to innumerable improvements. My heartiest thanks go to him.

E.L.S.

*Seattle*
*September 1970*

# CONTENTS

# THE THEORY OF UNIFORM ALGEBRAS

# *chapter one*/GENERAL THEORY OF COMMUTATIVE BANACH ALGEBRAS

In this chapter we present some of the general theory of commutative Banach algebras.

## *I* DEFINITION AND EXAMPLES

**1.1** ***Definition***

*A **complex Banach algebra** is a complex Banach space A in which there is defined a multiplication, to be denoted by juxtaposition, with respect to which A is an associative ring and in which scalar multiplication satisfies* $\alpha(ab) = (\alpha a)b = a(\alpha b)$ *for all* $\alpha \in \mathbb{C}$, $a,b \in A$.[1] *We require, in addition, that for all* $a,b \in A$, $\|ab\| \leq \|a\| \, \|b\|$ *and that if A has a multiplicative identity, this identity be of norm* 1.

Since in this book we shall consider only complex Banach algebras, we shall suppress the adjective *complex* and speak simply of *Banach algebras.*

In the event that a Banach algebra has an identity, we shall denote this identity by 1. Admittedly, this introduces an ambiguity since 1 denotes both the identity of $A$ and a certain complex number, but in practice this ambiguity will cause no difficulty. In the same way, if $\alpha$ is a complex number, we may regard it as an element of $A$, provided, of course, that $A$ has an identity.

Banach algebras abound in analysis. The following are some examples; later we shall introduce many more.

[1] Thus, $A$ is a linear associative algebra over the complex numbers.

**1.2** ***Example***

Let $X$ be a topological space, and let $\mathscr{C}^B(X)$ denote the algebra of all bounded, continuous, complex-valued functions on $X$. If we give $\mathscr{C}^B(X)$ the norm $\|\cdot\|_X$ defined by

$$\|f\|_X = \sup\{|f(x)| : x \in X\},$$

then $\mathscr{C}^B(X)$ is a commutative Banach algebra with identity. Without additional hypotheses on the space $X$, $\mathscr{C}^B(X)$ need not contain any nonconstant functions, but if $X$ is normal, Tietze's extension theorem implies that $\mathscr{C}^B(X)$ is large enough to be interesting. When $X$ is compact,[2] $\mathscr{C}^B(X)$ is the algebra $\mathscr{C}(X)$ of all continuous $\mathbb{C}$-valued functions on $X$.

**1.3** ***Example***

Let $K$ be a compact set in complex $N$-space, $\mathbb{C}^N$, and let $\mathscr{A}(K)$ denote the algebra of all functions continuous on $K$ and holomorphic on the interior of $K$. With respect to the uniform norm $\|\cdot\|_K$ defined as in the preceding example, $\mathscr{A}(K)$ is a Banach algebra, a closed subalgebra of $\mathscr{C}(K)$. We could also consider $\mathscr{P}(K)$, the smallest closed subalgebra of $\mathscr{C}(K)$ containing the polynomials. For certain sets we shall have $\mathscr{P}(K) = \mathscr{A}(K)$, but in general these two algebras will differ.

A very special but important case is that in which $K$ is $\bar{\Delta}$, the closure of the open unit disc $\Delta = \{z \in \mathbb{C} : |z| < 1\}$ in the complex plane. The algebra $\mathscr{A}(\bar{\Delta})$ is known as the *disc algebra.* It has been studied very thoroughly, and it serves as a model for a great deal of the theory which we shall develop. The algebra $\mathscr{A}(\bar{\Delta})$ will be used so frequently in the sequel that it will be useful to reserve a special symbol for it: *Henceforth* $\mathscr{A}$ *will denote the disc algebra.* It is easy to see that $\mathscr{A} = \mathscr{P}(\bar{\Delta})$, for if $f \in \mathscr{A}$, then the function $f_r, r \in (0, 1)$, defined by $f_r(z) = f(rz)$, is in $\mathscr{A}$ and is, in fact, holomorphic on a neighborhood of $\Delta$. As $r \uparrow 1$, $f_r$ converges uniformly on $\bar{\Delta}$ to $f$, and since, for fixed $r$, the power series for $f_r$ converges uniformly on $\bar{\Delta}$ to $f_r$, it follows that $f$ can be approximated uniformly on $\bar{\Delta}$ by polynomials. Thus, $f \in \mathscr{P}(\bar{\Delta})$, and we may conclude that $\mathscr{P}(\bar{\Delta}) = \mathscr{A}$.

**1.4** ***Example***

Let $K$ be a compact set in $\mathbb{C}^N$, and denote by $\mathscr{R}(K)$ the algebra of all those $f \in \mathscr{C}(K)$ which are uniformly approximable on $K$ by functions of the form $p/q$, where $p$ and $q$ are polynomials and $q$ is zero free on $K$. Thus, $\mathscr{A}(K) \supset \mathscr{R}(K) \supset \mathscr{P}(K)$.

[2] We shall use throughout this book the convention that a compact space is Hausdorff by definition.

1.5 *Example*

If $\Omega$ is an open set in $\mathbb{C}^N$, let $H^\infty(\Omega)$ denote the algebra of all functions that are holomorphic and bounded on $\Omega$. This is a closed subalgebra of $\mathscr{C}^B(\Omega)$.

1.6 *Example*

Let $A$ be the algebra of all formal power series

$$\alpha = \sum_{k=0}^{\infty} \alpha_k Z^k, \qquad \alpha_k \in \mathbb{C}$$

which satisfy

$$\|\alpha\| = \sum_{k=0}^{\infty} |\alpha_k| e^{-k^2} < \infty.$$

With this norm and the usual addition and multiplication of power series, $A$ is a commutative Banach algebra with identity. Note that as a Banach space, the algebra $A$ is just the $L^1$ space of the measure on the nonnegative integers which assigns to the set $\{k\}$ the mass $e^{-k^2}$. The fact that $\|\alpha\beta\| \leq \|\alpha\| \; \|\beta\|$ may be seen as follows. If $\alpha = \sum_{k=0}^{\infty} \alpha_k Z^k$, $\beta = \sum_{k=0}^{\infty} \beta_k Z^k$, then

$$\begin{aligned}
\|\alpha\beta\| &= \left\| \sum_{m=0}^{\infty} \left( \sum_{k+l=m} \alpha_k \beta_l \right) Z^m \right\| \\
&\leq \sum_{m=0}^{\infty} e^{-m^2} \left( \sum_{k+l=m} |\alpha_k| \, |\beta_l| \right) \\
&\leq \sum_{m=0}^{\infty} \sum_{k+l=m} |\alpha_k| e^{-k^2} |\beta_l| e^{-l^2} \\
&= \|\alpha\| \; \|\beta\|.
\end{aligned}$$

1.7 *Example*

Let $\mathscr{X}$ be a Banach space, and denote by $\mathscr{L}(\mathscr{X})$ the space of bounded linear operators on $\mathscr{X}$. Normed in the usual way, $\mathscr{L}(\mathscr{X})$ is a noncommutative (if $\dim \mathscr{X} > 1$) Banach algebra. In particular, if $\dim \mathscr{X} = N < \infty$, and if we choose a basis for $\mathscr{X}$, we obtain on the space of $N \times N$ matrices a Banach algebra structure by identifying the element of $\mathscr{L}(\mathscr{X})$ with the matrix which represents it with respect to the chosen basis.

1.8 *Example*

Let $\mathscr{C}^{(m)}(I)$ denote the space of functions on the open interval $I = (0, 1)$ which have $m$ derivatives all of which are uniformly continuous. With pointwise operations and the norm $\|\cdot\|$ given by

$$\|f\| = \sum_{k=0}^{m} \frac{1}{k!}\|f^{(k)}\|_I,$$

$\mathscr{C}^{(m)}(I)$ is a commutative Banach algebra with identity.

There is a canonical way of adjoining an identity to a Banach algebra which is useful in many situations. Suppose that, given a Banach algebra $A$, we denote by $A'$ the direct sum $\mathbb{C} \oplus A = \{(\lambda, a) : \lambda \in \mathbb{C}, a \in A\}$. If we define a norm $\|\cdot\|'$ in $A'$ by $\|(\lambda, a)\|' = |\lambda| + \|a\|$, $\|\cdot\|$ the norm of $A$, and if we define addition and multiplication in $A'$ by

$$(\lambda, a) + (\mu, b) = (\lambda + \mu, a + b)$$

and

$$(\lambda, a)(\mu, b) = (\lambda\mu, \lambda b + \mu a + ab),$$

respectively, then $A'$ becomes a Banach algebra. The algebra $A'$ has an identity, $(1, 0)$, and it contains $A$ as a maximal two-sided ideal. If $A$ has no identity, we have in $A'$ a canonical extension of $A$ to an algebra with identity. We shall refer to $A'$ as *the algebra obtained by adjoining an identity to A.*

In the definition of a Banach algebra, we have required that the norm satisfy $\|ab\| \leq \|a\|\,\|b\|$. This implies that multiplication is jointly continuous, i.e., that if $\lim a_n = a$, $\lim b_n = b$, then $\lim a_n b_n = ab$. In the opposite direction, we have the following result.

**1.9 *Theorem***

*Suppose given a complex Banach space A which is a complex linear associative algebra in which multiplication satisfies the continuity conditions that if* $\lim a_n = a$, *then* $\lim a_n b = ab$ *and* $\lim ba_n = ba$ *for all* $b \in A$. *Then there is a norm* $\|\cdot\|'$ *on A equivalent to the given norm* $\|\cdot\|$ *on A with respect to which A is a Banach algebra.*

PROOF

Consider first the case in which $A$ has an identity. For each $a \in A$, let $M_a \in \mathscr{L}(A)$ be defined by $M_a(b) = ab$. The map $a \mapsto M_a$ is an algebraic isomorphism of $A$ into $\mathscr{L}(A)$. The operators of form $M_a$ are easily characterized: $T \in \mathscr{L}(A)$ is an $M_a$ if and only if $T(bc) = T(b)c$ for all $b, c \in A$. It is clear that all the operators $M_a$ have this property. In the other direction, suppose that $T \in \mathscr{L}(A)$ has the indicated property. If $a = T(1)$, then for all $b \in A$, $T(b) = T(1 \cdot b) = T(1)b = ab$, so $T = M_a$. This characterization of the operators $M_a$ shows that the range $\tilde{A}$ of

$a \mapsto M_a$ is closed, for if $\{M_{a_n}\}$ converges in $\mathscr{L}(A)$ to $L$, then $L(bc) = \lim M_{a_n}(bc) = \lim M_{a_n}(b)c = L(b)c$. Thus, $\tilde{A}$ is not only (norm) closed, but is also strongly closed. We have the estimate

$$\begin{aligned} \|M_a\| &= \sup\{\|ab\| : b \in A, \|b\| \leq 1\} \\ &\geq \|a\|, \end{aligned}$$

so the map $M_a \mapsto a$ is continuous. The open mapping theorem now implies that $a \mapsto M_a$ is a topological isomorphism. Since $\|M_a M_b\| \leq \|M_a\| \, \|M_b\|$ and $\|M_1\| = 1$, it follows that if we define $\|a\|'$ to be $\|M_a\|$, then $A$ is a Banach algebra under the norm $\| \cdot \|'$.

In the case that $A$ has no identity, we can apply the result just obtained to the algebra $A'$ obtained by adjoining 1 to $A$. Then $A$ is a Banach algebra under the restriction to $A$ of the new norm $\| \cdot \|'$ obtained in $A'$.

We are dealing with complex Banach algebras, but at one point in the sequel it will be important for us to realize that if we make the obvious definition of real Banach algebras, then Theorem 1.9 holds for real Banach algebras, too. The argument does not depend on the fact that the scalars in question come from $\mathbb{C}$.

# 2 THE GROUP OF INVERTIBLE ELEMENTS

Let $A$ be a Banach algebra with identity.

2.1 ***Definition***

$\mathscr{G}(A) = \{a \in A : \textit{for some } b \in A,\ ab = ba = 1\}$.

Thus, $\mathscr{G}(A)$ is the group of invertible elements of the algebra $A$.

2.2 ***Theorem***

*The group $\mathscr{G}(A)$ is open in $A$, and the map $a \mapsto a^{-1}$ is a homeomorphism of $\mathscr{G}(A)$ onto itself.*

PROOF

We shall show that $\mathscr{G}(A)$ contains the ball $\{a \in A : \|1 - a\| < 1\}$. From this fact, the openness of $\mathscr{G}(A)$ follows, for suppose given $a \in \mathscr{G}(A)$. Let $b$ satisfy $\|a - b\| < \|a^{-1}\|^{-1}$. Then

$$\begin{aligned} \|1 - a^{-1}b\| &= \|a^{-1}(a - b)\| \\ &\leq \|a^{-1}\| \, \|a - b\| < 1. \end{aligned}$$

Thus, $a^{-1}b$ is invertible: For some $c \in A$, $ca^{-1}b = 1 = a^{-1}bc$. Multiplying the equation $1 = a^{-1}bc$ on the left by $a$ and on the right by $a^{-1}$, we find $1 = bca^{-1}$. Thus, $ca^{-1}$ is the inverse for $b$.

Now suppose that $\|1 - a\| < 1$. The series $\sum_{k=0}^{\infty}(1 - a)^k$ is majorized term by term by the convergent geometric series $\sum_{k=0}^{\infty}\|1 - a\|^k$ and so converges in $A$, say to $b$. We have

$$\begin{aligned} ab &= [1 - (1 - a)] \lim_{N\to\infty} \sum_{k=0}^{N} (1 - a)^k \\ &= \lim_{N\to\infty}\left[\sum_{k=0}^{N}(1 - a)^k - \sum_{k=0}^{N}(1 - a)^{k+1}\right] \\ &= \lim_{N\to\infty}[1 - (1 - a)^{N+1}]. \end{aligned}$$

Since $\|1 - a\| < 1$, $\lim_{N\to\infty}(1 - a)^{N+1} = 0$, so $ab = 1$. Similarly, $ba = 1$, and consequently $a$ is invertible. Thus, $\mathscr{G}(A)$ is open.

To prove that $a \mapsto a^{-1}$ is a homeomorphism of $\mathscr{G}(A)$, it suffices to prove it continuous since this map is its own inverse. We want to show that if $a \in \mathscr{G}(A)$, then as $h \to 0$, $(a + h)^{-1} \to a^{-1}$. By the calculations of the last paragraph, we know that if $\|h\| < \|a^{-1}\|^{-1}$, then $a + h \in \mathscr{G}(A)$. We can write

$$\begin{aligned} (a + h)^{-1} &= [a(1 + a^{-1}h)]^{-1} \\ &= (1 + a^{-1}h)^{-1}a^{-1} \\ &= \left[\sum_{k=0}^{\infty}(-1)^k(a^{-1}h)^k\right]a^{-1} \\ &= a^{-1} + \left[\sum_{k=1}^{\infty}(-1)^k(a^{-1}h)^k\right]a^{-1}. \end{aligned}$$

This infinite series is dominated term by term by the series

$$\sum_{k=1}^{\infty}\|a^{-1}\|^k\|h\|^k,$$

which has $\|a^{-1}\|\,\|h\|(1 - \|a^{-1}\|\,\|h\|)^{-1}$ as its sum. Thus, we may conclude that as $h \to 0$, $(a + h)^{-1} \to a^{-1}$; i.e., the map $a \mapsto a^{-1}$ is continuous.

**2.3** ***Definition***

*$\mathscr{G}_0(A)$ is the component of $\mathscr{G}(A)$ which contains the identity.*

We now consider the exponential function in a Banach algebra. Given an element $a \in A$, the series $\sum_{k=0}^{\infty} a^k/k!$ is majorized term by term by the series $\sum_{k=0}^{\infty} \|a\|^k/k!$ and so converges in $A$. We denote its sum by $e^a$ or $\exp a$. Elementary computations show that $1 = e^a e^{-a}$, and that if $a$ and $b$ commute, then $\exp(a + b) = (\exp a)(\exp b)$. Thus, if $A$ is commutative, the range of the exponential function is a subgroup of $\mathscr{G}(A)$. The mapping $a \mapsto e^a$ is continuous since

$$\begin{aligned}\|\exp(a + h) - \exp a\| &= \left\| \sum_{n=0}^{\infty} \left[\frac{1}{n!}(a + h)^n - \frac{1}{n!}a^n\right]\right\| \\ &\leq \sum_{n=1}^{\infty} \frac{1}{n!} \sum_{j=0}^{n-1} \binom{n}{j} \|a\|^j \|h\|^{n-j} \\ &= e^{\|a\| + \|h\|} - e^{\|a\|} \\ &= e^{\|a\|}(e^{\|h\|} - 1).\end{aligned}$$

As $h \to 0$, this last expression tends to 0.

**2.4** ***Theorem***
*If $A$ is commutative, then $\mathscr{G}_0(A) = \{e^a : a \in A\}$.*

PROOF
For the purposes of this demonstration, set $\mathscr{E}(A) = \{e^a : a \in A\}$. By the continuity of the exponential function, $\mathscr{E}(A)$ is connected. Since $1 = e^0 \in \mathscr{E}(A)$, it follows that $\mathscr{E}(A) \subset \mathscr{G}_0(A)$.

We shall prove that $\mathscr{E}(A)$ is both open and closed in $\mathscr{G}_0(A)$; from this fact we may conclude that $\mathscr{E}(A) = \mathscr{G}_0(A)$. To show that $\mathscr{E}(A)$ is open, it is clearly enough to show that it contains a ball around 1. An elementary though computational way to see this is as follows:

For $k = 1, 2, \ldots$, define $A(k, m)$ by

$$A(k, m) = \sum\left\{\frac{1}{m_1} \cdots \frac{1}{m_k} : 1 \leq m_1, \ldots, m_k \leq m, m_1 + m_2 + \cdots + m_k = m\right\}.$$

Thus, $A(k, m) = 0$ if $m < k$, for in this case the sum in question is empty. Also, define $A(0, 0) = 1$, $A(0, m) = 0$ if $m \geq 1$. For $\zeta \in \mathbb{C}$, $|1 - \zeta| < 1$, we have

$$\log \zeta = -\sum_{m=1}^{\infty} \frac{1}{m}(1 - \zeta)^m,$$

so

$$(\log \zeta)^k = (-1)^k \sum_{m=0}^{\infty} A(k, m)(1 - \zeta)^m.$$

For $\zeta$ with $|1 - \zeta| < e^{-1}$, consider the double series

$$\sum_{k,m=0}^{\infty} \frac{A(k, m)}{k!} |1 - \zeta|^m. \tag{1}$$

The terms of this series are all positive, and we have the obvious bound $A(k, m) < m^k$. For each $m$, $\sum_{k=0}^{\infty} m^k/k! = e^m$, and since $|1 - \zeta| < e^{-1}$, it follows that the repeated series

$$\sum_{m=0}^{\infty} |1 - \zeta|^m \sum_{k=0}^{\infty} \frac{A(k, m)}{k!}$$

is convergent. Thus, the double series (1) is convergent, so the series

$$\sum_{k,m=0}^{\infty} \frac{(-1)^k A(k, m)}{k!} (1 - \zeta)^m$$

is absolutely convergent as long as $|1 - \zeta| < e^{-1}$. Consequently, for all such values of $\zeta$ we may write

$$\begin{aligned}
\sum_{k,m=0}^{\infty} \frac{(-1)^k A(k, m)}{k!} (1 - \zeta)^m &= \sum_{k=0}^{\infty} \frac{(-1)^k}{k!} \sum_{m=0}^{\infty} A(k, m)(1 - \zeta)^m \\
&= \sum_{k=0}^{\infty} \frac{(\log \zeta)^k}{k!} \\
&= e^{\log \zeta} \\
&= \zeta.
\end{aligned}$$

But also,

$$\sum_{k,m=0}^{\infty} \frac{(-1)^k A(k, m)}{k!} (1 - \zeta)^m = \sum_{m=0}^{\infty} (1 - \zeta)^m \sum_{k=0}^{\infty} \frac{(-1)^k A(k, m)}{k!}.$$

Thus, since $\zeta = 1 - (1 - \zeta)$, we find that

$$\sum_{k=0}^{\infty} \frac{(-1)^k A(k, 0)}{k!} = 1, \tag{2}$$

$$\sum_{k=0}^{\infty} \frac{(-1)^k A(k, 1)}{k!} = -1, \tag{3}$$

$$\sum_{k=0}^{\infty} \frac{(-1)^k A(k, m)}{k!} = 0, \qquad m = 2, 3, \ldots. \tag{4}$$

Now let $a \in A$ satisfy $\|1 - a\| < e^{-1}$. The series

$$-\sum_{m=1}^{\infty} \frac{1}{m} (1 - a)^m$$

converges to an element $b$ of $A$, and we have that

$$e^b = \sum_{k=0}^{\infty} \frac{b^k}{k!}$$
$$= \sum_{m=0}^{\infty} (1-a)^m \sum_{k=0}^{\infty} \frac{(-1)^k A(k,m)}{k!}.$$

The identities (2), (3), and (4) show that this last sum is $a$. Thus, $\Sigma(A)$ contains the ball $\{a \in A : \|1-a\| < e^{-1}\}$ and so is open.

It is easy to see that $\mathscr{E}(A)$ is closed in $\mathscr{G}_0(A)$. Let $\{y_n\}$ be a sequence in $\mathscr{E}(A)$ which converges to $y_0 \in \mathscr{G}_0(A)$. We know that $\mathscr{E}(A)$ contains a neighborhood of 1, so for large $n$, $y_0 y_n^{-1} \in \mathscr{E}(A)$, say $y_0 y_n^{-1} = e^{b_n}$. If $y_n = e^{a_n}$, this yields $y_0 = e^{a_n + b_n}$, so $y_0 \in \mathscr{E}(A)$, and the theorem is proved.

There is a less computational though more sophisticated way to see that $\mathscr{E}$(A) contains a neighborhood of 1. It depends on the implicit-function theorem. (See Appendix C.) If we define $F: A \times A \to A$ by $F(x, y) = x - e^y$, then $F(1, 0) = 0$, and $F$ is continuously differentiable. If $D_2$ denotes differentiation with respect to $y$, then the relation

$$\lim_{y \to 0} \frac{\|F(1, y) + y\|}{\|y\|} = 0$$

shows that $-D_2F(1, 0)$ is the identity operator of $A$ onto itself. Thus, the implicit-function theorem provides a neighborhood $V$ of 1 and a continuous map $h: V \to A$ with $h(1) = 0$ and $F(x, h(x)) = 0$; i.e., $x = e^{h(x)}$. Consequently, each point of $V$ is in $\mathscr{E}(A)$, and it follows that $\mathscr{E}(A)$ is open.

2.5 ***Example***

Let $X$ be a compact space and consider $\mathscr{C}(X)$. We have $\mathscr{G}(\mathscr{C}(X)) = \{f : f \text{ has no zero on } X\}$. The group $\mathscr{G}_0(\mathscr{C}(X))$ can be described as the set of all those $f \in \mathscr{G}(\mathscr{C}(X))$ which, as maps into $\mathbb{C} \setminus \{0\}$, are homotopic to constants. That is, $f \in \mathscr{G}(\mathscr{C}(X))$ lies in $\mathscr{G}_0(\mathscr{C}(X))$ if and only if there exists a continuous map $F: X \times [0, 1] \to \mathbb{C} \setminus \{0\}$ such that $F(x, 0)$ is a constant, $c$, independent of $x$, and $F(x, 1) = f(x)$. To establish this fact, suppose first that such a homotopy exists. Then the map $[0, 1] \to \mathscr{G}(\mathscr{C}(X))$ given by $s \mapsto F(\cdot, s)$ is continuous by the continuity of $F$. Its range is connected and contains both $f$ and the constant $c \in \mathscr{C}(X)$. Since $c \in \mathscr{G}_0(\mathscr{C}(X))$, we must have $f \in \mathscr{G}_0(\mathscr{C}(X))$. Conversely, let $f \in \mathscr{G}_0(\mathscr{C}(X))$, say $f = e^g$. Define $F: X \times [0, 1] \to \mathbb{C} \setminus \{0\}$ by $F(x, s) = e^{sg(x)}$. The function $F$ is continuous, $F(x, 0) = 1$, and $F(x, 1) = f(x)$. Thus, $F$ is the desired homotopy.

As a consequence of these observations, the group $\mathcal{G}(\mathscr{C}(X))/\mathcal{G}_0(\mathscr{C}(X))$ can be identified with the group of homotopy classes of maps from $X$ into $\mathbb{C} \setminus \{0\}$. This is a very special case of a general fact about commutative Banach algebras which we shall establish in a later section.

Occasionally it is important to consider the elements of $A$ which lie in the topological boundary of $\mathcal{G}(A)$. They are *topological divisors* of zero in the following sense.

2.6 ***Definition***

*The element $x$ of a Banach algebra $A$ is a* ***left*** *(****right****)* ***topological divisor of zero*** *provided there is a sequence $\{y_n\}$ in $A$ such that $\|y_n\| = 1$ and such that the sequence $\{xy_n\}$ ($\{y_n x\}$) converges to zero.*

In the case of a commutative algebra, the distinction between left and right disappears, and we shall speak simply of a topological divisor of zero.

2.7 ***Theorem***

*If $A$ is a commutative Banach algebra with identity, and if $x \in A$ lies in the topological boundary of $\mathcal{G}(A)$, then $x$ is a topological divisor of zero.*

PROOF

Let $x = \lim x_n$, $x_n \in \mathcal{G}(A)$. If, for each $n$, we set

$$y_n = x_n^{-1}\|x_n^{-1}\|^{-1},$$

then $\|y_n\| = 1$, and

$$\begin{aligned}\|xy_n\| &= \|(x - x_n)y_n + x_n y_n\| \\ &\leq \|x - x_n\|\,\|y_n\| + \|x_n^{-1}\|^{-1}.\end{aligned}$$

The sequence $\{\|x_n^{-1}\|\}$ is unbounded, for otherwise

$$\begin{aligned}\|1 - xx_n^{-1}\| &= \|(x_n - x)x_n^{-1}\| \\ &\leq C\|x_n - x\|\end{aligned}$$

for some constant $C$. We have $\|x_n - x\| \to 0$, so if $n$ is large enough, $xx_n^{-1} \in \mathcal{G}(A)$ by Theorem 2.2. However, $xx_n^{-1} \in \mathcal{G}(A)$ and $x_n \in \mathcal{G}(A)$ imply that $x \in \mathcal{G}(A)$, contrary to hypothesis. Thus, $\|x_n^{-1}\|^{-1} \to 0$. Also $\|x - x_n\|\,\|y_n\| \to 0$, so we conclude that $\|xy_n\| \to 0$, and that $x$ is a zero divisor.

# 3 THE SPECTRUM OF AN ELEMENT

3.1 ***Definition***

(a) *If $A$ is a Banach algebra with identity and $a \in A$, then $\sigma_A(a)$, the* **spectrum of *a* with respect to *A***, *is the set $\{\lambda \in \mathbb{C} : \lambda - a$ is not invertible$\}$.*

(b) *If $A$ has no identity, $\sigma_A(a)$ is defined to be the spectrum of $a$ with respect to the algebra $A'$ obtained by adjoining to $A$ an identity.*

It is an interesting fact that it is possible to define $\sigma_A(a)$ for algebras lacking an identity without recourse to the extended algebra $A'$. For this purpose we make the following definition.

3.2 ***Definition***

(a) *If $A$ is a Banach algebra and $a,b \in A$, then $a \circ b = a + b - ab$.*

(b) *The element $a \in A$ is called* **quasiregular** *if there is $b \in A$ such that $0 = a \circ b = b \circ a$.*

The element $a \circ b$ is called the *circle product* of $a$ and $b$. The importance of the circle product for us is given by the following result.

3.3 ***Lemma***

*If $A$ has no identity, then $\sigma_A(a) = \{0\} \cup \{\lambda \in \mathbb{C} : \lambda^{-1}a$ is not quasiregular$\}$.*

PROOF

Fix $a \in A$, and suppose that $\lambda^{-1}a$ is quasiregular, say $b \circ (\lambda^{-1}a) = (\lambda^{-1}a) \circ b = 0$. Then in $A'$, $1 - b \circ (\lambda^{-1}a) = 1 - (\lambda^{-1}a) \circ b = 1$. But

$$\begin{aligned} 1 - b \circ (\lambda^{-1}a) &= (1 - b)(1 - \lambda^{-1}a) \\ &= \lambda^{-1}(1 - b)(\lambda - a), \end{aligned}$$

and similarly for $1 - (\lambda^{-1}a) \circ b$. Thus, $\lambda - a \in \mathscr{G}(A')$. Conversely, if $\lambda - a \in \mathscr{G}(A')$, write $(\lambda - a)^{-1}$ in the form $\lambda^{-1}(1 - b)$. The element $b$ is then in $A$, and $b \circ (\lambda^{-1}a) = 0 = (\lambda^{-1}a) \circ b$. Finally, observe that $0 \in \sigma_A(a)$ for all $a \in A$ if $A$ has no identity.

For the remainder of this section, let $A$ have an identity.

3.4 ***Theorem***

*The set $\sigma_A(a)$ is a nonvoid, compact subset of $\{\lambda \in \mathbb{C} : |\lambda| \leq \|a\|\}$.*

PROOF

We begin by showing that $\mathbb{C} \setminus \sigma_A(a)$ is open. If $\lambda - a \in \mathcal{G}(A)$, then

$$\|(\lambda - a) - (\mu - a)\| = |\lambda - \mu|,$$

and so the openness of $\mathcal{G}(A)$ implies that as soon as $\mu$ is sufficiently near $\lambda$, $\mu - a$ is also in $\mathcal{G}(A)$. Thus, $\mathbb{C} \setminus \sigma_A(a)$ is open, whence $\sigma_A(a)$ is closed.

Next, $\sigma_A(a) \subset \{\lambda \in \mathbb{C} : |\lambda| \leq \|a\|\}$, for if $|\lambda| > \|a\|$, then $\lambda - a = \lambda(1 - \lambda^{-1}a)$, and since $\|\lambda^{-1}a\| < 1$, this element is invertible.

It remains to show that $\sigma_A(a)$ is not the empty set. Suppose it is so that for all $\lambda$, $(\lambda - a)^{-1}$ exists in $A$. Given a continuous linear functional $\varphi$ on $A$, define a function $F_\varphi$ on $\mathbb{C}$ by

$$F_\varphi(\lambda) = \varphi[(\lambda - a)^{-1}].$$

The function $F_\varphi$ is continuous; we shall show it to be an entire function. For this it suffices to show that $F_\varphi$ is differentiable at every point of $\mathbb{C}$. We have

$$\begin{aligned}\frac{1}{h}[F_\varphi(\lambda + h) - F_\varphi(\lambda)] &= \frac{1}{h}\varphi[-h(\lambda + h - a)^{-1}(\lambda - a)^{-1}] \\ &= -\varphi[(\lambda + h - a)^{-1}(\lambda - a)^{-1}].\end{aligned}$$

The continuity of $\varphi$ and the continuity of the map $b \mapsto b^{-1}$ on $\mathcal{G}(A)$ imply that as $h \to 0$, $\varphi[(\lambda + h - a)^{-1}(\lambda - a)^{-1}] \to \varphi[(\lambda - a)^{-2}]$. Thus, $F_\varphi$ is an entire function. Moreover, as $\lambda \to \infty$, $F_\varphi(\lambda) \to 0$, for

$$\begin{aligned}|F_\varphi(\lambda)| &= \left|\frac{1}{\lambda}\varphi[(1 - \lambda^{-1}a)^{-1}]\right| \\ &\leq \frac{1}{|\lambda|}\|\varphi\| \, \|(1 - \lambda^{-1}a)^{-1}\|,\end{aligned}$$

and the latter expression tends to zero as $\lambda \to \infty$. Thus, $F_\varphi$ is the zero function. This holds for every choice of the continuous linear functional $\varphi$, so by the Hahn–Banach theorem, $a^{-1} = 0$, an obvious impossibility. Consequently, $\sigma_A(a)$ is nonvoid.

As a corollary to the proof just given, we have the following fact.

3·5 ***Corollary***

*Given $a \in A$ and a continuous linear functional $\varphi$ on $A$, the map*

$$\lambda \mapsto \varphi[(\lambda - a)^{-1}]$$

*is holomorphic in $\mathbb{C} \setminus \sigma_A(a)$ and vanishes at infinity.*

Another corollary is the Gelfand–Mazur theorem, which is of fundamental importance in the theory.

3·6 ***Theorem***

*A Banach algebra that is a division ring is isometrically isomorphic to the complex numbers.*

PROOF

(Recall that a division ring is a ring in which each nonzero element is invertible.) Let $A$ be such a Banach algebra. For each $a \in A$, $\sigma_A(a)$ contains exactly one element, because if $\lambda - a$ is not invertible, then $\lambda - a = 0$. Let $L: A \to \mathbb{C}$ be the map which carries each $a \in A$ to the unique element of $\sigma_A(a)$. Then $L$ is an isometry and so one to one. It is additive and multiplicative and so an isomorphism.

3·7 ***Theorem***

*If $a \in A$, then*

$$\sup\{|\mu| : \mu \in \sigma_A(a)\} = \lim_{n \to \infty} \|a^n\|^{1/n}.$$

*In particular, the latter limit exists.*

PROOF

For $|\lambda| > \|a\|$, $\lambda - a \in \mathcal{G}(A)$, and we can write

$$(\lambda - a)^{-1} = \sum_{k=0}^{\infty} \lambda^{-(k+1)} a^k.$$

Given $\varphi$, a continuous linear functional on $A$, we have that $\varphi[(\lambda - a)^{-1}]$ is a holomorphic function of $\lambda$ on the set $|\lambda| > \sup\{|\mu| : \mu \in \sigma_A(a)\}$, so on this set

$$\varphi(\lambda - a)^{-1} = \sum_{k=0}^{\infty} \varphi(a^k)\lambda^{-(k+1)}.$$

Thus, for any particular $\lambda$, we have that the set of numbers $\{\varphi(a^k)\lambda^{-k}\}_{k=0}^{\infty}$ is bounded. This holds for all choices of $\varphi$, so by the Banach–Steinhaus theorem, the set $\{a^k\lambda^{-k}\}_{k=0}^{\infty}$ is norm bounded in $A$, say $\|a^k\lambda^{-k}\| \leq C_\lambda$. Thus, $\|a^k\|^{1/k} \leq C_\lambda^{1/k}|\lambda|$, whence

$$\limsup_{k \to \infty} \|a^k\|^{1/k} \leq |\lambda|.$$

This inequality is valid for all $\lambda$ under consideration, so we get

(1) $$\limsup_{k \to \infty} \|a^k\|^{1/k} \leq \sup\{|\mu| : \mu \in \sigma_A(a)\}.$$

To establish the reversed implication, we start by noting that if $\lambda \in \sigma_A(a)$, then $\lambda^k \in \sigma_A(a^k)$. To see this, suppose that $\lambda^k - a^k \in \mathcal{G}(A)$ so that for some choice of $y \in A$, $y(\lambda^k - a^k) = 1 = (\lambda^k - a^k)y$. Then it follows from

$$\lambda^k - a^k = (\lambda - a)(\lambda^{k-1} + \lambda^{k-2}a + \cdots + a^{k-1})$$

that $(\lambda^{k-1} + \cdots + a^{k-1})y$ is a right inverse for $\lambda - a$. Similarly, $y(\lambda^{k-1} + \cdots + a^{k-1})$ is a left inverse for $\lambda - a$. These two inverses must be the same and so must be an inverse for $\lambda - a$.

We know that for any $b \in A$ and $\mu \in \sigma_A(b)$, the inequality $|\mu| \leq \|b\|$ is valid. Applied to $\lambda^k \in \sigma_A(a^k)$, this yields $|\lambda| \leq \|a^k\|^{1/k}$. Thus, for all $\lambda \in \sigma_A(a)$,

$$|\lambda| \leq \liminf_{k \to \infty} \|a^k\|^{1/k},$$

and so

$$\sup\{|\mu| : \mu \in \sigma_A(a)\} \leq \liminf_{k \to \infty} \|a^k\|^{1/k}.$$

Combined with inequality (1), this gives the theorem.

**3.8 *Definition***

*The number* $\lim_{k \to \infty} \|a^k\|^{1/k}$ *is called the* ***spectral radius*** *of* $a$; *we shall denote it by* $v(a)$.

**3.9 *Theorem***

*The functional* $v$ *on* $A$ *has the following properties:*

(a) $v(\lambda a) = |\lambda| v(a)$ *for* $\lambda \in \mathbb{C}$, $a \in A$.
(b) $v(ab) = v(ba)$ *and* $v(a^k) = v(a)^k$ *for* $a,b \in A$.
(c) *If* $ab = ba$, *then* $v(ab) \leq v(a)v(b)$ *and* $v(a + b) \leq v(a) + v(b)$.

PROOF

Parts (a) and (b) are immediate from the definitions. For the first part of (c), note that if $a$ and $b$ commute, then

$$\|(ab)^n\|^{1/n} \leq \|a^n\|^{1/n} \|b^n\|^{1/n}.$$

Since this holds for all $n$, we have $v(ab) \leq v(a)v(b)$.

The proof of the final part of (c) requires more care. Choose $\alpha > v(a)$, $\beta > v(b)$, and put $c = \alpha^{-1}a$, $d = \beta^{-1}b$. Since $a$ and $b$ commute, we have

$$\|(a + b)^n\|^{1/n} = \left\| \sum_{k=0}^{n} \binom{n}{k} a^k b^{n-k} \right\|^{1/n}$$
$$\leq \left\{ \sum_{k=0}^{n} \binom{n}{k} \alpha^k \beta^{n-k} \|c^k\| \, \|d^{n-k}\| \right\}^{1/n}.$$

Given $n$, choose $n'$ and $n''$ so that $n' + n'' = n$ and

$$\|c^{n'}\| \, \|d^{n''}\| = \max_{0 \le k \le n} \|c^k\| \, \|d^{n-k}\|.$$

For this choice of $n'$ and $n''$ we have

$$(2) \qquad \nu(a + b) \le (\alpha + \beta) \liminf_{n \to \infty} \|c^{n'}\|^{1/n} \|d^{n''}\|^{1/n}.$$

For some sequence $\{n_m\}$, the limit $\lim_{n \to \infty} n'_m/n_m$ exists; call it $\delta$. If $\delta \ne 0$, then $n'_m \to \infty$, and we have

$$\lim_{m \to \infty} \|c^{n'_m}\|^{1/n_m} = \nu(c)^{\delta} < 1,$$

while if $\delta = 0$, then

$$\limsup_{m \to \infty} \|c^{n'_m}\|^{1/n_m} \le \lim_{m \to \infty} \|c\|^{n'_m/n_m} \le 1.$$

In either case, $\limsup_{m \to \infty} \|c^{n'_m}\|^{1/n_m} \le 1$. Also $\limsup_{m \to \infty} \|d^{n''_m}\|^{1/n_m} \le 1$. From (2) it now follows that $\nu(a + b) \le \alpha + \beta$. As this holds for all $\alpha > \nu(a)$ and $\beta > \nu(b)$, we may conclude that $\nu(a + b) \le \nu(a) + \nu(b)$.

Thus, in case $A$ is commutative, $\nu$ is a seminorm on $A$. In general it is not a norm, and in the event that is it a norm, $A$ is usually not complete with respect to it. Of course, if $\upsilon$ is a norm and if $A$ is complete with respect to it, then $\nu$ is equivalent to the given norm in $A$ by virtue of the open mapping theorem.

3.10 ***Definition***

*The element $a$ is said to be **topologically nilpotent** if $\nu(a) = 0$.*

By Theorem 3.7, $a$ is topologically nilpotent if and only if $\sigma_A(a) = \{0\}$.

On the basis of Theorem 3.9, it is easy to see that if $A$ is commutative, then the set of topologically nilpotent elements is a closed ideal in $A$. It is immediate that it is an ideal. To see that it is closed, suppose that $\{a_n\}$ is a sequence which converges to $a$ and that for all $n$, $\nu(a_n) = 0$. If $\varepsilon > 0$, then for sufficiently large $n$, $\|a_n - a\| \le \varepsilon$. We have

$$\begin{aligned} \nu(a) &\le \nu(a - a_n) + \nu(a_n) \\ &\le \varepsilon. \end{aligned}$$

Thus, $\nu(a) = 0$.

It is easy to give examples of topologically nilpotent elements. Consider, for instance, the element $Z$ of the algebra in Example 1.6. We have $\|Z^k\| = e^{-k^2}$, so $\|Z^k\|^{1/k} = e^{-k} \to 0$. Thus, $Z$ is topologically nilpotent. It is not nilpotent, however.

As a less trivial example, define on $L^1(0, 1)$, the space of Lebesgue integrable functions on (0, 1), a multiplication $*$ by

$$f * g(x) = \int_0^x f(t)g(x - t)\,dt.$$

With respect to this multiplication and the usual $L^1$-norm, $L^1(0, 1)$ is a commutative Banach algebra. Let $I \in L^1(0, 1)$ be defined by $I(t) = t$. It is easy to verify that $I * I(t) = t^3/3!$, $I * I * I(t) = t^5/5!, \ldots$, i.e., that $I^n(t) = t^{2n-1}/(2n - 1)!$ and thus that $\|I^n\| = 1/(2n)!$. It follows that $v(I) = 0$, so $I$ is topologically nilpotent. Thus, every element in $S$, the linear span of the monomials $x, x^3, x^5, \ldots$, is topologically nilpotent. If we define $I_k$ by $I_k(x) = x^k$, then

$$\begin{aligned} I_k * I_k(x) &= \int_0^x t^k(x - t)^k\,dt \\ &= \sum_{j=0}^{k} (-1)^{k-j}\binom{k}{j} x^j \int_0^x t^{2k-j}\,dt \\ &= \sum_{j=0}^{k} (-1)^{k-j}\binom{k}{j} \frac{x^{2k+1}}{2k - j + 1} \end{aligned}$$

so that $I_k * I_k \in S$ for all $k$, whence $I_k * I_k$ is topologically nilpotent. Thus, $I_k$ itself is topologically nilpotent, and it follows that every polynomial is topologically nilpotent in $L^1(0, 1)$. Since the polynomials are dense in $L^1(0, 1)$, we may conclude that every element of $L^1(0, 1)$ is topologically nilpotent. An immediate consequence of this is that $L^1(0, 1)$ has no identity, a fact that could be established by other means.

# 4 IDEALS AND HOMOMORPHISMS

### 4.1 Theorem

*If $I$ is a closed two-sided ideal in the Banach algebra $A$, then $A/I$ is a Banach algebra under a norm equivalent to the quotient norm $\|\cdot\|_q$ given by*

$$\|a + I\|_q = \inf\{\|a + b\| : b \in I\}.$$

PROOF

If $I = A$, the theorem is true but uninteresting; we suppose therefore that $I \neq A$. The ideal $I$ is a closed subspace of $A$, so $A/I$ is a Banach space under the quotient norm. We have

$$\begin{aligned}\|ab + I\|_q &= \inf\{\|ab + c\| : c \in I\}\\ &\leq \inf\{\|ab + ad' + db + dd'\| : d,d' \in I\}\\ &\leq \inf\{\|a + d\|\,\|b + d'\| : d,d' \in I\}\\ &= \|a + I\|_q \|b + I\|_q.\end{aligned}$$

If $A$ has an identity, we have

$$\begin{aligned}\|1 + I\|_q &= \inf\{\|1 + a\| : a \in I\}\\ &\leq \|1\| = 1.\end{aligned}$$

Also, choosing an $a \notin I$ so that $\|a + I\| \neq 0$, we have

$$\begin{aligned}\|a + I\|_q &= \|(a + I)(1 + I)\|_q\\ &\leq \|a + I\|_q \|1 + I_q\|,\end{aligned}$$

whence $1 \leq \|1 + I\|_q$. Thus, if $A$ has an identity or if $A/I$ has no identity, it follows that $A/I$ is a Banach algebra under the quotient norm. It may happen that $A$ has no identity but that $A/I$ has one, say $e + I$. In this case, it need not be that $\|e + I\|_q = 1$, and thus to obtain a Banach algebra norm on $A/I$, we have to introduce a new norm, a norm equivalent to $\|\cdot\|_q$. That this is possible is shown by Theorem 1.9.

4.2 ***Example***

Let us give an example of the situation contemplated in the last paragraph. Let $A$ be the algebra of all $\mathbb{C}$-valued functions $f$ defined on $(-1, 1)$ which are continuously differentiable and which satisfy

$$\lim_{t \to \pm 1} f(t) = \lim_{t \to \pm 1} f'(t) = 0.$$

The operations are to be taken pointwise, and we define a norm in $A$ by

$$\|f\| = \sup\{|f(t)| : t \in (-1, 1)\} + \sup\{|f'(t)| : t \in (-1, 1)\}.$$

With this norm, $A$ is a Banach algebra without identity. Let $I \subset A$ be the ideal consisting of those functions which vanish on $[-\frac{1}{2}, \frac{1}{2}]$. Since there exist $f \in A$ which are identically 1 on $[-\frac{1}{2}, \frac{1}{2}]$, $A/I$ has an identity. Suppose though that $f$ is real and identically 1 on $[-\frac{1}{2}, \frac{1}{2}]$ so that $f + I$ is the identity of $A/I$. Then by the mean-value theorem, $1 = f(-\frac{1}{2}) - f(-1) = \frac{1}{2}f'(\xi)$ for some $\xi \in (-1, -\frac{1}{2})$, so $f'(\xi) = 2$. Thus, $\|f + I\|_q \geq 3$, and so with respect to the quotient norm, the identity of $A/I$ has norm greater than 1.

To proceed, we shall need certain elementary facts from ring theory.

**4·3** ***Definition***

*The ideal $I$ in a ring $R$ is said to be* **regular** *if there exists an identity modulo $I$ for $R$, i.e., if there exists $u \in R$ such that for all $a \in R$, $ua - a$ and $au - a$ both lie in $I$.*

**4·4** ***Lemma***

*Every regular ideal $I$ in a ring $R$ is contained in a maximal ideal.*

Note that the maximal ideal will be regular.

PROOF

Let $u$ be an identity modulo $I$. The element $u$ can lie in no proper ideal of $R$ which contains $I$, for if $u \in J$, $J$ an ideal in $R$ which contains $I$, then given $x \in R$, we have $x = xu + (x - xu)$. We have $xu \in J$ since $J$ is an ideal, and $x - xu \in I$ since $u$ is an identity modulo $I$. Thus, $x \in J$, and the ideal $J$ cannot be proper.

The collection $\mathscr{J}$ of all proper ideals that contain $I$ is partially ordered by inclusion, and if $\{J_\alpha\}$ is a subset of $\mathscr{J}$ which is linearly ordered by inclusion, then $\bigcup_\alpha J_\alpha$ is again in $\mathscr{J}$. Thus, Zorn's lemma applies; we may conclude that $\mathscr{J}$ contains a maximal element, $M$. The ideal $M$ is proper since it does not contain $u$, and it is maximal.

**4·5** ***Lemma***

*A regular ideal $I$ in a commutative ring $R$ is maximal if and only if $R/I$ is a field.*

PROOF

If $R/I$ is not a field, it has a proper ideal, $J$. Then $I_0 = \{a \in R : a + I \in J\}$ is an ideal in $R$ which properly contains $I$ and which is proper, for it cannot contain an element $u$ which is, modulo $I$, an identity for $R$. Thus, $I$ is not maximal. Conversely, if $R/I$ is a field, it can have no proper ideals, and so there are no proper ideals $I_0$ of $R$ that contain $I$ properly.

The following theorem shows that the openness of $\mathscr{G}(A)$ has importance for the ideal theory of Banach algebras.

**4·6** ***Theorem***

*If $I$ is a maximal, regular ideal in the Banach algebra $A$, then $I$ is closed.*

PROOF

It is easily verified that $I^-$, the closure of $I$ in $A$, is an ideal. We must show, under the hypothesis that $I$ is regular, that $I^- \neq A$. If $A$ has an

4.8 ***Corollary***

*If $A$ is a commutative Banach algebra, each of its complex homomorphisms is continuous and of norm no more than 1. If $A$ has an identity, the nonzero complex homomorphisms are of norm 1.*

PROOF

If $\varphi$ is not the zero homomorphism, then for each $a$ we have $\varphi(a) \in \sigma_A(a)$ and thus

$$|\varphi(a)| \le \sup\{|\lambda| : \lambda \in \sigma_A(a)\}.$$

By Theorem 3.4, this supremum does not exceed $\|a\|$. If $1 \in A$, then $\varphi(1) = 1$, so $\|\varphi\| = 1$.

4.9 ***Definition***

*The **radical** of a commutative Banach algebra is the intersection of all maximal regular ideals. Algebras with zero radical are said to be **semi-simple**.*

4.10 ***Theorem***

*The radical $R$ of a commutative Banach algebra $A$ is a closed ideal of $A$, and $R = \{a \in A : a$ is topologically nilpotent$\}$.*

PROOF

By Theorem 4.6, a maximal regular ideal is closed, so $R$, which is the intersection of all these ideals, is a closed ideal.

An element $a$ is topologically nilpotent if and only if $\sigma_A(a) = \{0\}$. Thus, if $a$ is topologically nilpotent, it is annihilated by each complex homomorphism and so is contained in each maximal regular ideal of $A$. Conversely, if $a$ is contained in every maximal regular ideal of $A$, it is annihilated by every complex homomorphism of $A$, so by part (b) of the last theorem, $\sigma_A(a) = \{0\}$, whence $a$ is topologically nilpotent.

4.11 ***Example***

Let us consider the algebra $A$ of Example 1.6. The element $Z$ of $A$, lies in the radical, $R$, of $A$, since, as we have already seen, $\lim_{n\to\infty} \|Z^n\|^{1/n} = 0$. Thus, the closed subalgebra of $A$ generated by $Z$ lies in $R$. But this closed subalgebra is the set of all $\sum_{k=0}^{\infty} \alpha_k Z^k$ with $a_0 = 0$. Call this set $S$.

It is easy to describe the complex homomorphisms of the algebra $A$. If $\varphi$ is such a homomorphism $\varphi \neq 0$, and if $a = \sum \alpha_k Z^k$, then $\varphi(a) = \varphi(a_0 + b)$, where $b = \sum_{k=1}^{\infty} a_k Z^k$. Since $b$ lies in the radical of $A$, $\varphi(a) = \varphi(a_0) = a_0$. Consequently, there is only one nonzero complex homomorphism of $A$. The set $S$ is therefore the unique maximal ideal of $A$, and thus it is the radical of $A$.

identity, in which case all ideals are regular, then $I^-$ must be for since $I$ is proper it can contain no element of $\mathcal{G}(A)$, so dist($I$ dist($I$, 1) $= \inf\{\|a - 1\| : a \in I\} \geq 1$.

In the case that $A$ has no identity, we are obliged to use quasii Let $u$ be an identity modulo $I$ for $A$. Then dist($I, u$) $\geq 1$. To see suppose that $a_0 \in I$ satisfies $\|a_0 - u\| < 1$. In the algebra $A'$ o by adjoining to $A$ an identity, the element $1 + (a_0 - u)$ is at less than 1 from $1 \in A'$, and is therefore in $\mathcal{G}(A')$. Thus, there i such that $(1 + b_0)[1 + (a_0 - u)] = 1 = [1 + (a_0 - u)](1 + b_0)$,

$$0 = a_0 - u + b_0 + a_0 b_0 - u b_0.$$

By hypothesis $u$ is an identity modulo $I$, so $b_0 - ub_0 \in I$. Als so $a_0 b_0 \in I$. Consequently, $u \in I$, and this is impossible since $I$ is Thus, dist($I, u$) $\geq 1$ as asserted, and it follows that $I^-$ is prope

4·7 ***Theorem***

*Let A be a commutative Banach algebra.*

(a) *Each maximal regular ideal of A is the kernel of a nonzero homomorphism of A.*

(b) *If A has an identity and if $a \in A$, then*

$\sigma_A(a) = \{\varphi(a) : \varphi$ *is a nonzero complex homomorphis*

*whereas if A has no identity,*

$\sigma_A(a) = \{0\} \cup \{\varphi(a) : \varphi$ *is a nonzero complex homomorphi*

PROOF

(a) If $I$ is a maximal regular ideal in $A$, it is closed by the last so $A/I$ is a Banach algebra under some norm equivale quotient norm. It is also a field since $I$ is maximal, and th Gelfand–Mazur theorem (Theorem 3.6), it is isomorphic $\psi : A/I \to \mathbb{C}$ be an isomorphism. If $\eta : A \to A/I$ is the natu $\psi \circ \eta$ is a nonzero complex homomorphism with kernel $I$

(b) Suppose that $A$ has an identity. If $\lambda \notin \sigma_A(a)$, then for $b(\lambda - a) = 1$, whence $1 = \varphi(b)[\lambda - \varphi(a)]$ for all nonzero homomorphisms $\varphi$. This implies that for no $\varphi$ is $\varphi(a) = \lambda$ therefore have

$\sigma_A(a) \supset \{\varphi(a) : \varphi$ is a nonzero complex homomorphis

Conversely, if $\lambda \in \sigma_A(a)$, then $\lambda - a$ is not invertible, so some maximal ideal which by (a) is the kernel of a nonzero homomorphism, say $\psi$. We have then that $\lambda = \psi(a)$.

If $A$ has no identity, adjoin one to get $A'$, apply the r derived to the algebra $A'$, and recall that

$$\sigma_A(a) = \sigma_{A'}(a).$$

In semisimple commutative Banach algebras, the algebraic and topological structures are related in an especially nice way. This is reflected, for instance, in the following result.

4.12 ***Theorem***

*If $A$ and $B$ are commutative Banach algebras, if $B$ is semisimple, and if $\Phi: A \to B$ is a homomorphism, then $\Phi$ is continuous.*

PROOF

We shall show that the graph of $\Phi$ is closed; the result is thus a consequence of the closed graph theorem. Let $\{(a_n, \Phi(a_n))\}_{n=1}^{\infty}$ be a sequence in the graph of $\Phi$, and suppose that $a_n \to a$, $\Phi(a_n) \to b$. If $\varphi$ is a complex homomorphism of $B$, $\varphi \circ \Phi$ is one of $A$ and thus is continuous. Thus, $\varphi(\Phi(a_n)) \to \varphi(\Phi(a))$. But also, $\varphi(\Phi(a_n)) \to \varphi(b)$, so for all $\varphi$, $\varphi(b) = \varphi(\Phi(a))$. We have assumed $B$ to be semisimple, so we may conclude that $b = \Phi(a)$, and the theorem is proved.

4.13 ***Corollary***

*If $A$ is a commutative semisimple Banach algebra with respect to a norm $\|\cdot\|$ and if $\|\cdot\|_1$ is a norm with respect to which $A$ is a Banach algebra, then $\|\cdot\|_1$ is equivalent to $\|\cdot\|$.*

PROOF

Apply the theorem to the map $a \mapsto a$ from $A$ with the norm $\|\cdot\|_1$ to $A$ with the norm $\|\cdot\|$.

4.14 ***Corollary***

*Every automorphism and every endomorphism of a commutative semisimple Banach algebra is continuous.*

# 5 *THE SPECTRUM OF A COMMUTATIVE BANACH ALGEBRA*

Every complex homomorphism of a commutative Banach algebra $A$ is, in particular, a linear functional and is continuous by Corollary 4.8. Thus, the set of complex homomorphisms of $A$ is a subset of $A^*$, the dual space of $A$, and so it is meaningful to speak of the weak $*$ topology on this set, i.e., the relative topology of this set as a subset of $A^*$ with the weak $*$ topology.

5.1 ***Definition***

*If $A$ is a commutative Banach algebra, the* **spectrum** *of $A$, $\Sigma(A)$, is the set of all nonzero complex homomorphisms of $A$ topologized with the weak * topology.*

Thus, a net $\{\varphi_i\}$ in $\Sigma(A)$ converges to $\varphi \in \Sigma(A)$ if and only if $\lim_i \varphi_i(a) = \varphi(a)$ for all $a \in A$.

In the literature of Banach algebra theory, the object $\Sigma(A)$ has been assigned several other names. Rickart [1] used the term *carrier space*, and Bourbaki [1] used the term *character space*. Recalling the one-to-one correspondence between nonzero complex homomorphisms of $A$ and the maximal regular ideals of $A$, we may use the topology on $\Sigma(A)$ to induce a topology on the set of all these ideals. The space of ideals obtained in this way is referred to as the *maximal ideal space of A*. Working with $\Sigma(A)$ is entirely equivalent to working with the maximal ideal space of $A$.

5.2 ***Theorem***

*If $A$ is a commutative Banach algebra, its spectrum is a locally compact space which is compact if $A$ has an identity.*

PROOF

For each $a \in A$, let $D_a = \{z \in \mathbb{C} : |z| \leq \|a\|\}$. Define $\Phi : \Sigma(A) \to \prod\{D_a : a \in A\}$ by requiring that $\pi_a(\Phi\varphi) = \varphi(a)$ for all $\varphi \in \Sigma(A)$, $a \in A$. Here $\pi_a$ is the natural projection of the product space onto $D_a$. We shall show that $\Phi$ is a homeomorphism of $\Sigma(A)$ onto $\Phi(\Sigma(A))$ and that $\{0\} \cup \Phi(\Sigma(A))$ is a closed subset in the product space if 0 is that point with $\pi_a(0) = 0$ for all $a$. Since the product space is compact by the Tychonoff theorem, it follows that $\Phi(\Sigma(A))$ is locally compact, and it will be compact if 0 is not a limit point of it. The point 0 cannot be a limit point of $\Sigma(A)$ if $A$ has an identity, for $\varphi(1) = 1$ for all $\varphi \in \Sigma(A)$.

The mapping $\Phi$ is one to one, for suppose that $\varphi$ and $\psi$ are distinct elements of $\Sigma(A)$. For some $a \in A$, $\varphi(a) \neq \psi(a)$, and consequently, $\pi_a(\Phi\varphi) \neq \pi_a(\Phi\psi)$. Thus, $\Phi$ is one to one. We have that a net $\{\varphi_i\}$ in $\Sigma(A)$ converges to $\varphi \in \Sigma(A)$ if and only if $\lim_i \varphi_i(a) = \varphi(a)$ for all $a \in A$, and this is equivalent to $\lim_i \pi_a(\Phi\varphi_i) = \pi_a(\Phi\varphi)$. This last formulation is simply the statement that $\lim_i \Phi\varphi_i = \Phi\varphi$. Thus, both $\Phi$ and $\Phi^{-1}$ are continuous, so $\Phi$ is a homeomorphism. Finally, the set $\{0\} \cup \Phi(\Sigma(A))$ is closed. Let $p$ be a limit point of $\Phi(\Sigma(A))$, say $p = \lim_i \Phi\varphi_i$, $p \neq 0$.

Define $\psi$ on $A$ by $\psi(a) = \pi_a(p)$. Then

$$\begin{aligned}\psi(a+b) &= \pi_{a+b}(p)\\ &= \lim_i \pi_{a+b}\Phi(\varphi_i) = \lim_i \varphi_i(a) + \lim_i \varphi_i(b)\\ &= \psi(a) + \psi(b),\end{aligned}$$

so $\psi$ is linear. Similarly, it is homogeneous and multiplicative. Since $p \neq 0$, $\psi \neq 0$, so $\psi \in \Sigma(A)$, and $\Phi\psi = p$. Thus, $\{0\} \cup \Phi(\Sigma(A))$ is closed and the proof is concluded.

5·3 ***Definition***

(a) *If $\Lambda$ is a set, $\mathbb{C}^\Lambda$ denotes the Cartesian product of $\Lambda$ copies of the complex plane given the usual product topology.*

(b) *If $\lambda \in \Lambda$, $\pi_\lambda$ is the projection of $\mathbb{C}^\Lambda$ onto the $\lambda$th coordinate.*

(c) *$\mathscr{P}_\Lambda$ is the smallest algebra of continuous $\mathbb{C}$-valued functions on $\mathbb{C}^\Lambda$ containing the constants and the projections $\pi_\lambda$.*

The elements of $\mathscr{P}_\Lambda$ will be called the *polynomials* in $\Lambda$ variables. An element of $\mathscr{P}_\Lambda$ is a function of the form $p = \sum_{j=1}^{k} p_j$, where each $p_j$ is a finite product of powers of the projections $\pi_\lambda$. Thus, we see that in case $\Lambda = \{1, \ldots, N\}$ so that $\mathbb{C}^\Lambda$ is the usual complex $N$-space, $\mathscr{P}_\Lambda$ is the ring of polynomials in $N$ complex variables.

5·4 ***Definition***

(a) *If $K$ is a compact set in $\mathbb{C}^\Lambda$, $\hat{K}$, the* ***polynomially convex hull*** *of $K$, is the set*

$$\{\mathfrak{z} \in \mathbb{C}^\Lambda : |p(\mathfrak{z})| \leq \|p\|_K \quad \text{for all } p \in \mathscr{P}_\Lambda\}.$$

*The set $K$ is* ***polynomial convex*** *if $K = \hat{K}$.*

(b) *The set $K$ is* ***rationally convex*** *if, given $\mathfrak{z}^0 \in \mathbb{C}^\Lambda \setminus K$, there exist $p, q \in \mathscr{P}_\Lambda$ such that $q$ has no zero on $K$ and such that*

$$|p(\mathfrak{z}^0)| > |q(\mathfrak{z}^0)| \sup\left\{\left|\frac{p(\mathfrak{z})}{q(\mathfrak{z})}\right| : \mathfrak{z} \in K\right\}.$$

5·5 ***Examples***

In case $\Lambda$ reduces to a singleton so that $\mathbb{C}^\Lambda$ is just the complex plane, the notions of polynomial and rational convexity are atypically simple. A compact set $K$ in $\mathbb{C}$ is polynomially convex if and only if $C \setminus K$ is connected, and every compact set is rationally convex.

The rational convexity of a compact set $K \subset \mathbb{C}$ is entirely elementary, for if $z^0 \notin K$, then the function $f_{z^0}$ given by $f_{z^0}(z) = (z - z^0)^{-1}$ is rational.

The functions $f_{z^0}$ obtained in this way suffice to show that $K$ is rationally convex.

Next, a compact set $K \subset \mathbb{C}$ is polynomially convex if and only if $\mathbb{C} \setminus K$ is connected. Suppose that $\mathbb{C} \setminus K$ is not connected. Since $K$ is compact, the set $\mathbb{C} \setminus K$ must have a bounded component, say $V$. Choose $z_0 \in V$, and let $p$ be a polynomial in one variable. Since $p$ is holomorphic on the whole plane, we have by the maximum modulus theorem that $|p(z_0)| \leq \|p\|_{\partial V} \leq \|p\|_K$. As this inequality holds for all polynomials $p$, we conclude that $z^0 \in \hat{K}$, so $K$ is not polynomially convex. Conversely, if $\mathbb{C} \setminus K$ is connected, let $z_0 \in \mathbb{C} \setminus K$. Let $D$ be a neighborhood of $z_0$, $D$ an open disc whose closure misses $K$, and let $D'$ be a domain in $\mathbb{C}$ bounded by a simple closed curve such that $\bar{D} \cap \bar{D}' = \varnothing$, and such that $D' \supset K$. Define $f$ on $D \cup D'$ by $f|D \equiv 1$, $f|D' \equiv 0$. Since $D \cup D'$ does not separate the plane, Runge's theorem provides a polynomial $p$ such that $|p(z)| < \frac{1}{2}$ if $z \in K$, $|p(z_0) - 1| < \frac{1}{2}$. Thus, the set $K$ is polynomially convex.

In higher dimensional spaces, the situation is much more complicated and is not thoroughly understood yet. Consider in $\mathbb{C}^2$ two circles given by

$$\Lambda_1 = \{(e^{i\theta}):\theta \text{ real}\}$$

and

$$\Lambda_2 = \{(e^{i\theta}, e^{-i\theta}):\theta \text{ real}\}.$$

The circle $\Lambda_1$ is not polynomially convex, but $\Lambda_2$ is. Consider $\Lambda_1$. If $p$ is a polynomial in two complex variables, the polynomial $\tilde{p}$ in one variable given by $\tilde{p}(z) = p(z, z)$ satisfies $|\tilde{p}(z_0)| \leq \sup\{|\tilde{p}(z)| : |z| = 1\}$ if $|z_0| < 1$. This implies that if $|z| < 1$, then $|p(z, z)| \leq \|p\|_{\Lambda_1}$, so

$$\hat{\Lambda}_1 \supset \{(z, z) \in \mathbb{C}^2 : |z| \leq 1\},$$

and thus $\Lambda_1 \neq \hat{\Lambda}_1$.

Consider now $\Lambda_2$ and a point $(z_1^0, z_2^0) \in \mathbb{C}^2 \setminus \Lambda_2$. If $|z_1^0| > 1$, then $\pi_1$, the projection onto the first coordinate, satisfies $|\pi_1(z_1^0, z_2^0)| = |z_1^0| > 1 = \|\pi_1\|_{\Lambda_2}$, so $(z_1^0, z_2^0) \notin \hat{\Lambda}_2$. Similarly, if $|z_2^0| > 1$, then $(z_1^0, z^0) \notin \hat{\Lambda}_2$. Now let $p$ be the polynomial $p(z_1, z_2) = \frac{1}{2}(1 + z_1 z_2)$. The polynomial $p$ assumes the value 1 on $\Lambda_2$ and nowhere else on the set $\{(z_1, z_2):|z_1|, |z_2| \leq 1\}$. Thus, if $(z_1^0, z_2^0) \notin \Lambda_2$ but $|z_1^0| \leq 1$, $|z_2^0| \leq 1$, then there is a polynomial in one variable, say $q$, such that $q(1) = 0$, $q(p(z_1^0, z_2^0)) = 1$. Since $q \circ p$ is a polynomial, $(z_1^0, z_2^0) \notin \hat{\Lambda}_2$, so $\Lambda_2$ must be polynomially convex.

5.6 ***Definition***

*Given a commutative Banach algebra with identity, $A$, and a set $E = \{a_\lambda\}_{\lambda \in \Lambda}$ of elements of $A$, the* ***joint spectrum*** *with respect to $A$ of $E$*

*is the set*

$$\sigma_A(E) = \{\mathfrak{z} \in \mathbb{C}^\Lambda : \textit{for some } \varphi \in \Sigma(A), \varphi(a_\lambda) = \pi_\lambda(\mathfrak{z}) \textit{ for all } \lambda \in \Lambda\}.$$

Thus, if $E$ reduces to a single element, say $E = \{a\}$, then $\sigma_A(E)$ is just the spectrum of $a$.

5·7 ***Definition***

*Let $A$ be a commutative Banach algebra with identity. A* ***set of generators*** *for $A$ is a set $E \subset A$ such that the smallest closed subalgebra with identity of $A$ containing $E$ is $A$ itself.*

5·8 ***Theorem***

*If $A$ is a commutative Banach algebra with identity and $E = \{a_\lambda\}_{\lambda\in\Lambda}$ is a set of generators for $A$, then $\sigma_A(E)$ is polynomially convex in $\mathbb{C}^\Lambda$, and the map $\Phi: \Sigma(A) \to \mathbb{C}^\Lambda$ determined by $\pi_\lambda(\Phi\varphi) = \varphi(a_\lambda)$ is a homeomorphism of $\Sigma(A)$ onto $\sigma_A(E)$.*

PROOF

The space $\Sigma(A)$ is compact, so in order to show that $\Phi$ is a homeomorphism, it is enough to prove that $\Phi$ is continuous and one to one. The continuity of $\Phi$ is immediate from the definition of the topologies on $\Sigma(A)$ and $\mathbb{C}^\Lambda$.

If $\varphi$ and $\psi$ are distinct elements of $\Sigma(A)$ at which $\Phi$ assumes the value $\mathfrak{z}^0 \in \mathbb{C}^\Lambda$, let $a \in A$ be such that $\hat{a}(\varphi) \neq \hat{a}(\psi)$. Since $E$ is a set of generators for $A$, given $\varepsilon > 0$, there is $b_\varepsilon \in A$, $b_\varepsilon$ of the form $b_\varepsilon = \sum_{j=1}^N \alpha_j c_j$, $c_j$ a finite product of the elements $a_\lambda$, and $\alpha_j$ a complex number, such that $\|a - b_\varepsilon\| < \varepsilon$. Since $\psi$ and $\varphi$ are of norm 1 we find that

$$\begin{aligned}|\varphi(a) - \psi(a)| &\leq |\varphi(a) - \varphi(b_\varepsilon)| + |\psi(a) - \psi(b_\varepsilon)| + |\varphi(b_\varepsilon) - \psi(b_\varepsilon)| \\ &\leq 2\varepsilon + |\varphi(b_\varepsilon) - \psi(b_\varepsilon)|.\end{aligned}$$

The hypothesis that $\Phi\psi = \Phi\varphi$ implies that $\varphi(b_\varepsilon) = \psi(b_\varepsilon)$, so $|\varphi(a) - \psi(a)| < 2\varepsilon$. As $\varepsilon$ is at our disposal, we are led to $\varphi(a) = \psi(a)$, contrary to our assumption. Thus, $\Phi$ is one to one and so a homeomorphism.

The range of $\Phi$ is the set $\sigma_A(E)$ so it remains to prove that $\sigma_A(E)$ is polynomially convex.

Suppose given a point $\mathfrak{z}^0 \in (\sigma_A(E))\hat{}$. The map $p \mapsto p(\mathfrak{z}^0)$ is a nonzero complex homomorphism of $\mathscr{P}_\Lambda$.

Define a map $h: \mathscr{P}_\Lambda \to A$ by setting $h(\pi_\lambda) = a_\lambda$ and demanding that $h$ be linear and multiplicative. Thus, if $\pi_{\lambda_1}^{m_1} \cdots \pi_{\lambda_j}^{m_j} \in \mathscr{P}_\Lambda$, then $h(\pi_{\lambda_1}^{m_1} \cdots \pi_{\lambda_j}^{m_j}) = a_{\lambda_1}^{m_1} \cdots a_{\lambda_j}^{m_j}$. The range of the homomorphism $h$ is the smallest subalgebra, not necessarily closed, of $A$ containing the constants and the set $\{a_\lambda\}_{\lambda\in\Lambda}$; call this range $B$. Thus, $B$ is a dense subalgebra of $A$.

Define $\psi$ on $B$ by setting $\psi(h(p)) = p(\mathfrak{z}^0)$ if $p \in P_\Lambda$. The functional $\psi$ is well defined, for if $h(p) = h(q)$, then for each $\varphi \in \Sigma(A)$, we have $0 = \varphi(h(p) - h(q))$, and by the definition of $h$, this implies that $0 = p(\Phi(\varphi)) - q(\Phi(\varphi))$. Thus, $0 = \sup\{|p(\mathfrak{z}) - q(\mathfrak{z})| : \mathfrak{z} \in \Phi(\Sigma(A))\}$, and since $\mathfrak{z}^0$ is in $(\sigma_A(E))\hat{} = (\Phi(\Sigma(A)))\hat{}$, we may conclude that $p(\mathfrak{z}^0) = q(\mathfrak{z}^0)$. Consequently, $\psi$ is well defined. It is clearly a nonzero complex homomorphism of $B$. If $b \in B$, say $b = h(p)$, then

$$\begin{aligned} |\psi(b)| &= |p(\mathfrak{z}^0)| \leq \sup\{|p(\mathfrak{z})| : \mathfrak{z} \in \Phi(\Sigma(A))\} \\ &= \sup\{|\varphi(b)| ; \varphi \in \Sigma(A)\} \\ &\leq \|b\|, \end{aligned}$$

so $\psi$ can be extended to an element, still to be denoted by $\psi$, of $\Sigma(A)$. But then $\mathfrak{z}^0 = \Phi\psi$, so $\mathfrak{z}^0 \in \Phi(\Sigma(A))$, and we may conclude that this set is, in fact, polynomially convex, and the proof is concluded.

Let us observe that as a consequence of this theorem, we may conclude that the spectrum of a Banach algebra with a single generator is, topologically, a subset of the plane which, by our characterization of the polynomially convex compacta in the plane, does not separate the plane.

Although Theorem 5.8 shows that the joint spectrum of a set of generators for a commutative Banach algebra with identity is polynomially convex, it is not true, as simple examples show, that the joint spectrum of an arbitrary set is polynomially convex.

5·9 ***Definition***

*If $I$ is an ideal in a commutative Banach algebra, then hull $I$,* ***the hull of the ideal*** *$I$, is the closed set*

$$\{\varphi \in \Sigma(A) : \varphi(a) = 0 \quad \textit{for all} \quad a \in I\}.$$

If $\Phi : A \to B$ is a homomorphism of commutative Banach algebras with identity, there is an induced map $\Phi^* : \Sigma(B) \to \Sigma(A)$ given by $\Phi^*\varphi(a) = \varphi(\Phi a)$ for all $\varphi \in \Sigma(B)$, all $a \in A$. The map $\Phi^*$ is continuous, even though $\Phi$ may not be, for if $\{\psi_\alpha\}$ is a net in $\Sigma(B)$ which converges to $\psi$, then since

$$\begin{aligned} \Phi^*\psi_\alpha(a) &= \psi_\alpha(\Phi a) \\ &\to \psi(\Phi a) \\ &= \Phi^*\psi(a), \end{aligned}$$

we see that $\{\Phi^*\psi_\alpha\}$ converges to $\Phi^*\psi$. If the range of $\Phi$ is dense in $B$, $\Phi^*$ is evidently one to one, and so a homeomorphism. This dual map is extremely useful as we shall see in the further development of the theory.

5.10 ***Theorem***

*If $A$ and $B$ are commutative Banach algebras, and if $\Phi: A \to B$ is a homomorphism of $A$ onto $B$, the dual map $\Phi^*: \Sigma(B) \to \Sigma(A)$ is a homeomorphism of $\Sigma(B)$ onto hull(ker $\Phi$).*

In this theorem, we do not require that $A$ and $B$ have identities.

PROOF OF THE THEOREM

Note to begin with that since $\Phi(A) = B$, each $\Phi^*\varphi$ is a nonzero complex homomorphism of $A$, so $\Phi^*$ does carry $\Sigma(B)$ to $\Sigma(A)$. As above, $\Phi^*$ is continuous and one to one. If $\varphi \in \Sigma(B)$ and $a \in \ker \Phi$, then $\Phi^*\varphi(a) = \varphi(\Phi a) = 0$, so $\Phi^*\varphi \in \text{hull}(\ker \Phi)$.

To show that the range of $\Phi^*$ is the whole of hull(ker $\Phi$), suppose $\psi$ to lie in this latter set. Define $\varphi$ on $B$ by $\varphi(b) = \psi(a)$ if $a \in \Phi^{-1}(b)$. The functional $\varphi$ is well defined, for if $\Phi(a) = \Phi(a')$, then $a - a' \in \ker \Phi$, so $\psi(a - a') = 0$; i.e., $\psi(a) = \psi(a')$. Since $\Phi$ is a homomorphism, $\varphi$ is one, and, thus, $\varphi \in \Sigma(B)$. It is clear that $\Phi^*\varphi = \psi$.

Finally, $\Phi^*$ is a homeomorphism. If $B$ has an identity so that $\Sigma(B)$ is compact, this fact is implied by what has already been said. If $B$ has no identity, neither does $A$. But in this case, regarding $\Sigma(A)$ and $\Sigma(B)$ as subsets of the dual spaces $A^*$ and $B^*$, respectively, we have that $\Phi^*$ extends to a continuous map from $\Sigma(B) \cup \{0\}$ to $\Sigma(A) \cup \{0\}$, and this extended map must be a homeomorphism. Thus, $\Phi$ takes $\Sigma(B)$ into $\Sigma(A)$ homeomorphically.

This theorem cannot be proved under the weaker hypothesis that the range of $\Phi$ is dense in $B$.

5.11 ***Corollary***

*If $\Phi: A \to B$ is an isomorphism, then $\Phi^*$ is a homeomorphism of $\Sigma(B)$ onto $\Sigma(A)$.*

5.12 ***Corollary***

*If $I \subset A$ is a closed ideal and $\eta: A \to A/I$ is the natural map, then $\eta^*: \Sigma(A/I) \to$ hull $I$ is a homeomorphism.*

Thus, we have identified the spectrum of a quotient algebra. It is important in many instances to know the spectrum of a closed ideal in a commutative Banach algebra. Such an ideal is again a commutative

Banach algebra under the induced norm, so it is meaningful to speak of its spectrum.

5.13 ***Theorem***

*Let $I$ be a closed ideal in $A$. The spectrum of $I$ may be identified with $\Sigma(A) \setminus \text{hull } I$.*

PROOF

It is clear that each element of $\Sigma(A) \setminus \text{hull } I$ defines an element of $\Sigma(I)$. We have to see that every element of $\Sigma(I)$ is of this form. If $\varphi \in \Sigma(I)$, there is $a_0 \in I$ with $\varphi(a_0) = 1$. Define $\psi$ on $A$ by $\psi(a) = \varphi(a_0 a)$ for all $a \in A$. The functional $\psi$ is defined on the whole of $A$ since $I$ is an ideal, and it is clearly linear. It is also multiplicative, for $\psi(b_1 b_2) = \varphi(a_0 b_1 b_2) = \varphi(a_0)\varphi(a_0 b_1 b_2) = \varphi(a_0^2 b_1 b_2) = \varphi(a_0 b_1)\varphi(a_0 b_2) = \psi(b_1)\psi(b_2)$. Thus, $\psi \in \Sigma(A)$. On $I$, $\psi$ and $\varphi$ obviously agree. Thus, as a set, $\Sigma(I)$ may be identified with $\Sigma(A) \setminus \text{hull } I$. Note that $\psi$ is the unique extension of $\varphi$ to an element of $\Sigma(A)$.

To conclude the proof, we have to verify that the topology on $\Sigma(I)$ agrees with the relative topology on $\Sigma(A) \setminus \text{hull } I$. This follows, however, from the fact that if $\{\varphi_i\}$ is a net in $\Sigma(I)$ and if $\psi_i$ denotes the extension of $\varphi_i$ to an element of $\Sigma(A)$, then $\{\varphi_i\}$ is convergent if and only if $\{\psi_i\}$ is.

5.14 ***Definition***

*If for $k = 1, 2, \ldots, N$, $A_k$ is a commutative Banach algebra, then $\sum_{k=1}^N A_k$, the* ***direct sum*** *of the $A_k$, is the set of all $N$-tuples $(a_1, \ldots, a_N)$, $a_k \in A_k$, with addition and multiplication defined coordinatewise, and with the norm $|||\cdot|||$ given by $|||(a_1, \ldots, a)||| = \max\{\|a_1\|, \ldots, \|a_N\|\}$.*

With this norm, $\sum_{k=1}^N A_k$ is again a commutative Banach algebra; it is sometimes denoted by $A_1 \oplus \cdots \oplus A_k$. The following theorem identifies the spectrum of $\sum_{k=1}^N A_k$.

5.15 ***Theorem***

*The spectrum of the direct sum $\sum_{k=1}^N A_k$ is the disjoint union of the spectra of the algebras $A_1, \ldots, A_k$.*

PROOF

By induction, it is enough to consider the case $N = 2$. Let $\psi \in \Sigma(A_1 \oplus A_2)$ and define $\psi_1$ and $\psi_2$ on $A_1 \oplus A_2$ by $\psi_1(a_1, a_2) = \psi(a_1, 0)$, $\psi_2(a_1, a_2) = \psi(0, a_2)$. Both of these functionals are complex homomorphisms of

$A_1 \oplus A_2$, and we have $\psi = \psi_1 + \psi_2$; so not both $\psi_1$ and $\psi_2$ are zero. Suppose that $\psi_1 \neq 0$. We shall show that $\psi = \psi_1$, i.e., that $\psi_2 = 0$. If not, choose $a \in A_1 \oplus A_2$ such that $\psi_1(a) \neq 0$, $\psi_2(a) = 0$. For any $b \in A_1 \oplus A_2$, we have $\psi(ab) = \psi_1(ab) + \psi_2(ab) = \psi_1(a)\psi_1(b) + \psi_2(a)\psi_2(b)$. But also

$$\begin{aligned}\psi(ab) &= \psi(a)\psi(b) \\ &= [\psi_1(a) + \psi_2(a)][\psi_1(b) + \psi_2(b)].\end{aligned}$$

It follows that $\psi_1(a)\psi_2(b) = 0$ since $\psi_2(a) = 0$. Since this holds for all choices of $b \in A_1 \oplus A_2$, we may conclude that $\psi_2 = 0$, so $\psi = \psi_1$ as asserted.

Thus, each element of $\Sigma(A_1 \oplus A_2)$ is of the form $\varphi \circ \pi_j$, $\pi_j : A_1 \oplus A_2 \to A_j$ the natural projection and $\varphi \in \Sigma(A_j)$. Conversely, it is clear that each $\varphi \circ \pi_j$ is an element of $\Sigma(A_1 \oplus A_2)$. The theorem now follows, for it is clear that the topology on $\Sigma(A_1 \oplus A_2)$ is that of the disjoint union of $\Sigma(A_1)$ and $\Sigma(A_2)$.

We now come to the important notion of *Gelfand transform.*

5.16 ***Definition***

*If $a \in A$, $A$ a commutative Banach algebra, then $\hat{a}$ is the function on $\Sigma(A)$ defined by $\hat{a}(\varphi) = \varphi(a)$. The function $\hat{a}$ is called the* ***Gelfand transform*** *of $a$. We denote by $\hat{A}$ the set of all Gelfand transforms of elements of $A$.*

The function $\hat{a}$ is continuous on $\Sigma(A)$, and in the event that $A$ has no identity, $\hat{a}$ tends to zero at infinity. The latter point follows from the fact that if $\{\varphi_i\}$ is a net in $\Sigma(A)$ which converges to infinity, then when we regard $\Sigma(A)$ as a subspace of the dual space of $A$, $\{\varphi_i\}$ converges weak $*$ to 0. The map $a \mapsto \hat{a}$ is a homomorphism of $A$ onto a point separating subalgebra of $\mathscr{C}_0(\Sigma(A))$.

In general the algebra $\hat{A}$ is not the whole of $\mathscr{C}_0(\Sigma)$ and it need not be dense. The kernel of the Gelfand transform is the set of those $a \in A$ which are annihilated by all complex homomorphisms of $A$, i.e., the radical of $A$. Thus, $\hat{}$ is one to one if $A$ is semisimple. The Gelfand transform is continuous since $\|\hat{a}\| = \sup\{|\hat{a}(\varphi)| : \varphi \in \Sigma(A)\}$ and for all $\varphi \in \Sigma(A)$, $|\hat{a}(\varphi)| \leq \|a\|$ by Corollary 4.8.

If $A$ is a Banach algebra of continuous functions on a topological space $X$, the operations being pointwise, and if $A$ separates points on $X$ in the strong sense that, given $x,y \in X$, there is $f \in A$ such that $0 \neq f(x) \neq f(y)$, then there is a natural map $i : X \to \Sigma(A)$ given by $(ix)(f) = f(x)$. The map $i$ is one to one and continuous, and the function $\hat{f}$ on $\Sigma(A)$ is, in a natural way, an extension of $f$ from the set

$i(X)$ to the whole of $\Sigma(A)$. In general $\Sigma(A)$ will be considerably larger than $X$. If $X$ is a compact space, then $i$ is necessarily a homeomorphism of $X$ into $\Sigma(A)$, but if $X$ is only locally compact, $i$ need not be a homeomorphism.

5.17 ***Example***

As an example of this latter phenomenon, let us consider the following algebra. Let $\lambda$ and $\mu$ be positive real numbers such that $\lambda/\mu$ is irrational. Let $A$ be the algebra of all functions on the real line of the form

$$f(t) = \sum_{m,n\in\mathbb{Z}} A_{mn}e^{i(m\lambda+n\mu)t} \tag{1}$$

with

$$\|f\| = \Sigma|A_{mn}| < \infty.$$

This is a commutative Banach algebra with identity, and its spectrum can be identified with the two-dimensional torus $T^2 = \{(e^{i\alpha}, e^{i\beta}) : \alpha, \beta$ real$\}$, as the following argument shows. Let $\varphi \in \Sigma(A)$. For $m,n \in \mathbb{Z}$, let $\chi_{m,n} \in A$ be defined by

$$\chi_{m,n}(t) = e^{i(m\lambda+n\mu)t}.$$

We have $\|\chi_{m,n}\| = 1$ and $\chi_{m,n}\chi_{-m,-n} = 1$, so

$$|\varphi(\chi_{m,n})| = 1 \qquad \text{and} \qquad \varphi(\chi_{m,n}) = \overline{\varphi(\chi_{-m,-n})}.$$

Consequently, there exist $\alpha,\beta \in \mathbb{R}$ such that if $f$ is given by (1), then

$$\varphi(f) = \Sigma A_{mn}\, e^{i(m\alpha+n\beta)}.$$

Conversely, each pair $\alpha,\beta \in \mathbb{R}$ determines such a homomorphism, $\varphi_{\alpha,\beta}$. Since we are considering absolutely convergent series, it follows that if $\alpha \to \alpha'$ and $\beta \to \beta'$, then $\varphi_{\alpha,\beta} \to \varphi_{\alpha',\beta'}$ in $\Sigma(A)$. Since $\varphi_{\alpha,\beta}$ and $\varphi_{\alpha',\beta'}$ agree if and only if $\alpha - \alpha' \equiv 0 (\text{mod } 2\pi)$ and $\beta - \beta' \equiv 0 (\text{mod } 2\pi)$, it follows that $\Sigma(A)$ can indeed be identified with $T^2$.

The map $i : \mathbb{R} \to \Sigma(A)$ considered above is given, in this instance, by $i(t) = (e^{i\lambda t}, e^{i\mu t})$ and is continuous and one to one. It is not, however, a homomorphism. It follows from a theorem of Kronecker that $i(\mathbb{R})$ is dense in $T^2$ since $\lambda/\mu$ is irrational. (See Appendix E.) Choose $t_0 \in \mathbb{R}$. The interval $I = [t_0 - 1, t_0 + 1]$ is mapped homeomorphically into $T^2$ by $i$. Let $\{V_n\}_{n=1}^{\infty}$ be a sequence of disjoint open sets in $T^2 \setminus i(I)$ such that if for all $n$, $s_n$ is a point of $V_n$, then $\{s_n\}$ converges to $i(t_0)$. The density of $i(\mathbb{R})$ in $T^2$ implies the existence of a point $t_n \in \mathbb{R}$ with $i(t_n) \in V_n$. Thus, the sequence $\{i(t_n)\}$ converges to $i(t_0)$, but since $t_n \notin I$, $\{t_n\}$ does not converge to $t_0$. Thus, $t$ is not a homeomorphism.

# 6 SOME EXAMPLES

In this short section we digress from the systematic development of the theory of commutative Banach algebras in order to give some examples of the theory developed so far. We begin with two applications to classical analysis. The first of these is a famous theorem of Wiener.

6.1 ***Theorem***

*If the function $f_0$ defined on the unit circle has an absolutely convergent Fourier series, and if $f$ does not assume the value zero, then $1/f_0$ also has an absolutely convergent Fourier series.*

PROOF

Denote by $A$ the algebra of all functions defined on $T$, the unit circle, which have absolutely convergent Fourier series. If $g \in A$, set

$$\|g\| = \Sigma|a_n|$$

if $g$ has as its Fourier series $\Sigma a_n e^{in\theta}$. With this norm, $A$ is a commutative Banach algebra with identity. To show that $1/f_0 \in A$, it is enough to show that $f_0$ lies in no maximal ideal of $A$ or, equivalently, that no complex homomorphism of $A$ annihilates $f_0$.

We shall determine the complex homomorphisms of $A$. Let $\varphi \in \Sigma(A)$. Define $h_0 \in A$ by $h_0(e^{i\theta}) = e^{i\theta}$ so that if $g \in A$, then $g = \Sigma b_n h_0^n$. Since this series converges in $A$ to $g$, the continuity of $\varphi$ implies that $\varphi(g) = \Sigma b_n \varphi(h_0^n) = \Sigma b_n[\varphi(h_0)]^n$. The functional $\varphi$ is of norm 1, and we have $1 = \varphi(1) = \varphi(h_0\bar{h}_0) = \varphi(h_0)\varphi(\bar{h}_0)$. Thus, $|\varphi(h_0)| = 1$, and we may conclude that if $f \in A$, then $\varphi(f) = f(\varphi(h_0))$, $\varphi(h_0) \in T$.

By hypothesis, the function $f_0$ does not assume the value zero at any point of $T$, so for no complex homomorphism $\varphi$ is $\varphi(f_0) = 0$. Thus, $f_0$ lies in no proper ideal of $A$, and so it is invertible.

Theorem 6.1 was proved by Wiener [1], who used entirely classical methods. The algebraic proof we have given here is from the fundamental paper of Gelfand [1].

Our second example concerns the disc algebra $\mathscr{A}$ introduced in Example 1.3.

6.2 ***Theorem***

*If $f_1, \ldots, f_n \in \mathscr{A}$ satisfy $|f_1(z)| + \cdots + |f_n(z)| > 0$ for all $z \in \bar{\Delta}$, the closed unit disc, then there exist $g_1, \ldots, g_n \in \mathscr{A}$ such that $f_1 g_1 + \cdots + f_n g_n = 1$.*

PROOF

We want to show that the identity of $\mathscr{A}$ lies in the ideal generated by $f_1, \ldots, f_n$. If not, this ideal lies in some maximal ideal and so there is some complex homomorphism $\varphi$ of $\mathscr{A}$ which annihilates each of the functions $f_1, \ldots, f_n$.

Denote by $Z$ the identity map of $\bar{\Delta}$ onto itself. Then $Z \in \mathscr{A}$, and $\|Z\| = 1$. Thus, if $\varphi \in \Sigma(A)$, then $|\varphi(Z)| \leq 1$. It follows that if $p \in \mathscr{A}$ is a polynomial, then $\varphi(p) = p(\varphi(Z))$. As we observed in Example 1.3, the polynomials are dense in $\mathscr{A}$, so the continuity of $\varphi$ implies that if $f \in \mathscr{A}$, then $\varphi(f) = f(\varphi(Z))$. Since $\varphi(Z) \in \bar{\Delta}$ and since $|f_1| + \cdots + |f_n| > 0$ on $\bar{\Delta}$, we may conclude that $\varphi$ does not annihilate all the functions $f_1, \ldots, f_n$, and consequently the desired functions $g_j$ exist and the theorem is proved.

Since $\mathscr{A}$ separates points on $\bar{\Delta}$, what we have proved establishes that, in a natural way, $\Sigma(A) = \bar{\Delta}$.

A classical proof of Theorem 6.2 was given by Carleson [1], and Cohen [1] has also given a more constructive proof. The methods of Cohen also suffice to prove Weiner's theorem.

Considering the ease with which we obtained Theorem 6.2, it is natural to try to extend the result to bounded functions: Suppose that $f_1, \ldots, f_n \in H^\infty(\Delta)$ satisfy $|f_1(z)| + \cdots + |f_n(z)| \geq \delta > 0$ for some fixed $\delta > 0$ and all $z \in \Delta$. Do there exist $g_1, \ldots, g_n \in H^\infty(\Delta)$ such that $f_1 g_1 + \cdots + f_n g_n = 1$? This question is the much discussed *corona problem*, and it too has been settled by Carleson [3]. The desired functions $g_j$ do exist. Carleson's proof uses only classical analysis. It would be of great interest to have a solution to the corona problem that draws less on classical methods and more on algebraic analysis, but to the best of my knowledge, no such proof has been discovered yet.

Our next example concerns the algebra $\mathscr{C}(X)$, $X$ a compact space. This is one of the very few Banach algebras for which it is possible to give a complete description of all the closed ideals.

6.3 ***Example***

If $X$ is a compact space, there is an obvious collection of closed ideals in $\mathscr{C}(X)$. If $E \subset X$ is a closed set, $I(E) = \{f \in \mathscr{C}(X) : f|E = 0\}$ is a closed ideal in $\mathscr{C}(X)$. We shall show that every closed ideal is of this form. Suppose that $I$ is a proper closed ideal of $\mathscr{C}(X)$, and set $E = \bigcap\{Z(f) : f \in I\}$. Here $Z(f)$ denotes the zero set of the function $f$. The set $E$ is nonempty, for if $E = \varnothing$, then by compactness there exist finitely many functions $f_1, \ldots, f_n \in I$ such that $|f_1(x)| + \cdots + |f_n(x)| > 0$ for all $x \in X$. Since $I$ is an ideal, $g = f_j \bar{f}_j = \Sigma |f_j|^2$ is an element of $I$

which is zero free on $X$. Thus, $1/g \in \mathscr{C}(X)$, so $1 = g(1/g) \in I$, an impossibility since $I$ is proper. Thus, the set $E$ is nonempty.

We now prove that if $h \in \mathscr{C}(X)$ vanishes on the set $E$, then $h \in I$. Let $\varepsilon > 0$ and put $V = \{x \in X : |h(x)| \geq \varepsilon\}$. By normality there is an open set $W$ such that $X \setminus V \supset W^- \supset W \supset E$. Let $p \in \mathscr{C}(X)$ satisfy $p(x) = 0$ if $x \in W$, $p(x) = 1$ if $x \in V$, and $0 \leq p(x) \leq 1$ for all $x$. The set $X \setminus W$ is compact, so there exist finitely many elements of $I$, say $f_1, \ldots, f_q$, such that $\Sigma |f_j(x)| > 0$ if $x \in X \setminus W$. Then $f_0 = \Sigma f_j \bar{f}_j \in I$ and $f_0$ is bounded away from zero on $X \setminus W$. Since $p$ vanishes identically on $W$, we can define a $g \in \mathscr{C}(X)$ by requiring that $g(x) = 0$ if $x \in W$, and $g(x) = h(x)p(x)f_0^{-1}(x)$ if $x \in X \setminus W$. The $g$ so defined is continuous and so $f_0 g \in I$. We have $|f_0(x)g(x) - h(x)| = |h(x)(p(x) - 1)|$. If $x \in V$, this is zero, and if $x \in X \setminus V$, it is no more than $2\varepsilon$ since for such $x$, $|h(x)| < \varepsilon$ and $|p(x) - 1| \leq 2$. Thus, $h$ is in the closure of $I$, and as $I$ is closed, $h \in I$. Thus, $I = I(E)$ as asserted.

In particular, if $I$ is a maximal ideal of $\mathscr{C}(X)$, $I$ must be of the form $\{f \in \mathscr{C}(X) : f(x_0) = 0\}$ for some fixed $x_0 \in X$. It follows that if $\varphi_x \in \Sigma(\mathscr{C}(X))$ is defined by $\varphi_x(f) = f(x)$, the map $x \mapsto \varphi_x$ is a continuous map of $X$ onto $\Sigma(\mathscr{C}(X))$. Since $\mathscr{C}(X)$ separates points on $X$ this map is a homeomorphism, and thus we have identified $\Sigma(\mathscr{C}(X))$ with $X$.

One consequence of this identification is the fact that $\mathscr{C}(X)$ and $\mathscr{C}(Y)$ are isomorphic, $X$ and $Y$ compact, if and only if $X$ and $Y$ are homeomorphic. Clearly if $\eta : X \to Y$ is a homeomorphism of $X$ onto $Y$, then the map $f \mapsto f \circ \eta$ is a homeomorphism of $\mathscr{C}(Y)$ onto $\mathscr{C}(X)$. Conversely, if $\Phi : \mathscr{C}(Y) \to \mathscr{C}(X)$ is an isomorphism, then $\Phi^* : \Sigma(\mathscr{C}(X)) \to \Sigma(\mathscr{C}(Y))$ is a homeomorphism and the result follows.

This last result raises an intriguing problem. Since $X$ and $Y$ are homeomorphic if and only if $\mathscr{C}(X)$ and $\mathscr{C}(Y)$ are algebraically isomorphic, it is reasonable to expect that every topological property of $X$ should have a faithful representation in some corresponding algebraic property of $\mathscr{C}(X)$. Some results in this direction are known. For example, it is clear that $\mathscr{C}(X)$ contains nontrivial idempotents if and only if $X$ is not connected. Other results are to be found in the thesis of Ahlberg [1]. A recent contribution is due to Watts [1], who shows how to compute the Čech cohomology with real coefficients of $X$ from $\mathscr{C}(X)$.

The Stone–Čech compactification of a completely regular space is an interesting application of the Banach algebra theory at our disposal.

## 6.4 *Example*

Let $X$ be a completely regular space and let $\mathscr{C}^B(X)$ be the Banach algebra of bounded complex-valued continuous functions on $X$ with the pointwise operations and the uniform norm $\|\cdot\|_X$.

We have a map $i:X \to \Sigma(\mathscr{C}^B(X))$ given in the usual way; we shall prove that it is a homeomorphism. It is one to one since $\mathscr{C}^B(X)$ separates points, and it is continuous. Suppose that $\{x_\alpha\}$ is a net in $X$ such that $\{ix_\alpha\}$ converges to $ix_0$, $x_0 \in X$. Thus, for all $f \in \mathscr{C}^B(X)$, $\lim_\alpha \hat{f}(ix_\alpha) = \hat{f}(ix_0)$, i.e., $\lim_\alpha f(x_\alpha) = f(x_0)$. Since this holds for all $f \in \mathscr{C}^B(X)$, it follows from the complete regularity of $X$ that $\lim_\alpha x_\alpha = x_0$, for if this latter limit relation does not hold, then there is a neighborhood $V$ of $x_0$ such that for all $\alpha$, there is $\beta > \alpha$ with $x_\beta \notin V$. Complete regularity provides a function $f_0 \in \mathscr{C}^B(X)$ with $f_0(x_0) = 0$, $f_0|(X \setminus V) = 1$. For this $f_0$, it is not true that $f(x_0) = \lim_\alpha f(x_\alpha)$ and we have a contradiction. Consequently, $i$ is a homeomorphism.

We next prove that $\Sigma(\mathscr{C}^B(X))$ is a compactification of $iX$, i.e., that $iX$ is dense in $\Sigma(\mathscr{C}^B(X))$. If it is not, there is $\varphi \in \Sigma(\mathscr{C}^B(X))$ not contained in $(iX)^-$. Since $(iX)^-$ is compact, there exists a finite family $\{f_1, \ldots, f_m\}$ in $\mathscr{C}^B(X)$, each $f_j$ nonnegative, such that $\varphi(f_j) = 0$ for all $j$ and such that $\sum_{j=1}^m f_j(x) \geq \varepsilon > 0$ for all $x \in X$ and some $\varepsilon$. Thus, if $g_p = f_p(\sum_{j=1}^m f_j)^{-1}$, then $\sum_{p=1}^m g_p = 1$ on $X$ but $\varphi(g_p) = 0$ for each $p$, and we have a contradiction. Thus, $iX$ is dense in $\Sigma(\mathscr{C}^B(X))$.

Now, suppose that $Y$ is a compact space and $\sigma: X \to Y$ is continuous. Then $f \mapsto f \circ \sigma$ is a continuous homomorphism from $\mathscr{C}(Y)$ to $\mathscr{C}^B(X)$, and so its adjoint is a continuous map $\sigma^*: \Sigma(\mathscr{C}^B(X)) \to \Sigma(\mathscr{C}(Y)) = Y$. We have $\sigma^* \circ i = \sigma$, and thus it follows from one of the well-known characterizations of the Stone–Čech compactification of a completely regular space (see Appendix A) that $\Sigma(\mathscr{C}^B(X))$ is the Stone–Čech compactification $\beta X$ of $X$.

As a further nontrivial example of the Gelfand transform, we shall consider $L^\infty$ as a Banach algebra.

## 6.5 *Example*

Let $X$ be a set, $\mathscr{F}$ a $\sigma$-algebra of subsets of $X$, and $\mu$ a positive, finite measure on $\mathscr{F}$. Thus, $(X, \mathscr{F}, \mu)$ is a totally finite measure space in the terminology of Halmos [1]. Denote by $L^\infty$ the space of equivalence classes of essentially bounded complex-valued measurable functions, and let $\|\cdot\|$ be the essential supremum norm on $L^\infty$. With this norm, $L^\infty$ is a commutative Banach algebra with identity.

We begin by noting that if $f \in L^\infty$, $\sigma(f)$ can be identified with the *essential range* of $f$:

$$\sigma(f) = \{z \in \mathbb{C} : \text{if } V \text{ is a neighborhood of } z, \mu(f^{-1}(V)) > 0\}.$$

This remark imples that the Gelfand transform carries $L^\infty$ isometrically into $\mathscr{C}(\Sigma(L^\infty))$, for if $f \in L^\infty$, then

$$\begin{aligned}\|f\| &= \text{essential supremum } |f| \\ &= \sup\{|z| : z \in \text{essential range of } f\} \\ &= \sup\{|z| : z \in \sigma(f)\} \\ &= \|\hat{f}\|_{\Sigma(L^\infty)}.\end{aligned}$$

Thus, $\hat{}$ is an isometry and consequently its range is closed. Next, $\hat{}$ preserves complex conjugations. To see this, let $f \in L^\infty$, $\varepsilon > 0$. There exist finitely many measurable sets $E_j \in \mathscr{F}$ such that if $\chi_{E_j}$ is the characteristic function of $E_j$, then for some constants $\alpha_j$, $\|f - \Sigma\alpha_j\chi_{Ej}\| < \varepsilon$. Thus, $\|\hat{f} - \Sigma\alpha_j\hat{\chi}_{E_j}\|_{\Sigma(L^\infty)} < \varepsilon$. Also $\|\bar{f} - \Sigma\bar{\alpha}_j\chi_{E_j}\| < \varepsilon$, so $\|\bar{f}\hat{} - \Sigma\bar{\alpha}_j\hat{\chi}_{E_j}\| < \varepsilon$. But $\Sigma\bar{\alpha}_j\hat{\chi}_{E_j} = (\Sigma\bar{\alpha}_j\chi_{E_j})\hat{}$, so we find that $\|\hat{f}^- - \bar{f}\hat{}\| < 2\varepsilon$. The number $\varepsilon$ is at our disposal, so we may conclude that $\hat{f}^- = \bar{f}\hat{}$, i.e., that $\hat{}$ preserves complex conjugations. Thus, $(L^\infty)\hat{}$ is a closed, conjugate closed point separating subalgebra of $\mathscr{C}(\Sigma(L^\infty))$, so, by the Stone–Weierstrass theorem, $(L^\infty)\hat{} = \mathscr{C}(\Sigma(L^\infty))$.

Since finite linear combinations of characteristic functions are dense in $L^\infty$, it follows that the functions $\chi_E$, $E \in \mathscr{F}$, constitute a set of generators for $L^\infty$. Consequently, the functions $\hat{\chi}_E$ separate points on $\Sigma(L^\infty)$, and we can prove that sets of the form

$$\{\varphi \in \Sigma(L^\infty) : \hat{\chi}_E(\varphi) = 0\} \tag{1}$$

form a basis for the topology of $\Sigma(L^\infty)$. To see this, let $\varphi_0 \in \Sigma(L^\infty)$, and let $V$ be an open neighborhood of $\varphi_0$. If $\psi \notin V$, there is $E$ such that $\hat{\chi}_E(\varphi_0) = 1$, $\hat{\chi}(\psi) = 0$. The essential range of $\chi_E$ is $\{0, 1\}$, so it follows that the sets $\{\varphi : \hat{\chi}_E(\varphi) = 0\}$ and $\{\varphi : \hat{\chi}_E(\varphi) = 1\}$ are both open and closed. The compactness of $\Sigma(L^\infty) \setminus V$ implies the existence of a finite collection $E_1, \ldots, E_n$ of elements of $\mathscr{F}$ such that if $F = E_1 \cap \cdots \cap E_n$, then

$$\hat{\chi}_F(\varphi_0) = 1 \qquad \text{and} \qquad \hat{\chi}_F|(\Sigma(L^\infty) \setminus V) = 0,$$

because $\chi_F = \Pi\{\chi_{E_j} : j = 1, \ldots, n\}$. Thus, $V$ contains the set $\{\varphi : \hat{\chi}_{X \setminus F}(\varphi) = 0\}$, and so sets of form (1) do constitute a basis for the topology of $\Sigma(L^\infty)$. Consequently, the space $\Sigma(L^\infty)$ is totally disconnected.

If $E \in \mathscr{F}$, the set $\hat{E}' = \{\varphi \in \Sigma(L^\infty) : \hat{\chi}_E(\varphi) = 1\}$ is an open and closed set in $\Sigma(L^\infty)$. The converse is also true, for if $S \subset \Sigma(L^\infty)$ is both open and closed, then $\chi_S$ is continuous on $\Sigma(L^\infty)$, and $\chi_S = \hat{f}$ for some $f \in L^\infty$. Since $\chi_S$ is idempotent, $f$ must be also, and thus $f = \chi_E$ for some $E \in \mathscr{F}$, and the assertion follows.

Call a compact space *extremally disconnected* if the closure of every open set is again open. The space $\Sigma(L^\infty)$ is extremally disconnected.

Let $S$ be open in $\Sigma(L^\infty)$. Then $S = \bigcup_\alpha \hat{E}_\alpha$ for some family $\{E_\alpha\}$ of measurable sets. Let

$$m = \inf\{\mu(F) : F \in \mathscr{F}, F \supset \bigcup_\alpha E_\alpha\}.$$

Since $\mathscr{F}$ is closed under the formation of countable intersections, it follows that for some $F \in \mathscr{F}$, $\mu(F) = m$, and $\hat{F} \supset \bigcup \hat{E}_\alpha$. We have that $\hat{F}$ is closed and that it contains $S$. If $\hat{F} \neq \bar{S}$, there is $\varphi \in \hat{F} \setminus \bar{S}$, and since $\hat{F}$ is also open, there is a measurable set $M$ with $\varphi \in \hat{M} \subset \hat{F} \setminus \bar{S}$. Since $\hat{M}$ is nonvoid, $\mu(M) > 0$, so $\mu(F \setminus M) < m$ and this contradicts the choice of $m$ and $F$. Thus, $\hat{F} = \bar{S}$, so $\bar{S}$ is both open and closed, and $\Sigma(L^\infty)$ is extremally disconnected.

Our discussion of $L^\infty$ has followed, more or less, that given by Hoffman [1]. See also Wang [1].

## 7 THE SHILOV BOUNDARY

In this section we shall develop some of the theory of certain *boundaries* which are associated with algebras of functions.

**7.1 *Definition***

*If $\mathscr{F}$ is a family of complex-valued functions on a set $X$, a subset $\Gamma$ of $X$ is called a* **boundary** *for $\mathscr{F}$ if for each $f \in \mathscr{F}$, there is $x \in \Gamma$ such that*

$$|f(x)| = \sup\{|f(y)| : y \in X\}.$$

As we shall see, the notion of boundary plays an important role in the theory of commutative Banach algebras.

**7.2 *Definitions***

*If $X$ is a locally compact space, a* **function algebra** *on $X$ is a subalgebra of $\mathscr{C}_0(X)$ which strongly separates the points of $X$ in the sense that if $x, y \in X$, $x \neq y$, then there exists an $f \in A$ with $0 \neq f(x) \neq f(y)$. If $X$ is compact, $A$ is assumed to contain the constants. A* **uniform algebra** *is a function algebra on a compact space that is uniformly closed.*

Since we shall treat uniform algebras in considerable detail, it is convenient to introduce special notation for them. We shall use expressions such as "$(A, X)$ is a uniform algebra" and understand thereby that $X$ is a compact space and $A$ a uniform algebra on $X$. This convention will serve to abbreviate the formulation of many theorems, etc. We do not require, in this notation, that $X$ be the spectrum of $A$.

The most obvious example of a function algebra is the algebra $\hat{A}$ of Gelfand transforms on the spectrum $\Sigma(A)$ of a commutative Banach algebra.

For future reference, let us note that if $(A, X)$ is a uniform algebra, so is $(\hat{A}, \Sigma(A))$, for since $A$ is uniformly closed in $\mathscr{C}(X)$ and is provided with the supremum norm, the Gelfand transform on $A$ is an isometry: If $f \in A$ and $z \in f(X)$, then $f - z$ is not invertible, so $z \in \sigma(f)$. Consequently, $\|f\|_X \leq \|\hat{f}\|_{\Sigma(A)}$. From general principles we know that $\|\hat{f}\|_{\Sigma(A)} \leq \|f\|_X$, so these two norms are, in fact, equal, and $\hat{A}$ is seen to be a uniform algebra on $\Sigma(A)$.

We shall need the following topological fact.

7·3 ***Lemma***

*If $A$ is a function algebra on the locally compact space $X$, then the weak topology on $X$ defined by $A$ agrees with the given topology on $X$.*

Recall that the weak topology on $X$ defined by $A$ is the topology in which a net $\{x_\alpha\}$ in $X$ converges to $x_0 \in X$ if and only if $\{f(x_\alpha)\}$ converges to $f(x_0)$ for all $f$ in $A$.

PROOF

It is clear that if the net $\{x_\alpha\}$ in $X$ converges in the given topology of $X$ to $x_0$, then it converges in the weak topology induced by $A$ to $x_0$, for the elements of $A$ are continuous.

Suppose, conversely, that $\{f(x_\alpha)\}$ converges, for all $f \in A$, to $f(x_0)$, $x_0 \in X$, but assume, to derive a contradiction, that $\{x_\alpha\}$ does not converge. There are two alternatives. Either $\{x_\alpha\}$ has a subnet which converges to a point $y_0 \in X$, $y_0 \neq x_0$, or else it has a subnet which converges to the point at infinity in the customary sense that it is eventually off any given compact set in $K$. If the subnet $\{x_{\alpha_i}\}$ converges to $y_0 \neq x_0$, and if $f \in A$ satisfies $f(x_0) \neq f(y_0)$, we have a contradiction to the assumption that $\lim f(x_\alpha) = f(x_0)$ for all $f \in A$. On the other hand, if the subnet $\{x_{\alpha_i}\}$ converges to infinity, and if $f \in A$ does not vanish at $x_0$, then since $f$ vanishes at infinity, we have $0 = \lim_i |f(x_{\alpha_i})| \neq |f(x_0)|$, which contradicts the assumption that $f(x_\alpha) \to f(x_0)$. This concludes the proof.

7·4 ***Theorem***

*If $A$ is a function algebra on the locally compact space $X$, then there exists a unique minimal closed boundary for $A$.*

There do exist closed boundaries for $A$; the set $X$ itself is one. We shall establish the theorem by showing that $\Gamma$, the intersection of all closed boundaries for $A$, is itself a boundary for $A$.

The proof depends on a lemma.

7·5 ***Lemma***

*Let $f_1, \ldots, f_m \in A$ and set $V = \{x \in X : |f_j(x)| < 1, 1 \leq j \leq m\}$. Then either $V$ meets every closed boundary for $A$ or else for every closed boundary $M_0$ for $A$, the set $M_0 \setminus V$ is a closed boundary.*

PROOF

For any closed set $E \subset X$, the set $E \setminus V$ is closed. Suppose that $\Gamma_0$ is a closed boundary but that $\Gamma_0 \setminus V$ is not one. Thus, there is $f \in A$ such that $\|f\|_X = 1$, and $\|f\|_{\Gamma_0 \setminus V} < 1$. Replacing $f$ by a high power if necessary, we may suppose that $\|f\|_{\Gamma_0 \setminus V} < \delta$, where $\delta$ is so small that $\delta \|f_j\|_X < 1$ for each $j$. By the choice of $f$ and the definition of $V$, we have $|ff_j| < 1$ on $\Gamma_0 \setminus V$ and also on $V$. Since $\Gamma_0$ is a boundary for $A$, it follows that $|ff_j| < 1$ on the whole of $X$. Consequently, every point at which $|f|$ assumes the value one must be contained in $V$, and therefore $V$ meets every boundary for $A$.

PROOF OF THE THEOREM

Assume, to begin with, that $A$ contains the constants so that $X$ is compact, and let $\Gamma$ be the intersection of all closed boundaries for $A$. Thus, $\Gamma$ is the set of all points $x \in X$ such that given an open set $V$ containing $x$, there is $f \in A$ with $\|f\|_{X \setminus V} < \|f\|_V$. Let $f \in A$ satisfy $\|f\|_\Gamma < 1$. If $\|f\|_X > 1$, the set $E = \{x \in X : |f(x)| \geq 1\}$ is a nonvoid compact set, and if $x_0 \in E$, the definition of $\Gamma$ implies the existence of a closed boundary $\Gamma_0$ not containing $x_0$. For some choice of functions $f_j$ in $A$, the neighborhood $V_{x_0} = \{x \in X : |f_1(x)| < 1, \ldots, |f_n(x)| < 1\}$ will contain $x_0$ and miss $\Gamma_0$—here we use Lemma 7.3. Finitely many of the neighborhoods $V_x$ cover $E$, say $V_1, \ldots, V_q$. But then by using Lemma 7.5 $q$ times, we see that $X \setminus \bigcup V_j$ is a closed boundary for $A$. Since $|f| < 1$ on $X \setminus \bigcup V_j$, it follows that $\|f\|_X < 1$, and so we have the result.

If $A$ does not contain the constants, adjoin to $A$ an identity to obtain $A'$ regarded in the natural way as a function algebra on $X^*$, the one-point compactification of $X$. We shall show that if $\Gamma'$ is the Shilov boundary for $A'$, then $\Gamma = \Gamma' \cap X$ is the Shilov boundary for $A$. The set $\Gamma$ is plainly a closed boundary for $A$. If $x \in \Gamma$, let $V$ be a neighborhood of $x$ in $X$. There is $f \in A'$ with $1 = f(y) = \|f\|_X$ for some $y \in V$ but with $|f| < \frac{1}{2}$ on $X \setminus V$. As $y \neq \infty$, there is $g \in A$ with $g(y) \neq 0$. Then $gf^N \in A$

for all $N$ since $A$ is an ideal in $A'$. For sufficiently large $N$, $gf^N$ achieves its maximum only on $V$, so $\Gamma$ is the minimal closed boundary for $A$.

The boundary $\Gamma$ of the theorem is called the *Shilov boundary for A* because it was Shilov who first demonstrated its existence. Shilov's proof depends on transfinite induction or a similar technique. The proof we have given is due to Hörmander [1]. Bear [3] has given an elementary proof of the existence of a Shilov boundary for a point separating linear space of continuous functions on a compact space, and Arens and Singer [1] have proved that certain multiplicative semigroups of continuous functions possess such boundaries. Another reference for related matters is Bauer [1].

If $A$ is a function algebra, $\Gamma(A)$ will denote its Shilov boundary.

There are certain points which are surely contained in the Shilov boundary for $A$, the *strong boundary points.*

**7.6** ***Definition***

*A point $x \in X$ is a* ***strong boundary point*** *for A if for every neighborhood $V$ of $x$ there is an $f \in A$ such that $\|f\|_X = f(x) = 1$, and $|f(y)| < 1$ if $y \in X \setminus V$. The point $x$ is a* ***peak point*** *for A if there is $f \in A$ such that $f(x) = 1$, and $|f(y)| < 1$ for $y \in X \setminus \{x\}$.*

In general there will be no peak points for a function algebra on a locally compact space $X$ because a peak point is necessarily a $G_\delta$, and in general $X$ need contain no points of type $G_\delta$. We shall see later that strong boundary points exist in great abundance, at least for uniform algebras.

Before taking up the theory of the Shilov boundary, let us consider some examples.

**7.7** ***Example***

The motivating example for the notion of Shilov boundary arises in connection with the disc algebra, $\mathscr{A}$. In this case the maximum modulus theorem shows that the Shilov boundary is contained in the boundary of the unit disc, and the function $f_0 \in \mathscr{A}$ given by

$$f_0 = \tfrac{1}{2}(1 + ze^{-i\theta})$$

peaks at $e^{i\theta}$, so the Shilov boundary must be the whole boundary of the disc. In this case, every point of the Shilov boundary is a strong boundary point. This is not the general phenomenon, as the next example shows.

7.8 ***Example***

Let $X = \{(z,t): z \in \mathbb{C},\ t \in \mathbb{R},\ |z|, |t| \leq 1\}$, and let $A = \{f \in \mathscr{C}(X): f(z,0)$ depends holomorphically on $z$, $|z| < 1\}$. In this case, the point $(z,t)$ of $X$ is a peak point for $A$ as long as $t \neq 0$, or if $t = 0$, as long as $|z| = 1$. However, the points $(z,0)$ with $|z| < 1$ are not peak points as follows from the maximum modulus principle. In this case, $X$ is the Shilov boundary for $A$.

7.9 ***Example***

Let $\mathbb{C}^*$ denote the Riemann sphere, and, if $E$ is a compact set in the plane, let

$$\mathscr{A}_E = \{f \in \mathscr{C}(\mathbb{C}^*): f \text{ is holomorphic off } E\}.$$

If the set $E$ is too small, $\mathscr{A}_E$ will reduce to the constants, but as soon as $\mathscr{A}_E$ contains a single nonconstant function, it contains three functions which separate points. To see this, let $\psi_1 \in \mathscr{A}_E$ be nonconstant on $\mathbb{C} \setminus E$. Since $\mathscr{A}_E$ contains the constants, we may suppose that $\psi_1(\infty) = 0$. Set $\psi_2(z) = z\psi_1(z)$. Finally, choose a point $c \in \mathbb{C} \setminus E$ at which $\psi_1$ does not vanish and set $\psi_3(z) = [\psi_1(z) - \psi_1(c)]/z - c$. This function is in $\mathscr{A}_E$ since $\psi_1$ is differentiable at $c$. The functions $\psi_1$ and $\psi_2$ separate points off the set $\psi_1^{-1}(0)$, and on this set $\psi_3$ separates points. Thus, $\mathscr{A}_E$ is a uniform algebra on $\mathbb{C}^*$ if it contains a single nonconstant function. Under the assumption that $\mathscr{A}_E$ contains nonconstant functions, the maximum modulus theorem shows that the Shilov boundary for $\mathscr{A}_E$ is contained in $E$. In this way, it is possible to construct nontrivial examples of function algebras with prescribed compact, planar sets as Shilov boundary.

Suppose that the set $E$ is of finite, positive two-dimensional Lebesgue measure, and define $\psi$ by

$$\psi(z) = \int_E \frac{du\,dv}{z - w} \qquad w = u + iv.$$

The function $\psi$ is holomorphic outside the set $E$, and we shall prove that it is continuous on $\mathbb{C}^*$. Observe that $\lim_{z \to \infty} z\psi(z)$ is just $\int_E dx\,dy \neq 0$, so $\psi$ is nonconstant. To prove that $\psi$ is continuous on $\mathbb{C}$, define $L$ by $L(z) = 1/z$ if $|z| \leq 2R$, $L(z) = 0$ if $|z| > 2R$, where $R$ is chosen so that $E \subset \{z: |z| < R\}$. The function $L$ is Lebesgue integrable on the plane, so for $\varepsilon > 0$ there exists a continuous function $M_\varepsilon$ with compact support so that

$$\int_{\mathbb{C}} |M_\varepsilon - L| < \varepsilon.$$

Consider $z_1, z_2 \in \mathbb{C}$ such that $|z_1|, |z_2| < R$. We have

$$|\psi(z_1) - \psi(z_2)| \leq \int_{\mathbb{C}} |L(z_1 - w) - L(z_2 - w)| \, du \, dv.$$

If $\varepsilon > 0$ is given, we have

$$\begin{aligned} |\psi_1(z_1) - \psi(z_2)| \leq & \int_{\mathbb{C}} |L(z_1 - w) - M_\varepsilon(z_1 - w)| \, du \, dv \\ & + \int_{\mathbb{C}} |M_\varepsilon(z_1 - w) - M_\varepsilon(z_2 - w)| \, du \, dv \\ & + \int_{\mathbb{C}} |M_\varepsilon(z_2 - w) - L(z_2 - w)| \, du \, dv \\ \leq & \, 2\varepsilon + \int_{\mathbb{C}} |M_\varepsilon(z_1 - w) - M_\varepsilon(z_2 - w)| \, du \, dv. \end{aligned}$$

Now $M_\varepsilon$ is uniformly continuous on $\mathbb{C}$ and has compact support, so if $|z_1 - z_2|$ is small enough, the last integral is no more than $\varepsilon$. Thus, for all $z_1$ near enough $z_2$, $|\psi(z_1) - \psi(z_2)| < 3\varepsilon$, i.e., $\psi$ is continuous in $\{z : |z| < R\}$, and, thus, $\psi$ is continuous. (Observe that this argument is just a special case of the fact that the convolution of a bounded function and an integrable function is continuous.)

Suppose now that $E$ is *locally* of positive measure in the sense that if $z \in E$, and $V$ is a neighborhood of $z$, then $V \cap E$ has positive planar measure. Then, if $z \in E$ and $V$ is a neighborhood of $z$, $\mathscr{A}_{\bar{V} \cap E}$ contains nonconstant functions and is contained in $\mathscr{A}_E$. Thus, $\mathscr{A}_E$ contains elements which achieve their maximum modulus only in $\bar{V}$, and it follows that the Shilov boundary for $\mathscr{A}_E$ coincides with the set $E$.

There exist, for example, sets of Cantor type that are locally of positive measure, and it is possible to construct arcs that have the same property. (See Appendix A.)

We can show that for every $f \in \mathscr{A}_E$, $f(E) = f(\mathbb{C}^*)$. To do this, it suffices to prove that if $f \in \mathscr{A}_E$ does not assume the value zero on $E$, then it does not assume the value zero on $\mathbb{C}^*$. By analyticity, an $f \in \mathscr{A}_E$ without zeros on $E$ can have only finitely many zeros in the finite plane, say $z_1, \ldots, z_q$. The function $h$ defined by $h(z) = f(z) \prod_{j=1}^{q} (z - z_j)^{-1}$ is in $\mathscr{A}_E$ and is zero free in the finite plane. It follows that there is a branch of $\log h$ defined in $\mathbb{C}$, but this is impossible since $h$ vanishes at infinity. Thus, if $E$ is an arc or a Cantor set, $\mathscr{A}_E$ is an algebra of dimension raising functions.

One further step in this direction is of interest. Let $E \subset \mathbb{C}$ be a Cantor set of locally positive measure, let $f_1$ be any nonconstant function in $\mathscr{A}_E$ which vanishes at infinity, and let $f_2(z) = zf_1(z)$. If $E_0 = \{z \in E : f_1(z) = 0\}$, the two functions $f_1$ and $f_2$ separate points on $E \setminus E_0$, and under the map $z \mapsto (f_1(z), f_2(z))$, $E$ goes onto a set $E'$ in $\mathbb{C}^2$ which contains $(0, 0)$ and which we can show to be a Cantor set. It is clear that $E'$ is compact and totally disconnected and that it has the property that all its points are limit points with the possible exception of the origin. The point $(0, 0)$ will fail to be a limit point of $E'$ if and only if $E \cap f_1^{-1}(0)$ is relatively open in $E$. But if $E \cap f_1^{-1}(0)$ is relatively open in $E$, there is connected open set $\Omega$ in $\mathbb{C}$ with $\Omega \cap E = f_1^{-1}(0) \cap E$. The function $f_1$ is continuous on $\Omega$ and holomorphic on $\Omega \setminus f_1^{-1}(0)$, so by a theorem of Radó which we shall prove later (Theorem 25.20), $f_1$ is holomorphic throughout $\Omega$. Since $E$ is of locally positive measure, $E \cap \Omega$ is uncountable, so $f_1|\Omega$ must be the zero function, and this leads to a contradiction to our choice of $f_1$. Thus, $E \cap f_1^{-1}(0)$ is not relatively open in $E$, and $(0, 0)$ is therefore a limit point of $E'$. It follows that $E'$ is a Cantor set. We may regard $E'$ as the decomposition space obtained from $E$ by collapsing $E_0$ to a point.

The subalgebra of $\mathscr{A}_E$ generated by $f_1$ and $f_2$ is isomorphic to $\mathscr{P}(E')$, the uniform closure of the polynomials in two variables on $E'$; it is a doubly generated algebra of dimension raising functions.

In connection with these examples, see the papers of Rudin [5] and Wermer [3].

It is true, though nontrivial, that if $\mathscr{A}_E$ contains nonconstant functions, then $\Sigma(\mathscr{A}_E) = C^*$. (See Theorem 24.6.)

The precise conditions on $E$ which ensure that $\mathscr{A}_E$ is nontrivial seem to be unknown. Arens [1] gives sufficient conditions weaker than being of positive area. For an exposition of several matters related to this question, see Zalcman [2].

## 7.10 *Example*

Let $B_N = \{\mathfrak{z} = (z_1, \ldots, z_N) \in \mathbb{C}^N : \|\mathfrak{z}\| = |z_1|^2 + \cdots + |z_N|^2 \leq 1\}$, and let $\mathscr{P}(B_N)$ be as in Example 1.3. The Shilov boundary for $\mathscr{P}(B_N)$ can be identified with the $(2N - 1)$-sphere which constitutes the boundary of $B_N$: If $\mathfrak{z}^0 = (z_1^0, \ldots, z_N^0)$ is a point of $\partial B_N$, then the function $f$ given by

$$f(z_1, \ldots, z_N) = \Sigma\, z_j \bar{z}_j^0$$

is an element of $\mathscr{P}(B_N)$ which is bounded in modulus by 1 and which assumes the value 1 only at the point $\mathfrak{z}^0$. Thus, $\mathfrak{z}^0$ is the peak point for $\frac{1}{2}(1 + f)$, so $\mathfrak{z}^0$ lies in the Shilov boundary for $\mathscr{P}(B_N)$. It follows from the maximum modulus theorem that the Shilov boundary for $\mathscr{P}(B_N)$ is in fact $\partial B_N$.

If instead of $B_N$, we consider $\bar{\Delta}^N$, the product of $N$ copies of the closed unit disc, and the associated algebra $\mathscr{P}(\bar{\Delta}^N)$, then the Shilov boundary is found not to be the topological boundary of $\bar{\Delta}^N$ but rather the torus $T^N = \partial\bar{\Delta} \times \cdots \times \partial\bar{\Delta}$. To see this, let $\mathfrak{z} = (e^{i\theta_1}, \ldots, e^{i\theta_N}) \in T^N$, and set

$$f(z_1, \ldots, z_N) = \prod_{j=1}^{N} \tfrac{1}{2}(1 + z_j e^{-i\theta_j}).$$

The function $f$ peaks at $\mathfrak{z}$, so $T^N$ is contained in the Shilov boundary for $\mathscr{P}(\bar{\Delta}^N)$. And, if $f \in \mathscr{P}(\bar{\Delta}^N)$, and if $(z_1, \ldots, z_N) \in \bar{\Delta}^N$, then since for fixed $z_2, \ldots, z_N, f(\cdot, z_2, \ldots, z_N)$ is an element of the disc algebra, there is a $\theta_1$ such that

$$|f(e^{i\theta_1}, z_2, \ldots, z_N)| \geq |f(z_1, \ldots, z_N)|.$$

Fixing $z_3, \ldots, z_N$, there is a $\theta_2$ such that

$$|f(e^{i\theta_1}, e^{i\theta_2}, z_3, \ldots, z_N)| \geq |f(e^{i\theta_1}, z_2, \ldots, z_N)|.$$

Iterating this process, we finally find $\theta_1, \ldots, \theta_N$ such that

$$|f(e^{i\theta_1}, \ldots, e^{i\theta_N})| \geq |f(\mathfrak{z})|.$$

Consequently, $T^N$ is the Shilov boundary for $\mathscr{P}(\bar{\Delta}^N)$.

For various other examples of Shilov boundaries we cite the following references: Gelfand, Raikov, and Shilov [1, 2]; Nelson [1]; Epe [1]; Hua [1]; Rickart [1]; and Fuks [1].

If $A$ is a commutative Banach algebra, then $\hat{A}$ is a function algebra on $\Sigma(A)$. We shall refer to the Shilov boundary for $\hat{A}$ as *the Shilov boundary for A*. We shall often denote it by $\Gamma(A)$.

We now turn to the investigation of some of the general properties of the Shilov boundary.

7.11 ***Theorem***

*Let $A$ be a commutative Banach algebra with identity, and let $\{a_\lambda\}_{\lambda\in\Lambda}$ be a set of generators for $A$. If $\Phi:\Sigma(A) \to \mathbb{C}^\Lambda$ is the canonical embedding of $\Sigma(A)$ into $\mathbb{C}^\Lambda$ determined by the given set of generators, then under $\Phi$, $\Gamma(A)$ goes onto the minimal closed set whose polynomially convex hull is $\Phi(\Sigma(A))$.*

PROOF

Let $\Gamma' = \Phi(\Gamma(A))$, $\Sigma' = \Phi(\Sigma(A))$. We are to prove that $(\Gamma')\hat{}$, the polynomially convex hull of $\Gamma'$, is $\Sigma'$, and that $\Gamma'$ is minimal with respect to this property. We know that $\Sigma'$ is polynomially convex, so it follows that $(\Gamma')\hat{} \subset \Sigma'$. Let $\mathfrak{z} \in \Sigma'$. We can define a functional $\varphi$ on $\mathscr{P}_\Lambda$ by $\varphi(p) = p(\mathfrak{z})$ for all polynomials $p$. We have the canonical homomorphism $\Psi : \mathscr{P}_\Lambda \to A$ determined by $\Psi(\pi_\lambda) = a_\lambda$ for each coordinate projection $\pi_\lambda$. Then $|\varphi(p)| \leq \|\Psi(p)\|_\Sigma \leq \|\Psi(p)\|_\Gamma = \|p\|_{\Gamma'}$. As this holds for all $p \in \mathscr{P}_\Lambda$, we have $\mathfrak{z} \in (\Gamma')\hat{}$, so $(\Gamma')\hat{} = \Sigma'$.

Suppose that $E'$ is a proper closed subset of $\Gamma'$. Choose $\mathfrak{z} \in \Gamma' \setminus E'$, and let $V'$ be a neighborhood of $\mathfrak{z}$ which misses $E'$. Then $V = \Phi^{-1}(V')$ is a neighborhood of $\varphi_0 = \Phi^{-1}(\mathfrak{z})$ which misses $E = \Phi^{-1}(E')$. Consequently, there is an $a \in A$ such that $\|\hat{a}\|_\Sigma = 1$, $\|\hat{a}\|_{\Sigma \setminus V} < 1$. Since $\{a_\lambda\}$ is a set of generators for $A$, we may suppose that $a = \Psi(p)$, $p \in \mathscr{P}_\Lambda$. But then $\|p\|_{E'} < 1$, $\|p\|_{V'} = 1$, so there are points of $\Sigma'$ which do not lie in $(E')\hat{}$.

Thus, we have established that $(\Gamma')\hat{} = \Sigma'$ and that $\Gamma'$ is minimal with respect to this property. The existence of a *unique* closed subset of $\Sigma'$ minimal with respect to this property follows from the fact that $\hat{A}$ has a unique minimal closed boundary.

7.12 ***Theorem***

*Let $A$ be a commutative Banach algebra with identity. If $a \in A$, then $\{\hat{a}(\varphi) : \varphi \in \Sigma(A)\}$ is contained in $\{\hat{a}(\varphi) : \varphi \in \Gamma(A)\}\hat{}$.*

PROOF

If $z_0 \in \mathbb{C} \setminus \{\hat{a}(\varphi) : \varphi \in \Gamma(A)\}\hat{}$, then there is a polynomial $p$ in a single variable such that $|p(z_0)| > |p(\hat{a}(\varphi))|$ for all $\varphi \in \Gamma(A)$. Suppose that $p(z) = \sum_{j=0}^{M} \beta_j z^j$. If $z_0 = \hat{a}(\varphi_0)$, and if $b = \Sigma \beta_j a^j$, then $\varphi_0(b) = p(z_0)$, and we find that $|\hat{b}(\varphi_0)| > |\hat{b}(\varphi)|$ for all $\varphi \in \Gamma(A)$. This is impossible, so $z_0 \neq \hat{a}(\varphi_0)$ for all $\varphi_0 \in \Sigma(A)$, and we have the result.

7.13 ***Theorem***

*Let $A$ be a function algebra on $X$, and let $I \subset A$ be an ideal in $A$. Set $E = \{x \in X : f(x) = 0$ for all $f \in I\}$. Then $\Gamma(I) = \Gamma(A) \setminus E$.*

PROOF

Since $I$ is an ideal in $A$, $I$ is a function algebra on the locally compact space $X \setminus E$, so our assertion is meaningful in a natural way.

Having noted this, we recognize that the argument used in the last part of the proof of Theorem 7.4 applies with only trivial modification to cover the present situation.

7.14 ***Corollary***

*With $A$, $I$, and $E$ as in the theorem,* $\Gamma(I) \subseteq \Gamma(A) \subseteq \Gamma(I) \cup E$.

Simple examples show that either of these inclusions can be strict.

No comparably simple characterization of the Shilov boundary for a quotient algebra exists.

One of the principal reasons for the importance of the Shilov boundary is contained in our next result.

**7.15 *Theorem***

*If $A$ is a function algebra on the locally compact space $X$, then for each $x \in X$, there exists a positive regular Borel measure $\mu$ on $\Gamma(A)$ which represents $x$ in the sense that*

$$f(x) = \int_\Gamma f\, d\mu$$

*for all $f \in A$. If $A$ contains the constants, $\mu$ has total mass 1.*

PROOF

Assume that $X$ is compact so that by standing assumption $A$ contains the constants. We may suppose that $A$ is uniformly closed. The map $f|\Gamma \mapsto f(x)$ is a continuous linear functional of norm 1 on the closed subalgebra $\{f|\Gamma : f \in A\}$ of $\mathscr{C}(\Gamma)$. By the Hahn–Banach theorem, there is a continuous linear functional of norm 1 on $\mathscr{C}(\Gamma)$ which extends this functional. By the Riesz representation theorem there is a regular Borel measure $\mu$ of total variation one on $\Gamma$ such that

$$\varphi(g) = \int_\Gamma g\, d\mu$$

for all $g \in \mathscr{C}(\Gamma)$. In particular, then,

$$f(x) = \int_\Gamma f\, d\mu$$

for all $f \in A$. Since $\mu$ is of total variation 1 and since $\int_\Gamma 1\, d\mu = 1$, $\mu$ must be a positive measure.

In the event that $A$ has no identity and so $X$ is noncompact, let $A'$ be the algebra obtained by adjoining to $A$ an identity so that $A'$ is, in a natural way, a function algebra on $X^*$, the one-point compactification of $X$. Given $x \in X$, the last paragraph provides a measure $\mu$ such that $f(x) = \int_{\Gamma(A')} f\, d\mu$ for all $f \in A'$. Since $\Gamma(A) \subseteq \Gamma(A') \subseteq \Gamma(A) \cup \{\infty\}$, and since if $f \in A$, then $f(\infty) = 0$, it follows that for $f \in A$,

$$f(x) = \int_{\Gamma(A)} f\, d\mu,$$

as asserted.

In the sequel we shall use the term *probability measure* to describe a positive measure of total mass 1.

Let us point out that our calculations do not prove that the measure $\mu$ is a probability measure if $A$ has no identity. It does not seem obvious that in the noncompact case, every point can be represented by a measure of total mass 1. In fact, however, such a representation is possible, as we shall see in Section 11.

An especially important case of the theorem is that in which $A$ is the algebra of Gelfand transforms of a Banach algebra:

**7.16** ***Corollary***

*Let $A$ be a commutative Banach algebra, and let $\varphi \in \Sigma(A)$. There exists a positive measure $\mu$ on $\Gamma(A)$ such that for all $a \in A$,*

$$\varphi(a) = \int_{\Gamma(A)} \hat{a}\, d\mu.$$

**7.17** ***Example***

An interesting example of this theorem occurs in connection with the disc algebra $\mathscr{A}$. If $f \in \mathscr{A}$ and $z = \rho e^{i\theta}$, $|\rho| < 1$, then by the Poisson integral formula,

$$f(\rho e^{i\theta}) = \frac{1}{2\pi} \int_{-\pi}^{\pi} f(e^{it}) \frac{1 - \rho^2}{1 - 2\rho \cos(t - \theta) + \rho^2}\, dt.$$

Thus, the measure $m_z$ given by

$$dm_z = \frac{1}{2\pi} \frac{1 - \rho^2}{1 - 2\rho \cos(t - \theta) + \rho^2}\, dt$$

is a representing measure for the point $\rho e^{i\theta}$. It is easy to see, on the basis of the uniqueness part of the Riesz representation theorem, that $m_z$ is the only representing measure for $z$ on the Shilov boundary for $\mathscr{A}$, i.e., on $T$, the unit circle. To see this, suppose that $\mu$ is a positive measure on $T$ such that

$$f(z) = \int_T f\, d\mu$$

for all $f \in \mathscr{A}$. Since $\mu$ is real, $\int \operatorname{Re} f\, d\mu = \operatorname{Re} f(z)$. The functions $\operatorname{Re} f$, $f \in \mathscr{A}$, are dense in $\mathscr{C}_{\mathbb{R}}(T)$, for if $g \in \mathscr{C}_{\mathbb{R}}(T)$, let $u_g$ be the real harmonic function on the disc which assumes the boundary values $g$ continuously, and let $v_g$ be its harmonic conjugate. If

$$h_r(z) = u_g(rz) + iv_g(rz), \qquad r \in (0, 1),$$

then $h_r \in \mathscr{A}$, and for $r$ near 1, $\operatorname{Re} h_r$ is uniformly near $g$ on $T$. Thus, the measures $\mu$ and $m_z$ agree on a dense subspace of $\mathscr{C}_{\mathbb{R}}(T)$, so $\mu = m_z$, and we have the uniqueness assertion.

The disc algebra is atypically simple in that every point admits a unique representing measure. In general a point in the spectrum of a uniform algebra will admit myriads of representing measures. Algebras for which representing measures are unique have many striking properties, as we shall see later.

It is of interest to point out that the Cauchy integral

$$f(z) = \frac{1}{2\pi i} \int_{|\zeta| = 1} f(\zeta) \frac{d\zeta}{\zeta - z}$$

does not yield a representing measure for the point $z$ in the sense of our definition, for the measure $(1/2\pi i)(\zeta - z)^{-1}\, d\zeta$ is not a *positive* measure.

Representing measures play an important role in the theory of strong boundary points, as our next result shows.

7.18 ***Theorem***

*If $(A, X)$ is a uniform algebra, and if $x \in X$, then the following conditions are equivalent:*

(a) *If $V$ is a neighborhood of $x$, there is $f \in A$ with $\|f\|_X = 1$, $|f(x)| > \frac{3}{4}$, and $\|f\|_{X \setminus V} < \frac{1}{4}$.*
(b) *The point $x$ is a strong boundary point for $A$.*
(c) *The only representing measure for $x$ is $\delta_x$, the unit point mass at $x$.*

PROOF

(a) implies (b). We let $V$ be a neighborhood of $x$ and we shall construct a function $g \in A$ such that $g(x) = 1 = \|g\|_X$, but $\|g\|_{X \setminus V} < 1$. For this purpose we begin by constructing inductively a sequence of neighborhoods $V_1 = V \supset V_2 \supset \cdots$ of $x$, and a sequence $g_1, g_2, \ldots$ of elements of $A$ such that

(1) $$g_k(x) = 1,$$

(2) $$\|g_k\| \leq \tfrac{4}{3},$$

(3) $$\|g_k\|_{X \setminus V_k} < \tfrac{1}{3},$$

and

(4) $$\|g_j\|_{V_k} < 1 + 4^{-2k} \qquad \text{if } j < k.$$

For this construction we take $V_1 = V$. By (a), there is $f_1 \in A$ such that $\|f_1\|_X \leq 1$, $\|f_1\|_{X \setminus V_1} < \frac{1}{4}$, and $f_1(x)$ is real and at least $\frac{3}{4}$. We take

$g_1 = (f_1(x))^{-1}f_1$. Suppose now that $V_1, \ldots, V_k$ and $g_1, \ldots, g_k$ have been constructed and satisfy conditions (1) through (4). Define $V_{k+1}$ by

$$V_{k+1} = \{y \in V_1 \cap \cdots \cap V_k : |g_j(y)| < 1 + 4^{-2(k+1)} \text{ for } 1 \leq j \leq k\}.$$

Condition (a) of the theorem provides a function $f_{k+1}$ with $\| f_{k+1} \|_X = 1$, $f_{k+1}(x) > \frac{3}{4}$, and $|f_{k+1}(y)| < \frac{1}{4}$ if $y \notin V_{k+1}$. Define $g_{k+1}$ to be $(f_{k+1}(x))^{-1}f_{k+1}$. In this way we obtain the desired sequences $\{V_k\}$ and $\{g_k\}$. We define the function $g$ by

$$g = \sum_{k=1}^{\infty} 2^{-k} g_k.$$

By (1), $g(x) = 1$, and by (3) and the fact that $\{V_k\}$ is a decreasing sequence with $V_1 = V$, it follows that $|g(y)| < 1$ if $y \notin V$. Item (4) implies that if $y \in \bigcap V_k$, then $|g(y)| \leq 1$. Finally, if $y$ is in $V_N$ but not $V_{N+1}$, then from

$$g(y) = \sum_{k=1}^{N-1} 2^{-k} g_k(y) + 2^{-N} g_N(y) + \sum_{k=N+1}^{\infty} 2^{-k} g_k(y)$$

we have the estimate

$$\begin{aligned} |g(y)| &\leq (1 + 4^{-2N})(1 - 2^{-N+1}) + 2^{-N} \cdot \tfrac{4}{3} + 2^{-N} \cdot \tfrac{1}{3} \\ &< 1. \end{aligned}$$

The function $g$ has the announced properties, and we may conclude that $x$ lies in the strong boundary for $A$.

Next, (b) implies (c). Let $x$ lie in the strong boundary for $A$, and let $\mu$ be a representing measure for $x$. Regularity implies that to show $\mu = \delta_x$, it suffices to show that if $V$ is a neighborhood of $x$, then $\mu(V) = 1$. If $V$ is a neighborhood of $x$, choose $f \in A$ such that $f(x) = 1 = \| f \|$ and $\| f \|_{X \setminus V} < 1$. We can suppose that $|f(y)| = 1$ implies $f(y) = 1$. Then for all $n$,

$$1 = f^n(x) = \int f^n \, d\mu.$$

By the dominated convergence theorem, $\lim_n \int f^n \, d\mu = \mu\{y : f(y) = 1\}$, so $\mu(V) = 1$.

The proof of the theorem is concluded by showing that (c) implies (a), a fact that follows from the following special case of a theorem of Bauer [1].

7.19 ***Lemma***

*If $(B, Y)$ is a uniform algebra and if $\varphi \in \Sigma(B)$, define $\underline{Q}_\varphi(u)$ and $\bar{Q}_\varphi(u)$ for every $u \in \mathscr{C}_\mathbb{R}(Y)$ by*

$$\underline{Q}_\varphi(u) = \sup\{\operatorname{Re} \varphi(f) : f \in B, \operatorname{Re} f \leq u\}$$

*and*

$$\bar{Q}_\varphi(u) = \inf\{\operatorname{Re} \varphi(f) : f \in B, \operatorname{Re} f \geq u\}.$$

*If $\mu \in \mathscr{M}_\varphi$, $\mathscr{M}_\varphi$ is the set of representing measures for $\varphi$, then for every $u \in \mathscr{C}_\mathbb{R}(Y)$,*

$$\underline{Q}_\varphi(u) \leq \int u\, d\mu \leq \bar{Q}_\varphi(u).$$

*Moreover, if $\gamma \in \mathbb{R}$ satisfies*

$$\underline{Q}_\varphi(u) \leq \gamma \leq \bar{Q}_\varphi(u),$$

*there is $\mu \in \mathscr{M}_\varphi$ such that $\gamma = \int u\, d\mu$. Consequently,*

$$\underline{Q}_\varphi(u) = \inf\left\{\int u\, d\mu : \mu \in \mathscr{M}_\varphi\right\}$$

*and*

$$\bar{Q}_\varphi(u) = \sup\left\{\int u\, d\mu : \mu \in \mathscr{M}_\varphi\right\}.$$

That this lemma suffices to prove the implication (c) implies (a) follows easily, for note that in the case that $\varphi$ has a unique representing measure, $\mu$, it implies that

$$\int u\, d\mu = \sup\{\operatorname{Re} \varphi(f) : f \in B, \operatorname{Re} f \leq u\}.$$

Let $\delta_x$ be the unique representing measure on $X$ for $x$, and let $V$ be a neighborhood of $x$. There is a nonpositive $f \in \mathscr{C}_\mathbb{R}(X)$ such that $f(x) = 0$ and $f \leq \log \frac{1}{4}$ on $X \setminus V$. Then

$$0 = f(x) = \sup\{\operatorname{Re} g(x) : \operatorname{Re} g \leq f, g \in A\},$$

so there is $g \in A$ with $\operatorname{Re} g \leq f$, $\operatorname{Re} g(x) > \log \frac{3}{4}$. If $\tilde{g} = e^g$, then $\tilde{g} \in A$, $\|\tilde{g}\| \leq 1$, $|\tilde{g}(x)| > \frac{3}{4}$, and $|\tilde{g}| < \frac{1}{4}$ on $X \setminus V$. Thus, it only remains to prove the lemma.

PROOF OF THE LEMMA

The first assertion is clear since, e.g., if $\operatorname{Re} f \le u, f \in B$, then

$$\operatorname{Re} \varphi(f) = \operatorname{Re} \int f \, d\mu$$
$$= \int \operatorname{Re} f \, d\mu \le \int u \, d\mu.$$

For the second assertion, we need to note that $\bar{Q}_\varphi$ is subadditive and positively homogeneous:

$$\bar{Q}_\varphi(u + v) \le \bar{Q}_\varphi(u) + \bar{Q}_\varphi(v)$$

and

$$\bar{Q}_\varphi(tu) = t\bar{Q}_\varphi(u)$$

if $t \ge 0$. These properties are easily verified directly from the definition. Let $E$ be the subspace

$$E = \{tu : t \in \mathbb{R}\}$$

of $\mathscr{C}_{\mathbb{R}}(Y)$ for some fixed $u$, and suppose that $\underline{Q}_\varphi(u) \le \gamma \le \bar{Q}_\varphi(u)$. The functional $tu \mapsto t\gamma$ is a well-defined linear functional $\psi$ on $E$, and it satisfies $\psi(h) \le \bar{Q}_\varphi(h)$ for all $h \in E$. Thus, by the Hahn–Banach theorem $\psi$ extends to a linear functional, still denoted by $\psi$, on $\mathscr{C}_{\mathbb{R}}(Y)$ which satisfies $\psi(h) \le \bar{Q}_\varphi(h)$ for all $h \in \mathscr{C}_{\mathbb{R}}(Y)$. From $\psi(-h) \le \bar{Q}_\varphi(-h)$ we find that

$$\psi(h) = -\psi(-h) \ge -\bar{Q}_\varphi(-h),$$

and as $-\bar{Q}_\varphi(-h) = \underline{Q}_\varphi(h)$, we have

$$\underline{Q}_\varphi(h) \le \psi(h) \le \bar{Q}_\varphi(h)$$

for all $h \in \mathscr{C}_{\mathbb{R}}(Y)$. The last inequalities yield that

$$0 = \underline{Q}_\varphi(0) \le \underline{Q}_\varphi(h)$$
$$\le \psi(h)$$

if $h \ge 0$, so $\psi$ is a nonnegative linear functional. There is a positive measure $\mu$ such that

$$\psi(h) = \int h \, d\mu$$

for all $h \in \mathscr{C}_{\mathbb{R}}(Y)$. As $\underline{Q}_\varphi(\operatorname{Re} f) = \bar{Q}_\varphi(\operatorname{Re} f)$ for all $f \in A$, it follows that

$$\operatorname{Re} \varphi(f) = \int \operatorname{Re} f \, d\mu,$$

which implies that $\mu \in \mathscr{M}_\varphi$. Since $\psi(u) = \gamma$, we have the lemma.

Theorem 7.18 was given by Bishop [1] and Bishop and deLeeuw [1].

There is another characterization of strong boundary points, the $(\alpha - \beta)$-criterion, which is due to Gonchar [1] and which generalizes condition (a) of the last theorem. The proof we give is from Zalcman [1].

7.20 ***Corollary***

*If $(A, X)$ is a uniform algebra, the point $x \in X$ is a strong boundary point for $A$ if and only if there exist $\alpha$ and $\beta$, $0 < \alpha < \beta < 1$, such that given a neighborhood $V$ of $x$, there exists $f \in A$ with $\|f\|_X \leq 1$, $f(x) = \beta$, $\|f\|_{X \setminus V} < \alpha$.*

PROOF

The necessity of the condition is clear. Suppose then that $x$ satisfies the $(\alpha - \beta)$-condition but that it is not a strong boundary point. Then by Theorem 7.18, there is a probability measure $\mu$, $\mu \neq \delta_x$, which represents $x$. For some $\varepsilon$ between 0 and 1, we have $\mu = \nu + \varepsilon\delta_x$, where $\nu(\{x\}) = 0$. Then, $(1 - \varepsilon)^{-1}\nu$ is a positive measure, and

$$(1 - \varepsilon)^{-1} \int f\, d\nu = (1 - \varepsilon)^{-1} \int f\, d(\mu - \varepsilon\delta_x)$$

$$= f(x);$$

i.e., $(1 - \varepsilon)^{-1}\nu$ is a representing measure for $x$, call it $\tilde{\mu}$. We have $\tilde{\mu}(\{x\}) = 0$. There exists a neighborhood $V$ of $x$ such that $\tilde{\mu}(V) < \beta - \alpha$, and there exists $g \in A$ with $\|g\|_X \leq 1$, $g(x) = \beta$, and $\|g\|_{X \setminus V} < \alpha$. Then

$$\begin{aligned} \beta = g(x) &= \int g\, d\tilde{\mu} \\ &= \int_V g\, d\tilde{\mu} + \int_{X \setminus V} g\, d\tilde{\mu} \\ &< (\beta - \alpha) + \alpha \\ &= \beta. \end{aligned}$$

We have reached a contradiction, so the corollary is proved.

Our next result establishes the fact that the set of strong boundary points is a boundary, the *strong boundary*.

7.21 ***Theorem***

*If $(A, X)$ is a uniform algebra, every $f \in A$ achieves its maximum modulus on the set $P$ of strong boundary points for $A$.*

We need a lemma in the proof of this fact. Let us say that a set $E \subset X$ is a *peak set for A* if there is $f \in A$ such that $E = f^{-1}(1)$, $\|f\|_X = 1$, and $|f(x)| = 1$ implies that $f(x) = 1$. Note that a finite or countable intersection of peak sets is a peak set: If $f_n$ peaks on $E_n$, then $\sum_{n=1}^{\infty} 2^{-n} f_n$ peaks on $\bigcap E_n$.

**7.22** ***Lemma***

*If $F \subset X$ is an intersection of peak sets and if $g \in A$, there exists $h \in A$ such that $\|h\|_X = \|g\|_F$, and $h|F = g|F$.*

PROOF

We may suppose that $\|g\|_F = 1$. Since $F$ is an intersection of peak sets, compactness implies the existence of a peak set $E_0$ with $F \subset E_0 \subset \{x : |g(x)| < \frac{3}{2}\}$. Let $f_0$ peak on $E_0$. Then for large integers $n$, $gf_0^n \in A$, $gf_0^n|F = g|F$, and $\|gf_0^n\|_X < \frac{3}{2}$. Thus, we may suppose that $\|g\|_X < \frac{3}{2}$.

Let $S_n = \{x \in X : |g(x)| < 1 + 2^{-n}\}$. Then for each $n$, there is an element $f_n \in A$ such that

$$f_n|F = 1 = \|f_n\|_X$$

and

$$\|f_n\|_{X \setminus S_n} < 1.$$

There exist integers $k(n)$ such that

$$\|f_n^{k(n)} g\|_{X \setminus S_n} < 2^{-n}.$$

Define $h \in A$ by

$$h(x) = \sum_{n=1}^{\infty} 2^{-n} f_n^{k(n)} g.$$

If $x \in F$, then $f_n^{k(n)}(x) = 1$, so $h(x) = g(x)$. If $|g(x)| \leq 1$, then $|h(x)| \leq 1$ since $\|f_n^{k(n)}\|_X = 1$. And if $g(x) > 1$, then for some $n_0$, $x \in S_{n_0}$, but for all $n > n_0$, $x \notin S_n$. For such an $x$, we have

$$\begin{aligned} |h(x)| &< (1 + 2^{-n_0}) \sum_{n=1}^{n_0} 2^{-n} + 2^{-n_0} \sum_{n=n_0+1}^{\infty} 2^{-n} \\ &= (1 + 2^{-n_0})(1 - 2^{-n_0}) + 2^{-n_0} 2^{-n_0} \\ &= 1 - 4^{-n_0} + 4^{-n_0} = 1. \end{aligned}$$

PROOF OF THE THEOREM

Let $g \in A$. We are to prove that $|g|$ achieves its maximum on $P$. We may suppose that $\|g\|_X = 1$ and that $g$ assumes the value one. Set $P_g = \{x : g(x) = 1\}$. Replacing $g$ by $\frac{1}{2}(1 + g)$ if necessary, we may suppose that $P_g = \{x : |g(x)| = 1\}$. Let $\mathscr{E}$ be the family of all nonvoid intersections of peak sets contained in $P_g$. Compactness together with Zorn's lemma implies that $\mathscr{E}$ contains a minimal element, say $E_0$. Assume that $E_0$ contains more than one point. There is $f \in A$ which is nonconstant on $E_0$, which assumes the value 1 on $E_0$, say at $x_1$, and which is bounded by 1 on $E_0$. Apply the lemma to $f$: There exists $h \in A$ with

$$h|E_0 = f$$

and

$$\|h\|_X = 1.$$

Then $\frac{1}{2}(1 + h)$ peaks on a set $P_h$ which contains $x_1$ but which does not contain $E_0$. This implies that $E_0 \cap P_h \in \mathscr{E}$ is properly contained in $E_0$, whence $E_0$ is not minimal. This argument establishes the fact that the minimal elements of $\mathscr{E}$ are singletons.

Finally, let $\{x_0\}$ be a minimal element of $\mathscr{E}$. If $V$ is a neighborhood of $x_0$, there is $f \in A$ which peaks on a set $E$, $V \supset E \supset \{x_0\}$, and thus $x_0 \in P$, so $P_g \cap P \neq 0$, as was to be shown.

In the metric case, this result is due to Bishop [1]; the general case is in Rickart [1].

For the sake of future reference, let us note the following immediate consequence of Lemma 7.22.

7.23 ***Corollary***

*If $E \subset X$ is an intersection of peak sets for $A$ and if $F \subset E$ is an intersection of peak sets for $\{f|E : f \in A\}$, a certain subalgebra of $\mathscr{C}(E)$, then $F$ is an intersection of peak sets for $A$.*

7.24 ***Corollary***

*The Shilov boundary for $A$ is the closure of the set of strong boundary points.*

PROOF

The Shilov boundary $\Gamma$ is the minimal closed boundary for $A$. It contains all the strong boundary points, so it must contain the closure $\bar{P}$ of the set of strong boundary points. By the theorem, $\bar{P}$ is a closed boundary, so $\bar{P} = \Gamma$.

The strong boundary includes every peak point, and the peak points can be characterized among the strong boundary points very easily:

7.25 ***Lemma***

*A point is a peak point if and only if it is a strong boundary point and a $G_\delta$.*

PROOF

A peak point is a $G_\delta$ and a strong boundary point. Suppose, conversely, that $x$ is a $G_\delta$ and a strong boundary point. Thus, there is a sequence $\{V_n\}$ of open sets with $\bigcap V_n = \{x\}$, and since $x$ is an intersection of peak sets, there is $f_n \in A$ with $f_n(x) = 1 = \|f_n\|_X$ and $\|f_n\|_{X \setminus V_n} < \frac{1}{2}$. Set $g = \sum_{n=1}^{\infty} 2^{-n} f_n$. The function $g$ peaks at the point $x$.

If we restrict attention to uniform algebras on compact metric spaces, the fact that in such a space every point is a $G_\delta$, taken with Lemma 7.25 and Theorem 7.21, yields the following important fact due to Bishop [1]:

7.26 ***Theorem***

*If $(A, X)$ is a uniform algebra, $X$ metric, then $P$, the set of peak points, constitutes a boundary for $A$ which is contained in every boundary for $A$.*

7.27 ***Definition***

*Let $A$ be a function algebra on a space $X$. The set $E \subset X$ is the* ***minimal boundary*** *for $A$ if $E$ is a boundary for $A$ and if, in addition, $E$ is contained in every boundary for $A$.*

Thus, by Theorem 7.26, a uniform algebra on a metrizable space has a minimal boundary. In general, no minimal boundary need exist.

7.28 ***Example***

Let $\Lambda$ be an uncountable index set, and for each $\lambda \in \Lambda$, let $I_\lambda = [0, 1]$. Let $I = \Pi\{I_\lambda : \lambda \in \Lambda\}$, a certain infinite-dimensional cube. The algebra $\mathscr{C}(I)$ has no minimal boundary. Consider in $I$ two sets $E_0$ and $E_1$. $E_0$ is the set of those points $\{x_\lambda\}_{\lambda \in \Lambda}$ of $I$ such that $x_\lambda = 0$ for all but countably many $\lambda$, and $E_1$ is the set of those $\{x_\lambda\}_{\lambda \in \Lambda}$ with $x_\lambda = 1$ for all but countably many $\lambda$. Then $E_0 \cap E_1 = \varnothing$, but both $E_0$ and $E_1$ are boundaries for $\mathscr{C}(I)$. This follows from the Stone–Weierstrass theorem. Let $\mathscr{P}$ be the minimal subalgebra of $\mathscr{C}(I)$ containing the constants and the coordinate projection $\pi_\lambda$. The Stone–Weierstrass theorem implies that $\mathscr{P}$ is dense in $\mathscr{C}(I)$. Thus, if $f \in \mathscr{C}(I)$, there is a sequence $\{p_n\}_{n=1}^{\infty}$ in $\mathscr{P}$ which converges uniformly to $f$ on $I$. There exists a set $\Lambda_0$ in $\Lambda$ with $\Lambda \setminus \Lambda_0$ countable such that for all $n$, $p_n(x) = p_n(y)$ if $\pi_\lambda(x) = \pi_\lambda(y)$ for all $\lambda \in \Lambda \setminus \Lambda_0$. Thus, $f(x) = f(y)$ as long as $\pi_\lambda(x) = \pi_\lambda(y)$ for all $\lambda \in \Lambda \setminus \Lambda_0$. If $x_0 = \{x_\lambda^0\}_{\lambda \in \Lambda}$ is a point of $X$ such that $|f(x_0)| = \|f\|_I$, then the points $x_0'$ and $x_0''$ also have this property if $\pi_\lambda(x_0') = \pi_\lambda(x_0'') = x_\lambda^0$

for $\lambda \in \Lambda \setminus \Lambda_0$, and $\pi_\lambda(x_0') = 0$, $\pi_\lambda(x_0'') = 1$ for $\lambda \in \Lambda_0$. Thus, both $E_0$ and $E_1$ are boundaries for $\mathscr{C}(I)$. Since they are disjoint, we may conclude that $\mathscr{C}(I)$ does not possess a minimal boundary.

A discussion of the minimal boundary for $\mathscr{C}(X)$ for various spaces $X$ is contained in a paper of Gulick [1].

The following result, which is also due to Bishop [1], gives a useful fact about the structure of the minimal boundary in the metric case.

7.29 ***Theorem***

*If $(A, X)$ is a uniform algebra with $X$ metric, then the minimal boundary for $A$ is a $G_\delta$.*

PROOF

We use the $(\alpha - \beta)$-condition: $x \in X$ is a peak point if and only if for every neighborhood $V$ of $x$, there is $f \in A$ with $\|f\|_X = 1$, $|f(x)| > \frac{3}{4}$, and $\|f\|_{X \setminus V} < \frac{1}{4}$.

Let $\rho$ be a metric on $X$ and define sets $D_n(x)$ and $U_n$ by

$$D_n(x) = \{y : \rho(x, y) > 1/n\}$$

and

$$U_n = \{x \in X : \text{for some } f \in A \text{ with } \|f\| \leq 1, |f(x)| > \tfrac{3}{4}, \text{ but } |f(y)| < \tfrac{1}{4} \text{ for all } y \in D_n(x)\}.$$

The set $U_n$ is open, and if $x \in \bigcap\{U_n : n = 1, 2, \ldots\}$, the $(\frac{1}{4} - \frac{3}{4})$-condition shows that $x$ is a peak point. Since every peak point is an $\bigcap_n U_n$, the result follows.

The next matter to which we turn our attention is the relation of the boundaries we have considered to the extremal structure of the unit ball in the space dual to the given algebra. The main result is the following fact established by Bishop and deLeeuw [1].

7.30 ***Theorem***

*Let $(A, X)$ be a uniform algebra, let $A^*$ be the space of continuous linear functionals on $A$ with the weak $*$ topology, and define $i: X \to A^*$ by $ix(f) = f(x)$. The map $i$ is a homeomorphism of $X$ into the unit sphere of $A^*$. If $K = \{\varphi \in A^* : \|\varphi\| = 1 = \varphi(1)\}$, then $K$ is the closed, convex hull of the range of $i$. Under $i$, the strong boundary for $A$ goes onto the set of extreme points of $K$, and consequently the Shilov boundary goes onto the closure of this set.*

PROOF

By definition, $A$ separates points on $X$, so it follows from the obvious continuity of $i$ and the compactness of $X$ that $i$ takes $X$ homeomorphically onto a closed set in $A^*$; and since $A$ contains the constants, $i(X)$ lies in the unit sphere of $A^*$.

The set $K$ is weak * compact and convex, so by the Krein–Milman theorem it is the weak * closure of the convex hull of the set of its extreme points. As the map $i$ takes $X$ into $K$, the proof of the theorem can be concluded by showing that the strong boundary goes onto the set of extreme points of $K$.

We first show that if $x$ is a strong boundary point, then $\varphi = ix$ is an extreme point of $K$. If not, $\varphi = \frac{1}{2}(\varphi_1 + \varphi_2)$, $\varphi_1$ and $\varphi_2$ distinct elements of $K$. By the Hahn–Banach and Riesz representation theorems, there are probability measures $\mu_1$ and $\mu_2$ such that

$$\varphi_j(f) = \int f \, d\mu_j.$$

One of $\mu_1$ and $\mu_2$ is not the point mass at $x$, so $\frac{1}{2}(\mu_1 + \mu_2)$ is a representing measure for $x$ other than point mass at $x$. By Theorem 7.18, $x$ is not a strong boundary point.

Conversely, let $\varphi$ be an extreme point of the set $K$, and let $\mu$ be a probability measure which represents $\varphi$. Let $S$ be a Borel set with $\mu(S) \neq 0, 1$. Then for all $f \in A$,

$$\varphi(f) = \mu(S)\varphi_1(f) + \mu(X \setminus S)\varphi_2(f),$$

where

$$\varphi_1(f) = \mu(S)^{-1} \int_S f \, d\mu$$

and

$$\varphi_2(f) = \mu(X \setminus S)^{-1} \int_{X \setminus S} f \, d\mu.$$

Since $\varphi$ is assumed to be extreme, we conclude that $\varphi = \varphi_1 = \varphi_2$, and therefore that

$$\int_S f \, d\mu = \mu(S) \int f \, d\mu$$

for all Borel sets $S$ with $0 < \mu(S) < 1$. This implies that $f$ is constant a.e. $[\mu]$, and as this holds for every $f \in A$ and $A$ separates points on $X$, we conclude that $\mu$ is a point mass, say at $x \in X$. Thus, $\varphi = ix$. Moreover, what we have just done proves that the only representing measure for

$x$ is the point mass at $x$; so by Theorem 7.18, $x$ lies in the strong boundary for $A$. This concludes the proof of the theorem.

7.31 ***Corollary***

*Let $(A, X)$ be a uniform algebra and let $T$ be a linear isometry of $A$ onto $A$. Then there exists an algebra automorphism $\Phi: A \to A$ and a function $\alpha \in A$ such that $|\alpha| \equiv 1$, $\alpha^{-1} \in A$, and such that $Tf = \alpha\Phi(f)$ for all $f \in A$.*

PROOF

Let $T^*: A^* \to A^*$ be the mapping dual to $T$ so that $T^*$ is also a linear isometry. If $E$ denotes the set of extreme points of the unit ball of $A^*$, then $T^*$ carries $E$ onto itself. Since the elements of $E$ are precisely the functionals of the form $\lambda\varphi, \varphi \in K, |\lambda| = 1$, $K$ as in the theorem, it follows that for each strong boundary point $x \in X$, there are defined a number $\lambda(x)$ of modulus 1 and a strong boundary point $\tau(x)$ such that

$$T^*(ix) = \lambda(x)i(\tau(x));$$

i.e.,

$$Tf(x) = \lambda(x)f(\tau(x)).$$

Taking for $f$ the function identically 1, we find that $\lambda(x) = (T1)(x)$, so $\lambda \in A$, and by its definition, it is constantly of modulus 1 on the strong boundary of $A$. For points $x$ of the strong boundary, we find that

$$\begin{aligned} T(fg)(x) &= \lambda(x)fg(\tau(x)) \\ &= \lambda(x)f(\tau(x))g(\tau(x)) \end{aligned}$$

and, thus, $\lambda(x)T(fg)(x) = (Tf)(x)(Tg)(x)$. Since this relation holds on the strong boundary, it must hold on the whole of $X$; i.e.,

$$\lambda T(fg) = T(f)T(g).$$

For $f = g = T^{-1}(1)$, this yields $\lambda^{-1} \in A$. Since $\lambda$ and $\lambda^{-1}$ both lie in $A$ and are of modulus 1 on the strong boundary, it follows that $|\lambda| \equiv 1$ on $X$, and we have the result.

For this corollary, we have followed Hoffman's discussion [1, pp. 146–147].

Much of the importance of Theorem 7.30 derives from the fact that it makes possible certain integral representations. The possibility of these representations depends on a portion of functional analysis which we shall not digress to develop. The results we need are summarized in Appendix C, and a very readable, detailed exposition of the whole theory, which goes far beyond our needs, has been given by Phelps [1].

**7.32** ***Theorem***

*Let $(A, X)$ be a uniform algebra, and let $P$ be the strong boundary for $A$. If $\varphi \in A^*$, there exists a finite regular Borel measure $\sigma$ on $X$ such that*

$$\varphi(f) = \int f \, d\sigma$$

*for all $f \in A$, and such that $\sigma(E) = 0$ for all Baire sets and all $G_\delta$'s disjoint from $P$.*

PROOF

By the Hahn–Banach and Riesz representation theorems, we can write

$$\varphi(f) = \int f \, d\mu$$

for some finite regular Borel measure $\mu$ on $X$. We have $\mu = \mu_1 - \mu_2 + i(\mu_3 - \mu_4)$, where the $\mu_j$ are positive measures. Set $\psi_j(f) = \int f \, d\mu_j$. We may suppose that $\|\mu_j\| = 1$; the functional $\psi_j$ then lies in the set

$$K = \{\varphi \in A^* : \varphi(1) = \|\varphi\| = 1\}.$$

Since $K$ is weak $*$ compact, the Choquet–Bishop–deLeeuw theorem [Appendix C] provides a regular Borel measure $\nu_j$ on $K$ such that for all linear functionals $L$ on $A^*$ which are continuous with respect to the weak $*$ topology, we have

$$L\psi_j = \int L(m) \, d\nu_j(m),$$

and $\nu_j(E) = 0$ if $E$ is a Baire set or a $G_\delta$ which misses the set of extreme points of $K$. Thus, if $f \in A$, we have

$$\int f \, d\mu_j = \psi_j(f) = \int m(f) \, d\nu_j(m). \tag{1}$$

Let $i : X \to K$ be as before; i.e., $ix(f) = f(x)$. Since the set $iX$ is compact and contains all the extreme points of $K$, the set $K \setminus iX$ is open and so a $G_\delta$, and it misses the set of extreme points of $K$. By (1), we have

$$\int f \, d\mu_j = \int_{iX} ix(f) \, d\nu_j(ix).$$

The map $i$ is a homeomorphism of $X$ into $K$ which carries $P$ onto the set of extreme points of $K$. If we define $\nu_j'$, a measure on $X$, by

$$\nu_j'(E) = \nu(iE)$$

for all Borel sets $E$ in $X$, then

$$\int f\, d\mu_j = \int f(x)\, d\nu'_j(x),$$

whence there exists a $\sigma$ of the desired kind with

$$\varphi(f) = \int f\, d\sigma.$$

We would like to be able to write

$$\varphi(f) = \int_P f(x)\, d\sigma, \tag{2}$$

but in general this is not possible, for the strong boundary $P$ need not be a Borel set. (An example is given in Section 14.) However, if $X$ is metrizable, then, as we have seen, $P$ is a $G_\delta$ and so measurable. In this case, we have $|\sigma|(X \setminus P) = 0$, and formula (2) is both meaningful and correct.

**7·33** ***Corollary***

*If $(A, X)$ is a uniform algebra with strong boundary $B$, then either $B$ is uncountable or else $X = B$.*

PROOF

Assume $B$ countable, say $B = \{x_1, x_2, \ldots\}$, and $X \setminus B$ nonempty. Since $B$ is countable, the set $X \setminus B = \bigcap_{n=1}^{\infty} (X \setminus \{x_n\})$ is a set of type $G_\delta$. It follows from the last theorem that if $x \in X \setminus B$, there is a probability measure $\mu$ supported on $B$ which represents $x : \int f\, d\mu = f(x)$ for all $f \in A$. Let $V$ be a neighborhood of $x_n$ which misses $x$. As $x_n$ lies in the strong boundary, there is $f \in A$ with $f(x_n) = 1 = \|f\| > |f(x)|$. We may suppose that $|f(y)| = 1$ implies $f(y) = 1$. Then

$$\begin{aligned} 0 = \lim_{n\to\infty} f^n(x) &= \lim_{n\to\infty} \int f^n\, d\mu \\ &= \mu(P_f), \end{aligned}$$

where we write $P_f$ for the set on which $f$ assumes the value 1. As $P_f \supseteq \{x_n\}$, it follows that $\mu(\{x_n\}) = 0$ since $\mu$ is a positive measure. This statement is independent of the choice of $n$, so since $B$ is the countable union of the sets $\{x_n\}$, we get $\mu(B) = 0$, which is obviously incompatible with the assumption that $\mu$ represents $x$.

It seems worthwhile to mention the fact that this corollary can be obtained without recourse to the Bishop–deLeeuw theory by using the

Shilov idempotent theorem established in the next section to reduce to the case that $X$ is connected, and then by using the easily established fact that the strong boundary for a uniform algebra on a perfect and *a fortiori* on a connected space has the property that each of its points is a limit point. (Since the term *perfect* is frequently reserved for *closed* sets, we do not say that the strong boundary is perfect.)

# 8 THE HOLOMORPHIC FUNCTIONAL CALCULUS

In Theorem 6.1 we saw that if a function $f$ on the unit circle has an absolutely convergent Fourier series, and if $f$ does not assume the value zero, then $1/f$ also has an absolutely convergent Fourier series. This is a very special case of a general phenomenon that we shall examine in this section. We begin with a simple but important special case.

**8.1** ***Theorem***

*Let $A$ be a commutative Banach algebra with identity. If $a \in A$ and $f$ is function holomorphic in a domain $\Omega$ containing $\sigma(a)$, then there exists $b \in A$ with $\hat{b} = f \circ \hat{a}$.*

The proof of this result is most easily effected by the use of a vector-valued integral, so we shall preface the proof with a few remarks about such integrals. Suppose that we have a continuous function $F$ defined on a domain $D$ in $\mathbb{C}$ which takes values in a Banach space $B$. If $\gamma$ is a rectifiable curve in $D$, i.e., a continuous function of bounded variation on $[0, 1]$ with values in $D$, we can form vector-valued Riemann sums

$$\sum_{j=0}^{N-1} (\gamma(t_{j+1}) - \gamma(t_j))F(\gamma(s_j))$$

for any partition $0 = t_0 < t_1 < \cdots < t_N = 1$ of $[0, 1]$ and any choice of $s_j \in [t_j, t_{j+1}]$. Just as in elementary calculus, as the mesh of the partition becomes finer and finer, these sums approach in norm an element $b$ of $B$ independently of the choice of the intermediate points $s_j$. This element is defined to be $\int_\gamma F(z)\, dz$.

PROOF OF THE THEOREM

From complex-variable theory, there is a system $\Gamma$ of rectifiable simple closed curves $\gamma_1, \ldots, \gamma_N$ which bound a domain $D, \Omega \supset \bar{D} \supset D \supset \sigma(a)$ such that for $z \in D$,

$$f(z) = \frac{1}{2\pi i}\int_\Gamma \frac{f(\zeta)}{\zeta - z}\, d\zeta.$$

Since for each $\zeta \in \gamma_j$, $\zeta \notin \sigma(a)$, it follows that $(\zeta - a)^{-1}$ is a well-defined element of $A$, and $f(\zeta)(\zeta - a)^{-1}$ is a continuous $A$-valued function. Thus, by our remarks on vector-valued integrals,

$$b = \frac{1}{2\pi i}\int_\Gamma f(\zeta)(\zeta - a)^{-1}\, d\zeta$$

is a well-defined element of $A$. If $\varphi \in \Sigma(A)$ and

$$S = \sum_{k=1}^{N} f(\gamma_j(s_k))(\gamma_j(s_k) - a)^{-1}(\gamma_j(t_{k-1}) - \gamma_j(t_k))$$

is a Riemann sum for $\int_{\gamma_j} f(\zeta)(\zeta - a)^{-1}\, d\zeta$, we have

$$\varphi(S) = \sum_{k=1}^{N} f(\gamma_j(s_k))(\gamma_j(s_k) - \varphi(a))^{-1}(\gamma_j(t_{k+1}) - \gamma_j(t_k)).$$

Since the Riemann sums for $\int_{\gamma_j} f(\zeta)(\zeta - a)^{-1}\, d\zeta$ approach this integral in norm and since $\varphi(a) \in \sigma(a)$, we may conclude that

$$\varphi(b) = \sum_{j=1}^{N} \frac{1}{2\pi i}\int_{\gamma_j} \frac{f(\zeta)}{\zeta - \varphi(a)}\, d\zeta.$$

This last integral is $f(\varphi(a))$, so $\hat{b}(\varphi) = f(\hat{a}(\varphi))$, as we wished to prove.

The theorem of Wiener, Theorem 6.1, is a special case of this result, the case that $f(z) = 1/z$.

Suppose now that $\Omega$, $\Gamma$, and $D$ are as above, and that $\Gamma'$ is a new system of curves in $D$ which bounds a domain $D'$, $D' \supset \sigma(a)$. We can form the integral

$$b' = \frac{1}{2\pi i}\int_{\Gamma'} f(\zeta)(\zeta - a)^{-1}\, d\zeta.$$

If $\Gamma''$ is the boundary of $D \setminus D'$, then

$$b - b' = \frac{1}{2\pi i}\int_{\Gamma''} f(\zeta)(\zeta - a)^{-1}\, d\zeta.$$

If $L$ is a continuous linear functional on $A$, then since $L((\zeta - a)^{-1} f(\zeta))$ is holomorphic, as a function of $\zeta$, in the domain $D \setminus D'$, the Cauchy integral theorem yields

$$0 = \frac{1}{2\pi i}\int_{\Gamma''} L(f(\zeta)(\zeta - a)^{-1})\, d\zeta$$
$$= L(b) - L(b').$$

Thus, $b = b'$, and consequently the integral $(1/2\pi i)\int_\Gamma f(\zeta)(\zeta - a)^{-1}\, d\zeta$ is independent of the choice of $\Gamma$.

Denote by $H$ the map $f \mapsto (1/2\pi i)\int_\Gamma f(\zeta)(\zeta - a)^{-1}\, d\zeta$ from the space of functions holomorphic on $\Omega$ into $A$. The map $H$ is clearly linear, and we shall prove that it is multiplicative. Let $f$ and $g$ be holomorphic in $\Omega$, let $\Gamma$ be chosen so that

$$H(f) = \frac{1}{2\pi i}\int_\Gamma f(\zeta)(\zeta - a)^{-1}\, d\zeta,$$

and let $D$ be the domain bounded by $\Gamma$. Choose $\Gamma' \subset D$ so that

$$H(g) = \frac{1}{2\pi i}\int_{\Gamma'} g(\zeta)(\zeta - a)^{-1}\, d\zeta.$$

We have then

$$H(f)H(g) = \left(\frac{1}{2\pi i}\right)^2 \int_{\Gamma'}\int_{\Gamma} f(\zeta)g(\tau)(\zeta - a)^{-1}(\tau - a)^{-1}\, d\tau\, d\zeta.$$

Since

$$(\zeta - a)^{-1} - (\tau - a)^{-1} = (\tau - \zeta)(\zeta - a)^{-1}(\tau - a)^{-1},$$

we have

$$\begin{aligned} H(f)H(g) = {} & \left(\frac{1}{2\pi i}\right)^2 \int_{\Gamma'} f(\zeta)(\zeta - a)^{-1}\left\{\int_\Gamma g(\tau)(\tau - \zeta)^{-1}\, d\tau\right\} d\zeta \\ & - \left(\frac{1}{2\pi i}\right)^2 \int_{\Gamma} g(\tau)(\tau - a)^{-1}\left\{\int_{\Gamma'} f(\zeta)(\tau - \zeta)^{-1}\, d\zeta\right\} d\tau. \end{aligned}$$

Since $\tau$ lies outside the domain bounded by $\Gamma'$, we have that

$$\int_{\Gamma'} f(\zeta)(\tau - \zeta)^{-1}\, d\zeta = 0,$$

so

$$\begin{aligned} H(f)H(g) &= \frac{1}{2\pi i}\int_{\Gamma'} f(\zeta)g(\zeta)(\zeta - a)^{-1}\, d\zeta \\ &= H(fg). \end{aligned}$$

Thus, $H$ is multiplicative and so an algebra homomorphism.

Theorem 8.1 is trivially false if we do not require $A$ to have an identity. However, there is a result in this case.

**8.2 *Theorem***

*Let $A$ be a commutative Banach algebra without identity. If $a \in A$, if $f$ is holomorphic on $\sigma(A)$, and if $f(0) = 0$, then there is $b \in A$ such that $\hat{b} = f \circ \hat{a}$.*

PROOF

Embed $A$ as a maximal ideal in the algebra $A'$ in the canonical way. The function $f$ is holomorphic on $\sigma^A(a) = \sigma_{A'}(a)$, so Theorem 8.1 applied to $A'$ yields $b \in A'$ such that $\hat{b} = f \circ \hat{a}$. Since $f(0) = 0$, we have $\hat{b}(\infty) = 0$ and therefore $b \in A$.

Our next object is an extension of Theorem 8.1 to an analogous result involving holomorphic functions of several complex variables. Our results depend on certain nontrivial facts from the theory of functions of several complex variables, results that will not be proved in this book but are summarized in Appendix D. Recall that if $\underline{a} = \{a_1, \ldots, a_N\}$ is a collection of elements of $A$, then $\sigma_A(\underline{a})$ denotes the joint spectrum of $\underline{a}$.

8.3 ***Theorem***

*If $A$ is a commutative Banach algebra with identity, if $\underline{a} = \{a_1, \ldots, a_N\}$ is a finite subset of $A$, and if $F$ is a function holomorphic in a neighborhood of $\sigma_A(\underline{a})$, then there is $b \in A$ such that*

$$\hat{b}(\varphi) = F(\hat{a}_1(\varphi), \ldots, \hat{a}_N(\varphi))$$

*for all $\varphi \in \Sigma(A)$.*

We shall obtain this theorem in the course of our development of a *holomorphic functional calculus*. For this functional calculus, certain preliminaries are necessary.

If $\mathscr{M}$ is a complex manifold, we use $\mathscr{O}(\mathscr{M})$ to denote the algebra of holomorphic functions on $\mathscr{M}$. If $K$ is a compact subset of $\mathscr{M}$, we let $\mathscr{V}(K)$ denote the collection of all open sets in $\mathscr{M}$ which contain $K$; and in $\bigcup\{\mathscr{O}(V): V \in \mathscr{V}(K)\}$ we define an equivalence relation $\sim$ by saying that if $F \in \mathscr{O}(V)$ and $F' \in \mathscr{O}(V')$, then $F \sim F'$ provided there is $V'' \in \mathscr{V}(K)$ such that $F|V'' = F'|V''$. The collection of equivalence classes defined in this way is denoted by $\mathscr{O}(K)$. Given $F \in \mathscr{O}(V)$, $V \in \mathscr{V}(K)$, $\tilde{F}$ will denote the element of $\mathscr{O}(K)$ containing it. The map $F \mapsto \tilde{F}$ from $\mathscr{O}(V)$ into $\mathscr{O}(K)$ is a homomorphism, but in general it is neither one to one nor onto. Observe that if $\mathfrak{z} \in K$, then $\tilde{F}(\mathfrak{z})$ is well defined for all $\tilde{F} \in \mathscr{O}(K)$. A word of warning is in order at this point: If $V, V' \in \mathscr{V}(K)$, if $F \in \mathscr{O}(V)$ and $G \in \mathscr{O}(V')$, and if $F(\mathfrak{z}) = G(\mathfrak{z})$ for all $\mathfrak{z} \in K$, it does not necessarily follow that $F \sim G$. Consider, e.g., the case that $K$ is a singleton. We shall prove the following fact.

8.4 ***Theorem***

*Let $A$ be a commutative Banach algebra with identity. There exists a family $H_{\underline{a}}$ of homomorphisms indexed by the finite subsets of $A$ such that if*

*$\underline{a} = \{a_1, \ldots, a_N\} \subset A$, then $H_{\underline{a}}: \mathcal{O}(\sigma_A(\underline{a})) \to A$ is a homomorphism with the following properties:*

(1) $H_{\underline{a}}(\tilde{Z}_j) = a_j$, $1 \le j \le N$.
(2) $\varphi(H_{\underline{a}}(\tilde{F})) = \tilde{F}(\hat{a}_1(\varphi), \ldots, \hat{a}_N(\varphi))$ *for* $\varphi \in \Sigma(A)$.
(3) *If* $\underline{b} = \{a_1, \ldots, a_N, b_1, \ldots, b_s\} \subset A$, *and if* $\pi: \mathbb{C}^{N+S} \to \mathbb{C}^N$ *is the natural projection, then* $H_{\underline{b}}((F \circ \pi)^\sim) = H_{\underline{a}}\tilde{F}$ *for all* $F \in \mathcal{O}(V)$, $V \in \mathscr{V}(\sigma_A\{\underline{a}\})$.

In the statement of the theorem we have used $Z_j$ to denote the $j$th coordinate function on $\mathbb{C}^N$. The proof of Theorem 8.4 will be in several steps; we shall obtain Theorem 8.3 at a rather early stage of the proof.

In proving Theorem 8.4, we shall need to deal with *polynomial polyhedra.* A polynomial polyhedron is a set of the form

$$P(\delta; p_1, \ldots, p_m) = \{\mathfrak{z} = (z_1, \ldots, z_N) \in \mathbb{C}^N : |z_1| < \delta_1, \ldots,$$
$$|z_N| < \delta_N, |p_1(\mathfrak{z})| < \delta_{N+1}, \ldots, |p_m(\mathfrak{z})| < \delta_{N+m}\},$$

where $\delta = (\delta_1, \ldots, \delta_{N+m})$ is a vector with strictly positive entries and $p_1, \ldots, p_m$ are polynomials. Suppose that $P = P(\delta; p_1, \ldots, p_m)$ is a polynomial polyhedron. If $q_1, \ldots, q_s$ are polynomials, let

$$\Omega = \{\mathfrak{z} \in P : |q_1(\mathfrak{z})| < r_1, \ldots, |q_s(\mathfrak{z})| < r_s\},$$

where $r_1, \ldots, r_s$ are positive numbers so that $\Omega$ is a polynomial polyhedron contained in $P$. If we let

$$\Delta(r_1, \ldots, r_s) = \{(z_1, \ldots, z_s) \in \mathbb{C}^s : |z_1| < r_1, \ldots, |z_s| < r_s\},$$

then we have an associated *Oka map*

$$\psi: \Omega \to P \times \Delta(r_1, \ldots, r_s)$$

given by

$$\psi(\mathfrak{z}) = \psi(z_1, \ldots, z_N) = (z_1, \ldots, z_N, q_1(\mathfrak{z}), \ldots, q_s(\mathfrak{z})).$$

The Oka map carries $\Omega$ onto a complex submanifold of $P \times \Delta(r_1, \ldots, r_s)$. An important result in function theory is the fact that if $f$ is holomorphic on $\Omega$, then there is a function $F$ holomorphic on $P \times \Delta(r_1, \ldots, r_s)$ with the property that $f = F \circ \psi$. (See Appendix D.) We shall also need the ideal theoretic fact contained in the following lemma.

**8.5** ***Lemma***

*If $F$ is holomorphic on $P \times \Delta(r_1, \ldots, r_s)$, and if $F \circ \psi$ is the zero function, then there exist functions $F_1, \ldots, F_s$ holomorphic on $P \times \Delta(r_1, \ldots, r_s)$ such that*

$$F(z_1, \ldots, z_{N+s}) = \sum_{j=1}^{s} F_j(z_1, \ldots, z_{N+s})(z_{N+j} - q_j(z_1, \ldots, z_N)).$$

In other words, the ideal in $\mathcal{O}(P \times \Delta(r_1, \ldots, r_s))$ consisting of those functions which vanish on the range of $\psi$ is generated by the polynomials $z_{N+j} - q_j(z_1, \ldots, z_N)$.

This is a special case of a more general result in the Oka–Cartan theory of ideals of holomorphic functions, but the following elementary proof is due to Allan [3].

PROOF

The proof is by induction on the number of polynomials $q_j$, i.e., on the integer $s$.

Consider first the case that $s = 1$. We can define a function on $(P \times \Delta(r_1)) \setminus \psi(\Omega)$ by setting

$$F_1(z_1, \ldots, z_{N+1}) = F(z_1, \ldots, z_{N+1})(z_{N+1} - q_1(z_1, \ldots, z_N))^{-1}.$$

The function $F_1$ is obviously holomorphic off $\psi(\Omega)$, and we shall show that it admits an extension to a function on the whole of $P \times \Delta(r_1)$. For this purpose, let $\mathfrak{z}^0 = (z_1^0, \ldots, z_N^0, q_1(z_1^0, \ldots, z_N^0))$ be a point of $\psi(\Omega)$. In a neighborhood of the point $\mathfrak{z}^0$, the map

$$(z_1, \ldots, z_{N+1}) \mapsto (z_1, \ldots, z_N, z_{N+1} - q_1(z_1, \ldots, z_N))$$

has nonvanishing Jacobian, so it carries a neighborhood of the point $\mathfrak{z}^0$ biholomorphically onto a neighborhood of the point $(z_1^0, \ldots, z_N^0, 0)$. Consequently, $F$ can be expanded in a power series in

$$z_1 - z_1^0, \ldots, z_N - z_N^0, z_{N+1} - q_1(z_1, \ldots, z_N),$$

and since $F$ vanishes on $\psi(\Omega)$, every term of this power series will contain the factor $z_{N+1} - q_1(z_1, \ldots, z_N)$. Thus, near $\mathfrak{z}^0$ we have

$$F(z_1, \ldots, z_N) = (z_{N+1} - q_1(z_1, \ldots, z_N))G(z_1, \ldots, z_N),$$

$G$ holomorphic near $\mathfrak{z}^0$, so it follows that $F_1$ is holomorphic near $\mathfrak{z}^0$.

Now assume that the lemma is true for all choices of $P$ and every choice of $\Omega$ obtained by using fewer than $s$ polynomials $q_j$. On the basis of this assumption, we shall prove that the assertion of the lemma is correct with $P$ and $\Omega$ as given. For this purpose, we define $\Omega'$ by

$$\Omega' = \{\mathfrak{z} \in P : |q_s(\mathfrak{z})| < r_s\},$$

and we define maps $\psi_1 : \mathbb{C}^N \to \mathbb{C}^{N+s-1}$ and $\psi_2 : \mathbb{C}^{N+s-1} \to \mathbb{C}^{N+s}$ by

$$\psi_1(z_1, \ldots, z_N) = (z_1, \ldots, z_N, q_1(\mathfrak{z}), \ldots, q_{s-1}(\mathfrak{z}))$$

and

$$\psi_2(z_1, \ldots, z_{N+s-1}) = (z_1, \ldots, z_{N+s-1}, q_s(z_1, \ldots, z_N)).$$

The composition $\psi_2 \circ \psi_1$ is the Oka map $\psi$.

We have that

$$\psi_2(\Omega' \times \Delta(r_1, \ldots, r_{s-1})) \subset P \times \Delta(r_1, \ldots, r_s),$$

so we can define $G$ on $\Omega' \times \Delta(r_1, \ldots, r_{s-1})$ by $G = F \circ \psi_2$. The function $G \circ \psi_1$ vanishes identically on $\Omega'$, so by the induction hypothesis, there exist $G_1, \ldots, G_{s-1}$ holomorphic on $\Omega' \times \Delta(r_1, \ldots, r_{s-1})$ such that

$$F \circ \psi_2 = G = \sum_{j=1}^{s-1} G_j(z_1, \ldots, z_{N+s-1})(z_{N+j} - q_j(z_1, \ldots, z_N)).$$

Now

$$\begin{aligned}\Omega' &\times \Delta(r_1, \ldots, r_{s-1}) \\ &= \{(z_1, \ldots, z_{N+s-1}) \in \Omega \times \Delta(r_1, \ldots, r_{s-1}) : |q_s(z_1, \ldots, z_N)| < r_s\},\end{aligned}$$

a polynomial polyhedron with associated Oka map $\psi_2$, so there exist functions $H_1, \ldots, H_{s-1}$ on $P \times \Delta(r_1, \ldots, r_s)$ such that for $(z_1, \ldots, z_{N+s-1}) \in \Omega' \times \Delta(r_1, \ldots, r_{s-1})$,

$$H_j(\psi_2(z_1, \ldots, z_{N+s-1})) = G_j(z_1, \ldots, z_{N+s-1}).$$

Define $H$ on $P \times \Delta(r_1, \ldots, r_s)$ by

$$\begin{aligned}H(&z_1, \ldots, z_{N+s}) \\ &= F(z_1, \ldots, z_{N+s}) - \sum_{j=1}^{s-1} H_j(z_1, \ldots, z_{N+s})(z_{N+j} - q_j(z_1, \ldots, z_N)).\end{aligned}$$

This function is holomorphic of $P \times \Delta(r_1, \ldots, r_s)$, and $H \circ \psi_2 = 0$; so by the case $s = 1$ of the lemma which we have already proved, there is a function $H_s$ holomorphic on $P \times \Delta(r_1, \ldots, r_s)$ such that

$$H(z_1, \ldots, z_{N+s}) = H_s(z_1, \ldots, z_{N+s})(z_{N+s} - q_s(z_1, \ldots, z_N)),$$

and this proves the lemma.

In addition to this algebraic result, we shall also need a geometric result which, roughly put, says that arbitrary polynomially convex sets can be approximated by polynomial polyhedra.

### 8.6 *Lemma*

*If $E$ is a compact, polynomially convex subset of the open set $\Omega$ in $\mathbb{C}^N$, there is a polynomial polyhedron $P$ with $E \subset P \subset \bar{P} \subset \Omega$.*

PROOF

The set $E$ is compact, so $E \subset \Delta_r^N = \{\mathfrak{z} \in \mathbb{C}^N : |z_1| < r, \ldots, |z_N| < r\}$ for some sufficiently large $r$. Given a point $\mathfrak{z} \in \bar{\Delta}_r^N \setminus E$, the polynomial convexity of $E$ implies the existence of a polynomial $p_\mathfrak{z}$ such that $p_\mathfrak{z}(\mathfrak{z}) > 1$, $\|p_\mathfrak{z}\|_E < 1$. Thus, for some neighborhood $V_\mathfrak{z}$ of $\mathfrak{z}$, we have $|p_\mathfrak{z}| > 1$ on $V_\mathfrak{z}$. Finitely many of the neighborhoods $V_\mathfrak{z}$, say $V_1, \ldots, V_M$, cover the compact set $\bar{\Delta}_r^N \setminus \Omega$. Let the associated polynomials be $p_1, \ldots, p_M$. The polynomial polyhedron

$$\{\mathfrak{z} \in \mathbb{C}^N : |z_1| < r, \ldots, |z_N| < r, |p_1(\mathfrak{z})| < 1, \ldots, |p_M(\mathfrak{z})| < 1\}$$

contains $E$ and is contained in $\Omega$.

Recall that $\hat{E}$ denotes the polynomially convex hull of the set $E$.

8.7 ***Lemma***

*Let $A$ be a commutative Banach algebra with identity, and let $\{a_1, \ldots, a_N\} \subset A$. Given an open set $\Omega$ in $\mathbb{C}^N$ which contains $\sigma_A(\{a_1, \ldots, a_N\})$, there exist elements $a_{N+1}, \ldots, a_M$ of $A$ such that if $\pi : \mathbb{C}^M \to \mathbb{C}^N$ is the natural projection of $\mathbb{C}^M$ onto its first $N$ coordinates, then*

$$\pi\hat{\sigma}_A(\{a_1, \ldots, a_M\}) \subset \Omega.$$

PROOF

If $\mathfrak{z}^0 = (z_1^0, \ldots, z_N^0) \in \hat{\sigma}_A(a_1, \ldots, a_N) \setminus \Omega$, the functions $\hat{a}_1 - z_1^0, \ldots, \hat{a}_N - z_N^0$ of $\hat{A}$ have no common zero on $\Sigma(A)$, so there exist $b_1, \ldots, b_N$ such that $1 = \Sigma b_j(a_j - z_j^0)$. If $\pi : \mathbb{C}^{2N} \to \mathbb{C}^N$ is the natural projection onto the first $N$ coordinates, then $\pi$ carries $\sigma_A(\{a_1, \ldots, a_N, b_1, \ldots, b_N\})$ onto $\sigma_A(\{a_1, \ldots, a_N\})$. The point $\mathfrak{z}^0$ does not lie in $\pi\hat{\sigma}_A(\{a_1, \ldots, a_N, b_1, \ldots, b_N\})$. To see this, denote by $B$ the closed subalgebra (with identity) generated by $a_1, \ldots, a_N, b_1, \ldots, b_N$. We have $\Sigma(B) = \hat{\sigma}_A(\{a_1, \ldots, a_N, b_1, \ldots, b_N\})$, and we have $1 = \Sigma b_j(a_j - z_j^0)$, whence $1 = \Sigma \hat{b}_j(\hat{a}_j - z_j^0)$. Thus, if $(w_1, \ldots, w_N, w_{N+1}, \ldots, w_{2N})$ is a point of $\hat{\sigma}_A(\{a_1, \ldots, a_N, b_1, \ldots, b_N\})$, we must have $\Sigma w_{N+j}(w_j - z_j^0) = 1$. This precludes the possibility that $w_j = z_j^0$, i.e., the possibility that $\mathfrak{z}^0 \in \pi\hat{\sigma}(\{a_1, \ldots, b_N\})$. Thus, $\pi\hat{\sigma}_A(\{a_1, \ldots, b_N\})$ misses a neighborhood $V_{\mathfrak{z}^0}$ of $\mathfrak{z}^0$. Finitely many of the neighborhoods $V_{\mathfrak{z}^0}$, say $V_1, \ldots, V_k$, cover the compact set $\hat{\sigma}_A(\{a_1, \ldots, a_N\}) \setminus \Omega$. Let the associated $b$'s be $b_1, \ldots, b_N, b_{N+1}, \ldots, b_{2N}, \ldots, \ldots, b_{kN}$. If then $\pi : \mathbb{C}^{N(1+k)} \to \mathbb{C}^N$ is the projection onto the first $N$ coordinates, our construction shows that

$$\pi\hat{\sigma}_A(\{a_1, \ldots, a_N, b_1, \ldots, b_{kN}\}) \subset \Omega.$$

PROOF OF THEOREM 8.4

We begin by considering a fixed neighborhood $\Omega$ of $\sigma_A(\underline{a})$, $\underline{a} = (a_1, \ldots, a_N)$, and constructing a certain homomorphism $H_{\underline{a}}^\Omega : \mathcal{O}(\Omega) \to A$. Invoke Lemma 8.7 to find $a_{N+1}, \ldots, a_M$ in $A$ such that if $\pi : \mathbb{C}^M \to \mathbb{C}^N$

is the natural projection, then $\pi(\hat{\sigma}_A(\{a_1,\ldots,a_M\})) \subset \Omega$, and then use Lemma 8.6 to find a polynomial polyhedron $P$ such that

$$\hat{\sigma}_A(\{a_1,\ldots,a_M\}) \subset P \subset \bar{P} \subset \pi^{-1}(\Omega).$$

In the notation of the definition of polynomial polyhedra, let $P = P(\delta; p_1,\ldots,p_L)$. The Oka mapping $\psi:\mathbb{C}^M \to \mathbb{C}^{M+L}$ given by

$$\psi(\mathfrak{z}) = (\mathfrak{z}, p_1(\mathfrak{z}),\ldots,p_L(\mathfrak{z}))$$

carries $P$ onto a closed analytic submanifold, $\mathscr{M}_P$, of the polycylinder

$$\Delta_\delta = \{(z_1,\ldots,z_{M+L}) \in \mathbb{C}^{M+L} : |z_1| < \delta_1,\ldots,|z_{M+L}| < \delta_{M+L}\}.$$

There is a natural homomorphism $\eta:\mathscr{O}(\Delta_\delta) \to A$ defined as follows: Let $G \in \mathscr{O}(\Delta_\delta)$ have the power-series expansion

$$G(\mathfrak{z}) = \Sigma\alpha(k)\mathfrak{z}^k,$$

where the summation extends over all $(M + L)$-tuples $k$ of nonnegative integers and where $\mathfrak{z}^k$ denotes $z_1^{k_1}\cdots z_{M+L}^{k_{M+L}}$. Let $c_j = p_j(a_1,\ldots,a_M)$, a certain element of $A$, and set

$$\eta G = \Sigma\alpha(k)a_1^{k_1}\cdots a_M^{k_M}c_1^{k_{M+1}}\cdots c_L^{k_{M+L}}.$$

We must verify, as a first step, that this prescription does define an element of $A$. To do this, we observe that under the mapping $\psi$, $\hat{\sigma}_A(\{a_1,\ldots,a_M\})$ goes to a compact set $K$ in $\Delta_\delta$, so we can find an $(M+L)$-tuple $(r_1,\ldots,r_{M+L})$ such that if $(z_1,\ldots,z_{M+L}) \in \psi(\hat{\sigma}_A(\{a_1,\ldots,a_M\}))$, then for $j = 1,\ldots,M + L$, $|z_j| < r_j < \delta_j$. Since for all $a \in A$, we have $\lim_{n\to\infty} \|a^n\|^{1/n} = \sup\{|\varphi(a)| : a \in \Sigma(A)\}$, it follows that if $k_1,\ldots,k_{N+L}$ are large enough, we shall have

$$\|a_1^{k_1}\| < r_1^{k_1},\ldots,\|a_M^{k_M}\| < r_M^{k_M}$$

and

$$\|c_1^{k_{M+1}}\| < r_{M+1}^{k_{M+1}},\ldots,\|c_L^{k_{M+L}}\| < r_{M+L}^{k_{M+L}}.$$

The series $\Sigma|\alpha(k)|r_1^{k_1}\cdots r_{M+L}^{k_{M+L}}$ converges, so our estimates imply that the series used to define $\eta G$ converges in $A$. Consequently, $\eta$ is a well-defined mapping. Clearly, it is a homomorphism.

If $\varphi \in \Sigma(A)$, we have

$$\begin{aligned}\varphi(\eta G) &= \Sigma\alpha(k)\varphi(a_1)^{k_1}\cdots\varphi(c_L)^{k_{M+L}}\\ &= G(\hat{a}_1(\varphi),\ldots,\hat{c}_L(\varphi))\\ &= G(\psi(\hat{a}_1(\varphi),\ldots,\hat{a}_M(\varphi))).\end{aligned}$$

The proof of Theorem 8.3 may be concluded now by appealing to the fact (Appendix D) that given $F \in \mathscr{O}(\Omega)$, there is $G \in \mathscr{O}(\Delta_\delta)$ such that if $\pi:\mathbb{C}^M \to \mathbb{C}^N$ is the natural projection onto the first $N$ coordinates, then $G \circ \psi = F \circ \pi$. If $F$ and $G$ are so related, our calculation shows that $\varphi(\eta G) = F(\hat{a}_1(\varphi), \ldots, \hat{a}_N(\varphi))$.

To continue with the proof of Theorem 8.4, observe that if $Q_j(z_1, \ldots, z_{M+L}) = z_{M+j} - p_j(z_1, \ldots, z_M)$, then $\psi(P)$ is the manifold

$$\mathscr{M}_P = \{(z_1, \ldots, z_{M+L}) \in \Delta_\delta : Q_j(z_1, \ldots, z_{M+L}) = 0, 1 \le j \le L\}$$

in $\Delta_\delta$. Denote by $I(\mathscr{M}_P)$ the ideal $\{F \in \mathscr{O}(\Delta_\delta) : F|\mathscr{M}_P = 0\}$. By Lemma 8.5 the polynomials $Q_j$ generate $I(\mathscr{M}_P)$, so if $F \in I(\mathscr{M}_P)$, then $F = \sum_{j=1}^{L} G_j Q_j$ for some choice of $G_j$ in $\mathscr{O}(\Delta_\delta)$. We have that $\eta(z_{M+j}) = p_j(a_1, \ldots, a_M)$ and $\eta p_j = p_j(a_1, \ldots, a_M)$; so $\eta Q_j = 0$ and, thus, $I(\mathscr{M}_P)$ is contained in the kernel of $\eta$. Consequently, $\eta$ induces a homomorphism $\tilde{\eta}:\mathscr{O}(\Delta_\delta)/I(\mathscr{M}_P) \to A$. Again, since each function holomorphic on $\mathscr{M}_P$ extends to a holomorphic function on $\Delta_\delta$, $\mathscr{O}(\Delta_\delta)/I(\mathscr{M}_P)$ may be identified with $\mathscr{O}(\mathscr{M}_P)$. The desired homomorphism $H_{\underline{a}}^{\Omega}$ can now be constructed. Given $F \in \mathscr{O}(\Omega)$, $F \circ \pi \circ \psi^{-1}$ is in $\mathscr{O}(\mathscr{M}_P)$, and we define

$$H_{\underline{a}}^{\Omega} F = \tilde{\eta}(F \circ \pi \circ \psi^{-1}).$$

The mapping $H_{\underline{a}}^{\Omega}$ so defined is a homomorphism of $\mathscr{O}(\Omega)$ into $A$, and we have seen that $\varphi(H_{\underline{a}}^{\Omega} F) = F(\hat{a}_1(\varphi), \ldots, \hat{a}_N(\varphi))$ for all $\varphi \in \Sigma(A)$. The definition of $H_{\underline{a}}^{\Omega}$ shows that $H_{\underline{a}}^{\Omega} Z_j = a_j$.

In the construction of the homomorphism $H_{\underline{a}}^{\Omega}$, we have made certain choices; we must verify that $H_{\underline{a}}^{\Omega}$ is, in fact, independent of these choices. The choices involved in the construction of $H_{\underline{a}}^{\Omega}$ were these:

(A) We chose elements $a_{N+1}, \ldots, a_M \in A$ such that

$$\pi\hat{\sigma}_A(\{a_1, \ldots, a_M\}) \subset \Omega.$$

(B) Having fixed $a_{N+1}, \ldots, a_M$, we chose a polynomial polyhedron $P = P(\delta; p_1, \ldots, p_L)$ with

$$\hat{\sigma}_A(\{a_1, \ldots, a_M\}) \subset P \subset \bar{P} \subset \pi^{-1}(\Omega).$$

Assume that we are given the homomorphism $H_{\underline{a}}^{\Omega}$ constructed with the given data (A) and (B). We construct an alternative homomorphism $H_{\underline{a}}'^{\Omega}$ as follows. Begin by choosing additional elements $a_{M+1}, \ldots, a_{M'}$ in $A$. If $\pi':\mathbb{C}^{M'} \to \mathbb{C}^N$ is the natural projection, then $\pi'\hat{\sigma}_A(\{a_1, \ldots, a_{M'}\}) \subset \Omega$. For $j = 1, \ldots, L$, the function $p_j' = p_j \circ \pi''$, where $\pi'':\mathbb{C}^{M'} \to \mathbb{C}^M$ is the natural projection, is a polynomial. If we choose polynomials $p_{L+1}', \ldots, p_{L'}'$ and numbers $\delta_{M+L+1}, \ldots, \delta_{M+L'}$, $\varepsilon_{M+1}, \ldots, \varepsilon_{M'}$ such that the polynomial polyhedron $P'$ defined by

$$P' = \{\mathfrak{z} = (z_1, \ldots, z_{M'}) : |z_1| < \delta_1, \ldots, |z_M| < \delta_M, |z_{M+1}| < \varepsilon_{M+1}, \ldots, |z_{M'}| < \varepsilon_{M'}, |p_j'(\mathfrak{z})| < \delta_{j+M}, 1 \le j \le L'\}$$

is a neighborhood of $\hat{\sigma}_A(a_1, \ldots, a_{M'})$ whose closure is contained in $(\pi')^{-1}(\Omega)$, then we can define the homomorphism $H_{\underline{a}}'^{\Omega}$ using $a_1, \ldots, a_{M'}$ and $P'$. We have $\psi : C^M \to \mathbb{C}^{M+L}$; define in addition $\psi' : \mathbb{C}^{M'} \to \mathbb{C}^{M'+L'}$ by $\psi'(\mathfrak{z}) = (\mathfrak{z}, p_1'(\mathfrak{z}), \ldots, p_{L'}'(\mathfrak{z}))$.

Similarly, we let

$$\Delta'_{\delta,\varepsilon} = \{(z_1, \ldots, z_{M'+L'}) : |z_1| < \delta_1, \ldots, |z_M| < \delta_M, |z_{M+1}| < \varepsilon_{M+1}, \ldots, \\ |z_{M'}| < \varepsilon_{M'}, |z_{M'+1}| < \delta_{M+1}, \ldots, |z_{M'+L'}| < \delta_{M+L'}\}.$$

If $f \in \mathcal{O}(\Omega)$, and $f \circ \pi = F \circ \psi$ for $F \in \mathcal{O}(\Delta_\delta)$ with

$$F(z_1, \ldots, z_{M+L}) = \Sigma\alpha(k) z_1^{k_1} \cdots z_{M+L}^{k_{M+L}},$$

then

$$H_{\underline{a}}^{\Omega} f = \Sigma\alpha(k) a_1^{k_1} \cdots a_M^{k_M} c_1^{k_{M+1}} \cdots c_{M+L}^{k_{M+L}}.$$

But now, notice that if $F$ and $f$ are related in this way, and if $F' \in \mathcal{O}(\Delta'_{\delta,\varepsilon})$ is defined by

$$F'(z_1, \ldots, z_{M'+L'}) = F(z_1, \ldots, z_M, z_{M'+1}, \ldots, z_{M'+L}),$$

then $F' \circ \psi' = f \circ \pi'$ and we may conclude that $H_{\underline{a}}'^{\Omega} f = H_{\underline{a}}^{\Omega}$.

Let us digress to point out that as soon as we have shown $H_{\underline{a}}^{\Omega}$ to be defined independently of the choices (A) and (B), this last fact will imply that if $\underline{b} = \{a_1, \ldots, a_N, b_1, \ldots, b_S\}$, then $H_{\underline{b}}^{\Omega'}(F \circ \pi) = H_{\underline{a}}^{\Omega} F$ for all $F \in \mathcal{O}(\Omega)$ provided $\pi(\Omega') \subset \Omega$, $\pi : \mathbb{C}^{N+S} \to \mathbb{C}^N$ the natural projection.

We can show now that our homomorphism $H_{\underline{a}}^{\Omega}$ is, in fact, independent of the choices (A) and (B). Let $H_{\underline{a}}^{\Omega}$ be defined using the data (A) and (B), and let $H_{\underline{a}}''^{\Omega}$ be constructed from the data

(A″) Elements $a''_{N+1}, \ldots, a''_{M''}$ of $A$ such that

$$\pi''(\hat{\sigma}_A(\{a_1, \ldots, a_N, a''_{N+1}, \ldots, a''_{M''}\})) \subset \Omega.$$

(B″) A polynomial polyhedron $P'' = P(\delta'', p''_1, \ldots, p''_{L''})$ with

$$\hat{\sigma}_A(\{a_1, \ldots, a_N, a''_{N+1}, \ldots, a''_{M''}\}) \subset P'' \subset \bar{P}'' \subset (\pi'')^{-1}(\Omega).$$

If we now construct $H_{\underline{a}}'^{\Omega}$ as in the next-to-last paragraph using $a''_{N+1}, \ldots, a''_{M''}$ in place of $a_{M+1}, \ldots, a_{M'}$ and $p''_1, \ldots, p''_L$ in place of $p'_{L+1}, \ldots, p'_{L'}$, we find that if $f \in \mathcal{O}(\Omega)$, then $H_{\underline{a}}^{\Omega} f = H_{\underline{a}}'^{\Omega} f$ and also $H_{\underline{a}}''^{\Omega} f = H_{\underline{a}}'^{\Omega} f$, whence $H_{\underline{a}}^{\Omega} f = H_{\underline{a}}''^{\Omega} f$. Thus, $H_{\underline{a}}^{\Omega}$ depends only on $\underline{a}$ and $\Omega$ and not on the choices (A) and (B).

Suppose now that $\Omega \supset \Omega' \supset \sigma_A(\{a_1, \ldots, a_N\})$. We have homomorphisms $H_{\underline{a}}^{\Omega} : \mathcal{O}(\Omega) \to A$ and $H_{\underline{a}}^{\Omega'} : \mathcal{O}(\Omega') \to A$. In addition, we have a natural inclusion $i : \mathcal{O}(\Omega) \to \mathcal{O}(\Omega')$ given by restriction. The fact that $H_{\underline{a}}^{\Omega}$ is defined independently of the choices (A) and (B) implies that

$H_{\underline{a}}^{\Omega} = H_{\underline{a}}^{\Omega'} \circ i$. This is clear, for if the choices (A) and (B) are made to construct $H_{\underline{a}}^{\Omega'}$, then since $\Omega \supset \Omega'$, the same choices will work in the construction of $H_{\underline{a}}^{\Omega}$ and our remark follows.

Therefore, the homomorphisms $H_{\underline{a}}^{\Omega}$, for fixed $\underline{a}$, fit together to yield a homomorphism $H_{\underline{a}}:\mathcal{O}(\sigma_A(\{a_1,\ldots,a_N\}))\to A$. Given $\tilde{F}\in\mathcal{O}(\sigma_A(\{a_1,\ldots,a_N\}))$, choose a neighborhood $\Omega$ of $\sigma_A(\{a_1,\ldots,a_N\})$ in which there is defined an $F \in \tilde{F}$. Define $H_{\underline{a}}\tilde{F}$ to be $H_{\underline{a}}^{\Omega}F$. The element of $A$ defined in this way is defined independently of the particular choice of $F \in \tilde{F}$, and the map $H_{\underline{a}}$ defined in this way is a homomorphism which has the desired properties (1) and (2). The fact that the homomorphisms $H_{\underline{a}}$ have the property (3) follows from our observation that if $\Omega' \in \mathscr{V}(\sigma_A(\{a_1,\ldots,a_N,b_1,\ldots,b_s\})$, and if $\pi(\Omega') \subset \Omega$, then $H_{\underline{b}}^{\Omega'}(\pi \circ F)^{\sim} = H_{\underline{a}}^{\Omega}\tilde{F}$ for all $F \in \mathcal{O}(\Omega)$.

Some remarks on the history of Theorems 8.3 and 8.4 are now in order. The use of the Cauchy integral in one variable to define functions of Banach algebra elements is found in the paper of Gelfand [1]. Before this, Beurling [2] had used the method in Fourier analysis. For finitely generated algebras, Theorem 8.3 was proved by Shilov [1], who used as an essential tool Weil's integral formula (Weil [1]). Theorem 8.3 for arbitrary commutative Banach algebras is due to Arens and Calderón [1], who also based their work on the Weil integral. It was Waelbroeck [1] who, working in a slightly different setting, realized the relevance of the Oka–Cartan theory for this sort of problem and who observed that it is possible to construct a *homomorphism* from $\mathcal{O}(\sigma_A(\{a_1,\ldots,a_N\}))$ into $\Sigma(A)$. Another exposition of the work of Waelbroeck is in Waelbroeck [2]. Arens [3] has shown that it is possible to base the construction of the homomorphisms $H_{\underline{a}}$ on an integral representation. Bourbaki [1] also bases this construction on an integral representation theorem.

It is possible to show that the homomorphisms $H_{\underline{a}}$ have an additional property of continuity. If we fix $\underline{a}$ and if $V$ is a neighborhood of $\sigma_A(\underline{a})$, then $\mathcal{O}(V)$ has a natural topology, that of uniform convergence on compacta. With this topology, we can induce a topology on $\mathcal{O}(\sigma_A(\underline{a}))$. $\mathcal{O}(\sigma_A(\underline{a}))$ is given the strongest topology which makes the maps $i_V:\mathcal{O}(V) \to \mathcal{O}(\sigma_A(\underline{a}))$ continuous. It is not hard to show, on the basis of our construction, that the homomorphisms $H_{\underline{a}}:\mathcal{O}(\sigma_A(\underline{a})) \to A$ are all continuous.

There is also a uniqueness assertion: There is only one family of continuous homomorphisms $H_{\underline{a}}:\mathcal{O}(\sigma(\underline{a})) \to A$ with the properties enunciated in Theorem 8.4. For this point, see Bourbaki [1].

It is interesting to see how the theorem just proved can be formulated in terms of certain functions on infinite-dimensional spaces. Let $\Lambda$ be an index set and consider the product space $\mathbb{C}^\Lambda$ equipped with the usual product topology. On $\mathbb{C}^\Lambda$ consider the ring $\mathscr{P}_\Lambda$ of polynomials in $\Lambda$ variables. (Recall Definition 5.3.)

Given $V$, an open set in $\mathbb{C}^\Lambda$, let $\mathcal{O}(V)$ be the collection of functions which admit the following approximation: $F \in \mathcal{O}(V)$ if for all $\mathfrak{z} \in V$ there is a neighborhood $N$ of $\mathfrak{z}$ on which $F$ is the uniform limit of a sequence of elements of $\mathscr{P}_\Lambda$. Since $\mathbb{C}^\Lambda$ has the product topology, this is a very stringent condition on $F$ as our next lemma shows. If $\Lambda_0 \subset \Lambda$, let $\pi_{\Lambda_0}$ denote the natural projection from $\mathbb{C}^\Lambda$ to $\mathbb{C}^{\Lambda_0}$.

**8.8** *Lemma* (Rickart)

*Let $\Lambda_0$ be a finite sunset of $\Lambda$, let $\Omega \subset \mathbb{C}^{\Lambda^0}$ πe open, and let $F$ be a uniform limit on $\pi_{\Lambda_0}^{-1}(\Omega)$ of a sequence in $\mathscr{P}_\Lambda$. For some finite set $\Lambda_0' \subset \Lambda$ that contains $\Lambda_0$, we have $F(\mathfrak{z}) = F(\mathfrak{z}')$ if $\pi_{\Lambda_0'}(\mathfrak{z}) = \pi_{\Lambda_0'}(\mathfrak{z}')$.*

Thus, roughly put, the elements of $\mathcal{O}(V)$ depend locally on only finitely many variables.

PROOF

Let $\|F - p_n\|_{\pi_{\Lambda_0}^{-1}(\Omega)} \to 0$ as $n \to \infty$. Thus, the sequence $\{p_n\}$ is uniformly Cauchy on $\pi_{\Lambda_0}^{-1}(\Omega)$. The set $\pi_{\Lambda_0}^{-1}(\Omega)$ is $\Omega \times \mathbb{C}^{\Lambda \setminus \Lambda_0}$, and since the polynomial $p_n - p_m$ is bounded on this set, it follows that it must be constant on each of the fibers $\pi_{\Lambda_0}^{-1}(\mathfrak{z})$, $\mathfrak{z} \in \Omega$. Consequently, $p_n - F$ must be constant on these fibers. If $p_n \in \mathscr{P}_{\Lambda_0'}$ for a subset $\Lambda_0'$ of $\Lambda$, which can be chosen to include $\Lambda_0$, then we have that $F(\mathfrak{z}) = F(\mathfrak{z}')$ provided $\pi_{\Lambda_0'}(\mathfrak{z}) = \pi_{\Lambda_0'}(\mathfrak{z}')$, and the lemma is proved.

In the situation of the lemma, the function $F$ admits a factorization $F = F_1 \circ \pi_{\Lambda_0'}$ for $F_1$ a function holomorphic on $\pi_{\Lambda_0'}(\Omega \times \mathbb{C}^{\Lambda \setminus \Lambda_0}) = \Omega \times \mathbb{C}^{\Lambda_0' \setminus \Lambda_0}$.

Let $K \subset \mathbb{C}^\Lambda$ be a compact set, and let $V$ be an open set which contains $K$. If $\mathfrak{z} \in K$ and $F \in \mathcal{O}(V)$, there is a neighborhood $W_\mathfrak{z}$ of $\mathfrak{z}$ contained in $V$ on which $F$ is a uniform limit of polynomials. Finitely many of the $W_\mathfrak{z}$, say $W_1, \ldots, W_q$, cover $K$, and by the lemma there is, for each $j$, a finite set $\Lambda_j \subset \Lambda$ such that if $\mathfrak{z}$ and $\mathfrak{z}'$ lie in $W_j$ and if $\pi_{\Lambda_j}(\mathfrak{z}) = \pi_{\Lambda_j}(\mathfrak{z}')$, then $F(\mathfrak{z}) = F(\mathfrak{z}')$. Thus, if we set $W = W_1 \cup \cdots \cup W_q$ and $\Lambda_0 = \Lambda_1 \cup \cdots \cup \Lambda_q$, then $K \subset W \subset V$; and given $\mathfrak{z}, \mathfrak{z}' \in W$ with $\pi_{\Lambda_0}(\mathfrak{z}) = \pi_{\Lambda_0}(\mathfrak{z}')$, we have $F(\mathfrak{z}) = F(\mathfrak{z}')$. Consequently, $F = F_1 \circ \pi_{\Lambda_0}$, where $F_1$ is a function holomorphic on a neighborhood of $\pi_{\Lambda_0}(K)$ in $\mathbb{C}^{\Lambda_0}$.

If we are given a compact $K$ in $\mathbb{C}^\Lambda$, we define $\mathcal{O}(K)$ to be the quotient of $\bigcup\{\mathcal{O}(V) : V$ is an open set containing $K\}$ modulo the equivalence relation $\sim$ defined by saying that $F \sim G$ for $F \in \mathcal{O}(V)$, $G \in \mathcal{O}(V')$ if there exists an open set $V''$ containing $K$ and contained in $V' \cap V$ on which $F$ and $G$ agree. The last paragraph shows that $\mathcal{O}(K)$ can be identified in a natural way with

$$\bigcup\{\mathcal{O}(\pi_{\Lambda_0}(K)) : \Lambda_0 \text{ a finite subset of } \Lambda\}$$

when this union is reduced modulo an evident equivalence relation.

Let $A$ be a commutative Banach algebra with identity, and in $A$ consider a set $\{a_\lambda : \lambda \in \Lambda\}$ of generators for $A$. Let $\Sigma_0 \subset \mathbb{C}^\Lambda$ be the canonical embedding of $\Sigma(A)$ in $\mathbb{C}^\Lambda$ defined by this set of generators; i.e., $\Sigma_0$ is the image of $\Sigma(A)$ under the map $i : \Sigma(A) \to \mathbb{C}^\Lambda$ determined by $\pi_\lambda(i\varphi) = \varphi(a_\lambda)$. We know that $\Sigma_0$ is a compact, polynomially convex subset of $\mathbb{C}^\Lambda$. Theorem 8.4 allows us to define a homomorphism $H : \mathcal{O}(\Sigma_0) \to A$. If $\tilde{F} \in \mathcal{O}(\Sigma_0)$, there is a neighborhood $W$ of $\Sigma_0$ and an $F \in \mathcal{O}(W)$ that represents $F$. On a smaller neighborhood $W'$ of $\Sigma_0$, $F$ has the form $F = f \circ \pi_{\Lambda_0}$ for some finite subset $\Lambda_0$ of $\Lambda$ and some $f$ holomorphic on a neighborhood of $\pi_{\Lambda_0}(\Sigma_0) = \sigma_A(\{a_\lambda\}_{\lambda \in \Lambda_0})$. If $\tilde{f}$ is the germ of $f$ on $\sigma_A(\{a_\lambda\}_{\lambda \in \Lambda_0})$, then, by Theorem 8.4, $H_{S(\Lambda_0)}(\tilde{f})$ is a well-defined element of $A$, where we write $S(\Lambda_0)$ for $\{a_\lambda\}_{\lambda \in \Lambda_0}$. We define $H(F)$ to be this element of $A$, and we see that $H(F)$ is well defined and that $H$ is the desired homomorphism.

For a further generalization of the holomorphic functional calculus, see the appendix to this section. Additional related material is contained in the papers of Arens [4], Allan [1], and Rickart [5].

We next turn to some of the immediate consequences of the preceding theory. In subsequent sections we shall see other applications.

The first corollary we shall derive is a relation between the algebraic structure of a commutative Banach algebra and the topological structure of its spectrum. The result is due to Shilov [1], and it is frequently referred to as the *Shilov idempotent theorem.*

### 8.9 *Theorem*

*Let $A$ be a commutative Banach algebra with identity. If $\Sigma(A) = \Sigma_0 \cup \Sigma_1$ is a decomposition of $\Sigma(A)$ into the union of two disjoint open and closed sets, then there exists an element $e \in A$ such that $e^2 = e$, $\hat{e}|\Sigma_0$ is identically zero, and $\hat{e}|\Sigma_1$ is identically* 1.

Observe that the converse of the theorem is trivial: If $e \in A$ is an idempotent, then $\Sigma(A) = \{\varphi : \hat{e}(\varphi) = 0\} \cup \{\varphi : \hat{e}(\varphi) = 1\}$ is a decomposition of $\Sigma(A)$ into a union of disjoint open, closed subsets.

PROOF OF THEOREM 8.9

Given a point $\psi \in \Sigma_0$ and a point $\varphi \in \Sigma_1$, there is $a_\varphi \in A$ with $\hat{a}_\varphi(\varphi) = 1$, $\hat{a}_\varphi(\psi) = 0$. Thus, on a neighborhood $V_\varphi$ of $\varphi$ we have $\operatorname{Re} \hat{a} > \frac{3}{4}$ and on a neighborhood of $\psi$ we have $\operatorname{Re} \hat{a} < \frac{1}{4}$. Leaving $\psi$ fixed, letting $\varphi$ vary over $\Sigma_1$, and using the compactness of $\Sigma_1$, we find a finite collection of $a$'s, say $a_{\psi,1}, \ldots, a_{\psi,n}$ such that if $\varphi \in \Sigma_1$, then $\operatorname{Re} \hat{a}_{\psi,j}(\varphi) > \frac{3}{4}$ for some $j$, and such that on some fixed neighborhood $W_\psi$ of $\psi$, all the $\hat{a}_{\psi,j}$ have real parts less than $\frac{1}{4}$. Again by compactness, finitely many of the $W_\psi$ cover $\Sigma_0$, say $W_{\psi_1}, \ldots, W_{\psi_m}$. If $\{b_1, \ldots, b_M\}$ is the collection of $a_{\psi_i,j}$ obtained in this way, so that given $\varphi \in \Sigma_1, \psi \in \Sigma_0$, $\operatorname{Re} \hat{b}_j(\varphi) > \frac{3}{4}$ and $\operatorname{Re} \hat{b}_j(\psi) < \frac{1}{4}$ for some choice of $j$, then the joint spectrum $\sigma_A(\{b_1, \ldots, b_M\})$ is decomposed into two disjoint open, closed subsets

$$\sigma_0 = \{(\hat{b}_1(\varphi), \ldots, \hat{b}_M(\varphi)) : \varphi \in \Sigma_0\}$$

and

$$\sigma_1 = \{(\hat{b}_1(\varphi), \ldots, \hat{b}_M(\varphi)) : \varphi \in \Sigma_1\}.$$

If we define a function $F$ by requiring it to be zero on a neighborhood of $\sigma_0$ and 1 on a neighborhood of $\sigma_1$, then $F$ is holomorphic on a neighborhood of $\sigma_A(\{b_1, \ldots, b_M\})$. If $\underline{b} = \{b_1, \ldots, b_M\}$, the element $H_{\underline{b}}\tilde{F}$ of $A$ has the desired property.

It seems worthwhile to observe that Theorem 8.3 suffices to prove the present theorem. By Theorem 8.3, there is an $a \in A$ with $\hat{a}|\Sigma_1 = 1$, $\hat{a}|\Sigma_0 = 0$. If the algebra $A$ is semisimple, our proof is concluded since $\hat{a}^2 = \hat{a}$ implies, in this case, that $a^2 = a$. In the nonsemisimple case we must argue further. We have that $\hat{a}^2 - \hat{a}$ vanishes identically, so $a^2 - a$ is in the radical of $A$. We set $r = a^2 - a$ and

$$e = (a - \tfrac{1}{2})(1 + 4r)^{-1/2} + \tfrac{1}{2},$$

where the square root is to be computed according to the binomial expansion

$$(1 + 4r)^{-1/2} = \sum_{n=0}^{\infty} \binom{-1/2}{n} (4r)^n.$$

The element $r$ is in the radical, so $\lim_{n\to\infty} \|(4r)^n\|^{1/n} = 0$ and it follows that the series does converge in $A$. A simple computation shows that $e(e - 1) = 0$, i.e., that $e^2 = e$. Since $e$ differs from $A$ by an element of the radical, $\hat{e} = \hat{a}$, and the theorem is established again.

So far, we have considered only algebras with identity. However, there are results valid for algebras lacking an identity. For this purpose, we need to give a definition of the joint spectrum of an $N$-tuple of elements in an algebra lacking an identity. There is only one reasonable choice: If $A$ is a commutative Banach algebra without identity, and if $a_1, \ldots, a_N \in A$, then $\sigma_A(\{a_1, \ldots, a_N\})$ is defined to be $\sigma_{A'}(\{a_1, \ldots, a_N\})$, where $A'$ is obtained by adjoining to $A$ an identity.

**8.10** ***Theorem***

*If $A$ is a commutative Banach algebra, if $a_1, \ldots, a_N$ are elements of $A$, if $F$ is holomorphic on a neighborhood of $\sigma_A(\{a_1, \ldots, a_N\})$, and if $F(0) = 0$, then there is $b \in A$ with $\hat{b}(\varphi) = F(\hat{a}_1(\varphi), \ldots, \hat{a}_N(\varphi))$ for all $\varphi \in \Sigma(A)$.*

This result is obtained from Theorem 8.3 just as Theorem 8.2 was derived from Theorem 8.1.

The following is a version of the Shilov idempotent theorem valid in algebras lacking an identity.

**8.11** ***Theorem***

*If $K$ is an open, compact subset of the spectrum of the commutative Banach algebra $A$, then there exists an idempotent element $e$ of $A$ such that $\hat{e}|K = 1$, $\hat{e}|(\Sigma(A) \setminus K) = 0$.*

PROOF

Since $K$ is compact and open, it follows that when we regard it as a subset of $\Sigma(A')$, $A'$ the algebra obtained by adjoining an identity to $A$, it is still an open and closed set. Thus, there is an idempotent $e$ in $A'$ such that $\hat{e}|K = 1$, $\hat{e}|(\Sigma(A') \setminus K) = 0$. Since $\hat{e}$ vanishes on $\Sigma(A') \setminus \Sigma(A)$, $e$ is in $A$, and the result follows.

Recall that a *hull* in $\Sigma(A)$ is a set of the form $\{\varphi \in \Sigma(A): \hat{a}(\varphi) = 0$ for all $a \in E\}$ for some subset $E$ of $A$.

**8.12** ***Corollary***

*If $K$ is a compact, open subset of the hull $H$ in $\Sigma(A)$, then $K$ itself is a hull.*

PROOF

Let $I$ be the closed ideal $\{a \in A; \hat{a}|H = 0\}$. The spectrum of $A/I$ can be identified with $H$, and Theorem 8.11 yields an element $\alpha \in A/I$ with $\hat{\alpha}|K = 1$, $\hat{\alpha}|(H \setminus K) = 0$. The element $\alpha$ is an equivalence class of elements of $A$; let $a_0 \in \alpha$. For each $\varphi \in H \setminus K$, let $b_\varphi \in A$ be such that $\hat{b}_\varphi(\varphi) \neq 0$. Then

$$K = \{\psi \in \Sigma(A): (b_\varphi - b_\varphi a_0)\hat{\ }(\psi) = 0 \text{ for all } \varphi \in H \setminus K\}$$
$$\cap \{\psi \in \Sigma(A): \hat{a}(\psi) = 0 \text{ for all } a \in I\},$$

and we have that $K$ is a hull.

There is an essentially stronger version of this corollary, a version due to Rossi [1]. It depends on the following fact from set theoretic topology.

**8.13 *Lemma***

*Let $X$ be a compact space, $X_0$ a ~~compact~~ component of $X$, and $Y$ a closed subset of $X$ which is disjoint from $X_0$. Then $X = U \cup V$, where $U$ and $V$ are disjoint open and closed subsets of $X$ such that $X_0 \subset U$, $Y \subset V$.*

PROOF

Let $X_1$ be the intersection of all open and closed subsets of $X$ which contain $X_0$. We shall show that $X_0 = X_1$. If this equality is false, then $X_1$ is not connected, so $X_1 = X_1' \cup X_1''$ where this is a disjoint union of relatively open subsets of $X_1$. Since $X_0$ is connected, it is contained in $X_1'$ or else in $X_1''$, say in $X_1'$. The set $X_1$ is closed in $X$, and, consequently, so are $X_1'$ and $X_1''$. Thus, by normality, there exist disjoint open sets $V_1'$ and $V_1''$ which contain $X_1'$ and $X_1''$, respectively. The definition of $X_1$ implies the existence of an open and closed set $S$ of $X$ such that $X_1 \subset S \subset V_1' \cup V_1''$. Then $S' = V_1 = V_1 \cap S$ is an open and closed subset of $X$ which contains $X_0$ but which is strictly contained in $X_1$. Consequently, $X_0 = X_1$.

Let $\mathscr{S}$ be the family of open and closed subsets of $X$ which contain $X_0$. The set $\mathscr{S}$ is closed under the formation of finite intersections. We have $X_0 = \bigcap \{S : S \in \mathscr{S}\}$, and since $X_0 \cap Y = \varnothing$, it follows that there is $S \in \mathscr{S}$ which misses $Y$. Then

$$X = S \cup (X \setminus S)$$

is a decomposition of $X$ of the desired kind.

I am indebted to Mrs. S. Negrepontis for showing me this very simple proof.

The result of Rossi is the following.

**8.14 *Corollary***

*If $E$ is a compact component of the hull $H$ in $\Sigma(A)$, then $E$ itself is a hull.*

PROOF

Let $F$ be a component of $H$ disjoint from $E$. An application of Lemma 8.13 to $H^*$, the one-point compactification of $H$, provides a decomposition $H = H_0 \cup H_1$, where $H_0$ and $H_1$ are disjoint open and closed subsets of $H$ such that $H_0 \supset E$, $H_1 \supset F$, and such that $H_0$ is compact. Then $H_0$ is a hull by Theorem 8.11. Thus, $E$ is an intersection of hulls and so a hull.

**8.15 *Corollary***

*If $A$ is semisimple and if $\Sigma(A)$ is compact, then $A$ has an identity.*

PROOF

Since $\Sigma(A)$ is compact, there is $e \in A$ such that $\hat{e}$ is identically 1, and since $A$ is semisimple, $e$ must be an identity for $A$: The Gelfand transform $\hat{}: A \to \hat{A}$ is an isomorphism and for all $a \in A$, $(ea)\hat{} = \hat{e}\hat{a} = \hat{a}$, so $ea = a$.

This corollary is obviously false without the hypothesis of semisimplicity: Let $B$ be the maximal ideal of the algebra in Example 1.6, and let $A$ be any algebra with identity. If $A_1 = A \oplus B$, then $\Sigma(A_1) = \Sigma(A)$ and so $\Sigma(A_1)$ is compact. However, $A_1$ has no identity.

**8.16** ***Corollary***

*If $E$ is an open and closed subset of $\Sigma(A)$, and if $A$ has an identity, then $E$ meets the Shilov boundary for $A$.*

This corollary is true without the hypothesis that $A$ have an identity, but the proof in this case seems to require techniques we have not developed yet. We shall return to this point in the next section (Corollary 9.9).

PROOF OF COROLLARY 8.16

Since $E$ is open and closed in $\Sigma(A)$, there exists $a \in A$ with $\hat{a}$ identically 1 on $E$, identically zero on $\Sigma(A) \setminus E$, and therefore $E$ must meet $\Gamma(A)$.

**8.17** ***Corollary***

*If $A$ has an identity and $\Gamma(A)$ is connected, then $\Sigma(A)$ is connected.*

**8.18** ***Corollary***

*If $\Sigma(A)$ is totally disconnected, then $\hat{A}$ is dense in $\mathscr{C}_0(\Sigma(A))$.*

PROOF

Since $\Sigma(A)$ is totally disconnected, each $f \in \mathscr{C}_0(\Sigma(A))$ can be approximated uniformly by functions of the form $s = \sum_{j=1}^{M} \alpha_j \chi_{Ej}$, $\alpha_j \in \mathbb{C}$, and $\chi_{E_j}$ the characteristic function of the open, compact set $E_j \subset \Sigma(A)$. Since $E_j$ is compact and open in $\Sigma(A)$, $\chi_{E_j} = \hat{b}_j$ for some $b_j \in A$. Thus, $s \in \hat{A}$, and the result follows.

There is a generalization of Theorem 8.3 which is of importance in dealing with multiple-valued functions of Banach algebra elements. To formulate it, we need a preliminary definition.

**8.19** ***Definition***

*If $\mathcal{M}$ and $\mathcal{N}$ are complex manifolds, a* **spread of $\mathcal{M}$ into $\mathcal{N}$** *is a holomorphic map $\eta : \mathcal{M} \to \mathcal{N}$ which is a local homeomorphism.*

Thus, for example, the exponential mapping gives a spread of $\mathbb{C}$ into $\mathbb{C} \setminus \{0\}$.

The proof of the following result is due to Arens and Calderón [1].

**8.20** ***Theorem***

*Let $A$ be a commutative Banach algebra with identity, let $\mathcal{M}$ be a complex manifold, and let $\eta = (\eta_1, \ldots, \eta_N) : \mathcal{M} \to \mathbb{C}^N$ be a spread of $\mathcal{M}$ into $\mathbb{C}^N$. If $a_1, \ldots, a_N$ are elements of $A$ and if $\Psi : \Sigma(A) \to \mathcal{M}$ is a continuous map such that for all $\varphi \in \Sigma(A)$, $\eta \circ \Psi(\varphi) = (\hat{a}_1(\varphi), \ldots, \hat{a}_N(\varphi))$, then there exists a homomorphism $H : \mathcal{O}(\Psi(\Sigma(A))) \to A$ such that $H(\eta_j) = a_j$, and such that $(Hf)\hat{}(\varphi) = f(\Psi(\varphi))$ for all $f \in \mathcal{O}(\Psi(\Sigma(A)))$.*

The proof of this result depends on a lemma.

**8.21** ***Lemma***

*If $\mathcal{M}$ is a complex manifold, and if $\eta : \mathcal{M} \to \mathbb{C}^N$ is a spread, then there exists a metric $\rho$ on $\mathcal{M}$ with respect to which $\eta$ is a local isometry when $\mathbb{C}^N$ is provided with its usual metric $d$.*

PROOF

A very simple construction for $\rho$ is as follows. Assume first that $\mathcal{M}$ is connected. Given $p,q \in \mathcal{M}$, define

$$\rho_{\mathcal{M}}(p,q) = \inf \sum_{j=1}^{M} d(\eta(s_{j+1}), \eta(s_j)),$$

where the infimum is taken over all chains of points $p = s_1, \ldots, s_{m+1} = q$ such that for each $j$, there is an open set $V$ in $\mathcal{M}$ which contains both $s_j$ and $s_{j+1}$ and which is mapped homeomorphically by $\eta$ onto a ball $\{\mathfrak{z} : d(\mathfrak{z}, \mathfrak{z}^0) < k\}$ in $\mathbb{C}^N$. Then $\rho_{\mathcal{M}}$ is a metric on $\mathcal{M}$, and with respect to this metric, $\eta$ is a local isometry. If $\mathcal{M}$ is not connected, define $\rho_\alpha$ on each component $\mathcal{M}_\alpha$ of $\mathcal{M}$ as $\rho_{\mathcal{M}}$ was defined above, and then if $p$ and $q$ lie in the component $\mathcal{M}_\alpha$, let $\tilde{\rho}_\alpha(p,q) = \min(\rho_\alpha(p,q), 1)$. Finally, define $\rho$ on $\mathcal{M}$ by

$$\rho(p,q) = \begin{cases} 1 & \text{if } p \text{ and } q \text{ lie in different components of } \mathcal{M}, \\ \tilde{\rho}_\alpha(p,q) & \text{if } p \text{ and } q \text{ lie in the component } \mathcal{M}_\alpha \text{ of } \mathcal{M}. \end{cases}$$

The function $\rho$ so defined is a metric on $\mathscr{M}$ with the desired properties.

PROOF OF THEOREM 8.20

Let $\rho$ be a metric on $\mathscr{M}$ such that with respect to $\rho$ and the metric $d$ on $\mathbb{C}^N$, $\eta$ is a local isometry. Since $\Psi(\Sigma(A))$ is compact, there exists an $\varepsilon > 0$ with the property that $\eta$ is a homeomorphism on every subset of $\Psi(\Sigma(A))$ of diameter less than $\varepsilon$. The continuity of $\Psi$ implies that there is a neighborhood $V$ of the diagonal of $\Sigma(A) \times \Sigma(A)$ such that if $(\varphi, \psi) \in V$, then $\rho(\Psi(\varphi), \Psi(\psi)) < \varepsilon$. If $\varphi,\psi \in \Sigma(A)$, $(\varphi,\psi) \notin V$, there exists $a \in A$ with $\hat{a}(\varphi) \neq \hat{a}(\psi)$. Consequently, for some choice of $\delta > 0$ and some choice of $a_{N+1}, \ldots, a_{N+M} \in A$,

$$\sum_{k=N+1}^{N+M} |\hat{a}_k(\varphi) - \hat{a}_k(\psi)| \geq \delta \tag{1}$$

for all $(\varphi, \psi) \notin V$.

Define $\Phi : \Sigma(A) \to \mathbb{C}^{N+M}$ by

$$\Phi(\varphi) = (\hat{a}_1(\varphi), \ldots, \hat{a}_{N+M}(\varphi)),$$

and let $\pi : \mathbb{C}^{N+M} \to \mathbb{C}^N$ be the natural projection onto the first $N$ coordinates. Finally, let $\theta : \Phi(\Sigma(A)) \to \Psi(\Sigma(A))$ be the mapping determined by $\theta \circ \Phi = \Psi$. We must show that $\theta$ is uniquely determined by this condition, and this amounts to proving that $\Psi$ is constant on the fibers $\Phi^{-1}(\mathfrak{z})$, $\mathfrak{z} \in \sigma_A(\{a_1, \ldots, a_{M+N}\})$. If $\Phi(\varphi) = \Phi(\psi)$, $\varphi,\psi \in \Sigma(A)$, then

$$\sum_{N+1}^{N+M} |\hat{a}_k(\varphi) - \hat{a}_k(\psi)| = 0,$$

so $(\varphi,\psi) \in V$. We know that $\eta \circ \Psi = \pi \circ \Phi$, so if $\Psi(\varphi) \neq \Psi(\psi)$, then $\eta(\Psi(\varphi)) = \eta(\Psi(\psi))$ implies that $\rho(\Psi(\varphi), \Psi(\psi)) \geq \varepsilon$ since $\eta$ is a homeomorphism on sets of diameter less than $\varepsilon$. But since $(\varphi, \psi) \in V$, we have $\rho(\Psi(\varphi), \Psi(\psi)) < \varepsilon$ by definition. This contradiction implies that $\Psi$ is constant on the fibers $\Phi^{-1}(\mathfrak{z})$, and thus that $\theta$ is well defined. The map $\theta$ is also continuous. To see this, let $\Phi(\varphi)$ and $\Phi(\psi)$ lie within $M^{-1/2}\delta$ of each other. Then again $(\varphi,\psi) \in V$, so $\rho(\Psi(\varphi), \Psi(\psi)) < \varepsilon$, and thus since $\pi \circ \Phi = \eta \circ \Psi$, we have

$$\begin{aligned}\rho(\Psi(\varphi), \Psi(\psi)) &= d(\eta \circ \Psi(\varphi), \eta \circ \Psi(\psi)) \\ &= d(\pi \circ \Phi(\varphi), \pi \circ \Phi(\psi)) \\ &\leq d'(\Phi(\varphi), \Phi(\psi)),\end{aligned}$$

$d'$ the metric on $\mathbb{C}^{N+M}$, whence $\theta$ is continuous.

We have obtained the following commutative diagram of maps and spaces.

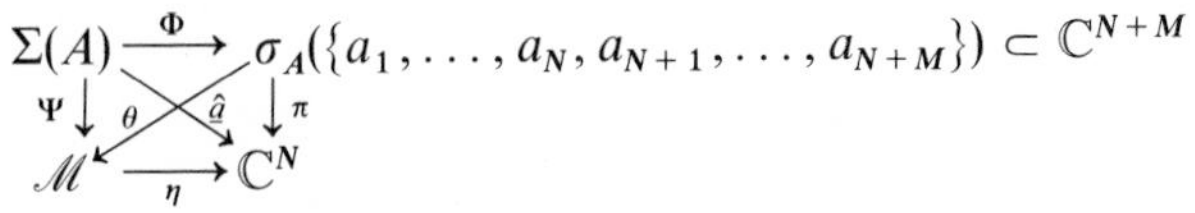

$$\Sigma(A) \xrightarrow{\ \Phi\ } \sigma_A(\{a_1, \ldots, a_N, a_{N+1}, \ldots, a_{N+M}\}) \subset \mathbb{C}^{N+M}$$

In this diagram $\underline{\hat{a}}$ is the map $\varphi \mapsto (\hat{a}_1(\varphi), \ldots, \hat{a}_N(\varphi))$, i.e., the canonical map from $\Sigma(A)$ onto $\sigma_A(\{a_1, \ldots, a_N\})$.

We can extend $\theta$ to a neighborhood of $\Phi(\Sigma(A))$ in $\mathbb{C}^{N+M}$. To do this note that if $\delta'$ is sufficiently small and if $\mathfrak{z} \in \mathbb{C}^{N+M}$ is at distance less than $\delta'$ from $\Phi(\psi)$, $\psi \in \Sigma(A)$, we may define $\theta(\mathfrak{z})$ to that point in $\mathscr{M}$ at $\rho$-distance less than $\varepsilon$ from $\Psi(\psi)$ which satisfies $\eta(\theta(\mathfrak{z})) = \pi(\mathfrak{z})$. This definition gives an extension of $\theta$ to a neighborhood $V$ of $\Phi(\Sigma(A))$. Since $\eta$ is an analytic local homeomorphism, the relation $\eta \circ \theta = \pi$ implies that $\theta$ is analytic in a neighborhood of $\Phi(\Sigma(A)) = \sigma_A(\{a_1, \ldots, a_{N+M}\})$.

Let $\tilde{H} : \mathscr{O}(\sigma_A(\{a_1, \ldots, a_{N+M}\})) \to A$ be a homomorphism with the properties (1) and (2) of Theorem 8.4. We can define the desired homomorphism $H : \mathscr{O}(\Psi(\Sigma(A))) \to A$ as follows: If $f$ is a holomorphic function germ on $\Psi(\Sigma(A))$, it is represented by a function $f$ holomorphic on a neighborhood of this compact set. The function $f \circ \theta$ is holomorphic on a neighborhood of $\sigma_A(\{a_1, \ldots, a_{N+M}\})$; let $(f \circ \theta)^\sim$ be its germ. We define $H(f)$ to be $\tilde{H}((f \circ \theta)^\sim)$, and the $H$ so defined is seen to have the desired properties.

It is, perhaps, worth pointing out that in the situation of Theorem 8.20, if we are given a function $F$ holomorphic on a neighborhood of $\Psi(\Sigma(A))$, then by invoking Theorem 8.3 rather than the more involved Theorem 8.4, we can obtain an $a \in A$ such that $\hat{a} = F \circ \Psi$. Of course, if we want $F \mapsto a$ to be a homomorphism, it is necessary to invoke the full force of Theorem 8.4. As an application of this weaker version of Theorem 8.20, we can complete our discussion of the exponential.

A very systematic treatment of the multiple-valued functional calculus has been given by Bonnard [1].

**8.22    *Corollary***

*If $A$ is a commutative Banach algebra with identity and if $a \in A$ is such that $\hat{a} = e^f$ for some $f \in \mathscr{C}(\Sigma(A))$, then there is $b \in A$ such that $a = e^b$.*

Thus, if $\hat{a}$ has a logarithm, $a$ is an exponential.

PROOF
Define $\eta : \mathbb{C} \to \mathbb{C} \setminus \{0\}$ by $\eta(z) = e^z$. Since $\eta \circ f = \hat{a}$, there is $c \in A$ with $\hat{c} = f$.

The element $d = ae^{-c}$ of $A$ has $\{1\}$ as its spectrum. Let $\gamma$ be a small positively oriented circle around 1 and consider the element $s$ of $A$ given by

$$s = \frac{1}{2\pi i}\int_\gamma (\zeta - d)^{-1} \log \zeta \, d\zeta.$$

By the simple remarks following Theorem 8.1, the map

$$f \mapsto \frac{1}{2\pi i}\int_\gamma f(\zeta)(\zeta - d)^{-1} \, d\zeta$$

is a homomorphism from the algebra of functions holomorphic on the closure of the interior of $\gamma$ into $A$. Thus,

$$s^k = \frac{1}{2\pi i}\int_\gamma (\zeta - d)^{-1}(\log \zeta)^k \, d\zeta,$$

and so

$$\sum_{k=0}^{\infty} \frac{s^k}{k!} = \frac{1}{2\pi i}\int_\gamma \sum_{k=0}^{\infty} \frac{1}{k!}(\zeta - d)^{-1}(\log \zeta)^k \, d\zeta;$$

i.e.,

$$e^s = \frac{1}{2\pi i}\int_\gamma (\zeta - d)^{-1}\zeta \, d\xi$$

$$= d.$$

Consequently, $a = e^c e^s = e^{(c+s)}$, and we have the result.

In the same way, we can show, for example, that if $a \in \mathcal{G}(A)$ and $\hat{a} = f^n$ for some continuous function $f$ on $\Sigma(A)$, then $a = b^n$ for some $b \in A$. In this instance, we need to assume that $a$ is invertible, for we have to consider the map $\eta : z \to z^n$ from $\mathbb{C}$ into $\mathbb{C}$, and this is not a spread of $\mathbb{C}$ into $\mathbb{C}$, but it is one of $\mathbb{C} \setminus \{0\}$ into $\mathbb{C} \setminus \{0\}$.

It is not possible to omit from the hypotheses of Theorem 8.20 the requirement that $\eta$ be a local homeomorphism. Difficulty arises in very simple cases.

As an application of Theorem 8.20, we have the following result on solving equations in Banach algebras.

8.23 ***Corollary***

*Let $A$ be a commutative Banach algebra with identity, let $a_1, \ldots, a_N \in A$, and let $h : \Sigma(A) \to \mathbb{C}$ be a continuous function. Let $G$ be a function of*

*$N+1$ complex variables defined and holomorphic on a neighborhood of the set*

$$S = \{(h(\varphi), \hat{a}_1(\varphi), \ldots, \hat{a}_N(\varphi)) : \varphi \in \Sigma(A)\} \subset \mathbb{C}^{N+1},$$

*and suppose that $G$ vanishes on the set $S$. If the derivative $G_1$ does not vanish on $S$, there exists $b \in A$ such that $\hat{b} = h$.*

PROOF

The proof of this fact depends on the implicit-function theorem, on Theorem 8.20, and on the use, in a very simple way, of the sheaf of germs of holomorphic functions on $\mathbb{C}^N$. Let $\mathcal{O}_{\mathbb{C}^N}$ denote this sheaf. Thus, it is a certain topological space and there is a mapping $\pi : \mathcal{O}_{\mathbb{C}^N} \to \mathbb{C}^N$ which is a local homeomorphism. Each component of $\mathcal{O}_{\mathbb{C}^N}$ has, in a natural way, the structure of a complex manifold and, with respect to this manifold structure, $\pi$ is holomorphic. Thus, $\pi$ is a spread of $\mathcal{O}_{\mathbb{C}^N}$ into $\mathbb{C}^N$. On $\mathcal{O}_{\mathbb{C}^N}$ there is an obvious holomorphic function $\mathscr{F}$ defined as follows: If $\alpha \in \mathcal{O}_{\mathbb{C}^N}$ and $\pi(\alpha) = \mathfrak{z}$, then $\alpha$ is the germ of a function $a$ holomorphic in a neighborhood of $\mathfrak{z}$. We define $\mathscr{F}(\alpha)$ to be $a(\mathfrak{z})$. The function $\mathscr{F}$ defined in this way is holomorphic on $\mathcal{O}_{\mathbb{C}^N}$.

To prove the corollary, we begin by defining a map $\tilde{h} : \Sigma(A) \to \mathcal{O}_{\mathbb{C}^N}$. Let $\varphi \in \Sigma(A)$ so that the point $(h(\varphi), \hat{a}_1(\varphi), \ldots, \hat{a}_N(\varphi))$ lies in $S$. By the implicit-function theorem there exists a neighborhood $V$ of the point $\mathfrak{z}(\varphi) = (\hat{a}_1(\varphi), \ldots, \hat{a}_N(\varphi))$ in $\mathbb{C}^N$, and in $V$ a unique holomorphic function $g_\varphi$ such that $g_\varphi(\hat{a}_1(\varphi), \ldots, \hat{a}_N(\varphi)) = h(\varphi)$, and

$$G(g_\varphi(z_1, \ldots, z_N), z_2, \ldots, z_N)) = 0$$

if $(z_1, \ldots, z_N) \in V$. We define $\tilde{h}(\varphi)$ to be the germ at the point $(\hat{a}_1(\varphi), \ldots, \hat{a}_N(\varphi))$ of the function $g_\varphi$. It is easily checked that the function $\tilde{h}$ defined in this way is continuous.

If we let $\hat{\underline{a}}$ be the map $\varphi \mapsto (\hat{a}_1(\varphi), \ldots, \hat{a}_N(\varphi))$ from $\Sigma(A)$ to $\mathbb{C}^N$, we have the commutativity relation $\pi \circ \tilde{h} = \hat{\underline{a}}$, and since $\pi$ is a spread, it follows from Theorem 8.20 that there exists an element $b \in A$ such that $\hat{b} = \mathscr{F} \circ \tilde{h} = h$. This concludes the proof of the corollary.

If the function $G$ is given by a power series

$$G(z_0, z_1, \ldots, z_N) = \sum_{0 \le k_0, \ldots, k_N} \alpha(k_0, \ldots, k_N) z_0^{k_0} \ldots z_N^{k_N}$$

which converges on a domain $D$ containing the set $S$, it is possible to strengthen Corollary 8.23 by proving the existence of $b \in A$ such that $G(b, a_1, \ldots, a_N) = 0$, i.e., such that the series

$$\Sigma \alpha(k_0, \ldots, k_N) b^{k_0} a_1^{k_1} \cdots a_N^{k_N}$$

converges in $A$ to zero and such that $\hat{b} = h$.

For the details of this result, we refer to the paper of Arens and Calderón [1]. For another result concerned with finding an $a \in A$ whose Gelfand transform is a given $f \in \mathscr{C}(\Sigma(A))$, see Shilov [2].

At this point a possible generalization suggests itself. Suppose $A$ to be a commutative Banach algebra with identity, and suppose that $f \in \mathscr{C}(\Sigma(A))$ has the property that given $\varphi \in \Sigma(A)$, there is a neighborhood $V$ of $\varphi$ and an $a \in A$ such that $f|V = \hat{a}|V$; i.e., *locally* $f$ belongs to $\hat{A}$. Might it be the case that $f = \hat{b}$ for some $b \in A$? In spite of certain assertions to be found in the literature, the function $f$ need not belong to $\hat{A}$. The following example was given by Kallin [1].

**8.24** ***Example***

In $\mathbb{C}^4$ let $X$ be the set $R \cup T_1 \cup T_2$ where $R$, $T_1$, and $T_2$ are given by

$$R = \{\mathfrak{z} \in \mathbb{C}^4 : z_1 z_2 = 2, 1 \leq |z_1| \leq 2, z_3 = z_4 = 0\},$$

$$T_1 = \{\mathfrak{z} \in \mathbb{C}^4 : z_1 z_2 = 2, |z_1| = 1, |z_3| \leq 1, z_4 = 0\},$$

and

$$T_2 = \{\mathfrak{z} \in \mathbb{C}^4 : z_1 z_2 = 2, |z_1| = 2, |z_3| \leq 1, z_4 = z_3^2\}.$$

The set $R$ is an annulus, $T_1$ and $T_2$ are solid tori, and $X$ is obtained by joining $T_1$ and $T_2$ to $R$ along its boundary. We have that

$$R \cap T_1 = \{\mathfrak{z} \in \mathbb{C}^4 : |z_1| = 1, z_2 = 2/z_1, z_3 = z_4 = 0\}$$

and

$$R \cap T_2 = \{\mathfrak{z} \in \mathbb{C}^4 : |z_1| = 2, z_2 = 2/z_1, z_3 = z_4 = 0\};$$

these sets are both circles.

We shall prove that the set $X$ is polynomially convex, so it is the spectrum of $\mathscr{P}(X)$, the uniform closure on $X$ of the polynomials. Also, we shall show that if $G \in \mathscr{C}(X)$ is defined by

$$G(\mathfrak{z}) = \begin{cases} 0 & \mathfrak{z} \in T_1 \cup R, \\ z_3 & \mathfrak{z} \in T_2, \end{cases}$$

then although $G$ belongs to $\mathscr{P}(X)$ locally, it does not belong to $\mathscr{P}(X)$.

First of all, $X$ is polynomially convex. To prove this fact, we shall show that $X$ is precisely the set of points $\mathfrak{z}$ in $\mathbb{C}^4$ which satisfy the following equations and inequalities:

(2) $$|z_1| \leq 2, |z_2| \leq 2, |z_3| \leq 1,$$

(3) $$|z_4 z_2^k| \leq \quad \text{and} \quad |z_1^k(z_4 - z_3^2)| \leq 1 \qquad \text{for } k = 1, 2, 3, \ldots,$$

and

(4) $$z_4(z_4 - z_3^2) = 0.$$

This observation will show that $X$ is polynomially convex. To prove our assertion, let $Y$ be the set of those points $\mathfrak{z}$ which satisfy the inequalities (2), the equality (4), and also the equation $z_1 z_2 = 2$. The set $Y$ is polynomially convex, and the points of $Y \setminus X$ fall into two classes: those for which $|z_1| < 2$ and $z_4 = z_3^2$, and those for which $|z_1| > 1$, $z_4 = 0$, and $z_4 - z_3^2 = -z_3^2 \neq 0$. If $\mathfrak{z}$ is a point of the first kind, then since $|z_1| < 2$ and $z_4 \neq 0$, we have $|z_2| > 1$, and thus $|z_4 z_2^k| > 1$ for all large positive integers $k$, while on $X$, either $z_4 = 0$ or else $z_2 = 1$ and in either case $|z_4 z_2^k| \leq 1$. Thus, the points of the first class do not lie in the polynomially convex hull of $X$. Neither do those of the second class, for if $\mathfrak{z}$ is of the second class, then $|z_1| > 1$, $z_4 = 0$, and $z_3 \neq 0$. Thus, $|z_1^k(z_4 - z_3^2)| > 1$ for large positive integers $k$, while on $X$ either $z_4 - z_3^2 = 0$ or else $|z_1| = 1$ and $|z_4 - z_3^2| = |z_3^2| \leq 1$; so in either case, on $X$, $|z_1^k(z_4 - z_3^2)| \leq 1$. Thus, $X$ is polynomially convex.

We conclude this example by showing that the function $G$ does not lie in $\mathscr{P}(X)$. This proof involves the consideration of measures orthogonal to $\mathscr{P}(X)$. Let $\mu$ be some measure on the two boundary circles $|z_1| = 1$ and $|z_1| = 2$ of $R$ which annihilates all continuous functions on $R$ that are analytic, as a function of $z_1$, on the annulus $\{\mathfrak{z} \in X : |z_1| \in (1, 2)\}$. Let the measure $\nu$ on $|z_3| = 1$ be given by $d\nu = (2\pi i z_3^2)^{-1}\, dz_3$. On $\Gamma = \{\mathfrak{z} \in X : |z_3| = 1 \text{ and } |z_1| = 1 \text{ or } |z_1| = 2\}$, consider the measure $\mu \times \nu$. The set $\Gamma$ is the disjoint union of two tori, and the measure $\mu \times \nu$ is orthogonol to $\mathscr{P}(X)$. To see this, suppose given a polynomial $p$, say

$$p(\mathfrak{z}) = q(\mathfrak{z}) + z_4 r(\mathfrak{z}).$$

Then, $\int q\, d\nu = \partial q/\partial z_3$, a polynomial, so

$$\int q\, d(\mu \times \nu) = \int \frac{\partial q}{\partial z_3}\, d\mu = 0.$$

Also

$$\int z_4 r(\mathfrak{z})\, d\nu = \frac{1}{2\pi i}\int_{|z_3|=1} z_4 r(\mathfrak{z}) z_3^{-2}\, dz_3,$$

and this is zero for all permissible values of $z_1$, $z_2$, and $z_4$, for either $z_4 = 0$ or else $|z_1| = 2$, in which case the integral becomes

$$\frac{1}{2\pi i}\int_{|z_3|=1} r(\mathfrak{z})\, dz_3 = 0.$$

Thus, $\int z_4 r(\mathfrak{z})\, d(\nu \times \mu) = 0$, and we conclude that $\mu \times \nu$ annihilates $\mathscr{P}(X)$.

On the other hand, for suitable choice of $\mu$, $\nu \times \mu$ does not annihilate the function $G$: Let $d\mu = z_1^{-1}\,dz_1$ on $|z_1| = 2$ and $-z_1^{-1}\,dz_1$ on $|z_1| = 1$. Then

$$\int G\,d(\mu \times \nu) = \int_{\substack{|z_1|=1\\|z_3|=1}} + \int_{\substack{|z_1|=2\\|z_3|=1}}.$$

The former integral is zero since $G$ is zero on the set $|z_1| = 1 = |z_3|$. The latter integral is

$$\frac{1}{2\pi i}\int_{\substack{|z_1|=2\\|z_3|=1}} z_3 \frac{dz_3}{z_3^2}\frac{dz_1}{z_1},$$

which, by the residue theorem, is $2\pi i$. Thus, for this choice of $\mu$, $\mu \times \nu$ does not annihilate $G$, so $G \notin \mathscr{P}(X)$.

In her paper [3], Kallin has constructed another example of a nonlocal algebra. In this case $X$ is a compact set in $\mathbf{C}^5$ which is the closure of its interior and again the algebra $\mathscr{P}(X)$ is not local.

Further material on the phenomenon of nonlocal algebras is contained in Sidney [2, 4] and Blumenthal [1]. See also Hurd [1].

The algebra $\mathscr{P}(X)$ considered above has several interesting properties in addition to being nonlocal. One of these is related to the *problem of division.* Suppose that $A$ is a commutative Banach algebra and $a,b \in A$ are such that $\hat{b}$ vanishes on a neighborhood of the zeros of $\hat{a}$. Does it follow that for some $c \in A$, $b = ac$? In general, such division is not possible. Let $f,g \in \mathscr{P}(X)$, $X$ the set in the example considered above, be given by $f(\mathfrak{z}) = z_3$, $g(\mathfrak{z}) = z_1 - \frac{3}{2}$. Then on $X$, $g$ vanishes only at $(\frac{3}{2}, \frac{4}{3}, 0, 0)$, while $f$ vanishes on the whole of $R$. If $f = gh$, $h$ continuous, then $h(\mathfrak{z}) = z_3(z_1 - \frac{3}{2})^{-1}$ on the support of the measure $\mu \times \nu$ which we constructed. This gives $\int h\,d(\mu \times \nu) = \frac{4}{3}\pi i \neq 0$. Thus, $h \notin \mathscr{P}(X)$.

We shall see in Theorem 14.9 that the spectrum of the algebra $A$ of all functions locally in $\mathscr{P}(X)$ is the set $X$, but this fact will depend on certain uniform algebra techniques that we have not developed yet.

## *APPENDIX: GENERALIZATIONS OF THE HOLOMORPHIC FUNCTIONAL CALCULUS*

In this brief appendix we shall consider certain extensions of the Shilov–Arens–Calderón–Waelbroeck theory. The amount of function theory necessary for our discussion is considerably greater than that used in the preceding section. We refer to Gunning and Rossi [1] for the necessary results and background material.

The main question to be considered is the following. Denote by $\Omega$ an open subset of $\mathbb{C}^N$, and let $V$ be an analytic subvariety of $\Omega$ so that $V$ is a closed subset of $\Omega$ which is, locally, the intersection of the zero sets of functions holomorphic on open subsets of $\Omega$. If $W \subset V$ is relatively open, we let $\mathcal{O}(W)$ denote the algebra of functions holomorphic on $W$ so that if $f$ is a $\mathbb{C}$-valued function on $W$, then $f \in \mathcal{O}(W)$ if and only if for each $\mathfrak{z} \in W$, there is an open subset $S$ of $\mathbb{C}^N$ containing $\mathfrak{z}$, and a function $F$ holomorphic on $S$ such that $F|(S \cap W) = f|(S \cap W)$. Let $A$ be a commutative Banach algebra with identity, and let $a_1, \dots, a_N \in A$. If the joint spectrum $\sigma_A(\{a_1, \dots, a_N\})$ is contained in $V$ and if $F \in \mathcal{O}(V)$, does there exist $b \in A$ with $\hat{b}(\varphi) = F(\hat{a}_1(\varphi), \dots, \hat{a}_N(\varphi))$ for all $\varphi \in \Sigma(A)$? If the desired element $b$ always exists, it is natural to ask whether the map taking $F$ to $b$ is an algebra homomorphism.

Questions of this kind were treated in papers [2] and [4] of Allan, and the results are somewhat surprising.

It is convenient to introduce certain additional notation. If $V$ is a variety in the open set $\Omega$ in $\mathbb{C}^N$, and if $K \subset V$ is a compact set, we denote by $\mathcal{O}_V(K)$ the ring of germs on $K$ of functions holomorphic on $V$: $\mathcal{O}_V(K)$ is the ring of equivalence classes in $\bigcup \{\mathcal{O}(W) : W \text{ a neighborhood in } V \text{ of } K\}$, where $f \in \mathcal{O}(W)$ and $f' \in \mathcal{O}(W)$ are equivalent if they agree on some neighborhood of $K$. If $F \in \mathcal{O}_V(K)$ and $\mathfrak{z} \in K$, then $F(\mathfrak{z})$ has an evident meaning even though $F$ is a germ and not a function. We can now state the main result of this appendix.

**8A.1** ***Theorem***

*Let $A$ be a commutative Banach algebra with identity, let $a_1, \dots, a_N \in A$, let $V$ be a variety in the open subset $\Omega$ of $\mathbb{C}^N$, and suppose that $\sigma_A(\{a_1, \dots, a_N\}) \subset V$.*

(a) *If $F \in \mathcal{O}_V(\sigma_A(\{a_1, \dots, a_N\}))$, there is $b \in A$ with*

$$\hat{b}(\varphi) = F(\hat{a}_1(\varphi), \dots, \hat{a}_N(\varphi))$$

*for all $\varphi \in \Sigma(A)$.*

(b) *If $V$ is a complex submanifold of $\Omega$, there is a homomorphism*

$$H : \mathcal{O}_V(\sigma_A(\{a_1, \dots, a_N\})) \to A$$

*such that for all $\varphi \in \Sigma(A)$ and all $F \in \mathcal{O}_V(\sigma_A(\{a_1, \dots, a_N\}))$,*

$$(HF)\hat{\ }(\varphi) = F(\hat{a}_1(\varphi), \dots, \hat{a}_N(\varphi)).$$

We shall see by example that the conclusion of (b) cannot be drawn for general varieties $V$.

PROOF

(a) Let $F \in \mathcal{O}_V(\sigma_A(\{a_1, \dots, a_N\}))$ and choose a neighborhood $U$ of $\sigma_A(\{a_1, \dots, a_N\})$ in $V$ such that in $U$ the holomorphic function $f$ represents the germ $F$. Let $\Omega_0$ be an open set in $\mathbb{C}^N$ whose intersection with $V$ is $U$. As $\Omega_0$ is a neighborhood (in $\mathbb{C}^N$) of $\sigma_A(\{a_1, \dots, a_N\})$, there exist $a_{N+1}, \dots, a_M \in A$ and a polynomial polyhedron $Q$ in $\mathbb{C}^M$ such that $Q \supset \hat{\sigma}_A(\{a_1, \dots, a_M\})$, and such that if $\pi : \mathbb{C}^M \to \mathbb{C}^N$ is the projection onto the first $N$ coordinates, then $\pi(Q) \subset \Omega_0$.

The function $f \circ \pi$ is holomorphic on the variety $\pi^{-1}(U) \cap Q$ in $Q$, so there exists a function $g \in \mathcal{O}(Q)$ with $g|(\pi^{-1}(U) \cap Q) = f \circ \pi$. As $g$ is holomorphic on a neighborhood (in $\mathbb{C}^M$) of $\sigma_A(\{a_1, \ldots, a_M\})$, there is, by the Arens–Calderón theorem, an element $b \in A$ with $\hat{b}(\varphi) = g(\hat{a}_1(\varphi), \ldots, \hat{a}_M(\varphi)) = f(\hat{a}_1(\varphi), \ldots, \hat{a}_N(\varphi)) = F(\hat{a}_1(\varphi), \ldots, \hat{a}_N(\varphi))$. This proves (a).

(b) To prove (b), choose $a_{N+1}, \ldots, a_M$ and $Q$ in $\mathbb{C}^M$ as above with $U = V$ and $\Omega_0 = \Omega$. As $V$ is a submanifold of $\Omega$, the set $V_0 = \pi^{-1}(V) \cap Q = (V \times \mathbb{C}^{(M-N)}) \cap Q$ is a submanifold of $Q$. Thus, by a minor modification of Theorem VIII.C.8 of Gunning and Rossi [1], there is a neighborhood $W$ of $V_0$ in $Q$ and a holomorphic retraction $\rho$ of $W$ onto $V_0 : \rho : W \to V_0$ is a holomorphic map which leaves $V_0$ fixed pointwise. If $f$ is holomorphic on a neighborhood (in $V$) of $\sigma_A(\{a_1, \ldots, a_N\})$, then $f \circ \pi \circ \rho$ is holomorphic on a neighborhood (in $\mathbb{C}^M$) of $\sigma_A(\{a_1, \ldots, a_M\})$. The map $f \mapsto f \circ \pi \circ \rho$ induces a homomorphism, $\eta$, from the ring of germs $\mathcal{O}_V(\sigma_A(\{a_1, \ldots, a_N\}))$ to the ring $\mathcal{O}(\sigma_A(\{a_1, \ldots, a_M\}))$. We know there to exist a homomorphism $H' : \mathcal{O}(\sigma_A(\{a_1, \ldots, a_M\})) \to A$ such that $(H'F)\hat{\ }(\varphi) = F(\hat{a}_1(\varphi), \ldots, \hat{a}_M(\varphi))$ for all $\varphi \in \Sigma(A)$. The map $F \mapsto H'(\eta F)$ is a homomorphism from $\mathcal{O}_V(\sigma_A(\{a_1, \ldots, a_N\})$ into $A$ with the desired properties.

**8A.2** ***Example*** (***Allan*** [**4**])

We shall show that if the variety $V$ is not a submanifold, then for $a_1, \ldots, a_N$ with $\sigma_A(\{a_1, \ldots, a_N\}) \subset V$ there need not exist a homomorphism with the properties of the homomorphism $H$ of part (b) of the theorem. By part (a) of the theorem, it is necessary to consider a nonsemisimple algebra to find an example of the kind we seek.

For this purpose let $\bar{\Delta}^2$ be the closed unit bicylinder in $\mathbb{C}^2$, and let $A$ be the uniform closure on $\bar{\Delta}^2$ of the polynomials in two complex variables. Denote by $Z$ and $W$ the coordinate functions on $\mathbb{C}^2$. In a natural way they are elements of $A$. Let $I$ be the principal ideal in $A$ generated by $Z^2W^2$. The ideal $I$ is closed, so the quotient algebra $A/I$ is a Banach algebra under the quotient norm. The spectrum of $A/I$ may be identified with the set $S = \{(z,w) \in \mathbb{C}^2 : z = 0 \text{ or } w = 0, |z| \leq 1, |w| \leq 1\}$, i.e., with $V \cap \bar{\Delta}^2$, where $V$ is the variety $\{(z, w) \in \mathbb{C}^2 : zw = 0\}$. If we denote by $\eta$ the canonical quotient map $A \to A/I$, then since $Z$ and $W$ constitute a set of generators for the algebra $A$, the elements $\eta Z$ and $\eta W$ generate $A/I$. Consequently, the joint spectrum $\sigma_{A/I}(\eta Z, \eta W)$ may be identified with the set $S$.

We shall show that there exists no homomorphism $h : \mathcal{O}(V) \to A/I$ such that

$$(1) \qquad (hF)\hat{\ }(\varphi) = F((\eta Z)\hat{\ }(\varphi), (\eta W)\hat{\ }(\varphi))$$

for $\varphi \in \Sigma(A/I)$. Assume, in order to derive a contradiction, that such a homomorphism $h$ exists. Apply (1) to the functions $Z$ and $W$ to obtain

$$(hZ)\hat{\ } = (\eta Z)\hat{\ }$$

and

$$(hW)\hat{\ } = (\eta W)\hat{\ }.$$

Thus,

$$hZ = \eta Z + r$$

and

$$hW = \eta W + s,$$

where $r$ and $s$ lie in the radical of $A/I$. We can determine the radical of $A/I$ quite easily. If $F \in A$, $(\eta F)\hat{}$ is simply the restriction of $F$ to the hull of the ideal $I$, so $(\eta F)\hat{}$ is the zero function if and only if $F$ vanishes on the subset $S$ of $\bar{\Delta}^2$. Such an $F$ admits a factorization $F = ZWG$, $G \in A$, so it follows that the radical of $A/I$ is the ideal generated by $\eta(ZW)$. Thus,

$$hZ = \eta(Z) + \eta(ZW)\eta(f)$$

and

$$hW = \eta(W) + \eta(ZW)\eta(g)$$

for some $f, g \in A$. Since $h$ is a homomorphism and since $ZW$ is the zero element of $\mathcal{O}(V)$, we conclude that

$$\begin{aligned} 0 &= h(ZW) \\ &= (\eta(Z) + \eta(ZW)\eta(f))(\eta(W) + \eta(ZW)\eta(g)) \\ &= \eta(ZW)(1 + \eta(W)\eta(f) + \eta(Z)\eta(g) + \eta(ZW)\eta(fg)). \end{aligned} \tag{2}$$

The equality (2) is one between elements of the quotient algebra $A/I$. When expressed explicitly in terms of elements of the algebra $A$, we find that there is an element $h \in A$ such that

$$ZW(1 + Wf + Zg + ZWfg) = Z^2W^2h. \tag{3}$$

This is an equality between elements of the algebra $A$, but it is contradictory, for it implies that

$$1 + Wf + Zg + ZWfg = ZWh, \tag{4}$$

and (4) is plainly impossible because the function on the right vanishes at the origin while that on the left takes there the value 1.

By invoking further results from several complex variables, it is possible to derive yet more general results in this direction. For the definition of *Stein space*, we refer to the book of Gunning and Rossi [1].

**8A.3** ***Corollary***

*Let $A$ be a commutative Banach algebra with identity, and let $X$ be a Stein space. If $\psi : \Sigma(A) \to X$ is a continuous map such that for each $F \in \mathcal{O}(X)$ there is $b \in A$ with $\hat{b} = F \circ \psi$, then, given $f$ holomorphic on a neighborhood of $\psi(\Sigma(A))$, there is $b \in A$ with $\hat{b} = f \circ \psi$. Moreover, if $X$ is a Stein manifold, there is a homomorphism $H : \mathcal{O}_X(\psi(\Sigma(A))) \to A$ such that $(HF)\hat{}(\varphi) = F(\psi(\varphi))$ for each $F \in \mathcal{O}_X(\psi(\Sigma(A)))$ and each $\varphi \in \Sigma(A)$.*

PROOF

By Corollary VII.A.4 of Gunning and Rossi, there exist $F_1, \ldots, F_M \in \mathcal{O}(X)$ such that the set

$$W = \{x \in X : |F_1(x)|, \ldots, |F_M(x)| < 1\}$$

contains $\psi(\Sigma(A))$, and such that under the map $\Phi$ defined by

$$\Phi(x) = (F_1(x), \ldots, F_M(x)),$$

$W$ goes biholomorphically onto an analytic subvariety $V$ of the unit polydisc $\Delta^M$ in $\mathbb{C}^M$. By hypothesis there exist elements $b_1, \ldots, b_M \in A$ with $\hat{b}_j = F_j \circ \psi$. It follows that $\sigma_A(\{b_1, \ldots, b_M\})$ is the set $\Phi(\psi(\Sigma(A)))$. If $f$ is holomorphic on a neighborhood of $\psi(\Sigma(A))$, then $f \circ \Phi^{-1}$ is holomorphic on a neighborhood (in $V$) of $\sigma_A(\{b_1, \ldots, b_M\})$. The theorem implies now that there is $a \in A$ with

$$\begin{aligned} \hat{a}(\varphi) &= f(\Phi^{-1}(\hat{b}_1(\varphi), \ldots, \hat{b}_M(\varphi)) \\ &= f(\psi(\varphi)). \end{aligned}$$

This establishes the first part of the corollary.

If $X$ is a manifold, then the same will be true of $V$. We know that in this case there is a homomorphism $H' : \mathcal{O}_V(\sigma_A(\{b_1, \ldots, b_M\})) \to A$ with $(H'F)(\varphi) = F(\hat{b}_1(\varphi), \ldots, \hat{b}_M(\varphi))$. The map $F \mapsto H'(F \circ \Phi^{-1})$ from $\mathcal{O}_X(\psi(\Sigma(A)))$ to $A$ is a homomorphism $H$ with the desired properties.

# 9 THE LOCAL MAXIMUM MODULUS THEOREM

We now turn to the circle of ideas surrounding Rossi's local maximum modulus theorem. We shall follow Rossi's original program (Rossi [1]) of deriving the local maximum modulus theorem from the local peak set theorem, but we shall establish a generalized version of the local peak set theorem given by Allan [5].

## 9.1 *Definition*

*Let $\mathcal{F}$ be a set of complex-valued functions on a topological space $X$. A set $E \subset X$ is **local peak set for** $\mathcal{F}$*

*If for the neighborhood $V$ we can take the whole of $X$, then $E$ is a **peak set for** $\mathcal{F}$.*

## 9.2 *Definition*

*If $\mathcal{F}$ is a family of continuous functions on the topological space $X$ and if $U$ is an open set in $X$, then $h \in \mathcal{C}(U)$ is **holomorphically approximable in** $U$ **by** $\mathcal{F}$ if there exist $f_1, \ldots, f_q \in \mathcal{F}$ and a function $H$ holomorphic on a* *neighborhood in* $\mathbb{C}^q$ *of* $\{(f_1(x), \ldots, f_q(x)) : x \in U\}$ *such that if* $x \in U$, *then*

$$h(x) = F(f_1(x), \ldots, f_q(x)).$$

The following is Allan's version of the local peak set theorem.

9·3 ***Theorem***

*Let A be a commutative Banach algebra, let E be a compact subset of* $\Sigma(A)$, *let U be a neighborhood of E, and let g be holomorphically approximable in U by* $\hat{A}$. *If* $g|E = 1$ *and* $|g| < 1$ *on* $U \setminus E$, *then E is a peak set for* $\hat{A}$.

The elements of $\hat{A}$ are holomorphically approximable on $\Sigma(A)$ by $\hat{A}$, so the theorem just stated entails the original version of the local peak set theorem:

9·4 ***Corollary***

*If* $E \subset \Sigma(A)$ *is a local peak set for* $\hat{A}$, *then it is a peak set for* $\hat{A}$.

Rossi's original proof of Corollary 9.4 has been simplified considerably, and the simplified version may be found in Gunning and Rossi [1]. (It is stated there for uniform algebras, but only trivial changes are required to obtain the general result by the given argument.) The proof of Allan's theorem is a refinement of the newer proof of Rossi's theorem.

The proof depends on a lemma.

9·5 ***Lemma***

*Let K be a compact set in* $\mathbb{C}^n$, *P a closed subset of K, and V a neighborhood in* $\mathbb{C}^n$ *of P. Let* $h \in \mathcal{O}(V)$ *vanish on P, and satisfy* $\operatorname{Re} h(x) < 0$ *for* $x \in (K \cap V) \setminus P$. *Let f be holomorphic on a neighborhood U of K, zero free in* $K \setminus V$. *If for some holomorphic function g we have* $f = he^{gh}$ *on* $V \cap U$, *then there is a function F holomorphic in a neighborhood of K such that* $F(x) = 1$, $x \in P$, *and* $|F(x)| < 1$ *if* $x \in K \setminus P$.

PROOF

We shall show that under the hypotheses of the lemma, the range of $f$ on $K$ is contained in the set $\{0\} \cup (\mathbb{C} \setminus \{z : |z - a| \leq a\})$ for some $a > 0$. Once we know this, we can take for $F$ the function $\varepsilon(\varepsilon - f)^{-1}$, $\varepsilon$ a small positive number.

Since $f$ is zero free on the compact set $K \setminus V$, there exists a neighborhood $V_0$ of $P$ whose closure is a compact subset of $V$ but which is large enough that on $K \setminus V_0$, $f$ is still zero free. Then the sets $f(K \setminus V_0)$ and

$\{z : |z - b| \le b\}$ are disjoint for all sufficiently small $b > 0$. Also, if $x \in K \cap V$, then $f = he^{gh}$, whence $h = fe^{-gh} = fe^{-\tilde{g}f}$, $\tilde{g} = ghf^{-1}$, a function holomorphic on $V \cap U$. From the power-series expansion of the exponential, it follows that

$$h = f + f^2 k$$

for a function $k$ holomorphic in $V \cap U$. Since $\operatorname{Re} h < 0$ off $P$, we have, on $(K \cap V_0) \setminus P$,

$$\begin{aligned} 0 &\ge \operatorname{Re} f + \operatorname{Re}(f^2 k) \\ &\ge \operatorname{Re} f - |f|^2 |k| \\ &\ge \operatorname{Re} f - M|f|^2, \end{aligned}$$

where $M$ is an upper bound for $|k|$ on $K \cap \overline{V}_0$. From $0 \ge \operatorname{Re} f - M|f|^2$, we get $(1/2M)^2 \le (1/2M)^2 - (\operatorname{Re} f/M) + |f|^2$, which reduces to $(1/2M)^2 \le |f - (1/2M)|^2$. Thus, the sets $f(K \cap V_0)$ and $\{z : |z - (1/2M)| < 1/2M\}$ are disjoint, and the lemma follows.

We now consider the theorem itself.

PROOF OF THE THEOREM

We deal first with the case of an algebra with identity. For notational ease, we introduce, for any set $S \subset \Sigma(A)$ and any $q$-tuple $b_1, \ldots, b_q$ of elements of $A$, the notation

$$\sigma_A(\{b_1, \ldots, b_q\}; S) = \{(\hat{b}_1(\varphi), \ldots, \hat{b}_q(\varphi)) : \varphi \in S\}.$$

In this notation, $\sigma_A(\{b_1, \ldots, b_q\}; \Sigma(A))$ is the usual joint spectrum $\sigma_A(\{b_1, \ldots, b_q\})$.

By hypothesis there exists $a_1, \ldots, a_k \in A$ and a function $H$ holomorphic on a neighborhood of $\sigma_A(\{a_1, \ldots, a_k\}; U)$ in $\mathbb{C}^k$ such that $g$ has the representation

$$g(\varphi) = H(\hat{a}_1(\varphi), \ldots, \hat{a}_k(\varphi)).$$

By shrinking $U$ if necessary we can suppose that $H$ is defined on an open set $\Omega$ containing the compact set $\sigma_A(\{a_1, \ldots, a_k\}; \overline{U})$. We have that $H(\hat{a}_1(\varphi), \ldots, \hat{a}_k(\varphi))$ is 1 if $\varphi \in E$, and is less than 1 in modulus if $\varphi \in U \setminus E$. Compactness implies that $E$ can be covered by a finite number of neighborhoods $N_1, \ldots, N_q$ of the form

$$N_j = \{\varphi \in \Sigma(A) : |\hat{a}_i(\varphi) - \alpha_{ij}| < r_{ij},\ 1 \le i \le k;\ |\hat{a}_m(\varphi)| < 1,\ k_{j-1} < m \le k_j\}.$$

Here each $\alpha_{ij}$ is a complex number, each $r_{ij}$ is a positive real number, $k_0 = k$, and $k_0 < k_1 < \cdots < k_q$ are positive integers. Moreover, writing $n$ for $k_q$, $a_{k+1}, \ldots, a_n$ are certain additional elements of the algebra $A$. We may choose $r_{ij}$, $\alpha_{ij}$, and $a_i$ so that each of the closed sets $\bar{N}_j$ is contained in $U$, and such that if

$$\Omega_j = \{\mathfrak{z} \in \mathbb{C}^k : |z_i - \alpha_{ij}| < r_{ij}, 1 \leq i \leq k\},$$

then $\bar{\Omega}_j \subset \Omega$. Write $M = 1 + \max\{\|a_1\|, \ldots, \|a_n\|\}$, and define sets $\Delta_j, j = 1, \ldots, q$, by

$$\Delta_j = \{\mathfrak{z} \in \mathbb{C}^n : (z_1, \ldots, z_k) \in \Omega_j, |z_i| < 1 \text{ for } k_{j-1} < i \leq k_j, \text{ and}$$
$$|z_i| < M \text{ for other values of } i\}.$$

The set $\Delta_j$ is thus the cartesian product of $\Omega_j$ and a certain polydisc, and as $\Omega_j$ is a polydisc, we find that $\Delta_j$ itself is a relatively compact polydisc in $\mathbb{C}^n$. If we let $\Delta = \bigcup_{j=1}^q \Delta_j$ and $N = \bigcup_{j=1}^q N_j$, then $E \subset N \subset \bar{N} \subset U$, and

$$\sigma_A(\{a_1, \ldots, a_n\}; N) = \sigma_A(\{a_1, \ldots, a_n\}) \cap \Delta.$$

We denote by $\pi$ the natural projection from $\mathbb{C}^n$ to $\mathbb{C}^k$, $\pi(z_1, \ldots, z_n) = (z_1, \ldots, z_k)$, and our definition of $\Delta$ shows that $\pi(\bar{\Delta}) \subset \Omega$. This fact enables us to define a function $H_1$ on a neighborhood $\Omega'$ of $\bar{\Delta}$ by $H_1 = H \circ \pi$. On $\Delta \cap \sigma_A(\{a_1, \ldots, a_n\})$ $|H_1| \leq 1$ with equality only on the set $\sigma_A(\{a_1, \ldots, a_n\}; E)$, where the value of $H_1$ is 1.

The set $\partial\Delta \cap \sigma_A(\{a_1, \ldots, a_n\})$ is a compact subset of $\{\mathfrak{z} \in \Omega' : |H_1(\mathfrak{z})| < 1\}$, so there exists a neighborhood $V$ of $\sigma_A(\{a_1, \ldots, a_n\}) \setminus \Delta$ with the property that $|H_1| < 1$ on $V \cap \Delta$. The set $V \cup \Delta$ is a neighborhood of $\sigma_A(\{a_1, \ldots, a_n\})$ in $\mathbb{C}^n$, so Lemma 8.7 provides elements $a_{n+1}, \ldots, a_p$ of $A$ such that under the canonical projection $\pi'$ from $C^p$ to $\mathbb{C}^n$, $\hat{\sigma}_A(\{a_1, \ldots, a_p\})$ goes into $V \cap \Delta$; then by Lemma 8.6 there exists a polynomial polyhedron $P$ with

$$\hat{\sigma}_A(\{a_1, \ldots, a_p\}) \subset P \subset \bar{P} \subset \pi'^{-1}(V \cup \Delta) = (V \cup \Delta) \times \mathbb{C}^{p-n}.$$

We write $P = P_0 \cup P_1$, $P_0 = P \cap \pi'^{-1}(V)$, $P_1 = P \cap \pi'^{-1}(\Delta)$, and on $P_1$, we define $h$ by

$$h(z_1, \ldots, z_p) = H_1(z_1, \ldots, z_n) - 1,$$

a function which is holomorphic and has negative real part on $P_0 \cap P_1$, so that on this set $\log h$ is defined and holomorphic. As $P$ is a polynomial polyhedron, the first problem of Cousin is solvable there (Appendix D), so there exist holomorphic functions $h_0$ on $P_0$ and $h_1$ on $P_1$ with the property that on $P_0 \cap P_1$

$$\log h = h_0 - h_1,$$

whence $h = e^{h_0 - h_1}$. Let $f_i = e^{h_i}$ on $P_i$, $i = 0, 1$, so that $f_i$ is invertible and $h = f_0/f_1$. Again, since we can solve the first Cousin problem on $P$, there exist holomorphic functions $h_0'$ and $h_1'$ on $P_0$ and $P_1$, respectively, such that on $P_0 \cap P_1$

$$f_0^{-1} \log h = h_0' - h_1'.$$

This leads to $\log h = f_0 h_0' - f_0 h_1'$, whence

$$h \exp(f_0 h_1') = \exp(f_0 h_0').$$

Since $f_0 = hf_1$, we obtain finally that

$$h \exp(f_1 h_1' h) = \exp(f_0 h_0').$$

It now follows that we can define a function $f$ holomorphic on $P$ by requiring that $f = \exp(f_0 h_0')$ on $P_0$, $f = h \exp(f_1 h_1' h)$ on $P_1$. The lemma proved above implies that there exists a function $F$ holomorphic on a neighborhood of $\sigma_A(\{a_1, \ldots, a_p\})$, which is 1 on $\sigma_A(\{a_1, \ldots, a_p\}; E)$ and which is less than 1 in modulus elsewhere on $\sigma_A(\{a_1, \ldots, a_p\})$. (To apply the lemma, take for the sets $K$, $P$, $V$, and $U$ of its statement the sets $\sigma_A(\{a_1, \ldots, a_p\})$, $\sigma_A(\{a_1, \ldots, a_p\}; E)$, $P_0 \cap P_1$, and $P$, respectively.) There is $b \in A$ with $\hat{b}(\varphi) = F(\hat{a}_1(\varphi), \ldots, \hat{a}_p(\varphi))$, and the theorem is proved, in the case of an algebra with identity.

If the algebra $A$ has no identity, apply the result proved so far to the algebra $A'$ obtained by adjoining to $A$ an identity to find an element $a \in A'$ which peaks on $E$. If $\Sigma(A') = \Sigma(A) \cup \{\infty\}$, then $a$ will lie in $A$ if and only if $\hat{a}(\infty) = 0$. If this equality does not hold, replace $a$ by the element $b$ given by

$$b = \frac{a - \hat{a}(\infty)}{1 - \bar{\hat{a}}(\infty)a} \frac{1 - \bar{\hat{a}}(\infty)}{1 - \hat{a}(\infty)},$$

which lies in $A$ and peaks on $E$. This concludes the proof.

The next two corollaries are also due to Allan.

### 9.6 *Corollary*

*Let $a, b \in A$, let $E \subset \Sigma(A)$ be compact, and let $U$ be a neighborhood of $E$. If $\hat{a}|E = \hat{b}|E$ and $\hat{a}$ is zero free on $E$, and if $|\hat{a}(\varphi)| < |\hat{b}(\varphi)|$ for $\varphi \in U \setminus E$, then $E$ is a peak set for $\hat{A}$.*

PROOF

The hypotheses imply that $\hat{b}$ is zero free in $U$, so if $h = \hat{a}\hat{b}^{-1}$, then $h$ is holomorphically approximable on $U$ by $\hat{A}$: If $H(z, w) = zw^{-1}$, then $h(\varphi) = H(\hat{a}(\varphi), \hat{b}(\varphi))$, and $H$ is holomorphic on a neighborhood of $\{(\hat{a}(\varphi), \hat{b}(\varphi)) : \varphi \in U\}$. Thus, the corollary follows directly from the theorem.

In addition to the local peak point theorem, there is a *local pit point theorem*, a result that does not seem to follow from Corollary 9.4.

**9.7** ***Corollary***

*If $E$ is a closed set in $\Sigma(A)$ and $U$ is a neighborhood of $E$, $E$ is a peak set for $\hat{A}$ if there is $a \in A$ with $\hat{a}(\varphi) = 1$, $\varphi \in E$, and $|\hat{a}(\varphi)| > 1$, $\varphi \in U \setminus E$.*

Another result we can derive now is the following generalized version of the local maximum modulus theorem.

**9.8** ***Theorem***

*Let $A$ be a commutative Banach algebra with identity. If $a, b \in A$, let $M(a, b)$ be the set $\{\varphi \in \Sigma(A) : |\hat{a}(\varphi)| \geq |\hat{b}(\varphi)|\}$. If $K$ is a component of $M(a, b)$ on which $\hat{b}$ is zero free, then $K$ meets the Shilov boundary for $A$.*

PROOF

Assume $K$ to be a component of $M(a, b)$ on which $b$ is zero free but which misses the Shilov boundary. We can invoke Lemma 8.13 to find an open and closed subset of $M(a, b)$ on which $b$ is zero free and which contains $K$; and as $K$ is compact and is the intersection of all the relatively open and closed subsets of $M(a, b)$ in which it lies, the fact that $K$ does not meet the Shilov boundary shows that our open and closed subset, call it $E$, can be chosen disjoint from the Shilov boundary. There is an open subset, $U$, of $\Sigma(A)$ such that $U \cap M(a, b) = E$, and if we shrink $U$ as necessary, we may suppose that $\hat{b}$ is zero free on $U$. On $U \setminus E$, we have $|\hat{a}| < |\hat{b}|$. Choose $\varphi_0 \in E$ such that

$$\begin{aligned} |\hat{a}(\varphi_0)\hat{b}(\varphi_0)^{-1}| &= \sup_{\varphi \in E} |\hat{a}(\varphi)\hat{b}(\varphi)^{-1}| \\ &= \sup_{\varphi \in U} |\hat{a}(\varphi)\hat{b}(\varphi)^{-1}|. \end{aligned}$$

Call this supremum $\lambda$; we have $\lambda \geq 1$. By replacing $a$ and $b$ by $\alpha a$ and $\beta b$, respectively, $\alpha$ and $\beta$ unimodular complex numbers, we may suppose that $\hat{a}(\varphi_0)$ and $\hat{b}(\varphi_0)$ are positive. The function $h$ defined on $U$ by

$$h(\varphi) = \tfrac{1}{2}(1 + \hat{a}(\varphi)\hat{b}(\varphi)^{-1}\lambda^{-1})$$

is continuous on $U$, and, relative to $U$, it peaks on a set $E_0$ containing $\varphi_0$. Moreover, the function $h$ is holomorphically approximable on $U$ by $A$, so it follows that the set $E_0$ is a peak set for $A$. By hypothesis the set $E$ is disjoint from the Shilov boundary, so we have reached a contradiction.

The theorem we have just proved is due to Allan.

If for a commutative Banach algebra $A$, an $a \in A$, and an $r > 0$, we set $M(a, r) = \{\varphi \in \Sigma(A) : |\hat{a}(\varphi)| \geq r\}$, then every component of $M(a, r)$ meets the Shilov boundary for $A$. This statement is what is usually called the *local maximum modulus theorem*, and it was proved originally by Rossi in the paper to which we have already referred. For algebras with identity, the local maximum modulus theorem is a special case of the theorem just proved, and for algebras lacking one, it follows from that theorem by the usual process of adjoining the constants, for in this case, every component of $M(a, r)$ is bounded away from the point at infinity.

We now turn to some of the immediate consequences of the local maximum modulus theorem. The first of these was promised in connection with Corollary 8.16.

9.9 ***Corollary***

*If $A$ is a commutative Banach algebra and if $S$ is an open and closed subset of $\Sigma(A)$, then $S$ meets the Shilov boundary for $A$.*

PROOF

We are concerned solely with the case that $A$ has no identity, for Corollary 8.16 covers the case of an algebra with identity. Let $a \in A$ have the property that $\hat{a}|S$ is not identically zero. If $t > 0$ is sufficiently small, $M(a, t) \cap S \neq \varnothing$. If $M_0$ is a component of $M(a, t)$ which meets $S$, then since $S$ is open and closed, $M_0 \subset S$, and the corollary follows from the local maximum modulus theorem.

Let $A$ be a uniform algebra on its spectrum $\Sigma$, and let $\Gamma$ be its Shilov boundary. If $X \subset \Sigma$ is a closed set, put

$$\hat{X} = \{x \in \Sigma : |f(x)| \leq \|f\|_X \text{ for all } f \in A\}.$$

9.10 ***Definition***

*The set $X$ is $A$-convex if $\hat{X} = X$.*

Given a compact set $X \subset \Sigma$, we denote by $A_X$ the uniform closure in $\mathscr{C}(X)$ of the algebra $\{f|X : f \in A\}$.

9.11 ***Lemma***

$\Sigma(A_X) = \hat{X}$.

PROOF

Let $\varphi \in \Sigma(A_X)$. The map $f \mapsto \varphi(f|X)$ is an element of $\Sigma(A)$, so for some $x_\varphi \in \Sigma$, $\varphi(f|X) = f(x_\varphi)$ for all $f \in A$. Since $|\varphi(f|X)| \leq \|f\|_X$, we have

$x_\varphi \in \hat{X}$. The density of $A$ in $A_X$ now implies that every $f \in A_X$ extends to a continuous function on $\hat{X}$ in such a way that $\varphi(f) = f(x_\varphi)$, and thus we conclude that $\Sigma(A_X) = \hat{X}$, as asserted.

**9.12** ***Theorem***

*Let $A$, $\Sigma$, and $\Gamma$ be as above. If $U$ is an open subset of $\Sigma \setminus \Gamma$, then $\Gamma(A_{\bar{U}})$ is contained in the topological boundary of $U$.*

PROOF

Clearly, $\Gamma(A_{\bar{U}}) \subset \bar{U}$. Suppose that $x \in \Gamma(A_{\bar{U}}) \cap U$. There is a neighborhood $V$ of $x$ such that $\bar{V} \subset U$, and there is $f \in A_{\bar{U}}$ such that $f|(\bar{U} \setminus V)$ has modulus no more than $\frac{1}{2}$ but $\|f\|_{\bar{U}} = 1$. Since $A|\bar{U}$ is dense in $A_{\bar{U}}$, it follows that $U$ contains a local peak set for $A$, and hence a global peak set for $A$. This contradicts $U \cap \Gamma = \varnothing$, so the theorem is proved.

**9.13** ***Corollary***

*If $K$ is a compact set in $\Sigma$, then*

$$\Gamma(A_K) \subset (\Gamma \cap K) \cup \partial K.$$

**9.14** ***Corollary***

*If $A$ is a uniform algebra with $\Sigma(A)$ the unit interval $[0, 1]$, then $\Gamma(A) = [0, 1]$.*

PROOF

If $t \in (0, 1) \setminus \Gamma(A)$, then for some $\varepsilon > 0$, $V = (t - \varepsilon, t + \varepsilon)$ is disjoint from $\Gamma(A)$. Then $\Gamma(A_{\bar{V}}) \subset \partial V$, a set of two points. This is impossible, since $A_{\bar{V}}$ has infinitely many maximal ideals. Thus, $\Gamma(A) \supset (0, 1)$, and hence, by closure, $\Gamma(A) = [0, 1]$.

It should be mentioned that apparently no examples of uniform algebras $A$ with $\Sigma(A) = [0, 1]$ are known except for the obvious case $A = \mathscr{C}([0, 1])$.

Our next application of the local maximum modulus theorem is a result that enables us to determine the spectrum of certain algebras.

For the statement of the result, certain notations are useful. Suppose that $X$ and $Y$ are compact spaces and $F \subset X$ is closed. Let $h$ be a homeomorphism of $F$ into $Y$. Then $X \cup_{F,h} Y$ is the space obtained by attaching $Y$ to $X$ along $F$ by way of $h$. Explicitly, $X \cup_{F,h} Y$ is the space obtained from $X \cup Y$, disjoint union, by identifying the point $x \in F$ with the point $h(x) \in Y$. When there is no chance of confusion, we shall suppress the $h$ and write simply $X \cup_F Y$.

The result we shall establish is due to Blumenthal [1].

9.15 ***Theorem***

*Let $A$ and $B$ be uniform algebras on the compact space $X$. Assume that $I$ is a closed ideal in $A$ which is also an ideal in $B$. Let $X = \Sigma(A)$ and $F = \{x \in X : f \in I \text{ implies } f(x) = 0\}$. Then*

$$\Sigma(B) = \Sigma(A) \bigcup_{F,i} \Sigma(B_F),$$

*where $i : F \to \Sigma(B_F)$ is the natural inclusion.*

PROOF

Since $B$ is a uniform algebra on $X$, the space $X$ is, in the usual way, a subset of $\Sigma(B)$. We know that $\Sigma(B_F) = \hat{F} \subset \Sigma(B)$, i.e.,

$$\Sigma(B_F) = \{\varphi \in \Sigma(B) : |\varphi(f)| \leq \|f\|_F \text{ for all } f \in B\}.$$

Since $F = \{x \in X : f(x) = 0 \text{ for all } f \in I\}$, it follows that $F = \hat{F} \cap X$. Consequently, our result will be proved if we show that $\varphi \in \Sigma(B) \setminus X$ implies that if $f \in B$, then

$$|\varphi(f)| \leq \|f\|_F. \tag{1}$$

Put $\text{hull}_B I = \{\varphi \in \Sigma(B) : f \in I \text{ implies } \varphi(f) = 0\}$, and assume that it is known that

$$\Sigma(B) \setminus X \subset \text{hull}_B I. \tag{2}$$

Then since $\Sigma(B) \setminus X$ is an open subset of $\Sigma(B)$ which misses the Shilov boundary for $B$, and since the topological boundary of $\Sigma(B) \setminus X$ lies in $F$ by (2), the local maximum modulus theorem yields (1).

Thus, it remains only to prove (2). Suppose then that $\varphi \in \Sigma(B) \setminus X$, $\varphi \notin \text{hull}_B I$. Then $\varphi$ is a nontrivial complex homomorphism of $I$, so there is $\psi \in \Sigma(A) = X$ with $\psi | I = \varphi$ since $I$ is an ideal of $A$. We have $\psi \in X \setminus F$ since $\psi$ does not annihilate $I$. This is impossible, however, for since $\psi \in X$, we have that $\psi \in \Sigma(B)$, and thus $\psi$ and $\varphi$ are distinct points of $\Sigma(B) \setminus \text{hull}_B I$ which agree on $I$. This contradiction concludes the proof of the theorem.

As an immediate corollary, we have a result of Bear [1]. We mention that Bear's proof of this corollary is more elementary than the one we give since it does not depend on the local maximum modulus theorem.

9.16 ***Corollary***

*Let $X$ be a compact space, let $E \subset X$ be closed, let $A$ be a uniform algebra with $\Sigma(A) = E$, and let $B = \{f \in \mathscr{C}(X) : f | E \in A\}$. Then $\Sigma(B) = X$.*

PROOF

Set $I = \{f \in \mathscr{C}(X) : f|E = 0\}$. Then $I$ is an ideal in $B$ and an ideal in $\mathscr{C}(X)$, and since $X = \Sigma(\mathscr{C}(X))$, the corollary follows directly from the theorem.

If $(A, X)$ is a uniform algebra, the *essential set* for $A$ is defined to be the complement of the hull of the largest closed ideal of $\mathscr{C}(X)$ contained in $I$. This is a well-defined set, for if $\{I_\alpha\}_{\alpha \in M}$ is the family of all closed ideals in $\mathscr{C}(X)$ which are contained in $A$, then $I$, the smallest closed ideal of $\mathscr{C}(X)$ containing $\bigcup I_\alpha$, is a closed ideal of $\mathscr{C}(X)$ contained in $A$. It is easy to see that if $E$ is the essential set for $(A, X)$, then $A|E$ is closed in $\mathscr{C}(E)$, for let $g \in \mathscr{C}(E)$ be a uniform limit of $A|E$. Thus, $g = \sum_{n=0}^{\infty} f_n$, $f_n \in A|E$. We may suppose that $\|f_n\|_E < 2^{-n}\|g\|$. But then there exist $F_n \in \mathscr{C}(X)$ with $\|F_n\|_X = \|f_n\|_E$ and $F_n|E = f_n$. The functions $F_n$ all lie in $A$, so the function $G = \sum_{n=0}^{\infty} F_n$ also lies $A$, and $G|E = g$. Thus, $A|E$ is closed.

The corollary says that in order to determine $\Sigma(A)$, it suffices to determine $\Sigma(A|E)$, $E$ the essential set for $A$.

# 10 FURTHER APPLICATIONS OF FUNCTION THEORY

In this section we shall consider some further applications of function theory to the theory of commutative Banach algebras. The main result we shall establish is the Arens–Royden theorem, a result on the cohomology of the spectrum of a commutative Banach algebra.

In deriving the Arens–Royden theorem, it is convenient to use the formalism of sheaf theory. For some information on sheaf theory and references to its literature, see Appendix A. We shall need the fact that if $\Delta$ is a polynomial polyhedron in $C^n$, then $H^1(\Delta, \mathscr{O}_\Delta) = 0$ if $\mathscr{O}_\Delta$ is the sheaf of germs of holomorphic functions on $\Delta$. This result is precisely the solvability of the first Cousin problem on ploynomial polyhedra, a result we have already used in connection with the local maximum modulus theorem. Since $\Delta$ is paracompact, the cohomology in question is that given by the Čech theory. If $\mathfrak{U} = \{U_\alpha\}$ is an open cover for $\Delta$ and $f_{\alpha\beta} \in \mathscr{O}(U_\alpha \cap U_\beta)$ has the property that $f_{\alpha\beta} + f_{\beta\alpha} = 0$ for all $\alpha$ and $\beta$ and $f_{\alpha\beta} + f_{\beta\gamma} + f_{\gamma\alpha} = 0$ on $U_\alpha \cap U_\beta \cap U_\gamma$ for all $\alpha$, $\beta$, and $\gamma$, then for each $\alpha$ there exists $f_\alpha \in \mathscr{O}(U_\alpha)$ such that on $U_\alpha \cap U_\beta$, $f_{\alpha\beta} = f_\alpha - f_\beta$. On the one hand, this is simply the solvability of the Cousin problem, and on the other hand, since it holds for all $\mathfrak{U}$, it is the statement that $H^1(\Delta, \mathscr{O}_\Delta) = 0$.

We also need the fact that if $X$ is compact or a polynomial polyhedron, then $H^1(X, \mathscr{C}_X) = 0$ if $\mathscr{C}_X$ is the sheaf of germs of continuous functions on $X$. This fact can be proved directly by a simple argument. Of course the result also follows from general principles: $\mathscr{C}_X$ is a fine sheaf, and the cohomology of a paracompact space with coefficients in a fine sheaf is trivial in degree greater than zero.

The following result is referred to as the Arens–Royden theorem as it was proved by Arens [5] and, independently, by Royden [1]. Recall that $\mathscr{G}(A)$ denotes the group of units in $A$ and $\mathscr{G}_0(A)$ its component of the identity.

10.1 ***Theorem***

*If $A$ is a commutative Banach algebra with identity, then $H^1(\Sigma(A), \mathbb{Z})$ is isomorphic to $\mathscr{G}(A)/\mathscr{G}_0(A)$.*

PROOF

We begin by considering the case in which $A = \mathscr{C}(\Sigma)$, where we write $\Sigma$ instead of $\Sigma(A)$. Denote by $\mathbb{Z}$ the constant sheaf of integers on $\Sigma$, denote by $\mathscr{C}_\Sigma$ the sheaf of germs of continuous functions on $\Sigma$, and denote by $\mathscr{C}_\Sigma^*$ the sheaf of germs of nonvanishing continuous functions on $\Sigma$. If $E:\mathscr{C}_\Sigma \to \mathscr{C}_\Sigma^*$ is the sheaf homomorphism determined by requiring that

$$E(f) = \exp f \qquad f \in \mathscr{C}_\Sigma(U)$$

for all open $U \subset X$, then we have the exact sequence of sheaves

$$0 \to \mathbb{Z} \to \mathscr{C}_\Sigma \xrightarrow{E} \mathscr{C}_\Sigma^* \to 0,$$

whence, on passing to the exact cohomology sequence, we have, in part,

$$\cdots \to H^0(\Sigma, \mathscr{C}_\Sigma) \xrightarrow{E^*} H^0(\Sigma, \mathscr{C}_\Sigma^*) \xrightarrow{\delta} H^1(\Sigma, \mathbb{Z}) \to H^1(\Sigma, \mathscr{C}_\Sigma) \to \cdots.$$

The group $H^0(\Sigma, \mathscr{C}_\Sigma)$ is $\mathscr{C}(\Sigma)$, and the group $H^0(\Sigma, \mathscr{C}_\Sigma^*)$ is $\mathscr{C}^*(\Sigma) = \mathscr{G}(\mathscr{C}(\Sigma))$. Also, since the sheaf $\mathscr{C}_\Sigma$ is fine, $H^1(\Sigma, \mathscr{C}_\Sigma) = 0$. By the definition of $E$, the range of $E^*$ is the set of continuous functions of form $e^f$, $f \in \mathscr{C}(X)$, i.e., $\mathscr{G}_0(\mathscr{C}(\Sigma))$. Therefore, $\delta$ is a homomorphism of $\mathscr{G}(\mathscr{C}(\Sigma))$ onto $H^1(\Sigma, \mathbb{Z})$ with kernel $\mathscr{G}_0(\mathscr{C}(\Sigma))$, and thus

$$H^1(\Sigma, \mathbb{Z}) \simeq \mathscr{G}(\mathscr{C}(\Sigma))/\mathscr{G}_0(\mathscr{C}(\Sigma)).$$

As a matter of notational convenience, we shall write $\mathscr{G}(\mathscr{C})$ and $\mathscr{G}_0(\mathscr{C})$ for $\mathscr{G}(\mathscr{C}(\Sigma))$ and $\mathscr{G}_0(\mathscr{C}(\Sigma))$, respectively.

Digressing from our proof, let us recall that in Example 2.5 we identified the group $\mathscr{G}(\mathscr{C})/\mathscr{G}_0(\mathscr{C})$ with the group of homotopy classes of maps from $\Sigma$ into $\mathbb{C}^* = \mathbb{C} \setminus \{0\}$. Thus, as a by-product of what we have done so far, we have the identification of $H^1(\Sigma, \mathbb{Z})$ with this group of homotopy classes, a standard result in topology.

Continuing with the proof, there is a canonical map,

$$\eta : \mathcal{G}(A)/\mathcal{G}_0(A) \to \mathcal{G}(\mathcal{C})/\mathcal{G}_0(\mathcal{C}),$$

which takes the coset $a + \mathcal{G}_0(A)$ to the coset $\hat{a} + \mathcal{G}_0(\mathcal{C})$. This map $\eta$ is well defined, for if $a \in \mathcal{G}(A)$, then $\hat{a} \in \mathcal{G}(\mathcal{C})$, and if $a \in \mathcal{G}_0(A)$, then $\hat{a} \in \mathcal{G}_0(\mathcal{C})$.

It follows from Corollary 8.22 that $\eta$ has a trivial kernel, for suppose that $\eta([a])$, $[a] \in \mathcal{G}(A)/\mathcal{G}_0(A)$, is the identity of $\mathcal{G}(\mathcal{C})/\mathcal{G}_0(\mathcal{C})$. This means that $\hat{a}$ is an exponential: $\hat{a} = e^f$ for some $f \in \mathcal{C}(\Sigma)$. Corollary 8.22 implies that $a = e^b$ for some $b \in A$. Thus, $[a]$ is the identity of $\mathcal{G}(A)/\mathcal{G}_0(A)$, so $\eta$ has trivial kernel.

The final step in the proof is to prove that the homomorphism $\eta$ has the whole of $\mathcal{G}(\mathcal{C})/\mathcal{G}_0(\mathcal{C})$ as its range, i.e., that every zero-free continuous function on $\Sigma(A)$ is homotopic to an $\hat{a}$, $a \in \mathcal{G}(A)$. Assume given $f_0 \in \mathcal{G}(\mathcal{C})$. Let $\{a_i\}_{i \in I}$ be a set of generators for $A$. The Stone–Weierstrass theorem implies that there is a function $f_1 \in \mathcal{C}(\Sigma(A))$ which is a polynomial in a finite number of the functions $\hat{a}_i$ and their complex conjugates, say $\hat{a}_1, \ldots, \hat{a}_M, (\hat{a}_1)^-, \ldots, (\hat{a}_M)^-$, and which satisfies

$$\| f_0 - f_1 \|_X < \tfrac{1}{2} \inf\{| f_0(x)| : x \in \Sigma(A)\}.$$

Then $\|1 - (f_1/f_0)\|_X < \frac{1}{2}$, so $f_0$ and $f_1$ differ by an exponential, and, consequently, it suffices to produce an $a \in A$ such that $\hat{a}$ is homotopic to $f_1$.

Let $\pi_M : \Sigma(A) \to \mathbb{C}^M$ be the map given by

$$\pi_M(x) = (\hat{a}_1(x), \ldots, \hat{a}_M(x))$$

so that $\pi_M(\Sigma(A))$ is $\sigma_A(\{a_1, \ldots, a_M\})$, the joint spectrum of $a_1, \ldots, a_M$. The function $f'_1 \in \mathcal{C}(\sigma_A(\{a_1, \ldots, a_M\}))$ defined by $f'_1 = f_1 \circ \pi_M^{-1}$ is well defined and zero free. Let $F'_1$ be an extension of $f'_1$ to a continuous function on $\mathbb{C}^M$, and let $\Omega$ be a neighborhood of $\sigma_A(\{a_1, \ldots, a_M\})$ on which $F'_1$ does not vanish.

By Lemma 8.7, there exist $a_{M+1}, \ldots, a_N \in A$ so that if $\pi_{NM} : C^N \to C^M$ is the natural map, then

$$\pi_{NM} \hat{\sigma}_A(\{a_1, \ldots, a_N\}) \subset \Omega.$$

Let $\Omega' = \pi_{NM}^{-1}(\Omega)$ and define $F''$ on $\Omega'$ by $F'' = F'_1 \circ \pi_{NM}$. Thus, $F''$ is a zero-free, continuous function on $\Omega'$. Let $\Delta$ be a polynomial polyhedron with

$$\Omega' \supset \Delta \supset \hat{\sigma}_A(\{a_1, \ldots, a_N\}).$$

Suppose it known that there exists a function $G$ holomorphic in $\Delta$ which is homotopic to $F''$. If $\pi_N : \Sigma(A) \to \mathbb{C}^N$ is given by $\pi_N(x) = (\hat{a}_1(x), \ldots, \hat{a}_N(x))$, then by Theorem 8.3, there is $b \in A$ with $\hat{b} = G \circ \pi_N$.

The function $\hat{b}$ is homotopic to $f_1$. To see this, let $h:[0, 1] \times \Delta \to \mathbb{C}^*$ be a continuous map with

$$h(0, \mathfrak{z}) = G(\mathfrak{z}), \qquad h(1, \mathfrak{z}) = F''.$$

If $\tilde{h}:[0, 1] \times \Sigma(A) \to \mathbb{C}^*$ is given by

$$\tilde{h}(t, x) = h(t, \pi_N(x)),$$

then

$$\begin{aligned}\tilde{h}(0, x) &= h(0, \pi_N(x)) \\ &= G(\pi_N(x)) = \hat{b}(x)\end{aligned}$$

and

$$\begin{aligned}\tilde{h}(1, x) &= F''(\pi_N(x)) \\ &= F_1'(\pi_{NM}(\pi_N(x)) \\ &= f_1(\pi_M^{-1}(\pi_M(x))) = f_1(x).\end{aligned}$$

Thus, $\hat{b}$ and $f_1$ are homotopic, and the proof will be concluded if we can establish the following lemma.

**10.2** ***Lemma***

*If $\Delta$ is a polynomial polyhedron in $\mathbb{C}^N$ and if $f$ is a zero-free continuous function on $\Delta$, then there exists a zero-free holomorphic function $G$ which is homotopic, as a map into $\mathbb{C} \setminus \{0\}$, to $f$.*

Although subsequent writers on Banach algebra theory seem to have missed the fact, this lemma is contained explicitly in the paper of Stein [1], published in 1951. (This is the paper in which Stein manifolds were introduced.) Stein's proof proceeds by entirely classical means, and, in particular, no sheaves or cohomology appear in the proof. The proof we shall give is sheaf theoretic and is, more or less, that given in the papers of Arens and Royden.

PROOF

Let $\mathbb{Z}$ be the constant sheaf of integers on $\Delta$. Let $\mathscr{C}_\Delta$ and $\mathscr{C}_\Delta^*$ be the sheaves on $\Delta$ as in the first part of the proof, and let $E:\mathscr{C}_\Delta \to \mathscr{C}_\Delta^*$ be the homomorphism induced by the exponential. Let $\mathcal{O}_\Delta$ and $\mathcal{O}_\Delta^*$ be, respectively, the sheaf of germs of holomorphic functions on $\Delta$ and the sheaf of germs of nonvanishing holomorphic function on $\Delta$. Then $\mathcal{O}_\Delta$ is a subsheaf of $\mathscr{C}_\Delta$, $\mathcal{O}_\Delta^*$ is a subsheaf of $\mathscr{C}_\Delta^*$, and $E$ carries $\mathcal{O}_\Delta$ onto $\mathcal{O}_\Delta^*$. If $i:\mathbb{Z} \to \mathbb{Z}$ is the identity map, and $j:\mathcal{O}_\Delta \to \mathscr{C}_\Delta$ and $k:\mathcal{O}_\Delta^* \to \mathscr{C}_\Delta^*$

are inclusion maps, then we have the commutative diagram of sheaves

$$\begin{array}{ccccccccc} 0 & \to & \mathbb{Z} & \to & \mathcal{O}_\Delta & \to & \mathcal{O}^*_\Delta & \to & 0 \\ & & \downarrow i & & \downarrow j & & \downarrow k & & \\ 0 & \to & \mathbb{Z} & \to & \mathscr{C}_\Delta & \to & \mathscr{C}^*_\Delta & \to & 0. \end{array}$$

Passing to the associated cohomology sequence and recalling, as already noted, that $H^1(\Delta, \mathscr{C}_\Delta) = 0$ and $H^1(\Delta, \mathcal{O}_\Delta) = 0$, we obtain the following commutative diagram of exact sequences:

$$\begin{array}{ccccccc} \cdots \to H^0(\Delta, \mathcal{O}_\Delta) & \overset{E^*}{\to} & H^0(\Delta, \mathcal{O}^*_\Delta) & \overset{\delta^*}{\to} & H^1(\Delta, \mathbb{Z}) & \to 0 \to \cdots \\ \downarrow j^* & & \downarrow k^* & & \downarrow i^* & \\ \cdots \to H^1(\Delta, \mathscr{C}_\Delta) & \overset{E^*}{\to} & H^0(\Delta, \mathscr{C}^*_\Delta) & \overset{\delta^*}{\to} & H^1(\Delta, \mathbb{Z}) & \to 0 \quad \cdots. \end{array}$$

We have that $H^0(\Delta, \mathscr{C}_\Delta)$ is the space of functions holomorphic on $\Delta$, and $H^0(\Delta, \mathcal{O}^*_\Delta)$ is the subset of zero-free elements. Similarly, $H^0(\Delta, \mathscr{C}_\Delta) = \mathscr{C}(\Delta)$ and $H^0(\Delta, \mathscr{C}^*_\Delta)$ is the space of zero-free elements of $\mathscr{C}(\Delta)$. The maps $j^*$ and $k^*$ are one to one and $i^*$ is the identity. Finally, $E^*f = e^f$. Since $i^*\delta^* = \delta^*k^*$ and both are onto, it follows that if $f \in H^0(\Delta, \mathscr{C}^*_\Delta)$, there is $g \in H^0(\Delta, \mathcal{O}^*_\Delta)$ with $\delta^*g = \delta^*f$. Thus, $f$ and $g$ differ only by an element of ker $\delta^*$, i.e., by a multiplicative factor of an exponential: $f = ge^h$ for some $h \in \mathscr{C}(\Delta)$; so $f$ and $g$ are homotopic and the lemma is proved.

This concludes the proof of the Arens–Royden theorem.

A different proof, one depending somewhat more on sheaf theoretic formalism, may be found in Rickart's paper [5].

Gamelin [3] has observed that for certain special algebras $A$, it is possible to express $H^1(\Sigma(A), \mathbb{Z})$ in terms of the real parts of the functions in $A$. Recall that Re $A$ denotes the space of real parts of function in $A$. Let $\mathscr{L}(\hat{A}) = \{\log|\hat{f}| : f \in \mathscr{G}(A)\}$.

10.3 ***Corollary***

*If $A$ contains no nonconstant real elements, then $H^1(\Sigma(A), \mathbb{Z}) \simeq \mathscr{L}(\hat{A})/\mathrm{Re}\, \hat{A}$.*

PROOF

Note that if $f \in \hat{A}$, then $e^f \in \hat{A}$, and $|e^f| = e^{\mathrm{Re}\, f}$; so Re $\hat{A} \subset \mathscr{L}(\hat{A})$, and the assertion is meaningful.

Define a map $\psi : \mathscr{G}(A)/\mathscr{G}_0(A) \to \mathscr{L}(\hat{A})/\mathrm{Re}\, \hat{A}$ by

$$\psi([a]) = \log|\hat{a}| + \mathrm{Re}\, \hat{A}.$$

The mapping $\psi$ is a homomorphism of groups, and it is also onto. The kernel of $\psi$ consists of those cosets $[a] \in \mathscr{G}(A)/\mathscr{G}_0(A)$ such that $\log|\hat{a}|$ is zero, i.e., of those cosets $[a]$ such that $|\hat{a}| = 1$. If $a \in A$ is such that

$|\hat{a}| = 1$, then $(a^{-1})\hat{} = (\hat{a})^-$ if $(\cdot)^-$ denotes complex conjugation. Thus, $b = a + a^{-1} \in A$ satisfies $\hat{b} = \hat{a} + (\hat{a})^-$ so that $\hat{b}$ is real and therefore constant, whence $\operatorname{Re} \hat{a}$ is constant. Similarly, $\operatorname{Im} \hat{a}$ is constant. Consequently, $\hat{a}$ is a constant of modulus 1, so $a = e^b$ by Corollary 8.22. Thus, $a \in \mathscr{G}_0(A)$, and $\psi$ has trivial kernel and so is an isomorphism. The result now follows from the Arens–Royden theorem.

Since $H^1(\Sigma(A), \mathbb{Z})$ contains no elements of finite order,[1] we obtain as a corollary of our result the following fact due to Lorch [1].

**10.4** ***Corollary***

*If $A$ is a commutative Banach algebra with identity, then the group $\mathscr{G}(A)$ is either connected or else has infinitely many components.*

Arens [6] has established a matrix analogue to Theorem 10.1. Denote by $GL(n)$ the group of $n \times n$ invertible matrices with complex entries. If $X$ is a topological space, let $\mathscr{C}(X, GL(n))$ be the space of continuous $GL(n)$-valued functions on $X$. Thus, each $F \in \mathscr{C}(X, GL(n))$ is an $n \times n$ matrix with entries that are continuous, complex-valued functions on $X$. If $X$ is compact, we can define a metric $\rho$ on $\mathscr{C}(X, GL(n))$ by setting

$$\rho(F, G) = \sup\{|F(x) - G(x)| : x \in X\},$$

where $|\cdot|$ denotes some fixed norm on the space of $n \times n$ matrices. The metric $\rho$ on $\mathscr{C}(X, GL(n))$ gives rise to a topology on this space; let $\mathscr{C}(X, GL(n))_0$ denote the component of the identity. Similarly, if $A$ is a commutative Banach algebra with spectrum $\Sigma$, we can consider $\mathscr{G}(A, n)$, the subset of $\mathscr{C}(\Sigma, GL(n))$ with entries of the form $\hat{a}$, $a \in A$. Let $\mathscr{G}(A, n)_0$ denote the component of the identity of this group. Arens's theorem is as follows:

**10.5** ***Theorem***

*If $A$ is a commutative Banach algebra with identity, then the groups $\mathscr{G}(A, n)/\mathscr{G}(A, n)_0$ and*

$$\mathscr{C}(\Sigma(A), GL(n))/\mathscr{C}(\Sigma(A), GL(n))_0$$

*are isomorphic.*

[1] The fact that $H^1(\Sigma(A), \mathbb{Z})$ has no elements of finite order (other than the identity) is a consequence of the $\mathscr{C}(X)$-case of the Arens–Royden theorem. If $X$ is any compact space, if $f \in \mathscr{C}(X)$ is zero free, and if $f^n$ has a continuous logarithm for some integer $n$, say $f^n = e^g$, then $(1/n)g$ is a logarithm for $f$.

Arens's proof of this fact depends on some very nontrivial results of Grauert in function theory, so we shall content ourselves with a reference to his paper for the proof.

Another result on the topology of $\Sigma(A)$ that depends on complex function theory is due to Browder [2].

10.6 ***Theorem***

*If $A$ is a commutative Banach algebra with identity which admits a system of $N$ generators, then $H^k(\Sigma(A), \mathbb{C}) = 0$ for $k \geq N$.*

Browder's proof is based on a theorem of Serre [1] according to which $H^k(\Omega, \mathbb{C}) = 0$ if $k \geq N$, and $\Omega \subset \mathbb{C}^N$ is a domain with the property that every $F \in \mathcal{O}(\Omega)$ can be approximated uniformly by polynomials on compacta in $\Omega$ (Runge domains) and on the continuity of Čech cohomology.

Browder's theorem has the following consequence, a consequence that does not seem to be very easy to establish by more direct methods. Denote by $S^N$ the $N$ sphere.

10.7 ***Corollary***

*$\mathscr{C}(S^N)$ admits no system of fewer than $N + 1$ generators.*

PROOF

This follows from Theorem 10.6 since $H^N(S^N, \mathbb{C}) = \mathbb{C}$.

In connection with the corollary, we should observe that it follows from the Stone–Weierstrass theorem that $\mathscr{C}(S^N)$ does admit a system of $N + 1$ generators.

For additional applications of several complex variables to the theory of Banach algebras, see Koppelman [1] and Royden [2].

# *chapter two*/UNIFORM ALGEBRAS

## II MEASURES ASSOCIATED WITH UNIFORM ALGEBRAS

In this chapter we specialize our considerations to the theory of uniform algebras. The first section is devoted to the discussion of certain measures naturally associated with a uniform algebra. We begin by setting forth some notations to be used systematically in the sequel.

If $X$ is a locally compact space, $\mathscr{M}(X)$ will denote the space of finite regular Borel measures on $X$. Recall that an element $\mu$ of $\mathscr{M}(X)$ is a *probability measure* if $\|\mu\| = 1 = \mu(X)$, i.e., if $\mu$ is a positive measure of norm 1. The *support* of $\mu \in \mathscr{M}(X)$ is the smallest closed set $S$ such that $\int f\, d\mu = 0$ for all $f \in \mathscr{C}_0(X)$ which vanish on $S$. If $E$ is a Borel set in $X$ and $\mu \in \mathscr{M}(X)$, then $\mu_E \in \mathscr{M}(X)$ is defined by requiring $\mu_E(S) = \mu(E \cap S)$ for all Borel sets $S$. Finally, if $\mu \in \mathscr{M}(X)$, $|\mu|$ is the unique positive measure such that $\mu = \theta|\mu|$, where $\theta$ is a Borel function of modulus 1 a.e. $[\mu]$.

II.1 ***Definition***

*If $(A, X)$ is a uniform algebra and $\varphi \in \Sigma(A)$, then $\mathscr{M}_\varphi$ denotes the set of all probability measures $\mu$ on $X$ which satisfy $\int f\, d\mu = \varphi(f)$ for all $f \in A$.*

Thus, the elements of $\mathscr{M}_\varphi$ are simply the representing measures for $\varphi$ which are supported by $X$. It is easy to see that $\mathscr{M}_\varphi$ is a weak * compact, convex subset of $\mathscr{M}(X)$.

By definition, an element of $\mathscr{M}_\varphi$ is positive, so as we noted in Example 7.17, the measure $\sigma_\zeta$ given by

$$d\sigma_\zeta = \frac{1}{2\pi i}\frac{z\, dz}{z - \zeta}, \qquad |\zeta| < 1,$$

on the unit circle is not an element of $\mathscr{M}_\zeta$ if we are considering the disc algebra. To get an element of $\mathscr{M}_\zeta$, we are obliged to use the Poisson measure $m_\zeta$ of Example 7.17. Note that $m_\zeta$ is absolutely continuous with respect to $\sigma_\zeta$. This is a special case of a general phenomenon.

II.2 ***Theorem***

*If $\sigma \in \mathscr{M}(X)$ is such that $\int f\, d\sigma = \varphi(f)$ for some $\varphi \in \Sigma(A)$ and all $f \in A$,* *then there exists a* $\mu \in \mathscr{M}_\varphi$ *which is absolutely continuous with respect to* $\sigma$.

This theorem was explicitly stated for the first time by Hoffman and Rossi [1]; for some remarks on its history, see Hoffman [3, p. 27].

PROOF OF THE THEOREM

Denote by $H^2(|\sigma|)$ the closure of the algebra $A$ in the Hilbert space $L^2(|\sigma|)$, and let $H^2_\varphi(|\sigma|)$ be the closure of $A_\varphi$, the kernel of $\varphi$. If $f \in A_\varphi$, we have

$$1 = \left| \int (1 - f) \frac{d\sigma}{d|\sigma|} \, d|\sigma| \right| \le \left( \int |1 - f|^2 \, d|\sigma| \right)^{1/2} \|\sigma\|^{1/2},$$

so $H^2(|\sigma|) \ne H^2_\varphi(|\sigma|)$. Let $h \in H^2(|\sigma|) \ominus H^2_\varphi(|\sigma|)$, $\|h\| = 1$. If $f \in A_\varphi$, we have that $fh \in H^2_\varphi(|\sigma|)$, for $h$ is the $L^2(|\sigma|)$ limit of a sequence $\{h_n\}$ in $A$, so $fh$ is the $L^2(|\sigma|)$ limit of $\{fh_n\}$, a sequence in $A_\varphi$. Consequently, $fh$ is orthogonal to $h$ for all $f \in A_\varphi$:

$$0 = \int f h \bar{h} \, d|\sigma| = \int f |h|^2 \, d|\sigma|.$$

The measure $\mu$ given by $d\mu = |h|^2 \, d|\sigma|$ is a positive measure of norm $\int |h|^2 \, d|\sigma| = 1$ which annihilates $A_\varphi$. If $f \in A$, then $\int f \, d\mu = \int (f - \varphi(f)) \, d\mu + \varphi(f) = \varphi(f)$, so $\mu \in \mathscr{M}_\varphi$, and the theorem is proved.

As an application of this result, we can settle a point left in abeyance in Section 7.

**11.3** ***Corollary***

*If* $A$ *is a function algebra on the locally compact space* $X$, *and if* $x \in X$, *there is a probability measure* $\mu$ *on* $\Gamma(A)$ *which represents* $x$.

We saw in Corollary 7.16 that a positive representing measure on $\Gamma(A)$ exists for $x$; it was not clear, however, that we could find a probability measure in the case of noncompact $X$.

PROOF OF THE COROLLARY

We may suppose $A$ to be uniformly closed. Let $X^*$ denote the one-point compactification of $X$, and regard $A'$, the algebra obtained by adjoining to $A$ an identity, as a function algebra on $X^*$. Let $\mu \in \mathscr{M}(\Gamma(A'))$ be a representing measure for the point $x$, and let $f \in A$ assume the value one at $x$. Then if $g \in A'$, we have

$$\int g f \, d\mu = g(x) f(x) = g(x),$$

so $f\,d\mu$ is a complex measure which represents $x$. By the theorem, there is a probability measure $\nu$ absolutely continuous with respect to $f\,d\mu$ which represents $x$. Since $f \in A$, we have $f(\infty) = 0$, so $\nu(\{\infty\}) = 0$. Thus, $\nu$ is, in fact, a probability measure on $\Gamma(A)$, and we have the result.

Another application of Theorem 11.2 concerns the geometry of the Shilov boundary. The result in the case of a separable algebra follows from the remarks after Corollary 7.33.

11.4 ***Corollary***

*If $A$ is a commutative Banach algebra with Shilov boundary $\Gamma$, and if $\Gamma \neq \Sigma(A)$, then $\Gamma$ contains a perfect set.*

PROOF

We may suppose that $A$ has an identity and that $A$ is uniformly closed on $\Sigma(A)$. If $\varphi \in \Sigma(A) \setminus \Gamma$, let $\mathscr{M}_\varphi(\Gamma)$ denote the set of representing measures for $\varphi$ supported by $\Gamma$, and if $\mu \in \mathscr{M}_\varphi(\Gamma)$, let $S(\mu)$ denote the support of the measure $\mu$. The set $\mathscr{S} = \{S(\mu) : \mu \in \mathscr{M}_\varphi(\Gamma)\}$ is partially ordered by inclusion, and if $\mathscr{S}_0$ is a subset of $\mathscr{S}$ linearly ordered by inclusion, then $\bigcap\{S(\mu) : S(\mu) \in \mathscr{S}_0)$ contains an element of $\mathscr{S}$. To see this, consider the net $\{\nu_E ; E \in \mathscr{S}_0, \leq\}$ where for each $E \in \mathscr{S}_0$, $\nu_E$ is some measure in $\mathscr{M}_\varphi(\Gamma)$ with support $E$, and where $E \leq F$ if and only if $E \supset F$. The set $\mathscr{M}_\varphi(\Gamma)$ is weak * compact, so this net has a weak * limit point, $\nu$. The measure $\nu$ is in $\mathscr{M}_\varphi(\Gamma)$, and its support is contained in every element of $\mathscr{S}_0$. Thus, Zorn's lemma implies the existence of a minimal element of $\mathscr{S}$, say $E_0$. Let $E_0$ be the support of $\mu_0 \in \mathscr{M}_\varphi(\Gamma)$.

Since $\mu_0$ is a representing measure for $\varphi$ and since $\varphi \in \Sigma(A) \setminus \Gamma$, it follows that $E_0$ cannot be a single point. If $x$ is an isolated point of $E_0$. there exists an $f \in \hat{A}$ with $f(x) = 0$, $f(\varphi) = 1$. Then the measure $f\,d\mu_0$ is a complex representing measure for the point $\varphi$ and, by the lemma, there is measure $\mu_1 \in \mathscr{M}_\varphi(\Gamma)$ with $\mu_1 \ll f\,d\mu_0$. Since $f(x) = 0$ and $x$ is an isolated point of $E_0$, we have $S(\mu_1) \subset E_0 \setminus \{x_0\}$, and this contradicts the minimality of $E_0$. Thus, $E_0$ can have no isolated points. As $E_0 \subset \Gamma$, we conclude that $\Gamma$ contains perfect sets.

Another corollary of the same kind is the following.

11.5 ***Corollary***

*If $A$ is a commutative Banach algebra and if $a \in A$ is such that for some $\varphi \in \Sigma(A) \setminus \Gamma$, $|\hat{a}(\varphi)| = \|\hat{a}\|_{\Sigma(A)}$, then $\hat{a}$ assumes the value $\hat{a}(\varphi)$ on a set which contains a perfect set in $\Gamma$.*

PROOF
Again we can suppose that $A$ has an identity and that $\hat{A}$ is uniformly closed. Moreover, we can suppose $\hat{a}(\varphi) = 1$. Replacing $a$ by $\frac{1}{2}(1 + a)$, we can suppose that for all $\psi \in \Sigma(A)$, $|\hat{a}(\psi)| < 1$ or else $\hat{a}(\psi) = 1$. If $\mu \in \mathcal{M}_\varphi(\Gamma)$, then

$$\begin{aligned} 1 = \hat{a}(\varphi)^n &= \int_\Gamma \hat{a}^n \, d\mu \\ &\to \mu(P_{\hat{a}}), \end{aligned}$$

where $P_{\hat{a}} = \{\psi \in \Gamma : \hat{a}(\psi) = 1\}$. Thus, the set $P_{\hat{a}}$ contains the support of an element of $\mathcal{M}_\varphi(\Gamma)$ and so, by the argument of the last corollary, a perfect set.

Corollary 11.5 is contained, e.g., in Rickart [1]; a weaker version is due to Holladay [1].

We now turn to the consideration of certain special representing measures.

11.6 ***Definition***

*Let $(A, X)$ be a uniform algebra, let $\varphi \in \Sigma(A)$, and let $\mu \in \mathcal{M}(X)$ be a positive measure. Then*

(a) *$\mu$ is an **Arens–Singer** measure for $\varphi$ if $\log|\varphi(f)| = \int \log|f| \, d\mu$ for all $f \in \mathcal{G}(A)$.*
(b) *$\mu$ is a **Jensen** measure for $\varphi$ if $\log|\varphi(f)| \le \int \log|f| \, d\mu$ for all $f \in A$.*

It is easy to see that a Jensen measure is necessarily an Arens–Singer measure. If $f \in \mathcal{G}(A)$ and $\mu$ is a Jensen measure, then

$$\log|\varphi(f)| \le \int \log|f| \, d\mu$$

and

$$-\log|\varphi(f)| \le -\int \log|f| \, d\mu$$

since $f$ and $f^{-1}$ are both in $A$. Thus, $\log|\varphi(f)| = \int \log|f| \, d\mu$. It is also clear if every $f \in \mathcal{G}(A)$ has a logarithm in $A$; i.e., if $H^1(\Sigma(A), \mathbb{Z}) = 0$, then every representing measure is an Arens–Singer measure, for if $f \in \mathcal{G}(A)$, say $f = e^g$, then

$$\begin{aligned} \log|\varphi(f)| &= \operatorname{Re} \varphi(g) \\ &= \int \log|f| \, d\mu. \end{aligned}$$

If $\mu \in \mathcal{M}(X)$ is an Arens–Singer measure for $\varphi$, then $\mu \in \mathcal{M}_\varphi$. This fact was noted by Arens and Singer in their paper [1]. To prove it, it suffices to show that $\int \operatorname{Re} f\, d\mu = \operatorname{Re} \varphi(f)$ for all $f$. But if $f \in A$, then $e^f \in A$, so by the Arens–Singer property,

$$\int \operatorname{Re} f\, d\mu = \int \log|e^f|\, d\mu = \log|\varphi(e^f)|$$

$$= \log|e^{\varphi(f)}| = \operatorname{Re} \varphi(f).$$

It is clear that the set of Arens–Singer measures for $\varphi \in \Sigma(A)$ is convex, as is the set of Jensen measures. The following additional facts were noted by Gamelin and Rossi [1].

11.7 ***Theorem***

*Let $(A, X)$ be a uniform algebra.*

(a) *The set of Arens–Singer measures for $\varphi \in \Sigma(A)$ is a weak * compact, convex subset of $\mathcal{M}_\varphi$, and so is the set of Jensen measures for $\varphi$.*

(b) *If $\mu, \nu \in \mathcal{M}_\varphi$ are Arens–Singer measures, so is every positive measure on the line through $\mu$ and $\nu$, and if $\varphi \in X$, and $\mu \in \mathcal{M}_\varphi$ is a Jensen measure, then every positive measure on the line through $\mu$ and $\delta_\varphi$, the unit point mass at $\varphi$, is a Jensen measure for $\varphi$.*

PROOF

(a) We need only prove compactness, and since $\mathcal{M}$ is weak * compact, we have only to show that the sets in question are weak * closed.

Suppose that $\{\mu_\alpha\}$ is a net of Jensen measures which converges in the weak * sense to $\mu$. Let $f \in A$, and choose $\varepsilon > 0$. Then $\log(|f| + \varepsilon)$ is continuous, and we have

$$\log|\varphi(f)| \leq \int \log(|f| + \varepsilon)\, d\mu_\alpha.$$

Take the limit over $\alpha$ and let $\varepsilon$ decrease to zero. The result is

$$\log|\varphi(f)| \leq \int \log|f|\, d\mu;$$

i.e., $\mu$ is a Jensen measure.

If the $\mu_\alpha$ are only Arens–Singer measures, we have only to consider $f \in \mathcal{G}(A)$. For such functions, $\log|f|$ is continuous, and we have, for all $\alpha$,

$$\log|\varphi(f)| = \int \log|f|\, d\mu_\alpha.$$

If we take the limit over $\alpha$, we find that $\mu$ is an Arens–Singer measure.

(b) Let $\varphi \in X$, and let $\mu \in \mathscr{M}_\varphi$ be a Jensen measure for $\varphi$. If $\sigma$ is a positive measure on the line through $\delta_\varphi$ and $\mu$, then $\sigma = \delta_\varphi + t(\mu - \delta_\varphi)$, for some $t > 0$. If $f \in A$, then

$$\log|\varphi(f)| \leq (1 - t)\log|(\varphi(f)| + t\int \log|f|\, d\mu$$

$$= \int \log|f|\, d\sigma,$$

and thus $\sigma$ is a Jensen measure for $\varphi$. An even simpler calculation provides the result for Arens–Singer measures.

Not every representing measure is an Arens–Singer measure, and not every Arens–Singer measure is a Jensen measure. We first give an example of the former occurrence.

11.8 ***Example*** (See Wermer [9])
Denote by $R$ the ring $\{z \in \mathbb{C} : \rho < |z| < 1\}$, $\rho \in (0, 1)$, and consider the algebra $\mathscr{A}(R)$ of all $f \in \mathscr{C}(\bar{R})$ which are holomorphic on $R$. The Laurent series expansion for functions holomorphic on $R$ shows that $\mathscr{A}(R)$ is generated by the functions $z$ and $z^{-1}$, and it follows from this that the spectrum of $\mathscr{A}(R)$ can be identified with the set $\bar{R}$. The Shilov boundary for $\mathscr{A}(R)$ is the topological boundary, $\partial R$, of $R$, and we regard $\mathscr{A}(R)$ as an algebra on $\partial R$. By Cauchy's theorem we know that if $f \in \mathscr{A}(R)$, then

$$0 = \frac{1}{i}\int_{\partial R} \zeta^{-1} f(\zeta)\, d\zeta$$

when $\partial R$ is properly oriented. Since $\partial R$ consists of the two circles $|z| = \rho$ and $|z| = 1$, this formula can be rewritten in the form

$$0 = \int_{\partial R} f(\zeta) H(\zeta)\, ds,$$

where $H(\zeta) = 1$ if $|\zeta| = 1$, $H(\zeta) = -1/\rho$ if $|\zeta| = \rho$ and where $ds$ denotes the differential of arclength on $\partial R$.

Given a point $z_0 \in R$, let $G(\cdot, z_0)$ be the Green's function of $R$ with singularity at $z_0$. Then from the elements of potential theory, we know that for any function $U$ that is continuous on $\bar{R}$ and harmonic in $R$,

$$U(z_0) = \frac{1}{2\pi}\int_{\partial R} U(\zeta)\frac{\partial G(\zeta, z_0)}{\partial n}\, ds \tag{1}$$

if $\partial/\partial n$ denotes the inner normal derivative. We know that $\partial G(\zeta, z_0)/\partial n > 0$ for $\zeta \in \partial R$. It follows that if $\varepsilon$ is small enough $(\partial G/\partial n) + \varepsilon H > 0$ on $\partial R$, and thus, all the measures

$$\left\{\frac{1}{2\pi}\frac{\partial G(\cdot, z_0)}{\partial n} + \varepsilon H(\cdot)\right\} ds$$

are representing measures for $z_0$.

We can show that the measure $(1/2\pi)[\partial G(\cdot, z_0)/\partial n]\, ds$ is the only Arens–Singer measure for the point $z_0$. This measure is an Arens–Singer measure, for if $f \in \mathcal{G}(\mathcal{A}(R))$, then $\log|f|$ is harmonic, so

$$\log|f(z_0)| = \frac{1}{2\pi}\int_{\partial R} \log|f(\zeta)|\frac{\partial G(\zeta, z_0)}{\partial n}\, ds.$$

Suppose that $\nu \in \mathcal{M}(\partial R)$ is an Arens–Singer measure for $z_0$. We have, since the function $z$ is an invertible element of $\mathcal{A}(R)$,

$$\log|z_0| = \int_{\partial R} \log|\zeta|\, d\nu(\zeta).$$

Let $U$ be a function harmonic on a neighborhood of $\bar{R}$. The function $U$ need not have a well-defined conjugate; but for a suitably chosen real constant $\beta$, $U(z) + \beta \log|z|$ will have one, say $V$, so that $g(z) = U(z) + \beta \log|z| + iV(z)$ is holomorphic on a neighborhood of $\bar{R}$. Since $e^g \in \mathcal{A}(R)$, and since $\nu$ is an Arens–Singer measure for $z_0$, we have

$$\begin{aligned}\int_{\partial R} \log|e^g|\, d\nu &= \log|e^{g(z_0)}| \\ &= U(z_0) + \beta \log|z_0|.\end{aligned}$$

But also,

$$\begin{aligned}\int_{\partial R} \log|e^g|\, d\nu &= \int_{\partial R} (U(\zeta) + \beta \log|\zeta|)\, d\nu(\zeta) \\ &= \int_{\partial R} U\, d\nu + \beta \log|z_0|.\end{aligned}$$

Consequently,

$$\int_{\partial R} U\, d\nu = U(z_0). \tag{2}$$

Since (2) holds for all functions $U$ harmonic on a neighborhood of $\bar{R}$, and since these functions are dense in the space of all functions harmonic on $R$ and continuous on $\bar{R}$, it follows that (2) holds for all functions of this latter class. Since every continuous function on $\partial R$ has a harmonic extension $U_f$ into the interior of $R$, and since the map $f \mapsto U_f(z_0)$ is a bounded linear functional on $\mathscr{C}(\partial R)$, the uniqueness portion of the Riesz representation theorem implies that only one measure can have property (2). Since (1) holds, we may conclude that $dv = [\partial G(\cdot, z_0)/\partial n]\, ds$.

Next, we give an example of an Arens–Singer measure that is not a Jensen measure.

11.9 ***Example***

Let $B$ be the algebra of functions continuous on the closed unit disc and holomorphic on the ring $R$ of the last example. Every invertible element of $B$ has a logarithm in $B$, so every representing measure is an Arens–Singer measure. Choose a point $z_0 \in R$, and let $\mu \in \mathscr{M}(\partial R)$ be a representing measure for $z_0$ for the algebra $\mathscr{A}(R)$ which is not a Jensen measure for $\mathscr{A}(R)$. Then $\mu$ is a representing measure, and hence an Arens–Singer measure for $B$, but it is not a Jensen measure for $B$.

Note that by Corollary 9.16, the spectrum of the algebra $B$ is the closed unit disc.

So far we have not shown that Arens–Singer measures or Jensen measures exist. Arens–Singer measures were shown to exist for general uniform algebras by Arens and Singer [1], and more generally Bishop [6] proved the following fact.

11.10 ***Theorem***

*If $A$ is a uniform algebra on the compact space $X$, and if $\varphi \in \Sigma(A)$, then there exists a Jensen measure in $\mathscr{M}_\varphi$ which is supported by the Shilov boundary for $A$.*

PROOF

Let $\Gamma$ be the Shilov boundary for $A$. Let $\mathscr{C}^-(\Gamma)$ be the set of all strictly negative continuous functions on $\Gamma$, and let $K$ be the set of all those $h \in \mathscr{C}(\Gamma)$ such that for some $f \in A$ with $|\varphi(f)| \geq 1$, $rh(x) > \log|f(x)|$ for all $x \in \Gamma$ and some $r > 0$. If any such $f$ and $r$ exist, we can find a suitable $r$ which is rational. If $rh(x) > \log|f(x)|$ for all $x \in \Gamma$, then $qrh(x) > \log|f^q(x)|$; so it follows that given $h_1$ and $h_2$ in $K$, we can find an $r_1$ and an $f_1$ for $h_1$, and an $r_2$ and $f_2$ for $h_2$, with $r_1 = r_2$. Then $r_1(h_1 + h_2) > \log|f_1 f_2|$ on $\Gamma$. Thus, $K$ is closed under addition. As it is also closed under multiplication by positive real numbers, it is convex.

We have $K \cap \mathscr{C}^-(\Gamma) = \varnothing$, and since these are disjoint open convex sets, it follows that for some continuous real linear functional $L$ on $\mathscr{C}_{\mathrm{R}}(\Gamma)$, $\|L\| = 1$, we have

$$\sup\{L(f): h \in \mathscr{C}^-(\Gamma)\} = 0 = \inf\{L(h): h \in K\}.$$

By the Riesz representation theorem, there is a real regular Borel measure $\mu$ on $\Gamma$ such that

$$L(g) = \int g\, d\mu$$

for all real continuous $g$. We have $L \leq 0$ on $\mathscr{C}^-(\Gamma)$, so $\mu$ is a positive measure, and $\|\mu\| = 1$.

If $f \in A$ satisfies $\varphi(f) = \alpha$, $\alpha \neq 0$, then for any $h \in \mathscr{C}_{\mathrm{R}}(\Gamma)$ with $h(x) > \log|\alpha^{-1}f(x)|$ for all $x \in \Gamma$, we have $h \in K$, and so $\int h\, d\mu \geq 0$. As $\inf\{\int h\, d\mu : h > \log|\alpha^{-1}f|\} = \int \log|\alpha^{-1}f|\, d\mu$, we have that

$$\begin{aligned} 0 &\leq \int \log|\alpha^{-1}f|\, d\mu \\ &= \int \log|f|\, d\mu - \log|\alpha|; \end{aligned}$$

i.e.,

$$\log|\varphi(f)| \leq \int \log|f|\, d\mu,$$

and $\mu$ is seen to be a Jensen measure for $\varphi$.

A reasonable next step would be to show that in the case of a uniform algebra on a metrizable space every complex homomorphism admits a Jensen measure which is supported on the minimal boundary. However, this step cannot be taken. We shall construct in a later section a uniform algebra $A$ which is *normal* on its spectrum in the sense that if $E$ and $F$ are disjoint closed sets in $\Sigma(A)$, there is $f \in A$ with $f|E \equiv 1$, $f|F \equiv 0$. The spectrum will be a planar set and so metrizable and $A$ will not be $\mathscr{C}(\Sigma(A))$. The minimal boundary for $A$ will differ from $\Sigma(A)$. For such an algebra $A$, it is clear that the only Jensen measures are the point masses: Let $\varphi \in \Sigma(A)$ and let $\mu \in \mathscr{M}(\Sigma(A))$ be a Jensen measure for $\varphi$. If $E$ is a closed set in $\Sigma(A) \setminus \{\varphi\}$ which has positive $\mu$ mass, and if $f \in A$ satisfies $\varphi(f) = 1, f|E \equiv 0$, we have

$$\begin{aligned} 0 = \log|\varphi(f)| &\leq \int \log|f|\, d\mu \\ &= -\infty. \end{aligned}$$

Thus, no such $E$ can exist and by regularity $\mu$ must be the unit point mass at $\varphi$.

As a first, rather elementary, application of the existence of Jensen measures, we consider the following situation. Let $A$ be a uniform algebra on $X$, and suppose that $X = \bigcup\{X_n : n = 1, 2, \ldots\}$, each $X_n$ a closed set. Then $X$ and $X_n$ are all contained in $\Sigma(A)$. We have that $X$ contains the Shilov boundary for $A$, so

$$\Sigma(A) = \hat{X} = \{\varphi \in \Sigma(A) : |\varphi(f)| \leq \|f\|_X \text{ for all } f \in A\}.$$

The following result is due to Gamelin and Wilken [1].

II.11 ***Theorem***

*If for all n, $A|X_n$ is closed in $\mathscr{C}(X_n)$, then $\Sigma(A) = \bigcup\{\hat{X}_n : n = 1, 2, \ldots\}$.*

PROOF

Let $\varphi \in \Sigma(A)$ and let $\mu \in \mathscr{M}(X)$ be a Jensen measure for $\varphi$. We have $X = \bigcup X_n$, so there is $n$ such that $\mu(X_n) \neq 0$. Since $A|X_n$ is closed, $\ker \varphi|X_n$ is a maximal ideal in $A|X_n$ or else $\ker \varphi|X_n = A|X_n$. In the latter case, there is $f \in \ker \varphi$ with $f|X_n = 1$. But then

$$\begin{aligned} 0 = \log|\varphi(1 - f)| &\leq \int \log|1 - f|\, d\mu \\ &= -\infty. \end{aligned}$$

Thus, in fact, $\ker \varphi|X_n$ is a maximal ideal of $A|X_n$ and so

$$\varphi \in \Sigma(A|X_n) = \hat{X}_n.$$

Simple examples show that the theorem fails without the hypothesis that $A|X_n$ be closed, even if there are only finitely many $X_n$.

II.12 ***Corollary***

*If $A$ is a uniform algebra on the compact space $X$, if $X = \bigcup\{X_n : n = 1, 2, \ldots\}$, $X_n$ closed, if $A|X_n$ is closed, and if $\Sigma(A|X_n) = X_n$, then $\Sigma(A) = X$.*

This corollary was established, in the case of finite closed covers, by Mullins [1].

# 12 THE GENERALIZED STONE–WEIERSTRASS THEOREM

The classical Weierstrass approximation theorem says that every function continuous on [0, 1] can be approximated uniformly on this interval by polynomials. The theorem was generalized by Stone [1] who established what is now known as the Stone–Weierstrass theorem. Stone's theorem has been extended in its turn by Shilov, and then by Bishop [3], who established a very general result concerning uniform algebras. It is this theorem of Bishop that we consider now.

Let $X$ be a compact space and let $A \subset \mathscr{C}(X)$ be a closed subalgebra which contains the constants. We do not assume that $A$ separates points, so $A$ is not necessarily a uniform algebra in our terminology. If $K \subset X$ has the property that every $f$ in $A$ which is real on $K$ is, in fact, constant on $K$, then $K$ will be called a set of *antisymmetry for A* or an *antisymmetric set for A*. If $X$ is itself a set of antisymmetry for $A$, $A$ will be called an *antisymmetric* algebra.

**12.1** ***Theorem***

*Every antisymmetric set for A is contained in a maximal antisymmetric set. The set $\mathscr{K}$ of maximal antisymmetric sets for A is a closed pairwise disjoint cover for X with the property that if I is a closed ideal in A, then*

(a) *$f \in \mathscr{C}(X)$ and $f|K \in I|K$ for all $K \in \mathscr{K}$ imply $f \in I$.*
(b) *$I|K$ is closed in $\mathscr{C}(K)$ for all $K \in \mathscr{K}$.*

The proof we give for this theorem is that given by Glicksberg [1], and it is based on a proof of the Stone–Weierstrass theorem given by deBranges [1]. The result was first proved by Bishop [3], at least in the case that $I = A$. Further results in the direction of Theorem 12.1 are contained in the papers of Arenson [1] and Sidney [3].

PROOF OF THE THEOREM

Zorn's lemma provides maximal antisymmetric sets for $A$, and the continuity of the elements of $A$ implies that these must be closed. Since each point of $X$ is a set of antisymmetry, it follows that the maximal antisymmetric sets for $A$ do constitute a closed cover for $X$. That this cover is pairwise disjoint is clear: If $K_1$ and $K_2$ are sets of antisymmetry and if $K_1 \cap K_2 \neq \varnothing$, then $K_1 \cup K_2$ is a set of antisymmetry. It remains to prove (a) and (b).

In the proof of (a) and (b), we shall use the following notation: If $S \subset \mathscr{C}(X)$, then $S^\perp = \{\mu \in \mathscr{M}(X): \int f\, d\mu = 0 \text{ for all } f \in S\}$. The elements of $S^\perp$ are referred to as the *measures orthogonal* to $S$. If $Y$ is a normed linear space, ball $Y$ will denote its closed unit ball.

We shall need several lemmas.

12.2 ***Lemma***

*If $\mu$ is an extreme point of ball $(I^\perp)$, then the support of $\mu$ is a set of antisymmetry for $A$.*

PROOF

By the Krein–Milman theorem, our assertion is not vacuous: Ball $(I^\perp)$ is weak $*$ compact and convex, so it does have extreme points.

Assume that $f \in A$ is real valued on $K$, the support of $\mu$. The algebra $A$ contains the constants, so replacing $f$ by $\frac{1}{2}[1 + (f/n)]$ for large $n$ if necessary, we can suppose that $0 < f(x) < 1$ for all $x \in K$. Then $f\mu$ and $(1 - f)\mu$ are nonzero measures, and we may write

$$\mu = \|f\mu\| \frac{f\mu}{\|f\mu\|} + \|(1-f)\mu\| \frac{(1-f)\mu}{\|(1-f)\mu\|}.$$

We have

$$\begin{aligned} \|f\mu\| + \|(1-f)\mu\| &= \int |f|\, d|\mu| + \int |1-f|\, d|\mu| \\ &= \int f\, d|\mu| + \int (1-f)\, d|\mu| \\ &= \int |\mu| = 1. \end{aligned}$$

The measures $f\mu$ and $(1-f)\mu$ both lie in $I^\perp$ since $I$ is an ideal, and since the measures $f\mu/\|f\mu\|$ and $(1-f)\mu/\|(1-f)\mu\|$ are both of norm 1, the hypothesis that $\mu$ is extreme implies that $\mu = f\mu/\|f\mu\|$, which, in turn, implies that $f$ is constant on the support of $\mu$.

From this lemma, part (a) of the theorem follows immediately. Suppose that $f \in \mathscr{C}(X)$ and $f \in I|K$ for all $K \in \mathscr{K}$. If $f \notin I$, there exists an extreme point of ball $(I^\perp)$, say $\mu$, such that $\int f\, d\mu \neq 0$. By the lemma, the support of $\mu$ is a set of antisymmetry for $A$, so some $K \in \mathscr{K}$ contains this support. Consequently, there is $g \in I$ with $f|K = g|K$. We have then that $\int f\, d\mu = \int g\, d\mu = 0$ since $\mu \in I^\perp$. This contradicts the choice of $\mu$, so part (a) is proved.

We turn now to part (b).

12.3 **Lemma**

*If $E \subset X$ is an intersection of peak sets for $A$, then $I/E$ is closed in $\mathscr{C}(E)$ and is isometrically isomorphic to the quotient algebra $I/(I \cap k(E))$, $k(E) = \{f \in A : f|E = 0\}$.*

Let us remark that the norm $\|\cdot\|_E$ on $I|E$ is the supremum over $E$, and that on $I/(I \cap k(E))$ is the usual quotient norm.

PROOF OF THE LEMMA

The lemma follows from Lemma 7.22: If $g \in I$, then, by Lemma 7.22, there is $h \in A$ which, by the construction given in the proof of the cited lemma, actually lies in $I$, such that $h|E = g|E$ and $\|h\|_X = \|g\|_E$. This implies that

$$\|g\|_E \geq \|g + (I \cap k(E)\|.$$

As the reversed inequality is obvious, we have the stated isometry. Since the quotient algebra is complete, so is $A|E$, and consequently $A|E$ is closed.

12.4 **Lemma**

*Every maximal antisymmetric set $K$ for $A$ is an intersection of peak sets.*

PROOF

Let $E$ be the intersection of all the peak sets for $A$ which contain $K$, and assume that $E \neq K$. Since $K$ is a maximal set of antisymmetry for $A$, there is $f \in A$ which is real and nonconstant on $E$. The function $f$ is constant on $K$; let is assume there the value $s$.

The compact set $f(E)$ is a $G_\delta$ in the plane, say

$$f(E) = \cap\{V_n : n = 1, 2, \ldots\},$$

each $V_n$ an open set. The set $f^{-1}(V_n)$ is open in $X$ and contains $E$, and consequently there is a set $F_n$ which is a finite intersection of peak sets and which satisfies $E \subset F_n \subset f^{-1}(V_n)$. We have

$$E \subset F = \bigcap_n F_n \subset \bigcap_n f^{-1}(V_n).$$

Since $\cap V_n = f(E)$, it follows that $f(E) = f(F)$. Moreover, since each of the sets $F_n$ is actually a peak set, so also is the set $F$, for if $g_n \in A$ peaks on $F_n$, the function $g = \sum_{n=1}^{\infty} 2^{-n} g_n$ peaks on $F$.

The set $f(F)$ is contained in the real line, so there is a polynomial $p$ such that $p(s) = 1$ and $|p(t)| < 1$ if $t \in f(F) \setminus \{s\}$. Then $p \circ f \in A$, and *on the set $F$*, $p \circ f$ peaks on $F_0 = f^{-1}(s)$, a set which contains $K$ but not the whole of $E$. (Of course, $|p \circ f|$ may be greater than 1 at points of $X \setminus F$.) Corollary 7.23 now applies, and we may conclude that the set $F_0$ is an intersection of peak sets for $A$. Since $E$ contains $F_0$ as a proper subset, we have reached a contradiction to the choice of $E$, so the lemma is proved.

The last two lemmas combine to give part (b) of the theorem.

The theorem contains as a very special case the Stone–Weierstrass theorem.

12.5 ***Corollary***

*Let $(A, X)$ be a uniform algebra. If $A$ is conjugate closed in the sense that $f \in A$ implies $\bar{f} \in A$, then $A = \mathscr{C}(X)$.*

PROOF

For the ideal $I$ of the theorem, take the whole algebra $A$. Since $A$ separates points on $X$ and is conjugate closed, it follows that every maximal antisymmetric set for $A$ is a single point and the corollary follows immediately.

Another corollary of this theorem is the fact that ideals are relatively conjugate closed.

12.6 ***Corollary***

*With $I$ and $A$ as in the theorem, if $f \in I$ and $\bar{f} \in A$, then $\bar{f} \in I$.*

PROOF

If $K \in \mathscr{K}$, then $f|K$ is constant, so $\bar{f}|K \in I|K$.

12.7 ***Remarks*** (See Glicksberg [1] and Tomiyama [2].)

Let $(A, X)$ be a uniform algebra. We can consider the maximal antisymmetric decomposition $\mathscr{K}$ of $X$ with respect to $A$, and we can also consider the maximal antisymmetric decomposition $\mathscr{K}_\Sigma$ of $\Sigma(A)$ with respect to $\hat{A}$. How are $\mathscr{K}$ and $\mathscr{K}_\Sigma$ related? This relation can be obtained as follows. Let $P$ be the collection of one-point elements of $\mathscr{K}_\Sigma$. Then

$$\mathscr{K} = P \cup \{K \cap X : K \in \mathscr{K}_\Sigma\}.$$

To prove this, consider $K \in \mathscr{K}_\Sigma$. We know that $\hat{A}|K$ is closed; denote by $\Gamma_K$ its Shilov boundary. Then $\Gamma_K \subset X$, for $\Gamma_K \cap X$ is closed, and since $K$ is an intersection of peak sets for $\hat{A}$, each strong boundary point for $\hat{A}|K$ must be a strong boundary point for $\hat{A}$ (Corollary 7.23) and so must lie in $\Gamma$, the Shilov boundary for $A$ which is contained in $X$.
for $\hat{A}|K$ must be a strong boundary point for $\hat{A}$ (Corollary 7.23) and so
$P \subset \mathscr{K}$. If $K \in \mathscr{K}_\Sigma$, and if $f$ is real on $K \cap X$, then $\hat{f}$ is real on the Shilov boundary for $\hat{A}|K$, whence $\hat{f}$ is real on $K$. Thus, $f|K$ is constant, so

$f|(K \cap X)$ is constant, and it follows that $K \cap X$ is a set of antisymmetry for $A$. The maximality of $K$ implies that of $K \cap X$, and our assertion is proved.

The decomposition $X = \bigcup\{K : K \in \mathscr{K}\}$ into a disjoint union of closed sets induces a corresponding decomposition of $\Sigma(A)$: We have

$$\Sigma(A) = \bigcup\{\Sigma(A|K) : K \in \mathscr{K}\},$$

and this is a disjoint union of closed sets. This fact is a consequence of Theorem 11.11, for we know that $A|K$ is closed in $\mathscr{C}(K)$ for each $K \in \mathscr{K}$, and we know that $\Sigma(A|K) = \hat{K}$. Thus, we need only prove that for distinct $K, K' \in \mathscr{K}$, $\hat{K}$ and $\hat{K}'$ are disjoint. This is so, for since $K \neq K'$, there is $f \in A$ with $f|K = 1$ and $f|K' = 0$. It follows that $\hat{K} \subset \{\varphi \in \Sigma(A) : \varphi(f) = 1\}$ and $\hat{K}' \subset \{\varphi \in \Sigma(A) : \varphi(f) = 0\}$, whence the sets $\hat{K}$ and $\hat{K}'$ are disjoint.

The Shilov idempotent theorem implies that each $K \in \mathscr{K}_\Sigma$ is connected. If $K = K' \cup K''$ with $K', K''$ disjoint open, closed subsets of $K$, then there is $f \in A$ with $\hat{f}|K' = 0, \hat{f}|K'' = 1$, because $K$ is $\Sigma(A|K)$. This is in violation of the fact that $K$ is a set of antisymmetry for $A$. If $X \neq \Sigma(A)$, then a maximal antisymmetric set in $X$ for $A$ need not be connected. However, if $K$ is such a set, then either $K$ consists of a single point or else $K$ contains a perfect set; for if $K$ does not reduce to a single point and if $K = K_\Sigma \cap X$, $K_\Sigma \in \mathscr{K}_\Sigma$, then as $K_\Sigma$ is connected, we may restrict attention to the case that $K_\Sigma \neq K$. The set $K$ contains the Shilov boundary for $A|K_\Sigma$, and this algebra has $K_\Sigma$ as its spectrum, so it follows by Corollary 11.4 that $K$ contains a perfect set.

As a consequence of this last remark, we have the following fact noted by Rudin [7].

**12.8** ***Corollary***

*If $(A, X)$ is a uniform algebra and if $X$ contains no perfect sets, then $A = \mathscr{C}(X)$.*

Various examples of compact spaces without perfect subsets are known:

1. Countable compact sets in $\mathbb{C}^N$.
2. The one-point compactification of discrete spaces.
3. The space of ordinal numbers not exceeding a given limit ordinal taken with the order topology.
4. The continuous image of a compact space containing no perfect sets.

The validity of the last example can be established by the following simple argument. Let $X$ and $Y$ be compact spaces, let $Y$ be the continuous image of $X$ under the map $f$, and suppose that $X$ contains no perfect subsets. If $E$ is a perfect set contained in $Y$, let $E'$ be a closed subset of $X$ minimal with respect to the property of being carried onto $E$ by $f$. As $X$ contains no perfect sets, there is a point $x$ in $E'$ which is not a limit point of $E'$. But then the set $E'' = E' \setminus \{x\}$ is closed and is carried onto $E$ by $f$. This contradicts the assumed minimality of $E'$, so $E$ cannot exist.

12.9 ***Corollary***

*If $E \subset \mathbb{C}^N$ is a countable, compact set, every continuous function on $E$ can be approximated uniformly by polynomials.*

# 13 *CHARACTERIZATIONS OF $\mathscr{C}(X)$ AS A UNIFORM ALGEBRA*

In this section we shall study $\mathscr{C}(X)$ as a uniform algebra. It is our intention to prove that many of the properties which $\mathscr{C}(X)$ possesses suffice to characterize it within the class of uniform algebras. The first result we consider is due to Wermer [10].

13.1 ***Theorem***

*If $(A, X)$ is a uniform algebra, and if the space* Re $A$ *is closed under multiplication, then $A = \mathscr{C}(X)$.*

PROOF

By the generalized Stone–Weierstrass theorem, it is enough to consider the case that $A$ is antisymmetric.

If $x_0 \in X$, the antisymmetry of $A$ implies that if $u \in \text{Re}\, A$, then there exists a unique $v$ such that $v(x_0) = 0$ and $u + iv \in A$. The map $u \mapsto u + iv$ from Re $A$ into $A$ is real linear and one to one. Thus, we can define a norm $N$ on Re $A$ by setting

$$N(u) = \|u + iv\|_X.$$

The completeness of $A$ implies that Re $A$ is complete in the norm $N$. If $u_0 \in \text{Re}\, A$, the map $u \mapsto uu_0$ is a real linear map of Re $A$ into itself, and its graph is closed. Thus, it is continuous, so there is $K(u_0)$, a constant, such that

$$(1) \qquad N(uu_0) \le K(u_0)N(u).$$

Consequently, if $\{u_n\}$ is a sequence in Re $A$, which converges to $\tilde{u}$ in the $N$-norm, then $\{u_n u_0\}$ converges to $\tilde{u}u_0$. It follows from Theorem

1.9 and the remarks following it that for some norm $N_0$ equivalent to $N$, we have

$$N_0(uu') \leq N_0(u)N_0(u')$$

for all $u, u' \in \operatorname{Re} A$. Since $N_0$ is equivalent to $N$, this leads to

$$N(uu') \leq CN(u)N(u')$$

for some absolute constant $C$.

We next show that if $p \in \operatorname{Re} A$, and $p > 0$ on $X$, then $\log p \in \operatorname{Re} A$. To see this, consider the space $S$ of functions $f = u + iv$, $u, v \in \operatorname{Re} A$. Since $\operatorname{Re} A$ is an algebra, so is $S$. The algebra $S$ contains $A$ and is closed under complex conjugation. If $u + iv \in S$, define $N(u + iv)$ to be $N(u) + N(v)$, and then set

$$N_1(u + iv) = \sup\{N(e^{i\theta}(u + iv)) : \theta \text{ real}\}.$$

With respect to this norm, $S$ is a complex Banach space, and we have $N_1(fg) \leq CN_1(f)N_1(g)$ for $f, g \in S$, so $S$ is a Banach algebra under some norm equivalent to $N_1$. If $\varphi \in \Sigma(S)$, then $f \mapsto \overline{\varphi(\bar{f})}$ is a well-defined element $\psi$ of $\Sigma(A)$. Since $p \in \operatorname{Re} A$, we have that $p = \frac{1}{2}(f + \bar{f})$ for some $f \in A$, so

$$\varphi(p) = \tfrac{1}{2}(\varphi(f) + \overline{\psi(f)}).$$

Now $p = \operatorname{Re} f$ is strictly positive on $X$, and since $X$ contains the Shilov boundary for $A$, it follows that if $\chi \in \Sigma(A)$, then $\operatorname{Re} \chi(f) > 0$. Applied to $\varphi$ and $\psi$, this implies that $\operatorname{Re} \varphi(p) > 0$. Thus, $\sigma_S(p)$ is contained in the right half-plane, so $\log p \in S$, and, consequently, $\log p \in \operatorname{Re} A$.

Now we can obtain the desired contradiction. Let $c$ be a positive real number. Let $g \in A$, $g(x_0) = 0$, $\|g\|_X = 1$, and let $y \in X$ be such that $|g(y)| = 1$. Let $q$ be a polynomial such that $\operatorname{Im} q(0) = 0$, $\operatorname{Im} q(g(y)) \geq c$, and $0 < \operatorname{Re} q(z) \leq 1$ if $|z| \leq 1$. Set $f = q \circ g \in A$. We have $0 < \operatorname{Re} f \leq 1$ on $X$, so $\log f \in \operatorname{Re} A$ by the last paragraph. Let $\log \operatorname{Re} f = \operatorname{Re} F$, $F \in A$, and put $V = e^{1/2F}$. The function $V$ is in $A$, and $|V|^2 = \operatorname{Re} f$. We have $\|V\|_X \leq 1$. The identity $(\operatorname{Re} z)^2 = \frac{1}{2}(\operatorname{Re} z^2 + |z|^2)$ applied to $V$ yields

$$(\operatorname{Re} V)^2 = \operatorname{Re} \tfrac{1}{2}(V^2 + f)$$

since $|V|^2 = f$. If $h$ is any element of $A$, we have

$$N(\operatorname{Re} h) \geq \|h\|_X - |h(x_0)|,$$

so

$$\begin{aligned} N((\operatorname{Re} V)^2) &\geq \tfrac{1}{2}\|V^2 + f\|_X - \tfrac{1}{2}|V^2(x_0) + f(x_0)| \\ &\geq \tfrac{1}{2}(\|f\|_X - \|V^2\|_X) - \tfrac{1}{2}(|V^2(x_0)| + |f(x_0)|) \\ &\geq \tfrac{1}{2}(c - 1) - 1, \end{aligned}$$

because $\|f\|_X \geq c$. As $N((\text{Re } V)^2) \leq C(N(\text{Re } V))^2 \leq C(2\|V\|_X)^2 \leq 4C$, we have a contradiction, for $c$ was chosen arbitrarily. Thus, we must conclude that $X$ consists of a single point at most, and the result follows.

Applied to the disc algebra, this result has the following consequence: If $u$ is a continuous function, let $u^{\sim}$ be the function harmonic in the unit disc which assumes continuously the boundary values $u$. There exists a $u$ such that the harmonic conjugate of $u^{\sim}$ has continuous boundary values but the harmonic conjugate of $(u^2)^{\sim}$ does not. This follows from the above theorem and the fact that not every $u$ is the boundary value function of the real part of an element of the disc algebra. Wermer attributes this observation to J.-P. Kahane.

Our second characterization of $\mathscr{C}(X)$ also involves the real parts of the algebra. It was published by Hoffman and Wermer [1], who credit both Rossi and Bear with having discovered the result independently. Other proofs have been given by Browder [3] and Arenson [1].

**13.2** ***Theorem***

*If $(A, X)$ is a uniform algebra and if* Re $A$ *is uniformly closed in* $\mathscr{C}_{\mathbf{R}}(X)$, *then* $A = \mathscr{C}(X)$.

The proof we shall give is that of Browder. It depends on a lemma from Banach space theory obtained by Reiter [1].

**13.3** ***Lemma***

*Let $B$ be a Banach space with dual $B^*$*, and let $V_1$ and $V_2$ be closed subspaces of $B$. Set

$$V_j^{\perp} = \{\varphi \in B^* : \varphi(x) = 0 \text{ if } x \in V_j\}.$$

*Then*

(a) $V_1 + V_2$ *is closed only if* $V_1^{\perp} + V_2^{\perp}$ is norm closed.

(b) *If* $V_1 + V_2$ *is closed, then*

(2) $$(V_1 \cap V_2)^{\perp} = V_1^{\perp} + V_2^{\perp}.$$

PROOF

(a) Let $\eta : B \to B/V_1$ be the quotient map. Then $V_1 + V_2$ is closed in $B$ if and only if $\eta(V_2)$ is closed in $B/V_1$. Let $\eta_1$ denote the restriction of $\eta$ to $V_2$. The adjoint $\eta_1^*$ maps $(B/V_1)^*$ to $V_2^*$, and its range is closed if and only if the range of $\eta_1$ is closed. We have $V_2^* = B^*/V_2^{\perp}$, and $(B/V_1)^* = V_1^{\perp}$. The range of $\eta_1^*$ is the set of cosets $\{\varphi + V_2^{\perp} : \varphi \in V_1^{\perp}\}$, and this is closed in $B/V_2^{\perp}$ if and only if $V_1^{\perp} + V_2^{\perp}$ is closed in $B^*$. Thus we have $(a)$.

(b) In any case, we have $(V_1 \cap V_2)^\perp \supset V_1^\perp + V_2^\perp$. Suppose, as a first case, that $V_1 + V_2 = B$ and $V_1 \cap V_2 = 0$. Then (2) surely holds. Then, considering $B/V_1 \cap V_2)$, we find that (2) holds whenever $V_1 + V_2 = B$. From this, the general case follows by the Hahn–Banach theorem. Let $\varphi \in (V_1 \cap V_2)^\perp$, and apply the last observation to $\varphi_1$, the restriction of $\varphi$ to the closed subspace $V_1 + V_2$ of $B: \varphi_1 = \psi_1 + \psi_2$, $\psi_j \in \{\psi \in (V_1 + V_2)^* : \psi | V_j = 0\}$. By the Hahn–Banach theorem, $\psi_j$ extends to an element $\psi_j'$ of $V_j^\perp$, and we have then that on $V_1 + V_2$, $\varphi = \psi_1' + \psi_2'$. Thus, $\varphi - \psi_1' - \psi_2' \in (V_1 + V_2)^\perp = V_1^\perp \cap V_2^\perp$, so since $V_1^\perp \cap V_2^\perp \subset V_1^\perp + V_2^\perp$, we have $\varphi - \psi_1' - \psi_2' = \psi_1'' + \psi_2''$, $\psi_j'' \in V_j^\perp$, whence $\varphi = \psi_1' + \psi_1'' + \psi_2' + \psi_2'' \in V_1^\perp + V_2^\perp$, and we have part (b).

PROOF OF THE THEOREM

If $\mathscr{F}$ is a set of complex-valued functions, let $\bar{\mathscr{F}}$ denote $\{\bar{f} : f \in \mathscr{F}\}$. Set $A_R = A \cap \bar{A}$ so that $A_R$ is the maximal conjugate closed subalgebra of $A$. We are supposing that Re $A$ is closed. From this it follows that $A + \bar{A}$ is closed, for if $f_n + \bar{g}_n \to u + iv$, then $\operatorname{Re} f_n + \operatorname{Re} g_n \to u$, so $u = \operatorname{Re} h_1, h_1 \in A$. Similarly, $v = \operatorname{Im} h_2, h_2 \in A$. But if $h \in A$, then $\operatorname{Re} h$, $\operatorname{Im} h \in A + \bar{A}$, so $u + iv \in A + \bar{A}$. Since $A + \bar{A}$ is closed, the lemma shows that $A^\perp + \bar{A}^\perp = A_R^\perp$, and consequently every real measure on $X$ which annihilates $A_R$ is the real part of a measure which annihilates $A$. Given $x \in X$, let $K_x = \{t \in X : h(t) = h(x) \text{ for all } h \in A_R\}$. If $K_x$ contains more than one point, there is a $g \in A$ which is nonconstant on $K_x$. Let $y,z \in K_x$ satisfy $|g(y)| = \|g\|_{K_x}$, $|g(z)| < |g(y)|$, and put $f = \frac{1}{2}(1 + g/g(y))$. Then $f \in A$, $f(y) = 1$, $|f(z)| < 1$, and also, if $|f(t)| \geq 1$ and $t \in K_x$, then $f(t) = 1$. Let $\delta_s$ denote the point mass at the point $s$. Then $\delta_y - \delta_z$ annihilates $A_R$ and, thus, for some measure $v$, $\mu = \delta_y - \delta_z + iv \in A^\perp$. If $A_R$ consists of constants only, then $K_x = X$, and we can take, say, $y$ to be a strong boundary point for $A$. Then by considering $\int h \, d\mu$ for functions $h$ in $A$ which peak on small sets containing $y$, we find that $0 = 1 + iv(\{y\})$, a contradiction, and we are finished. Assume, therefore, that $A_R$ is nontrivial.

If $h \in A_R$ is real and nonconstant, let

$$\tilde{h} = 1 - (M - m)^{-2}[h - h(y)]^2,$$

where $M = \max\{h(x) : x \in X\}$, $m = \min\{h(x) : x \in X\}$. Then $\tilde{h} \in A_R$, $0 \leq \tilde{h} \leq 1$, and $h(s) = 1$ if and only if $s \in K(h) = \{t \in X : h(t) = h(y)\}$. Now, $\int f^n \tilde{h}^k \, d\mu = 0$ for all $h \in A_R$ and all positive integers $n$ and $k$. If we let $k \to \infty$, we find $\int_{K(h)} f^n \, d\mu = 0$ since $\tilde{h}^k$ tends boundedly and pointwise, as $k \to \infty$, to the characteristic function of $K(h)$. The sets $K(h)$ all include $K_x$, they are directed by inclusion because $K(\tilde{h}) = K(h)$ and $K(\tilde{h}_1 \tilde{h}_2) = K(\tilde{h}_1) \cap K(\tilde{h}_2)$, and their intersection is $K_x$. Consequently, we may conclude that for every positive integer $n$, $\int_{K_x} f^n \, d\mu = 0$. We now let $n \to \infty$ to find that $\mu(E) = 0$ if $E = \{s \in K_x : f(s) = 1\}$, for if $s \in K_x$, $f^n(s)$ converges, as $n \to \infty$, to 0 or 1 according as $s \in E$ or $s \in K_x \setminus E$. Since $y \in E$, $z \notin E$, we have $1 + iv(E) = 0$, and this is not possible. Consequently, $K_x$ must reduce to a single point. Thus, $A_R$ separates points on $X$, and so by the Stone–Weierstrass theorem, $A_R$ is $\mathscr{C}_R(X)$, and the theorem follows.

13.4 ***Remarks***

Let us mention without proof a variation of the theorem of Hoffman and Wermer.

***Theorem*** (Glicksberg [6])

*If $(A, X)$ is a uniform algebra with $X$ metrizable, and if $I \subset A$ is a closed ideal in $A$ such that $\bar{I} + A$ is closed in $\mathscr{C}(X)$, then $I = \bar{I}$.*

The case that $I = A$ is a trivial consequence of the Hoffman–Wermer theorem.

Another characteristic property of $\mathscr{C}(X)$ is the fact that all its restriction algebras are closed.

13.5 ***Theorem*** (Glicksberg [2])

*If $(A, X)$ is a uniform algebra, and if for every closed set $E \subset X$, $A|E$ is closed in $\mathscr{C}(E)$, then $A = \mathscr{C}(X)$.*

PROOF

If $E \subset X$ is closed, then $A|E$ is closed in $\mathscr{C}(E)$, and if

$$k(E) = \{f \in A : f|E = 0\},$$

then the map

$$f + k(E) \mapsto f|E$$

from $A/kE$ to $A|E$ is a continuous linear map with range the closed-space $A|E$. Thus, by the open mapping theorem, there is a constant $C_E$ such that

$$\|f + k(E)\| \leq C_E \|f\|_E. \tag{3}$$

We shall show that if $E_0$ and $E_1$ are closed disjoint subsets of $X$, then there exists an $f \in A$ with $f|E_0 = 0$, $f|E_1 = 1$. To see this, let $x \in E_0$, $y \in E_1$. Since $A$ separates points on $X$, there is $f \in A$ with $f(x) = 0$, $f(y) = 1$. Let $W_y = \{w \in X : |f(w)| \leq \frac{1}{4}\}$, $V_y = \{w \in X : |f(w) - 1| \leq \frac{1}{4}\}$. Thus, $W_y$ and $V_y$ are disjoint closed subsets of $X$. There exists a sequence $\{p_n\}$ of polynomials which converges uniformly to zero on

$\{z \in \mathbb{C} : |z| \leq \frac{1}{4}\}$ and uniformly to 1 on $\{z \in \mathbb{C} : |z - 1| \leq \frac{1}{4}\}$. For all $n$, $p_n \circ f \in A$, and since $A|(V_y \cup W_y)$ is closed in $\mathscr{C}(W_y \cup V_y)$, it follows that for some $g_y \in A$ we have $g_y|V_y = 1$, $g_y|W_y = 0$. Finitely many of the sets $V_y$ cover $E_1$, say $V_1, \ldots, V_k$. Let $W_1, \ldots, W_k$ be the associated $W$'s, and let $g_1, \ldots, g_k$ be the associated $g$'s. The function $h_x$ given by

$$h_x = 1 - (1 - g_1) \cdots (1 - g_k)$$

is in $A$, vanishes on the neighborhood $U_x = \bigcap\{W_j : j = 1, \ldots, n\}$ of $x$, and is identically 1 on the neighborhood $\bigcup\{V_j : j = 1, \ldots, n\}$ of $E_1$. Finitely many of the neighborhoods $U_x$ cover $E_0$, say $U_1, \ldots, U_m$. Let $h_1, \ldots, h_m$ be the associated $h_x$'s, and set $f = h_1 \cdots h_m$. The function $f$ is in $A$, vanishes on a neighborhood of $E_0$, and is identically 1 on a neighborhood of $E_1$.

Thus, we have constructed an idempotent in $A/k(E_0 \cup E_1)$ which separates $E_0$ and $E_1$. We need to obtain a bound on such idempotents. The following lemmas of Katznelson [1] provide the desired bound. We say that *$A$ is bounded on the subset $V$ of $X$* if there exists a constant $C_V$ such that whenever $E \subset V$ is a closed subset of $X$, every idempotent of $A/k(E)$ is of norm no more than $C_V$; and we say that $A$ is *bounded at the point $x \in X$ if $A$* is bounded on some neighborhood of $x$.

**13.6** ***Lemma***

*If $A$ is bounded on the open sets $V_1$ and $V_2$ of $X$, it is bounded on every closed set contained in $V_1 \cup V_2$.*

PROOF

Let $E \subset V_1 \cup V_2$ be closed. The set $E \setminus V_2$ is closed in $X$, so there is an open set $W_1$ containing $E \setminus V_2$ with $\overline{W}_1 \subset V_1$.

Similarly, there is an open $W_2$ with $V_2 \supset \overline{W}_2 \supset W_2 \supset E \setminus W_1$. The closed sets $E \setminus W_1$ and $E \setminus W_2$ are disjoint, so there is $g \in A$ with $g|(E \setminus W_1) = 0$, $g|(E \setminus W_2) = 1$.

Let $E_0$ be a closed set in $E$, and let $f + k(E_0)$ be an idempotent in $A/k(E_0)$ so that $f$ is zero or 1 at every point of $E_0$. Then $f + k(E_0 \cap \overline{W}_j)$ is idempotent in $A/k(E_0 \cap \overline{W}_j)$, so there is $f_j \in f + k(E_0 \cap \overline{W}_j)$ with $\|f_j\| \leq C_{V_j} + 1$, for $A$ is bounded on $V_j$. On $E_0$ we have

$$f = gf_1 + (1 - g)f_2$$

since $E_0 \subset E \subset W_1 \cup W_2$. Thus

$$\begin{aligned} \|f + k(E_0)\| &\leq \|gf_1 + (1 - g)f_2\| \\ &\leq \|g\|(C_{V_1} + 1) + (1 + \|g\|)(C_{V_2} + 1), \end{aligned}$$

and the right side is a suitable bound for $C_E$.

As an immediate consequence, we have the following fact.

13.7 **Lemma**

*If $F \subset X$ is closed and if $A$ is bounded at each point of $F$, then there is an open set $V \supset F$ on which $A$ is bounded.*

13.8 **Lemma**

*There exist only finitely many points at which $A$ is not bounded.*

PROOF

If the lemma is false, there exists a sequence $\{x_n\}$ of distinct points in $X$ and a corresponding sequence $\{V_n\}$ of neighborhoods such that $x_n \in V_n$, $\bar{V}_n \cap (\bigcup\{V_m : m \neq n\})^- = \varnothing$, and such that $A$ is not bounded on $V_n$. Since $A$ is unbounded on each $V_n$, there is a closed set $E_n \subset V_n$ and an idempotent $f_n + k(E_n)$ in $A|k(E_n)$ which is of norm at least $n$. If $E = (\bigcup E_n)^-$, then $E = E_n \cup (\bigcup_{m \neq n} E_n)^-$, and $E_n \cap (\bigcup_{m \neq n} E_m)^- = \varnothing$. Then $E = E_n \cup (E \setminus E_n)$ is a decomposition of $E$ into a disjoint union of closed sets, so there is $g \in A$ with $g|E_n = 1, g|(E \setminus E_n) = 0$. Then $gf_n + k(E)$ is an idempotent in $A/k(E)$, and since

$$\|gf_n + k(E)\| \geq \|gf_n + k(E_n)\|$$

$$= \|f_n + k(E_n)\| \geq n,$$

the idempotents of $A/k(E)$ are not bounded. However, this contradicts (3) above, so the lemma is proved.

With these lemmas, we can conclude the proof of the theorem. Let $T = \{x_1, \ldots, x_m\}$ be the finite set of points at which $A$ is not bounded, and let $E$ be a closed set in $X \setminus T$. By Lemma 13.7, $A$ is bounded on a neighborhood $V$ of $E$ with $\bar{V} \cap T = \varnothing$; i.e., there is a constant $C_V$ so that whenever $K$ is a closed set contained in $V$, every idempotent of $A/k(K)$ is of norm no more than $C_V$.

Let $\mu$ be a regular Borel measure on $E$ which annihilates $A|E$, let $K \subset E$ be closed, and let $\varepsilon > 0$. Choose a closed set $K_0$ in $E \setminus K$ for which $|\mu|(E \setminus (K \cup K_0)) < \varepsilon(2C_V)^{-1}$, and choose $F \in A$ which is 1 on $K$, zero on $K_0$, and bounded by $2C_V$. Then

$$0 = \int F\,d\mu = \int_K F\,d\mu + \int_{E \setminus (K \cup K_0)} F\,d\mu,$$

and we have that

$$|\mu(K)| = \int_{E \setminus (K \cup K_0)} F\,d\mu < \varepsilon.$$

Since $\varepsilon$ is at our disposal, we conclude that $|\mu(K)| = 0$, and since $K$ is an arbitrary closed subset of $E$, we conclude that $|\mu|(E) = 0$. Thus, $A|E$ is dense in $\mathscr{C}(E)$, so $A|E = \mathscr{C}(E)$.

By regularity, this implies that any measure orthogonal to $A$ is carried on $T$, for if not, then $\mu(E) \neq 0$ for some closed set contained in $X \setminus T$ and some $\mu$ orthogonal to $A$. Let $V$ be an open set which contains $E$ and whose closure misses $T$. The sets $E$ and $V^c$ are disjoint and closed, so there is $f \in A$ with $F|E = 1, f|V^c = 0$. Then the measure $f\mu$ annihilates $A$ and is supported by $\bar{V}$. Since $\bar{V} \cap T = \varnothing$, $A|\bar{V} = \mathscr{C}(\bar{V})$, so $f\mu$ is orthogonal to $\mathscr{C}(\bar{V})$ and, thus, $f\mu = 0$. However, $f|E = 1$, so $f\mu(E) = \mu(E) \neq 0$, and we have a contradiction.

Consequently, every measure orthogonal to $A$ is supported on the finite set $T$. Since $A$ separates points on $X$, we may conclude that the only measure orthogonal to $A$ is the zero measure, so $A = \mathscr{C}(X)$, as desired.

There is a version of Glicksberg's theorem valid in locally compact spaces.

13.9 ***Corollary***

*If $A$ is a uniformly closed function algebra on the locally compact space $X$, and if $A|E$ is closed in $\mathscr{C}(E)$ for every compact $E \subset X$, then $A = \mathscr{C}_0(X)$.*

PROOF

Let $E \subset X$ be compact. The algebra $A|E$ is closed in $\mathscr{C}(E)$, and it separates points on $E$.

We shall show that it is $\mathscr{C}(E)$. Note at the outset that it is not obvious that $A|E$ contains the constants. Consider therefore the algebra $\mathbb{C} + A|E$. This too is a closed subalgebra of $\mathscr{C}(E)$, and if $F \subset E$ is a closed set, then by hypothesis $A|F$ is closed, so $(\mathbb{C} + A|E)|F = \mathbb{C} + A|F$ is closed in $\mathscr{C}(F)$. Consequently, $\mathbb{C} + A|E = \mathscr{C}(E)$, and so if $g \in \mathscr{C}(E)$, there is $f \in A$ and $\alpha \in \mathbb{C}$ such that on $E$, $g = f + \alpha$. Apply this in particular to $\bar{g}$, for $g \in A : \bar{g}|E = f|E + \alpha, f \in A$, so on $E$

$$g\bar{g} = g(f + \alpha) = gf + g\alpha,$$

and thus $g\bar{g} \in A|E$. A finite sum of functions $g\bar{g}$, $g \in A$, say $h$, exists such that $h|E$ has range bounded away from zero on $E$. But then $h^{-1} \in \mathscr{C}(E)$, so $h^{-1} = f + \alpha$ on $E$. Thus, on $E$,

$$1 = hh^{-1} = h(f + \alpha) \in A|E,$$

and we find that $\mathbb{C} + A|E = A|E$. Therefore, $A|E = \mathscr{C}(E)$ for all compact sets $E$ in $X$.

This implies that $A = \mathscr{C}_0(X)$. Let $f \in \mathscr{C}_0(X)$ have compact support, and choose a compact set $K \subset X$ off which $F$ vanishes. There exists $h \in \mathscr{C}_0(X)$ with range in $[0, 1]$, $h = 1$ on $K$, and $h = 0$ outside a compact neighborhood $U$ of $K$. Let $g \in A$ agree with $h$ on a neighborhood $V$ of

$U$, and set $G = \{x \in X \setminus V : |g(x)| \geq \frac{1}{2}\}$ so that $G$ is a certain compact set. There exists $k \in A$ such that $k = f$ on $U$, $k = 0$ on $G$. The functions $g^n k$ all lie in $A$ and we have that as $n \to \infty$, $g^n k \to f$ uniformly on $X$. Since $A$ is closed, $f \in A$, and thus $A$ contains each function in $\mathscr{C}_0(X)$, with compact support. As these functions are dense in $\mathscr{C}_0(X)$, $A = \mathscr{C}_0(X)$.

The problem of showing that if $A \subset \mathscr{C}_0(X)$ is closed and if $A|E = \mathscr{C}(E)$ for all compact sets $E$, then $A = \mathscr{C}_0(X)$ was proposed by R. E. Mullins at the Tulane Symposium on Function Algebras. Various solutions have been given; the argument used in the proof above is in Badé and Curtis [1], where it is attributed to Katznelson.

Since if $E$ is an intersection of peak sets, then $A|E$ is closed in $\mathscr{C}(E)$, Theorem 13.5 immediately implies the following fact. (Of course, the corollary can be proved more directly by considering measures which annihilate $A$.)

13.10 ***Corollary***

*If each closed set $E \subset X$ is an intersection of peak sets for $A$, then $A = \mathscr{C}(X)$.*

The algebras $\mathscr{C}(X)$ can also be characterized as those uniform algebras that are locally $\mathscr{C}(X)$. Even more is true.

13.11 ***Theorem*** (Gamelin and Wilkin [1])

*Let $(A, X)$ be a uniform algebra. If $X = \bigcup\{X_n : n = 1, 2, \ldots\}$ where each $X_n$ is closed, and if $A|X_n = \mathscr{C}(X_n)$, then $A = \mathscr{C}(X)$.*

PROOF

By the generalized Stone–Weierstrass theorem, we can suppose that $A$ is antisymmetric. Since $A|X_n = \mathscr{C}(X_n)$, it follows from Theorem 11.11 that $X = \Sigma(A)$. By the Baire category theorem there is an $n$ for which $X_n$ has a nonvoid interior. If $E$ is a closed $G_\delta$ in the interior of $X_n$, then since $A|X_n = \mathscr{C}(X_n)$, it follows that $E$ is a local peak set for $A$ and consequently, by the local peak set theorem, it is a peak set for $A$. This implies that if $\mu \in \mathscr{M}(X)$ is orthogonal to $A$, then $\mu_E$ is also orthogonal to $A$. Again, since $A|X_n = \mathscr{C}(X_n)$, we conclude that $\mu_E = 0$. By regularity, it follows that if $\mu \in A^\perp$, then $\mu$ restricts to the zero measure on the interior, $X_n^0$, of $X_n$.

There is a real $f \in \mathscr{C}(X)$ such that $f|(X \setminus X_n^0) = 0$ but $f$ is not identically zero. By what we have just said $\int f\,d\mu = 0$ for all $\mu \in A^\perp$. Consequently, $f \in A$. Since $A$ is antisymmetric, $f$ is constant, and since $f$ is not identically zero, we have $X \setminus X_n^0 = \emptyset$. This yields the result.

There is a locally compact version of this result.

13.12 ***Corollary***

*Let $X$ be a locally compact space and $A$ a uniformly closed function algebra on $X$. Let $\{X_\alpha\}$ be a family of closed subsets of $X$ such that $A|X_\alpha = \mathscr{C}_0(X_\alpha)$ and $X = \bigcup X_\alpha$. If every $x \in X$ has a neighborhood which meets only countably many of the sets $X_\alpha$, then $A = \mathscr{C}_0(X)$.*

PROOF

Let $E \subset X$ be a compact set. By compactness, there is a finite family of functions $\{f_1, \ldots, f_N\}$ in $A$ such that if $Y = \{y \in X : |f_1(y)| + \cdots + |f_N(y)| \neq 0\}$, then $Y \supset E$. The set $Y$ is an open, $\sigma$-compact set in $X$. Let $Y = \bigcup Y_n$, $Y_n$ compact. If $y \in Y_n$, there is a neighborhood $V(y)$ of $y$ which meets only countably many of the sets $X_\alpha$. Thus, by compactness, only countably many of the sets $X_\alpha$ (denote them by $X_{n,j}$, $j = 1, 2, \ldots$) meet $Y_n$. Let $I = \{f \in A : f(y) = 0 \text{ if } y \notin Y\}$, a certain closed ideal in $A$, and let $I'$ be the algebra on $Y^*$, the one point compactification of $Y$, obtained by adjoining to $I$ an identity. If $\infty$ denotes the point at infinity of $Y^*$, then $\{\infty\} \cup \{Y_n \cap X_{m,j} : j, m, n = 1, 2, \ldots\}$ is a countable closed cover for $Y^*$, call it $\{F_k\}_{k=1}^{\infty}$, such that $I'|F_k = \mathscr{C}(F_k)$. Consequently, the theorem implies that $I' = \mathscr{C}(Y^*)$, and thus $I|E = \mathscr{C}(E)$.

Thus, for all compact sets $E \subset X$, $A|E = \mathscr{C}(E)$, and the corollary now follows from Corollary 13.9.

So far we have proceeded in complete generality. If we impose topological conditions on the space $X$, then additional results can be obtained. The first of these which we shall consider is due to Gorin [1].

13.13 ***Theorem***

*If $(A, X)$ is a uniform algebra, $X$ metric, and if every strictly positive function on $X$ is the modulus of an invertible element of $A$, then $A = \mathscr{C}(X)$.*

PROOF

We may suppose without loss of generality that $A$ is antisymmetric. We shall begin by showing that the group $\mathscr{G}(A)$ of invertible elements of $A$ is connected. Suppose, to the contrary, that $\mathscr{G}(A)$ differs from $\mathscr{G}_0(A)$. Fix an $x_0 \in X$, and let $f \in \mathscr{G}(A) \setminus \mathscr{G}_0(A)$ assume the value 1 at $x_0$. Since $f \notin \mathscr{G}_0(A)$, $f$ is not constant. Given $\sigma \in \mathbb{R}$, let $f_\sigma \in \mathscr{G}(A)$ be determined by $f_\sigma(x_0) = 1$, and $|f_\sigma(x)| = |f(x)|^\sigma$ for all $x \in X$. By hypothesis such an $f_\sigma$ exists in $A$, and the antisymmetry of $A$ implies that only one such $f_\sigma$ can exist. To see this suppose $g_1, g_2 \in \mathscr{G}(A)$, $g_1(x_0) = g_2(x_0)$, and $|g_1| = |g_2|$. Then $h = g_1 g_2^{-1} \in A$ satisfies $|h| \equiv 1$ on $X$, and $h(x_0) = 1$. We have that $h + h^{-1} = h + \bar{h}$, a real element of $A$, so $h + \bar{h}$ is constant, whence $\operatorname{Re} h$ is constant. Similarly, $\operatorname{Im} h$ is constant, so $h$ is constant, and thus $g_1(x) = g_2(x)$ for all $x \in X$. Therefore, $f_\sigma$ is uniquely determined by $\sigma$. Next, we show that if $\sigma \neq \tau$, then $f_\sigma$ and $f_\tau$ lie in different cosets of $\mathscr{G}_0(A)$ in $\mathscr{G}(A)$. To see this, note that since $|f_{\sigma+\tau}| = |f|^{\sigma+\tau} = |f_\sigma||f_\tau|$, the map $\sigma \mapsto f_\sigma$ is a homomorphism of $\mathbb{R}$ into $\mathscr{G}(A)$. Then if $f_\sigma$ and $f_\tau$ lie in the same coset, $f_\sigma f_\tau^{-1} \in \mathscr{G}_0(A)$, whence $f_{\sigma-\tau} = e^h$. This implies that $f = \exp\{h/\sigma - \tau\}$, so $f \in \mathscr{G}_0(A)$ contrary to hypothesis.

Thus, if $[f_\sigma]$ denotes the element of $\mathscr{G}(A)/\mathscr{G}_0(A)$ which contains $f_\sigma$, the map $\sigma \mapsto [f_\sigma]$ is a one-to-one map of $\mathbb{R}$ into $\mathscr{G}(A)/\mathscr{G}_0(A)$, so this latter group must be uncountable. Since $\mathscr{G}_0(A)$ is an open set in $A$, and since distinct cosets are disjoint, it follows that $A$ contains uncountably many disjoint open sets. This is not possible, though, for since $X$ is a compact metric space, $\mathscr{C}(X)$, and hence $A$, is separable. Thus, $\mathscr{G}(A) = \mathscr{G}_0(A)$.

Now, let $u$ be a real-valued continuous function on $X$. By hypothesis, $e^u = |f|$ for some $f \in \mathscr{G}(A) = G_0(A)$. Therefore, $f = e^h$, so we have that $u = \operatorname{Re} h$. We have shown therefore that every real continuous function on $X$ is the real part of an element of $A$, so the theorem follows the Theorem 13.1 or Theorem 13.2.

The hypothesis of metrizability in Gorin's theorem cannot be abandoned completely.

13.14 ***Example***

Throughout this example, let us understand by $L^\infty$ the $L^\infty$-algebra associated with Lebesgue measure on the unit circle. If $u$ is a bounded harmonic function on the open unit disc, then for almost all $e^{it}$, the nontangential limit $\lim_{z \to e^{it}} u(z) = u^*(e^{it})$ exists, and the function $u^*$ so defined lies in $L^\infty$. Conversely, given $u^* \in L^\infty$, there is a bounded harmonic function $u$ with $u^*$ as its boundary-value function. In particular, if $f \in H^\infty(\Delta)$, i.e., if $f$ is a bounded holomorphic function in $\Delta$, then the boundary values $f^*$ exist almost everywhere, and the map $i: f \to f^*$ is an isometric isomorphism from $H^\infty(\Delta)$ into $L^\infty$. Let $B$ be the range of this isomorphism. Since $i$ is an isometry and the Gelfand transform on $L^\infty$ is an isometry, $\hat{B}$ is a closed subalgebra of $(L^\infty)\hat{} = \mathscr{C}(\Sigma(L^\infty))$. Let $h \in \mathscr{C}(\Sigma(L^\infty))$ be strictly positive. Then $\log h \in \mathscr{C}(\Sigma(L^\infty))$, so $\log h = u^*\hat{}$ for some bounded harmonic function $u$. Then if $v$ is a function conjugate to $u$, and if $f = e^{u+iv}$, then $f \in H^\infty(\Delta)$, and $|f^*| = e^{u^*}$, whence $|(if)\hat{}\,| = h$. Thus, $\{|\hat{g}| : g \in B\}$ contains every strictly positive function conjugate to $u$, and if $f = e^{u+iv}$, then $f \in H^\infty(\Delta)$, and $|f^*| = e^{u^*}$, bounded measurable function on the unit circle is the boundary-value function of an $f \in H^\infty(\Delta)$. This example is due to Schark [1]. One may also consult Hoffman [1] for this and related results.

Another relevant example is due to Hoffman and Ramsey [1], who construct, using the continuum hypothesis, a closed subalgebra of

$l_\infty$, the bounded sequences, such that each positive $\alpha \in l_\infty$ is the modulus of an element of $A$. Negrepontis [1] has shown it possible to avoid the continuum hypothesis in this construction.

Our final characterization of $\mathscr{C}(X)$, this time for $X$ compact and locally connected, was stated by Cirka [1].

13.15 ***Theorem***

*If $(A, X)$ is a uniform algebra with $X$ locally connected, and if for each $f \in A$, there is $g \in A$ with $g^2 = f$, then $A = \mathscr{C}(X)$.*

This theorem, under the additional hypothesis that each positive $h \in \mathscr{C}(X)$ is the modulus of an $f \in A$ was established by Gorin [1]. The proof we shall give makes explicit use of the hypothesis of local connectedness, and, as we shall see in a subsequent section, the theorem cannot be proved without some hypothesis in addition to the assumption that $A$ be closed under the extraction of square roots. The proof we give is based on ideas of Cirka, but it also incorporates a suggestion of Sidney.

PROOF OF THE THEOREM

Let $f \in A$, $\alpha > 0$. We shall show that if $E$ and $F$ are compact, connected sets in $X$ such that

$$E \subset \{x : |f(x)| > \alpha\} \qquad \text{and} \qquad F \subset \{x : |f(x)| < \alpha\},$$

then given $\varepsilon > 0$, there exists $g \in A$ with

$$\|g\|_X < 1 + \varepsilon, \qquad \|g\|_F < \varepsilon, \qquad \|1 - g\|_E < \varepsilon.$$

Choose a fixed point $x_0 \in F$, and for each $m = 1, 2, \ldots$, let $f_m \in A$ satisfy $f_m^{2m} = f^{2m} + \alpha^{2m}$. The function $f_m$ is not uniquely determined by this condition, but the value at $x_0$ is determined up to a multiplicative factor of a $2^m$th root of unity. We choose $f_m$ so that the sequence $\{f_m(x_0)\}$ converges to $\alpha$; the choice of this sequence determines uniquely the values of $f_m$ on the connected set $F$. For $m, N = 1, 2, \ldots$, let $f_{N,m} \in A$ satisfy $f_{N,m}^{2N} = f_m - \alpha$. We have that

$$\|f_m\|_X \leq 2\|f\|_X$$

for all $m$, so, given $\varepsilon > 0$, there is $N(\varepsilon)$ such that $N \geq N(\varepsilon)$ implies $\|f_{N,m}\|_X \leq 1 + \varepsilon$ for all $m$. The set $F$ is compact so for some $\delta > 0$, $\|f\|_F < \alpha - \delta$, and this implies that uniformly on $F$

$$\lim_{m \to \infty} f_m = \alpha,$$

as follows from the convergence of $\{f_m(x_0)\}$ to $\alpha$ and the connectedness of $F$.

We shall show that for all large $m$ and infinitely many $N$

$$\| f_{N,m} - 1 \|_E < \varepsilon.$$

For this purpose, let $V_1, \ldots, V_q$ be connected open sets with

$$E \subset V_1 \cup \cdots \cup V_q \subset \{x : |f(x)| > \alpha + \delta\}$$

for some $\delta > 0$. Require that each $V_j$ meet $E$, and let the $V_j$ be so small that the set $f(V_j)$ is contained in a sector with vertex the origin and with opening no more than $\pi/4$. There exists $L_j$, a branch of $\log f$ on $V_j$, and for some $t_j$ we have, for all $x$ in $V_i$, $\operatorname{Im} L_j(x) \in [t_j, t_j + \pi/4]$. Since on $V_j$, $|f| \geq \alpha + \delta$, the range of the function $1 + \alpha^{2m} f^{-2m}$ on this set is contained in the right half-plane. On the set $V_j$,

$$f_m^{2m} = f^{2m}(1 + \alpha^{2m} f^{-2m}),$$

so our remarks show that on this set there is a branch of $\log f_m$ whose imaginary part lies in an interval of length $(\pi/4) + 2^{-m}\pi$. Thus, for $m > 1$, the range of $f_m$ on $V_j$ is contained in a half-plane bounded by a line through the origin; and since $|f_m| > \alpha$ on $V_j$, we see that the function $f_m - \alpha$ omits from its range on $V_j$ a ray emanating from the origin. For $j = 1, \ldots, q$, let $x_j$ be a point of $V_j$. A diagonal process provides a sequence $\{N_k\}_{k=1}^{\infty}$, $\lim_k N_k = \infty$, such that for all $m$ and $j$,

$$\lim_{k \to \infty} f_{N_k,m}(x_j) = \lambda(j, m)$$

exists. We have $|\lambda(j, m)| = 1$ for all choices of $m$ and $j$. By choosing properly the branches of the square roots in question, we may assume that $\lambda(1, m) = 1$ for all $m$. With this normalization we have

$$\lim_{k \to \infty} f_{N_k,m}(x) = \lambda(j, m)$$

uniformly in $m$ and $x \in V_j$. This is so, for on $V_j$, $|f_m - \alpha|$ is bounded away from zero uniformly in $m$, so $\lim_k |f_{N_k,m}| = 1$ uniformly in $m$ and $x \in V_j$; moreover, on $V_j$, the range of $f_m - \alpha$ omits a ray, so the range of $f_{N_k,m}$ is contained in a sector of opening no more than $2\pi 2^{-N_k}$. Since $f_{N_k,m}(x_j) \to \lambda(j, m)$ it follows that, in fact, $f_{N_k,m} \to \lambda(j, m)$ uniformly on $V_j$ and moreover that the convergence is uniform in $m$. The connectedness of $E$ now implies that $\lambda(j, m) = 1$ for all $m$ and all $j$. We know that $\lambda(1, m) = 1$ so $f_{N_k,m} \to 1$ on $V_1$. Connectedness implies that one of $V_2, \ldots, V_q$, say $V_2$, meets $V_1$. Since $f_{N_k,m} \to \lambda(2, m)$ on $V_2$, we get that

$\lambda(2, m) = 1$. Proceeding in this way, we find successively that $\lambda(3, m) = 1, \ldots, \lambda(q, m) = 1$. The conclusion is that

$$\lim_{k\to\infty} f_{N_k,m}(x) = 1,$$

the convergence uniform in $m$ and in $x \in E$.

Thus, if $\varepsilon > 0$, we can choose $k$ so that

$$\| f_{N_k,m} - 1 \|_E < \varepsilon$$

for all $m$ and also so that

$$\| f_{N_k,m} \|_X < 1 + \varepsilon.$$

If $m$ is large enough, we shall have, in addition,

$$\| f_{N_k,m} \|_F < \varepsilon.$$

This concludes the first step in the proof.

Our next step is to prove that if $E$ and $F$ are disjoint closed subsets of $X$ and if $\varepsilon > 0$, then there exists $h \in A$ with

$$\|h\|_X \leq 1, \qquad \|h\|_F < \varepsilon, \qquad \|h - 1\|_E < \varepsilon.$$

To begin the construction of $h$, use the compactness and local connectedness of $X$ and the fact that $A$ separates points on $X$ to find a finite set of connected open sets $U_1, \ldots, U_p$ and $V_1, \ldots, V_q$ such that $E \subset U_1 \cup \cdots \cup U_p$, $F \subset V_1 \cup \cdots \cup V_q$, and such that for some choice of functions $f_{ij} \in A$,

$$U_i \subset \{x : |f_{ij}(x)| < \tfrac{1}{8}\}$$

and

$$V_j \subset \{x : |f_{ij}(x) - 1| < \tfrac{1}{8}\}.$$

Let $\delta \in (0, 1)$ and let $f$ be a conformal mapping of the open unit disc onto itself with the property that $f(1) = 1$ and $|f(0) + 1| < \delta$. Let $n > 0$ be so small that $|z| < \eta$ implies that $|f(z) + 1| < 2\delta$ and $|z - 1| < 2^p\eta$ implies that $|f(z) = 1| < 2\delta$. By the first step in the proof, there are functions $g_{ij} \in A$ with

$$\|g_{ij}\|_X < 1, \qquad \|g_{ij}\|_{\bar{U}_i} < \eta, \qquad \|g_{ij} - 1\|_{\bar{V}_j} < \eta.$$

If $g_j = \prod_{i=1}^p g_{ij}$, then $\|g_j\|_X < 1$, and on $\bigcup\{U_i : i = 1, \ldots, p\}$ we have $|g_j| < \eta$, so $\|g_j\|_E < \eta$. On the set $V_j$, $g_{ij} = 1 + \varepsilon_{ij}$, $|\varepsilon_{ij}| < \eta$, so there

$$g_j = \prod_{i=1}^{p} (1 + \varepsilon_{ij}),$$

whence

$$\|g_j - 1\|_{\bar{V}_j} < 2^p \eta.$$

Define $h \in A$ by

$$h = \prod_{j=1}^{p} \tfrac{1}{2}(1 - f \circ g_j).$$

It is clear that $\|h\|_X \leq 1$. If $x \in V_j$, then $|g_j(x) - 1| < 2^p\eta$, so by the choice of $\eta$, $|1 - f(g_j(x))| < 2\delta$, whence $|h(x)| < \delta$ and we have, therefore, that $\|h\|_F < \delta$. If $x \in E$, then $|g_j(x)| < \eta$, so $f \circ g_j(x)$ is within $2\delta$ of $-1$, and it follows that $|h(x) - 1| < 2^q\delta$. Thus, if $\delta$ is chosen so that $2^q\delta < \varepsilon$, the function $h$ has the desired properties.

Now suppose that $A \neq \mathscr{C}(X)$ so that there exists a nonzero measure $\mu$ on $X$ which annihilates $A$. By regularity there is a closed set $E$ with $\mu(E) \neq 0$ and a sequence $\{F_n\}$ of compact sets in $X \setminus E$ with $|\mu|(F_n) \to |\mu|(X \setminus E)$. For each $n$ let $h_n \in A$ satisfy

$$\|h_n\|_X < 1 + 1/n, \qquad \|h_n\|_{F_n} < 1/n, \qquad \|h_n - 1\|_E < 1/n.$$

By the dominated convergence theorem,

$$0 = \lim \int h_n \, d\mu = \mu(E) \neq 0,$$

a contradiction, and the theorem is proved.

It is possible to weaken slightly the hypotheses of the preceding theorem: The theorem is true if we assume only that for a dense set of $g$'s in $A$, $g = f^2$, $f \in A$. That this is so follows from the theorem just proved by way of the next lemma, another result which was stated by Cirka.

**13.16 *Lemma***

*Let $X$ be a compact, locally connected space, and let $\mathscr{F}$ and $\mathscr{G}$ be subsets of $\mathscr{C}(X)$. If every $g \in \mathscr{G}$ is of the form $g = f^2$ for some $f \in \mathscr{F}$, then every $g \in \overline{\mathscr{G}}$ is of the form $g = f^2$ for some $f \in \overline{\mathscr{F}}$.*

Here we use $\overline{\mathscr{F}}$ and $\overline{\mathscr{G}}$ to denote the closures in $\mathscr{C}(X)$ of $\mathscr{F}$ and $\mathscr{G}$, respectively.

PROOF

It suffices to prove that if $\{g_k\}$ is a sequence in $\mathscr{G}$ which converges to $g \in \mathscr{C}(X)$, and if, for each $k$, $g_k = f_k^2$ for some $f_k \in \mathscr{F}$, then some subsequence of $\{f_k\}$ is uniformly Cauchy. To do this, it is enough to show that for some sequence $\{k_j\}$ of integers both $\{\operatorname{Re} f_{k_j}\}$ and $\{\operatorname{Im} f_{k_j}\}$ are uniformly Cauchy.

We have that $f_k^2 \to g$, so $|f_k|^2 \to |g|$. If $f_k = u_k = u_k + iv_k$, $u_k$ and $v_k$ real, then it follows that $\{u_k^2 + v_k^2\}$ and $\{u_k^2 - v_k^2\}$ are both uniformly convergent, and hence that $\{u_k^2\}$ and $\{v_k^2\}$ are, too. This observation permits us to consider only real sequences: It suffices to prove that if $\{\varphi_k\}$ is a sequence in $\mathscr{C}_{\mathbf{R}}(X)$ such that $\{\varphi_k^2\}$ converges uniformly, say to $\psi$, then some subsequence of $\{\varphi_k\}$ is uniformly convergent.

For $N = 1, 2, \ldots$, let

$$E(N) = \{x \in X : \psi(x) > 1/N\}.$$

By local connectedness and compactness, it follows that for each $N$, the set $E(N)$ is contained in the union of a finite number of components of the set $E(2N)$, say

$$E(N) \subset \bigcup_{j=1}^{q(N)} C(N, j),$$

$C(N, j)$ a component of $E(2N)$. There is a countable set $S \subset X$ which meets each of the components $C(N, j)$, and a diagonal process yields a subsequence of $\{\varphi_k\}$ which converges at every point of $S$. We assume henceforth that the sequence $\{\varphi_k\}$ itself has this property, and we shall show, under this hypothesis, that $\{\varphi_k\}$ converges uniformly on $X$.

Let $N \geq 2$ be a positive integer. There is a $K$ such that if $k, k' > K$, then

$$\|\varphi_k^2 - \varphi_{k'}^2\|_X \leq (1/N)^2,$$

whence also

$$\|\varphi_k^2 - \psi\|_X \leq (1/N)^2.$$

For $j = 1, 2, \ldots, q(N)$, let $x_j$ be a point of $C(N, j) \cap S$ so that for each $j$, the sequence $\{\varphi_k(x_j)\}_{k=1}^{\infty}$ converges, say to $\lambda_j, |\lambda_j| \geq 1/N$. There is a $K' \geq K$ with the property that if $k > K'$, then for each $j$,

$$|\varphi_k(x_j) - \lambda_j| < 1/4N.$$

It follows that for each $j$ all the functions $\varphi_k$, $k > K'$, have the same sign over the whole of $C(N, j)$. Suppose not, so that, say, $\lambda_j > 0$ but that $\varphi_k(y) < 0$ for some $y \in C(N, j)$, $k > K'$. By connectedness there is $y_0 \in C(N, j)$ at which $\varphi_k$ vanishes. But then

$$|\varphi_k^2(y_0) - \psi(y_0)| = |\psi(y_0)| > 1/2N,$$

while we have assumed that $\|\varphi_k^2 - \psi\|_X \leq (1/N)^2$. Thus, for given $j$ the functions $\varphi_k$, $k > K'$, are all positive on $C(N, j)$ or else they are all negative there.

Consider now an $x \in X$ and $k,k' > K'$. If $x \in X \setminus E(N)$, then $|\psi(x)| \leq 1/N$, $\|\varphi_k^2 - \psi\|_X \leq (1/N)^2$, and $\|\varphi_{k'}^2 - \psi\| < (1/N)^2$, so

$$|\varphi_k(x) - \varphi_{k'}(x)| \leq |\varphi_k(x)| + |\varphi_{k'}(x)|$$
$$\leq 2[(1/N) + (1/N)^2]^{1/2}.$$

If $x \in E(N)$, then

$$|\varphi_k(x) - \varphi_{k'}(x)| = \left|\frac{\varphi_k^2(x) - \varphi_{k'}^2(x)}{\varphi_k(x) + \varphi_{k'}(x)}\right|.$$

The numerator is no more than $(1/N)^2$, and the denominator is no less that $2[(1/N) - (1/N^2)]^{1/2}$. These estimates suffice to prove the uniform convergence of $\{\varphi_k\}$.

# 14 SOME CONSTRUCTIONS

There are certain standard ways of constructing new uniform algebras from old, constructions which are quite useful in devising examples. In this section we shall consider some of these constructions together with certain applications of them.

The first of these constructions is the tensor product of a family of uniform algebras. Suppose given an index set $\Lambda$ and for each $\lambda \in \Lambda$, a uniform algebra $(A_\lambda, X_\lambda)$.

**14.1** ***Definition***

*The tensor product of the algebras $A_\lambda$, denoted by $\otimes\{A_\lambda : \lambda \in \Lambda\}$, is the smallest closed subalgebra of $\mathscr{C}(\Pi\{X_\lambda : \lambda \in \Lambda\})$ containing the functions $f \circ \pi_\lambda$, where for each $\lambda$, $\pi_\lambda$ is the natural projection of $\Pi\{X_\lambda : \lambda \in \Lambda\}$ onto $X_\lambda$ and where $f$ ranges over the algebra $A_\lambda$.*

If we are given two uniform algebras $A_1$ and $A_2$, their tensor product will be denoted by $A_1 \otimes A_2$.

**14.2** ***Examples***

Perhaps the simplest example of a tensor product is $\mathscr{P}(\bar{\Delta}^2)$, the uniform closure of the polynomials in two complex variables on $\bar{\Delta}^2$, the closed bicylinder in $\mathbb{C}^2$. We have that $\mathscr{P}(\bar{\Delta}^2) = \mathscr{A} \otimes \mathscr{A}$. (Recall that by standing convention, $\mathscr{A}$ denotes the disc algebra.)

In connection with this example, it is of interest to note that the corresponding result for $H^\infty(\Delta^2)$ is false: It is not true that $H^\infty(\Delta^2)$ is the tensor product of two copies of $H^\infty(\Delta)$. [To be consistent with our definition, we should regard $H^\infty(\Delta)$ and $H^\infty(\Delta^2)$ as algebras on their spectra. The product $\Sigma(H^\infty(\Delta)) \times \Sigma(H^\infty(\Delta))$ contains $\Delta^2$ as an open subset, so the following argument yields the result.] Suppose given $F \in H^\infty(\Delta) \otimes H^\infty(\Delta)$ so that there exists a sequence $\{F_k\}$ of functions of the form

$$F_k(z, w) = \sum_{j=1}^{N(k)} g_j^{(k)}(z) h_j^{(k)}(w),$$

$g_j^{(k)}, h_j^{(k)} \in H^\infty(\Delta)$, which converges to $F$ uniformly on $\Delta^2$. Since $\lim_{r\to 1^-} f(re^{i\theta})$ exists for almost all $\theta$ if $f \in H^\infty(\Delta)$, it follows that for some set $E \subset [0, 2\pi]$ of measure $2\pi$, we have that the limits

$$\lim_{r\to 1^-} f_j^{(k)}(re^{i\theta}) \qquad \text{and} \qquad \lim_{r\to 1^-} g_j^{(k)}(re^{i\theta})$$

exist for all $k$ and $j$ if $\theta \in E$. The set $E$ is of measure $2\pi$, so by altering it by a null set if necessary, we can suppose that $\theta \in E$ implies $2\pi - \theta \in E$. As the limit $\lim_k F_k = F$ is uniform on $\Delta^2$, we have for all $\theta \in E$,

$$\begin{aligned}\lim_{r\to 1^-} F(re^{i\theta}, re^{-i\theta}) &= \lim_{r\to 1^-} \lim_{k\to\infty} F_k(re^{i\theta}, re^{-i\theta}) \\ &= \lim_{k\to\infty} \lim_{r\to 1^-} F_k(re^{i\theta}, re^{-i\theta}),\end{aligned}$$

and this latter limit exists. Thus, if $F \in H^\infty(\Delta) \otimes H^\infty(\Delta)$, the limit $\lim_{r\to 1^-} F(re^{i\theta}, re^{-i\theta})$ exists for almost all $\theta$. If we define $G$ by

$$G(z, w) = f(\tfrac{1}{2}(1 + zw)),$$

$f \in H^\infty(\Delta)$, then $G(re^{i\theta}, re^{-i\theta}) = f(\frac{1}{2}(1 + r^2))$. If we take $f$ to be a function for which $\lim_{r\to 1^-} f(r)$ does not exist, then our remarks show that $G \notin H^\infty(\Delta) \otimes H^\infty(\Delta)$, though it is clear that $G \in H^\infty(\Delta^2)$.

Many proofs of this last observation have been given. See Birtel and Dubinski [1]. The very simple proof given above was shown me by S. J. Sidney.

In this same general vein, let us note that a theorem of Schatten [1] implies that if we denote by $L^\infty(I)$ the $L^\infty$-algebra associated with Lebesgue measure on $I = [0, 1]$, and regard it as a uniform algebra on its spectrum, then $L^\infty(I \times I) \neq L^\infty(I) \otimes L^\infty(I)$ if we take the product measure on $I \times I$.

Our first theorem establishes some of the general properties of the tensor products which make them useful in the theory of uniform algebras. It is difficult to assign priority of discovery to these results: In more or less general form they have been published by several authors. Probably the most systematic treatment from our point of view is that given by Forster [1], but relevant material is also found in Gelbaum [1], Tomiyama [1], and other sources.

14·3 ***Theorem***

*For each $\lambda \in \Lambda$, let $A_\lambda$ be a uniform algebra with spectrum $\Sigma_\lambda$, Shilov boundary $\Gamma_\lambda$, and strong boundary $B_\lambda$, respectively.*

(a) *The spectrum of $\otimes\{A_\lambda : \lambda \in \Lambda\}$ is $\Pi\{\Sigma_\lambda : \lambda \in \Lambda\}$.*
(b) *The Shilov boundary for $\otimes\{A_\lambda : \lambda \in \Lambda\}$ is $\Pi\{\Gamma_\lambda : \lambda \in \Lambda\}$.*
(c) *The strong boundary for $\otimes\{A_\lambda : \lambda \in \Lambda\}$ is $\Pi\{B_\lambda : \lambda \in \Lambda\}$.*

PROOF

The tensor product $\otimes A_\lambda$ is a uniform algebra on $\Pi\Sigma_\lambda$ in an evident way, so to prove (a) we need only show that each complex homomorphism $\varphi$ of $\otimes A_\lambda$ is of the form $\varphi(f) = f(p)$ for some point $p \in \Pi\Sigma_\lambda$. For each $\mu \in \Lambda$, we have a natural injection $i_\mu : A_\mu \to \otimes A_\lambda$, and the composition $\varphi \circ i_\mu$ is an element $p_\mu$ of $\Sigma_\mu$. Thus, if $g \in A_\mu$, $g(p_\mu) = \varphi \circ i_\mu(g)$. The point $p \in \Pi\Sigma_\lambda$ whose $\mu$th coordinate is $p_\mu$ then satisfies $\varphi f = f(p)$ for all $f \in \otimes A_\lambda$, and we have part (a).

To prove (c), let $p \in \Pi\Sigma_\lambda$ have $\mu$th coordinate $p_\mu$, and suppose that $p_\mu \in B_\mu$ for all $\mu \in \Lambda$. Then $p$ lies in $B$, the strong boundary for $\otimes A_\lambda$. To see this, let $V$ be a neighborhood of $p$. We can suppose $V$ to be of the form $V = \Pi\{V_\lambda : \lambda \in \Lambda_0\} \times \Pi\{\Sigma_\lambda : \lambda \in \Lambda \setminus \Lambda_0\}$, where $\Lambda_0$ is a finite subset of $\Lambda$ and where, for $\lambda \in \Lambda_0$, $V_\lambda$ is a neighborhood $p_\lambda$. Since $p_\lambda$ is a strong boundary point for $A_\lambda$, there is an $f_\lambda \in A_\lambda$ for each $\lambda \in \Lambda_0$ with $f_\lambda(p_\lambda) = 1 = \|f_\lambda\|_{\Sigma_\lambda} > \|f_\lambda\|_{\Sigma_\lambda \setminus V_\lambda}$. The product $F = \Pi_{\lambda \in \Lambda_0} f_\lambda \circ \pi_\lambda$ is an element of $\otimes A_\lambda$ which is of norm 1, which is 1 at $p$, and which is strictly less than 1 in modulus off $V$. Thus, $p \in B$. Conversely, suppose that $p \in B$. Let $\mu \in \Lambda$, and let $V_\mu$ be a neighborhood of $p_\mu$, the $\mu$th coordinate of $p$. Then $V_\mu \times \Pi\{\Sigma_\lambda : \lambda \in \Lambda \setminus \{\mu\}\}$ is a neighborhood, $W$, of $p$, so there is $f \in \otimes A_\lambda$ with $|f(p)| = 1 = \|f\|$ and $|f| < 1$ off $W$. For each $\lambda \in \Lambda \setminus \{\mu\}$, let $p_\lambda$ be a point of $\Sigma_\lambda$, and define $g$ on $\Sigma_\mu$ by requiring that for $s \in \Sigma_\mu$, $g(s) = f(j(s))$, where $j(s) \in \Pi\Sigma_\lambda$ is the point with $\mu$th coordinate $s$, $\lambda$th coordinate $p_\lambda$ if $\lambda \neq \mu$. Then $g \in A_\mu$, and $g(p_\mu) = 1 = \|g\|$, but $|g| < 1$ off $V$. Therefore, $p_\mu \in B_\mu$, and we have established part (c).

Part (b) is a consequence of part (c) and the fact that $\Gamma_\lambda$ is the closure of $B_\lambda$.

As a first application of the tensor product, let us give an example to which we alluded in the remarks following Theorem 7.32.

14·4 ***Example***

The strong boundary for a uniform algebra need not be a Borel set. Let $\Lambda$ be an uncountable index set, and for each $\lambda$, let $A_\lambda$ be the subalgebra of the disc algebra consisting of those functions such that

$f(0) = f(1)$. Then $\Sigma_\lambda = \Sigma(A_\lambda)$ is the space obtained from the closed unit disc by identifying the points 0 and 1. The Shilov boundary for $A_\lambda$ is the set in $\Sigma_\lambda$ which corresponds to the unit circle, and the minimal boundary is the set $B_\lambda$ obtained by deleting from the Shilov boundary the point 1. Consequently, by the theorem just proved, $B$, the strong boundary for $\mathfrak{A} = \otimes\{A_\lambda : \lambda \in \Lambda\}$, is the set $\Pi\{B_\lambda : \lambda \in \Lambda\}$, and this is not a Borel set since $\Lambda$ is uncountable. (See Appendix B.)

Our next construction, based on tensor products, is due to Sidney [1].

14.5 ***Theorem***

*Let $A'$ and $B$ be uniform algebras on $X = \Sigma(A')$ and $Y = \Sigma(B)$, respectively, and let $A \subset A'$ be a second uniform algebra on $X$. If $I$ is a closed ideal in $B$, let $\mathfrak{A}$ be the uniform closure in $\mathscr{C}(X \times Y)$ of the algebra*

$$A \otimes B + A' \otimes I. \tag{1}$$

*Then*

$$\Sigma(\mathfrak{A}) = (X \times Y) \bigcup_F (\Sigma(A) \times \text{hull } I), \tag{2}$$

*where $F$ is the set $X \times$ hull $I$.*

For the notation used in (2), recall Theorem 9.15 and the remarks that precede it.

PROOF

(Blumenthal [1].) We have $\Sigma(A' \otimes B) = X \times Y$. The algebra $A \otimes B$ is a closed subalgebra of $A' \otimes B$, and $A' \otimes I$ is an ideal in $A' \otimes B$. Thus, $\mathfrak{A}$ is a subalgebra of $A' \otimes B$ which contains the ideal $A' \otimes I$ of $A' \otimes B$. Now,

$$\{(x, y) \in X \times Y : f(x, y) = 0 \text{ for all } f \in A' \otimes I\}$$
$$= X \times \text{hull } I = F,$$

and $\Sigma((\mathfrak{A}|F)^-) = \Sigma(A) \times$ hull $I$. Our result is therefore a corollary of Theorem 9.15.

We now consider the direct limit of a family of uniform algebras. Our treatment will make essential use of the tensor product we have just constructed. Let us recall that a relation $<$ is a *quasiorder* on a set $D$ if it is reflexive and transitive. A *directed set* is a set $D$ together with a quasiorder such that if $\lambda, \mu \in D$, then there is $\nu \in D$ with $\lambda < \nu$, $\mu < \nu$. Let there be given a directed set $D$ which is fixed for the remainder of the discussion. For each $\lambda \in D$, let $(A_\lambda, X_\lambda)$ be a uniform algebra with

$X_\lambda$ the spectrum of $A_\lambda$. For $\lambda, \mu \in D$, $\lambda < \mu$, let $\Phi_{\lambda\mu}: A_\lambda \to A_\mu$ be a homomorphism and suppose that these homomorphisms satisfy the following compatibility conditions: $\Phi_{\lambda\lambda}$ is the identity, and $\Phi_{\mu\nu} \circ \Phi_{\lambda\mu} = \Phi_{\lambda\nu}$ if $\lambda < \mu < \nu$. Let $\Phi^*_{\lambda\mu}: X_\mu \to X_\lambda$ be the adjoint of $\Phi_{\lambda\mu}$. We then have that $\Phi^*_{\lambda\lambda}$ is the identity and that $\Phi^*_{\lambda\mu} \circ \Phi^*_{\mu\nu} = \Phi^*_{\lambda\nu}$. Then $\{A_\lambda, \Phi_{\lambda\mu}\}$ is a *directed system of uniform algebras*, and the system $\{X_\lambda, \Phi^*_{\lambda\mu}\}$ is an inverse system of compact spaces. (See Eilenberg and Steenrod [1].) The inverse limit of the system $\{X_\lambda, \Phi^*_{\lambda\mu}\}$ is the set

$$\mathfrak{Z} = \left\{ p \in \prod_{\lambda \in D} X_\lambda : \Phi^*_{\lambda\mu}(\pi_\mu(p)) = \pi_\lambda(p) \text{ for all } \lambda, \mu \in D, \lambda < \mu \right\}.$$

As usual, $\pi_\lambda$ is the projection onto the $\lambda$th coordinate. The set $\mathfrak{Z}$ is a closed subset of the product space and it is nonempty. It is therefore compact.

**14.6** ***Definition***
*With $\{A_\lambda, \Phi_{\lambda\mu}\}$, $\{X_\lambda, \Phi^*_{\lambda\mu}\}$, and $\mathfrak{Z}$ as above, the* **direct limit** *of $\{A_\lambda, \Phi_{\lambda\mu}\}$ is the uniform closure in $\mathscr{C}(\mathfrak{Z})$ of the set $\{f|\mathfrak{Z} : f \in \otimes \{A_\lambda : \lambda \in D\}\}$.*

The principal fact we shall need concerning the direct limit of uniform algebras is contained in our next theorem.

**14.7** ***Theorem***
*The spectrum of the direct limit of $\{A_\lambda, \Phi_{\lambda\mu}\}$ is the set $\mathfrak{Z}$.*

PROOF
Denote by $I$ the ideal in $\otimes A_\lambda$ consisting of those functions which vanish on $\mathfrak{Z}$. The quotient algebra $(\otimes A_\lambda)/I$ is isomorphic to $\{f|\mathfrak{Z} : f \in \otimes A_\lambda\}$, and we have that $\Sigma((\otimes A_\lambda)/I) = \text{hull } I$; we shall identify this hull with $\mathfrak{Z}$. Suppose given $p \in \Sigma(\otimes A_\lambda) = \Pi X_\lambda, p \notin \mathfrak{Z}$. For some $\lambda, \mu \in D$, $\Phi^*_{\lambda\mu}(\pi_\mu(p)) \neq \pi_\lambda(p)$. The algebra $A_\lambda$ separates points on $X_\lambda$, so there is $f \in A_\lambda$ with

$$f(\pi_\lambda(p)) \neq f(\Phi^*_{\lambda\mu}(\pi_\mu(p))).$$

The function $g = f \circ \pi_\lambda - (\Phi_{\lambda\mu}(f)) \circ \pi_\mu$ is in $\otimes A_\lambda$ and we have

$$\begin{aligned} g(p) &= f(\pi_\lambda(p)) - \Phi_{\lambda\mu} f(\pi_\mu(p)) \\ &= f(\pi_\lambda(p)) - f(\Phi^*_{\lambda\mu}(\pi_\mu(p))) \neq 0. \end{aligned}$$

But if $q \in \mathfrak{Z}$ so that $\Phi^*_{\lambda\mu}(\pi_\mu(q)) = \pi_\lambda(q)$, then $g(q) = 0$. Thus, $g \in I$, and $g(p) \neq 0$. Consequently, $\mathfrak{Z}$ is the hull of $I$, and we have that $\Sigma((\otimes A_\lambda)/I) = \mathfrak{Z}$. From this it follows that the spectrum of the direct limit of $\{A_\lambda, \Phi_{\lambda\mu}\}$ is $\mathfrak{Z}$.

This theorem combines with the local maximum modulus theorem to yield a result on the spectrum of certain algebras of functions locally in a given algebra.

14.8 ***Definition***

*Let $A$ be a uniform algebra. The function $f \in \mathscr{C}(\Sigma(A))$ is* ***locally A-approximable*** *if given $p \in \Sigma(A)$, there is a neighborhood $V$ of $p$ on which $f$ can be approximated uniformly by elements of $A$.*

The example of a nonlocal uniform algebra given in Example 8.24 shows that a function can be locally $A$-approximable without belonging to $A$. Our next theorem shows, however, that adjoining to $A$ functions which are locally $A$-approximable does not change the spectrum.

14.9 ***Theorem***

*Let $A$ be a uniform algebra, and let $\mathscr{F} \subset \mathscr{C}(\Sigma(A))$ be a family of locally $A$-approximable functions. If $B$ is the closed subalgebra of $\mathscr{C}(\Sigma(A))$ generated by $A$ and $\mathscr{F}$, then $\Sigma(B) = \Sigma(A)$.*

PROOF

By the local maximum modulus theorem, it is evident that $A$ and $B$ have the same Shilov boundary.

We shall prove the theorem first for the case that $\mathscr{F}$ is a singleton and then obtain the general case by a limiting process. Let $\mathscr{F} = \{f\}$, and denote by $[A, f]$ the closed subalgebra of $\mathscr{C}(\Sigma(A))$ generated by $f$ and $A$. There is a natural projection $\pi : \Sigma([A, f]) \to \Sigma(A)$ given by $\pi\varphi = \varphi | A$. If $g \in A$ and if $\hat{}$ denotes the Gelfand transform in $[A, f]$, then $\hat{g} = g \circ \pi$. The assumption that $f$ is locally $A$-approximable implies that $f \circ \pi$ is locally approximable on $\Sigma([A, f])$ by functions of the form $g \circ \pi$, $g \in A$.

Consider now the algebra $[A, \hat{f}, f \circ \pi]$ on $\Sigma([A, f])$ generated by $\hat{f}$, $f \circ \pi$, and $\{g \circ \pi : g \in A\}$. The Shilov boundary, $\Gamma$, for $[A, \hat{f}, f \circ \pi]$ can be identified with that for $A$. We have that $\hat{f} | \Gamma = f \circ \pi | \Gamma$, so $\hat{f} = f \circ \pi$ on $\Gamma$. Consequently, $\hat{f} = f \circ \pi$, and thus $\hat{f}$ is constant on each fiber $\pi^{-1}(p)$, $p \in \Sigma(A)$. It follows that the fibers $\pi^{-1}(p)$ all consist of a single point, i.e., that $\pi$ is one to one. Thus, $\Sigma([A, f]) = \Sigma(A)$, as we wished to show.

Now, denote by $\mathscr{S}(\mathscr{F})$ the collection of finite subsets of $\mathscr{F}$, and for each $F \in \mathscr{S}(\mathscr{F})$, let $[A, F]$ denote the closed subalgebra of $\mathscr{C}(\Sigma(A))$ generated by $A$ and $F$. It follows immediately from what we have said so far that $\Sigma([A, F]) = \Sigma(A)$. If $F, G \in \mathscr{S}(\mathscr{F})$, $F \subset G$, there is a natural inclusion $i_{F,G} : [A, F] \to [A, G]$, and there is the corresponding adjoint map $i^*_{F,G} : \Sigma([A, G]) \to \Sigma([A, F])$. The system $\{[A, F], i_{F,G}\}$ for $F, G \in \mathscr{S}(\mathscr{F})$ is a direct system of uniform algebras, and it follows from Theorem 14.7

that if $\mathfrak{A}$ is the direct limit of this system, then $\Sigma(\mathfrak{A})$ is the inverse limit of the system $\{\Sigma([A, F]), i^*_{FG}\}$ of compact spaces. The algebra $\mathfrak{A}$ is the algebra $[A, \mathscr{F}]$, and since each of the maps $i^*_{FG}$ is a homeomorphism onto, the inverse limit of the system $\{\Sigma([A, F]), i^*_{FG}\}$ may be identified with $\Sigma(A)$, and we have the desired result.

This theorem is originally due to Rickart [2]. The proof given above that $\Sigma([A, f]) = \Sigma(A)$ is attributed to Quigley (unpublished) by Gamelin and Wilken [1]. A result closely related to Rickart's is due to Stolzenberg [1], who proved that *if* $f \in \mathscr{C}(\Sigma(A))$ *has the property that for some finite decomposition* $\Sigma(A) = E_1 \cup \cdots \cup E_q$, $f|E_j \in A|E_j$ *for each* $j$, *then* $\Sigma([A, f]) = \Sigma(A)$. In particular, then, if $f$ is locally in $A$, then Stoltzenberg's theorem implies that $\Sigma([A, f]) = \Sigma(A)$. However, as the sets $E_j$ need not be open, Stolzenberg's result is not an obvious consequence of Rickart's. Gamelin and Wilken in the paper just cited have extended Stolzenberg's result to the case of countable rather than finite closed covers of $\Sigma(A)$.

# *chapter three*/ANALYTIC STRUCTURE IN SPECTRA

## 15 *SUFFICIENT CONDITIONS FOR ANALYTIC STRUCTURE*

How can a uniform algebra $A$ fail to be $\mathscr{C}(X)$? The most obvious answer is that $A$ may consist of functions holomorphic on a suitable subset of $\Sigma(A)$. In recent years, a good deal of effort has gone into the search for the suitable subsets of $\Sigma(A)$, and in this chapter we shall consider some of the results that have been obtained.

Recall that if $\Omega$ is an open set in $\mathbb{C}^N$, an *analytic variety* $V$ in $\Omega$ is a closed subset of $\Omega$ such that if $\mathfrak{z} \in V$, there is a neighborhood $U$ of $\mathfrak{z}$ and a family $\mathscr{F}$ of functions holomorphic in $U$ such that

$$V = \{\mathfrak{z}' \in U : f(\mathfrak{z}') = 0 \text{ for all } f \in \mathscr{F}\}.$$

A continuous function $f$ on $V$ is said to be holomorphic on $V$ if for each $\mathfrak{z} \in V$, there is a neighborhood $U$ of $\mathfrak{z}$ and a function $F$ holomorphic on $U$ such that $f|(V \cap U) = F|(V \cap U)$. The theory of analytic varieties and holomorphic functions on them is developed in detail, e.g., in Gunning and Rossi [1].

Given a commutative Banach algebra $A$ and an element $\varphi$ of $\Sigma(A)$, we shall seek conditions that guarantee the presence of an analytic variety in $\Sigma(A)$ through $\varphi$ in the sense that for some analytic variety $V$ in a neighborhood $\Omega$ of the origin in $\mathbb{C}^N$, $0 \in V$, there is a nonconstant continuous map $\Phi : V \to \Sigma(A)$, with $\Phi(0) = \varphi$ and $\hat{a} \circ \Phi$ holomorphic on $V$ for all $a \in A$. In this section we shall obtain certain sufficient conditions for the existence of such varieties in $\Sigma(A)$, and in subsequent sections of this chapter, we shall consider the related theory of Gleason parts for uniform algebras.

Our first result is due to Gleason [2], and it has to do with maximal ideals which are algebraically finitely generated in the sense of the following definition.

**15.1** ***Definition***

*The ideal $I$ in the Banach algebra $A$ is **algebraically generated** by $g_1, \ldots, g_N \in I$ if each $a \in I$ is of the form $\Sigma g_j b_j$ for some choice of $b_1, \ldots, b_N$ in $A$.*

15.2 ***Theorem***

*Let $A$ be a commutative Banach algebra, and let $\varphi_0 \in \Sigma(A)$. If* $\ker \varphi_0$ *is algebraically generated by $g_1, \ldots, g_N$, and if $\gamma : \Sigma(A) \to \mathbb{C}^N$ is the map given by $\gamma(\varphi) = (\varphi(g_1), \ldots, \varphi(g_N))$, then there is a neighborhood $\Omega$ of 0 in $\mathbb{C}^N$ such that*

(a) *$\gamma$ is a homeomorphism of $\gamma^{-1}(\Omega)$ onto an analytic variety $E$ in $\Omega$.*
(b) *For every $a \in A$, there is a holomorphic function $F$ on $\Omega$ such that on $\gamma^{-1}(\Omega)$, $\hat{a} = F \circ \gamma$.*

*Moreover, if $\psi \in \gamma^{-1}(\Omega)$, then* $\ker \psi$ *is algebraically generated by*

$$g_1 - \psi(g_1), \ldots, g_N - \psi(g_N).$$

The original proof Gleason gave for this theorem is quite elementary, but it does depend on the consideration of certain Banach space-valued analytic functions. It has developed that there is an even simpler, more direct proof of the theorem. This proof has evidently occurred to various people. I learned of it from some work of Read [1, 2].

PROOF OF THE THEOREM

Assume the theorem known for algebras with identity, and suppose that $A$ has no identity. Let $A'$ be the algebra obtained by adjoining to $A$ an identity, and let $\varphi_0'$ be the unique extension of $\varphi_0$ to an element of $\Sigma(A')$. Choose $a \in A$ with $\varphi_0'(a) = 1$. Then $\ker \varphi_0'$ is algebraically generated by $g_1, \ldots, g_N$, and $1 - a$, and the result for $A$ follows from the result for $A'$. We assume henceforth that the algebra $A$ has an identity.

The proof of the theorem is effected by constructing a homomorphism from $A$ to an algebra of holomorphic functions on a certain analytic variety and then considering the dual map. For notational convenience, set $I = \ker \varphi_0$. By hypothesis, the map

$$L : A \oplus \cdots \oplus A \oplus \mathbb{C} \to A$$

given by

$$L(a_1, \ldots, a_N, \lambda) = a_1 g_1 + \cdots + a_N g_N + \lambda$$

is onto; consequently, by the open mapping theorem there are constants $C'$ such that if $a \in A$, then $a = L(a_1, \ldots, a_N, \lambda)$ with $C'\|a\| \geq \max\{\|a_1\|, \ldots, \|a_N\|, |\lambda|\}$. Let $C_0$ be the infimum of all such $C'$, and let $C > C_0$. Thus, if $a_0 \in A$, we have

$$a_0 = \hat{a}_0(\varphi_0) + \sum_{i_1=1}^{N} a_{i_1} g_{i_1} \tag{1}$$

with $\|a_{i_1}\| \leq C\|a_0\|$. Apply this remark to each $a_{i_1}$ to get

$$a_{i_1} = \hat{a}_{i_1}(\varphi_0) + \sum_{i_2=1}^{N} a_{i_1 i_2} g_{i_2}$$

with $\|a_{i_1 i_2}\| \leq C^2\|a_0\|$, and thus

$$a_0 = \hat{a}_0(\varphi_0) + \sum_{i_1=1}^{N} \hat{a}_{i_1}(\varphi_0) g_{i_1} + \sum_{i_1, 1_2=1}^{N} a_{i_1 i_2} g_{i_1} g_{i_2}.$$

Iterating this process we find at the $(p+1)$st step,

$$(2) \qquad a_0 = \hat{a}_0(\varphi_0) + \sum_{i_1=1}^{N} \hat{a}_{i_1}(\varphi_0) g_{i_1} + \cdots + \sum_{i_1, \ldots, i_p=1}^{N} \hat{a}_{i_1, \ldots, i_p}(\varphi_0) g_{i_1} \cdots g_{1_p} + \sum_{i_1, \ldots, i_{p+1}=1}^{N} a_{i_1, \ldots, i_{p+1}} g_{i_1} \cdots g_{i_{p+1}},$$

and we have the estimate

$$(3) \qquad \|a_{i_1 \cdots i_k}\| \leq C^k \|a_0\|, \qquad 1 \leq k \leq p+1.$$

In this way, we generate from $a_0$ a certain formal series of elements from $A$:

$$(4) \qquad \hat{a}_0(\varphi_0) + \sum_{i_1=1}^{N} \hat{a}_{i_1}(\varphi_0) g_{i_1} + \cdots + \sum_{i_1, \ldots, i_p=1}^{N} \hat{a}_{i_1, \ldots, i_p}(\varphi_0) g_{i_1} \cdots g_{i_p} + \cdots.$$

It should be pointed out that since the linear operator $L$ is not necessarily one to one, the expansion (1) is not unique, i.e., the $a_{i_1}$ in (1) are not unique, so the series (4) is not at all uniquely determined by $a_0$; many choices are made in its construction. Also, we should observe that it is not convergent in any obvious topology on $A$. It does, however, have the following convergence property. If $\varphi \in \Sigma(A)$ satisfies $|\varphi(g_i)| < \alpha/CN$, $\alpha \in (0, 1)$, $1 \leq i \leq N$, then from (2) and (3),

$$\left| \varphi(a_0) - \hat{a}_0(\varphi_0) - \cdots - \sum_{i_1, \ldots, i_p=1}^{N} \hat{a}_{i_1, \ldots, i_p}(\varphi_0) \varphi(g_{i_1}) \cdots \varphi(g_{i_p}) \right| \leq N^{p+1} C^{p+1} \|a_0\| \left( \frac{\alpha}{CN} \right)^{p+1} = \|a_0\| \alpha^{p+1}.$$

As $p \to \infty$, this converges to zero, so the series (4) converges pointwise on the neighborhood

$$W = \{\varphi \in \Sigma(A) : |\varphi(g_i)| < 1/CN,\ i = 1, \ldots, N\},$$

and the convergence is uniform on the sets $W_\alpha$ given by

$$W_\alpha = \{\varphi \in \Sigma(A) : |\varphi(g_i)| < \alpha/CN, i = 1, \ldots, N\}$$

for $0 < \alpha < 1$.

Consider now the formal series in $Z_1, \ldots, Z_N$

$$\hat{a}_0(\varphi) + \sum_{i_1=1}^{N} \hat{a}_{i_1}(\varphi_0)Z_{i_1}$$
$$+ \cdots + \sum_{i_1,\ldots,i_p=1}^{N} \hat{a}_{i_1,\ldots,i_p}(\varphi_0)Z_{i_1}\cdots Z_{i_p} + \cdots.$$

Our remarks on the convergence of the series (4) imply that this series converges, at least in the polycylinder

$$V = \{(z_1, \ldots, z_N) \in \mathbb{C}^N : |z_1|, \ldots, |z_N| < 1/CN\}.$$

Denote by $f_{a_0}$ the function represented in $V$ by this series so $f_{a_0} \in \mathcal{O}(V)$. In light of the great freedom we enjoy in the construction of the functions $f_{a_0}$, we may suppose that for $j = 1, \ldots, N$, $f_{g_j}$ is the $j$th coordinate function on $\mathbb{C}^N$.

Thus, we have constructed a map $\tilde{\Phi} : A \to \mathcal{O}(V)$, viz., $\tilde{\Phi}(a) = f_a$ for all $a \in A$. The map $\tilde{\Phi}$ is well defined but it has no apparent properties of regularity, either algebraic or topological. However, we can obtain a homomorphism $\Phi$ from $\tilde{\Phi}$ as follows. In $\mathcal{O}(V)$, let $J$ be the ideal generated by the functions

$$f_{\lambda a} - \lambda f_a,$$
$$f_{a+b} - f_a - f_b,$$
$$f_{ab} - f_a f_b$$

for $a,b \in A$, $\lambda \in \mathbb{C}$. If $\eta : \mathcal{O}(V) \to \mathcal{O}(V)/J$ is the canonical quotient map, and if $\Phi : A \to \mathcal{O}(V)/J$ is the composition, $\eta \circ \tilde{\Phi}$, then by the very definition of the ideal $J$, it follows that $\Phi$ is an algebra homomorphism. Let $E \subset V$ be the analytic variety consisting of the common zeros of the elements of the ideal $J$.

A point $\mathfrak{z} \in E$ gives rise to a well-defined complex homomorphism of $\mathcal{O}(V)/J$. If $[f] \in \mathcal{O}(V)/J$ is the coset that contains $f$, then the complex number $f(\mathfrak{z})$ depends only on $[f]$ and not on the particular choice of representative of $[f]$, so $[f] \mapsto f(\mathfrak{z})$ is well defined and not identically zero. Thus, $a \mapsto f_a(\mathfrak{z})$ is a well-defined element of $\Sigma(A)$, call it $\psi_{\mathfrak{z}}$. Define

$\Phi^*: E \to \Sigma(A)$ by $\Phi^*(\mathfrak{z}) = \psi_{\mathfrak{z}}$. The map $\Phi^*$ defined in this way is continuous, for if $\mathfrak{z}_n \to \mathfrak{z}_0$, then for all $f \in \mathcal{O}(V)$, $f(\mathfrak{z}_n) \to f(\mathfrak{z}_0)$, so in particular $\psi_{\mathfrak{z}_n}(a) = f_a(\mathfrak{z}_n) \to f_a(\mathfrak{z}_0) = \psi_{\mathfrak{z}_0}(a)$. Under $\Phi^*$, the set $E$ goes into the neighborhood $W$ of $\varphi_0$ defined above, for $(\Phi g_j)(\mathfrak{z})$ is the $j$th coordinate of $\mathfrak{z}$ and so is no more than $1/CN$ in modulus.

If $\gamma$ is the map defined in the statement of the theorem, then all $\mathfrak{z} \in E$,

$$\begin{aligned}\gamma(\Phi^*(\mathfrak{z})) &= (\hat{g}_1(\Phi^*(\mathfrak{z})), \ldots, \hat{g}_N(\Phi^*(\mathfrak{z})))\\ &= (f_{g_1}(\mathfrak{z}), \ldots, f_{g_N}(\mathfrak{z}))\\ &= (z_1, \ldots, z_N) = \mathfrak{z},\end{aligned}$$

and thus, $\gamma \circ \Phi^*$ is the identity on $E$, and it follows that $\gamma$ carries $W$ onto a set containing $E$. We have that $\gamma(W) \subset E$, for if $\varphi \in W$ and $\gamma(\varphi) = \mathfrak{z} = (z_1, \ldots, z_N)$, then $\mathfrak{z} \in V$ since $|\varphi(g_i)| < 1/CN$. Also, since for all $a_0 \in A$, the series (4) converges pointwise on $W$ to $\varphi(a_0)$, it follows that the generators of the ideal $J$ must all vanish at $\mathfrak{z}$. Thus, $\mathfrak{z} \in E$. Finally, we note that $\Phi^* \circ \gamma$ is the identity on $W$, for if $\varphi \in W$, $\gamma(\varphi) = \mathfrak{z}$, and if $a \in A$, then

$$\begin{aligned}\Phi^*(\gamma(\varphi))(a) &= \gamma(\varphi)(\Phi a)\\ &= f_a(\mathfrak{z})\\ &= \varphi(a).\end{aligned}$$

The last equality is valid because of the convergence of (4) to $\varphi(a_0)$. Thus, $\Phi^* \circ \gamma$ is the identity on $W$.

So far we have shown that $\gamma$ is a homeomorphism of $W = \gamma^{-1}(V)$ onto the analytic variety $E$ and that if $a \in A$, then $\hat{a}(\varphi) = f_a \circ \gamma(\varphi)$ if $\varphi \in W$.

It remains only to prove that if $\varphi \in W$, then $\ker \varphi$ is generated algebraically by $g_i - \hat{g}_i(\varphi)$, $1 \le i \le N$. Define a map $L_1: A \oplus \cdots \oplus A \oplus \mathbb{C} \to A$ by

$$L_1(a_1, \ldots, a_N, \lambda) = a_1(g_1 - \hat{g}_1(\varphi)) + \cdots + a_N(g_N - \hat{g}_N(\varphi)) + \lambda.$$

If we give $A \oplus \cdots \oplus A \oplus \mathbb{C}$ the norm $\|\!| \cdot |\!\|$ defined by

$$\|\!|(a_1, \ldots, a_N, \lambda)|\!\| = \max(\|a_1\|, \ldots, \|a_N\|, |\lambda|),$$

then $L_1$ is a continuous linear operator, and we have that

$$\begin{aligned}\|L - L_1\| &= \sup\{\|L(x) - L_1(x)\| : x \in A \oplus \cdots \oplus A \oplus \mathbb{C}, \|\!|x|\!\| \le 1\}\\ &= \sup\left\{\left\|\sum_{j=1}^{N} a_j g_j(\varphi)\right\| : a_1, \ldots, a_N \in A, \|a_1\|, \ldots, \|a_N\| \le 1\right.\\ &\le 1/C.\end{aligned}$$

Thus, $\|L - L_1\| < C_0^{-1}$. Our next lemma establishes the fact that this implies that $L_1$ has the whole of $A$ as its range, and thus our proof will be complete.

If $X$ and $Y$ are Banach spaces, $\mathscr{L}(X, Y)$ denotes the space of bounded linear operators from $X$ to $Y$. If $S \in \mathscr{L}(X, Y)$ is surjective, it follows that for some constant $K$ and for all $y \in Y$, there is $x \in S^{-1}(y)$ such that $\|x\| \leq K\|y\|$. We denote by $K(S)$ the infimum of all such $K$. Thus, in the work above, $C_0 = K(L)$. The next lemma was given by Gleason [2], but the result had been discovered already by Dieudonné [1].

15·3 ***Lemma***

*If $S,T \in \mathscr{L}(X, Y)$, if $S$ is surjective, and if $\|S - T\| < K(S)^{-1}$, then $T$ is surjective. In particular, the set of surjective elements of $\mathscr{L}(X, Y)$ is open.*

PROOF

Let $U = S - T$ so $\|U\| < K(S)$. Choose a constant $k$ with $k\|U\| < 1$, $k > K(S)$, and let $y \in Y$. Since $k > K(S)$, we can find a sequence $\{x_n\}_{n=0}^{\infty}$ which satisfies $Sx_0 = y$, $Sx_1 = Ux_0, \ldots, Sx_n = Ux_{n-1}, \ldots$, and $\|x_0\| \leq k\|y\|$, $\|x_1\| \leq k\|Ux_0\|, \ldots, \|x_n\| \leq k\|Ux_{n-1}\|, \ldots$. We have then that $\|x_n\| \leq (k\|U\|)^n\|x_0\|$ for all $n$, so the series $\sum_{n=0}^{\infty} x_n$ converges in $X$, say to $x$, and we have that $T(x) = S(x) - U(x) = y$. Thus, $T$ is surjective, and the lemma is proved.

As a very special case of the theorem, we have the following result.

15·4 ***Corollary***

*If $A$ is a commutative Banach algebra, if $\varphi \in \Sigma(a)$ is not an isolated point, and if* ker $\varphi$ *is a principal ideal, say generated by $a$, then $\hat{a}$ carries some neighborhood $W$ of $\varphi$ homeomorphically onto a disc $D$ in $\mathbb{C}$ centered at 0, and on $W$ each $\hat{b} \in \hat{A}$ is of the form $\hat{b} = f \circ \hat{a}, f \in \mathcal{O}(D)$.*

As a first application of Gleason's theorem, we derive a result of Kra [1], who obtained it by entirely classical means.

15·5 ***Theorem***

*Let $X$ be a locally compact space, and let $A$ be a point separating algebra of continuous, complex-valued functions on $X$ containing the constants and satisfying the following conditions:*

(a) *If $f \in A$ is nonconstant, then $f$ is an open mapping.*
(b) *If $x \in X$, the ideal $A_x = \{f \in A : f(x) = 0\}$ is principal.*

*Then $X$ is an orientable, two-dimensional manifold which can be given the structure of a Riemann surface on which all the functions in $A$ are holomorphic.*

PROOF

Let $x \in X$, and denote by $f_x$ a generator for the ideal $A_x$. Since $A$ separates points, $f_x$ can vanish only at $x$. Let $V_x$ be a neighborhood of $x$ with compact closure, and consider the uniform algebra $(A|\bar{V}_x)^-$ on $\bar{V}_x$. There is a natural homeomorphism $i: \bar{V}_x \to \Sigma = \Sigma((A|\bar{V}_x)^-)$ by taking $y \in \bar{V}_x$ to the point evaluation at $y$. Since every $f \in A$ is an open map, it follows that $\Gamma$, the Shilov boundary for $(A|\bar{V}_x)^-$, is contained in $\partial V_x$. In particular, $\delta = \inf\{|f_x(y)| : y \in \Gamma\} > 0$. Let $g \in (A|\bar{V}_x)^-$ vanish at $x$. Thus, there exists a sequence $\{g_n\}$ in $A$ such that $\|g - g_n\|_{\bar{V}_x} \to 0$; we may suppose $g_n(x) = 0$. But then $g_n = h_n f_x$. Since $|f_x| > \delta$ on $\Gamma$, we find that $\|h_n - h_m\|_{\partial V_x} = \|g_n f_x^{-1} - g_n f_x^{-1}\| < 1/\delta \|g_n - g_n\|$, so $\{h_n\}$ is uniformly Cauchy on $\Gamma$. Since $h_n \in A$, it follows that for some $h \in (A|\bar{V}_x)^-$, we have $h = \lim h_n$ on $\Gamma$. But then $g = f_x h$. Thus, the ideal $\{g \in (A|\bar{V}_x)^- : g(x) = 0\}$ is a principal ideal in $(A|\bar{V}_x)^-$, and it is generated by $f_x$. Thus, by Gleason's theorem, there is $W$, a neighborhood of $x$ in $\Sigma$ which is carried homeomorphically onto a neighborhood $U$ of 0 by $\hat{f}_x$, and, moreover, if $g \in (A|\bar{V}_x)^-$, then on $W$, $g = G \circ \hat{f}_x$ for some $G$ holomorphic on $U$.

We have that $\hat{f}_x$ is a homeomorphism of $W$ onto $U$. Since by hypothesis $f_x$ is continuous and open, it follows that for some neighborhood $V'_x$ of $x$ in $X$, $V'_x \subset V_x$, we have that $f(V'_x)$ is a neighborhood of zero. Consequently, $f_x$ must carry $V'_x$ homeomorphically onto a neighborhood $U'$ of 0.

Thus, we obtain a collection $\{(V'_x, f_x) : x \in X\}$ of pairs consisting of an open set $V'_x$ and a function $f_x$ which carries $V_x$ homeomorphically into $\mathbb{C}$. The $V'_x$ constitute an open cover for $X$, and for all $x$ and $y$, the function $f_x \circ (f_y^{-1}|f_y(V_x \cap V'_y))$ is holomorphic. Thus, we have provided $X$ with the structure of a Riemann surface. Every function $g \in A$ is holomorphic, for in $V'_x$, $g = G \circ f_x$, $G$ holomorphic on $f_x(V'_x)$. This concludes the proof.

It seems interesting that there is evidently no analogue of Gleason's theorem valid for general topological algebras. That is, there are reasonable examples of topological algebras of functions in which maximal ideals are finitely generated but in which there is no analyticity present.

15.6 ***Example***

Consider $\mathscr{C}^\infty(\mathbb{R}^N)$, the space of all $\mathbb{C}$-valued, infinitely differentiable functions on $\mathbb{R}^N$. If $f \in \mathscr{C}^\infty(\mathbb{R}^N)$, $f(0) = 0$, set $F(t) = f(tx_1, \ldots, tx_N)$,

$t \in (-\infty, \infty)$. The function $F$ so defined is of class $\mathscr{C}^\infty(\mathbb{R})$, and by the fundamental theorem of calculus, since $f(0) = 0$, we have

$$\begin{aligned} F(1) &= \int_0^1 F'(t)\,dt \\ &= \int_0^1 \sum_{j=1}^N x_j f_j(tx_1, \ldots, tx_N)\,dt \\ &= \sum_{j=1}^N x_j g_j(x_1, \ldots, x_N). \end{aligned}$$

Here we have written $f_j$ for the partial derivative of $f$ with respect to $x_j$. Since $F(1) = f(x_1, \ldots, x_N)$, we have

$$f(x_1, \ldots, x_N) = \sum_{j=1}^N x_j g_j(x_1, \ldots, x_N).$$

The functions $g_j$ lie in $\mathscr{C}^\infty(\mathbb{R}^N)$, and therefore the ideal $\mathscr{C}_0^\infty(\mathbb{R}^N) = \{f \in \mathscr{C}^\infty(\mathbb{R}^N) : f(0) = 0\}$ is finitely generated.

The ideals $\mathscr{C}_x^\infty(\mathbb{R}^N) = \{f \in \mathscr{C}^\infty(\mathbb{R}^N) : f(x) = 0\}$ are the only maximal ideals in $\mathscr{C}^\infty(\mathbb{R}^N)$ which arise from complex homomorphisms. To see this, let $\varphi$ be a nonzero complex homomorphism of $\mathscr{C}^\infty(\mathbb{R}^N)$, and set $\alpha_i = \varphi(x_i)$. Given $f \in \mathscr{C}^\infty(\mathbb{R}^N)$ such that $f(\alpha_1, \ldots, \alpha_N) = 0$, we can write

$$f(x) = (x_1 - \alpha_1)g_1(x) + \cdots + (x_n - \alpha_n)g_N(x)$$

by a slight extension of the calculations of the last paragraph. Since $\varphi(x_j - \alpha_j) = 0$, we find $\varphi(f) = 0$, so if $g \in \mathscr{C}^\infty(\mathbb{R}^N)$, $\varphi(g) = g(\alpha_1, \ldots, \alpha_N)$.

There are maximal ideals in $\mathscr{C}^\infty(\mathbb{R}^N)$ which are not of the form $\mathscr{C}^\infty(\mathbb{R}^N)$: Let $\{y_k\}_{k=1}^\infty$ be a sequence in $\mathbb{R}^N$ without limit points, and let $I = \{f \in \mathscr{C}^\infty(\mathbb{R}^N) : f(y_k) = 0$ for all large $k\}$. The ideal $I$ is contained in some maximal ideal, but this maximal ideal is not a $\mathscr{C}_x^\infty(\mathbb{R}^N)$ and so is not the kernel of a complex homomorphism of $\mathscr{C}^\infty(\mathbb{R}^N)$.

In general it is difficult to determine that a maximal ideal in a Banach algebra is finitely generated, but Gleason [2] established a result which, roughly speaking, says that if the maximal ideal $m$ in a finitely generated Banach algebra is algebraically finitely generated, then it is generated by the right elements of the algebra.

15.7 ***Theorem***

*Let the commutative Banach algebra $A$ with identity be generated as a Banach algebra by $1, z_1, \ldots, z_M$. If $\varphi \in \Sigma(A)$ and if* ker *$\varphi$ is algebraically finitely generated, then* ker *$\varphi$ is algebraically generated by $z_1 - \varphi(z_1), \ldots, z_M - \varphi(z_M)$.*

PROOF

Let $J$ be the ideal generated by $z_1 - \varphi(z_1), \ldots, z_M - \varphi(z_m)$, and let $P$ be the subalgebra of $A$ generated by $1, z_1, \ldots, z_M$, so that $A$ is the closure of $P$. We have $J \subset \ker \varphi$. Since $P \cap J$ and $P \cap \ker \varphi$ both have codimension 1 in $P$, we have $P \cap J = P \cap \ker \varphi$. Also, $P \cap J$ is dense in $J$.

Let $\ker \varphi$ be generated by $g_1, \ldots, g_n$, and define $T: A \oplus \cdots \oplus A + \mathbb{C} \to A$ by

$$T(a_1, \ldots, a_n, \lambda) = a_1 g_1 + \cdots + a_n g_n + \lambda$$

so that $T$ is surjective. By Lemma 15.3, there is $\varepsilon > 0$ so that if $\|g_i - h_i\| < \varepsilon$, then the map $U$ given by

$$U(a_1, \ldots, a_n \lambda) = a_1 h_1 + \cdots + a_n h_n + \lambda$$

is also surjective. In this case, the ideal in $A$ generated by the $h$'s has codimension at most one. We can choose the $h$'s to lie in $P \cap \ker \varphi$. Since the $h$'s lie in $J$, the ideal $J$ must have codimension at most 1 in $A$, and since we have $J \subset \ker \varphi$, we have that $J = \ker \varphi$.

It is natural to wonder if there is a converse to Gleason's theorem. For instance, if $M$ is a complex manifold contained as an open set in $\Sigma(A)$ and if each $\hat{a}$ is holomorphic on $M$, does it follow that the maximal ideals corresponding to points of $M$ are finitely generated? As we shall see below, no such general converse exists. There is, however, the following example.

15.8 ***Example***

Let $B_2 = \{(z, w) \in \mathbb{C}^2 : |z|^2 + |w|^2 \leq 1\}$, and as usual let $\mathscr{P}(B_2)$ be the closure in $\mathscr{C}(B_2)$ of the polynomials in two complex variables. Let $I = \{f \in \mathscr{P}(B_2) : f(0) = 0\}$. In his paper [2], Gleason asked whether $I$ is finitely generated. We shall show that it is. The argument was shown me by Garnett, who attributed it indirectly to Leibnitzen.

If $f \in I$, and if $(z_0, w_0)$ lies in $B_2^0$, the interior of $B_2$, we have

$$\begin{aligned} f(z_0, w_0) &= \int_0^1 \frac{d}{dt} f(tz_0, tw_0)\, dt \\ &= z_0 \int_0^1 \frac{\partial f}{\partial z}(tz_0 tw_0)\, dt + w_0 \int_0^1 \frac{\partial f}{\partial w}(tz_0, tw_0)\, dt. \end{aligned}$$

We define $T_1 f$ and $T_2 f$ for $f \in \mathscr{P}(B_2)$ by

$$T_1 f(z_0, w_0) = \int_0^1 \frac{\partial f}{\partial z}(tz_0, tw_0)\, dt$$

and

$$T_2 f(z_0, w_0) = \int_0^1 \frac{\partial f}{\partial w}(tz_0, tw_0)\, dt.$$

For any $f \in \mathscr{P}(B_2)$, $T_1 f$ and $T_2 f$ are holomorphic in $B_1^0$; we shall show that they lie in $\mathscr{P}(B_2)$. If $f$ happens to be a polynomial, then $T_1 f$ and $T_2 f$ are polynomials. Since the operators $T_1$ and $T_2$ are linear, we may draw the desired conclusion if we can establish bounds of the form $\|T_j f\| \leq C\|f\|$ valid for some absolute constant $C$ and all polynomials $f$. The desired bound can be established as follows.

Consider a fixed point $p_0 \in B_2^0$ with coordinates $(z_0, w_0) \neq (0, 0)$. Define new coordinates $\xi, \eta$ on $\mathbb{C}^2$ by

$$\xi = \frac{\bar{z}_0}{N} z + \frac{\bar{w}_0}{N} w$$

and

$$\eta = \frac{w_0}{N} z - \frac{z_0}{N} w,$$

where $N^2 = |z_0|^2 + |w_0|^2$. Note that this is a unitary change of coordinates. We consider $T_1 f$, $f$ a polynomial in $I$. By the chain rule we have

$$\frac{\partial f}{\partial z} = \frac{\bar{z}_0}{N} \frac{\partial f}{\partial \xi} + \frac{w_0}{N} \frac{\partial f}{\partial \eta},$$

and thus

$$(T_1 f)(p_0) = \frac{\bar{z}_0}{N} \int_0^1 \frac{\partial f}{\partial \xi}(tp_0)\, dt + \frac{w_0}{N} \int_0^1 \frac{\partial f}{\partial \eta}(tp_0)\, dt.$$

If $f$ has, with respect to the $(\xi, \eta)$-coordinate system, the power-series expansion $\sum_{m,n \geq 0} c(m, n)\xi^m \eta^n$, then since $f(0) = 0$, $c(0, 0) = 0$, and we have

$$\frac{\partial f}{\partial \xi} = \sum_{m,n \geq 0} c(m, n) m \xi^{m-1} \eta^n.$$

The point $p_0$ has coordinates $(N, 0)$ with respect to the $(\xi, \eta)$-coordinate system, so

$$\frac{\partial f}{\partial \xi}(tp_0) = \sum_{m \geq 1} c(m, 0) m t^{m-1} N^{m-1}.$$

Thus,

$$\int_0^1 \frac{\partial f}{\partial \xi}(tp_0)\, dt = \sum_{m \geq 1} c(m, 0) N^{m-1}.$$

If $h(\xi) = \sum_{m \geq 0} c(m, 0)\xi^m$, then $h$ is holomorphic in $|\xi| < 1$ and for such values of $\xi$, we have $|h(\xi)| \leq \|f\|_{B_2}$. Moreover, $h(0) = 0$, so by Schwarz's lemma $|h(\xi)/\xi| \leq \|f\|_{B_2}$ if $|\xi| < 1$. Thus,

$$\left| \int_0^1 \frac{\partial f}{\partial \xi}(tp_0)\, dt \right| \leq \|f\|_{B_2}.$$

We can obtain a bound on the modulus of $\int_0^1 (\partial f/\partial \eta)(tp_0)\, dt$ by noting that if $|\zeta|^2 \leq 1 - t^2N^2$, then the point with $(\xi, \eta)$-coordinates $(N, \zeta)$ lies in $B_2$; so we can write by Cauchy's formula

$$\int_0^1 \frac{\partial f}{\partial \eta}(tp_0)\, dt = \int_0^1 \frac{1}{2\pi i} \int_\gamma \frac{f(tN, \zeta)}{\zeta^2}\, d\zeta\, dt,$$

where $\gamma$ is the circle $|\zeta|^2 = 1 - t^2N^2$. Thus, we have the estimate

$$\begin{aligned} \left| \int_0^1 \frac{\partial f}{\partial \eta}(tp_0)\, dt \right| &\leq \|f\|_{B_2} \int_0^1 (1 - t^2N^2)^{-1/2}\, dt \\ &\leq \|f\|_{B_2} \int_0^1 (1 - t^2)^{-1/2}\, dt \\ &= \pi/2 \|f\|_{B_2}, \end{aligned}$$

and, consequently,

$$|(T_1 f)(p_0)| \leq \pi/2 \|f\|_{B_2} \left( \frac{|z_0| + |w_0|}{N} \right) \leq \pi \|f\|_{B_2}.$$

It follows that $T_1 f \in \mathscr{P}(B_2)$ for all $f \in \mathscr{P}(B_2)$. Similarly, $T_2 f \in \mathscr{P}(B_2)$, and we conclude that the ideal $I$ is generated by the coordinate functions.

It also follows that if we let $J = \{f \in H^\infty(B_2) : f(0) = 0\}$, then $J$ is generated by the coordinate functions.

Kerzman and Nagel [1] have obtained a sheaf-theoretic proof of the fact that the ideal $I$ of this example is generated by the coordinate functions. Their methods are sufficiently powerful to yield the corresponding result for certain more general domains.

We conclude these considerations with an example, to which we have already alluded, that shows that no general converse to Theorem 15.2 exists. The example is due to T. T. Read [1, 2].

15.9 ***Example***

We shall produce a uniform algebra $A$ and an open set $V \subset \Sigma(A)$ such that for some continuous, one-to-one map $\Phi$ of the open unit disc onto $V$, each $f \in A$ has the property that $f \circ \Phi$ is holomorphic on the unit disc but for no $x \in V$ will the ideal $\{f \in A : \hat{f}(x) = 0\}$ be finitely generated.

Let $S$ be an infinite index set, and, for $s \in S$, let $\bar{\Delta}_s$ be a copy of the closed unit disc. Let $Y = \Pi\{\bar{\Delta}_s : s \in S\}$, and let $\pi_s : Y \to \bar{\Delta}_s$ be the natural projection. Define $B$ to be the uniform algebra on $Y$ generated by the functions $\pi_s$ so that $B$ is the tensor product of $S$ copies of the disc algebra and $Y = \Sigma(B)$. Let $I$ be the closed ideal in $B$ generated by the functions $fg$, where $f, g \in B$ vanish at the point 0, i.e., the point in $Y$ carried to 0 by all the $\pi_s$. The hull of the ideal $I$ is the point 0. Let $T$ denote the unit circle, and let $\mathscr{A}$ be the disc algebra regarded as a subalgebra of $\mathscr{C}(T)$. Set

$$A = (\mathscr{C}(T) \otimes I + \mathscr{A} \otimes B)^-,$$

an algebra on $T \times Y$. By Theorem 14.5,

$$\Sigma(A) = (T \times Y) \bigcup_F (\bar{\Delta} \times \{0\}),$$

where $F = T \times \{0\}$. The set $\Delta \times \{0\}$ is an open subset of $\Sigma(A)$ which is homeomorphic to $\Delta$ and on which the functions $\hat{f}$, $f \in A$, are holomorphic.

Let $J$ be the maximal ideal of $A$ consisting of those $f \in A$ which vanish at $(0,0) \in \bar{\Delta} \times \{0\}$. We shall prove that $J$ is not algebraically finitely generated. To prove this, it suffices to show that the quotient space $J/(J^2)^-$ is infinite dimensional, as we shall show below [$J^2$ denotes the ideal generated by products $fg, f, g \in J$. It is easy to see that $J^2$ is also generated by elements of the form $f^2, f \in J$. Although $J$ is closed, $J^2$ need not be; $(J^2)^-$ denotes its closure.]

We denote by $\mu$ the normalized Haar measure on $T^S$, the product of the boundaries of the $\bar{\Delta}_s$, and we let $m$ be that on $T$. If $s \in S$, it is evident that the measure $\bar{\pi}_s \, d\mu$ on $T^S$ annihilates the ideal $I$, as $I$ is generated by functions of the form $f^2$, $f(0) = 0$. We shall prove that for each $s \in S$, the measure $\bar{\pi}_s \, d(m \times \mu)$ on $T \times T^S$ annihilates $J^2$, i.e., that

$$\int f_1 f_2 \bar{\pi}_s \, d(m \times \mu) = 0$$

if $f_1, f_2 \in J$. By density, we may suppose that for $j = 1, 2$, $f_j \in \mathscr{C}(T) \otimes I + \mathscr{A} \otimes B$, say $f_j = g_j h_j + g_j' h_j'$, where $g_j \in \mathscr{C}(T)$, $h_j \in I$, $g_j' \in \mathscr{A}$, $h_j' \in B$, and where $g_j' h_j'$ vanishes at $(0,0) \in \Delta \times \{0\}$. Thus, we have to prove that $\bar{\pi}_s \, d(m \times \mu)$ annihilates functions of any of the following three forms: $g_1 g_2 h_1 h_2$, $g_1 g_2' h_1 h_2'$, and $g_1' g_2' h_1' h_2'$. The measure $\bar{\pi}_s \, d(m \times \mu)$ annihilates $g_1 g_2 h_1 h_2$, for $h_1$ and $h_2$ lie in $I$ and, as we noted above, $\bar{\pi}_s \, d\mu$ annihilates $I$. It annihilates $g_1 g_2' h_1 h_2'$ since $h_1 h_2' \in I$. Finally, it annihilates $g_1' g_2' h_1' h_2'$, for either $g_1'(0) g_2'(0) = 0$ in which case $m$ annihilates $g_1' g_2'$, or else $g_1'(0) \neq 0$ and $g_2'(0) \neq 0$, in which case $h_1'(0) = h_2'(0) = 0$ so that $h_1' h_2' \in I$ and the desired conclusion follows.

Since the measures $\bar{\pi}_s\, d(m \times \mu)$ annihilate $(J^2)^-$ but

$$\int \pi_t \bar{\pi}_s\, d(m \times \mu) = 0 \text{ or } 1$$

according as $t \neq s$ or $t = s$, it follows that the elements $\pi_s + (J^2)^-$ of $J/(J^2)^-$ are linearly independent. This implies that $J$ cannot be algebraically finitely generated, for if $g_1, \ldots, g_p \in J$ are such that the operator

$$L: A \oplus \cdots \oplus A \to J$$

given by $L(f_1, \ldots, f_p) = \Sigma f_j g_j$ is onto, then by the first isomorphism theorem

$$(A \oplus \cdots \oplus A/L^{-1}((J^2)^-) \simeq J/(J^2)^-.$$

If $\psi_j$ is defined on $A \oplus \cdots \oplus A$ by $\psi_j(f_1, \ldots, f_p) = \varphi(f_j)$, where $J = \ker \varphi$, then each element $(f_1, \ldots, f_p)$ which lies in the kernel of all the $\psi_j$ lies also in $L^{-1}((J^2)^-)$; for if $L(f_1, \ldots, f_p) = f = \Sigma f_i g_i$, then also

$$f = \Sigma \hat{f}_i(\varphi) g_i + \sum_{i,j} f_{ij} g_i g_j$$

for some $f_{ij} \in A$, and since each $\psi_j(f_1, \ldots, f_p) = 0$, we get that $f \in J^2$. Thus, dim $[(A \oplus \cdots \oplus A)/L^{-1}(J^2)] \leq p$, and we have a contradiction.

Another sufficient condition for the existence of analytic structure in the spectrum is based upon the existence of certain derivations of the algebra.

15.10 ***Definition***

*If $A$ is a commutative Banach algebra with identity and if $\varphi \in \Sigma(A)$, a bounded* ***derivation of type*** *$(I, \varphi)$ is a bounded linear operator $T: A \to A$ such that*

$$T(fg) = (Tf)g + \varphi(f)Tg$$

*and such that $\varphi \circ T$ is not the zero functional.*

15.11 ***Examples***

(a) If $\mathscr{A}$ is the usual disc algebra and if $T: \mathscr{A} \to \mathscr{A}$ is given by $Tf(z) = [f(z) - f(0)]/z$, then $T$ is a bounded derivation of type $(I, 0)$.
(b) More generally, let $\varphi \in \Sigma(A)$ have a principal ideal as kernel, say, $\ker \varphi = (a_0)$, and define $T: A \to A$ by $Ta = b$, where $b \in A$ satisfies $ba_0 = a - \hat{a}(\varphi)$. If $A$ is semisimple and if $\varphi$ is a limit point of $\Sigma(A)$, then $Ta$ is well defined, for our definition implies that $(Ta)\hat{} = (\hat{a} - \hat{a}(\varphi))\hat{a}_0^{-1}$ except at the point $\varphi$ where $\hat{a}_0$ vanishes.

The operator $T$ satisfies the required product rule, and the hypothesis of semisimplicity enables us to show that $T$ is continuous. Suppose that $a_n \to c$ and $Ta_n \to d$. Let $Ta_n = b_n$. Thus, $\hat{b}_n = (\hat{a}_n - \hat{a}_n(\varphi))\hat{a}_0^{-1} \to (\hat{c} - \hat{c}(\varphi))\hat{a}_0^{-1}$, whence $\hat{d} = (Tc)\hat{}$, and as $A$ is semisimple, it follows that $d = Tc$. The closed graph theorem now implies that $T$ is continuous. Note finally that since $Ta_0 = 1$, $\varphi \circ T \neq 0$, and we have shown that $T$ is a bounded derivation of type $(I, \varphi)$.

(c) It is not true that if there exists a bounded derivation of type $(I, \varphi)$, then $\ker \varphi$ is principal. To construct an example, let $\mathbb{C}^*$ denote the one-point compactification of $\mathbb{C}$, i.e., the Riemann sphere. Let $A = \{f \in \mathscr{C}(\mathbb{C}^*) : f \text{ is holomorphic in } |z| < 1\}$. Then $\Sigma(A) = \mathbb{C}^*$, and the Shilov boundary for $A$, call it $\Gamma$, is the complement in $\mathbb{C}^*$ of the open unit disc. Define $\varphi \in \Sigma(A)$ by $\varphi f = f(0)$, and let $k$ be a function continuous on $\Gamma$ with the property that $k(z) = \bar{z} = 1/z$ if $|z| = 1$. If $f \in A$, then $f - \varphi(f) \in \ker \varphi$, and there is a unique function $Tf$ in $A$ which agrees on $\Gamma$ with $k(f - \varphi(f))$. Since $\Gamma$ is the Shilov boundary for $A$, our definition of $Tf$ implies that the operator $T$ is continuous. We have, on $\Gamma$,

$$\begin{aligned} T(fg) &= T(fg - \varphi(f)\varphi(g)) \\ &= k(fg - \varphi(f)g) + k\varphi(f)(g - \varphi(g)) \\ &= gT(f) + \varphi(f)T(g). \end{aligned}$$

Since $\Gamma$ is the Shilov boundary, this relation persists over the whole of $\Sigma(A)$, and thus $T$ is a bounded derivation of type $(I, \varphi)$.

We now show that $\ker \varphi$ is not a principal ideal. If $\ker \varphi$ were the principal ideal generated by $f_0 \in A$, then $f_0$ could assume the value zero only at the point zero, and $f_0$ would have at zero a simple zero. Thus, if $\gamma$ is a small circle about zero,

$$\int_\gamma \frac{df}{f} = 2\pi i. \tag{5}$$

Denote by $D$ the component of $\mathbb{C}^* \setminus \gamma$ which contains the point at infinity. The set $D \cup \gamma$ is a closed disc, and since $f_0$ is zero free on this disc, there is a continuous branch of $\log f_0$ defined there. This implies that $\Delta_\gamma \log f = 0$, a contradiction to (5). Thus, $\ker \varphi$ is not principal.

It is worth noting that $\ker \varphi$ is finitely generated. Let $F \in A$ vanish on a neighborhood of infinity and be identically 1 on a neighborhood of the open unit disc. Then $1 - F \in \ker \varphi$, and if $f \in \ker \varphi$, $f = (1 - F)f + Ff$. Since $F(z)f(z) = zg(z)$ for some choice of $g \in A$, it follows that if $F_1 \in A$ vanishes on a neighborhood of infinity and satisfies $F_1(z) = z$ for those $z$ at which $F$ does not vanish, then $1 - F$ and $F_1$ generate the ideal $\ker \varphi$.

The following theorem was proved in a somewhat different form by Bishop [5]. We shall follow the treatment given by Banaschewski [1].

15.12 ***Theorem***

*Let $A$ be a commutative Banach algebra with identity, let $\varphi \in \Sigma(A)$, and let $T$ be a bounded derivation of type $(I, \varphi)$ on $A$. There exists a homeomorphism $\Phi$ of the open unit disc in $\mathbb{C}$ onto an open set in $\Sigma(A)$ with $\Phi(0) = \varphi$ which has the properties that for all $a \in A$, $\hat{a} \circ \Phi$ is holomorphic, and that if $|z_0| < 1$, there is $a \in A$ such that $d/dz(\hat{a} \circ \Phi)(z_0) \neq 0$.*

PROOF

If $\alpha$ is a nonzero complex number, then $\alpha T$ is also a bounded derivation of type $(I, \varphi)$, so we can suppose $\|T\| = 1$. Consequently, if $I$ denotes the identity operator and if $|z| < 1$, $I - zT$ is invertible in $\mathscr{L}(A)$; let its inverse be denoted by $S_z$.

The operators $S_z$ obey a certain product rule:

$$S_z(fg) = (S_z f - \varphi(S_z f))g + \varphi(S_z f)S_z g. \tag{6}$$

To see this, it is enough to show that

$$(I - zT)[(S_z f - \varphi(S_z f))g + \varphi(S_z f)S_z g] = fg.$$

The last equation is valid, for since $S_z$ is inverse to $I - zT$, the left side is

$$(S_z f)g - zT((S_z f)g) + z\varphi(S_z f)T(g). \tag{7}$$

We have $T((S_z f)g) = gT(S_z f) + \varphi(S_z f)T(g)$, so (7) is

$$\begin{aligned}(S_z f)g - zgT(S_z f) &= ((I - zT)S_z f)g \\ &= fg.\end{aligned}$$

From the product rule for the operators $S_z$, it follows that

$$\varphi \circ S_z(fg) = \varphi \circ S_z(f)\varphi \circ S_z(g).$$

Since the range of $S_z$ is the whole of $A$, $\varphi \circ S_z$ is not the zero functional, so for all $|z| < 1$, $\varphi \circ S_z \in \Sigma(A)$. We define $\Phi : \Delta \to \Sigma(A)$ by $(\Phi z)(a) = \varphi(S_z(a))$.

If $a \in A$, then

$$\begin{aligned}\hat{a}(\Phi z) &= \varphi(S_z a)\\ &= \varphi\left(\sum_{n=0}^{\infty} z^n T^n a\right)\\ &= \sum_{n=0}^{\infty} z^n \varphi(T^n a).\end{aligned}$$

Since $\|T\| = 1$ and $\|\varphi\| = 1$, this latter series is convergent for all $|z| < 1$, and its sum is a holomorphic function of $z$. The definition of the topology in $\Sigma(A)$ shows that the map $\Phi$ is continuous.

Next, $\Phi$ is one to one, for if $\Phi z = \Phi z'$, then

$$\varphi \circ (I - zT)^{-1} = \varphi \circ (I - z'T)^{-1}.$$

If we compose both sides with

$$(I - zT) \circ (I - z'T) = (I - z'T) \circ (I - zT),$$

we find

$$\varphi \circ (I - z'T) = \varphi(I - zT),$$

whence $(z - z')\varphi \circ T = 0$. We have assumed that $\varphi \circ T$ is not the zero functional, so $z - z' = 0$, whence $\Phi$ is one to one.

If $a \in A$ and $|z_0| < 1$, then

$$\begin{aligned}\frac{d}{dz}(\hat{a} \circ \Phi)(z_0) &= \sum_{n=1}^{\infty} n z_0^{n-1} \varphi(T^n a)\\ &= \varphi(T S_{z_0}^2(a)).\end{aligned}$$

Since $\varphi \circ T$ is not the zero functional and since $S_{z_0}$ is surjective, this last term does not vanish identically as a function of $a$.

It remains only to prove that the range of $\Phi$ is an open set and that $\Phi$ has a continuous inverse. For this we need a consequence of the product rule for the operator $T$. We want to show that if $\alpha \in \Delta$, if $a,b \in A$, and $\hat{b}(\varphi) = 0$, then

(8) $$(Tb)(a - \Phi(\alpha)a) = (b - \alpha Tb)T(S_\alpha(a)).$$

To establish this fact, use the fact that $ab = ba$ and the product rule for $T$ to find

$$(Ta)b + \varphi(a)Tb = (Tb)a + \varphi(b)Ta.$$

Since $\varphi(b) = 0$, this implies that

$$(Tb)(a - \varphi(a)) = bT(a).$$

Using this fact, we have

$$\begin{aligned}(b - \alpha Tb)T(S_\alpha a) &= bT(S_\alpha a) - \alpha TbT(S_\alpha a)\\ &= Tb(S_\alpha a - \varphi \circ S_\alpha(a) - \alpha TS_\alpha a)\\ &= Tb(a - \Phi(\alpha)(a)),\end{aligned}$$

and thus (8) is established.

Choose $b \in A$ with $\hat{b}(\varphi) = 0$, and set

$$U_b = \{\psi \in \Sigma(A) : (Tb)\hat{\ }(\psi) \neq 0\}.$$

Define $\chi_b : U_b \to \mathbb{C}$ by

$$\chi_b(\psi) = \hat{b}(\psi)((Tb)\hat{\ }(\psi))^{-1}.$$

If $\psi \in U_b$ and $\chi_b(\psi) = \alpha$, $|\alpha| < 1$, we have

$$\hat{b}(\psi) - \alpha(Tb)\hat{\ }\psi = 0,$$

so for all $a \in A$, we have by (8) that

$$(Tb)\hat{\ }(\psi)(\hat{a}(\psi) - \Phi(\alpha)(a)) = 0.$$

As $(Tb)\hat{\ }(\psi) \neq 0$ and this last equation is valid for all $a \in A$, we conclude that $\psi = \Phi(\alpha)$. Consequently, $\Phi \circ \chi_b$ is the identity on the open set $\{\psi \in U_b : |\chi_b(\psi)| < 1\}$ in $\Sigma(A)$. Also, $\chi_b \circ \Phi$ is the identity on $\Phi^{-1}(U_b) \subset \Delta$. This is immediate from the definitions, for

$$\begin{aligned}\chi_b \circ \Phi(\alpha) &= \hat{b}(\varphi \circ S_\alpha)((Tb)\hat{\ }(\varphi \circ S_\alpha))^{-1}\\ &= \left(\sum_{n=0}^{\infty} \alpha^n \varphi(T^n b)\right)\left(\sum_{n=0}^{\infty} \alpha^n \varphi(T^{n+1} b)\right)^{-1},\end{aligned}$$

and since $\varphi(b) = 0$, this is $\alpha$. Thus, to conclude the proof, we need only prove that for every $\alpha \in \Delta$, there is $b \in A$ such that $\hat{b}(\varphi) = 0$ but $(Tb)\hat{\ }(\Phi\alpha) \neq 0$. Since $\Phi(0) = \varphi$ and $\varphi \circ T \neq 0$, it follows that such a $b$ exists for $\alpha = 0$. For $\alpha \neq 0$ and $b \in \ker \varphi$, we have $\Phi(\alpha)(Tb) = 1/\alpha\Phi(\alpha)(b)$ by the definition of $\Phi$, and since $\Phi(\alpha) \neq \varphi$, this is not zero for some $b \in \ker \varphi$. This concludes the proof of the theorem.

As a corollary to Theorem 15.12, we shall establish a fact which is "obvious" but which does not seem so easy to prove by simpler means.

15.13 ***Corollary***

*If $\Delta$ is a planar domain such that $H^\infty(\Delta)$ contains a nonconstant function, then under the map $z \mapsto \varphi_z$, $\varphi_z(f) = f(z)$, $\Delta$ goes homeomorphically onto an open subset of $\Sigma(H^\infty(\Delta))$.*

PROOF

Since $\Delta$ is a planar domain on which there is a nonconstant bounded holomorphic function, it follows, as in Example 7.9, that $H^\infty(\Delta)$ contains three functions which separate points on $\Delta$; so the map $z \mapsto \varphi_z$, call it $i$, is one to one. It is plainly continuous, and it is easily verified that $i$ is a homeomorphism. The only question that remains is whether $i\Delta$ is *open* in $\Sigma(H^\infty(\Delta))$.

If $z_0 \in \Delta$, define $T : H^\infty(\Delta) \to H^\infty(\Delta)$ by

$$Tf(z) = \frac{f(z) - f(z_0)}{z - z_0}.$$

The mapping $T$ does take bounded functions to bounded functions, and is a bounded derivation of type $(I, \varphi_{z_0})$ on $H^\infty(\Delta)$. Thus, by Theorem 15.12, there is a neighborhood of $\varphi_{z_0}$ which is a disc. If $D$ is a disc around $z_0$, $D \subset \Delta$, then $i(D)$ is a disc in $\Sigma(H^\infty(\Delta))$ which contains $\varphi_{z_0}$, and it follows that $i(D)$ is, in fact, a neighborhood of $\varphi_{z_0}$. Thus, $i(\Delta)$ is open in $\Sigma(H^\infty(\Delta))$, as asserted.

If $\Delta$ is conformally equivalent to a bounded domain, then it can be established by essentially easier means that $\Delta$ is an open subset of $\Sigma(H^\infty(\Delta))$; but for general domains, the result is not so evident.

# 16 GLEASON PARTS

There is a general approach to the problem of finding analytic structure in the spectrum of a uniform algebra which has been remarkably successful in certain special cases. This approach involves the notion of *part* or *Gleason part* as defined by Gleason in [1].

Denote by $\rho$ the hyperbolic metric in the open unit disc in $\mathbb{C}$. (See Appendix D.)

**16.1** ***Lemma***

*If $A$ is a uniform algebra and if $\varphi, \psi \in \Sigma(A)$, then the following conditions are equivalent:*

(a) $\|\varphi - \psi\| < 2$.

(b) *For some constant $c \in (0, 1)$, $|\varphi(f)| \le c\|f\|$ for all $f \in A$ such that $\psi(f) = 0$.*

(c) *For some constant $M$, $\rho(\varphi(f), \psi(f)) \le M$ if $\|f\| \le 1$.*

PROOF

If $\psi(f) = 0$, then $\rho(\varphi(f), \psi(f)) = \rho(\varphi(f), 0)$, and if this is uniformly bounded in $f$, $\|f\| \leq 1$, then $|\varphi(f)|$ must be bounded away from 1; i.e., $|\varphi(f)| \leq c < 1$. Thus, if $f \in A$ and $\psi(f) = 0$, then $|\varphi(f)| \leq c\|f\|$, and therefore (c) implies (b).

If (c) fails, there is a sequence $\{f_n\}$ in the unit ball of $A$ with $\rho(\varphi(f_n), \psi(f_n)) \to \infty$. Since $\rho$ is invariant under conformal mappings of the unit disc onto itself, it follows that if $g_n \in A$ is defined to be $b_n \circ f_n$, where $b_n(z) = [z - \psi(f_n)]/[1 - \overline{\psi(f_n)}z]$, then since $b_n$ is a conformal map of the unit disc onto itself, we have

$$\begin{aligned} \rho(\varphi(g_n), \psi(g_n)) &= \rho(b_n(\varphi(f_n)), b_n(\psi(f_n))) \\ &= \rho(\varphi(f_n), \psi(f_n)) \\ &\to \infty. \end{aligned}$$

Since $\psi(g_n) = 0$, this yields $\rho(\varphi(g_n), 0) \to \infty$, which can only happen if $|\varphi(g_n)| \to 1$. Thus, (b) and (c) are equivalent.

It is clear that (c) implies (a), for if $\{f_n\}$ is a sequence in $A$ with $\|f_n\| < 1$ and $|\varphi(f_n) - \psi(f_n)| \to 2$, then $\rho(\varphi(f_n), \psi(f_n)) \to \infty$.

Finally, (a) implies (b). To see this, assume (b) false so that for some sequence $\{f_n\}$ in $A$ with $\|f_n\| < 1$ and $\psi(f_n) = 0$, we have $|\varphi(f_n)| \to 1$. Since $\varphi$ is complex linear, we can suppose $0 < \varphi(f_n) < 1$. Determine $\alpha_n \in (0, \varphi(f_n))$ by $\rho(0, \alpha_n) = \rho(\alpha_n, \varphi(f_n))$, and set

$$b_n(z) = \frac{z - \alpha_n}{1 - \alpha_n z}.$$

Then, if $g_n = b_n \circ f_n$, we have $g_n \in A$, and $\|g_n\| < 1$. Also,

$$\begin{aligned} \rho(\psi(g_n), 0) &= \rho(b_n(0), 0) \\ &= \rho(\alpha_n, 0) \end{aligned}$$

and

$$\begin{aligned} \rho(\varphi(g_n), 0) &= \rho(b_n(\varphi(f_n)), 0) \\ &= \rho(\varphi(f_n), \alpha_n). \end{aligned}$$

Thus, $\rho(\psi(g_n), 0)$ and $\rho(\varphi(g_n), 0)$ both tend to infinity, and since $\psi(g_n) < 0$, $\varphi(g_n) > 0$, this implies that

$$|\psi(g_n) - \varphi(g_n) \to 2.$$

We have that $\|g_n\| < 1$, so $\|\psi - \varphi\| = 2$, and (a) is false.

As a consequence of the equivalence of (a) and (c), it is clear that if we define a relation $\sim$ on $\Sigma(A)$ by agreeing that $\psi \sim \varphi$ if and only if $\|\varphi - \psi\| < 2$, then $\sim$ is an equivalence relation.

**16.2** ***Definition***

*The equivalence classes of $\Sigma(A)$ defined by the relation $\sim$ are the* ***parts*** *or* ***Gleason parts*** *for $A$.*

The relevance of the notion of part for the study of analytic structure in $\Sigma(A)$ is suggested by the fact that any connected analytic variety contained in $\Sigma(A)$ is contained within a single part: Let $V$ be a connected analytic variety in a domain $\Omega \subset \mathbb{C}^N$, and let $\Phi: V \to \Sigma(A)$ be a nonconstant map such that if $f \in A$, then $f \circ \Phi$ is holomorphic on $V$. If $\varphi, \psi \in \Phi(V)$ are not equivalent under the relation $\sim$, there is a sequence $\{f_n\}$ in $A$ with $\|f_n\| < 1$, $|\varphi(f_n)| \to 1$, and $\psi(f_n) = 0$. Then $\{f_n \circ \Phi\}$ is a normal family, so there is a function $g$ holomorphic on $V$ which is the pointwise limit of a subsequence $\{f_{n_j} \circ \Phi\}$ of $\{f_n \circ \Phi\}$. If $\Phi(\mathfrak{z}) = \varphi$, $\Phi(\mathfrak{z}') = \psi$, then $g(\mathfrak{z}') = 0$; but $|g(\mathfrak{z})| = 1 = \|g\|_V$, and we have a contradiction. Thus, connected varieties in $\Sigma(A)$ are subsets of parts.

The original hope was that parts would turn out to be varieties, at least in some generalized form. This hope has not been realized, for, as we shall see, parts can be very pathological, but in the case of certain special algebras, parts do, in fact, have the structure of varieties. We shall obtain results of this kind in the next section. In the remainder of this section, we shall consider some of the general properties of parts and give some examples.

The simplest nontrivial example of a Gleason part arises in connection with the disc algebra, $\mathscr{A}$.

**16.3** ***Example***

If $f \in \mathscr{A}$, if $f(0) = 0$, and if $|z| < 1$, then $|f(z)| \le |z| \, \|f\|$ by Schwarz's lemma, so $z$ and 0 lie in the same Gleason part for $\mathscr{A}$. If, for $|\alpha| = 1$, we set $f_\alpha(z) = \frac{1}{2}(z - \alpha)$, then for all $n$, $f_\alpha^{1/n} \in \mathscr{A}$, and $f_\alpha^{1/n}(\alpha) = 0$. Since $\|f_\alpha^{1/n}\| = 1$ and $|f_\alpha^{1/n}(\beta)| \to 1$ if $\alpha \ne \beta$, $|\beta| \le 1$, we have that $\alpha$ is equivalent, in the sense of Gleason, to no other point of the unit disc. Thus, for the disc algebra, there is one nontrivial part, the open unit disc, and many one-point parts, all the points of the unit circle.

The Gleason parts for $H^\infty(\Delta)$ have been described by Hoffman [5]; this description is quite complicated.

It is clear that every strong boundary point for $A$ constitutes a single part, but simple examples show that nontrivial parts can be contained completely in the Shilov boundary.

In the proof of Lemma 16.1, we made explicit use of the hypothesis that $A$ is an algebra, but each of the conditions is meaningful if $A$ is simply a complex Banach space. If $A$ is a complex Banach space, denote by $S(A)$ its unit sphere and by ball $(A)$ its unit ball. Thus, if $A^*$ is the dual of $A$, then $S(A^*) = \{\varphi \in A^* : \|\varphi\| = 1\}$ and ball $(A^*) = \{\varphi \in A^* : \|\varphi\| \leq 1\}$. If $A$ is a commutative Banach algebra with identity, then $\Sigma(A) \subset S(A^*)$. We define an equivalence relation $\approx$ on ball $(A^*)$ by requiring $\varphi \approx \psi$ if and only if for some $M$, $\rho(\varphi(a), \psi(a)) \leq M$ for all $a \in A$ with $\|a\| < 1$. Thus, if $A$ is a uniform algebra, $\approx$ is an extension of the relation $\sim$ from $\Sigma(A)$ to the whole of ball $(A^*)$. In this setting Lewittes [1] has shown that $\varphi \approx \psi$ is not generally equivalent to $\|\varphi - \psi\| < 2$. We call the equivalence classes in ball $(A^*)$ defined by $\approx$ the *hyperbolic parts* for $A$. Our interest in these hyperbolic parts stems from the fact that they contain structures on which the elements of $A$ are analytic as is shown by the following theorem of Lewittes.

16.4 ***Theorem***

*If $A$ is a complex Banach space, and if $\varphi, \psi \in S(A^*)$ lie in the same hyperbolic part for $A$, then there is a continuous map $\Phi$ from the unit disc in $\mathbb{C}$ into $S(A^*)$ such that if $a \in A$, then $\Phi(z)(a)$ depends holomorphically on $z$, and such that both $\varphi$ and $\psi$ lie in the range of $\Phi$.*

The proof depends on a simple lemma. If $z, w \in \mathbb{C}$, set $[z, w] = |(z - w)/(1 - \bar{w}z)|$. Thus, $\rho(z, w) = \frac{1}{2}\log\{(1 + [z, w])/(1 - [z, w])\}$. If $\varepsilon > 0$, let $D_\varepsilon = \{z \in \mathbb{C} : |z - t| < \varepsilon$ for some $t$ with $0 \leq t \leq 1\}$.

16.5 ***Lemma***

*Given $\theta < k < k + \theta < 1$, let $\varepsilon = \theta/k(1 + k + \theta)$. If $|x|, |y| < 1$ and $[x, y] < k$, then for $z \in D_\varepsilon$, $|y + z(x - y)| < 1$, and $[y + z(x - y), x] < k + \theta$.*

PROOF

We are given $[x, y] < k$ and $z \in D_\varepsilon$, and we are to prove that $[y + z(x - y), x] < k + \theta$. This follows by a direct calculation. We have $(1 + \varepsilon)/(k^{-1} - \varepsilon) = k + \theta < 1$, so as a function of $t$, $(1 + \varepsilon - t)/(k^{-1} - \varepsilon - t)$ decreases on $[0, 1]$ and thus, if $\theta \leq t \leq 1$, then

$$0 < \frac{1 + \varepsilon - t}{k^{-1} - \varepsilon - t} < \frac{1 + \varepsilon}{k^{-1} - \varepsilon} = k + \theta.$$

Given $z \in D_\varepsilon$, choose $t$, $0 \leq t \leq 1$, so that $|z - t| < \varepsilon$, whence $|z| < \varepsilon + t$

and $|1 - z| < 1 - t + \varepsilon$. Then

$$[y + z(x - y), x] = \left| \frac{y + z(x - y) - x}{1 - \bar{x}(y + z(x - y))} \right|$$
$$= \left| \frac{1 - z}{(1 - y\bar{x})(y - x)^{-1} + z\bar{x}} \right|$$
$$\leq \frac{1 - t + \varepsilon}{[y, x]^{-1} - |z\bar{x}|}$$
$$\leq \frac{1 - t + \varepsilon}{k^{-1} - \varepsilon - t}$$

since $[y, x] < k$, $|x| < 1$, and $|z| < t + \varepsilon$. Thus,

$$[y + z(x - y), x] < k + \theta.$$

This is half the lemma; the other half follows from this and the fact that if $|s| < 1$ and $[t, s] < 1$, then $|t| < 1$.

PROOF OF THE THEOREM

Assume that $\varphi \approx \psi$. Thus, for some $k < 1$,

$$\sup\{[\varphi(a), \psi(a)] : a \in A, \|a\| < 1\} = k.$$

Let $\varepsilon$ and $\theta$ be associated with $k$ as in the lemma, and let $f$ be a conformal map of the unit disc onto $D_\varepsilon$. Define, for $|z| < 1$, $\Phi(z) \in A^*$ by

$$\Phi(z) = \varphi + f(z)(\psi - \varphi).$$

Then if $a \in A$, $\Phi(z)(a)$ is holomorphic as a function of $z$. Since the range of $f$ contains both 0 and 1, that of $\Phi$ contains both $\varphi$ and $\psi$. We have $\|\Phi(z)\| \leq 1$ for all $z$, for, if not, there is a sequence $\{a_n\}$ in $A$ with $\|a_n\| \leq c < 1$ but $|\Phi(z)(a_n)| \to 1$. This implies that

$$[\Phi(z)(a_n), \psi(a_n)] \to 1$$

since $|\psi(a_n)| \leq c$. However, the lemma implies that $[\Phi(z)(a_n), \psi(a_n)] \leq k + \theta < 1$ and we have a contradiction. Therefore, the range of $\Phi$ is contained within the unit ball of $A^*$.

The lemma also implies that the range of $\Phi$ is contained within a single hyperbolic part for $A$. It is easy to see that no hyperbolic part for $A$ can contain points of $S(A^*)$ and points in ball$(A^*)$ of norm less than 1, for if $\chi, \chi' \in A^*$, $\|\chi'\| < 1$, $\|\chi\| = 1$, then there is a sequence $\{a_n\}$ in $A$ with $\|a_n\| < 1$, $|\chi(a_n)| \to 1$, and $|\chi'(a_n)| < \delta < 1$. We have $\rho(\chi(a_n),$

$\chi'(a_n)) \to \infty$, so $\chi$ and $\chi'$ are inequivalent. Since the range of $\Phi$ is contained in a hyperbolic part and contains points in $S(A^*)$, it must lie in $S(A^*)$, and our theorem is proved.

Thus, in the setting of Banach space theory, the hyperbolic parts do in fact give rise to analytic structure. Such a result does not obtain for the spectra of uniform algebras as we shall see later. If $\varphi,\psi \in \Sigma(A)$ lie in the same Gleason part, the range of the map $\Phi$ of Lewittes' theorem need not be contained in $\Sigma(A)$. This range can meet $\Sigma(A)$ only in the set $\{\varphi, \psi\}$.

If $A$ is a uniform algebra and $\varphi,\psi \in \Sigma(A)$, there are several alternative ways of expressing the fact that $\varphi$ and $\psi$ lie in the same Gleason part.

**16.6** ***Theorem*** (Bishop [7])
*If $(A, X)$ is a uniform algebra and if $\varphi,\psi \in \Sigma(A)$, then the following are equivalent.*

(a) $\|\varphi - \psi\| < 2$, *i.e., $\varphi$ and $\psi$ lie in the same Gleason part.*
(b) *There exist $\mu_\varphi \in \mathscr{M}_\varphi$ and $\mu_\psi \in \mathscr{M}_\psi$ which are mutually absolutely continuous and which have the property that the Radon–Nikodym derivatives $d\mu_\varphi/d\mu_\psi$ and $d\mu_\psi/d\mu_\varphi$ are both bounded.*
(c) *If $\mu_\varphi \in \mathscr{M}_\varphi$, there exists $\mu_\psi \in \mathscr{M}_\psi$ with respect to which $\mu_\varphi$ is absolutely continuous.*
(d) *There exist $\mu_\varphi \in \mathscr{M}_\varphi$ and $\mu_\psi \in \mathscr{M}_\psi$ which are not mutually singular.*

PROOF
(a) implies (b). Begin by observing that although $\varphi$ and $\psi$ are complex homomorphisms of $A$, it is meaningful to speak of $\varphi(u)$ and $\psi(u)$ for $u \in \operatorname{Re} A$: If $f = u + iv \in A$, then $\varphi(u) = \operatorname{Re} \varphi(f)$. Choose $c \in (0, 1)$ such that for some positive $u \in \operatorname{Re} A$, we have $\psi(u) = 1$, $\varphi(u) \le c$. If $f = u + iv \in A$, then $e^{-f} \in A$, call it $h$, and we have $\|h\| \le 1$, $|\psi(h)| = e^{-1}$, and $|\varphi(h)| \ge e^{-c}$. If we could choose $c$ arbitrarily near zero, then for some sequence $\{h_n\}$ in $A$ we would have $\|h_n\| \le 1$, $|\psi(h_n)| = e^{-1}$, and $|\varphi(h_n)| \to 1$. Since $\|\varphi - \psi\| < 2$, this is impossible by the equivalence of (a) and (c) in Lemma 16.1. Thus, there exists a $c$ such that $u \in \operatorname{Re} A$ and $u \ge 0$ imply $\psi(u) \ge c\varphi(u)$. By symmetry, reducing $c$ if necessary, $u \in \operatorname{Re} A$ and $u \ge 0$ also imply $\varphi(u) \ge c\psi(u)$. The functional $u \mapsto \psi(u) - c\varphi(u)$ is a positive linear functional on $\operatorname{Re} A$, and consequently there is a positive measure $\alpha$ on $X$ such that

(1) $$\psi(u) - c\varphi(u) = \int u \, d\alpha.$$

The representation (1) is valid not only for $u \in \operatorname{Re} A$ but also for $u \in A$

since $\alpha$ is a real measure. Similarly, there is a positive measure $\beta$ on $X$ such that

$$\varphi(f) - c\psi(f) = \int f\, d\beta \tag{2}$$

for all $f \in A$. Thus,

$$\begin{aligned}\psi(f) &= \int f\, d\alpha + c\varphi(f) \\ &= \int f\, d\alpha + c\int f\, d\beta + c^2\psi(f),\end{aligned}$$

and we find that

$$\psi(f) = (1 - c^2)^{-1} \int f\, d(\alpha + c\beta).$$

Similarly,

$$\varphi(f) = (1 - c^2)^{-1} \int f\, d(\beta + c\alpha).$$

Therefore, if $\mu_\psi = (1 - c^2)^{-1}(\alpha + c\beta)$ and $\mu_\varphi = (1 - c^2)^{-1}(\beta + c\alpha)$, then $\mu_\psi \in \mathscr{M}_\psi$ and $\mu_\varphi \in \mathscr{M}_\varphi$. We have

$$\begin{aligned}c\mu_\varphi &= (1 - c^2)^{-1}(c\beta + c^2\alpha) \\ &\leq (1 - c^2)^{-1}(\alpha + c\beta) \\ &= \mu_\psi,\end{aligned}$$

and also $c\mu_\psi \leq \mu_\varphi$. Thus, $\mu_\varphi$ and $\mu_\psi$ satisfy (b).

Next we prove that (b) implies (c). Let $\mu_\varphi \in \mathscr{M}_\varphi$. Choose by (b) $\nu_\varphi \in \mathscr{M}_\varphi$ and $\nu_\psi \in \mathscr{M}_\psi$ such that for some $c \in (0, 1)$, $c\nu_\varphi \leq \nu_\psi \leq c^{-1}\nu_\varphi$. If $\mu_\psi$ is defined by

$$\mu_\psi = \nu_\psi - c\nu_\varphi + c\mu_\varphi,$$

then $\mu_\psi \in \mathscr{M}_\psi$, and $\mu_\psi(E) = 0$ implies $\mu_\varphi(E) = 0$; i.e., $\mu_\varphi \ll \mu_\psi$.

It is clear that (c) implies (d).

To see that (d) implies (a), suppose that $\|\varphi - \psi\| = 2$ and let $\mu_\varphi \in \mathscr{M}_\varphi, \mu_\psi \in \mathscr{M}_\psi$. From $\|\varphi - \psi\| = 2$, we find that

$$\sup\left\{\left|\int f\, d(\mu_\varphi - \mu_\psi)\right| : f \in A, \|f\| \leq 1\right\} = 2,$$

whence

$$\sup\left\{\left|\int u\, d(\mu_\varphi - \mu_\psi)\right| : u \in \operatorname{Re} A, \|u\| \leq 1\right\} = 2,$$

so $\|\mu_\varphi - \mu_\psi\| = 2$. Let $d\mu_\psi = h\,d\mu_\varphi + d\sigma$ be the Lebesgue decomposition of $\mu_\psi$ with respect to $\mu_\varphi$. Thus, $h \geq 0$ a.e. $[\mu_\varphi]$, and $\int h\,d\mu_\varphi + \|\sigma\| = \|\mu_\psi\| = 1$. If $E^+ = \{x : h(x) \geq 1\}$, $E^- = \{x : h(x) < 1\}$, we have

$$\begin{aligned} 2 &= \int |1 - h|\,d\mu_\varphi + \|\sigma\| \\ &= \int_{E^-} (1 - h)\,d\mu_\varphi + \int_{E^+} (h - 1)\,d\mu_\varphi + \|\sigma\| \\ &= \mu_\varphi(E^-) - \mu_\varphi(E^+) + \int_{E^+} h\,d\mu_\varphi - \int_{E^-} h\,d\mu_\varphi + \|\sigma\|. \end{aligned}$$

Since $\mu_\varphi(E^-) + \mu_\varphi(E^+) = 1$, and $\int h\,d\mu_\varphi + \|\sigma\| = 1$, it follows that $\mu_\varphi(E^+) = 0$ and $\int_{E^-} h\,d\mu_\varphi = 0$. Thus, $h = 0$ a.e. $[\mu_\varphi]$, and it follows that $\mu_\psi = \sigma$; i.e., $\mu_\psi$ and $\mu_\varphi$ are mutually singular. This concludes the proof of the theorem.

The argument used to derive (c) from (b) yields the following interesting fact.

16.7 ***Corollary***

*If $\varphi$ and $\psi$ lie in the same Gleason part and if $\mathcal{M}_\psi$ has a unique element, then so does $\mathcal{M}_\varphi$.*

PROOF

Let $\mathcal{M}_\psi = \{\mu_\psi\}$ and let $\nu_\varphi \in \mathcal{M}_\varphi$ be boundedly mutually absolutely continuous with respect to $\mu_\psi$, say, $\nu_\varphi \leq (1/c)\mu_\psi$ for some $c \in (0, 1)$. If $\mu_\varphi \in \mathcal{M}_\varphi$, then

$$\mu_\psi - c\nu_\varphi + c\mu_\varphi$$

is an element of $\mathcal{M}_\psi$, so since $\mathcal{M}_\psi = \{\mu_\psi\}$, we have that $\nu_\varphi = \mu_\varphi$. Thus, $\mathcal{M}_\varphi = \{\nu_\varphi\}$ and we have the corollary.

As a second corollary, we have a maximum principle.

16.8 ***Corollary***

*Let $(A, X)$ be a uniform algebra, let $\Pi$ be a Gleason part for $A$, and let $f \in A$. If $|\varphi(f)| = \|f\|$ for some $\varphi \in \Pi$, then $f|\Pi$ is constant.*

PROOF

Let $\psi \in \Pi$. If $\mu_\varphi \in \mathcal{M}_\varphi$ and $\mu_\psi$ are mutually absolutely continuous, then

$$\begin{aligned} \|f\| = |\varphi(f)| &= \left|\int f\,d\mu_\varphi\right| \\ &\leq \|f\|, \end{aligned}$$

whence $f$ assumes the value $\varphi(f)$ a.e. $[\mu_\varphi]$. Thus, $f$ assumes the value $\varphi(f)$ a.e $[\mu_\psi]$, and from

$$\psi(f) = \int f \, d\mu_\psi,$$

we get $\psi(f) = \varphi(f)$.

A word of warning is in order at this point. The results developed in Section 18 will show that from $\varphi \in \Pi$ and $|\varphi(f)| = \|f\|_\Pi$ it does not necessarily follow that $f$ is constant on $\Pi$.

The following observation of S. J. Sidney shows that parts have certain convexity properties.

**16.9** ***Theorem***

*If $\Pi$ is a Gleason part for the uniform algebra $A$ and if $K \subset \Pi$ is compact, then the set*

$$\hat{K} = \{\varphi \in \Sigma(A) : |\varphi(f)| \leq \|f\|_K \text{ for all } f \in A\}$$

*is contained in $\Pi$.*

PROOF

If $\psi \in \hat{K} \setminus \Pi$, there is a probability measure $\mu$ on $K$ such that $\int_K f d\mu = \psi(f)$. Let $\varphi \in \Pi$. Since $\psi \notin \Pi$, there is a sequence $\{f_n\}$ in $A$ with $\|f_n\| \leq 1$, $\psi(f_n) \to 0$, and $\varphi(f_n) \to 1$. Since $\varphi(f_n) \to 1$ and $\|f_n\| \leq 1$, it follows that $\{f_n\}$ converges pointwise to 1 on the whole of $\Pi$. Thus, by the dominated convergence theorem, $\int_K f_n \, d\mu \to 1$. However, $\int_K f_n \, d\mu = \psi(f_n) \to 0$. This contradiction establishes the result.

We conclude this section with another example of a Gleason part, an example that hints at the rather general form which Gleason parts can assume.

**16.10** ***Example***

Let $E$ be a compact subset of the real axis which is of Lebesgue (one-dimensional) measure zero. Let $\{\lambda_j\}_{j=0}^{\infty}$ be a sequence of positive numbers, $\lambda_0 < 1$, $\lambda_j \to \infty$, such that

$$\sum_{k=1}^{\infty} (k\lambda_k)^{-1} = \infty.$$

Let $z_n = x_n + iy_n$, $n = 1, 2, 3, \ldots$, be points in the upper half-plane located so that the set $\{z_n : n = 1, 2, \ldots\}$ has the set $E$ as its set of limit points. If

$$\alpha_n = \inf\{\lambda_0, y_n\lambda_1, (y_n\lambda_3)^3, \ldots\},$$

then, since $\lambda_j \to \infty$, it follows that $\alpha_n > 0$, and we can choose $r_n$ so that

$$0 < r_n < 2^{-n} y_n \alpha_n,$$

and so that the closed discs $D_n$ with radius $r_n$ and center $z_n + ir_n$ are disjoint. The definitions of $\alpha_n$ and of $r_n$ imply that

$$\sum_{n=1}^{\infty} y_n^{-1} r_n \leq \lambda_0 < 1. \tag{3}$$

If $\mathbb{C}^*$ denotes the Riemann sphere, let $D = \mathbb{C}^* \setminus (E \cup \bigcup_n D_n)$ so that $D$ is an open set in $\mathbb{C}^*$, $\infty \in D$. The algebra $\mathscr{A}(\bar{D})$ has $\bar{D}$ as its spectrum, as we shall see in Section 24. We shall show that $D \cup E$ is a Gleason part for $\mathscr{A}(\bar{D})$. It is clear that $D$ is contained within a single Gleason part for $\mathscr{A}(\bar{D})$. To show that the set $E$ lies in the same part, we must derive a version of the Cauchy integral formula valid for functions in $\mathscr{A}(\bar{D})$. Denote by $\gamma_n$ the boundary of $D_n$ so that $\gamma_n$ is a certain circle. If $z$ is a point in $D$, then we can write, by the Cauchy integral formula,

$$f(z) = \frac{1}{2\pi i}\int_{C_N} \frac{f(\zeta)}{\zeta - z}\, d\zeta - \sum_{j=1}^{N} \frac{1}{2\pi i}\int_{\gamma_j} \frac{f(\zeta)}{\zeta - z}\, d\zeta + \frac{1}{2\pi i}\int_{C'_N} \frac{f(\zeta)}{\zeta - z}\, d\zeta$$

if $f \in \mathscr{A}(\bar{D})$. Here $C_N$ is a large circle centered, say, at the origin which contains $z$ in its interior, and $C'_N$ is a system of simple closed curves in $D$ which have the property that if $V_N$ is the union of the interiors of the curves in the system $C'_N$, then $V_N$ does not contain $\gamma_j$ for $1 \leq j \leq N$, but it does contain both $\gamma_j$, $j > N$, and $E$. As $C_N$ increases in radius, $(1/2\pi i)\int_{C_N} [f(\zeta)/(\zeta - z)]\, d\zeta$ converges to $f(\infty)$, and since the set $E$ has one-dimensional measure zero, the curves in the system $C'_N$, for large $N$, can be chosen to have arbitrarily small total length. Thus, for large $N$, $\int_{C'_N} [f(\zeta)/(\zeta - z)]\, d\zeta$ is small. The conclusion is that for $z \in D$,

$$f(z) = f(\infty) - \sum_{j=1}^{\infty} \frac{1}{2\pi i}\int_{\gamma_j} \frac{f(\zeta)}{\zeta - z}\, d\zeta.$$

If $z$ is a point in the lower half-plane, then since $\gamma_n$ is centered at $z_n + ir_n$, we have the estimate

$$\left| \frac{1}{2\pi i}\int_{\gamma_n} \frac{f(\zeta)}{\zeta - z}\, d\zeta \right| \leq y_n^{-1} r_n \| f \|_D.$$

By the estimate (3), we find that if $f \in \mathscr{A}(\bar{D})$ and $f(\infty) = 0$, then for $z$ in the lower half-plane

$$|f(z)| \leq \lambda_0 \| f \|_D.$$

Since $\lambda_0 < 1$, this implies that the closed lower half-plane is contained in the same Gleason part as the point $\infty$, and thus $D \cup E$ is a Gleason part for $\mathscr{A}(\bar{D})$.

This example was considered by Rudin [9] in a somewhat different connection. Note that although the set $E$ lies in the Shilov boundary for $\mathscr{A}(\bar{D})$, Corollary 16.8 implies that no point of $E$ is a peak point for $\mathscr{A}(\bar{D})$. Rudin also remarked that the set $E$ has the following curious property: Any function in $\mathscr{A}(\bar{D})$ that vanishes at infinitely many points of $E$ vanishes identically. This remark depends on the theory of quasi-analytic functions. For additional material on the theory of parts, see Bear [4].

# 17 ALGEBRAS WITH FEW REPRESENTING MEASURES

If $A$ is a uniform algebra, if $\varphi \in \Sigma(A)$, and if $\mathscr{M}_\varphi$, the set of representing measures for $\varphi$, is not too large, then through $\varphi$ there passes certain analytic structure. In this section, we shall treat in detail one case of this phenomenon, the case in which $\mathscr{M}_\varphi$ contains a single measure, and we shall consider without proof certain additional cases. The main result we shall establish is the following.

17.1 ***Theorem***

*Let $A$ be a uniform algebra with Shilov boundary $\Gamma$, and let $\varphi \in \Sigma(A)$. If $\Pi$, the Gleason part for $A$ which contains $\varphi$, consists of more than one point, and if there is a unique representing measure for $\varphi$ supported on $\Gamma$, then $\Pi$ is an analytic disc in the sense that there is a one-to-one continuous map $\Phi$ from $\Delta$, the open unit disc in $\mathbb{C}$, onto $\Pi$ such that if $f \in A$, then $\hat{f} \circ \Phi$ is holomorphic on $\Delta$.*

There are certain important classes of algebras to which this theorem applies.

17.2 ***Definitions***

*Let $A$ be a uniform algebra with Shilov boundary $\Gamma$.*

(a) *$A$ is a **Dirichlet algebra** if* Re *$A$ is dense in $\mathscr{C}_{\mathbb{R}}(\Gamma)$.*
(b) *$R$ is **logmodular** if the functions* $\log|f|$, *$f$ invertible in $A$, are dense in $\mathscr{C}_{\mathbb{R}}(\Gamma)$.*

With these two definitions, we can indicate briefly the history of Theorem 17.1. The first result of this kind is due to Wermer [8], who proved that if $A$ is a Dirichlet algebra, then every Gleason part for $A$ is either trivial or an analytic disc. Hoffman [2] showed that the same result obtains if $A$ is merely logmodular. Finally, in the paper [1] of Lumer, the full result is obtained.

It is clear that if $A$ is a Dirichlet algebra, then there is only one probability measure on $\Gamma$ which represents $\varphi \in \Sigma(A)$. It is less clear, though true, that the same result holds for logmodular algebras.

17.3 ***Theorem***

*If $A$ is a logmodular algebra, and if $\varphi \in \Sigma(A)$, there is a unique probability measure supported on $\Gamma$ which represents $\varphi$.*

PROOF

Let $\mu$ and $\nu$ be representing measures for $\varphi$ supported by $\Gamma$. If $f$ is an invertible element of $A$, then

$$\varphi(f) = \int f \, d\mu$$

and

$$\varphi(f)^{-1} = \int f^{-1} \, d\nu.$$

Thus,

$$|\varphi(f)| \leq \int |f| \, d\mu$$

and

$$|\varphi(f)|^{-1} \leq \int |f|^{-1} \, d\nu,$$

whence

$$1 \leq \int |f| \, d\mu \int |f|^{-1} \, d\nu.$$

Since the functions of form $\log|f|$, $f$ invertible in $A$, are dense in $\mathscr{C}_{\mathbb{R}}(\Gamma)$, it follows that if $u$ is a real continuous function on $\Gamma$, then

$$1 \leq \int e^{u} \, d\mu \int e^{-u} \, d\nu.$$

Let $u \in \mathscr{C}_{\mathbb{R}}(\Gamma)$ be fixed, and define $p$ on $\mathbb{R}$ by

$$p(t) = \int e^{tu}\, d\mu \int e^{-tu}\, d\nu.$$

The function $p$ is real valued and real analytic on $\mathbb{R}$. We have $p(t) \geq 1$ and $p(0) = 1$. Thus, $p'(0) = 0$. Since

$$p'(0) = \int u\, d\mu - \int u\, d\nu,$$

it follows that

$$\int u\, d\mu = \int u\, d\nu.$$

As this holds for all choices of $u \in \mathscr{C}_{\mathbb{R}}(\Gamma)$, $\mu = \nu$, as was to be proved.

Thus, Theorem 17.1 is applicable to both Dirichlet and logmodular algebras.

Let us give some examples of Dirichlet and logmodular algebras.

17.4 ***Examples***

(a) The most obvious example of a Dirichlet algebra is the disc algebra $\mathscr{A}$. The fact that $\mathscr{A}$ is Dirichlet was noted, implicitly, in Example 7.17.

(b) A group theoretic generalization of the preceding example may be obtained as follows. Let $G$ be a compact abelian group with dual group $\Gamma$, and suppose that $\Gamma$ can be ordered; i.e., suppose that there exist subsemigroups $\Gamma_+$ and $\Gamma_-$ of $\Gamma$ such that $\Gamma_+ \cup \Gamma_- = \Gamma$ and $\Gamma_+ \cap \Gamma_- = \{0\}$. Let $A$ be the algebra of continuous functions on $G$ such that if $m$ denotes the normalized Haar measure on $G$, then

$$\hat{f}(\gamma) = \int_G f(x)(-x, \gamma)\, dm(x) = 0$$

if $\gamma \notin \Gamma_+$. Since $\Gamma_+$ is a semigroup, $A$ is an algebra. It is, moreover, a Dirichlet algebra. To see this, recall first that the trigonometric polynomials are dense in $\mathscr{C}(G)$ so that if $u \in \mathscr{C}_{\mathbb{R}}(G)$ and $\varepsilon > 0$, then for some trigonometric polynomial $p$, say $p(x) = \sum_{j=1}^{N} \alpha_j(x, \gamma_j)$, we have $\|u - p\|_G < \varepsilon$. Since $u$ is real, we have $\|u - \operatorname{Re} p\| < \varepsilon$. We can write $p = p_+ + p_-$, where $p_+(x)$ is the sum of the terms in $\sum_{j=1}^{N} \alpha_j(x, \gamma_j)$ for which $\gamma_j \in \Gamma_+$, and $p_-(x)$ is the sum of the terms for which $\gamma_j \in \Gamma_- \setminus \{0\}$. We have $\overline{(x, \gamma_j)} = (x, -\gamma_j)$, so $\bar{p}_- \in A$. Since $\operatorname{Re}(p_+ + p_-) = \operatorname{Re}(p_+ + \bar{p}_-)$, it follows that

$$\|u - \operatorname{Re}(p_+ + \bar{p}_-)\| < \varepsilon.$$

The polynomial $p_+ + \bar{p}_-$ is in $A$, so $A$ is Dirichlet.

Algebras of this kind are treated in some detail from the point of view of harmonic analysis by Rudin [1].

(c) We shall see later that if $X$ is a compact set in the plane such that $\mathbb{C} \setminus X$ is connected, then $\mathscr{P}(X)$, the uniform closure in $\mathbb{C}(X)$ of the polynomials in a complex variable, is a Dirichlet algebra.

(d) The algebra $H^\infty(\Delta)$ of bounded holomorphic functions on the open unit disc is logmodular but not Dirichlet. In Example 13.14 we saw that every continuous, strictly positive function on the Shilov boundary for $H^\infty(\Delta)$ is the modulus of some invertible element of $H^\infty(\Delta)$. Thus, $H^\infty(\Delta)$ is logmodular. However, it is not a Dirichlet algebra. We saw in Example 13.14 that the Shilov boundary for $H^\infty(\Delta)$ is the spectrum of $L^\infty$, the $L^\infty$-algebra of Lebesgue measure on the unit circle, and this spectrum we know to be totally disconnected. (Recall Example 6.5.) We can now invoke an observation of Gleason [1] according to which there are no Dirichlet algebras with totally disconnected Shilov boundaries [except, of course, for the algebras $\mathscr{C}(X)$]. If $A$ is a Dirichlet algebra with totally disconnected Shilov boundary $\Gamma$, then given an open and closed subset $E$ of $\Gamma$, the Dirichlet property implies the existence of $f \in A$ with $|\operatorname{Re} f| < \frac{1}{4}$ on $E$, $|\operatorname{Re} f - 1| < \frac{1}{4}$ on $\Gamma \setminus E$. Since $\Gamma$ is the Shilov boundary for $A$, it follows, e.g., from Theorem 7.12, that there is a decomposition

$$\Sigma(A) = \Sigma_0 \cup \Sigma_1$$

of $\Sigma(A)$ into a disjoint union of open and closed sets such that $\Sigma_0 \supset E$, $\Sigma_1 \supset \Gamma \setminus E$. By the Shilov idempotent theorem, there is $g \in A$ such that $\hat{g}|\Sigma_1 = 0$, $\hat{g}|\Sigma_0 = 1$. Thus, $A$ contains the characteristic function of every open, closed subset of $\Gamma$. Since $\Gamma$ is totally disconnected, finite linear combinations of such functions are dense in $\mathscr{C}(\Gamma)$, and thus $A = \mathscr{C}(\Gamma)$.

(e) A method for producing some rather bizarre examples of logmodular algebras is as follows. Let $\Lambda$ be an index set, and for each $\lambda$, let $X_\lambda$ be a compact space which is the Shilov boundary for the logmodular algebra $A_\lambda$ so that if $h \in \mathscr{C}_{\mathrm{R}}(X_\lambda)$ and if $\varepsilon > 0$, there is $f \in \mathscr{G}(A_\lambda)$ such that $\|\log|f| - h\|_{X_\lambda} < \varepsilon$. Let $X$ be the disjoint union of the $X_\lambda$, and let $A$ be the subalgebra of $\mathscr{C}^B(X)$ consisting of those functions $F$ which agree on $X_\lambda$ with an element $f_\lambda$ of $A_\lambda$. It is clear that if $H$ is a bounded, real-valued function on $X$, then given $\varepsilon > 0$, there is $F \in \mathscr{G}(A)$ for which $\|\log|F| - H\|_X < \varepsilon$. Consequently, if we extend the elements of $A$ in the canonical way to functions on $\beta(X)$, the Čech compactification of $X$, to obtain a uniform algebra $A^*$ on $\beta(X)$, then $A^*$ is logmodular on $\beta(X)$.

The set $S = \bigcup\{\Sigma(A_\lambda):\lambda \in \Lambda\}$ is, in a natural way, an open subset of $\Sigma(A^*)$; it would be of considerable interest to determine whether $S$ is dense in $\Sigma(A^*)$.

We now turn from these preliminary remarks to begin developing the results needed to prove Theorem 17.1. Let $\Pi$ and $\varphi$ be as in the statement of the theorem. We denote by $\mu$ the unique probability measure on $\Gamma$ which represents $\varphi$, and we define $H^2(\mu)$ to be the closure of $A$ in $L^2(\mu)$. If $\psi$ is another point of $\Pi$, and if $\nu$ is the representing measure for $\psi$—$\nu$ is unique by Corollary 16.7—then $\mu$ and $\nu$ are boundedly mutually absolutely continuous; so, as sets, $L^2(\mu)$ and $L^2(\nu)$ are the same, and the norms induced by $\mu$ and $\nu$ are equivalent. Similarly, $H^2(\mu)$ and $H^2(\nu)$ are the same set. A similar remark applies to $L^\infty(\mu)$ and $L^\infty(\nu)$: Not only are they the same set, but the two norms are the same. We define $H^\infty(\mu)$ to be $H^2(\mu) \cap L^\infty(\mu)$. It is easy to see that $H^\infty(\mu)$ is an algebra. The only thing to be shown is that $H^\infty(\mu)$ is closed under multiplication. Suppose then that $f, g \in H^2(\mu) \cap L^\infty(\mu)$ so that there exist sequences $\{f_n\}$ and $\{g_n\}$ in $A$ such that

$$\int |f_n - f|^2 \, d\mu \to 0$$

and

$$\int |g_n - g|^2 \, d\mu \to 0.$$

By choosing $\{g_n\}$ first and then $\{f_n\}$, we can assume that

$$\int |f_n - f|^2 \, d\mu < 2^{-2n}(\|g_n\|_\infty + 1)^{-2},$$

$\|g_n\|_\infty$ the $L^\infty(\mu)$ norm of $g_n$. With this choice of $\{f_n\}$, we have

$$\begin{aligned} \|f_n g_n - fg\|_2 &\le \|f_n g_n - f g_n\|_2 + \|f g_n - fg\|_2 \\ &\le 2^{-n}\|g_n\|_\infty(\|g_n\|_\infty + 1)^{-1} + \|f\|_\infty \|g_n - g\|_2. \end{aligned}$$

[Here we use $\|\cdot\|_2$ to denote the norm in $L^2(\mu)$.] Since this last quantity tends to zero as $n \to \infty$, it follows that $fg$ is in $H^\infty(\mu)$ and so that $H^\infty(\mu)$ is an algebra. An argument similar to the one just given shows that $\mu$ is multiplicative on $H^\infty(\mu)$. The algebra $H^\infty(\mu)$ is plainly norm closed in $L^\infty(\mu)$, and it is also weak * closed, for if $f_n \in H^\infty(\mu)$ and $\int f_n g \, d\mu \to \int fg \, d\mu$ for all $g \in L^1(\mu)$, $f$ some element of $L^\infty(\mu)$, then given $g \in L^2(\mu) \ominus H^2(\mu)$, $0 = \int f_n g \, d\mu \to \int fg \, d\mu$, whence $f \in H^2(\mu)$ and therefore $f \in H^\infty(\mu)$.

[For future reference, we point out that these arguments do not depend on the special context of unique representing measures. They show quite generally that if $B$ is a subalgebra of an $L^\infty(m)$, $m$ a probability measure, on which $m$ is multiplicative, and if $H^\infty(B)$ is the intersection of $L^\infty(m)$ and $H^2(B)$, the closure of $B$ in $L^2(m)$, then $H^\infty(B)$ is norm and weak * closed in $L^\infty(m)$, and $m$ is multiplicative on $B$.]

If $\mu$ and $\nu$ are as above, we have that $H^\infty(\mu)$ and $H^\infty(\nu)$ are the same algebra and that they have the same norm. We shall use $H^\infty(\Pi)$ to denote this algebra so that $H^\infty(\Pi)$ is canonically associated with the part $\Pi$.

If $\psi \in \Pi$, we denote by $H^2_\psi(\mu)$ the set of all $f \in H^2(\mu)$ for which $0 = \int f\, d\nu$, $\nu$ the representing measure for $\psi$, and we set $H^\infty_\psi(\Pi) = H^\infty(\Pi) \cap H^2_\psi(\mu)$. Thus, $H^2_\psi(\mu)$ is a closed subspace of $H^2(\mu)$, and $H^\infty_\psi(\Pi)$ is a closed ideal in $H^\infty(\Pi)$. We need to prove that if $\psi \in \Pi$, then

$$L^2(\mu) = H^2_\psi(\mu) \oplus \overline{H^2(\mu)}.$$

The next three lemmas, which are purely measure theoretic, lead up to this fact. The development follows Hoffman [2] and Lumer [1].

**17.5** ***Lemma***

*If $m$ is a regular Borel probability measure on the compact space $X$, and if $f$ is a nonnegative element of $L^1(m)$, then*

$$\text{(1)} \qquad \exp \int \log f\, dm = \inf\left\{\int fe^g\, dm : g \in K, \int g\, dm = 0\right\}$$

*if $K$ is any one of the following spaces:*

(a) $L^1_{\mathbb{R}}(m)$.
(b) *The space of real bounded Borel functions on $X$.*
(c) $\mathscr{C}_{\mathbb{R}}(X)$.

PROOF

We have $f \in L^1_R(m)$ and $\log f \le f$, so either $\log f$ is integrable or $\int \log f\, dm = -\infty$. In the latter case, we define $\exp \int \log f\, dm$ to be 0. If $\log f$ is integrable and if $g \in L^1_{\mathbb{R}}(m)$ satisfies $\int g\, dm = 0$, then we have (Appendix B)

$$\int fe^g\, dm \ge \exp \int (g + \log f)\, dm$$

$$= \exp \int \log f\, dm.$$

By our convention in the case that $\int \log f\, dm = -\infty$, this latter inequality is also valid in that case, too. Thus, we have

$$(2) \qquad \exp \int \log f\, dm \leq \inf\left\{\int fe^g\, dm : g \in L^1_{\mathbf{R}}(m), \int g\, dm = 0\right\}.$$

If $\log f$ is integrable, then the function $g_0 = -\log f + \int \log f\, dm$ is in $L^1_{\mathbf{R}}(m)$ and has zero integral. Since

$$\int fe^{g_0}\, dm = \exp \int \log f\, dm,$$

we have equality in (2). If $\log f$ is not integrable, then given $\varepsilon > 0$, $\log(f + \varepsilon)$ is integrable, and by letting $\varepsilon$ decrease to zero in the inequality we have already established, we obtain the lemma, at least in case (a).

Cases (b) and (c) follow by approximation arguments. Consider case (b). If $g \in L^1_{\mathbf{R}}(dm)$ satisfies $\int g\, dm = 0$, there is a sequence $\{g_n\}$ of bounded Borel functions such that $\int g_n\, dm = 0$ and such that $\max(g_n, 0)$ increases to $\max(g, 0)$ and $\min(g_n, 0)$ decreases to $\min(g, 0)$. By the monotone convergence theorem

$$\lim_{n \to \infty} \int fe^{g_n}\, dm = \int fe^g\, dm,$$

and this implies the desired equality in case (b).

Finally, if $g$ is a bounded, real-valued Borel function, there exists a bounded sequence $\{g_n\}$ in $\mathscr{C}_{\mathbf{R}}(X)$ which converges pointwise a.e. $[dm]$ to $g$. If $\int g\, dm = 0$, then the $g_n$ can be chosen to have zero integral, and case (c) of the lemma follows from the dominated convergence theorem.

**17.6** ***Lemma***

*Let $m$ be a regular Borel probability measure on the compact space $X$, and let $g \in L^1_{\mathbf{R}}(m)$. If*

$$\int \log|1 - \xi g|\, dm \geq 0$$

*for all $\xi$ in an interval $(-\delta, \delta)$, $\delta > 0$, then $g = 0$ a.e. $[dm]$.*

PROOF

The proof of this lemma depends on classical function theory. Define an auxiliary function $u$ by

$$u(\zeta) = \int \log|1 - \zeta g(x)|\, dm(x).$$

This function is well defined in the half-planes $\operatorname{Im} \zeta > 0$ and $\operatorname{Im} \zeta < 0$, and it is harmonic in these half-planes, for it is the real part of the holomorphic function

$$\int \log(1 - \zeta g(x))\, dm(x).$$

Let $\zeta = \xi + i\eta$; we shall prove that

$$\lim_{\eta \to 0} \frac{1}{\eta} u(i\eta) = 0. \tag{3}$$

We have

$$\begin{aligned} \frac{1}{\eta} u(i\eta) &= \frac{1}{2} \int \frac{1}{\eta} \log[1 + \eta^2 g^2(x)]\, dm(x) \\ &= \frac{1}{2} \int \frac{1}{\eta g(x)} \log[1 + \eta^2 g^2(x)] g(x)\, dm(x), \end{aligned}$$

where we define $1/\eta g(t) \log[1 + \eta^2 g^2(t)]$ to be zero at those points where $g$ vanishes. The function $1/x \log(1 + x^2)$ is bounded on $(0, \infty)$ and on $(-\infty, 0)$, and

$$\lim_{x \to 0} \frac{1}{x} \log(1 + x^2) = 0.$$

Thus, the dominated convergence theorem applies and yields the limit relation (3).

By hypothesis $u(\xi) \geq 0$ for $\xi \in (-\delta, \delta)$, and, thus, $u$ is nonnegative in the strip $-\delta < \operatorname{Re} \zeta < \delta$, as follows from the fact that

$$\begin{aligned} u(\xi + i\eta) &= \int \log|1 - (\xi + i\eta) g(x)|\, dm(x) \\ &\geq \int \log|1 - \xi g(x)|\, dm(x). \end{aligned}$$

This fact, taken with (3), implies that $u$ vanishes identically. In order to prove this assertion, define $\tilde{u}$ by

$$\tilde{u}(z) = u(i\delta(1 - z)), \qquad |z| < 1.$$

The function $\tilde{u}$ is harmonic in the unit disc, and from (3) we conclude that

$$\lim_{r \to 1} \frac{\tilde{u}(r)}{1 - r} = 0. \tag{4}$$

By the theorem of Herglotz (Appendix D), $\tilde{u}$ is of the form

$$\tilde{u}(z) = \int P_z(\theta)\, d\mu(\theta),$$

where $P_z$ is the Poisson kernel for $z$ and $\mu$ is a nonnegative measure on the unit circle. We have that $P_r(\theta) \geq (1 - r)/(1 + r)$, so

$$\tilde{u}(r) \geq \frac{1 - r}{1 + r} \int d\mu,$$

and this, in conjunction with (4), yields $\mu = 0$. Thus, $\tilde{u}$ vanishes identically, and consequently so does $u$. In particular then, $u(i) = 0$, i.e.,

$$\frac{1}{2} \int \log(1 + g^2)\, dm = 0,$$

so $g = 0$ a.e. $[dm]$, and we have the lemma.

Next, consider a uniform algebra $(A, X)$ and a $\varphi \in \Sigma(A)$.

17.7 ***Lemma***

*If $\varphi \in \Sigma(A)$ admits a unique representing measure $\mu$, and if $h \in L^1_{\mathbb{R}}(\mu)$ satisfies $\int fh\, d\mu = 0$ for all $f \in A$, then $\int \log|1 - h|\, d\mu \geq 0$.*

PROOF

If $f \in \mathcal{G}(A)$, we have

$$\int f\, d\mu = \int f(1 - h)\, d\mu,$$

whence

$$\log\left|\int f\, d\mu\right| \leq \log \int |f||1 - h|\, d\mu.$$

Since $\mu$ is an Arens–Singer measure, we may conclude that

$$\exp \int \log|f|\, d\mu \leq \int |f||1 - h|\, d\mu.$$

Suppose now that $g \in \mathcal{C}_{\mathbb{R}}(X)$ satisfies $\int g\, d\mu = 0$. By Lemma 7.19, there is a sequence $\{f_n\}$ in $A$ such that $\int \operatorname{Re} f_n\, d\mu \to \int g\, d\mu$ and $\operatorname{Re} f_n \leq g$. Adding an imaginary constant if necessary, we can suppose that $\int \operatorname{Im} f_n\, d\mu = 0$. For a suitably chosen sequence $\{d_n\}$ of real numbers,

$\varphi(f_n + d_n) = 0$. Set $g_n = e^{f_n + d_n}$. The function $g_n$ lies in $\mathcal{G}(A)$, so

$$1 = \exp \int \operatorname{Re}(f_n + d_n)\, d\mu$$

$$= \exp \int \log|g_n|\, d\mu$$

$$\leq \int |\exp(f_n + d_n)|\, |1 - h|\, d\mu$$

$$= e^{d_n} \int e^{\operatorname{Re} f_n}|1 - h|\, d\mu$$

$$\leq e^{d_n} \int e^{g}|1 - h|\, d\mu.$$

Since $d_n \to 0$, we have

$$1 \leq \int e^{g}|1 - h|\, d\mu.$$

Since this inequality is valid for all $g \in \mathscr{C}_{\mathbb{R}}(X)$ with zero integral, Lemma 17.5 yields

$$1 \leq \exp \int \log|1 - h|\, d\mu,$$

whence

$$0 \leq \int \log|1 - h|\, d\mu,$$

as was to be established.

We can establish now the following crucial fact.

**17.8 *Lemma***

*If $(A, X)$ is a uniform algebra and if $\varphi \in \Sigma(A)$ admits a unique representing measure $\mu$ supported on the Shilov boundary, then*

$$L^2(\mu) = H_\varphi^2(\mu) \oplus \overline{H^2(\mu)}.$$

PROOF

It is clear that $H_\varphi^2(\mu)$ and $\overline{H^2(\mu)}$ are orthogonal in $L^2(\mu)$. To show that their sum is $L^2(\mu)$, let us consider an $h \in L^2(\mu)$ which is orthogonal to both $H_\varphi^2(\mu)$ and $\overline{H^2(\mu)}$. We need consider only the case of real $h$ because

$H^2_\varphi(\mu) + \overline{H^2(\mu)}$ is closed under complex conjugation. For such a function $h$,

$$0 = \int fth \, d\mu$$

for all $f \in A$ and $t \in \mathbb{R}$. Lemmas 17.6 and 17.7 yield the desired conclusion that $h$ is the zero function.

The final preparatory facts we shall need come not from the theory of uniform algebras but rather they are elementary points in the theory of Banach spaces. If $Y_0$ is a subspace of the Banach space $Y$, we use $\operatorname{codim}_Y Y_0$ to denote the codimension of $Y_0$ in $Y$.

**17.9** ***Lemma***

*Let $X$ and $Y$ be Banach spaces.*

(a) *If $T \in \mathscr{L}(X, Y)$ is one to one and has closed range, and if $\operatorname{codim}_Y T(X) = 1$, then $\operatorname{codim}_Y (T + S)(X) = 1$ for all $S \in \mathscr{L}(X, Y)$ with sufficiently small norm.*

(b) *If $\{T_n\}$ is a sequence of isometries in $\mathscr{L}(X, Y)$, if $\{T_n\}$ converges in norm to $T$, and if $\operatorname{codim}_Y T_n(X) = 1$ for all $n$, then $\operatorname{codim}_Y T(X) = 1$.*

PROOF

(a) The hypotheses that $T(X)$ is closed and $\operatorname{codim}_Y T(X) = 1$ imply the existence of a one-dimensional subspace $G$ of $Y$ such that $Y$ admits the topological direct sum decomposition

$$Y = T(X) \oplus G.$$

Let $\pi : Y \to G$ and $\eta : Y \to T(X)$ be the natural projections so that $\pi + \eta$ is the identity $I$ of $\mathscr{L}(Y)$, and define $J \in \mathscr{L}(Y)$ by

$$J(y) = \pi(y) + (T + S)(T^{-1}(\eta(y))).$$

The operator $J$ is continuous, for $T$ is a topological isomorphism of $X$ onto $T(X) = \eta(Y)$.

We have

$$\|I - J\| = \sup_{\|y\| \leq 1} \|ST^{-1}(\eta(y))\|,$$

and if $S$ is small enough in norm, this norm is less than 1 whence $J$ is an invertible element of the Banach algebra $\mathscr{L}(Y)$, i.e., a topological automorphism of $Y$. As automorphisms preserve direct sums, we have the direct-sum decomposition $Y = J(T(X)) \oplus J(G)$, and the definition of $J$ shows that this direct sum is just $Y = (T + S)(X) \oplus G$.

(b) As $\{T_n\}$ converges in norm to $T$, the fact that each $T_n$ is an isometry implies that $T$ also is an isometry, and from this it follows that $T(X)$ is a closed subspace of $Y$. If the assertion is false, $\operatorname{codim}_Y T(X) \geq 2$, so there are $y_1$ and $y_2$ in $Y$ such that $(y_1)$ and $(y_2)$, the cosets of $y_1$ and $y_2$, respectively, in $Y/T(X)$, are linearly independent. By the uniqueness of norm topology on finite-dimensional Banach spaces, it follows that if $\|\cdot\|_q$ is the quotient norm on $Y/T(X)$, then for some $\beta > 0$ and all $a,b \in \mathscr{C}$,

$$\|(ay_1 + by_2)\|_q \geq \beta(|a| + |b|).$$

If

$$M = \sup\{\|ay_1 + by_2\| : |a| + |b| = 1\},$$

choose an $n$ so large that

$$\|T_n - T\| < \tfrac{1}{2}\beta(M + \beta/2)^{-1}.$$

We shall show that if for $z \in Y$, $(z)_n$ denotes the element of $Y/T_n(X)$ which contains $z$, and if $\|\cdot\|_n$ denotes the quotient norm on $Y/T_n(X)$, then for all complex numbers $a$ and $b$

$$(5) \qquad \|(ay_1 + by_2)_n\|_n \geq \tfrac{1}{2}\beta(|a| + |b|).$$

If not, the definition of the quotient norm implies the existence of complex numbers $a$ and $b$ and an element $x$ in $X$ such that $|a| + |b| = 1$ and

$$\|ay_1 + by_2 - T_n(x)\| < \tfrac{1}{2}\beta.$$

The operator $T_n$ is an isometry so

$$\begin{aligned} \|x\| &= \|T_n(x)\| \\ &< \|ay_1 + by_2\| + \beta/2 \\ &\leq M + \beta/2. \end{aligned}$$

Thus,

$$\begin{aligned} \|(ay_1 + by_2)\|_q &\leq \|ay_1 + by_2 - T(x)\| \\ &< \|ay_1 + by_2 - T_n(x)\| + \|T_n(x) - T(x)\| \\ &< \beta/2 + \beta/2. \end{aligned}$$

This last inequality contradicts the choice of $\beta$, so we conclude that the inequality (5) must hold. However, (5) implies that $\dim Y/T_n(X) \geq 2$ contrary to hypothesis, and the lemma is proved.

Part (a) of the lemma is contained in paper [1] of Dieudonné, and I learned part (b) from Sidney.

**17.10** ***Proof of Theorem* 17.1**

We assume the notation of the statement of the theorem. Let $\psi$ be a point of $\Pi$, $\psi \neq \varphi$, and define $\gamma$ by

$$\gamma = \sup\{|\psi(f)| : f \in \ker \varphi, \|f\| \leq 1\}$$

so that $\gamma$ is the norm of the restriction of $\psi$ to $\ker \varphi$. By the Hahn–Banach and Riesz representation theorems, there exists a measure $\eta$ on $\Gamma$ with $\|\eta\| = \gamma$, and $\int f\, d\eta = \psi(f)$ for all $f \in \ker \psi$. Choose a sequence $\{f_n\}$ in $\ker \varphi$ such that $\|f_n\| \leq 1$ and $\psi(f_n) \to \gamma$. By passing to a subsequence if necessary, we may assume that $\{f_n\}$ converges weakly in $L^2(|\eta|)$, say to $E$. Then $|E| \leq 1$ a.e. $[|\eta|]$, and $\int E\, d\eta = \gamma = \|\eta\|$; so we may conclude that $|E| = 1$ a.e. $[|\eta|]$. We have that $E^{-1}$ is the Radon–Nikodym derivative $d\eta/d|\eta|$. Since $\gamma^{-1}\, d|\eta| = \gamma^{-1}E\, d\eta$ is a probability measure, and since for all $g \in A$, the function $gf_n$ lies in $\ker \varphi$, we have

$$\begin{aligned} \int g\gamma^{-1}E\, d\eta &= \lim \gamma^{-1} \int gf_n\, d\eta \\ &= \lim \gamma^{-1}\psi(g)\psi(f_n) \\ &= \psi(g), \end{aligned}$$

and it follows that $\gamma^{-1}E\, d\eta$ is the unique representing measure $\mu_\psi$ for $\psi$. (Recall Corollary 16.7.)

Since $E$ is the weak limit in $L^2\,(|\eta|)$ of a sequence in $A$, and since $|\eta|$ is boundedly mutually absolutely continuous with respect to $\mu_\psi$, it follows that $E \in H^2(\mu_\psi)$, and as $|E| = 1$ a.e., $E \in H^\infty(\Pi)$. It is meaningful, therefore, to consider the principal ideal $EH^\infty(\Pi)$. The fact that $|E| = 1$ a.e. implies that this ideal is closed, for if $\{Ef_n\}$ converges to $g$, then $\{f_n\} = \{\bar{E}Ef_n\}$ converges to $\bar{E}g$, so $g = E(\bar{E}g) \in EH^\infty(\Pi)$. Also, $1 \notin EH^\infty(\Pi)$, for $\int E\, d\mu_\varphi = 0$, so the ideal $EH^\infty(\Pi)$ is a proper ideal. We shall prove it to be maximal. To see this, it is convenient first to observe that $EH^2(\mu_\psi)$ is of codimension 1 in $H^2(\mu_\psi)$. To prove this, suppose $g \in H^2(\mu_\psi)$, $g \perp EH^2(\mu_\psi)$. Then

$$E\bar{g} \in [(H^2_\varphi(\mu_\psi) \cap H^2_\psi(\mu_\psi)) \oplus \overline{H^2(\mu_\psi)}]^\perp. \tag{6}$$

It is clear that $E\bar{g} \perp \overline{H^2(\mu_\psi)}$. Also, if $f \in H^2_\varphi(\mu_\psi) \cap H^2_\psi(\mu_\psi)$, then

$$\begin{aligned} \int \overline{fgE}\, d\mu_\psi &= \left(\int fg\bar{E}\, d\mu_\psi\right)^- \\ &= \gamma^{-1}\left(\int fg\, d\eta\right)^- \\ &= 0 \end{aligned}$$

since $\int h\, d\eta = 0$ for all $h \in (\ker \varphi \cap \ker \psi)$. By Lemma 17.8, the right side of (6) has codimension 1 in $L^2(\mu_\psi)$, so the space

$$\{E\bar{g} : g \in H^2(\mu_\psi) \ominus EH^2(\mu_\psi)\}$$

is one dimensional, and thus the fact that $|E| = 1$ a.e. implies that $H^2(\mu_\psi) - EH^2(\mu_\psi)$ has dimension 1. Now $H^2(\mu_\psi) \supseteq H^2_\varphi(\mu_\psi) \supseteq EH^2(\mu_\psi)$, and the first inclusion is proper. From the fact that $EH^2(\mu_\psi)$ has codimension 1 in $H^2(\mu_\psi)$, we may conclude that $EH^2(\mu_\psi) = H^2_\varphi(\mu_\psi)$. If $f \in H^\infty_\varphi(\Pi)$, then $f \in H^2_\varphi(\mu_\psi)$, so $f = Eg$, $g \in H^2(\mu_\psi)$, and indeed $g \in H^\infty(\Pi)$ because $f \in H^\infty(\Pi)$. Consequently, $EH^\infty(\Pi)$ is the maximal ideal $H^\infty_\varphi(\Pi)$ in $H^\infty(\Pi)$.

If $|\zeta| < 1$, define $E_\zeta \in H^\infty(\Pi)$ by

$$E_\zeta = \frac{E - \zeta}{1 - \bar{\zeta}E}.$$

The function $E_0$ evidently coincides with $E$. If $L(\zeta)$ denotes the operator $f \mapsto E_\zeta f$ on $H^\infty(\Pi)$, then since for each $\zeta$, $|E_\zeta| = 1$ a.e., the operators $L(\zeta)$ are all isometries. They vary continuously with the parameter $\zeta$ from the unit disc, so it follows from Lemma 17.9 that the codimension of $E_\zeta H^\infty(\Pi)$ is 1 for every $\zeta$ in the unit disc. Consequently, each of the ideals $E_\zeta H^\infty(\Pi)$ is a maximal ideal in $H^\infty(\Pi)$.

It follows that the Gelfand transform, $\hat{E}$, of $E$ carries $\Sigma(H^\infty(\Pi))$ onto the closed unit disc: The range of $\hat{E}$ is closed and contained in the closed unit disc since $\|E\| = 1$, and since each $E_\zeta H^\infty(\Pi)$ is a maximal ideal, $\hat{E}$ assumes every value in the open unit disc. Set

$$D = \{\theta \in \Sigma(H^\infty(\Pi)) : |\hat{E}(\theta)| < 1\}.$$

Under $\hat{E}$, $D$ is mapped continuously onto the open unit disc $\Delta$ in $\mathbb{C}$, and, moreover, $\hat{E}$ is one to one on this set; for if $\hat{E}(\theta) = \zeta$, $|\zeta| < 1$, then the kernel of $\theta$ contains $E_\zeta H^\infty(\Pi)$, so it is, in fact, equal to this maximal ideal. Since the kernel of each $\theta \in D$ is a maximal principal ideal, Gleason's theorem, Theorem 15.2, implies that $D$ admits the structure of an analytic variety on which each $\hat{f}$, $f \in H^\infty(\Pi)$, is holomorphic. Since $\hat{E}$ is a holomorphic, one-to-one map of $D$ onto $\Delta$, the variety $D$ is conformally the disc. (Observe that we could equally well invoke Theorem 15.12 to find that $D$ has the structure of a Riemann surface.) If $\chi : \Delta \to D$ is the inverse of $\hat{E}$, then for $f \in H^\infty(\Pi)$, $\hat{f} \circ \chi$ is holomorphic.

So far we have succeeded in obtaining analytic structure in $\Sigma(H^\infty(\Pi))$; we want to find it in $\Sigma(A)$. For this purpose, let $i : A \to H^\infty(\Pi)$ be the natural inclusion, and let $i^* : \Sigma(H^\infty(\Pi)) \to \Sigma(A)$ be its adjoint. We shall prove that $i^*$ carries $D$ onto $\Pi$ in a one-to-one, continuous fashion. Once we have this, we are finished, for if $\Phi = i^* \circ \chi$, then for all $f \in A$, $\hat{f} \circ \Phi = \hat{f} \circ i^* \circ \chi = (if)\hat{\ } \circ \chi$ is holomorphic on $\Delta$. The continuity of $i^*$ is clear; we need only prove it to be one to one and onto.

First, $i^*(D) = \Pi$, for if $\theta \in \Pi$, the map $\tilde{\theta}$ defined by

$$\tilde{\theta}f = \int f \, d\mu_\theta$$

is a well-defined element of $\Sigma(H^\infty(\Pi))$ which is carried onto $\theta$ by $i^*$. (Here, $\mu_\theta$ is the representing measure for $\theta$.) Since $\theta \in \Pi$, it is easily checked that $|\hat{E}(\tilde{\theta})| < 1$, i.e., that $\tilde{\theta} \in D$. It remains only to prove that $i^*$ is one to one. We shall accomplish this by showing that if $\{f_n\}$ is the sequence in $A$ which we used to define $E$, then the functions $(if_n)\hat{}\ n = 1, 2, \ldots,$ together separate points on $D$. We have that $\{(if_n)\hat{} \circ \chi\}$ is a sequence of holomorphic functions on the unit disc $\Delta$ which is uniformly bounded by 1 in modulus and which converges pointwise on $\Delta$ to $\hat{E} \circ \chi$, which is the identity map. This implies that for all $\zeta, \zeta' \in \Delta$, $\zeta \neq \zeta'$, $(if_n)\hat{}(\chi(\zeta)) \neq (if_n)\hat{}(\chi(\zeta'))$ for some choice of $n$, and thus that the functions $(if_n)\hat{}$ separate points on $\Pi$. This concludes the proof of Theorem 17.1.

A few words may be in order concerning the techniques used in this proof. They are not the ones originally used in this connection. The theorem was proved originally by means of certain Hilbert-space methods. We refer to the paper of Hoffman [2] for this proof. The proof we have given is based on ideas found in the paper of Gamelin [3], where a more general situation is considered. The idea of using the solution to a certain extremal problem to find analytic structure in $\Sigma(A)$ is due to Bishop [8] and has been used by several authors. Of course, it is reminiscent of one of the proofs of the Riemann mapping theorem.

More general results on analytic structure in $\Sigma(A)$ are due to Wermer [11], O'Neill [1], O'Neill and Wermer [1], and Gamelin [3]. Although we do not propose to develop in detail the results of these authors, we shall briefly summarize, without proofs, some of it. Wermer and O'Neill were concerned with hypo-Dirichlet algebras in the following sense.

17.11 ***Definition***

*If $A$ is a uniform algebra with Shilov boundary $\Gamma$, then $A$ is **hypo-Dirichlet** if there exist finitely many invertible elements $g_1, \ldots, g_N$ in $A$ such that the linear span of* Re $A$ *and* $\log|g_1|, \ldots, \log|g_N|$ *is dense in* $\mathscr{C}_{\mathbb{R}}(\Gamma)$.

The prototypic hypo-Dirichlet algebra is the algebra $\mathscr{A}(R)$ of all functions continuous on $\bar{R}$ and holomorphic on $R$, $R$ a domain in the plane bounded by finitely many mutually disjoint simple closed curves.

A somewhat more general class of examples is obtained by replacing the planar region $R$ by a proper open subset of $R$ of a compact Riemann surface, $\partial R$ again a finite collection of mutually disjoint simple closed curves. The theory of hypo-Dirichlet algebras is due largely to Ahern and Sarason [1, 2].

In the papers Wermer [11], O'Neill [1], and O'Neill and Wermer [1], the parts for hypo-Dirichlet algebras were investigated. To state the culmination of this work, let us introduce certain notations similar to those we have already used. The definition of hypo-Dirichlet algebras implies immediately that if $\varphi \in \Sigma(A)$, there exists a single Arens–Singer measure for $\varphi$. If $\mu_\varphi$ is this Arens–Singer measure, let $H^2(\mu_\varphi)$ be the closure in $L^2(\mu_\varphi)$ of $A$, and let $H^\infty(\mu_\varphi) = H^2(\mu_\varphi) \cap L^\infty(\mu_\varphi)$. If $\psi \in \Sigma(A)$ lies in the same part as $\varphi$, it can be shown that $H^\infty(\mu_\varphi) = H^\infty(\mu_\psi)$, and thus that the functional $\tilde{\psi}$ defined on $H^\infty(\mu_\varphi)$ by

$$\tilde{\psi}(f) = \int f \, d\mu_\psi$$

is a well-defined element of $\Sigma(H^\infty(\mu_\varphi))$. The work of Wermer and O'Neill concludes with the following result.

17.12 ***Theorem***

*Let $A$ be a hypo-Dirichlet algebra, let $\varphi \in \Sigma(A)$, and let $\Pi$ be the Gleason part for $A$ which contains $\varphi$. If $\Pi$ is nontrivial, then the set $\tilde{\Pi} = \{\tilde{\psi} : \psi \in \Sigma(A)\}$, with the relative topology inherited from $\Sigma(H^\infty(\mu_\varphi))$, can be given the structure of a one-dimensional analytic space on which the functions $\hat{f}$, $f \in H^\infty(\mu_\varphi)$, are holomorphic. Moreover, there is an $E \in H^\infty(\mu_\varphi)$ such that $\hat{E}$ maps $\Pi$ onto the unit disc in an $n$-to-one fashion for some positive integer $n$.*

The hypothesis that $A$ be a hypo-Dirichlet algebra is a global hypothesis in that it imposes conditions on $A$ over the whole of its Shilov boundary. It would be desirable to obtain results on analytic structure under *local hypotheses*, i.e., under hypotheses bearing on a single element of $\Sigma(A)$. We have one example of such a result: If $\varphi \in \Sigma(A)$ admits a unique representing measure and if the part $\Pi$ containing $\varphi$ is nontrivial, then $\Pi$ is an analytic disc. Taking this fact together with the observation that if $A$ is hypo-Dirichlet, then each $\varphi \in \Sigma(A)$ admits at most a finite-dimensional family of representing measures, it becomes natural to attempt to find analytic structure at $\varphi \in \Sigma(A)$ as soon as $\varphi$ admits only a finite-dimensional set of representing measures. This is an essentially more inclusive hypothesis than the hypothesis that the algebra be hypo-Dirichlet, as the following example shows.

17.13 ***Example***

Let $A$ consist of those functions $f$ in the disc algebra with the property that $f'(0) = 0$. It is easily verified that $\Sigma(A)$ is $\bar{\Delta}$, the closed unit disc, and that if $\varphi \in \Sigma(A)$ is given by $\varphi(f) = f(0)$, then $\varphi$ admits only a finite-dimensional set of representing measures on $T$, the unit circle which is the Shilov boundary for $A$: If $\mu$ is such a representing measure and if

$$\hat{\mu}(n) = \int_T e^{-inx}\, d\mu(x),$$

then we have $\hat{\mu}(n) = (\hat{\mu}(-n))^-$ since $\mu$ is real, and for $n = 2, 3, \ldots$, $\hat{\mu}(-n) = \varphi(z^n) = 0$ since $z^n \in A$ and $z^n$ vanishes at 0. Thus, $\hat{\mu}(n) = 0$ if $|n| \geq 2$. Also, $\hat{\mu}(0) = \varphi(1) = 1$. It follows that if $\lambda$ is the normalized Haar measure on $T$, then

$$\begin{aligned} d\mu &= (1 + \varepsilon e^{it} + \varepsilon e^{-it})\, d\lambda \\ &= (1 + 2\varepsilon \cos t)\, d\lambda \end{aligned}$$

for some choice of real $\varepsilon$ so small that $1 + 2\varepsilon \cos t$ is positive for all $t$, i.e., some $\varepsilon$ in $[-\frac{1}{2}, \frac{1}{2}]$. Thus, the set of representing measures for $\varphi$ is one dimensional.

The algebra $A$ is not hypo-Dirichlet, however, for it is clearly not a Dirichlet algebra, and we have the following fact.

17.14 ***Lemma***

*If $A$ is a hypo-Dirichlet algebra with $H^1(\Sigma(A), \mathbb{Z}) = 0$, then $A$ is in fact Dirichlet.*

PROOF

By the hypothesis that $A$ is hypo-Dirichlet there exist $g_1, \ldots, g_N$, invertible elements in $A$, such that the linear span of $\operatorname{Re} A$ and $\log|g_1|, \ldots, \log|g_N|$ is dense in $\mathscr{C}_{\mathbb{R}}(\Gamma)$. Since $H^1(\Sigma(A), \mathbb{Z}) = 0$, the Arens–Royden theorem (Theorem 10.1) implies that $g_j = e^{h_j}$, $h_j \in A$, so $\log|g_j| = \operatorname{Re} h_j$. Thus, $\operatorname{Re} A$ is itself dense in $\mathscr{C}_{\mathbb{R}}(\Gamma)$, and the lemma is established.

Since the spectrum of the algebra $A$ in our example is the disc $\bar{\Delta}$, we have $H^1(\Sigma(A), \mathbb{Z}) = 0$, and so since $A$ is not Dirichlet, it is not even hypo-Dirichlet.

The search for analytic structure in $\Sigma(A)$ at $\varphi$ under the hypothesis that $\varphi$ admit a finite-dimensional set of representing measures, was begun implicitly by O'Neill [1] and continued by Gamelin [3]. These authors establish the fact that if $\varphi \in \Sigma(A)$ admits a finite-dimensional set of representing measures, then there is a one-dimensional analytic variety in $\Sigma(A)$ which passes through $\varphi$, and Gamelin investigates in considerable detail the structure of this variety. For the particulars of these investigations, we refer to the original papers and to Gamelin [1].

# 18 A CHARACTERIZATION OF PARTS

Parts were introduced in the hope that they would provide analytic structure in the spectra of commutative Banach algebras. Impressive results have been obtained in this direction as we saw in the last section, but work of Garnett [1, 2,] has shown that in general parts can be quite pathological. It is Garnett's work that we consider in this section.

Before passing to general results, we need to consider in some detail a particular algebra, $\mathfrak{A}_\alpha$, which will be used in our constructions. Denote by $T^2$ the torus $\{(z, w) \in \mathbb{C}^2 : |z|, |w| = 1\}$. For a fixed positive, irrational number $\alpha$, let $S_\alpha = \{(m, n) : m + \alpha n \geq 0\}$ so that $S_\alpha$ is an additive semigroup in $\mathbb{Z}^2 = \mathbb{Z} \oplus \mathbb{Z}$. Since $S_\alpha$ is an additive semigroup, the set of functions $z^m w^n$, $(m, n) \in S_\alpha$, is a multiplicative semigroup in $\mathscr{C}(T^2)$, and consequently its closed linear span is an algebra which we shall denote by $\mathfrak{A}_\alpha$. The following theorem summarizes the results we need about $\mathfrak{A}_\alpha$.

**18.1** ***Theorem***

(a) *$\Sigma(\mathfrak{A}_\alpha)$ can be identified with the space obtained from $T^2 \times [0, 1]$ by collapsing the set $T^2 \times \{0\}$ to a single point.*

(b) *The Shilov boundary for $\mathfrak{A}_\alpha$ is $T^2$.*

(c) *The point in $\Sigma(\mathfrak{A}_\alpha)$ corresponding to $T^2 \times \{0\}$ constitutes a Gleason part for $\mathfrak{A}_\alpha$.*

(d) *If $K$ is a closed, proper subset of $T^2$, then $\mathfrak{A}_\alpha|K$ is dense in $\mathscr{C}(K)$.*

PROOF

Part (b) is clear.

To prove part (a), we suppose that $\alpha \in (0, 1)$; the case that $\alpha \in (1, \infty)$ is a consequence of this case, for $m + n\alpha > 0$ is equivalent, for $\alpha > 0$, to $n + m\alpha^{-1} > 0$. Denote by $\mathscr{B}$ the algebra of finite linear combinations of the monomials $z^m w^n$, $(m, n) \in S_\alpha$.

Define a map $\chi : T^2 \times [0, 1] \to \Sigma(\mathfrak{A}_\alpha)$ by requiring that if $(e^{i\theta_1}, e^{i\theta_2}) \in T^2$ and $R \in [0, 1]$, then the value of $\chi$ at $((e^{i\theta_1}, e^{i\theta_2}), R)$ is the element $\psi$ of $\Sigma(\mathfrak{A}_\alpha)$ determined by requiring that for $m, n$ with $m + n\alpha > 0$,

$$(1) \qquad \psi(z^m w^n) = R^{m + n\alpha} e^{i(m\theta_1 + n\alpha\theta_2)}.$$

We must verify that this condition does determine an element of

$\Sigma(\mathfrak{A}_\alpha)$. It is clear that the $\psi$ defined by (1) is well defined and multiplicative on $\mathscr{B}$. It has to be shown that if $P \in \mathscr{B}$, then $|\psi(P)| \le \|P\|_{T^2}$; the density of $\mathscr{B}$ in $\mathfrak{A}_\alpha$ will then imply that $\psi$ extends from $\mathscr{B}$ to a well-defined element of $\Sigma(\mathfrak{A}_\alpha)$ so that $\chi$ is a well-defined map. If

$$P(z, w) = \sum_{m+\alpha n \ge 0} A(m, n) z^m w^n$$

is a finite sum, then the function $f$ defined on $\mathbb{C}$ by

$$\begin{aligned} f(\zeta) &= P(e^{i(\zeta+\theta_1)}, e^{i(\zeta+\theta_2)}) \\ &= \Sigma A(m, n) \exp i[m(\zeta + \theta_1) + n\alpha(\zeta + \theta_2)] \end{aligned}$$

is an entire function. If $\zeta = \xi + i\eta$, we have

$$f(\zeta) = \Sigma A(m, n) e^{-\eta(m+\alpha n)} \exp i[m(\xi + \theta_1) + n\alpha(\xi + \theta_2)],$$

which is evidently bounded on the half plane $\eta \ge 0$; so by the Phragmen–Lindelöf theorem, we have

$$\sup\{|f(\zeta)| : \operatorname{Im} \zeta \ge 0\} = \sup\{|f(\zeta)| : \operatorname{Im} \zeta = 0\}.$$

In particular, therefore, we have, since $\log R < 0$,

$$|f(-i \log R)| \le \sup\{|f(\zeta)| : \operatorname{Im} \zeta = 0\}.$$

We have that $f(-i \log R) = \psi(P)$, and $\sup\{|f(\zeta)| : \operatorname{Im} \zeta = 0\} \le \|P\|_{T^2}$. (Equality actually holds, but inequality, which is obvious, suffices for our present purposes.) Thus, $|\psi(P)| \le \|P\|_{T^2}$, and it follows that $\chi$ is a well-defined map into $\Sigma(\mathfrak{A}_\alpha)$. It is easily checked that $\chi$ is one to one on $T^2 \times (0, 1]$ and that $\chi$ is constant on $T^2 \times \{0\}$. Also, $\chi$ is continuous.

We may conclude the proof of (a) by showing that $\chi$ is onto. Suppose, therefore, that $\varphi \in \Sigma(\mathfrak{A}_\alpha)$. If $\varphi(z) = 0$, then $\varphi(w^m z^n) = 0$ for all $m$ and $n$ with $m + n\alpha > 0$; for given such a pair of integers, there is an integer $k$ with $k(m + n\alpha) > 1$, and from

$$\varphi(w^m z^n)^k = \varphi(w^{km} z^{kn-1})\varphi(z) = 0$$

we get $\varphi(w^m z^n) = 0$. Similarly, if $\varphi(w) = 0$, then $\varphi(w^m z^n) = 0$ when $m + n\alpha > 0$. Thus, if $\varphi(z)\varphi(w) = 0$, $\varphi$ is the image under $\chi$ of the set $T^2 \times \{0\}$. Suppose, therefore, that $\varphi(z)\varphi(w) \ne 0$ so that $\varphi(z) = Re^{i\theta_1}$ for some $R \in [0, 1]$. We shall prove that $|\varphi(w)| = R^\alpha$. Consider first the case that $R < 1$. Since $\|\varphi\| = 1$ and $\|w\| = 1$, we have $|\varphi(w)| = R^\beta$ for some $\beta \ge 0$. We have to show that $\beta = \alpha$. By assumption $\alpha < 1$, so the function $zw^{-1}$ lies in $\mathfrak{A}_\alpha$. From $z = zw^{-1}w$, we find

$$|\varphi(zw^{-1})| = R^{\alpha-\beta}. \tag{2}$$

We have assumed that $R \in (0, 1)$, so (2) together with $\|zw^{-1}\| = 1 = \|\varphi\|$ implies that $\beta \le \alpha$. If $\beta < \alpha$, then there exists a pair of integers

$(m, n)$ such that $m + \alpha n > 0$, $m + \beta n < 0$. But then $z^m w^n \in \mathfrak{A}_\alpha$ and

$$|\varphi(z^m w^n)| = R^{m+\beta n}$$
$$> 1,$$

which is impossible. Thus, $\beta = \alpha$. If $R = 1$, i.e., $|\varphi(z)| = 1$, then $|\varphi(w)| = 1$, for otherwise we have the contradiction that

$$1 = |\varphi(z)| = |\varphi(zw^{-1})\varphi(w)|$$
$$< 1.$$

Consequently, if $\varphi(z)\varphi(w) \neq 0$, then $\varphi(z^m w^n) = R^{m+n\alpha}e^{i(m\theta_1 + n\alpha\theta_2)}$ for some $R \in (0, 1]$ and some $\theta_1$ and $\theta_2$. It follows that $\varphi$ lies in the range of $\chi$, and we have established part (a).

To prove (c), let $\varphi \in \Sigma(\mathfrak{A}_\alpha)$ be determined by

$$\varphi(z^m w^n) = R^{m+n\alpha}e^{i(m\theta_1 + n\alpha\theta_2)}, \qquad R \neq 0,$$

and denote by $\varphi_0$ the complex homomorphism determined by $\varphi(P) = A(0, 0)$ if $P(z, w) = \Sigma A(m, n)z^m w^n$ is a finite sum in $\mathfrak{A}_\alpha$. Thus, $\varphi_0$ is the point in $\Sigma(\mathfrak{A}_\alpha)$ corresponding to the set $T^2 \times \{0\}$ under $\chi$. If $I$ is the kernel of $\varphi_0$, $I$ contains $z^m w^n$ as long as $(m, n) \in S_\alpha$, $(m, n) \neq (0, 0)$, and it follows that

$$\sup\{|\varphi(f)| : f \in I, \|f\| \leq 1\} \geq R^{m+n\alpha} \tag{3}$$

for all $(m, n) \in S_\alpha$. By a theorem of diophantine analysis (Appendix E), we can choose $(m, n) \in S_\alpha$ with $m + n\alpha$ arbitrarily near zero. It follows that the supremum (3) is 1, so $\varphi$ and $\varphi_0$ lie in different Gleason parts for $\mathfrak{A}_\alpha$. Consequently, $\varphi_0$ constitutes a part.

It remains to establish part (d). Part (d) is a consequence of the following lemma, the proof of which depends on certain elementary facts from harmonic analysis.

**18.2** ***Lemma***

*The algebra $\mathfrak{A}_\alpha$ is a maximal subalgebra of $\mathscr{C}(T^2)$: If $B$ is a closed subalgebra of $\mathscr{C}(T^2)$ and if $\mathfrak{A}_\alpha \subseteq B \subseteq \mathscr{C}(T^2)$, then either $B = \mathfrak{A}_\alpha$ or else $B = \mathscr{C}(T^2)$.*

PROOF

Let $\mathfrak{A}_\alpha \subseteq B \subseteq \mathscr{C}(T^2)$, $B$ a closed subalgebra. Consider again the complex homomorphism $\varphi_0$ used in the proof of part (c). If $\mu$ is the normalized Haar measure on $T^2$, then $\mu$ is a representing measure for $\varphi_0$. Two cases are possible for $\varphi_0$. It may be that $\varphi_0$ does not extend to an element of $\Sigma(B)$. In this case $z$ and $w$ are both invertible in $B$, for as we noted above, if $\psi \in \Sigma(\mathfrak{A}_\alpha)$, then $\psi(z) = 0$ if and only if $\psi(w) = 0$, and this happens only if $\psi = \varphi_0$. Thus, $z^{-1}$ and $w^{-1}$ both lie in $B$, and since on $T^2$, $z^{-1} = \bar{z}$ and $w^{-1} = \bar{w}$, $B$ is closed under complex conjugation. By the Stone–Weierstrass theorem $B = \mathscr{C}(T^2)$.

On the other hand, it may be that $\varphi_0$ does extend to $\psi \in \Sigma(B)$. Let $\sigma$ be a representing measure for $\psi$. Since $\psi$ is an extension of $\varphi_0$, and since $\varphi(z^m w^n) = 0$ if $m + n\alpha > 0$, we find that

(4) $$\int z^m w^n \, d\sigma = 0 \qquad m + n\alpha > 0.$$

By conjugation, we conclude that the integral (4) vanishes for all $m$ and $n$ subject only to $m + n\alpha \neq 0$, i.e., for all $(m, n) \neq (0, 0)$. Thus, we have that for all $m$ and $n$,

$$\int z^m w^n \, d\sigma = \int z^m w^n \, d\mu.$$

As the closed linear span in $\mathscr{C}(T^2)$ of the monomials $z^m w^n$ is dense in $\mathscr{C}(T^2)$, the uniqueness part of the Riesz representation theorem implies that $\sigma = \mu$.

If $f \in B$, then $z^m w^n f \in \ker \psi$ if $m + n\alpha > 0$, since $\psi$ is an extension of $\varphi_0$. Thus,

$$\int z^m w^n f \, d\mu = 0 \qquad m + n\alpha > 0.$$

Consequently, the Fourier series for $f$ is of the form

$$f \sim \sum_{m + n\alpha \geq 0} A(m, n) e^{i(m\theta_1 + n\theta_2)},$$

and from this it follows that $f$ can be approximated uniformly on $T^2$ by finite linear combinations of the monomials $z^m w^n$ for $m + n\alpha \geq 0$. We may conclude that $f \in \mathfrak{A}_\alpha$, so $B = \mathfrak{A}_\alpha$. This established the lemma.

We conclude the proof of the theorem by showing how the lemma implies part (d). Let $K$ be a proper closed subset of $T^2$. For convenience, suppose that $(1, 1) \notin K$. If $F \in \mathscr{C}(T^2)$ has range in $[0, 1]$, assumes the value 1 at $(1, 1)$, and vanishes off $\{(e^{i\theta_1}, e^{i\theta_2}) : |\theta_1|, |\theta_2| < \delta\}$ for small $\delta > 0$, then

$$\int_{-\pi}^{\pi} \int_{-\pi}^{\pi} e^{i\theta_1} F(e^{i\theta_1}, e^{i\theta_2}) \, d\theta_1 \, d\theta_2 \neq 0,$$

so $F \notin \mathfrak{A}$. Consequently, $B$, the closed subalgebra of $\mathscr{C}(T^2)$ generated by $\mathfrak{A}_\alpha$ and $F$, is the whole of $\mathscr{C}(T^2)$. We have $F|K = 0$, so $B|K = \mathfrak{A}_\alpha|K$, whence $\mathfrak{A}_\alpha|K$ is dense in $\mathscr{C}(K)$ as asserted.

18.3 ***Corollary***

*If $K$ is a proper closed subset of $T^2$ and if $x \in \Sigma(\mathfrak{A}_\alpha) \setminus K$, there is $f \in \mathfrak{A}_\alpha$ such that $\|f\|_K < |f(x)|$.*

PROOF

If not, $f|K \mapsto f(x)$ is a well-defined, nonzero complex homomorphism of $\mathfrak{A}_\alpha|K$ which is of norm 1. Thus, it gives rise to an element of $\Sigma(\mathscr{C}(K))$, and so there is $y \in K$ such that $f(y) = f(x)$ for all $f \in \mathfrak{A}_\alpha$. The algebra $\mathfrak{A}_\alpha$ separates points on $\Sigma(\mathfrak{A}_\alpha)$, so we have obtained a contradiction.

With our study of $\mathfrak{A}_\alpha$ concluded, we can turn to Garnett's theory. Recall that if $E \subset \Sigma(A)$, $A$ a commutative Banach algebra, then $E$ is a hull if

$$E = \{\varphi \in \Sigma(A) : \hat{a}(\varphi) = 0 \text{ for all } a \in A \text{ with } \hat{a}|E = 0\}.$$

Our next theorem contains the basic construction.

18.4 ***Theorem***

*Let $A$ be a uniform algebra, let $S$ be a hull in $\Sigma(A)$, and let $\Pi$ be a part for $A$ with $S \cap \Pi \neq \varnothing$. There exists a uniform algebra $B$ with a part $\Pi_1$ homeomorphic to $S \cap \Pi$, and with the property that $B|\Pi_1$ is isometrically isomorphic to an algebra lying between $A|(S \cap \Pi)$ and its uniform closure in $\mathscr{C}^B(S \cap \Pi)$.*

PROOF

Let $\alpha$ be a positive, irrational number, and let $\mathfrak{A}_\alpha$ be the algebra considered above. Then $\Sigma(\mathfrak{A}_\alpha \otimes A) = \Sigma(\mathfrak{A}_\alpha) \times \Sigma(A)$.

Let $J = \{\mathfrak{z} \in T^2 : \mathfrak{z} = (z_1, z_2), \operatorname{Re} z_1 \leq 0\}$, and set $X = (J \times \Sigma(A)) \cup (\Sigma(\mathfrak{A}_\alpha) \times S)$ so that $X$ is a certain closed set in $\Sigma(\mathfrak{A}_\alpha) \times \Sigma(A)$. Let $B$ be the closure in $\mathscr{C}(X)$ of the restriction algebra $\{f|X : f \in \mathfrak{A}_\alpha \otimes A\}$. The spectrum of $B$ can be identified with the $\mathfrak{A}_\alpha \otimes A$-convex hull of $X$, i.e., with the set

$$\hat{X} = \{x \in \Sigma(\mathfrak{A}_\alpha) \times \Sigma(A) : |f(x)| \leq \|f\|_X \text{ for all } f \in \mathfrak{A}_\alpha \otimes A\}.$$

We shall prove that $\hat{X} = X$ so that $X = \Sigma(B)$. To see this, consider a point $x^0 = (x_1^0, x_2^0) \in (\Sigma(\mathfrak{A}_\alpha) \times \Sigma(A)) \setminus X$. Thus, $x_1^0 \notin J$ and $x_2^0 \notin S$. Since $S$ is a hull, there is $g \in A$ with $g|S = 0$, $g(x_2^0) = 1$, and since $x_1^0 \notin J$, there is, as noted in our discussion of the algebra $\mathfrak{A}_\alpha$, an $f \in \mathfrak{A}_\alpha$ with $f(x_1^0) = 1$, $\|f\|_J < 1$. Replacing $f$ by a high power of $f$ if necessary, we may suppose that $\|f\|_J < \|g\|^{-1}$, $\|\cdot\|$ the supremum over $\Sigma(A)$. If $h$ is given by $h(x_1, x_2) = f(x_1)g(x_2)$, then $h \in \mathfrak{A}_\alpha \otimes A$, $h(x^0) = 1$, and $\|h\|_X < 1$. Thus, $x^0 \notin \hat{X}$, so $X = \hat{X}$ and $\Sigma(B) = X$.

We shall show that the set $Q = \{0\} \times (\Pi \cap S)$ is a part for $B$. If $s \in \Pi \cap S$, let $p_s = (0, s) \in Q$. If the point $x = (x_1, x_2)$ lies in $X \setminus Q$, and if $x_1 \neq 0$, then since $\{0\}$ is a part for $\mathfrak{A}_\alpha$, the functions in the subalgebra $\mathfrak{A}_\alpha$ of $\mathfrak{A}_\alpha \otimes A$ suffice to show that $p_s$ and $x$ are not equivalent in the sense of Gleason. Similarly, if $x_2 \notin \Pi$, then $p_s$ and $x$ are inequivalent. Finally, if $x_2 \notin S$, then the choice of $X$ shows that $x_1 \neq 0$, so $p_s$ and $x$ are inequivalent. Thus, $Q$ is a union of parts for $B$.

We now show that $Q$ is, in fact, a part. If $s \in \Pi \cap S$, $g \in B$, then

$$g(p_s) = g(0, s) = \int_{T^2} g(x, s)\, d\mu(x),$$

where $\mu$ is the normalized Haar measure on $T^2$; for $\mu$ is a representing measure for 0 for the algebra $\mathfrak{A}_\alpha$. If $s, t \in \Pi \cap S$, and if $g \in B$ satisfies $\|g\| \leq 1$, then

$$|g(p_s) - g(p_t)| \leq \int_{T^2} |g(x, s) - g(x, t)|\, d\mu(x)$$

$$\leq 2\mu(T^2 \setminus J) + \int_J |g(x, s) - g(x, t)| d\mu(x).$$

If $x \in J$, then $\{x\} \times \Sigma(A) \subset X$, and thus the map $y \mapsto g(x, y)$ is in $A$ and is of norm no more than 1. Since $s$ and $t$ lie in the part $\Pi$, it follows that for some $c \in (0, 2)$ which is independent of $x$, $|g(x, s) - g(x, t)| < c$. Thus, $\int_J |g(x, s) - g(x, t)|\, d\mu(x) \leq c\mu(J)$. Since $\mu(T^2) = 1$, we are led to $|g(p_s) - g(p_t)| \leq 2\mu(T^2 \setminus J) + c\mu(J) < 2$; i.e., $p_s$ and $p_t$ lie in the same part. Thus, $Q$ is a part.

It is clear that $Q$ is homeomorphic to $\Pi \cap S$. Finally, $B|Q = \{f(0, \cdot): f \in B\}$ and this lies between $A|(\Pi \cap S)$ and its uniform closure in $\mathscr{C}^B(\Pi \cap S)$. Therefore, for the part $\Pi_1$ of the theorem, we can take $Q$ and our theorem is proved.

Now suppose that we are given a uniform algebra $A$ and suppose $\Pi$ to be a part for $A$. What can we say about the topological structure of $\Pi$? [We are concerned with the topology induced on $\Pi$ from $\Sigma(A)$, which has the weak * topology.] Two observations are immediate: $\Pi$ is completely regular, and $\Pi$ is $\sigma$-compact. The complete regularity of $\Pi$ is clear since $\Pi$ is a subset of the compact space $\Sigma(A)$, and the $\sigma$-compactness follows from the observation that if $y \in \Pi$, then $\Pi = \bigcup\{P_n : n = 1, 2, \ldots\}$, where for all $n$, $P_n$ is the compact set

$$\{x \in \Sigma(A) : \|y - x\| \leq 2 - 1/n\}.$$

It happens that these are the only two topological restrictions on parts.

18.5 ***Theorem***

*Let $X$ be a $\sigma$-compact, completely regular space. There exists a uniform algebra $B$ and a part $\Pi$ for $B$ homeomorphic to $X$ such that $B|\Pi$ is* *isometrically isomorphic to $\mathscr{C}^B(X)$. If $X$ is compact and if $C$ is a uniform algebra with $\Sigma(C) = X$, there exists a uniform algebra $B$ with a part homeomorphic to $X$ such that $B|\Pi$ is isomorphic to $C$.*

PROOF

If $I$ is an index set, let $Y_I = \Pi\{\Delta_i : i \in I\}$, where each $\bar{\Delta}_i$ is a copy of the closed unit disc, and let $A_I$ be the tensor product of $I$ copies of the disc algebra $\mathscr{A}$ so that $\Sigma(A_I) = Y_I$. Denote by $\theta$ the point in $Y_I$ all of whose coordinates are zero, and let $\Pi_\theta$ be the part containing $\theta$. If $\mathfrak{z} \in Y_I$, then $\mathfrak{z} \in \Pi_\theta$ if and only if for some $a < 1$, $|z_i| < a$ for each coordinate $z_i$ of $\mathfrak{z}$. To see this, suppose first that such an $a$ exists, and define $\rho : \Delta \to Y_I$, $\Delta$ the open unit disc in $\mathbb{C}$, by requiring the $i$th coordinate of $\rho(\zeta)$ to be $\zeta z_i a^{-1}$. Then $\rho(0) = \theta$, and if $f \in A_I$, then $f \circ \rho \in \mathscr{A}(\Delta)$ and $\|f \circ \rho\| \leq \|f\|$. Since 0 and $a$ lie in the same Gleason part for $\mathscr{A}(\Delta)$, it follows that $\theta$ and $\mathfrak{z}$ are equivalent. Conversely, if no such $a$ exists, then for some sequence $\{g_n\}$ of coordinate functions, i.e., $g_n(\mathfrak{z}) = z_{i(n)}$ for some $i(n)$, we have $\|g_n\| = 1$, $g_n(\theta) = 0$, $|g_n(\mathfrak{z})| \to 1$, so $\theta$ and $\mathfrak{z}$ are inequivalent.

Now, let us suppose that the space $X$ is not compact. Let $\beta X$ be its Čech compactification. Take for the index set $I$ the set $\beta X \setminus X$, and let $A = \{f \in \mathscr{C}(\beta X \times Y_I) : f|(\{x\} \times Y_I) \in A_I$ for all $x \in \beta X$ and $f|(\beta X \times \{\theta\})$ is constant$\}$. Then $\Sigma(A)$ is the space obtained from $\beta X \times Y_I$ by collapsing the set $\beta X \times \{\theta\}$ to a point, and $\Pi = \{(x, \mathfrak{z}) \in \Sigma(A) : \mathfrak{z} \in \Pi_\theta\}$ is a part for $A$.

Write $X = \bigcup_n X_n$ where $X_1 \subseteq X_2 \subseteq \cdots$ and each $X_n$ is compact. Given $x \in \beta X \setminus X$, there exists a function $h_x \in \mathbb{C}(\beta X)$ with range contained in $[\frac{1}{2}, 1]$, with $h_x(x) = 1$ and $\|h_x\|_{X_n} \leq 1 - 2^{-n}$. Define a map $\rho : \beta X \to \Sigma(A)$ by $\rho(y) = (y, H(y))$, where the $x$th coordinate of $H(y)$ is $h_x(y)$ for all $x \in \beta X \setminus X$. The map $\rho$ is a homeomorphism of $\beta X$ into $\Sigma(A)$; set $S - \rho(\beta X)$. We have that $\Pi \cap S = \rho(X)$ by the observation in the first paragraph of the proof. Moreover, $S$ is a hull in $\Sigma(A)$; for if for each $x \in \beta X \setminus X$, we set

$$g_x(y, \mathfrak{z}) = (h_x(y) - z_x)(3h_x(y) - z_x)^{-1},$$

then $S = \bigcap\{g_x^{-1}(0) : x \in \beta X \setminus X\}$, and each $g_x$ is in $A$. (Here $z_x$ denotes the $x$th coordinate of $\mathfrak{z}$.) Next, we observe that $A|(S \cap \Pi)$ is isomorphic to $\mathscr{C}^B(X)$: If $f \in \mathscr{C}^B(X)$, it has a unique extension, $F$, to an element of $\mathscr{C}(\beta X)$. If $x \in \beta X \setminus X$, the function $g$ given by

$$g(y, \mathfrak{z}) = z_x F(y)/h_x(y)$$

lies in $A$, and we have $f = (g \circ \rho)|X$. The first half of our theorem now follows.

The proof of the second half of the theorem follows essentially the same line as the first half, but it is simpler. Given $C$ with $\Sigma(C) = X$, define $A$ to be $\{f \in C \otimes \mathscr{A} : f(x, 0) \text{ is constant}\}$. Then $\Sigma(A)$ is the space obtained from $X \times \bar{\Delta}$ by collapsing the set $X \times \{0\}$ to a point, and the set $\Pi = \{(x, z) : x \in X, |z| < 1\} \subset \Sigma(A)$ is a part for $A$. Let $S = \{(x, \frac{1}{2}) : x \in X\}$. The set is homeomorphic to $X$, and it is a hull since it is the zero set of the function $f \in A$ given by $f(x, z) = 1 - 2z$. Let $\rho : X \to S$ be given by $\rho(x) = (x, \frac{1}{2})$. If $F \in C$, define $f \in A$ by $f(x, z) = 2zF(x)$. Then $f|S = F \circ \rho^{-1}$, and it follows that $A|S$ is isomorphic to $C$. The theorem now follows from the previous one.

## 19 *ALGEBRAS WITH ONE-POINT PARTS*

This section is a continuation of our study of the pathology of Gleason parts. We shall construct uniform algebras $(A, X)$, $A \neq \mathscr{C}(X)$, with the property that every Gleason part for $A$ consists of a single point. We shall even obtain algebras with the property that although $A \neq \mathscr{C}(X)$, every point of $\Sigma(A)$ is a peak point for $\hat{A}$. These refute certain conjectures that had been made. It had been conjectured that if every point of $\Sigma(A)$ is a peak point for $A$, then $\hat{A} = \mathscr{C}(\Sigma(A))$, and perhaps that the same conclusion could be drawn provided only that $A$ has trivial parts. Using a remarkably simple idea, Cole [1] was able to show that both of these conjectures are false. It is to this work of Cole that we now turn.

Consider a fixed uniform algebra $(A_0, X_0)$, $X_0 = \Sigma(A_0)$. We begin by constructing a new uniform algebra $(A'_0, X'_0)$ such that $A'_0$ contains $A_0$ and each $f \in A_0$ has square roots in $A'_0$. On the product space $\mathbb{C}^{A_0}$ we have the usual topology, and for each $f \in A_0$, we have the projection $p_f : \mathbb{C}^{A_0} \to \mathbb{C}$, which takes $\mathfrak{z} \in \mathbb{C}^{A_0}$ onto $z_f$, its $f$th coordinate. We set

$$X'_0 = \{(x, \mathfrak{z}) \in X_0 \times \mathbb{C}^{A_0} : f(x) = p_f^2(\mathfrak{z})\}.$$

Since it is a closed subspace of the space

$$X_0 \times \{\mathfrak{z} \in \mathbb{C}^{A_0} : |p_f(\mathfrak{z})| \leq (\|f\|_{X_0})^{1/2}\},$$

which is compact by the Tychonoff theorem, $X'_0$ is compact. We define $\pi : X'_0 \to X_0$ by $\pi(x, \mathfrak{z}) = x$. This map takes $X'_0$ onto $X_0$, so the map $f \mapsto f \circ \pi$, call it $\pi^*$, is an isometric isomorphism of $\mathscr{C}(X_0)$ into $\mathscr{C}(X'_0)$. We define $A'_0$ to be the closed subalgebra of $\mathscr{C}(X'_0)$ generated by the projections $p_f$ and the functions $f \circ \pi$, $f \in A'_0$, and thus $A'_0$ contains $\pi^* A_0$, an isometric copy of $A_0$. If we denote by $\mathbb{Z}_2$ the multiplicative group consisting of 1 and $-1$, then the product group $\mathbb{Z}_2^{A_0}$ acts on $X'_0$ as follows: If $\gamma \in \mathbb{Z}_2^{A_0}$ and $(x, \mathfrak{z}) \in X'_0$, we let $\gamma(x, \mathfrak{z})$ be the point $(x, \mathfrak{z}')$ where the $f$th coordinate $p_f(\mathfrak{z}')$ of $\mathfrak{z}'$ is $\gamma_f \mathfrak{z}_f$. (For each $f \in A_0$, $\gamma_f = 1$ or $-1$.) The map

$$\mathbb{Z}_2^{A_0} \times X_0' \to X_0'$$

given by $(\gamma, (x, \mathfrak{z})) \mapsto \gamma(x, \mathfrak{z})$ is continuous, so $\mathbb{Z}_2^{A_0}$ acts as a topological transformation group on $X_0'$. The group $\mathbb{Z}_2^{A_0}$ is a compact abelian group, and we denote by $\mu$ its normalized Haar measure. Finally, let $\Gamma(A_0)$ and $\Gamma(A_0')$ be, respectively, the Shilov boundaries for $A_0$ and $A_0'$. Among these several entities at least the following relations hold.

**19.1** ***Theorem***

(a) *If for* $f \in \mathscr{C}(X_0')$ *we define* $Sf$ *by*

$$Sf(x) = \int_{\mathbb{Z}_2^{A_0}} f(\gamma(x))\, d\mu(\gamma),$$

*then* $S$ *is a projection of norm* 1 *of* $\mathscr{C}(X_0')$ *onto* $\pi^*\mathscr{C}(X_0)$ *which carries* $A_0'$ *onto* $\pi^* A_0$.

(b) $\Sigma(A_0') = X_0'$.

(c) $A_0'|\pi^{-1}(x)$ *is dense in* $\mathscr{C}(\pi^{-1}(x))$ *for all* $x \in X$.

PROOF

(a) We must show that the integral defining $Sf$ is a continuous function of $x$. In the case that $X$ is metric so that we need only consider sequences, the continuity of $Sf$ follows immediately from the dominated convergence theorem. As this theorem fails for nets, we must devise a different argument for the general case. Let $\varepsilon > 0$. We shall find a neighborhood $V$ of $x_0$, $x_0$ a given point of $X$, such that if $y \in V$, then $|Sf(x_0) - Sf(y)| < \varepsilon$. If $x \in X$, $\gamma \in \mathbb{Z}_2^{A_0}$, let $U(\gamma)$ be a neighborhood of $\gamma$ and $V(x)$ a neighborhood of $x$ such that if $\gamma' \in U(\gamma)$ and $x' \in V(x)$, then $|f(\gamma(x)) - f(\gamma'(x'))| < \varepsilon/2$. Thus, the range of $f$ on $\{\gamma'(x') : \gamma' \in U(\gamma), x' \in V(x)\}$ has diameter less than $\varepsilon$. By compactness, finitely many of the set $U(\gamma) \times V(x)$ cover $\mathbb{Z}_2^{A_0} \times X$, say $U_1 \times V_1, \ldots, U_q \times V_q$. We may assume that $x_0 \in V_j$ if $1 \le j \le p$, $x_0 \notin V_j$ if $p < j \le q$. Put $V = V_1 \cap \cdots \cap V_p$, and let $U_1' = U_1$, $U_2' = U_2 \setminus U_1, \ldots, U_p' = U_p \setminus (U_1 \cup \cdots \cup U_{p-1})$. If $y \in V$, we have

$$\int_{\mathbb{Z}_2^{A_0}} f(\gamma(y)) - f(\gamma(x))\, d\mu(\gamma) \le \sum_{j=1}^{p} \int_{U_j'} |f(\gamma(y)) - f(\gamma(x))|\, d\mu(\gamma),$$

and since the range of $f$ on $\{\gamma'(x') : x' \in V, \gamma' \in U_j'\}$ has diameter less than $\varepsilon$, this latter sum is no more than $\varepsilon \int d\mu = \varepsilon$. This establishes the continuity of $Sf$.

The fact that $\|\mu\| = 1$ implies that $\|Sf\|_{X_0'} \leq \|f\|_{X_0'}$ and that $S1 = 1$, so $S$ is of norm 1. It is also a projection, for from the translation invariance of $\mu$ we have

$$\begin{aligned}\int_{\mathbb{Z}_2^{A_0}} Sf(\gamma(x))\, d\mu(\gamma) &= \int_{\mathbb{Z}_2^{A_0}} \left\{ \int_{\mathbb{Z}_2^{A_0}} f(\gamma'(\gamma(x)))\, d\mu(\gamma') \right\} d\mu(\gamma) \\ &= \int_{\mathbb{Z}_2^{A_0}} Sf(x)\, d\mu(\gamma) \\ &= Sf(x);\end{aligned}$$

i.e., $S^2 = S$. Also, the translation invariance of $\mu$ implies that if $\gamma \in \mathbb{Z}_2^{A_0}$ and $f \in A_0$, then $S(f \circ \gamma) = Sf$, whence it follows that the range of $S$ is the set of all $f \in \mathscr{C}(X_0')$ which are constant on the orbits $\{\gamma(x) : \gamma \in \mathbb{Z}_2^{A_0}\}$, $x$ a fixed point of $X_0'$. The definitions of $X_0'$ and of the action of $\mathbb{Z}_2^{A_0}$ on $X_0'$ show that these orbits are the fibers $\pi^{-1}(y)$, $y \in X$. Any $f \in \mathscr{C}(X_0')$ constant on the fibers $\pi^{-1}(y)$ is of the form $\tilde{f} \circ \pi$, $\tilde{f} \in \mathscr{C}(X_0)$. Thus, the range of the projection $S$ is $\pi^*\mathscr{C}(X_0)$.

Since $\pi^*A_0$ is carried onto itself by $S$, the proof that $S$ takes $A_0'$ onto $\pi^*A_0$ involves only proving that $S(A_0') \subset \pi^*A_0$. To prove this, it is enough to show that $S$ carries every polynomial in $p_{f_1}, \ldots, p_{f_n}$ with coefficients from $\pi^*A_0$ into $\pi^*A_0$ for all choices of $f_1, \ldots, f_n \in A_0$, and to do this we use induction on $n$. The case $n = 0$ is clear since $S$ leaves $\pi^*A_0$ fixed pointwise. Suppose now that we know the result for all polynomials involving no more than $k - 1$ of the $p_f$. Let $h$ be a polynomial in $p_{f_1}, \ldots, p_{f_k}$. Since $p_{f_k}^2 \in \pi^*A_0$, we can suppose the polynomial to be of the form $h_1 + h_2 p_{f_k}$, where $h_1$ and $h_2$ are polynomials which involve only $p_{f_1}, \ldots, p_{f_{k-1}}$. Choose $\gamma \in \mathbb{Z}_2^{A_0}$ such that $\gamma_{f_j} = 1$, $1 \leq j < k$, $\gamma_{f_k} = -1$. Then

$$\begin{aligned}h \circ \gamma &= h_1 \circ \gamma + (h_2 \circ \gamma)(p_{f_k} \circ \gamma) \\ &= h_1 \circ \gamma - (h_2 \circ \gamma) p_{f_k}.\end{aligned}$$

The definition of the operator $S$ and the translation invariance of $\mu$ imply that $S(h_2 p_{f_k} - (h_2 \circ \gamma) p_{f_k}) = 0$, so we find that

$$S(h + h \circ \gamma) = S(h_1 + h_1 \circ \gamma).$$

Again invoking the translation invariance of $\mu$, we have that $S(h) = S(h_1)$. By the induction hypothesis, $S(h_1) \in \pi^*A_0$, so $S(h) \in \pi^*(A_0)$, as we wished to prove. This concludes the proof of (a).

(b) Let $\varphi \in \Sigma(A_0')$. The map $f \mapsto \varphi(\pi^*f)$ is an element of $\Sigma(A_0)$, so there is $x_0 \in X_0$ with $\varphi(\pi^*f) = f(x_0)$ for all $f \in A_0$. Since for all $f \in A_0$,

$p_f^2 = \pi^* f$, it follows that $\varphi(p_f)$ is one of the square roots of $f(x_0)$, and consequently, for some point $x_0'$ in the fiber $\pi^{-1}(x_0)$, $\varphi f = f(x_0')$ for all $f \in A_0'$.

(c) If $x_0 \in X_0$, then $A_0'|\pi^{-1}(x_0)$ is generated by the functions $p_f|\pi^{-1}(x_0)$, and each of them assumes only two values on this fiber. If $p_f$ assumes on $\pi^{-1}(x_0)$ the values $\alpha$ and $\beta$, there is a polynomial $q$ in one complex variable such that $q(\alpha) = \bar{\alpha}$, $q(\beta) = \bar{\beta}$. Then $q \circ p_f \in A_0'|\pi^{-1}(x_0)$, and as $q \circ p_f = \bar{p}_f$, we conclude that $A_0'|\pi^{-1}(x_0)$ is closed under complex conjugation. Thus, part (c) is a consequence of the Stone–Weierstrass theorem.

**19.2** ***Corollary***

*$A_0' = \mathscr{C}(X_0)$ if and only if $A_0 = \mathscr{C}(X_0)$.*

So far we have obtained an algebra $A_0'$ which contains $A_0$ isometrically and in which each element of $A_0$ has a square root. We call $A_0'$ the *root algebra* of $A_0$. We shall now iterate this process inductively to obtain an algebra $\tilde{A}$ which contains $A_0$ isometrically and which has the property that every element of $\tilde{A}$ has, in $\tilde{A}$, a square root. Let $\Omega$ denote the first uncountable ordinal.

We construct the following system. For each $\alpha \le \Omega$, $(A_\alpha, X_\alpha)$ is a uniform algebra with $X_\alpha = \Sigma(A_\alpha)$, and for each pair $\alpha \le \beta \le \Omega$, $\pi_{\alpha,\beta}: X_\beta \to X_\alpha$ is a continuous map of $X_\beta$ onto $X_\alpha$. We require that the following conditions be met:

(a) $(A_0, X_0)$ is arbitrary.
(b) The homomorphism $\pi_{\alpha,\beta}^*: \mathscr{C}(X_\alpha) \to \mathscr{C}(X_\beta)$ given by $\pi_{\alpha,\beta}^*(f) = f \circ \pi_{\alpha,\beta}$ carries $A_\alpha$ into $A_\beta$.
(c) For $\alpha \le \beta \le \gamma \le \Omega$, $\pi_{\alpha,\beta} \circ \pi_{\beta\gamma} = \pi_{\alpha,\gamma}$, and $\pi_{\alpha,\alpha}$ is the identity map.
(d) If $\beta = \alpha + 1$, $A_\beta$ is the root algebra of $A_\alpha$, $X_\beta$ is $\Sigma(A_\beta)$, and $\pi_{\alpha,\beta}: X_\beta \to X_\alpha$ is the map considered in Theorem 19.1.
(e) If $\beta$ is a limit ordinal, then $(A_\beta, X_\beta)$ is the direct limit of the system $\{A_\alpha, \pi_{\alpha,\gamma}^*\}_{\alpha,\gamma<\beta}$ of uniform algebras as defined in Section 14.

In addition to this system of algebras, we shall need certain operators.

**19.3** ***Lemma***

*There exists a system $\{T_\alpha\}_{\alpha\le\Omega}$ of linear operators, $T_\alpha$, taking $\mathscr{C}(X_\alpha)$ onto $\mathscr{C}(X_0)$ such that*

(a) *$T_0$ is the identity operator on $\mathscr{C}(X_0)$;*
(b) *$T_\alpha$ and $T_\alpha|A_\alpha$ have norm one, and $T_\alpha$ maps $A_\alpha$ onto $A_0$;*
(c) *$T_\beta \circ \pi_{\alpha,\beta}^* = T_\alpha$ for all $\alpha \le \beta$.*

PROOF

We have specified $T_0$. Suppose that $\beta \leq \Omega$ is given, that $T_\alpha$ is defined for all $\alpha < \beta$, and that the operators $T_\alpha$ which are defined satisfy the conditions (b) and (c). Two cases are to be considered. It may be that $\beta = \gamma + 1$ for some $\gamma$. In this case, $A_\beta$ is the root algebra of $A_\gamma$. We have, as in Theorem 19.1, an operator $S:\mathscr{C}(X_\beta) \to \pi^*_{\gamma,\beta}\mathscr{C}(X)$, and if we set $T_\alpha = T_\gamma \circ (\pi^*_{\gamma,\beta})^{-1} \circ S$, we have the desired map. On the other hand, it may be that $\beta$ is a limit ordinal. In this case $X_\beta$ is the inverse limit of the system $\{X_\alpha, \pi_{\alpha,\gamma}\}_{\alpha,\gamma<\beta}$ of compact spaces, and $A_\beta$ is the direct limit of the system $\{A_\alpha, \pi^*_{\alpha,\gamma}\}$ of uniform algebras. For each $\alpha < \beta$, $\pi^*_{\alpha,\beta}\mathscr{C}(X_\alpha)$ is a closed subspace of $\mathscr{C}(X_\beta)$ on which $T^0_\alpha = T_\alpha \circ \pi^{*\,-1}_{\alpha,\beta}$ is a well-defined linear operator onto $\mathscr{C}(X_0)$. The operator $T^0_\alpha$ has norm 1 since $\pi^{*\,-1}_{\alpha,\beta}$ is an isometry. Moreover, if $\gamma < \alpha$, then on $\pi^*_{\gamma\beta}\mathscr{C}(X_\gamma)$,

$$\begin{aligned} T^0_\alpha &= T_\alpha \circ \pi^{*\,-1}_{\alpha,\beta} \\ &= T_\alpha \circ \pi^*_{\gamma,\alpha} \circ \pi^{*\,-1}_{\gamma,\beta} \\ &= T_\gamma \circ \pi^{*\,-1}_{\gamma,\beta} \\ &= T^0_\gamma. \end{aligned}$$

Thus, the operators $\{T^0_\alpha\}$ agree, pairwise, on their common domains of definition; so taken together, they define on the union of their domains, i.e., on $\bigcup\{\pi^*_{\alpha,\beta}\mathscr{C}(X_\alpha) : \alpha < \beta\}$, a well-defined linear operator, $\tilde{T}_\beta$, which is of norm 1. Since $\bigcup\{\pi^*_{\alpha,\beta}\mathscr{C}(X_\alpha) : \alpha < \beta\}$ is dense in $\mathscr{C}(X_\beta)$, $\tilde{T}_\beta$ extends to $T_\beta : \mathscr{C}(X_\beta) \to \mathscr{C}(X_0)$, and $T_\beta$ has the desired properties.

We now consider the algebra $(A_\Omega, X_\Omega)$, which we call the *universal root algebra* of $A_0$. The following theorem summarizes what we have done.

19.4 ***Theorem***

*If $(A_0, X_0)$, $X_0 = \Sigma(A_0)$, is a uniform algebra, and $(A_\Omega, X_\Omega)$ is its universal root algebra, then for every $f \in A_\Omega$, there is $g \in A_\Omega$ with $f = g^2$. Moreover, $A_\Omega = \mathscr{C}(X_\Omega)$ if and only if $A_0 = \mathscr{C}(X_0)$.*

PROOF

If $f \in A_\Omega$, then since $\Omega$ is the first uncountable ordinal, it follows that for some $\beta < \Omega$, $f \in \pi^*_{\beta,\Omega}A_\beta$ for some $\beta < \Omega$. Since $A_{\beta+1}$ is the root algebra for $A_\beta$, there is $g \in \pi^*_{\beta+1,\Omega}A_{\beta+1}$ with $f = g^2$.

Since the operator $T_\Omega : \mathscr{C}(X_\Omega) \to \mathscr{C}(X_0)$ carries $\mathscr{C}(X_\Omega)$ onto $\mathscr{C}(X_0)$ and $A_\Omega$ onto $A_0$, it follows that if $A_\Omega = \mathscr{C}(X_\Omega)$, then $A_0 = \mathscr{C}(X_0)$. Conversely, if $A_0 = \mathscr{C}(X_0)$, then by induction, using Corollary 19.2, it follows that $A_\Omega = \mathscr{C}(X_\Omega)$. To see this, consider an ordinal $\gamma \leq \Omega$, and suppose that if $\alpha < \gamma$, then $A_\alpha = \mathscr{C}(X_\alpha)$. If $\gamma = \beta + 1$, then, by Corollary 19.2, $A_\gamma = \mathscr{C}(X_\gamma)$. If $\gamma$ is a limit ordinal, then since $A_\gamma$ is the closure of the subalgebra of $\mathscr{C}(X_\gamma)$ generated by the functions $\pi_{\alpha,\gamma}{}^*f$, $f \in \mathscr{C}(X_\alpha)$, $\alpha < \gamma$, it follows that $A_\gamma$ is conjugate closed and so by the Stone–Weierstrass theorem, $A_\gamma = \mathscr{C}(X_\gamma)$.

In particular, then, there exist uniform algebras $(A, X)$, $A \neq \mathscr{C}(X)$, such that each $f \in A$ has a square root in $A$. Suppose that $\varphi$ and $\psi$ are distinct elements of $\Sigma(A)$. There is $f \in A$ with $\varphi(f) = 0$, $\psi(f) \neq 0$, and $\|f\| < 1$. If $g_1, g_2, \ldots$ are elements of $A$ such that $g_1^2 = f$, $g_2^2 = g_1, \ldots$, then $\|g_n\| < 1$, $\varphi(g_n) = 0$, but $|\psi(g_n)| = |\psi(f)|^{2^{-n}} \to 1$; so $\varphi$ and $\psi$ lie in different Gleason parts.

By construction, we have the map $\pi_{\Omega,0}: X_\Omega \to X_0$, which carries $X_\Omega$ onto $X_0$.

19.5 ***Lemma***

*$A_\Omega|\pi_{\Omega,0}^{-1}(x_0)$ is dense in $\mathscr{C}(\pi_{\Omega,0}^{-1}(x_0))$ for all $x_0 \in X_0$.*

PROOF

This lemma follows by induction, using part (c) of Theorem 19.1. If $\gamma$ is the first ordinal $\leq \Omega$ for which $A_\gamma|\pi_{\gamma,0}^{-1}(x_0)$ is not dense in $\mathscr{C}(\pi_{\gamma,0}^{-1}(x_0))$, then by Theorem 19.1 and the definition of the algebras $A_\alpha$, $\gamma$ must be a limit ordinal. Then $\mathscr{C}(\pi_{\gamma,0}^{-1}(x_0))$ is generated by functions of the form $f \circ \pi_{\gamma,\alpha}$, $f \in \mathscr{C}(\pi_{\alpha,0}^{-1}(x_0))$, $\alpha < \gamma$. By hypothesis, each $f \in \mathscr{C}(\pi_{\alpha,0}^{-1}(x_0))$ can be approximated uniformly by functions of the form $g|\pi_{\alpha,0}^{-1}(x_0)$, $g \in A_\alpha$; and it follows that $\mathscr{C}(\pi_{\gamma,0}^{-1}(x_0))$ also is generated by $\{g \circ \pi_{\gamma,\alpha} : g \in A_\alpha\}$, i.e., that $A_\gamma|\pi_{\gamma,0}^{-1}(x_0)$ is dense in $\mathscr{C}(\pi_{\gamma,0}^{-1}(x_0))$.

19.6 ***Theorem***

*If $(A_0, X_0)$ is a uniform algebra with universal root algebra $(A_\Omega, X_\Omega)$, and if $A_0$ has the property that every Jensen measure is a point mass, then every point of $X_\Omega$ is a strong boundary point for $A_\Omega$.*

PROOF

Let $\mu$ be a representing measure for $\varphi \in \Sigma(A_\Omega)$. If $f \in A_\Omega$ and $\{g_n\}$ is a sequence in $A_\Omega$ with $g_1^2 = f$, $g_2^2 = g_1, \ldots$, we have

$$\left| \int g_n \, d\mu \right| = |\varphi(f)|^{2^{-n}},$$

whence

$$\left( \int |f|^{2^{-n}} \, d\mu \right)^{2^n} \geq |\varphi(f)|.$$

As this holds for all $n$,

$$|\varphi(f)| \leq \lim_{n\to\infty} \left( \int |f|^{1/2^n} \, d\mu \right)^{2^n}$$

$$= \exp \int \log |f| \, d\mu.$$

(For the last equality, see Appendix B.) Thus,

$$\log |\varphi(f)| \leq \int \log |f| \, d\mu,$$

and $\mu$ is a Jensen measure, i.e., every representing measure for $A_\Omega$ is a Jensen measure.

Define $\sigma$ on $X_0$ by $\sigma(E) = \mu(\pi_{\Omega,0}^{-1}(E))$. The measure $\sigma$ is a probability measure, and if $f \in A_0$, then

$$\int_{X_0} f \, d\sigma = \int f \circ \pi_{\Omega,0} \, d\mu = \varphi(f \circ \pi_{\Omega,0}),$$

so $\sigma$ is a representing measure for $\tilde{\varphi} \in \Sigma(A_0)$ given by $\tilde{\varphi}(f) = \varphi(f \circ \pi_{\Omega,0})$. Moreover, $\sigma$ is a Jensen measure for $\tilde{\varphi}$ since $\mu$ is one for $\varphi$. By hypothesis this implies that $\sigma$ is a point mass at some point, $x_0$, of $X_0$. Consequently, $\mu$ is supported by the fiber $\pi_{\Omega,0}^{-1}(x_0)$, and since $A_\Omega | \pi_{\Omega,0}^{-1}(x_0)$ is dense in $\mathscr{C}(\pi_{\Omega,0}^{-1}(x_0))$, $\mu$ itself must be a point mass. Thus, every representing measure for the algebra $A_\Omega$ is a point mass, and so by Theorem 7.18, every point of $X_\Omega = \Sigma(A_\Omega)$ is a strong boundary point for $A_\Omega$.

Combining Theorems 19.4 and 19.6 and the fact to be proved in Section 27 that there are algebras, even algebras on metrizable spaces, other than $\mathscr{C}(X)$ for which the only Jensen measures are the point masses, we arrive at the following fact.

**19.7** ***Corollary***

*There exists uniform algebras $(A, X)$, $X = \Sigma(A)$ and $A \neq \mathscr{C}(X)$, for which every point of $X$ is a strong boundary point.*

In connection with the universal root algebra $(A_\Omega, X_\Omega)$ of the algebra $(A_0, X_0)$, we should refer to the theorem of Cirka, Theorem 13.15: The space $X_\Omega$ can never be locally connected if $A_0 \neq \mathscr{C}(X_0)$. It is not at all evident what topological properties $X_\Omega$ has.

The construction we have just given for the universal root algebra is natural in the sense that it provides a canonical way of embedding a given uniform algebra in a new uniform algebra, the spectrum of which consists exclusively of one-point parts. For certain purposes it is undesirable, though, for the universal root algebra is enormous. Given a uniform algebra $(A, X)$ with $X$ metric, it is reasonable to seek to embed $A$ in a uniform algebra $B$ with a metric spectrum that consists of one-point parts. The process used above can be modified to provide nontrivial examples of this kind but at the cost of naturality. The new procedure is not at all canonical. We can omit safely most of the details as they parallel the details of the general construction just given.

We consider a uniform algebra $(A_0, X_0)$ with $X_0$ metrizable and hence $A_0$ separable, and we let $\mathscr{F}_0$ be some countable, dense subset of $A_0$. Let $X_1$ be the set

$$\{(x, \mathfrak{z}) \in X_0 \times \mathbb{C}^{\mathscr{F}_0} : p_f^2(\mathfrak{z}) = f(x) \text{ for all } f \in \mathscr{F}_0\},$$

and let $A_1$ be the uniform closure in $\mathscr{C}(X_1)$ of the polynomials in the coordinate projections $p_f$ and the functions $f \circ \pi$, where $\pi : X_0 \times \mathscr{C}^{\mathscr{F}_0} \to X_0$ is the natural projection. Let $\mathscr{F}_1 \subset A_1$ be the collection of polynomials in $f \circ \pi$ and $p_f$, $f \in \mathscr{F}$, polynomials with coefficients of the form $\alpha + i\beta$, $\alpha$ and $\beta$ rational. Then $\mathscr{F}_1$ is a countable dense subset of $A_1$ which contains $\mathscr{F}_0$ naturally. We iterate this construction inductively, stopping this time at $\omega$, the first infinite ordinal. For each $n < \omega$, we have a uniform algebra $(A_n, X_n)$, $X_n$ metric, and in $A_n$ we have a countable dense set $\mathscr{F}_n$. Each $A_n$ is naturally contained in $A_{n+1}$, and each $\mathscr{F}_n$ in $\mathscr{F}_{n+1}$. Moreover, if $f \in \mathscr{F}_n$, it has a square root in $\mathscr{F}_{n+1}$. $(A_\omega, X_\omega)$ is the direct limit of the system $\{A_n\}$, and if we set $\mathscr{F}_\omega = \bigcup \mathscr{F}_n$, regarding $A_n$ and $\mathscr{F}_n$ as subsets of $A_\omega$ in the natural way, then $\mathscr{F}_\omega$ is dense in $A_\omega$, and $\mathscr{F}_\omega$ is closed under the extraction of square roots: Every $f \in \mathscr{F}_\omega$ has a square root in $\mathscr{F}_\omega$.

If $\varphi$ and $\psi$ are complex homomorphisms, and if $f \in A_\omega$ is of norm no more than 1 and is annihilated by $\varphi$ but not by $\psi$, let $f_n \in \mathscr{F}_\omega$ satisfy $\|f_n - f\| < e^{-4^n}$, $\|f_n\| \leq 1$, and let $g_n \in \mathscr{F}_\omega$ satisfy $g_n^{2^n} = f_n$. Then $\|g_n\| \leq 1$, and

$$\begin{aligned} |\varphi(g_n)| &= |\varphi(f_n)|^{2^{-n}} = |\varphi(f_n - f)|^{2^{-n}} \\ &< e^{-2^n} \to 0, \end{aligned}$$

and also

$$\begin{aligned} |\psi(g_n)| &= |\psi(f) + \psi(f_n - f)|^{2^{-n}} \\ &> (|\psi(f)| - e^{-4^n})^{2^{-n}}. \end{aligned}$$

For large values of $n$, this last term is at least $(\frac{1}{2}|\psi(f)|)^{2^{-n}}$, and as $n \to \infty$, this tends to 1. Thus, $\varphi$ and $\psi$ lie in different Gleason parts. In the same way, it follows that if every Jensen measure for the algebra $A_0$ is a point mass, then every point of $X_\omega = \Sigma(A_\omega)$ is a strong boundary point for $A_\omega$. Since $X_\omega$ is metrizable, this implies that each point is actually a peak point.

After developing the general theory discussed above, Cole was able to construct finitely generated uniform algebras with the property that each point of the spectrum is a peak point. An extremely simple example of this kind has been found by Basener.

19.8 ***Example*** (Basener [1])
Let $E$ be a compact subset of the open unit disc for which the algebra $\mathscr{R}(E)$ differs from $\mathscr{C}(E)$ but has the property that the only Jensen measures are point masses. A suitable $E$ is constructed in Section 27. Define $X_E$ by

$$X_E = \{(z, w) \in \mathscr{C}^2 : z \in E, |z|^2 + |w|^2 = 1\}.$$

Basener observed that the algebra $\mathscr{R}(X_E)$ has $X_E$ as its spectrum but that it is not $\mathscr{C}(X_E)$. From this it follows that $\mathscr{R}(X_E)$ is an example of the kind we seek, for given a point of $X_E$ there is a polynomial which peaks there.

It is convenient to observe that $\mathscr{R}(X_E)$ is the subalgebra $B$ of $\mathscr{C}(X_E)$ generated by the functions $w$, $w^{-1}$, and $f \circ \pi$, where $f \in \mathscr{R}(E)$ and $\pi : \mathscr{C}^2 \to \mathscr{C}$ is the projection onto the $z$ coordinate. The assumption that $E$ is a compact subset of the open unit disc implies that if $(z, w) \in X_E$, then $w \neq 0$, so $w^{-1}$ is in $\mathscr{R}(X_E)$. Assume for the moment that $\Sigma(B) = X_E$. The set $X_E$ is the joint spectrum of the set $\{z, w\}$ of elements of $B$, so if $h$ is a function holomorphic on a neighborhood of $X_E$, in particular, if $h$ is a rational function which is holomorphic on $X_E$, then $h|X_E$ lies in $B$. It follows that $\mathscr{R}(X_E)$ is contained in $B$, so these two algebras are identical.

To prove that $X_E$ is the spectrum of $B$, let $\varphi$ lie in $\Sigma(B)$, and let $\mu$ be a Jensen measure for $\varphi$. If we denote by $\varphi^*$ the element of $\Sigma(\mathscr{R}(E))$ defined by $\varphi^*(f) = \varphi(f \circ \pi)$ for $f \in \mathscr{R}(E)$ and by $\mu^*$ the measure on $E$ defined by $\int_E g \, d\mu^* = \int_{X_E} g \circ \pi \, d\mu$, then $\mu^*$ is a Jensen measure for $\varphi^*$. Our hypothesis on $E$ implies that $\mu^*$ is the unit point mass at a point, say $z_0$, of $E$. It follows that $\mu$ is concentrated on the circle $K = \{(z_0, \sqrt{1 - |z_0|^2}\, e^{i\theta}) : \theta \text{ real}\}$. We have

$$\varphi(w) = \int w \, d\mu = \int_K w \, d\mu$$
$$\leq \|w\|_K = \sqrt{1 - |z_0|^2}.$$

Also $\varphi(w^{-1}) \leq \|w^{-1}\|_K = (\sqrt{1 - |z_0|^2})^{-1}$. As $1 = ww^{-1}$, we see that both inequalities must, in fact, be equalities, and thus $|\varphi(w)| = \sqrt{1 - |z_0|^2}$. This implies that for some point $p = (z_0, w_0) \in K$, $\varphi(g) = g(p)$ for all $g \in B$, and therefore $\Sigma(B) = X_E$, as we wished to show.

Finally, $\mathscr{R}(X_E) \neq \mathscr{C}(X_E)$. This follows from the fact that $\mathscr{R}(E) \neq \mathscr{C}(E)$, as we shall show by lifting a measure on $E$ which annihilates $\mathscr{R}(E)$

to one on $X_E$ which annihilates $\mathscr{R}(X_E)$. Suppose, to this end, that $\nu$ is a measure on $E$, not the zero measure, which annihilates $\mathscr{R}(E)$. Let $\nu$ be the weak* limit of the sequence $\{\nu_n\}$,

$$\nu_n = \sum_{k=1}^{m(n)} a_{n,k}\delta_{n,k},$$

where each $\delta_{n,k}$ is the unit point mass at the point $z_{n,k} \in E$. For each $n$, define a measure $\tilde{\nu}_n$ on $X_E$ by

$$\tilde{\nu}_n(f) = \sum_{k=1}^{m(n)} \frac{1}{2\pi} \int_{-\pi}^{\pi} f(z_{n,k}, \sqrt{1 - |z_{n,k}|^2}\, e^{i\theta})\, d\theta,$$

and let $\tilde{\nu}$ be a weak* limit point of $\{\tilde{\nu}_n\}$. The measure $\tilde{\nu}$ annihilates the functions $w^p$, $p = 1, 2, 3, \ldots$, and the functions $f \circ \pi$, $f \in \mathscr{R}(E)$. But if $g \in \mathscr{C}(E)$ is not annihilated by $\nu$, then $g \circ \pi$ is not annihilated by $\tilde{\nu}$.

# *chapter four*/INTERPOLATION

## 20 *GENERAL RESULTS ON INTERPOLATION*

If $(A, X)$ is a uniform algebra, call the closed subset $E$ of $X$ an *interpolation set for* $A$ if $A|E = \mathscr{C}(E)$, where $A|E$ denotes, as usual, the algebra of restrictions to $E$ of functions in $A$. In this chapter we shall address ourselves to certain problems connected with the general question of interpolation. In the present section we shall establish various conditions which ensure that a set $E$ is an interpolation set. In the next section we shall consider the problem of interpolation for the polydisc algebras, and we shall devote the section following that to a theorem to the effect that, under certain hypotheses, $\mathscr{C}(E)$ is a Banach space direct summand of $A$ if $E$ is an interpolation set for $A$. In the final section we shall establish some generalizations of the well-known F. and M. Riesz theorem of classical function theory.

In addition to interpolation sets, we shall consider *peak interpolation sets*, i.e., sets $E$ with the property that if $f \in \mathscr{C}(E)$ does not vanish identically, there is $F \in A$ with $F|E = f$, $|F(x)| < \|f\|_E$ for all $x \in X \setminus E$. Applying the definition of peak interpolation set to the function in $\mathscr{C}(E)$ which is identically 1, it is clear that a peak interpolation set is a peak set for $A$. There is a converse to this observation, for it follows from Lemma 7.22 that if $E$ is a peak set for $A$ and an interpolation set, then $E$ is a peak interpolation set for $A$. The following lemma records this simple observation.

**20.1** ***Lemma***

*The set E is a peak interpolation for* $(A, X)$ *if and only if it is simultaneously a peak set and an interpolation set.*

Aside from the Tietze extension theorem, it seems that the first nontrivial result on interpolation was proved by Rudin [6] and, in a slightly weaker form, by Carleson [2]. The result is frequently referred to as the *Rudin–Carleson theorem*, and it refers to the disc algebra:

**20.2** ***Theorem***

*A closed subset of the unit circle is a peak interpolation set for* $\mathscr{A}$ *if and only if it has zero Lebesgue measure.*

This result has motivated much of the subsequent research on the general theory of interpolation. In its proof we shall denote by $T$ the unit circle in the complex plane and by $\lambda$ Lebesgue measure on $T$, normalized so that $\lambda(T) = 1$. Thus, $\lambda$ is the representing measure for $0: f \in \mathscr{A}$ implies that $f(0) = \int_T f\, d\lambda$. Our proof of the Rudin–Carleson theorem will depend on a function theoretic lemma.

20.3 ***Lemma***

*Every closed set of T which has Lebesgue measure zero is a peak set for A.*

PROOF

Begin by observing that if $f_\varepsilon(x) = \varepsilon \log(1 - x^2)$ then $f_\varepsilon \leq 0, f_\varepsilon$ is real analytic on $(-1, 1)$, and it has limit $-\infty$ at $\pm 1$. Moreover, it is integrable on $(-1, 1)$ and by making $\varepsilon$ small, its integral can be made as small in modulus as we wish.

Let $E \subset T$ have zero Lebesgue measure, $E$ closed. If we write $T \setminus E = \bigcup\{L_n : n = 1, \ldots\}$, where the $L_n$ are pairwise disjoint open arcs, then $\Sigma\lambda(L_n) = 1$. There exists a sequence $\{c_n\}$ of positive numbers, $\lim c_n = \infty$, such that $\Sigma c_n \lambda(L_n) < \infty$. By the last paragraph there exists a function $u_n$ on $L_n$ with the following properties:

(a) $u_n \leq -c_n$ and $u_n \to -\infty$ at the end points of $L_n$.
(b) $u_n$ is real analytic.
(c) $|\int_{L_n} u_n\, d\lambda| \leq 2c_n \lambda(L_n)$.

The function $u$ on $T \setminus E$ that agrees on $L_n$ with $u_n$ is integrable with respect to $\lambda$ and is real analytic on $T \setminus E$. Moreover, $u(p) \to -\infty$ as $p \to E, p \in T \setminus E$. Let $\tilde{u}$ be the function harmonic in the unit disc which assumes the boundary values $u$ on $T$, and let $\tilde{v}$ be its conjugate. As $u$ is real analytic on $T \setminus E$, $u$ can be continued harmonically across $T \setminus E$, and it follows that the same is true of $\tilde{v}$. The function $f_0 = \tilde{u} + i\tilde{v}$ is continuous on $\bar{\Delta} \setminus E$, $\bar{\Delta}$ the closed unit disc, and $\lim_{\zeta \to \zeta_0} \operatorname{Re} f(\zeta) = -\infty$ if $\zeta$ approaches $\zeta_0 \in E$ through the set $\bar{\Delta} \setminus E$. By construction, $\operatorname{Re} f_0 \leq 0$, so the function $f$ given by

$$f = f_0(f_0 - 1)^{-1}$$

lies in $\mathscr{A}$. It clearly peaks on $E$, so $E$ is a peak set, as we wished to show.

PROOF OF THE THEOREM

The necessity of the condition is clear. If $f \in \mathscr{A}$ peaks on the set $E \subset T$, then $1 - f$ vanishes on $E$, and $1 - f(0) \neq 0$. The measure $\lambda$ is a Jensen measure for 0, so

$$-\infty < \log|1 - f(0)| \leq \int_T \log|1 - f|\, d\lambda,$$

and this implies that $\lambda(E) = 0$.

That sets of zero measure are peak interpolation sets is an immediate consequence of the lemma. If $\lambda(E) = 0$, $E$ closed, then $E$ is a peak set for $\mathscr{A}$, and so, by Lemma 12.3, $\mathscr{A}|E$ is closed in $\mathscr{C}(E)$. It remains only to prove that $\mathscr{A}|E$ is dense in $\mathscr{C}(E)$. There are many ways to see this. A very simple way is to show that each $\mu \in \mathscr{M}(E)$ which annihilates $\mathscr{A}|E$ is the zero measure. Suppose that $\mu \in \mathscr{M}(E)$ and $0 = \int f\,d\mu$ for all $f \in \mathscr{A}$. If $F \subset E$ is a closed set, $\lambda(F) = 0$, so there is $g \in \mathscr{A}$ which peaks on $F$. This leads to

(1) $$0 = \int g^n\,d\mu,$$

and by letting $n \to \infty$ we find from the dominated convergence theorem that $0 = \int_F d\mu$. As this holds for all closed $F \subset E$, we may conclude by regularity that $\mu$ is the zero measure.

**20.4** ***Corollary*** (Fatou [1])
*If $E \subset T$ is a closed set of measure zero, there is a function $f \in \mathscr{A}$ with $E$ as its zero set.*

**20.5** ***Corollary*** (F. and M. Riesz Theorem)
*If $\mu \in \mathscr{M}(T)$ annihilates $\mathscr{A}$, then $\mu$ is absolutely continuous with respect to Lebesgue measure.*

PROOF
If $E \subset T$ is a closed Lebesgue null set, there is $g \in \mathscr{A}$ which peaks on $E$. Then again $0 = \int g^n\,d\mu$, whence $\mu(E) = 0$. This implies by regularity that $\mu \ll \lambda$.

**20.6** ***Corollary***
*The set $E \subset T$ is a peak interpolation set for $\mathscr{A}$ if and only if it is the zero set for an element of $\mathscr{A}$.*

We now pass to more general considerations, but much of what we do will be motivated by these concrete results from classical function theory. Many of our general results are due to Glicksberg [1].

As the proof given above for the Rudin–Carleson theorem suggests, a general theory of interpolation will necessarily include criteria for the closure of a restriction algebra. Already we have established one such criterion: By Lemma 12.3, if $E$ is a peak set for $A$ or even if it is only an intersection of peak sets, then $A|E$ is closed in $\mathscr{C}(E)$. In the paper of Glicksberg cited above another useful criterion was established, a criterion which is valid not only for subalgebras of $\mathscr{C}(X)$ but also for subspaces.

**20.7** ***Theorem***

*If $X$ is a compact space, if $E$ is a closed subset of $X$, and if $A$ is a closed subspace of $\mathscr{C}(X)$, then $A|E$ is closed in $\mathscr{C}(E)$ if and only if for some $c \geq 1$,*

$$\|v + (A|E)^{\perp}\| \leq c\|v + A^{\perp}\| \tag{1}$$

*for all $v \in \mathscr{M}(E)$. When (1) holds, $A|E$ is topologically isomorphic to $A/\ker E$, and the isomorphism $f|E \mapsto f + \ker E$ from $A|E$ to $A/\ker E$ has norm no more than $c$.*

A remark on our notation may be in order. We have already used the notation $A^{\perp}$ for the set $\{\mu \in \mathscr{M}(X) : \int f\, d\mu = 0 \text{ for all } f \in A\}$. By $(A|E)^{\perp}$, we denote $A^{\perp} \cap \mathscr{M}(E)$. The norms in (1) are the quotient norms

$$\|v + (A|E)^{\perp}\| = \inf\{\|v + \mu\| : \mu \in (A|E)^{\perp}\}$$

and

$$\|v + A^{\perp}\| = \inf\{\|v + \mu\| : \mu \in A^{\perp}\}.$$

Finally, $\ker E = \{f \in A : f|E = 0\}$.

PROOF OF THE THEOREM

The proof depends on the fact that a linear operator between Banach spaces has closed range if and only if its adjoint does.

Denote by $\rho : A \to (A|E)^{-}$ the restriction operator given by $\rho f = f|E$, and let $\rho^*$ be its adjoint. The dual space of $A$ can be identified with $\mathscr{M}(X)/A^{\perp}$ and that of $(A|E)^{-}$ with $\mathscr{M}(E)/(A|E)^{\perp}$. The map $\rho^*$ carries the coset $\mu + (A|E)^{\perp}$ to the coset $\mu + A^{\perp}$, and since

$$\|v + A^{\perp}\| \leq v + (A|E)^{\perp}\|, \tag{2}$$

$\rho^*$ has norm no more than 1.

If $T : B_1 \to B_2$ is a continuous, one-to-one linear operator from the Banach space $B_1$ to the Banach space $B_2$, the range of $T$ is closed if and only if $\|Tx\| \geq k\|x\|$ for some fixed $k > 0$ and all $x \in B_1$. Applying this to $\rho^*$, we see that (1) is satisfied if and only if $\rho^*$ has closed range, and this latter condition is equivalent to the closure of $\rho(A)$. Thus, we have the first part of the theorem.

Assume now that condition (1) is satisfied so that $\rho$ has closed range, and consequently, by the open mapping theorem, $\rho$ is an open mapping. As $\ker E$ is the kernel of $\rho$, the induced map

$$\tilde{\rho} : A/\ker E \to A|E$$

is an isomorphism of Banach spaces. We can identify $(A/\ker E)^*$ with a subspace of $A^*$, and when this identification is made, we find that

$\rho^*$ and $\tilde{\rho}^*$ coincide. Thus, bounds on $\rho^* = \tilde{\rho}^*$ and $(\rho^*)^{-1} = (\tilde{\rho}^*)^{-1} = (\tilde{\rho}^{-1})^*$ imply bounds on $\tilde{\rho}$ and $\tilde{\rho}^{-1}$, respectively. By (1), we have that $\|(\tilde{\rho}^{-1})^*\| \le c$, and, by (2), $\|\tilde{\rho}^*\| \le 1$. Thus, $\|\tilde{\rho}^{-1}\| \le c$ and $\|\tilde{\rho}\| \le 1$.
In particular, $\tilde{\rho}$ is an isometry if and only if $c = 1$.

**20.8** ***Corollary*** (Glicksberg [I])
(a) *If $E \subset X$ is a closed set, then $A|E = \mathscr{C}(E)$ if and only if for some $\alpha \ge 1$,*

$$\|\mu_E\| \le \alpha\|\mu_{E^c}\| \tag{3}$$

*for all $\mu \in A^\perp$. When (3) holds, the norm of the isomorphism $f|E \mapsto f + \ker E$ from $A|E$ onto $A/\ker E$ is no more than $\alpha$.*
(b) *$A|E = \mathscr{C}(E)$ if and only if there is a constant $\alpha < 1$ such that $\|\mu_E\| \le \alpha\|\mu\|$ for all $\mu \in A^\perp$.*
(c) *$A/\ker E$ is isometrically isomorphic to $\mathscr{C}(E)$ if and only if $\|\mu_E\| \le \frac{1}{2}\|\mu\|$ for all $\mu \in A^\perp$.*

PROOF
(a) If $A|E = \mathscr{C}(E)$, then $(A|E)^\perp = 0$, so $\|\nu\| \le \alpha\|\nu + A^\perp\|$ for some $\alpha \ge 1$ and all $\nu$, as the last theorem asserts. If $\mu \in A^\perp$, this inequality implies that

$$\|\mu_E\| \le \alpha\|\mu_{E^c}\|, \tag{4}$$

so (3) holds.

Conversely, if (3) holds, then $(A|E)^\perp = 0$, for if $\mu \in (A|E)^\perp$, then (3) yields

$$\begin{aligned}\|\mu\| = \|\mu_E\| &\le \alpha\|\mu_{E^c}\| \\ &= 0.\end{aligned}$$

Consequently, (3) implies that $A|E$ is dense in $\mathscr{C}(E)$. As $(A|E)^\perp = 0$, it follows that (1) is equivalent to

$$\|\nu\| \le c\|\nu + A^\perp\| \tag{5}$$

for $\nu \in \mathscr{M}(E)$. For such a $\nu$ and for a $\mu \in A^\perp$, we have by (4) that

$$\begin{aligned}\|\nu + \mu\| &= \|(\mu + \nu)_E\| + \|(\mu + \nu)_{E^c}\| \\ &\ge 1/\alpha\|\mu_E + \nu\| + \|\mu_{E^c}\| \\ &\ge 1/\alpha\|\nu\| - 1/\alpha\|\mu_E\| + \|\mu_{E^c}\| \\ &\ge 1/\alpha\|\nu\|.\end{aligned}$$

Thus, (5) holds with $c = \alpha$, so the inequality (1) holds with $c = \alpha$, and part (a) is seen to follow from the theorem.

Parts (b) and (c) constitute a rephrasing of (a). Consider (b). If $\|\mu_E\| \leq \alpha\|\mu\|$, $\alpha < 1$, then $\|\mu_E\| \leq \alpha(\|\mu_E\| + \|\mu_{E^c}\|)$, whence $\|\mu_E\| \leq \alpha(1 - \alpha)^{-1}\|\mu_{E^c}\|$, so (3) is satisfied. Conversely, if $\|\mu_E\| \leq \alpha\|\mu_{E^c}\|$ for some $\alpha \geq 1$ and all $\mu \in A^\perp$, then $\|\mu\| \geq \|\mu_E\| + 1/\alpha\|\mu_E\|$, whence $\|\mu_E\| \leq \alpha(1 + \alpha)^{-1}\|\mu\|$. This gives (b). Part (c) is proved in a similar way.

As a consequence of this corollary, we are able to give a very simple proof of the following theorem.

20.9 ***Theorem***

*Let $(A, X)$ be a uniform algebra, let $E \subset X$ be a closed set, and assume that for every $u \in \mathscr{C}_{\mathbf{R}}(E)$, there is $f \in A$ with $\operatorname{Re} f|E = u$. Then $E$ is an interpolation set for $A$.*

This result, which may be regarded as a local version of the theorem of Hoffman and Wermer, Theorem 13.2, was proved originally by Sidney and Stout [1]. An alternative, simpler, proof was given by Glicksberg [6], and yet a simpler proof was discovered by Natzitz [1]. It is the proof of Natzitz which we shall present. A similar theorem has been established in a somewhat more general context by Bernard [2].

PROOF

By the open mapping theorem, there is a constant $M$ such that if $u \in \mathscr{C}_{\mathbf{R}}(E)$, then $u = \operatorname{Re} f|E$ with $\|f\|_X \leq M\|u\|_E$. In particular, then, $-M\|u\|_E \leq \operatorname{Re} f$, $\operatorname{Im} f \leq M\|u\|_E$. If $\mu \in A^\perp$, there is a real Borel function $\psi$ on $X$ with range in $[0, 2\pi]$ such that $d|\mu|/d\mu = e^{i\psi}$. By regularity of $|\mu|$, it follows that if $\varepsilon > 0$, there is a real continuous function $\psi_0$ on $X$ with range in $[0, 2\pi]$ such that

$$\int_E |e^{i\psi} - e^{i\psi_0}| d|\mu| < \varepsilon.$$

The function $\psi_0$ agrees on $E$ with $\operatorname{Re} f, f = u + iv \in A$ with $\|u\|_X, \|v\|_X \leq 2\pi M$. As $\mu \in A^\perp$, we have

$$0 = \int e^{i(u+iv)}\, d\mu,$$

so

$$\left|\int_E e^{iu-v}\, d\mu\right| = \left|\int_{E^c} e^{iu-v}\, d\mu\right|.$$

The right-hand side of this does not exceed $e^{2\pi M}\|\mu\|_{E^c}$, and for the left-hand side we have the estimate

$$\left|\int_E e^{iu-v}\,d\mu\right| = \left|\int_E e^{-v}(e^{i\psi_0} - e^{i\psi})\,d\mu + \int_E e^{-v}\,d|\mu|\right|$$
$$\geq e^{-2\pi M}\left(\int_E d|\mu| - \varepsilon\right)$$
$$= e^{-2\pi M}(\|\mu\|_E - \varepsilon).$$

Since $\varepsilon$ is at our disposal, we are led to

$$\|\mu\|_E \leq e^{4\pi M}\|\mu\|_{E^c},$$

and the result follows from the last theorem.

By invoking the theorem of Hoffman and Wermer, it is possible to draw a slightly stronger conclusion: $A|E = \mathscr{C}(E)$ if $\operatorname{Re} A$ is closed in $\mathscr{C}_{\mathbb{R}}(E)$, for then $\operatorname{Re} A|E = \operatorname{Re}(A|E)^-$; and by the theorem of Hoffman and Wermer, it follows that $(A|E)^- = \mathscr{C}(E)$. Thus, $\operatorname{Re} A|E = \mathscr{C}_{\mathbb{R}}(E)$, and our assertion follows from the theorem just proved.

Our next result gives another criterion for the closure of a restriction algebra, a criterion that may be regarded as a generalization of the implication (c) implies (b) of Theorem 7.18. The result was proved by Glicksberg [1].

**20.10** ***Theorem***

*If $(A, X)$ is a uniform algebra and if $E \subset X$ is a closed set such that $\mu \in A^\perp$ implies $\mu_E \in A^\perp$, then $A|E$ is closed in $\mathscr{C}(E)$ and in fact $E$ is an intersection of peak sets for $A$. Consequently, if $f \in A|E$, there is $F \in A$ with $\|F\|_X = \|f\|_E$ and $F|E = f$. If in addition $\mu_E = 0$ for all $\mu \in A^\perp$, then $A|E = \mathscr{C}(E)$.*

Thus, if $E$ is a $G_\delta$, then $E$ itself is a peak set.

PROOF

We shall show that $A|E$ is closed; from this, the last statement of the theorem follows immediately. Using the closure of $A|E$, we shall prove that $E$ is an intersection of peak sets for $A$. The rest of the theorem then follows directly from Lemma 7.22.

To prove that $A|E$ is closed, it is enough to prove that it is isometrically isomorphic to the complete space $A/\ker E$. This isometry follows from Theorem 20.7, for if $\mu \in A^\perp$, then $\mu_E \in A^\perp$; so if $\nu \in \mathscr{M}(E)$, and if we set $(A^\perp)_E = \{\mu_E : \mu \in A^\perp\}$, then

$$\begin{aligned} \|\nu + (A|E)^\perp\| &= \|\nu + (A^\perp)_E\| \\ &= \inf\{\|\nu + \mu_E\| : \mu \in A^\perp\} \\ &\le \inf\{\|\nu + \mu\| : \mu \in A^\perp\}. \end{aligned}$$

We shall need the following lemma to finish the proof of the theorem.

20.11 ***Lemma***

$(\ker E)^\perp = A^\perp + \mathscr{M}(E)$.

PROOF

If $\mu = \sigma + \tau$, $\sigma \in A^\perp$, $\tau \in \mathscr{M}(E)$, and if $f \in A$, $f|E = 0$, then $\int f\, d\mu = 0$, so $A^\perp + \mathscr{M}(E) \subset (\ker E)^\perp$. In the other direction, any $f \in \mathscr{C}(X)$ that is annihilated by $A^\perp + \mathscr{M}(E)$ is in $A$ since it is annihilated by $A^\perp$, and it vanishes on $E$ since it is annihilated by $\mathscr{M}(E)$. Thus, in the weak $*$ topology of $\mathscr{M}(X)$, $A^\perp + \mathscr{M}(E)$ is dense in $(\ker E)^\perp$. Consequently, the proof of the lemma can be concluded by showing $A^\perp + \mathscr{M}(E)$ to be weak $*$ closed, and by the Krein–Šmulian theorem, $\mathscr{A}^\perp + \mathscr{M}(E)$ is weak $*$ closed if its intersection with the unit ball of $\mathscr{M}(X)$ is.

Suppose, then, that $\{\sigma_\alpha + \tau_\alpha\}$ is a net in the unit ball of $\mathscr{M}(X)$ which converges in the weak $*$ sense to $\mu$, $\sigma_\alpha \in A^\perp$, $\tau_\alpha \in \mathscr{M}(E)$. We have $\|\mu\| \le 1$. As we have already observed, if $\nu \in \mathscr{M}(E)$, then

$$\|\nu + (A|E)^\perp\| = \|\nu + A^\perp\|.$$

Consequently, there is $\lambda_\alpha \in (A|E)^\perp$ such that

$$\|\tau_\alpha - \lambda_\alpha\| \le \|\tau_\alpha + \sigma_\alpha\| + 1 \le 2.$$

Then $\{\tau_\alpha - \lambda_\alpha\}$ has a weak $*$ limit point $\nu$ which lies in $\mathscr{M}(E)$ since both $\tau_\alpha$ and $\lambda_\alpha$ lie in $\mathscr{M}(E)$, and $\mathscr{M}(E)$ is a weak $*$ closed subset of $\mathscr{M}(X)$. It follows that $\{\sigma_\alpha + \lambda_\alpha\} = \{(\sigma_\alpha + \tau_\alpha) - (\tau_\alpha - \lambda_\alpha)\}$ has $\mu - \nu$ as a weak $*$ cluster point. The measures $\sigma_\alpha$ and $\lambda_\alpha$ both lie in $A^\perp$, so $\mu - \nu \in A^\perp$, and thus $\mu \in A^\perp + \mathscr{M}(E)$. As $\|\mu\| \le 1$, the lemma is proved.

To continue with the proof of the theorem itself, denote by $Y$ the space obtained from $X$ by allowing the set $E$ to collapse to a single point, and let $\rho : X \to Y$ be the natural projection. Set $y_0 = \rho(E)$. Let $\tilde\rho : \mathscr{C}(Y) \to \mathscr{C}(X)$ be given by $\tilde\rho f = f \circ \rho$, and let $\tilde\rho^* : \mathscr{M}(X) \to \mathscr{M}(Y)$ be the adjoint mapping. Define $B \subset \mathscr{C}(Y)$ to be the algebra

$$\{f \in \mathscr{C}(Y) : f \circ \rho \in A\},$$

and let $\ker\{y_0\} = \{f \in B : f(y_0) = 0\}$. Proving that $E$ is an intersection of peak sets is equivalent to proving that the point $y_0$ lies in the strong boundary for the algebra $B$.

We need to know that $B$ is a uniform algebra on $Y$. It is clear that $B$ is uniformly closed in $\mathscr{C}(Y)$, but it is less clear that $B$ separates points on $Y$. To see that it does, suppose it known that if $x \in X \setminus E$, there is $f \in \ker E$ which does not vanish at $x$. Consider then a pair of points $y, y' \in Y$, $y \neq y'$. If, e.g., $y' = y_0$, choose $f \in \ker E$ which does not vanish at $\rho^{-1}(y)$. Then $g = f \circ \rho^{-1} \in B$, and $g(y) \neq g(y')$. If neither $y$ nor $y'$ is the point $y_0$, choose $f \in \ker E$ which does not vanish at $\rho^{-1}(y)$, and $g \in A$ such that $g(\rho^{-1}(y)) \neq 0$, $g(\rho^{-1}(y')) = 0$. Then $fg \in \ker E$ and this function separates the point $\rho^{-1}(y)$ from $\rho^{-1}(y')$. If $h \in B$ is defined to be $(fg) \circ \rho^{-1}$, then $h$ separates $y$ from $y'$. Thus, to see that $B$ separates points on $Y$, it suffices to prove that if $x \in X \setminus E$, there is $f \in \ker E$ such that $f(x) \neq 0$.

If this latter statement is false, there is a point $x_0 \in X \setminus E$ such that $f \in \ker E$ implies that $f(x_0) = 0$. Let $\delta_0$ be the unit point mass at $x_0$ so that $\delta_0 \in (\ker E)^\perp$. By the lemma proved above, $\delta_0 = \mu + \nu$, $\mu \in A^\perp$, $\nu \in \mathscr{M}(E)$, and therefore $\delta_0 - \nu \in A^\perp$. By hypothesis $\mu_E \in A^\perp$, so it follows that $\delta_0 \in A^\perp$, an obvious contradiction. Thus, $B$ does separate points on $Y$.

Next we show that $(\ker\{y_0\})^\perp = \tilde{\rho}^*((A^\perp)_{X \setminus E}) + \{\lambda\delta_0 : \lambda \in \mathbb{C}\}$, where $\delta_0$ denotes the unit point mass at $y_0$ and $(A^\perp)_{X \setminus E} = A^\perp \cap \mathscr{M}(X \setminus E)$. We need to know, first, that $(\ker\{y_0\})^\perp = \tilde{\rho}^*((\ker E)^\perp)$. We know that $\tilde{\rho}$ is one to one, so $\tilde{\rho}^*$ is onto. Thus, each $\nu \in (\ker\{y_0\})^\perp$ is $\tilde{\rho}^*\mu$ for some $\mu \in \mathscr{M}(X)$. To see that $\mu \in (\ker E)^\perp$, let $f \in \ker E$, say $f = \tilde{\rho}g$, $g \in \ker\{y_0\}$. We have

$$\begin{aligned} \int f\,d\mu &= \int \tilde{\rho}g\,d\mu \\ &= \int g\,d(\tilde{\rho}^*\mu) = 0, \end{aligned}$$

so $\mu \in (\ker E)^\perp$ as desired. Now if $\mu \in (\ker E)^\perp$, then by the lemma, $\mu = m + \nu$, $\nu \in \mathscr{M}(E)$ and $m \in A^\perp$. By hypothesis $m_E \in A^\perp$, so $m_{X \setminus E} \in A^\perp$, and thus we have a decomposition $\mu = m' + \nu'$, $m' \in (A^\perp)_{X \setminus E}$ and $\nu' \in \mathscr{M}(E)$. It follows that $(\ker\{y_0\})^\perp = \tilde{\rho}^*(A^\perp)_{X \setminus E} + \{\lambda\delta_0 : \lambda \in \mathbb{C}\}$.

Suppose now that $\mu \in \mathscr{M}(Y)$ is a representing measure for $y_0 : \int f\,d\mu = f(y_0)$ for all $f \in B$. Then $\mu \in (\ker\{y_0\})^\perp$, so $\mu = \tilde{\rho}^*\nu + \lambda\delta_0$ by the last paragraph. As $\nu \in (A^\perp)_{X \setminus E}$, and as $\mu$ is a probability measure, it follows that $\nu = 0$, for zero is the only nonnegative element of $A^\perp$. Thus, $\mu = \lambda\delta_0$ and we find that $\lambda = 1$, whence $\mu = \delta_0$. As the unit point mass at $y_0$ is the only representing measure for $y_0$, Theorem 7.18

implies that $y_0$ is a strong boundary point for $B$. Thus, we finally have that $E$ is an intersection of peak sets for $A$.

Our next result concerns interpolation sets for ideals, interpolation with prescribed majorization of the interpolating function over the space in question.

20.12 ***Theorem***

*Let $(A, X)$ be a uniform algebra, let $E \subset X$ be an interpolation set for $A$ which is an intersection of peak sets, and let $I$ be a closed ideal in $A$ whose hull misses $E$. Then $I|E = \mathscr{C}(E)$. Moreover, if $p$ is a strictly positive, continuous function on $X$, and if $f \in \mathscr{C}(E)$ satisfies $|f(x)| \leq p(x)$ for $x \in E$, then there is $F \in I$ such that $F|E = f$ and $|F(x)| \leq p(x)$ for all $x \in X$.*

PROOF

First we prove that $I|E = \mathscr{C}(E)$, and to do this we start by showing that there is $f_0 \in I$ which is strictly positive on $E$. If $x \in E$, there is $f \in I$ such that $f(x) \neq 0$. Thus, by compactness there exist finitely many elements of $I$, say $f_1, \ldots, f_q$, such that if $x \in E$, one of the $f_j$ does not vanish at $x$. As $E$ is an interpolation set for $A$, there exist $g_1, \ldots, g_q \in A$ such that $g_j|E = \bar{f}_j|E$. Then $f_0 = \Sigma f_j g_j$ is in $I$ and is strictly positive on $E$. If $h \in \mathscr{C}(E)$, then $(f_0|E)^{-1}h \in \mathscr{C}(E)$, so it has an extension to an element $g$ in $A$. The function $f_0 g = h$ is in $I$ and $f_0 g|E = h$.

To complete the proof, we establish the possibility of the stated majorization. Let $f \in \mathscr{C}(E)$. We may suppose with no loss of generality that $\|f\|_E = 1$. Assume that $p$ is a strictly positive function on $X$ such that $p(x) \geq |f(x)|$ if $x \in E$. There exists $F \in I$ such that $F(x) = f(x)$ if $x \in E$ by the first paragraph of the proof. The set

$$V = \{x \in X : |F(x)| \geq \tfrac{5}{4}p(x)\}$$

is closed and disjoint from $E$, so there is a function $h \in A$ which peaks on a set containing $E$ and contained in $X \setminus V$. Replacing $F$ by $h^N F$ for sufficiently large $N$ provides an element, $F_0$, of $I$ which interpolates $f$ on $E$ and which satisfies $|F_0(x)| < \frac{5}{4}p(x)$ for all $x \in X$.

For $n = 1, 2, \ldots,$ let

$$K_n = \{x \in X : (1 + 2^{-(n+1)})p(x) < |F_0(x)| \leq (1 + 2^{-n})p(x)\}.$$

We have $\bigcup_{n=1}^{\infty} K_n = \{x : |F_0(x)| > p(x)\}$, and for $N = 1, 2, \ldots, K_N$ is a set whose closure does not intersect $E$. Thus, for $N = 1, 2, \ldots,$ there is a function $h_N \in A$ which peaks on a set containing $E$ and contained in $X \setminus \bar{K}_N$. For each $N$, there is an integer $m(N)$ such that if $x \in K_N$, then

$$|F_0(x)h_N^{m(N)}(x)| < 4^{-N} \inf_{x \in K_N} p(x).$$

Define $F_1$ by

$$F_1 = \sum_{N=1}^{\infty} 2^{-N} F_0 h_N^{m(N)},$$

a certain function in $I$. It is clear that $F_1|E = f$ and that if $x \in X \setminus \bigcup K_n$, then $|F_1(x)| \le p(x)$. Finally, if $x \in K_M$, then

$$\begin{aligned} |F_1(x)| &\le \sum_{\substack{N=1 \\ N \ne M}}^{\infty} 2^{-N}(1 + 2^{-M})p(x) + 2^{-M}4^{-M}p(x) \\ &= \{(1 + 2^{-M})(1 - 2^{-M}) + 2^{-M}4^{-M}\}p(x) \\ &< p(x). \end{aligned}$$

The introduction of the majorizing function $p$ in the setting of the last theorem is due to Bishop [4]. A result on interpolation by ideals was given by Gamelin [2] in the case of the disc algebra.

In connection with the question of the closure of a restriction algebra, it is natural to ask whether $A|(E_1 \cup E_2)$ is closed, granted that $A|E_1$ and $A|E_2$ are both closed. In general, such a conclusion cannot be drawn, but Glicksberg [1] has established the following facts.

**20.13 *Theorem***

*Let $(A, X)$ be a uniform algebra and let $E_1, E_2$, and $E$ be closed subsets of $X$.*

(a) *If $\mu_{E_1} \in A^\perp$ for each $\mu \in A^\perp$, and if $A|E_2$ is closed in $\mathscr{C}(E_2)$, then $A|(E_1 \cup E_2)$ is closed in $\mathscr{C}(E_1 \cup E_2)$.*
(b) *If $A$ is logmodular on $X$, then $A|E$ is closed in $\mathscr{C}(E)$ if and only if $\mu \in A^\perp$ implies $\mu_E \in A^\perp$.*
(c) *If $A$ is logmodular on $X$ and if $A|E_j$ is closed in $\mathscr{C}(E_j)$, then $A|(E_1 \cup E_2)$ is closed in $\mathscr{C}(E_1 \cup E_2)$.*

PROOF

It is clear that part (c) follows immediately from (b) and (a).

Part (a) is a corollary of Theorem 20.7. By that theorem there is a constant $c \ge 1$ for which

$$\|\nu + (A|E_j)^\perp\| \le c\|\nu + A^\perp\|$$

for all $\nu \in \mathscr{M}(E_j)$, $j = 1, 2$. If $\nu \in \mathscr{M}(E_1 \cup E_2)$ and $\mu \in A^\perp$, then

$$\begin{aligned} \|\nu - \mu\| &= \|\nu_{E_1} - \mu_{E_1}\| + \|\nu_{E_2 \setminus E_1} - \mu_{X \setminus E_1}\| \\ &\ge \frac{1}{c}\|\nu_{E_1} + (A|E_1)^\perp\| + \frac{1}{c}\|\nu_{E_2 \setminus E_1} + (A|E_2)^\perp\| \end{aligned}$$

because both $\mu_{E_1}$ and $\mu_{X\setminus E_1} = \mu - \mu_{E_1}$ lie in $A^\perp$. It follows that

$$\|\nu - \mu\| \geq \frac{1}{c}\{\|\nu_{E_1} + (A|(E_1 \cup E_2))^\perp\| + \|\nu_{E_2\setminus E_1} + (A|(E_1 \cup E_2))^\perp\|\}$$

$$\geq \frac{1}{c}\|\nu + (A|(E_1 \cup E_2))^\perp\|,$$

whence $A|(E_1 \cup E_2)$ is closed.

To prove (b), it suffices, in light of Theorem 20.10, to show that if $A$ is logmodular on $X$ and $A|E$ is closed, then $\mu \in A^\perp$ implies that $\mu_E \in A^\perp$. If $A|E$ is closed, then

$$\|\nu + (A|E)^\perp\| \leq c\|\nu + A^\perp\|$$

for some $c$ and all $\nu \in \mathscr{M}(E)$. Thus, for $\mu \in A^\perp$,

$$\text{(6)} \qquad \|\mu_E + (A|E)^\perp \leq c\|\mu_E + A^\perp\| \leq c\|\mu - \mu_E\| = c\|\mu_{E^c}\|.$$

Consider a $\mu \in A^\perp$ with $\|\mu\| = 1$. If $\varepsilon > 0$ is given and if a compact set $K \subset E^c$ is such that $\|\mu_{E^c\setminus K}\| < \varepsilon$, then since $\{\log|f| : f \text{ is invertible in } A\}$ is dense in $\mathscr{C}(X)$, it follows that there is an invertible $f \in A$ such that $\|f\|_X \leq 1$, $|f| > 1 - \varepsilon$ on $E$, and $|f| < \varepsilon$ on $K$. The measure $f\mu$ is in $A^\perp$, so, by (6), we have

$$\|\mu_E + (A|E)^\perp\| = \|f^{-1}\{(f\mu)_E + (A|E)^\perp\}\|$$
$$\leq \|f^{-1}\|_E\|(f\mu)_E + (A|E)^\perp\|$$
$$\leq \frac{1}{1-\varepsilon}2c\varepsilon.$$

As this estimate is valid for all positive $\varepsilon$, we conclude that $\mu_E \in (A|E)^\perp$, as was to be shown.

We know from Theorem 20.10 that if $\mu_E = 0$ for all $\mu \in A^\perp$, then $A|E = \mathscr{C}(E)$. In general the converse of this statement does not hold, but there is an important class of algebras for which the converse is true. These are the algebras which *approximate in modulus* in the following sense.

**20.14** ***Definition***

*The uniform algebra $(A, X)$ approximates in modulus if, given a positive continuous function $f$ on $X$ and $\varepsilon < 0$, there is $F \in A$ such that $\||F| - f\|_X < \varepsilon$.*

It is clear that a logmodular algebra approximates in modulus on its Shilov boundary, but the class of algebras that approximate in modulus is much more extensive than the class of logmodular algebras, as the following result shows.

20.15 ***Theorem*** (Glicksberg [1])
*If $(A, X)$ is a uniform algebra and if the collection $\mathscr{U}$ of elements of $A$ of constant modulus 1 separates points on $X$, then $A$ approximates in modulus on $X$.*

PROOF
Note that the collection $\overline{\mathscr{U}}A = \{\bar{u}f : u \in \mathscr{U}, f \in A\}$ is an algebra. It is closed under multiplication and also under addition since

$$(\bar{u}_1 f_1 + \bar{u}_2 f_2) = \bar{u}_1\bar{u}_2(u_2 f_1 + u_1 f_2)$$

if $u_1, u_2 \in \mathscr{U}$. Thus, by the Stone–Weierstrass theorem, $\overline{\mathscr{U}}A$ is dense in $\mathscr{C}(X)$ since it contains the point separating conjugate closed subalgebra generated by $\mathscr{U}$ and $\overline{\mathscr{U}}$. It follows that if $h$ is a positive continuous function on $X$, and if $\varepsilon > 0$, there exist $u \in \mathscr{U}$ and $f \in A$ such that $\|h - \bar{u}f\|_X < \varepsilon$. But this implies that $\|h - |\bar{u}f|\|_X < \varepsilon$, i.e., $\|h - |f|\|_X < \varepsilon$, so $A$ approximates in modulus.

As an example of an algebra that approximates in modulus but which is not logmodular, consider $\mathscr{A}(\bar{\Delta}^N)$, the uniform closure of the polynomials in $N$ complex variables on $\bar{\Delta}^N$, the closed unit polycylinder in $\mathbb{C}^N$. As an algebra on its Shilov boundary

$$T^N = \{(e^{i\theta_1}, \ldots, e^{i\theta_N}) : \theta_1, \ldots, \theta_N \text{ real}\},$$

$\mathscr{A}(\bar{\Delta}^N)$ approximates in modulus by the last theorem since, e.g., the monomials separate points on $T^N$. However, $\mathscr{A}(\bar{\Delta}^N)$ is not logmodular.

For algebras that approximate in modulus, there is the following interpolation theorem.

20.16 ***Theorem*** (Glicksberg [1])
*If $(A, X)$ is a uniform algebra that approximates in modulus, then the following conditions on a closed set $E \subset X$ are equivalent:*

(a) $A|E = \mathscr{C}(E)$.
(b) $\mu \in A^{\perp}$ *implies* $\mu_E = 0$.
(c) *Each* $f \in \mathscr{C}(E)$ *has an extension* $f$ *in* $A$ *such that* $\|f\|_X = \|f\|_E$.

*If these equivalent conditions are satisfied, then $E$ is an intersection of peak sets for $A$.*

PROOF

Theorem 20.10 shows that condition (b) implies that $E$ is an intersection of peak sets and that $A|E = \mathscr{C}(E)$. Thus, (b) implies (a) and thus also (c) by the same theorem. Consequently, we need only prove that (a) implies (b).

Assume that $A|E = \mathscr{C}(E)$ so that for some constant $\alpha$ and all $\mu \in A^\perp$, $\|\mu_E\| \leq \alpha\|\mu_{E^c}\|$ by Corollary 20.8. Consider a $\mu \in A^\perp$ with $\|\mu\| = 1$. If $\varepsilon > 0$, there is a compact set $K \subset E^c$ with $|\mu|(E^c \setminus K) < \varepsilon$. There exists an element $h \in A$ with $\|h\| \leq 1$, $|h| > 1 - \varepsilon$ on $E$, and $|h| < \varepsilon$ on $K$ since $A$ approximates in modulus on $X$. We have $h\mu \in A^\perp$ and thus

$$\begin{aligned}(1 - \varepsilon)\|\mu_E\| &\leq \|(h\mu)_E\| \\ &\leq \alpha\|(h\mu)_{E^c}\| \\ &\leq \alpha\|(h\mu)_{E^c \setminus K}\| + \alpha\varepsilon\|\mu_K\| \\ &< 2\alpha\varepsilon.\end{aligned}$$

It follows that $\mu_E = 0$, as was to be shown.

So far in this section we have not considered an obvious question: Do there exist nontrivial interpolation sets for a given uniform algebra? We now turn to this question. We would like to show that if $(A, X)$ is a uniform algebra with $X$ infinite, then there exists an infinite set $E \subset X$ which is an interpolation set or even a peak interpolation set for $A$. However, the existence of such a set seems to be an open question. The result is known only under various topological restrictions on the space $X$. For instance, Bernard [1] has shown that if $X$ is an infinite metric space, then there exists an infinite peak interpolation set for $A$, and Heard and Wells [1] have observed that if $x$ is a limit point of $X$ which is a peak point for $A$, then there is an infinite interpolation set for $A$. The result of Bernard is an easy consequence of the theory now at our disposal as the following remarks show.

Let $A$ be a uniform algebra on $X$, $X$ compact metric, and let $B$ be the minimal boundary for $A$. If $B$ is countable, then by Corollary 7.33, $B = X$, whence, by Corollary 12.8, $A = \mathscr{C}(X)$ and we are done. If $B$ is uncountable, it contains a countably infinite set $E = \{x_0, x_1, \ldots, x_n, \ldots\}$ with $x_0 = \lim_{n\to\infty} x_n$. If $\mu \in A^\perp$, then since each of the points $x_n$ is a peak point for $A$, $\mu(\{x_n\}) = 0$, so $\mu(E) = 0$. Thus, $E$ is a peak interpolation set for $A$.

An essentially better result is due to Pełczyński [3].

**20.17** ***Theorem***

*If $(A, X)$ is a uniform algebra with $X$ an uncountable compact metric* *space, there exists a set E in X which is homeomorphic to the Cantor set and which is a peak interpolation set for A.*

The proof we give is not that given originally by Pełczyński but is somewhat less computational.

PROOF

For the proof we shall assume a special case of the result, the case that $X$ is a planar set and $A$ is the uniform closure in $\mathscr{C}(X)$ of the polynomials in a complex variable. This special case of the result is contained in our general discussion of algebras on planar sets in the next chapter and is, of course, independent of the present result. For the special case, see Corollary 25.10.

We begin the proof of the theorem with several simple reductions. First, since each uncountable closed subset of a compact metric space contains a set homeomorphic to a Cantor set, it suffices to prove the existence of an uncountable peak interpolation set. Second, we may suppose that $X$ is the spectrum of $A$, for in any case $X \subset \Sigma(A)$ and $X$ contains the Shilov boundary for $A$. As it is clear that a peak interpolation set for $A$ will lie in the Shilov boundary, we see that every such set must be contained in $X$. It now follows that we can assume $X$ connected, for if $X$ is totally disconnected, then since $A$ is uniformly closed, it follows easily from the Shilov idempotent theorem that $A$ is $\mathscr{C}(X)$, and we are done. (Recall Corollary 8.18.) Thus, if $X$ has no nontrivial components, then the result is immediate. On the other hand, if $K$ is a nontrivial component of $X$, then $K$ is the countable intersection of open and closed subsets $K_n$ of $X$, and again by the Shilov idempotent theorem, each $K_n$ is a peak set, so $K$ itself is a peak set. Thus, $A$ is closed in $\mathscr{C}(K)$, and it suffices to produce an uncountable peak interpolation set for $A|K$. This shows that we may suppose $X$ connected. Next it is enough to prove the existence of an uncountable, totally disconnected peak set. To see this, observe that if $E$ is such a set, then the restriction algebra $A|E$ is closed in $\mathscr{C}(E)$ and, as it can be identified with the quotient algebra $A/I(E)$, $I(E) = \{f \in A : f|E = 0\}$, which has $E$ as its spectrum, it follows that $A|E = \mathscr{C}(E)$ because $E$ is totally disconnected.

As $X$ is a compact metric space, $\mathscr{C}(X)$ is separable and so $A$ is separable, and thus it admits a countable set of generators, say, $\{f_1, f_2, \ldots\}$. Let $\mathbb{N} = \{1, 2, 3, \ldots\}$, let $\mathbb{C}^{\mathbb{N}}$ be the usual cartesian product, and for $n \in \mathbb{N}$, let $\pi_n : \mathbb{C}^{\mathbb{N}} \to \mathbb{C}$ be given by $\pi_n(\mathfrak{z}) = z_n$ if $z_n$ is the $n$th coordinate of $\mathfrak{z}$. Let $\rho$ denote the metric on $\mathbb{C}^{\mathbb{N}}$ given by

$$\rho(\mathfrak{z}, \mathfrak{w}) = \sum_{n=1}^{\infty} 2^{-n} \frac{|\pi_n(\mathfrak{z}) - \pi_n(\mathfrak{w})|}{1 + |\pi_n(\mathfrak{z}) - \pi_n(\mathfrak{w})|}.$$

The $\rho$-diameter of $\mathbb{C}^N$ is 1. If $i: X \to \mathbb{C}^N$ is the continuous map given by requiring that $\pi_n(ix) = f_n(x)$, then under $i$, $X$ goes onto a compact, polynomially convex subset of $\mathbb{C}^N$. (Recall Theorem 5.8.) We identify $X$ with $iX$ so that $X$ itself is a polynomially convex set in $\mathbb{C}^N$, and $A$ is then identified via the Gelfand transform with the algebra $\mathscr{P}(X)$, the uniform closure in $\mathscr{C}(X)$ of the algebra generated by the projections $\pi_n$.

Consider now the smallest integer $n_1 \in \mathbb{N}$ with the property that $\pi_{n_1}(X)$ is not a point. Thus, $\pi_{n_1}(X)$ is a certain continuum in $\mathbb{C}$, so there exists a totally disconnected uncountable peak interpolation set $E_{n_1}$ for the algebra $\mathscr{P}(\pi_{n_1}(X))$. If $g \in \mathscr{P}(\pi_{n_1}(X))$, then since $A$ is uniformly closed, $g \circ \pi_{n_1} \in A$ because we are assuming that $X$ is the spectrum of $A$. Consequently, the set $\check{E}_{n_1} = \pi_{n_1}^{-1}(E_{n_1}) \cap X$ is an uncountable peak set for $A$. The set $E_{n_1}$ is totally disconnected, so either $\check{E}_{n_1}$ contains an open and closed subset which is uncountable and totally disconnected, or else $\check{E}_{n_1}$ contains two nontrivial continua, $X_{1,0}$ and $X_{1,1}$. In the former case we are finished, for an open and closed subset of a peak set for $A$ is again a peak set. In the latter case, note that since $E_{n_1}$ is totally disconnected, $X_{1,0}$ and $X_{1,1}$ both project onto points under $\pi_{n_1}$ and hence have $\rho$-diameter no more than $2^{-n_1} \leq 2^{-1}$. Each of the sets $X_{1,0}$ and $X_{1,1}$ is a peak set for $A$. Define $X_1$ to be $X_{1,0} \cup X_{1,1}$, a peak set for $A$. Let $n_2$ be the smallest element of $\mathbb{N}$ such that $\pi_{n_2}$ is nonconstant on one of $X_{1,0}, X_{1,1}$, say on $X_{1,0}$. Arguing as above, we either obtain an uncountable, totally disconnected peak set for $A$ contained in $X_{1,0}$, or else we find nontrivial disjoint connected sets $X_{2,0}$ and $X_{2,1}$ in $X_{1,0}$ which are peak sets for $A$ and which project under $\pi_{n_2}$ onto points. Thus, $X_{2,0}$ and $X_{2,1}$ have $\rho$-diameter no more than $2^{-n_2}$. If $\pi_{n_2}$ is nonconstant on $X_{1,1}$, construct similar sets $X_{2,2}$ and $X_{2,3}$ in $X_{1,1}$, and define $X_2$ to be the set $X_{2,0} \cup \cdots \cup X_{2,3}$. If $\pi_{n_2}$ is constant on $X_{1,1}$, then $X_{1,1}$ has $\rho$-diameter no more than $2^{-n_2}$, and we define $X_2$ to be $X_{2,0} \cup X_{2,1} \cup X_{1,1}$. In either case, $X_2$ is a peak set with at least three components each of which has $\rho$-diameter no more than $2^{-2}$.

We iterate this process and either we obtain a set of the desired kind at some step, or else we obtain a sequence $\{X_n\}_{n=1}^{\infty}$ of compact subsets of $X$ with the following properties:

(a) $X_1 \supset X_2 \supset \cdots$.

(b) $X_n$ has at least $n+1$ components each of which has $\rho$-diameter no more than $2^{-n}$.

(c) $X_n$ is a peak set for $A$.

Let $E = \bigcap_{n=1}^{\infty} X_n$. The set $E$ is a totally disconnected, uncountable peak set for $A$, and the theorem is proved.

# 21 INTERPOLATION WITH POLYDISC ALGEBRAS

We began this chapter on interpolation with the Rudin–Carleson theorem, a result that completely settles the problem of interpolation for the disc algebra. It is natural to consider the interpolation problem for various $N$-dimensional analogues of the disc algebra, and it is one such generalization that we shall consider in this section. The theory with which we shall be concerned has not yet attained the degree of completeness of the one-dimensional theory; it presents yet another manifestation of the familiar fact that the transition from the study of functions of a single complex variable to the study of functions of several complex variables involves one in many essentially new problems.

The setting for our considerations will be the *polydisc algebras.* Recall that $\Delta^N$ is the unit polydisc in $\mathbb{C}^N$:

$$\Delta^N = \{(z_1, \ldots, z_N) \in \mathbb{C}^N : |z_1|, \ldots, |z_N| < 1\},$$

and that $T^N$ is the torus

$$T^N = \{(e^{i\theta_1}, \ldots, e^{i\theta_N})\theta_1, \ldots, \theta_N \text{ real}\}.$$

Throughout this section, we shall denote by $\mathscr{A}_N$ the uniform closure of the polynomials on the set $\bar{\Delta}^N$, and we shall refer to the algebras $\mathscr{A}_N$ as the *polydisc algebras.* Thus, $\mathscr{A}_1$ is the disc algebra that we have considered frequently in previous sections. The spectrum of $\mathscr{A}_N$ can be identified with $\bar{\Delta}^N$, and the Shilov boundary is the torus $T^N$.

The algebra $\mathscr{A}_N$ is generated by the coordinate functions on $\mathbb{C}^N$, and as these functions assume on $T^N$ values of modulus 1, it follows from Theorem 20.15 that the algebra $\mathscr{A}_N$ approximates in modulus on $T^N$. Consequently, Theorem 20.16 applies; it yields the following facts.

21.1 ***Theorem***

*The following conditions on a closed set $E \subset T^N$ are equivalent:*

(a) $\mu \in \mathscr{A}_N^\perp$ *implies* $\mu_E = 0$.
(b) $\mathscr{A}_N | E = \mathscr{C}(E)$.
(c) *$E$ is a peak interpolation set for $\mathscr{A}_N$.*

This list of equivalences can be extended as the following result shows.

21.2 ***Theorem*** (Stout [1])

*The set $E \subset T^N$ is a peak interpolation set for $\mathscr{A}_N$ if and only if there is $f \in \mathscr{A}_N$ such that*

$$E = \{\mathfrak{z} \in \bar{\Delta}^N : f(\mathfrak{z}) = 0\}.$$

PROOF

One implication is clear: A peak interpolation set is surely a zero set.

In the other direction, note that if $E \subset T^N$ is the zero set of an element of $\mathscr{A}_N$ and if $I_E = \{f \in \mathscr{A}_N : f|E = 0\}$, then $\mathscr{A}_N/I_E$ has $E$ as its spectrum, and it is generated by elements $\pi_j + I_E$, if we use $\pi_j$ to denote the projection which takes $\mathfrak{z} \in \mathbb{C}^N$ onto its $j$th coordinate. The set $E$ is in $T^N$, so the spectrum of the element $\pi_j + I_E$ of $\mathscr{A}_N/I_E$ is a subset of the unit circle, and this implies that $\pi_j + I_E$ is invertible. Let $f_j + I_E$ be its inverse. On $E$, we have $f_j = 1/\pi_j = \bar{\pi}_j$, so $\mathscr{A}_N|E$ contains the conjugate closed subalgebra of $\mathscr{C}(E)$ generated by $\pi_j$ and $\bar{\pi}_j, j = 1, \ldots, N$. It follows from the Stone–Weierstrass theorem that $\mathscr{A}_N|E$ is dense in $\mathscr{C}(E)$.

To prove that $\mathscr{A}_N|E$ is closed in $\mathscr{C}(E)$, we shall prove that since $E$ is the zero set of an element $f \in \mathscr{A}_N$, $E$ is a peak set for $\mathscr{A}_N$. We may suppose that $\|f\| < 1$. The function $g$ given by

$$g = \frac{\log f}{-1 + \log f}$$

lies in $\mathscr{A}_N$ and it peaks on $E$. Thus, $\mathscr{A}_N|E$ is closed, and the theorem is proved.

There are various corollaries of this theorem. The first we mention is a fact about zero sets, a fact that does not seem to admit a more direct proof. (The result in the one-dimensional case is trivial because of the simple characterization of the subsets of the circle which are zero sets of elements of the disc algebra as the closed sets of measure zero.)

21.3 ***Corollary***

*If $E \subset T^N$ is the zero set of an element of $\mathscr{A}_N$, so is every closed subset of $E$.*

PROOF

Every closed subset of a peak interpolation set for $\mathscr{A}_N$ is again such a set.

21.4 ***Corollary***

*If $E \subset T^N$ is a closed set which is a union $\bigcup\{E_n : n = 1, 2, \ldots\}$ of sets $E_n$ each of which is the zero set of an element of $\mathscr{A}_N$, then $E$ is the zero set of an element of $\mathscr{A}_N$.*

PROOF

It follows from the equivalence of (a) and (c) in Theorem 21.1 that the countable closed union of peak interpolation sets is a peak interpolation set and hence a zero set.

We can use Theorem 21.2 to give some examples of peak interpolation sets for $\mathscr{A}_N$. These examples are from the paper of Rudin and Stout [1].

21.5 ***Theorem***

*A closed subset $E$ of $T^N$ is a peak interpolation set for $\mathscr{A}_N$ if it satisfies either of the following conditions:*

(a) *$E$ has zero linear measure.*

(b) *There exists a closed subset $S$ of the unit circle which has Lebesgue measure zero, and there exist strictly positive integers $k_1, \ldots, k_N$ such that*

$$E \subset \{(z_1, \ldots, z_N) \in T^N : z_1^{k_1} \cdots z_N^{k_N} \in S\}.$$

The linear (or one-dimensional Hausdorff) measure of a set in $\mathbb{C}^N$ is to be computed with respect to the usual metric on $\mathbb{C}^N$. (For the definition of Hausdorff measures, see Appendix B.)

PROOF

(a) Again we use $\pi_j$ to denote the projection from $\mathbb{C}^N$ onto the $j$th coordinate. Since $E$ has zero linear measure, the set $\pi_j(E)$ is a Lebesgue null set in the circle, so there is $f_j \in \mathscr{A}_1$ which peaks on $E_j = \pi_j(E)$. If $\varphi \in \mathscr{A}_N$ is defined by

$$\varphi(z_1, \ldots, z_N) = f_1(z_1) \cdots f_N(z_N),$$

the function $\varphi$ peaks on the set $F = E_1 \times \cdots \times E_N$ in $T^N$, and as $E \subset F$, the result follows from the results already established.

(b) Let $f \in \mathscr{A}_1$ vanish on the set $S$ and nowhere else, and define $\varphi \in \mathscr{A}_N$ by

$$\varphi(z_1, \ldots, z_N) = f(z_1^{k_1} \cdots z_N^{k_N}).$$

The function $\varphi$ vanishes exactly on the set

$$\{(z_1, \ldots, z_N) \in T^N : z_1^{k_1} \cdots z_N^{k_N} \in S\}, \tag{1}$$

so the result follows from Theorem 21.2.

The condition (b) has a group theoretic significance which should be mentioned. Define $\chi$ on $T^N$ by

$$\chi(e^{i\theta_1}, \ldots, e^{i\theta_N}) = e^{i(k_1\theta_1 + \cdots + k_N\theta_N)}.$$

The function $\chi$ is a character of the group $T^N$, and if $G$ is its kernel, then the set (1) is a certain union of cosets of $G$.

For $N = 1$, the Rudin–Carleson theorem shows that there is a very simple measure theoretic condition which characterizes the peak interpolation sets for the disc algebra. There does not seem to be a comparably simple metric characterization of peak interpolation sets for $\mathscr{A}_N$. Part (a) of the preceding theorem gives a metric condition sufficient for a set to be a peak interpolation set, and part (b) provides examples of peak interpolation sets which are not even if $\sigma$-finite $(N - 1)$-dimensional measure. The following example indicates the complexity of the situation and suggests that metric criteria may not by sufficient to characterize the peak interpolation sets for $\mathscr{A}_N$.

21.6 ***Theorem*** (Rudin and Stout [1])
*If $\varepsilon > 0$, there exist closed sets $E \subset T^N$ which are of linear measure no more than $\varepsilon$ but which have the property that if $f \in \mathscr{A}_N$ vanishes on $E$, then $f$ is the zero function.*

PROOF
Let $\mu_1, \ldots, \mu_N$ be positive numbers which are rationally independent in that $q_k\mu_1 + \cdots + q_N\mu_N = 0$, with integers $q_j$, implies $0 = q_1 = \cdots = q_N$. Let $Q = \{\zeta \in \mathbb{C} : \operatorname{Re} \zeta < 0\}$, and let $\tilde{E}$ be a compact set of positive Lebesgue measure in the imaginary axis. If we define $\varphi : \bar{Q} \to \bar{\Delta}^N$ by

$$\varphi(\zeta) = (e^{\mu_1\zeta}, \ldots, e^{\mu_N\zeta}),$$

and $E$ by $E = \varphi(\tilde{E})$, then $E$ will have one-dimensional measure less than $\varepsilon$ if $\tilde{E}$ has small enough Lebesgue measure.

If $f \in \mathscr{A}_N$ vanishes on $E$, then $f \circ \varphi$, a function continuous on $\bar{Q}$ and holomorphic on $Q_1$, vanishes on the set $\tilde{E}$. As we have assumed $\tilde{E}$ to have positive measure, this implies that $f \circ \varphi$ must vanish identically, so $f$ vanishes on the range of $\varphi$.

We have assumed that the numbers $\mu_j$ are rationally independent, so Kronecker's theorem (Appendix *E*) implies that $\varphi(\partial Q)$ is dense in $T^N$. Thus, $f$ vanishes identically, and the theorem is proved.

We next turn to a localization theorem for peak interpolation sets for $\mathscr{A}_N$ which generalizes a very simple observation about the disc algebra. Let $D_1, \ldots, D_q$ be a finite family of open discs in $\mathbb{C}$ which cover the unit circle, and let $E$ be a closed subset of the unit circle. Using the Rudin–Carleson theorem, it is easy to see that $E$ is a peak interpolation set for $\mathscr{A}_1$ provided that $E \cap (D_j \cap \Delta)^-$ is a peak interpolation set for the algebra $\mathscr{A}(\bar{D}_j \cap \bar{\Delta})$. Such a localization result obtains for peak interpolation sets for $\mathscr{A}_N$, but it requires somewhat more effort to prove. It is based on the solution of a Cousin II-like problem on $\bar{\Delta}^N$, a problem in which Cousin data are required to have continuous boundary values, as is the solution. Explicitly, we shall prove the following result.

21.7 ***Theorem***

*Let $\mathscr{V} = \{V_\alpha\}_{\alpha\in I}$ be an open cover for $\bar{\Delta}^N$, and for each $\alpha$ let $f_\alpha \in \mathscr{C}(\bar{V}_\alpha)$ be holomorphic on the interior (with respect to $\mathbb{C}^N$) of $V_\alpha$. If for each pair $\alpha,\beta \in I$ there exists a zero-free $f_{\alpha\beta} \in \mathscr{C}(\overline{V_\alpha \cap V_\alpha})$ such that*

$$f_\alpha|(\overline{V_\alpha \cap V_\beta}) = (f_\beta|(\overline{V_\alpha \cap V_\beta}))f_{\alpha\beta},$$

*then there is $F \in \mathscr{A}_N$ such that for each $\alpha$ and some zero-free function $g_\alpha \in \mathscr{C}(\bar{V}_\alpha)$, $F|\bar{V}_\alpha = g_\alpha f_\alpha$.*

The sets $V_\alpha$ are relatively open in $\bar{\Delta}^N$, but, of course, they need not be open in $\mathbb{C}^N$. Note also that if no $f_\alpha$ vanishes on an open set, then the functions $g_\alpha$ are necessarily holomorphic on the interior of $V_\alpha$. Similarly, $f_{\alpha\beta}$ is necessarily holomorphic on each component of the interior of $V_\alpha \cap V_\beta$ where $f_\beta$ does not vanish identically.

Before proving the theorem, let us see that it does imply a localization result of the announced kind.

21.8 ***Corollary***

*Let $E$ be a closed subset of $T^N$, and let $\mathscr{V} = \{V_\alpha\}$ be a collection of relatively open subsets of $\bar{\Delta}^N$ whose union contains $E$ and which have the property that for each $\alpha$, the set $E \cap \bar{V}_\alpha$ is a peak interpolation set for the algebra $\mathscr{A}(\bar{V}_\alpha)$ of all $f \in \mathscr{C}(\bar{V}_\alpha)$ which are holomorphic on the interior of $V_\alpha$. Then $E$ is a peak interpolation set for $\mathscr{A}_N$.*

PROOF

If $\mathfrak{z} = (z_1, \ldots, z_N) \in E$, the openness of the elements of $\mathscr{V}$ implies that for some $\delta(\mathfrak{z}) > 0$, the neighborhood

$$N(\mathfrak{z}, \delta(\mathfrak{z})) = \{(w_1, \ldots, w_N) \in \bar{\Delta}^N : |w_1 - z_1|, \ldots, |w_N - z_N| < \delta(\mathfrak{z})\}$$

is contained in one of the elements of $\mathscr{V}$. Compactness implies the existence of finitely many points $\mathfrak{z}_j \in E$ such that

$$E \subset \bigcup N(\mathfrak{z}_j, \tfrac{1}{2}\delta(\mathfrak{z}_j)).$$

By hypothesis, $E \cap \bar{V}_\alpha$ is a peak interpolation set for $\mathscr{A}(V_\alpha)$, so it follows that $E \cap \overline{N(\mathfrak{z}_j, \delta(\mathfrak{z}_j))}$ is a peak interpolation set for $\mathscr{A}(\overline{N(\mathfrak{z}_j, \delta(\mathfrak{z}_j))})$. Consequently, the set $E_j = E \cap \overline{N(\mathfrak{z}_j, \frac{1}{2}\delta(\mathfrak{z}_j))}$ is also a peak interpolation set for $\mathscr{A}(\overline{N(\mathfrak{z}_j, \delta(\mathfrak{z}_j))})$, so there is a function $h_j$ in this latter algebra which has $E_j$ as its zero set.

Let $\mathfrak{A}$ be the cover $\{U_1, U_2\}$ for $\bar{\Delta}_N$ given by

$$U_1 = N(\mathfrak{z}_j, \delta(\mathfrak{z}_j))$$

and

$$U_2 = \bar{\Delta}_N \setminus \overline{N(\mathfrak{z}_j, \tfrac{3}{4}\delta(\mathfrak{z}_j))},$$

so if $(w_1, \dots, w_N) \in U_2$, at least one of $|w_1 - z_1|, \dots, |w_N - z_N|$ is at least $\frac{3}{4}\delta(\mathfrak{z}_j)$. Associate with this cover the set of Cousin data $h_j \in \mathscr{A}(\overline{U}_1)$ and $g_j \in A(\overline{U}_2)$, where $g_j$ is the function identically 1 on $\overline{U}_2$. Since $\overline{U}_1 \cap \overline{U}_2$ is disjoint from the zero set of $h_j$, the theorem stated above applies, and we find that there is an element $F_j \in \mathscr{A}_N$ which has the set $E_j$ as its zero set. If we let $F$ be the product of the finitely many functions $F_j$ constructed in this way, then $F$ is an element of $\mathscr{A}_N$ which has $E$ as its zero set. The corollary now follows from Theorem 21.2.

Before passing to the proof of Theorem 21.7, it is necessary to give certain analytic preliminaries. To begin with, we shall need a very simple estimate on the size of the partial sums of the power series of a bounded holomorphic function. Suppose that $f$ is bounded and holomorphic in the unit disc, say $|f(z)| \leq M$. If $f(z) = \sum_{n=0}^{\infty} a_n z^n$ and if $s_n(z) = a_0 + a_1 z + \cdots + a_n z^n$, then each of the coefficients $a_j$ satisfies $|a_j| \leq M$, so we have the estimate that $|s_n(z)| \leq M(n+1)$. [It is not true that $s_n(z)$ is bounded uniformly in $n$ and $z$, but our very simple estimate is not nearly the best estimate possible: It is not hard to prove that $|s_n(z)| \leq CM \log n$ for some absolute constant $C$. These questions are discussed in Titchmarsh [1].]

The second preliminary fact we need is a simple estimate on the Fourier coefficients of smooth functions. Suppose that $f$ is a $\mathscr{C}^\infty$ function on the circle so that $f$ has a Fourier series expansion

$$f(e^{i\theta}) = \sum_{n=-\infty}^{\infty} a_n e^{in\theta}.$$

The coefficients $a_n$ are given by

$$a_n = \frac{1}{2\pi} \int_{-\pi}^{\pi} f(e^{i\varphi}) e^{-in\varphi}\, d\varphi.$$

If we integrate by parts repeatedly, we find that for $k = 1, 2, \dots$, there exist constants $C_k$ such that

$$|a_n| < C_k(|n| + 1)^{-k}.$$

21.9 ***Lemma***

*If $F \in \mathscr{A}_N$ and if $h \in \mathscr{C}^\infty(T)$, the function $G$ defined by*

$$G(z_1, \dots, z_N) = \frac{1}{2\pi i} \int_{|\zeta|=1} \frac{F(\zeta, z_2, \dots, z_N)}{\zeta - z_1} h(\zeta)\, d\zeta$$

*is an element of $\mathscr{A}_N$.*

PROOF

It is clear that $G$ is holomorphic in $\Delta^N$; we must prove continuity on $\bar{\Delta}^N$. For this purpose, let $h$ have the Fourier expansion

$$h(e^{i\theta}) = \sum_{n=-\infty}^{\infty} a_n e^{in\theta}.$$

If

$$g_n(z_1, \ldots, z_N) = \frac{1}{2\pi i} \int_{|\zeta|=1} \frac{F(\zeta, z_2, \ldots, z_N)}{\zeta - z_1} \zeta^n \, d\zeta,$$

we have the expansion

$$G(z_1, \ldots, z_N) = \sum_{n=-\infty}^{\infty} a_n g_n(z_1, \ldots, z_N) \tag{2}$$

valid in $\Delta^N$. For $n \geq 0$, the Cauchy integral theorem yields that

$$g_n(z_1, \ldots, z_N) = z_1^n F(z_1, z_2, \ldots, z_N),$$

so for such values of $n$, $g_n \in \mathscr{A}_N$, and $\|g_n\|_{\Delta^N} = \|F\|_{\Delta^N}$. For $n < 0$, we have that if $s_q(z_1, \ldots, z_N)$ denotes the $q$th partial sum of the power-series expansion of the function $F(\cdot, z_2, \ldots, z_N)$, then

$$g_n(z_1, \ldots, z_N) = z_1^n(F(z_1, \ldots, z_N) - s_{|n|-1}(z_1, \ldots, z_N)).$$

We have that $|s_{|n|-1}(z_1, \ldots, z_N)| \leq |n| \|F\|$, and we know that $|a_n| \leq C(|n| + 1)^{-3}$, so it follows that the series (2) converges uniformly on $\bar{\Delta}^N$, and thus $G \in \mathscr{A}_N$, as was to be shown.

We now come to the lemma which is basic for our proof of Theorem 21.7. It is more convenient to work on cubes than on polydiscs, so we introduce the following notation:

$$K = \{x + iy : |x|, |y| < 1\},$$

and $K^N$ is the product of $N$ copies of $K$. Let

$$K_+ = \{x + iy \in K : x > -\tfrac{1}{2}\}$$

and

$$K_- = \{x + iy \in K : x < \tfrac{1}{2}\}.$$

21.10 ***Lemma***

*If $F \in \mathscr{A}((K_+ \cap K_-) \times K^{N-1})$ is zero free, then there are zero-free elements $F_+ \in \mathscr{A}(K_+ \times K^{N-1})$ and $F_- \in \mathscr{A}(K_- \times K^{N-1})$ such that on $(K_+ \cap K_-) \times K^{N-1}$, $F = F_+ F_-$.*

PROOF

Since $F$ is zero free, the topological simplicity of $(K_+ \cap K_-) \times K^{N-1}$ implies that $G = \log F$ is a well-defined member of $\mathscr{A}((K_+ \cap K_-) \times K^{N-1})$. To prove the lemma, it is sufficient to show that on $(K_+ \cap K_-) \times K^{N-1}$,

$$G = G_+ + G_-, \qquad G_+ \in \mathscr{A}(K_+ \times K^{N-1}), \qquad G_- \in \mathscr{A}(K_- \times K^{N-1}).$$

Let $h$ be a function defined on $\partial(K_+ \cap K_1)$, call this set $\gamma$, which is identically 1 on the part of $\gamma$ to the left of the line $x = -\frac{1}{4}$ and is identically zero on the part of $\gamma$ to the right of the line $x = \frac{1}{4}$. Require also that $h$ be infinitely differentiable on the intervals $\{t + i : -\frac{1}{2} < t < \frac{1}{2}\}$ and $\{t - i : -\frac{1}{2} < t < \frac{1}{2}\}$. By the Cauchy integral formula, we have that if $(z_1, \ldots, z_N) \in (K_+ \cap K_-) \times K^{N-1}$, then

$$\begin{aligned} G(z_1, \ldots, z_N) &= \frac{1}{2\pi i}\int_\gamma \frac{G(\zeta, z_2, \ldots, z_N)}{\zeta - z_1}\, d\zeta \\ &= \frac{1}{2\pi i}\int_\gamma \frac{G(\zeta, z_2, \ldots, z_N)}{\zeta - z_1} h(\zeta)\, d\zeta \\ &\quad + \frac{1}{2\pi i}\int \frac{G(\zeta, z_2, \ldots, z_N)}{\zeta - z_1}(1 - h(\zeta))\, d\zeta. \end{aligned}$$

Call the first of these integrals $G_+$, the second $G_-$. The functions $G_+$ and $G_-$ are holomorphic in $K_+ \times K^{N-1}$ and $K_- \times K^{N-1}$, respectively, and it follows easily from the last lemma that they actually have continuous boundary values. Thus, $G = G_+ + G_-$ is the desired decomposition.

Douady [1] has established a somewhat more general version of this lemma.

With these preliminaries at our disposal, we can now establish the theorem by a subdivision process familiar in this sort of problem.

## 21.11 *Proof of Theorem* 21.7

Again it is convenient to work on cubes rather than on polydiscs. For the purposes of our problem, cubes are entirely equivalent to polydiscs, for if $K$ is the open unit square and if $\varphi$ is a conformal mapping of the unit disc onto $K$, then $\varphi$ extends to a homeomorphism of the closed unit disc onto the closure of $K$; and if we define $\Phi : \bar{\Delta}^N \to \bar{K}^N$ by $\Phi(z_1, \ldots, z_N) = (\varphi(z_1), \ldots, \varphi(z_N))$, then $\Phi$ is a homeomorphism which takes $\Delta^N$ biholomorphically onto $K^N$.

Assume given an open covering $\mathscr{V} = \{V_\alpha\}$ of $\bar{K}^N$, and, for each $\alpha$, an $f_\alpha \in \mathscr{A}(V_\alpha)$ such that on $V_\alpha \cap V_\beta$, $f_\alpha = f_{\alpha\beta} f_\beta$, where $f_{\alpha\beta}$ is an invertible element of $\mathscr{A}(V_\alpha \cap V_\beta)$. We assume that none of the functions $f_\alpha$ is the zero function, for if one is, all are, and the problem is trivial. Let us assume that our problem is not solvable, i.e., that there does not exist $F \in \mathscr{A}(K^N)$ such that for some invertible $h_\alpha \in \mathscr{A}(V_\alpha)$, $F = h_\alpha f_\alpha$ on $V_\alpha$. Let

$$X_{1+} = \bar{K}_+ \times \bar{K}^{N-1}$$

and

$$X_{1-} = \bar{K}_- \times \bar{K}^{N-1},$$

and assume the problem solvable on both $X_{1+}$ and $X_{1-}$; i.e., assume that there are $F_+ \in \mathscr{A}(X_{1+})$ and $F_- \in \mathscr{A}(X_{1-})$ such that for all indices $\alpha$, there exist invertible functions $g_{+\alpha} \in \mathscr{A}(X_{1+} \cap V_\alpha)$ and $g_{-\alpha} \in \mathscr{A}(X_{1-} \cap V_\alpha)$ with the property that

$$F_+ = g_{+\alpha} f_\alpha \qquad \text{on } X_{1+} \cap V_\alpha$$

and

$$F_- = g_{-\alpha} f_\alpha \qquad \text{on } X_{1-} \cap V_\alpha.$$

From the fact that on $X_{1+} \cap X_{1-} \cap V_\alpha$ we have

(3) $$F_- g_{+\alpha} g_{-\alpha}^{-1} = F_+,$$

it follows that we can obtain a well-defined $h \in \mathscr{A}(X_{1+} \cap X_{1-})$ by requiring that on $X_{1+} \cap X_{1-} \cap V_\alpha$, $h$ be given by

(4) $$h = g_{+\alpha} g_{-\alpha}^{-1}.$$

The function $h$ obtained in this way is zero free, so by the preceding lemma, there is a decomposition, valid on $X_{1+} \cap X_{1-}$,

(5) $$h = h_+^{-1} h_-,$$

where $h_+$ is an invertible element of $\mathscr{A}(X_{1+})$ and $h_-$ an invertible element of $\mathscr{A}(X_{1-})$. Define a function $F$ on $\bar{K}^N$ by

$$F = \begin{cases} F_- h_- & \text{on } X_{1-}, \\ F_+ h_+ & \text{on } X_{1+}. \end{cases}$$

On $V_\alpha \cap X_{1+} \cap X_{1-}$ we have, because of (3), (4), and (5), that

$$\begin{aligned} F_- h_- &= F_- h_- h_+^{-1} h_+ \\ &= F_- h h_+ \\ &= F_+ h_+, \end{aligned}$$

so $F$ is a well-defined element of $\mathscr{A}(\bar{K}^N)$.

If $V_\alpha \in \mathscr{V}$ is contained within $X_{1+}$, then we have that on $V_\alpha$,

$$F = F_+ h_+$$
$$= h_+ g_{+\alpha} f_\alpha,$$

so $F$ differs from $f_\alpha$ by a multiplicative factor of an invertible element of $\mathscr{A}(V_\alpha)$. A similar observation is valid if $V_\alpha$ is contained within $X_{1-}$. We shall prove next that even if $V_\alpha$ is not contained entirely within one of $X_{1+}$ or $X_{1-}$, $F$ still differs from $f_\alpha$ by an invertible element of $\mathscr{A}(V_\alpha)$. This is so, for, as above, we have that on $V_\alpha \cap X_{1+}$,

$$F = h_+ g_{+\alpha} f_\alpha, \tag{6}$$

while on $V_\alpha \cap X_{1-}$,

$$F = h_- g_{-\alpha} f_\alpha. \tag{7}$$

We have on $X_{1-} \cap X_{1+}$, $h = h_+^{-1} h_-$, and on $X_{1-} \cap X_{1+} \cap V_\alpha$, $h = g_{+\alpha} g_{-\alpha}^{-1}$, so on $V_\alpha \cap X_{1-} \cap X_{1+}$, $h_- g_{-\alpha} = h_+ g_{+\alpha}$. Thus, (6) and (7) show that on $V_\alpha$, $F$ differs from $f_\alpha$ by a well-defined invertible element of $\mathscr{A}(V_\alpha)$.

We assumed at the outset that no such function $F$ exists, so we are led to conclude that either $F_-$ or $F_+$ does not exist. Suppose, for the sake of definiteness, that $F_+$ does not exist. If we set

$$X_{2+} = \{(z_1, \ldots, z_N) \in X_{1+} : z_2 = x_2 + iy_2 \text{ with } -\tfrac{1}{2} \leq y_2 \leq 1\}$$

and

$$X_{2-} = \{(z_1, \ldots, z_N) \in X_{1+} : -1 \leq y_2 \leq \tfrac{1}{2}\},$$

then, arguing as above, we find that our problem is not solvable on one of $X_{2+}$, $X_{2-}$. We iterate this process, proceeding cyclicly through the real coordinates, and we obtain thereby a decreasing sequence $\{Y_j\}_{j=1}^{\infty}$ of closed subsets of $\bar{K}^N$ such that on no $Y_j$ is our problem solvable and such that, moreover, the diameter of $Y_j$ tends to zero. As $\mathscr{V}$ is an open cover for $\bar{K}^N$, it follows that if $j$ is large enough, $Y_j$ is contained in some $V_\alpha \in \mathscr{V}$, and we have the desired contradiction; for on $V_\alpha$ and therefore on any subset of $V_\alpha$, the function $f_\alpha$ is a solution to the problem.

The subdivision process we have used in the proof of this theorem is well known in connection with the second Cousin problem. It can be used to establish a result similar to Theorem 21.7 but valid for bounded functions. For this result, see the paper of Stout [3].

For some applications of the theorem just proved to interpolation problems, it is convenient to introduce certain additional maps and spaces. Define $\Psi:\mathbb{C}^N \to \mathbb{C}^N$ by

$$\Psi(z_1, \ldots, z_N) = (e^{iz_1}, \ldots, e^{iz_N}).$$

The map $\Psi$ is a local homeomorphism of $\mathbb{C}^N$ onto a dense open subset of $\mathbb{C}^N$, and it carries $\mathbb{R}^N$ onto the torus $T^N$. We shall denote by $Q_+$ the set $\{z \in \mathbb{C}: \text{Im } z \geq 0\}$, and by $Q_+^N$ the $N$-fold cartesian product of $Q_+$ with itself, so that $Q_+^N$ is a certain closed subset of $\mathbb{C}^N$ which, under the map $\Psi$, goes onto a dense subset of $\bar{\Delta}^N$. We shall consider the subalgebra $\mathscr{B}_N$ of $\mathscr{C}_0(Q_+^N)$ consisting of the functions which are holomorphic on the interior of $Q_+^N$. The algebra $\mathscr{B}_N$ is not a uniform algebra in our sense of this term, for the constants do not belong to it. It is a uniformly closed function algebra, and by the maximum modulus theorem, the Shilov boundary for $\mathscr{B}_N$ can be identified with $\mathbb{R}^N$. It is quite easy to relate peak interpolation sets for $\mathscr{B}_N$ to those for $\mathscr{A}_N$ by way of Theorem 21.7. Observe that since the elements of $\mathscr{B}_N$ vanish at infinity, a peak interpolation set for $\mathscr{B}_N$ is necessarily a compact subset of $\mathbb{R}^N$.

As a Banach algebra, $\mathscr{B}_N$ may be identified with a certain ideal in $\mathscr{A}_N$. To achieve this identification, recall that if

$$\varphi(z) = \frac{1}{i}\frac{z+1}{z-1},$$

then $\varphi$ carries the open unit disc onto the interior of the half-plane $Q_+$ and takes the point 1 to the point at infinity. If we define $\Phi:\Delta^N \to Q_+^N$ by

$$\Phi(z_1, \ldots, z_N) = (\varphi(z_1), \ldots, \varphi(z_N)), \tag{8}$$

then $\Phi$ carries $\Delta^N$ biholomorphically onto the interior of $Q_+^N$; and it is clear that if $E_0$ is the closed set

$$\{(z_1, \ldots, z_N) \in \bar{\Delta}^N : z_j = 1 \text{ for some } j\},$$

and if $I(E_0) = \{f \in \mathscr{A}_N : f|E_0 = 0\}$, then the map

$$f \mapsto f \circ \Phi$$

is an isometric isomorphism from $\mathscr{B}_N$ onto $I(E_0)$. The set $E_0$ is the zero set of the element $f \in \mathscr{A}_N$ given by

$$f(z_1, \ldots, z_N) = (z_1 - 1)\cdots(z_N - 1).$$

For $\mathscr{A}_N$, zero sets (in $T^N$) and peak interpolation sets coincide. It is clear that for $\mathscr{B}_N$ peak interpolation sets are zero sets. Also, if $F \subset \mathbb{R}^N$ is a compact zero set, then $E_0 \cup \Phi^{-1}(F)$ is a zero set for $\mathscr{A}_N$, so $\Phi^{-1}(F)$ is a zero set for $\mathscr{A}_N$ by Corollary 21.8. Thus, $\Phi^{-1}(F)$ is a peak interpolation set for $\mathscr{A}_N$, and Theorem 20.12 implies that $F$ is a peak interpolation set for $\mathscr{B}_N$. In addition, $F$ will be a peak interpolation set for

$\mathscr{B}_N$ if and only if $\mu_F = 0$ for every finite Borel measure on $\mathbb{R}^N$ which annihilates $\mathscr{B}_N$.

**21.12** ***Theorem***

*The compact subset $F$ of $\mathbb{R}^N$ is a peak interpolation set for $\mathscr{B}_N$ if and only if $\Psi(F)$ is a peak interpolation set for $\mathscr{A}_N$.*

PROOF

Assume $F$ to be a peak interpolation set for $\mathscr{B}_N$, and let $f \in \mathscr{B}_N$ be a function which has $F$ as its zero set. If $\mathfrak{z} \in F$, let $N(\mathfrak{z})$ be a closed neighborhood of $\mathfrak{z}$ on which the map $\Psi$ is one to one. If $\tilde{\mathfrak{z}} = \Psi(\mathfrak{z})$, there exists a closed neighborhood $\tilde{N}$ of $\tilde{\mathfrak{z}}$ in $\mathbb{C}^N$ such that

(a) $\tilde{N} \cap \bar{\Delta}^N$ is equivalent to a polydisc under a homeomorphism which is holomorphic on the interior of this set, and
(b) there exists a map $\psi_{\mathfrak{z}}$ from $\tilde{N}$ onto $N(\mathfrak{z})$ which carries $\tilde{\mathfrak{z}}$ onto $\mathfrak{z}$ and which is inverse to $\Psi$.

The function $f \circ \psi_{\mathfrak{z}}$ lies in $\mathscr{A}(\tilde{N} \cap \bar{\Delta}^N)$, and it has $\Psi(N(\mathfrak{z}) \cap F)$ as its zero set. By condition (a), we know that $\Psi(N(\mathfrak{z}) \cap F)$ is a peak interpolation set for $\mathscr{A}(\tilde{N} \cap \bar{\Delta}^N)$ since it is a zero set for this algebra, and it follows from Corollary 21.8 that $\Psi(F)$ is a peak interpolation set for $\mathscr{A}_N$.

In the opposite direction, let us suppose that the compact set $F \subset \mathbb{R}^N$ has the property that $\Psi(F)$ is a peak interpolation set for $\mathscr{A}_N$. We can assume that $\Psi$ is a homeomorphism on $F$ because $\Psi$ is a local homeomorphism, and because a finite union of peak interpolation sets for $\mathscr{B}_N$ is again a peak interpolation set. Let $g \in \mathscr{B}_N$ be zero free on the set $F$ and let $h \in \mathscr{C}(F)$. If $\psi : \Psi(F) \to F$ is the map inverse to $\Psi$, then there is $H \in \mathscr{A}_N$ with $H|\Psi(F) = (h \circ \psi)(g \circ \psi)^{-1}$, so the element $(H \circ \Psi)g$ of $\mathscr{B}_N$ interpolates $h$ on $F$. Thus $F$ is an interpolation set for $\mathscr{B}_N$. Using our identification of $\mathscr{B}_N$ with an ideal in $\mathscr{A}_N$, using the fact that interpolation sets for $\mathscr{A}_N$ are peak interpolation sets, and invoking Theorem 20.12 we see that $F$ is a peak interpolation set for $\mathscr{B}_N$.

An alternative proof for this theorem is possible using a technique of Forelli [3] (see also Rudin [3]) which enables one to compare measures in $\mathscr{A}_N^{\perp}$ with measures in $\mathscr{B}_N^{\perp}$.

The theorem shows that to obtain examples of peak interpolation sets for $\mathscr{A}_N$, it suffices to construct peak interpolation sets for $\mathscr{B}_N$ and project them into $T^N$ by the covering map $\Psi$. Using this idea, Forelli has shown that certain interesting sets are peak interpolation sets for $\mathscr{A}_N$. It is to this result of Forelli that we now turn our attention.

Recall that if $\mu$ is a finite Borel measure on $\mathbb{R}^N$, its Fourier–Stieltjes transform, $\hat{\mu}$, is the function on $\mathbb{R}^N$ defined by

$$\hat{\mu}(x) = \int_{\mathbb{R}^N} e^{-ix\cdot y}\, d\mu(y),$$

where for $x,y \in \mathbb{R}^N$, we write $x \cdot y = x_1y_1 + \cdots + x_Ny_N$. We shall use $\mathbb{R}^N_+$ to denote the open cone $\{(x_1, \ldots, x_N) \in \mathbb{R}^N : x_1, \ldots, x_N > 0\}$, and $\mathbb{R}^N_-$ will be the set $\{x \in \mathbb{R}^N : -x \in \mathbb{R}^N_+\}$. Forelli's theorem deals with the null sets of measures $\mu \in \mathcal{M}(\mathbb{R}^N)$ with the property that $\hat{\mu}$ vanishes on $\mathbb{R}^N_-$. The relevance of these considerations for our problem stems from the fact that if $\mu \in \mathcal{M}(\mathbb{R}^N)$ annihilates the algebra $\mathcal{B}_N$, then $\hat{\mu}$ vanishes on $\mathbb{R}^N_-$. To see this, let $\varphi : \Delta \to Q_+$ be the conformal map considered above, and let $\psi$ be its inverse. We define $f_k$ by

$$f_k(z_1, \ldots, z_N) = \prod_{j=1}^{N} (1 - \psi(z_j))^{1/k}$$

where for the $k$th root in question we take that branch which is real for real values of the argument. The functions $f_k$ all lie in $\mathcal{B}_N$, and for fixed $(z_1, \ldots, z_N)$ we have that

$$f_k(z_1, \ldots, z_N) \to 1$$

as $k \to \infty$. If we let

$$g_k(z_1, \ldots, z_N) = f_k(z_1, \ldots, z_N)e^{-i(x_1z_1 + \cdots + x_Nz_N)}$$

for all $(x_1, \ldots, x_N) \in \mathbb{R}^N_-$, then $g_k$ also lies in $\mathcal{B}_N$; so if $\mu \in \mathcal{B}_N^\perp$, then

(9) $$0 = \int g_k\, d\mu.$$

We may apply the dominated convergence theorem to take the limit as $k \to \infty$ in (9) and conclude that $\hat{\mu}(x_1, \ldots, x_N) = 0$. Note that by continuity, $\hat{\mu}$ vanishes on the closure of $\mathbb{R}^N_-$.

We come now to the formulation and proof of Forelli's theorem. If $x = (x_1, \ldots, x_N)$ is a point in $\mathbb{R}^N$, and if $E$ is a set in $\mathbb{R}$, we shall denote by $E^x$ the set

$$\{y \in \mathbb{R}^N : x \cdot y \in E\},$$

and if $S \subset \mathbb{R}^N$, we put

$$E^S = \bigcup\{E^x : x \in S\}.$$

Given a set $S \subset \mathbb{R}^N$, the subset $F$ of $\mathbb{R}^N$ is said to have *null S-width*, provided that for all $\varepsilon > 0$, there is a collection of pairs $(I_\alpha, w_\alpha)$, $w_\alpha \in S$ and $I_\alpha$ an open interval in $\mathbb{R}$, such that

(a) $F \subset \bigcup_\alpha I_\alpha^{w_\alpha}$
(b) $\sum_\alpha (\text{length } I_\alpha) < \varepsilon$.

21.13 ***Theorem***

*If S is a compact set of unit vectors in $\mathbb{R}^N_+$, and if the Borel set $E \subset \mathbb{R}^N$ has null S-width, then E is a null set for every $\mu \in \mathscr{M}(\mathbb{R}^N)$ for which $\hat{\mu}$ vanishes on $\mathbb{R}^N_-$.*

The proof we shall give for this theorem is not Forelli's original proof but rather a considerable simplification of it which he subsequently found. It was presented in his Warwick lectures (Forelli [3]). We shall need two lemmas, the first of which is a fact from harmonic analysis, the second a rather special result about certain Hilbert spaces.

21.14 ***Lemma***

*If $\lambda$ is a finite Borel measure on $\mathbb{R}$ with Fourier–Stieltjes transform $\hat{\lambda}$, and if $\hat{\lambda} \in L^1(\mathbb{R})$, then the function F given by*

$$F(x) = \int e^{ixy}\hat{\lambda}(y)\,dy$$

*is in $L^1(\mathbb{R})$, and $\lambda = (1/2\pi)F$ in the sense that for all Borel sets $E \subset \mathbb{R}$,*

$$\lambda(E) = \frac{1}{2\pi}\int_E F(x)\,dx.$$

PROOF

For the proof of this, we need to evaluate a definite integral. Define $\varphi(y)$, $y \in \mathbb{R}$, by

$$\varphi(y) = \int_{-\infty}^{\infty} \exp\left(-\frac{x^2}{2} - ixy\right) dx.$$

Differentiation under the integral shows that

$$\frac{d}{dy}\varphi(y) = i\int_{\mathbb{R}} \exp(-ixy)\frac{d}{dx}\exp\left(-\frac{x^2}{2}\right) dx,$$

and if we integrate by parts, we are led to

$$\frac{d}{dy}\varphi(y) = -y\varphi(y),$$

whence

$$\varphi(y) = \varphi(0)\exp\left(-\frac{y^2}{2}\right).$$

As $\varphi(0)$ is well known to be $\sqrt{2\pi}$, we find finally that

$$\varphi(y) = \sqrt{2\pi}\exp\left(-\frac{y^2}{2}\right). \tag{10}$$

The knowledge of this special integral may be used to derive the lemma in the following way. The function $F$ is bounded, so $\varphi F$ lies in $L^1(\mathbb{R})$. If we can prove that

$$\int_E \varphi(x)\, d\lambda(x) = \frac{1}{2\pi}\int_E \varphi(x)F(x)\, dx \tag{11}$$

for all Borel sets $E$, then since $\varphi$ is zero free, it will follow that $\lambda$ is absolutely continuous with respect to Lebesgue measure on the real line; then by the Radon–Nikodym theorem we shall be able to conclude that $F \in L^1(\mathbb{R})$, and that $\lambda = (1/2\pi)F$ in the sense of the lemma. To prove (11), it is enough to show that the Fourier–Stieltjes transform of the measure $\varphi\lambda$ and the Fourier transform of $(1/2\pi)\varphi F$ are the same function. From (10), we have, interchanging $x$ and $y$ and taking complex conjugates of both sides in the definition of $\varphi$,

$$\varphi(x) = \frac{1}{\sqrt{2\pi}}\int e^{ixy}\varphi(y)\, dy,$$

and from this it follows, using Fubini's theorem, that

$$\begin{aligned}(\varphi\lambda)\hat{}\,(v) &= \int e^{-ixv}\varphi(x)\, d\lambda(x)\\ &= \int e^{-ixv}\frac{1}{\sqrt{2\pi}}\int e^{ixy}\varphi(y)\, dy\, d\lambda(x)\\ &= \frac{1}{\sqrt{2\pi}}\int \varphi(y)\int e^{-ix(v-y)}\, d\lambda(x)\, dy\\ &= \frac{1}{\sqrt{2\pi}}\int \hat{\lambda}(v-y)\varphi(y)\, dy.\end{aligned}$$

Also, the definition of the function $F$ implies that

$$\begin{aligned}(\varphi F)\hat{}\,(v) &= \int e^{-ixv}\varphi(x)\int e^{ixy}\hat{\lambda}(y)\, dy\, dx\\ &= \int \hat{\lambda}(y)\int e^{-ix(v-y)}\varphi(x)\, dx\, dy. \qquad (12)\end{aligned}$$

We know that

$$\sqrt{2\pi}\varphi(v-y) = \int e^{-ix(v-y)}\varphi(x)\, dx,$$

so from (12) we may conclude that

$$(\varphi F)\hat{\ }(v) = \sqrt{2\pi}\int \hat{\lambda}(y)\varphi(v-y)\,dy,$$

and thus $((1/2\pi)\varphi F)\hat{\ } = (\varphi\lambda)\hat{\ }$. This establishes the lemma.

Before stating our second lemma, we have to introduce certain notations. If $v \in \mathbb{R}^N$, we denote by $\chi_v$ the character on $\mathbb{R}^N$ given by

$$\chi_v(t) = e^{it\cdot v}.$$

For a space $M$ of functions on $\mathbb{R}^N$ and for $v \in \mathbb{R}^N$, we let

$$M_v = \chi_v M = \{\chi_v f : f \in M\}.$$

21.15 ***Lemma***

*Let $\mu$ be a positive finite measure on $\mathbb{R}^N$ and let $M$ be a closed subspace of $L^2(\mu)$ with the following properties:*

(a) *If $v \in \mathbb{R}^N_+$, then $M \subset \chi_v M$.*
(b) *If $w$ is a unit vector in $\mathbb{R}^N_+$, then $\vee\{M_{tw} : t \in \mathbb{R}\} = L^2(\mu)$ and $\wedge\{M_{tw} : t \in \mathbb{R}\} = 0$.*

*Then for $w$, a unit vector in $\mathbb{R}^N_+$, for $t > 0$, and for $g \in M_{-tw} \wedge M^{\perp}_{tw}$,*

$$\int_{Ew} |g|^2\,d\mu \le t\left(\int |g|^2\,d\mu\right) m(E)$$

*for every Borel set $E$ in $\mathbb{R}$. Here $m(E)$ denotes the Lebesgue measure of $E$.*

In the statement of this lemma we have used the customary Hilbert-space notation that if $\{H_\alpha\}$ is a family of closed subspaces of the Hilbert space $\mathscr{H}$, then $\wedge H_\alpha$ is the largest closed subspace contained in all the $H_\alpha$ and $\vee H_\alpha$ is the least closed subspace containing all the $H_\alpha$.

PROOF OF THE LEMMA

Since $w \in \mathbb{R}^N_+$, we have that if $s > 2t$, then $(s-2t)w \in \mathbb{R}^N_+$; so by condition (a), $M \supset M_{(s-2t)w}$, whence

$$M_{tw} \supset \chi_{sw} M_{-tw}$$

if $s > 2t$. As $g \in M_{-tw} \wedge M^{\perp}_{tw}$, it follows that

$$\int \chi_{sw} g\bar{g}\,d\mu = \int \chi_{sw}|g|^2\,d\mu = 0$$

if $s > 2t$, and by conjugation, we draw the same conclusion if $|s| > 2t$.

We define the Borel measure $\lambda$ on $\mathbb{R}$ by

$$\lambda(E) = \int_{E^w} |g|^2 \, d\mu$$

for every Borel set $E$, and it follows that for a Borel function $f$ on $\mathbb{R}$,

$$\int_{\mathbb{R}} f \, d\lambda = \int_{\mathbb{R}^N} f(w \cdot x)|g(x)|^2 \, d\mu(x).$$

(The definition of $\lambda$ implies the validity of this last formula if $f$ is a characteristic function, and the general case follows from this.) Our hypotheses on $g$ imply that $\hat{\lambda}$ vanishes off the interval $[-2t, 2t]$, for if $s \in \mathbb{R}$, then

$$\begin{aligned} \hat{\lambda}(s) &= \int_{\mathbb{R}} e^{-isy} \, d\lambda(y) \\ &= \int_{\mathbb{R}^N} e^{-is(w \cdot x)}|g(x)|^2 \, d\mu(x) \\ &= \int_{\mathbb{R}^N} \chi_{-sw}(x)|g(x)|^2 \, d\mu(x), \end{aligned}$$

and we know this latter integral to vanish if $|s| > 2t$. The function $\hat{\lambda}$ is continuous, so it lies in $L^1(\mathbb{R})$; so the last lemma yields the fact that if

$$F(r) = \int_{\mathbb{R}} e^{irs}\hat{\lambda}(s) \, ds,$$

then

$$\lambda(E) = \frac{1}{2\pi} \int_E F(r) \, dr$$

for $E$ a Borel set in $\mathbb{R}$. Now

$$\begin{aligned} |F(r)| &\leq \int_{-2t}^{2t} |\hat{\lambda}(s)| \, ds \\ &\leq 4t \int |g(x)|^2 \, d\mu(x). \end{aligned}$$

Thus, if $E$ is a Borel set in $\mathbb{R}$, we have

$$\lambda(E) \leq \frac{2t}{\pi}\left(\int |g(x)|^2 \, d\mu(x)\right) m(E),$$

and the lemma is proved.

### 21.16 *Proof of Forelli's Theorem*

Let $\mu \in \mathscr{M}(\mathbb{R}^N)$ satisfy $\hat{\mu} = 0$ on $\mathbb{R}^N_-$, and let $\lambda$ be the total variation of $\mu : \lambda = |\mu|$. Choose for $M$ the $L^2(\lambda)$ closure of the linear span of the characters $\chi_y$ with $y \in \mathbb{R}^N_+$. Plainly, $M$ has property (a) of the preceding lemma. It also has property (b). That for a unit vector $w \in \mathbb{R}^N_+$,

$$\vee \{M_{tw} : t \in \mathbb{R}\} = L^2(\lambda)$$

follows from the fact that the linear span of the characters $\chi_y$, $y \in \mathbb{R}^N$, is dense in $L^2(\lambda)$. That

$$\wedge \{M_{tw} : t \in \mathbb{R}\} = 0$$

requires a bit more care. Let $f \in M_{tw}$ for all $t \in \mathbb{R}$, and let $\gamma$ be the Radon–Nikodym derivative $d\mu/d\lambda$ so that $\gamma$ is a certain unimodular Borel function. We shall show that

$$\int f\gamma\chi_y \, d\lambda = 0$$

for all $y \in \mathbb{R}^N$ so that $f\gamma = 0$ a.e. $[\lambda]$ and thus $f = 0$ a.e. $[\lambda]$. Choose a $t \in \mathbb{R}$ such that $tw - y \in \mathbb{R}^N_+$, and let $\varepsilon > 0$. As $f \in M_{tw}$, there exists a trigonometric polynomial $g \in M_{tw}$ such that

$$\|\lambda\| \int |f - g|^2 \, d\lambda \leq \varepsilon^2.$$

Then

$$\left| \int f\gamma\chi_y \, dy \right| \leq \|\lambda\|^{1/2} \left( \int |f - g|^2 \, d\lambda \right)^{1/2} + \left| \int g\gamma\chi_y \, d\lambda \right|.$$

The second integral on the right is zero, for $\gamma = d\mu/d\lambda$, $g\chi_y$ is a finite sum of characters $\chi_z$, $z \in \mathbb{R}^N_+$, and $\hat{\mu}$ vanishes on $\mathbb{R}^N_-$.

Let $v$ be a unit vector in $\mathbb{R}^N_+$, and put

$$c = \operatorname{lub}\left\{ 1 + \frac{v_1}{w_1} + \cdots + \frac{v_N}{w_N} : (w_1, \ldots, w_N) \in S \right\}.$$

The compactness of $S$ implies that $c$ is finite. By the definition of $c$, we have $cw - v \in \mathbb{R}^N_+$ for each $w \in S$. If $r > 0$ and $s = cr$, then for any $w \in S$, $sw - rv \in \mathbb{R}^N_+$. Whenever $y - x \in \mathbb{R}^N_+$, we have $M_x \supset M_y$, so

$$M_{sw} \subset M_{rv} \subset M_{-rv} \subset M_{-sw},$$

whence

$$M_{-rv} \wedge M_{rv}^{\perp} \subset M_{-sw} \wedge M_{sw}^{\perp}.$$

If $g \in M_{-rv} \wedge M_{rv}^{\perp}$, then for each interval $I$ the last lemma yields

$$\int_{I^w} |g|^2 \, d\lambda \leq s\left(\int |g|^2 \, d\lambda\right)(\text{length } I).$$

Thus, if $r > 0$ and $g \in M_{-rv} \wedge M_{rv}^{\perp}$, then

$$\int_{I^w} |g|^2 \, d\lambda \leq cr\left(\int |g|^2 \, d\mu\right)(\text{length } I) \tag{13}$$

for every $w \in S$ and every open interval $I$.

Consider now a Borel set $E \subset \mathbb{R}^N$ which has zero $S$-width. The definition of zero $S$-width provides a set of pairs $(I_\alpha, w_\alpha)$, $I_\alpha$ an interval in $\mathbb{R}$ and $w_\alpha \in S$, with $E \subset \bigcap I_\alpha^{w_\alpha}$ and $\sum_\alpha (\text{length } I_\alpha) < \varepsilon$. If $r > 0$, and $g \in M_{-rv} \wedge M_{rv}^{\perp}$, then, by (13),

$$\int_E |g|^2 \, d\lambda \leq \text{cr}\left(\int |g|^2 \, d\mu\right)\varepsilon,$$

and thus we may conclude that

$$\int_E |g|^2 \, d\lambda = 0. \tag{14}$$

Equation (14) holds for all $g \in M_{-rv} \wedge M_{rv}^{\perp}$ and all $r > 0$.

To conclude the proof we show that

$$\vee \{M_{-rv} \wedge M_{rv}^{\perp} : r > 0\} = L^2(\mu). \tag{15}$$

This is easily seen, for it is clear that if $s > r > 0$, then

$$M_{-sv} \supset M_{-rv} \quad \text{and} \quad M_{sv}^{\perp} \supset M_{rv}^{\perp},$$

so

$$M_{-rv} \wedge M_{rv}^{\perp} \subset M_{-sv} \wedge M_{sv}^{\perp}.$$

Moreover, if $\chi$ is any character of $\mathbb{R}^N$, then for sufficiently large $s$, $\chi \in M_{-sv}$ and $\chi \in M_{sv}^{\perp}$, whence $\chi \in M_{-sv} \wedge M_{sv}^{\perp}$. It follows that every character of $\mathbb{R}^N$ is contained in $\vee (M_{-rv} \wedge M_{rv}^{\perp})$, and thus (15) is correct. This finishes the proof of Forelli's theorem.

As an immediate corollary of the theorem just proved, we can show that certain subsets of $T^N$ are peak interpolation sets for the polydisc algebra $\mathscr{A}_N$. We continue to use $\Psi$ to denote the covering map from $\mathbb{R}^N$ to $T^N$ considered in Theorem 21.12.

21.17 ***Theorem***

*If the closed set $E$ in $T^N$ is contained in $\Psi(F)$, where the set $F$ is a union of a sequence $\{F_n\}$ of closed sets in $\mathbb{R}^N$, $F_n$ of null $S_n$-width for some* *compact set $S_n$ of unit vectors in $\mathbb{R}^N_+$, then E is a peak interpolation set for $\mathscr{A}_N$.*

PROOF

$E = \bigcup \{E \cap \Psi(F_n): n = 1, 2, \ldots\}$, and as $F_n$ is closed it is a countable union of compacta, $F_{n,m}$. Thus, it is enough to prove that $E \cap \Psi(F_{n,m})$ is a peak interpolation set for $\mathscr{A}_N$. But by Forelli's theorem, the set $F_{n,m}$ is a peak interpolation set for the algebra $\mathscr{B}_N$, and thus the set $\Psi(F_{n,m})$ is one for $\mathscr{A}_N$, whence $E \cap \Psi(F_{n,m})$ is also one for $\mathscr{A}_N$.

As a concrete instance of this theorem, we have the following example.

21.18 ***Corollary***

*Let E be a closed subset of $\mathbb{R}$ of Lebesgue measure zero, and let $\beta \in \mathbb{R}^N_+$. Every closed subset of $\Psi(E^\beta)$ is a peak interpolation set for $\mathscr{A}_N$.*

The case that $E$ is a singleton is due to Rudin and Stout [1], and the general case was proved by Forelli [2] and Stout [3].

In his paper [2], Forelli has given some explicit examples of curves in $\mathbb{R}^2$ which are peak interpolation sets for $\mathscr{B}_N$. His construction is based on the following elementary result.

21.19 ***Theorem***

*Let $\gamma(t) = (f(t), g(t))$, $a < t < b$, be a curve of class $\mathscr{C}^2$ in $\mathbb{R}^2$ such that the tangent vector $(f'(t), g'(t))$ never vanishes. If K is a compact interval in $(a, b)$, then $\gamma(K)$ has null S-width if S is the (compact) set of unit vectors whose directions are those of the normal vectors $(g'(t), -f'(t))$, $t \in K$.*

PROOF

Using the Taylor theorem of order 2, we see that the hypotheses that $f$ and $g$ have two continuous derivatives and that the tangent vector to $\gamma$ not vanish imply that there is an absolute constant $C$ such that if $t, t_0 \in K$ and $|t - t_0| < 1$, then the distance from the point $\gamma(t)$ to the line tangent to $\gamma$ at $t_0$ is no more than $C|t - t_0|^2$. It follows that if, for $N >$ (length of $K$), we choose $N + 1$ points of subdivision,

$$t_1 < t_2 < \cdots < t_{N+1},$$

with $|t_{j+1} - t_j| = 1/N$ (length of $K$), then the set $\gamma(K)$ is covered by $N$ strips $I_j^{w_j}$, where $w_j$ is the unit vector in the direction of $(g'(t_j), -f'(t_j))$, and $I_j$ has length no more than $CN^{-2}$(length of $K)^2$. Thus,

$$\Sigma(\text{length } I_j) \leq CN^{-1}(\text{length of } K)^2,$$

and by taking $N$ large, this can be made smaller than any preassigned $\varepsilon > 0$. This proves the theorem.

As an example of this last theorem, let $\gamma$ be the curve in $\mathbb{R}^2$ defined by

$$\gamma(t) = (t^{-1}, t), \qquad 0 < t < \infty,$$

i.e., one branch of the hyperbola $x_1 x_2 = 1$. Every compact subset of this curve is a peak interpolation set for $\mathscr{B}_2$ as follows from the corollary and our last theorem. However, the corresponding statement for the curve $x_1 x_2 = -1$, $x_1 < 0$, is false. To see this, note that under the map

$$z \mapsto (-z^{-1}, z),$$

$Q_+$ goes into $Q_+^2$, and $\mathbb{R}$ goes into the curve under question. If $F \in \mathscr{B}_2$ vanishes on an interval of the curve $x_1 x_2 = -1$, $x_1 < 0$, then the element $f$ of $\mathscr{B}_1$ given by

$$f(z) = F(-z^{-1}, z)$$

vanishes on an interval of $\mathbb{R}$ and so is the zero function. Thus, $F$ vanishes identically on the curve in question, and we have the result. These hyperbolas were considered by Forelli.

In the same way we see that the quarter circle $(\cos t, \sin t)$, $0 \leq t \leq \pi/2$, is a peak interpolation set for $\mathscr{B}_2$, but it can be shown that the quarter circle $(\cos t, \sin t)$, $-\pi/2 \leq t \leq 0$, is not one.

We conclude this discussion of interpolation with one final example, an example in which we do not interpolate $\mathscr{C}(E)$ but rather a different algebra.

21.20 ***Theorem*** (Rudin and Stout [1])
*Let $E \subset T^N$ be a peak interpolation set for $\mathscr{A}_N$. If $f \in \mathscr{C}(E \times \bar{\Delta}^M)$ has the property that for each $\mathfrak{z} \in E$, the function $f(\mathfrak{z}, \cdot) \in \mathscr{C}(\bar{\Delta}^M)$ is in $\mathscr{A}_M$, then there is $F \in \mathscr{A}_{N+M}$ such that $F|E \times \bar{\Delta}^M = f$.*

PROOF

If $A$ denotes the algebra $\mathscr{A}_{N+M}|(E \times \bar{\Delta}^M)$, then since $E$ is a peak set for $\mathscr{A}_N$, it is clear that $A$ is uniformly closed in $\mathscr{C}(E \times \bar{\Delta}^M)$. The fact that $E$ is an interpolation set for $\mathscr{A}_N$ implies that the maximal sets of antisymmetry for $A$ are sets of the form $\{\mathfrak{z}\} \times \bar{\Delta}^M$. Thus, by hypothesis, $f$ agrees on every one of these maximal sets of antisymmetry with an element of $A$, so by the generalized Stone–Weierstrass theorem, $f \in A$, and we have the result.

For more on the theory of interpolation for the polydisc algebras $\mathscr{A}_N$, we refer the reader to the paper of Forelli already cited, and to the book of Rudin [3]. Certain additional results are known, but the theory is far from complete. It seems to be a difficult problem to characterize exactly the peak interpolation sets for $\mathscr{A}_N$.

One final word is in order. In this section we have dealt exclusively with problems on $\bar{\Delta}^N$, but there are various other domains in $\mathbb{C}^N$ which could be considered. We have fixed attention on $\bar{\Delta}^N$ for a very simple reason: At this time, there are, apparently, no results known concerning interpolation on other sets in $\mathbb{C}^N$.

# 22 *LINEAR EXTENSIONS*

If $E$ is a closed subset of the space $X$ which is an interpolation set for the uniform algebra $A$ on $X$, it happens in some situations that essentially more is true, viz., there exists a continuous linear operator $L:\mathscr{C}(E) \to A$ such that $(Lf)|E = f$ for all $f \in \mathscr{C}(E)$. We shall call such an operator a *linear extension*, and in this section we shall give a sufficient condition for the existence of a linear extension. This theory is due to Pełczyński [1, 2] and to Michael and Pełczyński [1, 2], who have obtained results beyond those which we shall present. Observe that if a linear extension $L:\mathscr{C}(E) \to A$ exists, then the space $\mathscr{C}(E)$ is a direct summand of the Banach space $A$.

Let $X$ be a topological space, let $E$ be a closed set in $X$, and let $A$ be a closed *subspace* of $\mathscr{C}^B(X)$, the bounded, continuous functions on $X$.

22.1 ***Definition***

*If $A'$ is a closed subspace of $\mathscr{C}^B(E)$, the pair $(A', A)$ has the* ***bounded extension property*** *if given $\varepsilon > 0$, every $f \in A'$ has a bounded family of extensions*

$$\{F_{\varepsilon,W} : \varepsilon > 0, W \text{ a neighborhood of } E\}$$

*in $A$ such that $|F_{\varepsilon,W}(x)| < \varepsilon$ if $x \in X \setminus W$.*

The main result of this section will contain as a special case the following theorem.

22.2 ***Theorem***

*If $X$ is a compact metric space and if $(\mathscr{C}(E), A)$ has the bounded extension property, $A$ a closed subspace of $\mathscr{C}(X)$, then there exists a linear extension $L:\mathscr{C}(E) \to A$ which has norm 1.*

The proof of our main result will depend on several preliminaries. Let us preface these preliminaries with a few remarks.

First, we note that the theorem just stated is rather generally applicable, for if $A$ is a uniform algebra on a compact space $X$, and if $E$ is a peak interpolation set for $A$, then the pair $(\mathscr{C}(E), A)$ plainly has the bounded extension property.

Second, granted the existence of a linear extension $L:\mathscr{C}(E) \to A$ of norm 1, it is natural to ask if $L(1_E) = 1_X$. Here $1_E$ denotes the function identically 1 on $E$, $1_X$ the corresponding function on $X$. The question presupposes that $1_X \in A$, a condition satisfied, e.g., if $A$ is a uniform algebra on $X$. Granted $1_X \in A$, the answer to our question is no, as an example shows.

22.3 *Example*

Let $E$ be a closed set of Lebesgue measure zero in the unit circle so that by the Rudin–Carleson theorem $E$ is a peak interpolation set for the disc algebra, $\mathscr{A}$. Thus, by the theorem we have stated, there exists a linear extension $L:\mathscr{C}(E) \to \mathscr{A}$ with $\|L\| = 1$. However, if $E$ contains more than one point, then for no choice of $L$ is $L(1) = 1$.

To see this, choose $z_1$ and $z_2$, distinct points of $E$, and choose $f_1 \in \mathscr{A}$ so that $1 = f_1(z_1) \geq f(z) \geq f_1(z_2) = 0$ for all $z \in E$. Set $f_2 = 1 - f_1$. Then $\|f_i\|_E = f_i(z_i) = 1$, $i = 1, 2$, and $f_i(z_j) = 0$ if $i \neq j$. Also, $f_1 + f_2 = 1$. Thus, if $\alpha_1$ and $\alpha_2$ are complex numbers

$$\begin{aligned} \max(|\alpha_1|, |\alpha_2|) &\geq \|\alpha_1 f_1 + \alpha_2 f_2\|_E \\ &\geq \max_{i=1,2} |(\alpha_1 f_1 + \alpha_2 f_2)(z_i)| \\ &= \max(|\alpha_1|, |\alpha_2|). \end{aligned}$$

The operator $L$ is of norm 1 by hypothesis, so it is an isometry of $\mathscr{C}(E)$ into $\mathscr{A}$, and consequently

$$\begin{aligned} \|\alpha_1 Lf_1 + \alpha_2 Lf_2\|_X &= \|\alpha_1 f_1 + \alpha_2 f_2\|_E \\ &= \max(|\alpha_1|, |\alpha_2|). \end{aligned}$$

If we fix a point $z$ in the unit disc and take $\alpha_i = 0$ if $Lf_i(z) = 0$, $\bar{\alpha}_i = Lf_i(z)|Lf_i(z)|^{-1}$ otherwise, we are led to the fact that for all $z$, $|z| < 1$,

$$|Lf_1(z)| + |Lf_2(z)| \leq 1. \tag{1}$$

If $L1$ were 1, we would have $Lf_1 + Lf_2 = 1$, and combining this with (1), $Lf_1$ and $Lf_2$ would have to be nonconstant, positive elements of $\mathscr{A}$. As no such functions exist, we conclude that $L(1) \neq 1$.

Next let us observe that if $A$ is a uniform algebra on $X$, and if $L:\mathscr{C}(E) \to A$ is a linear extension, then generally $L$ will not be multiplicative, i.e., generally $L$ will not be an algebra homomorphism. If $L$ is an algebra homomorphism and if $L^*:\Sigma(A) \to E$ is the adjoint, then $L^*|X$ is a continuous map of $X$ onto $E$ which leaves $E$ fixed pointwise. On purely topological grounds it is clear that such a map can exist only in very special cases.

Finally, the theorem should be compared with an earlier result due to Bartle and Graves [1], according to which if $T$ is any linear operator from a Banach space $S_1$ onto a Banach space $S_2$, then there exists a continuous, though not necessarily linear, map $\mu:S_2 \to S_1$ such that $T \circ \mu$ is the identity. (For some generalizations of this result, see Michael [1, 2, 3].)

In order to establish Theorem 22.2, we need to know that $\mathscr{C}(X)$, $X$ compact metric, enjoys a certain exhaustion property.

22.4 ***Definition***

*The finite set* $\Phi = \{\varphi_1, \ldots, \varphi_m\}$ *in* $\mathscr{C}(X)$ *is a* ***peaked partition of unity*** *on* $X$ *if*

(a) $0 \leq \varphi_j \leq 1$.
(b) $\|\varphi_j\|_X = 1$.
(c) $\Sigma\varphi_j(x) = 1$ *for all* $x \in X$.

We shall denote by $[\Phi]$ the linear span in $\mathscr{C}(X)$ of the set $\Phi$. This is a finite-dimensional subspace and is therefore closed in $\mathscr{C}(X)$.

22.5 ***Theorem***

*If $X$ is a compact metric space, then there exists a sequence $\{S_n\}$ of subspaces of the form $[\Phi]$, $\Phi$ a peaked partititon of unity on $X$, such that*

(a) $S_1 \subset S_2 \subset \cdots$.
(b) $\bigcup S_n$ *is dense in* $\mathscr{C}(X)$.

For certain special spaces, e.g., the Cantor set, the existence of such a sequence of subspaces is obvious, but the general result is not obvious. The proof of Theorem 22.5 will depend on three lemmas.

22.6 ***Lemma***

*Given an finite open covering $\{U_1, \ldots, U_n\}$ for $X$, and for each $i$ a point $x_i \in U_i$, $x_i \neq x_j$ unless $i = j$, there exists a peaked partition of unity $\{\varphi_1, \ldots, \varphi_n\}$ on $X$ such that $\varphi_i(x_i) = 1$ and $\varphi_i(x) = 0$ if $x \in X \setminus U_i$.*

PROOF

Set $V_i = U_i \setminus \{x_1, \dots, x_{i-1}, x_{i+1}, \dots, x_n\}$. The set $\{V_i\}$ is an open covering for $X$, and if $\{\varphi_1, \dots, \varphi_n\}$ is a paritition of unity on $X$ subordinate to $\{V_i\}$, then $\varphi_i(x_i) = 1$.

22.7 ***Definition***

*If $\Phi = \{\varphi_1, \dots, \varphi_n\}$ is a partition of unity on $X$ and if $\{U_\alpha\}$ is an open cover for $X$, then $\Phi$ is said to be $\varepsilon$-**subordinate** to $\{U_\alpha\}$ if there exist $U_1, \dots, U_n \in \{U_\alpha\}$ such that $\Sigma\varphi_i(x) \leq \varepsilon$, where the sum is over those $i \in \{1, \dots, n\}$ for which $U_i$ does not contain $x$.*

22.8 ***Lemma***

*If $\Phi = \{\varphi_1, \dots, \varphi_n\}$ is a peaked partition of unity on $X$, if $\{U_\alpha\}$ is an open cover for $X$, and if $\varepsilon > 0$, then there exists a peaked partition of unity $\Psi = \{\psi_1, \dots, \psi_m\}$ which is* $\varepsilon$-subordinate to $\{U_\alpha\}$ *and which satisfies $[\Phi] \subset [\Psi]$.*

PROOF

We may suppose $0 < \varepsilon < 1$ so that $0 < 1 - \varepsilon < 1$ also holds. Let $x_1, \dots, x_n \in X$ be such that $\varphi_i(x_i) = 1$ for $i = 1, 2, \dots, n$. For each $z \in X$, choose $U_z \in \{U_\alpha\}$ such that $z \in U_z$, and put

$$(2) \qquad U_{z,i} = \begin{cases} U_z & \text{if } \varphi_i(z) = 0, \\ U_z \cap \{x \in X : \varphi_i(x) > (1 - \varepsilon)\varphi_i(z)\} & \text{if } \varphi_i(z) > 0. \end{cases}$$

If we set

$$U(z) = \bigcap_i U_{z,i},$$

then $\{U(z)\}_{z \in X}$ is an open cover for $X$, so by compactness, there exist $z_1, \dots, z_m \in X$ with the property that $\{U(z_i) : i = 1, \dots, m\}$ is an open cover for $X$. We may suppose that $m \geq n$ and that for $j = 1, \dots, n$, $z_j = x_j$.

By Lemma 22.6, there is a peaked partition of unity $\Psi^* = \{\psi_1^*, \dots, \psi_m^*\}$ on $X$ such that $\psi_j^*$ vanishes off $U(z_j)$ and $\psi_j^*(z_j) = 1$, $j = 1, \dots, m$. If we define $\varphi_i^*$ by

$$(3) \qquad \varphi_i^*(x) = \sum_{j=1}^m \varphi_i(z_j)\psi_j^*(x),$$

then

$$(4) \qquad \varphi_i^*(z_k) = \varphi_i(z_k),$$

for $\psi_j^*(z_k)$ is zero unless $j = k$, in which case it is 1. We also have

$$(5) \qquad \varphi_i(x) \geq (1 - \varepsilon)\varphi_i^*(x)$$

for all $x$, for if $\psi_j^*(x) > 0$ so that $x \in U(z_j)$, then $\varphi_i(x) \geq (1 - \varepsilon)\varphi_i(z_j)$ by the definition of $U(z_i)$. From this it follows that

$$\sum_{j=1}^{m} \varphi_i(x)\psi_j^*(x) \geq (1 - \varepsilon) \sum_{j=1}^{m} \varphi_i(z_j)\psi_j^*(x)$$

$$= (1 - \varepsilon)\varphi_i^*(x).$$

For all $x \in X$ and each $i = 1, 2, \ldots, n$, let

$$\gamma_i(x) = \sup\{\gamma \leq 1 : \varphi_i(x) \geq \gamma\varphi_i^*(x)\}$$

and

$$\gamma(x) = \min\{\gamma_i(x) : i = 1, \ldots, n\}.$$

Then $\gamma(x) \geq 1 - \varepsilon$ by (5), and $\gamma(z_j) = 1$ by (4). We have

$$\varphi_i(x) \geq \gamma(x)\varphi_i^*(x). \tag{6}$$

We shall prove that the function $\gamma$ is continuous. For this it is sufficient to show that the functions $\gamma_1, \ldots, \gamma_n$ are continuous. On the set $\{x \in X : \varphi_i^*(x) > 0\}$, the continuity of $\gamma_i$ is clear, for there

$$\gamma_i(x) = \min\{1, \varphi_i(x)\varphi_i^*(x)^{-1}\}.$$

Consider now an $x_0$ at which $\varphi_i^*$ vanishes. At such an $x_0$, $\gamma_i$ assumes the value 1, and we shall prove that $\gamma_i(x) = 1$ for all $x$ near $x_0$. If $\varphi_i(x_0) > 0$, this is clear, so suppose that $\varphi_i(x_0) = 0$. If we put

$$J_i = \{j \leq m : \varphi_i(z_j) > 0\}$$

and

$$W_i = \{x \in X : \varphi_i(x) < (1 - \varepsilon)\varphi_i(z_j) \text{ for all } j \in J_i\},$$

then $W_i$ is open and $x_0 \in W_i$ since $\varphi_i(x_0) = 0$. If $x \in W_i$, we shall see that $\varphi_i^*(x) = 0$, whence, as above, $\gamma_i(x) = 0$. To prove that $\varphi_i^*(x) = 0$ for $x \in W_i$, it suffices, by (3), to show that if $\varphi_i(z_j) > 0$ for some $j < m$, then $\psi_j^*(x) = 1$. But if $\varphi_i(z_j) > 0$, then $j \in J_i$, so by the definition of $W_i$ and (2), $x \notin U_{z_j,i}$. But $U_{z_j,i} \supset U(z_j)$ and hence $\psi_j^*(x) = 0$. Thus, $\gamma$ is continuous.

We define $\psi_j$ by

$$\psi_j(x) = \begin{cases} \gamma(x)\psi_j^*(x) + \varphi_j(x) - \gamma(x)\varphi_j^*(x), & 1 \leq j \leq n, \\ \gamma(x)\psi_j^*(x), & n < j \leq m, \end{cases}$$

and we set $\Psi = \{\psi_1, \ldots, \psi_m\}$. The set $\Psi$ is the desired peak partition of unity on $X$.

We have $\psi_j(x) \geq \gamma(x)\psi_j^* \geq 0$, and $\psi_j(z_j) = 1$ since $\gamma(z_j) = 1$ and $\psi_j^*(z_j) = 1$. By (3) we have

$$\begin{aligned}\sum_{j=1}^{m} \varphi_i(z_j)\psi_j(x) &= \gamma(x) \sum_{j=1}^{n} \psi_j^*(x)\varphi_i(z_j) + \sum_{j=1}^{n} \varphi_j(x)\varphi_i(z_j) \\ &\qquad - \gamma(x) \sum_{j=1}^{n} \varphi_i(z_j)\varphi_j^*(x) + \gamma(x) \sum_{j=n+1}^{m} \psi_j^*(x)\varphi_i(z_j) \\ &= \gamma(x)\varphi_i^*(x) + \sum_{j=1}^{n} \varphi_j(x)\varphi_i(z_j) - \gamma(x) \sum_{j=1}^{n} \varphi_i(z_j)\varphi_j^*(x).\end{aligned}$$

As $\Sigma\varphi_i = 1$ and $\varphi_i(x_j) = 0$ or 1 according as $i = j$ or not, it follows that the first of the sums in the last term is $\varphi_i(x)$ and that the second is $\gamma(x)\varphi_i^*(x)$. Consequently,

$$\sum_{j=1}^{m} \varphi_i(z_j)\psi_j(x) = \varphi_i(x),$$

which implies that $[\Psi] \supset [\Phi]$. We also have that $\Psi$ is a partition of unity since

$$\begin{aligned}\sum_{j=1}^{m} \psi_j(x) &= \sum_{j=1}^{m} \sum_{i=1}^{n} \varphi_i(z_j)\psi_j(x) \\ &= \sum_{i=1}^{n} \sum_{j=1}^{m} \varphi_i(z_j)\psi_j(x) \\ &= \sum_{i=1}^{n} \varphi_i(x) = 1.\end{aligned}$$

Finally, $\Psi$ is $\varepsilon$-subordinate to $\{U_\alpha\}$. To see this, choose $U_j, j = 1, \ldots, m$, in $\{U_\alpha\}$ such that $U(z_j) \subset U_j$. If $x \in X$, let $J(x) = \{j \leq m : x \notin U_j\}$ and $K(x) = \{j \leq m : x \in U_j\}$. From $\psi_j(x) \geq \gamma(x)\psi_j^*(x)$, we have

$$\begin{aligned}\sum_{j\in J(x)} \psi_j(x) &= 1 - \sum_{j\in K(x)} \psi_j(x) \\ &\leq 1 - \gamma(x) \sum_{j\in K(x)} \psi_j^*(x) \\ &= 1 - \gamma(x) \sum_{j=1}^{m} \psi_j^*(x) = 1 - \gamma(x) \\ &\geq 1 - (1 - \varepsilon) = \varepsilon.\end{aligned}$$

This proves Lemma 22.8.

We need one more lemma before we can give the proof of Theorem 22.5.

22.9 ***Lemma***

*If* $f \in \mathscr{C}(X)$ *and* $\varepsilon > 0$, *there exists a* $\delta > 0$ *and an open covering* $\mathscr{U}$ *of* $X$ *such that if* $\Phi$ *is a partition of unity on* $X$ *which is* $\delta$*-subordinate to* $\mathscr{U}$, *then* $\|f - g\| < \varepsilon$ *for some* $g \in [\Phi]$.

PROOF

If $f$ is the zero function, there is nothing to prove. If $f$ does not vanish identically, set $\delta = \varepsilon/4\|f\|^{-1}$. For $z \in \mathbb{C}$, $|z| \le \|f\|$, let $U(z) = \{x \in X : |f(x) - z| < \varepsilon/2\}$, let $\mathscr{U} = \{U(z) : |z| \le \|f\|\}$, and let $\Phi = \{\varphi_i\}_{i=1}^m$ be $\delta$-subordinate to $\mathscr{U}$. By the definition of $\delta$-subordinate, there exist complex numbers $z_1, \ldots, z_m$ of modulus no more than $\|f\|$ such that if $I(x) = \{i \le m : x \notin U(z_i)\}$, then

$$\sum\{\varphi_i(x) : i \in I(x)\} < \delta.$$

If we set

$$g(x) = \Sigma z_i \varphi_i(x),$$

then for any $x \in X$ we have

$$\begin{aligned} |f(x) - g(x)| &= |\Sigma\varphi_i(x)(f(x) - z_i)| \\ &= \sum_{i \in I(x)} \varphi_i(x)|f(x) - z_i| + \sum_{i \notin I(x)} \varphi_i(x)|f(x) - z_i| \\ &\le 2\|f\| \sum_{i \in I(x)} \varphi_i(x) + \varepsilon/2 \\ &\le 2\|f\|\delta + \varepsilon/2 \\ &= \varepsilon. \end{aligned}$$

These lemmas enable us to prove Theorem 22.5 as follows.

22.10 ***Proof of Theorem* 22.5**

The space $X$ is a compact metric space, so if $\rho$ is a metric for $X$, there is a sequence $\{\mathscr{U}_n\}$ of finite open covers of $X$ such that each element of $\mathscr{U}_n$ has diameter less than $1/n$. Using this sequence, we can construct the desired sequence $\{S_n\}$ inductively. Let $\Phi_1$ be an arbitrary peaked partition of unity on $X$, and use Lemmas 22.6 and 22.8 to construct peaked partitions of unity on $X$, $\Phi_2, \Phi_3, \ldots$, so that $[\Phi_n] \subset [\Phi_{n+1}]$, and $\Phi_n$ is $(1/n)$-subordinate to $\mathscr{U}_n$ for $n > 1$. We take $S_n = [\Phi_n]$, and it follows from Lemma 22.9 that $\bigcup S_n$ is dense in $\mathscr{C}(X)$.

We now turn to the formulation and proof of a theorem which includes Theorem 22.2 as a special case, as follows from Theorem 22.5.

22.11 ***Definition***

*A separable Banach space $B$ is a $\pi_1$-**space** if it has an increasing sequence $B_1 \subset B_2 \subset \cdots$ of finite-dimensional subspaces whose union is dense in $B$ and which have the additional property that there exists, for each $n$, a projection of norm 1 from $B$ onto $B_n$.*

The result we have just derived implies that $\mathscr{C}(X)$, $X$ a compact metric space, is a $\pi_1$-space; for let $\{S_n\}$ be a sequence of peaked partition subspaces of $\mathscr{C}(X)$ as in Theorem 22.5, say, $S_n = [\Phi_n]$, where $\Phi_n = \{\varphi_1^{(n)}, \ldots, \varphi_N^{(n)}\}$ is a peaked partition of unity on $X$. Let $x_1, \ldots, x_N \in X$ be points such that $\varphi_j^{(n)}(x_j) = 1$. If we define $P:\mathscr{C}(X) \to [\Phi_n]$ by

$$P(f) = \Sigma f(x_i)\varphi_i^{(h)},$$

then $P$ is a projection of norm 1 with range $[\Phi_n]$. Thus the space $\mathscr{C}(X)$ is a $\pi_1$-space.

We will establish the following theorem.

22.12 ***Theorem***

*Let $X$ be a topological space and let $E$ be a closed subspace. If $S \subset \mathscr{C}^B(E)$ and $A \subset \mathscr{C}^B(X)$ are closed subspaces, $S$ a separable $\pi_1$-space, and if $(S, A)$ has the bounded extension property, then there exists a linear extension $L:S \to A$ of norm 1.*

The proof of this result depends on further lemmas.

22.13 ***Lemma***

*Let $S_0 \subset S$ be a finite-dimensional space. If $\delta > 0$, there exists a bounded family $\{L_W : W$ a neighborhood of $E\}$ of linear extensions $L_W:S_0 \to A$ such that $|L_W f(x)| \leq \delta$ if $x \in X \setminus W$ and $\|f\| \leq 1$.*

PROOF

Let $s^1, \ldots, s^n$ be a base for $S_0$ so that if $f \in S_0$, then $f = \Sigma f(i)s^i$ for certain uniquely determined complex numbers $f(i)$. Denote by $\alpha_i$ the norm of the projection $f \mapsto f(i)s^i$ on $S_0$, let $\alpha = \Sigma\alpha_i$, and set $\varepsilon = \delta\alpha^{-1}$. For each neighborhood $W$ of $E$, let $s_W^i$ be an extension of $s^i$ to an element of $A$ which is of modulus less than $\varepsilon$ on $X \setminus W$. If we define $L_W:S_0 \to A$ to be the unique linear operator which takes $s^i$ to $s_W^i$, the family $\{L_W\}$ has the desired properties.

22.14 ***Lemma***

*If $S_0$ and $S_1$ are finite-dimensional subspaces of $S$, if $L_0:S_0 \to A$ is a linear extension, and if $\gamma > 0$, then there exists a linear extension*

*$L_1 : S_1 \to A$ such that*

$$\|L_0 f + L_1 g\| \le 1 + \gamma$$

*whenever $f \in S_0$, $g \in S_1$, $\|f + g\|_E \le 1$, and $\|L_0 f\| \le 1$.*

PROOF

For each linear extension operator $L : S_1 \to A$, let

$$K_L = \{L_0 f + Lg : f \in S_0, g \in S_1, \|f + g\|_E \le 1, \|L_0 f\| \le 1\}$$

and

$$\alpha_L(x) = \sup\{|h(x)| : h \in K_L\}.$$

We shall prove that $\alpha_L$ is a continuous function on $X$. For this purpose, we note that $K_L$ is compact, for if we set

$$\tilde{K}_L = \{(f, g) \in S_0 \times S_1 : \|f + g\|_E \le 1, \|L_0 f\| \le 1\},$$

then $\tilde{K}_L$ is a closed subset of $S_0 \times S_1$, and it is also bounded. This is so, for $L_0$ is linear and one to one since it is an extension operator, and the range of $L_0$ is finite dimensional. Consequently, there is a constant $c$ such that $\|f\|_E \le \|L_0 f\|_X \le c\|f\|_E$ for all $f \in S_0$. Thus, $\|f + g\|_E \le 1$ and $\|L_0 f\| \le 1$ imply that neither $\|f\|_E$ nor $\|g\|_E$ exceeds some absolute constant $C$. It follows that $\tilde{K}_L$ is compact, and since $K_L$ is the image of $\tilde{K}_L$ under the map $(f, g) \mapsto L_0 f + Lg$, $K_L$ is also compact. Define a map $\varphi : X \to \mathscr{C}(K_L)$ by $\varphi(x)(h) = h(x)$. The map $\varphi$ is continuous, for if $h_1, \ldots, h_q \in K_L$ constitute a basis for the space $M$ spanned by $K_L$, the compactness of $K_L$ implies the existence of a constant $C'$ such that if $f \in K_L$, then $f = \Sigma \alpha_j h_j$ with $\max|\alpha_j| \le C'$. If $\{x_\gamma\}$ is a net in $X$ which converges to $x_0$, we have

$$\begin{aligned} \|\varphi(x_0) - \varphi(x_\gamma)\| &= \sup\{|f(x_0) - f(x_\gamma)| : f \in K_L\} \\ &\le C'\Sigma|h_j(x_0) - h_j(x_\gamma)|, \end{aligned}$$

and by the continuity of $h_1, \ldots, h_q$, this last sum is small if $x_\gamma$ is near $x_0$. Thus, $\varphi$ is continuous. Since $\alpha_L(x) = \|\varphi(x)\|$, $\alpha_L$ is also continuous.

Let $\{L_W\}$ be a bounded family of linear extensions from $S_1$ as in the last lemma, say $\|L_W\| \le M$, where the $\delta$ of that lemma is taken to be $\gamma/4$. We shall write $\alpha_W$ in place of the more cumbersome $\alpha_{L_W}$. If $\|f + g\|_E \le 1$ and $\|L_0 f\|_X \le 1$, then $\|f\|_E \le 1$, so $\|g\| \le 2$. It follows that

$$\alpha_W(x) \le \begin{cases} 1 & \text{if } x \in E, \\ 1 + 2\delta & \text{if } x \in X \setminus W, \\ 1 + 2M & \text{if } x \in X. \end{cases}$$

Let $W_1 = X$ and define inductively sets $W_2, W_3, \ldots \supset E$, by

$$W_{n+1} = W_n \cap \{x \in X : \alpha_{W_n}(x) < 1 + 2\delta\},$$

and then for notational ease, write $L_n$ and $\alpha_n$ in place of $L_{W_n}$ and $\alpha_{W_n}$, respectively.

By our definition, if $x \in X$, there is at most one $n$ for which $\alpha_n(x) > 1 + 2\delta$, for if there are any such $n$, let $n_0$ be the smallest. Then $x \notin W_n$ for any larger $n$, so $\alpha_n(x) \leq 1 + 2\delta$ for all $n > n_0$.

If $N$ is an integer such that $(1 + 2M)N^{-1} \leq \delta$, then $x \in X$ implies that

$$\begin{aligned}\frac{1}{N}\sum_{n=1}^{N} \alpha_n(x) &\leq \frac{1}{N}\{(N-1)(1+2\delta) + (1+2M)\}\\ &\leq 1 + 3\delta.\end{aligned}$$

We define $L$ to be the operator $1/N \sum_{n=1}^{N} L_n$. If $f \in S_0$ and $g \in S_1$ satisfy $\|f + g\|_E \leq 1$ and $\|L_0 f\|_X \leq 1$, then for all $x \in X$,

$$\begin{aligned}|(L_0 f + Lg)(x)| &= \frac{1}{N}\left|\sum_{n=1}^{N} (L_0 f + Lg)(x)\right|\\ &\leq \frac{1}{N}\sum_{n=1}^{N} |L_0 f + L_n g|(x)\\ &\leq 1 + 3\delta,\end{aligned}$$

so $\|L_0 f + Lg\|_X \leq 1 + 3\delta$. It follows that for the $L_1$ of the lemma, we can take the operator $L$.

22.15 ***Corollary***

*Let $S_0$ and $S_1$ be finite-dimensional subspaces of $S$, $S_0 \subset S_1$, and suppose that there exists a linear projection $\pi : S_1 \to S_0$ of norm 1. If $\varepsilon > 0$ and $L_0 : S_0 \to A$ is a linear extension with $\|L_0\| < 1 + \varepsilon$, then there exists a linear extension $L' : S_1 \to A$ which satisfies $\|L'\| < 1 + \varepsilon$, and $L'|S_0 = L_0$.*

PROOF

Let $\gamma > 0$ be such that $(1 + \gamma)\|L_0\| < 1 + \varepsilon$, pick a linear extension $L_1 : S_1 \to A$ in accordance with the lemma, and set

$$L'f = L_0(\pi f) + L_1(f - \pi f).$$

Since $L_0$ and $L_1$ are both extension operators, so is $L'$. We have to prove that if $g \in S_1$, $\|g\|_E \leq 1$, then $\|L'g\|_X < 1 + \varepsilon$. Since $\|L_0\| \geq 1$, it suffices to show that if $g' = g\|L_0\|^{-1}$, then $\|L'g'\|_X \leq 1 + \gamma$. For this purpose, set $h = \pi g'$, $h' = g' - \pi g'$, so that $L'g' = L_0 h + L_1 h'$. Then

$$\|h + h'\| = \|g'\| \leq \|g\| \leq 1$$

and

$$\begin{aligned}\|L_0 h\| &= \|L_0 \pi g'\| \leq \|L_0\| \, \|\pi\| \, \|g'\|_E \\ &= \|\pi\| \, \|g\|_E \leq 1;\end{aligned}$$

so the choice of $L_1$ implies that $\|L'g'\| \leq 1 + \gamma$ and we have the result.

22.16 *Corollary*

*If $S_0$ is a finite-dimensional subspace of $S$, if $\varepsilon > 0$, and if $L$ is a linear extension from $S_0$ to $A$ with $\|L\| < 1 + \varepsilon$, then there is a linear extension $L' : S_0 \to A$ with $\|L'\| = 1$ and $\|L - L'\| < \varepsilon$.*

PROOF

It will suffice to construct $L''$, a linear extension from $S_0$ to $A$, such that for some positive $\gamma < \frac{1}{2}\varepsilon$ we have $\|L''\| < 1 + \gamma$ and $\|L'' - L\| < \varepsilon - \gamma$; for then we can inductively repeat the process and obtain a convergent sequence $L_n$ of linear extensions whose limit has the property we seek.

Let $\nu = \|L\|$ and let $\gamma < \frac{1}{2}\varepsilon$ be such that

$$(\nu - 1)(\gamma + 1) < \varepsilon - \gamma.$$

Choose $L_1$ in accordance with Lemma 22.14, taking $S_0 = S_1$, and let

$$L'' = \nu^{-1} L + (\nu - 1)\nu^{-1} L_1.$$

To prove $\|L''\| < 1 + \gamma$, let $h \in S_0$, $\|h\| \leq 1$. If $f = \nu^{-1} h$ and $g = (\nu - 1)\nu^{-1} h$ so that

$$L''h = Lf + L_1 g,$$

then since

$$\|f + g\|_E = \nu^{-1}(1 + (\nu - 1))\|h\| \leq 1$$

and

$$\|Lf\|_X \leq \|L\| \, \|f\|_E = \nu\nu^{-1}\|h\| \leq 1,$$

it follows that $\|L''h\|_X \leq 1 + \gamma$, whence $\|L''\| \leq 1 + \gamma$.

We complete the proof by showing that $\|L - L''\| < \varepsilon - \gamma$. For this purpose, consider an $h \in S_0$ with $\|h\|_E \leq 1$. Let $f = \nu^{-1} h$,

$g = -v^{-1}h$. We shall show that

$$\|Lf + L_1 g\|_X \leq 1 + \gamma; \tag{7}$$

for if this is so, then

$$\begin{aligned}\|(L - L'')h\|_X &= (v - 1)\|Lf + L_1 g\|_X \\ &\leq (v - 1)(1 + \gamma) < \varepsilon - \gamma\end{aligned}$$

since

$$L - L'' = (v - 1)v^{-1}(L - L_1).$$

To establish (7), note that $\|f + g\|_E = 0$ and

$$\begin{aligned}\|Lf\|_X \leq \|L\| \, \|f\| &= vv^{-1}\|h\| \\ &\leq 1,\end{aligned}$$

so the choice of $L_1$ implies that $\|Lf + L_1 g\|_X \leq 1 + \gamma$. This establishes (7).

We are at last in a position to prove Theorem 22.12 and obtain therewith Theorem 22.2.

22.17 ***Proof of Theorem 22.12***

Let $S_1 \subset S_2 \subset \cdots$ be a sequence of finite-dimensional subspaces of $S$ whose union is dense in $S$ and which have the property that for each $n$, there is a projection of norm 1 from $S$ onto $S_n$. Let $L_1$ be a linear extension of norm 1 from $S_1$ to $A$—one exists by Corollary 22.16 and the obvious fact that there exist some linear extensions from $S_1$ to $A$. By Corollaries 22.15 and 22.16 we can find inductively a sequence $\{L_n\}$, $L_n : S_n \to A$ a linear extension of norm 1 such that

$$\|L_{n+1}|S_n - L_n\| < 2^{-n}.$$

For each fixed $n$, the sequence $\{L_m|S_n\}_{m=n}^{\infty}$ is Cauchy in $\mathscr{L}(S_n, A)$, and therefore it has a limit, $L'_n$, in $\mathscr{L}(S_n, A)$, $\|L'_n\| = 1$. Since $L'_{n+1}|S_n = L'_n$, the operators $L'_n$ jointly yield an operator $L' : \bigcup S_n \to A$ which has norm 1 and which is an extension operator. By the continuity of $L'$ and the density of $\bigcup S_n$ in $S$, $L'$ can be extended to the desired extension operator, and the theorem is proved.

# 23 *THE STRUCTURE OF ORTHOGONAL MEASURES*

At many points in our study of uniform algebras, but particularly in our consideration of interpolation, we have had to deal with measures that annihilate a given uniform algebra. In this section we shall study the structure of such measures in some detail and obtain a rather explicit formula for annihilating measures, a formula that exhibits clearly the relation between annihilating measures and representing measures. Our treatment is based on an extension of the Hahn–Banach theorem due to König.

Recall that if $E$ is a real vector space and $\theta$ is a real-valued sublinear functional on $E$, then given $x \in E$, there is a real linear functional $\varphi$ on $E$ such that $\varphi(x) = \theta(x)$ and $\varphi \leq \theta$; i.e., $\varphi(y) \leq \theta(y)$ for all $y \in E$. The following fact may be regarded as a generalization of this formulation of the Hahn–Banach theorem.

23.1 ***Theorem*** (König [1])
*If $M$ is a nonempty convex subset of the real linear space $E$, and if $\theta$ is a real sublinear functional on $E$, then there is a real linear functional $\varphi$ on $E$ dominated by $\theta$ and satisfying*

$$\inf_{x \in M} \varphi(x) = \inf_{x \in M} \theta(x).$$

PROOF
Set $I = \inf_{x \in M} \theta(x)$. If $I = -\infty$, then we may choose for $\varphi$ any linear functional dominated by $\theta$, so we suppose that $I > -\infty$. Define $\eta : E \to \mathbb{R}$ by

$$\eta(x) = \inf\{\theta(x + ty) - tI : y \in M, t \geq 0\}.$$

The functional $\eta$ is dominated by $\theta$ since for any $t \geq 0$ and any $y \in M$,

$$\begin{aligned}\eta(x) &\leq \theta(x + ty) - tI \\ &\leq \theta(x) + t(\theta(y) - I) \leq \theta(x),\end{aligned}$$

and $\eta$ is sublinear since

$$\begin{aligned}\eta(x + x') &= \inf\{\theta(x + x' + ty) - tI : y \in M, t \geq 0\} \\ &= \inf\{\theta(x + x' + \tfrac{1}{2}ty + \tfrac{1}{2}ty') - tI : y, y' \in M, t \geq 0\} \\ &\leq \eta(x) + \eta(x').\end{aligned}$$

By the Hahn–Banach theorem, there exists a real linear functional $\varphi$ on $E$ which is dominated by $\eta$. We have then that if $x \in M$, then

$$\begin{aligned}-\varphi(x) &= \varphi(-x) \leq \eta(-x) \\ &\leq \theta(-x + x) - I \\ &= -I,\end{aligned}$$

so $\varphi(x) \geq I$. Thus,

$$\inf_{x \in M} \varphi(x) \geq \inf_{x \in M} \theta(x),$$

and, as $\varphi \leq \theta$, this gives the result.

The application of this result to uniform algebras arises in the following way. Let $(A, X)$ be a uniform algebra, let $\varphi \in \Sigma(A)$, and define $\theta$ on $\mathscr{C}_{\mathbb{R}}(X)$ by

$$\theta(u) = \inf\{\operatorname{Re} \varphi(f) : f \in A, \operatorname{Re} f \geq u\}.$$

The functional $\theta$ is sublinear on $\mathscr{C}_{\mathbb{R}}(X)$, and we have that for all $u \in \mathscr{C}_{\mathbb{R}}(X)$,

$$\inf_{x \in X} u(x) \leq \theta(u) \leq \sup_{x \in X} u(x).$$

If $\sigma$ is a measure on $X$ such that $\int u \, d\sigma \leq \theta(u)$ for all $u \in \mathscr{C}_{\mathbb{R}}(X)$, then $\int \operatorname{Re} f \, d\sigma = \operatorname{Re} \varphi(f)$ for every $f \in A$ and $\|\sigma\| \leq 1$, so $\sigma$ is a positive representing measure for $\varphi$, i.e., $\sigma \in \mathscr{M}_\varphi$.

We need the following fact, which we shall refer to as the *formal modification lemma.*

23.2 ***Lemma***

*If* $\{u_n\}_{n=1}^{\infty}$ *is a sequence of nonnegative elements of* $\mathscr{C}_{\mathbb{R}}(X)$ *with the property that* $\theta(u_n) \to 0$, *there exist elements* $f_n \in A$ *with* $|f_n| \leq 1$, $\theta(|f_n - 1|) \to 0$, *and* $\|u_n f_n\|_X \to 0$.

PROOF

Define $\alpha_n$ to be the number $\theta(u_n)$ so that $\alpha_n \to 0$, and let $\{t_n\}$ be a sequence of reals, $t_n \geq 1$ for all $n$, such that $t_n \to \infty$ but $t_n(\alpha_n + (1/n)) \to 0$. Choose $g_n \in A$ with $\operatorname{Re} g_n \geq u_n$, $\operatorname{Re} \varphi(g_n) \leq \alpha_n + (1/n)$, and $\operatorname{Im} \varphi(g_n) = 0$. If $f_n = \exp(-t_n g_n)$, then $f_n \in A$, and the sequence $\{f_n\}$ has the desired properties. To see that $\|u_n f_n\|_X \to 0$, let $\varepsilon > 0$, and write, for each $n$, $X = X_+^n \cup X_-^n$, where

$$X_-^n = \{x \in X : u_n(x) < \varepsilon\}$$

and

$$X_+^n = \{x \in X : u_n(x) \geq \varepsilon\}.$$

For all $n$, $|u_n(x) f_n(x)| < \varepsilon$ if $x \in X_-^n$ because $\|f_n\|_X \leq 1$, and if $x \in X_+^n$, then

$$\begin{aligned} |u_n(x) f_n(x)| &\leq |u_n(x)| \exp(-t_n u_n(x)) \\ &\leq |t_n u_n(x)| \exp(-t_n u_n(x)) \\ &\leq \exp(-t_n \varepsilon + \log t_n \varepsilon). \end{aligned}$$

For large values of $n$ this last term is small, so we have that $\|u_n f_n\|_X \to 0$.

Finally, we show that $\theta(|f_n - 1|) \to 0$. By the Hahn–Banach and Riesz representation theorems, we know that for given $n$, there is a measure $\sigma$ such that $\theta$ dominates the functional $u \mapsto \int u\, d\sigma$ on $\mathscr{C}_{\mathbf{R}}(X)$, and such that

$$\theta(|f_n - 1|) = \int |f_n - 1|\, d\sigma \leq \left(\int |f_n - 1|^2\, d\sigma\right)^{1/2}.$$

The measure $\sigma$ lies in $\mathscr{M}_\varphi$, so we have the estimate

$$\begin{aligned}\int |f_n - 1|^2\, d\sigma &= \int (|f_n|^2 + 1 - 2\operatorname{Re} f_n)\, d\sigma \\ &\leq 2 - 2\operatorname{Re}\int f_n\, d\sigma \\ &\leq 2 - 2\exp(-t_n(\alpha_n + 1/n)).\end{aligned}$$

Since $t_n(\alpha_n + (1/n)) \to 0$, we find that $\theta(|f_n - 1|) \to 0$, and we have the lemma.

The next lemma contains our first application of the generalized Hahn–Banach theorem.

23.3 **_Lemma_**

*If $\lambda$ is a nonnegative measure on $X$ which is singular with respect to each element of $\mathscr{M}_\varphi$, then there are $f_n \in A$ with $\|f_n\|_X \leq 1$, $f_n \to 1$ a.e. $[\mu]$ for every $\mu \in \mathscr{M}_\varphi$, and $f_n \to 0$ a.e. $[\lambda]$.*

PROOF

Apply Theorem 23.1 to the functional $L$ given by $L(u) = \theta(u) - \int u\, d\lambda$, $\theta$ as above, and the convex set $M = \{u \in \mathscr{C}_{\mathbf{R}}(X) : 0 \leq u \leq 1\}$ to find a linear functional $L_1$ dominated by $L$ such that $\inf_{0 \leq u \leq 1} L_1(u) = \inf_{0 \leq u \leq 1} L(u)$. If $L_2(u) = L_1(u) + \int u\, d\lambda$, then $L_2$ is dominated by $\theta$ and so is continuous, and it admits a representing measure $\sigma$ which lies in $\mathscr{M}_\varphi$. For this $\sigma$ we have, by construction,

$$\text{(1)} \qquad \inf_{0 \leq u \leq 1}\left(\int u\, d\sigma - \int u\, d\lambda\right) = \inf_{0 \leq u \leq 1}\left(\theta(u) - \int u\, d\lambda\right).$$

Since $\sigma$ and $\lambda$ are mutually singular, the infimum on the left is $-\lambda(X)$, for there exist sequences $\{E_n\}$ and $\{F_n\}$ of compact sets such that $E_n \cap F_n = \varnothing$, $\sigma(E_n) \to \|\sigma\| = 1$, and $\lambda(F_n) \to \|\lambda\|$. By Tietze's extension theorem, there is a function $u_n \in \mathscr{C}_{\mathbf{R}}(X)$ with values in $[0, 1]$ which vanishes on $E_n$ and which is constantly 1 on $F_n$. The functions $u_n$ show the left side of (1) to be $-\lambda(X)$. Thus,

$$\inf_{0 \le u \le 1} \theta(u) + \int (1 - u)\, d\lambda = 0. \tag{2}$$

Consequently, there is a sequence $\{u_n\}$ in $\mathscr{C}_{\mathbf{R}}(X)$, $0 \le u_n \le 1$, such that

$$\theta(u_n) \to 0 \qquad \text{and} \qquad u_n \to 1 \text{ a.e. } [\lambda].$$

Applying the formal modification lemma, we find a sequence $\{f_n\}$ in $A$ with $\|f_n\|_X \le 1$, $\|f_n u_n\|_X \to 0$, and $\theta(|f_n - 1|) \to 0$. Passing to subsequences if necessary, we may suppose that $\theta(|f_n - 1|) < 1/4^n$. Since $f_n u_n \to 0$ uniformly and $u_n \to 1$ a.e. $[\lambda]$, if follows that $f_n \to 0$ a.e. $[\lambda]$. Also, since $\theta(|f_n - 1|) < 1/4^n$, the definition of $\theta$ provides elements $g_n \in A$ with $\operatorname{Re} g_n \ge |f_n - 1|$, $\operatorname{Re} \varphi(g_n) < 1/2^n$. Consequently, if $\mu \in \mathscr{M}_\varphi$, we have

$$\begin{aligned} \frac{1}{2^n} > \operatorname{Re} \varphi(g_n) &= \operatorname{Re} \int g_n \, d\mu \\ &\ge \int |f_n - 1| \, d\mu. \end{aligned}$$

The convergence of the series $\Sigma \int |f_n - 1|\, d\mu$ implies that $\Sigma |f_n - 1|$ converges a.e. $[\mu]$, and therefore that $f_n \to 1$ a.e. $[\mu]$.

The next lemma will enable us to obtain the desired decomposition formula for elements of $A^\perp$.

## 23.4 Lemma

*If $\lambda$ is a measure on $X$ which annihilates $A$, there is a generalized Lebesgue decomposition $\lambda = \lambda_c + \lambda_s$, where*

(a) *$\lambda_c$ and $\lambda_s$ annihilate $A$.*
(b) *$\lambda_c$ is absolutely continuous with respect to some element of $\mathscr{M}_\varphi$.*
(c) *$\lambda_s$ is concentrated on a set $E$ of type $F_\sigma$ which is a null set for every $\mu \in \mathscr{M}_\varphi$.*

PROOF

Let us call a set $F$ *$\mathscr{M}_\varphi$-null* if $\mu(F) = 0$ for each $\mu \in \mathscr{M}_\varphi$. Choose a sequence $\{E_n\}$ of closed $\mathscr{M}_\varphi$-null sets such that $|\lambda|(E_n) \to \sup\{|\lambda|(F) : F$ is $\mathscr{M}_\varphi$-null$\} = S$, and let $E = \bigcup E_n$. The set $E$ is an $\mathscr{M}_\varphi$-null set of type $F_\sigma$, and if $\lambda_E$ denotes the restriction of $\lambda$ to $E$, and $\lambda_{E^c}$ the restriction to $E^c = X \setminus E$, then $\lambda_{E^c}$ vanishes on every set which is $\mathscr{M}_\varphi$-null; for if $F$

is an $\mathcal{M}_\varphi$-null set with $|\lambda_{E^c}| > 0$, then $E \cup F$ is $\mathcal{M}_\varphi$-null, and $|\lambda|(E \cup F) > S$, contradicting the definition of $S$.

To prove that $\lambda_{E^c}$ is absolutely continuous with respect to some element of $\mathcal{M}_\varphi$, we consider, somewhat more generally, an arbitrary positive measure $m$ which vanishes on all $\mathcal{M}_\varphi$-null sets. If $\mu \in \mathcal{M}_\varphi$, let $m_\mu$ denote the component of $m$ which is absolutely continuous with respect to $\mu$, and let $m'_\mu$ be the singular part: If $m = \tau + \sigma$ is the Lebesgue decomposition of $m$ with respect to $\mu$, $\tau \ll \mu$, $\sigma \perp \mu$, then $m_\mu = \tau$ and $m'_\mu = \sigma$. Let $\gamma = \sup\{\|m_\mu\| : \mu \in \mathcal{M}_\varphi\}$, choose $\mu_n \in \mathcal{M}_\varphi$, $n = 1, 2, \ldots$, with $\|m_{\mu_n}\| \to \gamma$, and let $\mu_0 = \sum_{n=1}^\infty 2^{-n}\mu_n$. The convexity and weak * compactness of $\mathcal{M}_\varphi$ imply that $\mu_0 \in \mathcal{M}_\varphi$, and we shall prove that $m \ll \mu_0$. For each $n$, there are complementary Borel sets $S_n$ and $T_n$ in $X$ such that

$$m_{\mu_n}(E) = m(E \cap S_n), \qquad m'_{\mu_n}(E) = m(E \cap T_n),$$

for all $E$. Let $S = \bigcup S_n$, $T = \bigcap T_n$, and consider the restrictions $m_S$ and $m_T$: $m_S(E) = m(E \cap S)$ and $m_T(E) = m(E \cap T)$. If $F$ is a Borel set such that $\mu_0(F) = 0$, then $\mu_n(F) = 0$ for all $n$; so $\mu_n(S_n \cap F) = 0$, and thus $m_{\mu_n}(S_n \cap F) = 0$; i.e., $m(S_n \cap F) = 0$. This is valid for all $n$, so $m(\bigcup(S_n \cap F)) = 0$, whence $m(S \cap F) = 0$. This yields $m_S(F) = 0$, and we have that $m_S \ll \mu_0$. Also, $m_T \perp \mu_0$, for

$$\mu_0(T) = \sum_{n=1}^{\infty} 2^{-n}\mu_n(T) \le \sum_{n=1}^{\infty} 2^{-n}\mu_n(T_n)$$
$$= 0.$$

The Lebesgue decomposition of $m$ with respect to $\mu$ is, therefore,

$$m = m_S + m_T.$$

We have that $\|m_S\| = m(S) \ge m(S_n) \to \gamma$, so $\|m_S\| = \gamma$. It follows that $T$ is $\mathcal{M}_\varphi$-null, and as $m$ vanishes on $\mathcal{M}_\varphi$-null sets, $m_T = 0$. Thus, $m \ll \mu_0$. Apply this to the measure $|\lambda_{E^c}|$: It is absolutely continuous with respect to some $\mu \in \mathcal{M}_\varphi$, so the same is true of $\lambda_{E^c}$ itself.

Now we apply Lemma 23.3, with $|\lambda_E|$ replacing the $\lambda$ of that lemma, to find a sequence $\{f_n\}$ in $A$ with $\|f_n\|_X \le 1$, $f_n \to 1$ a.e. $[\mu]$ for all $\mu \in \mathcal{M}_\varphi$, and $f_n \to 0$ a.e. $[|\lambda_E|]$. If $f \in A$, then

$$(3) \qquad 0 = \int ff_n\, d\lambda = \int ff_n\, d\lambda_E + \int ff_n\, d\lambda_{E^c}.$$

Since $f_n \to 0$ a.e. $[|\lambda_E|]$, the first of the integrals in (3) is small for large $n$, and since $\lambda_{E^c}$ is absolutely continuous with respect to some element of $\mathcal{M}_\varphi$, the choice of $\{f_n\}$ shows that the second integral is, for large $n$, nearly $\int f\, d\lambda_{E^c}$. Thus, $\lambda_{E^c} \in A^\perp$, so the same is true of $\lambda_E$. This finishes the proof.

Before formulating our main result, whose proof is quite simple on the basis of the preceding preliminaries, it is desirable to introduce a definition.

23.5 ***Definition***

*If $(A, Y)$ is a uniform algebra, the measure $\sigma$ on $X$ is said to be* ***completely singular*** *if it is singular with respect to every representing measure for every nonzero complex homomorphism of $A$.*

23.6 ***Theorem***

*Let $(A, X)$ be a uniform algebra, and let $\lambda \in \mathcal{M}(X)$ annihilate $A$. There exists a sequence $\varphi_n$, $n = 1, \ldots,$ of elements of $\Sigma(A)$ which lie in distinct Gleason parts such that for some representing measure $\mu_n$ for $\varphi_n$ and some Borel set $E$,*

$$\lambda = \sum_n \lambda_n + \lambda_E,$$

*where $\lambda_n \ll \mu_n$, $\lambda_E$ is completely singular, and the measures $\lambda_n$ and $\lambda_E$ all annihilate $A$. The series converges in the sense of the norm on $\mathcal{M}(X)$.*

PROOF

Let $\varphi, \psi \in \Sigma(A)$ lie in different Gleason parts, and let

$$\lambda = \lambda_\varphi + \lambda_{E_\varphi} = \lambda_\psi + \lambda_{E_\psi}$$

be decompositions of $\lambda$ with respect to $\mathcal{M}_\varphi$ and $\mathcal{M}_\psi$, respectively, as in Lemma 23.4, so that $\lambda_\varphi, \lambda_{E_\varphi} \in A^\perp$, $\lambda_\varphi \ll \mu$ for some $\mu \in \mathcal{M}_\varphi$, $E_\varphi$ is $\mathcal{M}_\varphi$-null, and similar conditions hold on $\lambda_\psi$ and $\lambda_{E_\psi}$. We see that $\lambda_\varphi$ and $\lambda_\psi$ are mutually singular because $\varphi$ and $\psi$ lie in different Gleason parts, and so if $\mu \in \mathcal{M}_\varphi$ and $v \in \mathcal{M}_\psi$, then $\mu$ and $v$ are mutually singular.

The fact that $\lambda$ is a finite measure implies the existence of a sequence $\Pi_n$, $n = 1, \ldots,$ of Gleason parts for $A$ such that if $\Pi$ is a Gleason part for $A$ not one of the $\Pi_n$, and if $v$ is a representing measure for a point of $\Pi$, then $\lambda$ and $v$ are mutually singular.

For each $n$, let $\varphi_n \in \Pi_n$. We write

$$\lambda = \lambda_{\mu_1} + \lambda_{E_1},$$

where $\lambda_{\mu_1}$ is the component of $\lambda$ absolutely continuous with respect to the element $\mu_1$ of $\mathcal{M}_{\varphi_1}$ which satisfies $\|\lambda_{\mu_1}\| = \sup\{\|\lambda_\mu\| : \mu \in \mathcal{M}_{\varphi_1}\}$, as in Lemma 23.4, $E_1$ is $\mathscr{M}_{\varphi_1}$-null, and $\lambda_{\mu_1}, \lambda_{E_1} \in A^\perp$. Next decompose $\lambda_{E_1}$ with respect to $\mathscr{M}_{\varphi_2}$:

$$\lambda_{E_1} = \lambda_{\mu_2} + \lambda_{E_2}.$$

We have that $E_2 \subset E_1$ and that $E_2$ is $\mathscr{M}_{\varphi_2}$-null. Generally, for $n = 2, 3, \ldots$, write

$$\lambda_{E_n} = \lambda_{\mu_{n+1}} + \lambda_{E_{n+1}},$$

decomposing with respect to $\mathscr{M}_{\varphi_{n+1}}$. The measures $\lambda_{\mu_n}$ generated in this way are all mutually singular, and the sets $E_n$ constitute a decreasing sequence. If we set $E = \bigcap E_n$, then

$$\lambda = \sum_{n=1}^{\infty} \lambda_{\mu_n} + \lambda_E$$

is the decomposition we seek.

The theorem we have just proved is the culmination of a long series of generalizations of the classical F. and M. Riesz theorem, Corollary 20.5. Let us indicate briefly some of the steps in this development. We begin by mentioning the proof given by Helson [1] for the classical F. and M. Riesz theorem. Helson's point of view was to consider a measure $\mu$ on the unit circle which is orthogonal to the disc algebra, to consider the Lebesgue decomposition of $\mu$ with respect to Lebesgue measure on the circle, say, $\mu = \mu_c + \mu_s$, $\mu_c$ the absolutely continuous part of $\mu$ and $\mu_s$ the singular part, and to show that both $\mu_c$ and $\mu_s$ annihilate the disc algebra. Finally, he showed that $\mu_s = 0$. Working in a more general context involving compact abelian groups with ordered duals, Helson and Lowdenslager [1] showed, in effect, that if $(A, X)$ is a Dirichlet algebra, if $\varphi \in \Sigma(A)$, and if $\mu$ is the representing measure for $\varphi$, then given any $\lambda \in A^\perp$ with Lebesgue decomposition $\lambda = \lambda_\mu + \lambda'_\mu$, $\lambda_\mu \ll \mu$, $\lambda'_\mu \perp \mu$, then $\lambda_\mu, \lambda'_\mu \in A^\perp$. In this generality, of course, it is not possible to prove that $\lambda'_\mu = 0$ as it is in the classical case. Basing his work on a proof given by Forelli [1] for some of the Helson–Lowdenslager results, Ahern [1] proved that if $(A, X)$ is an arbitrary uniform algebra, and if $\varphi \in \Sigma(A)$ has the property that there is $\mu \in \mathscr{M}_\varphi$ with respect to which every element of $\mathscr{M}_\varphi$ is absolutely continuous, and if $\lambda \in A^\perp$ has the Lebesgue decomposition $\lambda = \lambda_\mu + \lambda'_\mu$ with respect to $\mu$, then both $\lambda_\mu$ and $\lambda'_\mu$ lie in $A^\perp$. Generalizing the Helson–Lowdenslager results in a somewhat different direction, Glicksberg and Wermer [1] proved the version of Theorem 23.6 valid for Dirichlet algebras. In [5], Glicksberg showed that a decomposition slightly weaker than that of Theorem 23.6 could be obtained. König and Seever [1] obtained a decomposition formally different from that given by Glicksberg.

Finally, Rainwater [1] showed that the decomposition given by Glicksberg coincided with that given by König and Seever, and thus the full result, Theorem 23.6, was obtained. The development given by Glicksberg depended on a version of the minimax theorem of von Neuman and was somewhat more involved than that presented above. The idea of basing the whole theory on the Hahn–Banach theorem as in the treatment just given is due to König [1].

We now turn to some further details in connection with the decomposition of orthogonal measures. We observe, first of all, that certain choices are implicit in the decomposition. The Gleason parts $\Pi_n$ of the theorem are determined uniquely by the measure $\lambda$, but we have made a choice of a point $\varphi_n$ in each $\Pi_n$. Because of general results on parts, the decomposition obtained for $\lambda$ is independent of this choice. To see this, let $\psi_n$ be another point of $\Pi_n$, and let

$$\lambda = \Sigma\lambda_{\nu_n} + \lambda_F$$

be the decomposition of $\lambda$ obtained by using the homomorphism $\psi_n$ rather than $\varphi_n$. Thus, $\nu_n \in \mathcal{M}_{\psi_n}$ and $\lambda_{\nu_n} \ll \nu_n$. By Theorem 16.6, there is a measure $\mu_n' \in \mathcal{M}_{\varphi_n}$ such that $\nu_n \ll \mu_n'$. Since $\mu_n$ is an element of $\mathcal{M}_{\varphi_n}$ such that

$$\|\lambda_{\mu_n}\| = \sup\{\|\lambda_\mu\| : \mu \in \mathcal{M}_{\varphi_n}\},$$

it follows that $\lambda_{\nu_n} \ll \mu_n$, for otherwise, with $\tau = \frac{1}{2}(\mu_n + \mu_n')$, we have

$$\|\lambda_\tau\| > \|\lambda_{\mu_n}\|,$$

a contradiction. As $\lambda_{\nu_k} \perp \mu_n$ for all $k \neq n$, and as $\lambda_F \perp \mu_n$, it follows that $\lambda_{\nu_n} = \lambda_{\mu_n}$, so the decomposition does not depend on the particular choice of $\varphi_n \in \Pi_n$.

As a consequence of our decomposition for elements of $A^\perp$, we have the following interpolation result.

**23.7** ***Corollary***

*Let $(A, X)$ be a uniform algebra with the property that no element of $A^\perp$ is completely singular. Then the closed set $E$ in $X$ is a peak interpolation set for $A$ if and only if it is a $G_\delta$, and $\mu(E) = 0$ for every representing measure $\mu$ for every $\varphi \in \Sigma(A)$.*

As Glicksberg [5] noted, it is possible to obtain a sharper version of this result by using the Bishop–deLeeuw theorem: The conclusion of the corollary is correct under the weakened hypothesis that no extreme point of the unit ball of $A^\perp$ be completely singular. We refer to Glicksberg's paper for the details of this result.

In general there will be completely singular measures among the extreme points of the unit ball of $A^{\perp}$, but in certain important special cases there are no completely singular measures at all in $A^{\perp}$. As we shall see in the next chapter, this is true of the algebra $\mathscr{R}(X)$, $X$ a compact set in the plane.

Our next result is interesting in its own right, and it will enable us to give a nontrivial application of our decomposition for elements of $A^{\perp}$. In the case of Dirichlet algebras, it goes back to a result of Hoffmann and Wermer (see Wermer [12]), and in its present form it was given by Glicksberg [5]. For further results in this direction, see Lumer [2] and Gamelin and Lumer [1].

**23.8** ***Theorem***

*If $f$ is a bounded Borel function on $X$ for which there exists a sequence $\{f_n\}$ in $A$ with*

$$\sup_{\mu \in \mathscr{M}_\varphi} \int |f - f_n|^2 \, d\mu \to 0,$$

*then there is a sequence $\{g_n\}$ in $A$ with $\|g_n\| \leq \|f\|$ and $g_n \to f$ a.e. $[\mu]$ for each $\mu \in \mathscr{M}_\varphi$.*

PROOF

We assume that $\|f\| = 1$ and that

$$\sup_{\mu \in \mathscr{M}_\varphi} \int |f - f_n|^2 \, d\mu \leq n^{-4}$$

so that $f_n \to f$ a.e. $[\mu]$ for all $\mu \in \mathscr{M}_\varphi$. If

$$E_n = \{x : |f_n(x)| \geq 1 + \varepsilon\},$$

then $\sup_{\mu \in \mathscr{M}_\varphi} \mu(E_n) \to 0$, and

$$\begin{aligned} \int_{E_n} \log |f_n| \, d\mu &\leq \int_{E_n} |f_n| \, d\mu \\ &\leq \int_{E_n} |f - f_n| \, d\mu + \int_{E_n} |f| \, d\mu \\ &\leq \left( \int |f - f_n|^2 \, d\mu \right)^{1/2} + \int_{E_n} |f| \, d\mu. \end{aligned}$$

This implies that if we set

$$\eta_n = \sup_{\mu \in \mathscr{M}_\varphi} \int_{E_n} \log |f_n| \, d\mu,$$

then $\eta_n \to 0$. From this we may infer that if

$$\eta'_n = \sup_{\mu \in \mathcal{M}_\varphi} \int \log^+|f_n|\, d\mu,$$

then $\eta'_n \to 0$, for if $\mu \in \mathcal{M}_\varphi$, then

$$\begin{aligned}\int \log^+|f_n|\, d\mu &\le \int_{E_n} \log|f_n|\, d\mu + \log(1 + \varepsilon) \\ &\le \eta_n + \varepsilon.\end{aligned}$$

When $n$ is large, this is less than $2\varepsilon$ independently of $\mu$, and as $\varepsilon$ is at our disposal, $\eta'_n$ must approach zero.

The function $u_n = -\log^+|f_n|$ lies in $\mathscr{C}_{\mathbf{R}}(X)$, and since

$$-\eta'_n = \inf\left\{\int u_n\, d\mu : \in \mathcal{M}_\varphi\right\},$$

it follows from Lemma 7.19 that there exists a function $h_n \in A$ with

$$\operatorname{Re} h_n \le u_n$$

and

$$0 \ge \operatorname{Re} \varphi(h_n) \ge -\eta'_n - \frac{1}{n}.$$

As $\eta'_n \to 0$, we may suppose by passing to a subsequence, if necessary, that

$$\operatorname{Re} \varphi(h_n) > -n^{-4}.$$

Finally, we define $g_n$ by

$$g_n = f_n e^{h_n} \overline{\operatorname{sgn} e^{\varphi(h_n)}}.$$

The function $g_n$ lies in $A$, and we have

$$\begin{aligned}|g_n| &= |f_n| \exp \operatorname{Re} h_n \\ &\le |f_n| \exp(-\log^+|f_n|) \le 1.\end{aligned}$$

If $h'_n = e^{h_n} \operatorname{sgn} e^{\varphi(h_n)}$, then $h'_n \in A$, and

$$|h'_n| = \exp \operatorname{Re} h_n \le 1,$$

so

$$\begin{aligned}1 \ge \varphi(h'_n) &= |e^{\varphi(h_n)}| \\ &= \exp(\operatorname{Re} \varphi(h_n)) > e^{-n^{-4}} \\ &> 1 - n^{-4}.\end{aligned}$$

This yields the fact that $h_n' \to 1$ a.e. $[\mu]$ for every $\mu \in \mathscr{M}_\varphi$, and therefore that $g_n = f_n h_n' \to f$ a.e. $[\mu]$ for every $\mu \in \mathscr{M}_\varphi$.

If we combine the theorem on modified convergence with our decomposition of orthogonal measures, we obtain in the case of algebras with unique representing measures the following result established by Glicksberg and Wermer [1] in the case of Dirichlet algebras.

23.9 ***Theorem***

*If $(A, X)$ is a uniform algebra with the property that each $\varphi \in \Sigma(A)$ admits a unique representing measure and if $\lambda \in A^\perp$, then there exists a sequence $\{\mu_n\}$ of representing measures for points of $\Sigma(A)$ lying in distinct parts, there exists for each n an $f_n \in H^1_{\varphi n}(\mu_n)$, and there exists a Borel set E which is a null set for every representing measure for A, such that $\lambda = \Sigma f_n \mu_n + \lambda_E$, the series convergent in the sense of the norm on $\mathscr{M}(X)$ and $\lambda_E \in A^\perp$.*

PROOF

We know that there is a decomposition of the form $\lambda = \Sigma f_n \mu_n + \lambda_E$ with $f_n \in L^1(\mu_n)$, $f_n \mu_n \in A^\perp$, and $\lambda_E \in A^\perp$. The problem is to show that $f_n \in H^1(\mu_n)$. To prove this, it is enough to show that if $\psi$ is a continuous linear functional on $L^1(\mu_n)$ which annihilates $H^1(\mu_n)$, then $\psi(f_n) = 0$. Such a functional $\psi$ is of the form

$$\psi(f) = \int fm \, d\mu_n$$

for some fixed $m \in L^\infty(\mu_n)$. As $\psi$ annihilates $H^1(\mu_n)$ which contains $H^2(\mu_n)$, and as $L^2(\mu_n) = H^2(\mu_n) \oplus \overline{H^2_{\varphi n}(\mu_n)}$, it follows that $m \in H^2_{\varphi n}(\mu_n)$. As $m$ is bounded, there exists a bounded sequence $\{h_k\}$ in $A$ which converges a.e. $[\mu_n]$ to $m$. We have $f_n \mu_n \in A^\perp$, so for all $k$,

$$\int f_n h_k \, d\mu_n = 0,$$

and passing to the limit as $k \to \infty$, we find

$$\int f_n m \, d\mu_n = 0,$$

whence $\psi(f_n) = 0$.

There are certain situations more general than the unique representing measure setting in which the modified convergence theorem is available. Probably the simplest of these is that involving *strongly dominant representing measures.*

**23.10** ***Definition***

*Let $(A, X)$ be a uniform algebra, let $\varphi \in \Sigma(A)$, and let $\mu \in \mathcal{M}_\varphi$. The measure $\mu$ is a* ***dominant*** *representing measure for $\varphi$ if each $\nu \in \mathcal{M}_\varphi$ is absolutely continuous with respect to $\mu$, and it is* ***strongly dominant*** *if it is dominant and if, in addition, there is a constant $C$ such that for each $\nu \in \mathcal{M}_\varphi$, the derivative $d\nu/d\mu$ is bounded by $C$.*

These notions were introduced by Glicksberg [7], who made the following observations. If $\mathcal{M}_\varphi$ is norm separable, then a dominant representing measure exists, for if $\{\mu_n\}_{n=1}^{\infty}$ is a countable dense subset of $\mathcal{M}_\varphi$, then $\sum_{n=1}^{\infty} 2^{-n}\mu_n$ is an element of $\mathcal{M}_\varphi$, and it is clearly dominant. More generally, $\varphi$ admits a dominant representing measure if $\mathcal{M}_\varphi$ is contained in $L^1(m)$ for any measure $m$. To see this, let $m = m_\mu + m_\mu'$ be the Lebesgue decomposition of $m$ with respect to $\mu$, $\mu \in \mathcal{M}_\varphi$, $m_\mu \ll \mu$, and let $\mu_0 \in \mathcal{M}_\varphi$ be such that

$$\|m_{\mu_0}\| = \sup\{\|m_\mu\| : \mu \in \mathcal{M}_\varphi\}.$$

(Recall the similar situation in Lemma 23.4.) As in that earlier corollary, each $\mu \in \mathcal{M}_\varphi$ is absolutely continuous with respect to $\mu_0$. In particular, if $(\mathrm{Re}\, A)^\perp$ is norm separable, then dominant representing measures exist; for if $\mu, \nu \in \mathcal{M}_\varphi$, then $\mu - \nu \in (\mathrm{Re}\, A)^\perp$, and the separability of $(\mathrm{Re}\, A)^\perp$ implies that of $\mathcal{M}_\varphi$. In the event that $(\mathrm{Re}\, A)^\perp$ is finite dimensional, it is clear that strongly dominant representing measures exist, for then $\mathcal{M}_\varphi$ is a finite-dimensional compact convex set, and thus there is a point $\mu \in \mathcal{M}_\varphi$ with the property that $\mu + \varepsilon(\mu - \nu) \in \mathcal{M}_\varphi$ for all $\nu \in \mathcal{M}_\varphi$ and some fixed $\varepsilon > 0$. From this it follows that $(1 + \varepsilon)\mu - \varepsilon\nu \geq 0$ for all $\nu \in \mathcal{M}_\varphi$, so $\nu \ll \mu$ and the Randon–Nikodym derivative $d\nu/d\mu$ is bounded in modulus uniformly in $\nu$.

The modified convergence theorem is applicable whenever there is a strongly dominant representing measure.

**23.11** ***Corollary***

*If $\mathcal{M}_\varphi$ contains a strongly dominant representing measure, $\mu$, and if $f$ is a bounded function which lies in the $L^2(\mu)$-closure of $A$, then there is a sequence $\{g_n\}$ in $A$, $\|g_n\| \leq \|f\|$, such that $g_n \to f$ a.e. $[\nu]$ for all $\nu \in \mathcal{M}_\varphi$.*

PROOF

If $\nu \in \mathcal{M}_\varphi$, then $d\nu/d\mu \leq C$ for some absolute constant $C$; so if $\{f_n\}$ is a sequence in $A$ such that

$$\int |f_n - f|^2 \, d\mu \to 0,$$

then

$$\int |f_n - f|^2 \, d\nu = \int |f_n - f|^2 \frac{d\nu}{d\mu} \, d\mu$$
$$\leq C \int |f_n - f|^2 \, d\mu.$$

Thus,

$$\sup_{\nu \in \mathscr{M}_\varphi} \int |f_n - f|^2 \, d\nu \to 0,$$

so the corollary follows from Theorem 23.8.

We conclude this section by introducing conditions sufficient for a representing measure to be dominant or strongly dominant.

23.12 ***Definition***

*A representing measure $\mu$ for the uniform algebra $(A, X)$ is* ***enveloped*** *if, given a sequence $\{u_n\}$ in $\mathscr{C}_{\mathbf{R}}(X)$, $u_n \geq 0$, such that $\int u_n \, d\mu \to 0$, there is a subsequence $\{u_{n_k}\}$ and a sequence $f_k \in A$ with $|f_k| \leq \exp(-u_{n_k})$ and $\int f_k \, d\mu \to 1$.*

The following result was given by Gamelin and Lumer [1]; the proof is essentially due to Forelli [1].

23.13 ***Theorem***

*If $\mu$ is an enveloped representing measure for $\varphi$, then $\mu$ is dominant.*

PROOF

It is enough to show that if $\lambda \in A^\perp$ has the Lebesgue decomposition $\lambda = \lambda_\mu + \lambda'_\mu$, $\lambda_\mu \ll \mu$, $\lambda'_\mu \perp \mu$ with respect to $\mu$, then $\lambda_\mu, \lambda'_\mu \in A^\perp$; for then if $\nu \in \mathscr{M}_\varphi$, we have $\nu - \mu \in A^\perp$, and if $\nu - \mu$ has the Lebesgue decomposition $\tau + \sigma$ with respect to $\mu$, $\sigma \perp \mu$, then $\sigma \in A^\perp$; but $\sigma$ is nonnegative since $\nu$ is. It follows that $\sigma = 0$, so $\nu \ll \mu$, as was to be shown.

Now consider $\lambda \in A^\perp$ with Lebesgue decomposition $\lambda = \lambda_\mu + \lambda'_\mu$ with respect to $\mu$. There is an increasing sequence of compact sets, $\{K_n\}$, such that $|\lambda'_\mu|(\bigcup K_n) = \|\lambda'_\mu\|$, $\mu(K_n) = 0$. Let $u_n \in \mathscr{C}_{\mathbf{R}}(X)$ be nonnegative and satisfy $u_n \geq n$ on $K_n$ and $\int u_n \, d\mu \to 0$. Since $\mu$ is enveloped, it follows that for some subsequence $\{u_{n_k}\}$ of $\{u_n\}$, there is a corresponding sequence $\{f_k\}$ in $A$ with $|f_k| \leq \exp(-u_{n_k})$ and $\int f_k \, d\mu \to 1$. We have that $f_k \to 0$ pointwise on $K = \bigcup K_n$, and by passing to a subsequence of $\{f_k\}$ if necessary, we may assume that $f_k \to 1$ a.e. $[\mu]$.

If $g \in A$, we have that

$$\begin{aligned}\int g\, d\lambda'_\mu &= \int g\chi_K\, d\lambda \\ &= \lim_{k\to\infty} \int g(1 - f_k)\, d\lambda \\ &= 0\end{aligned}$$

since $\lambda \in A^\perp$. Thus, $\lambda'_\mu \in A^\perp$ and consequently $\lambda_\mu$ also lies in $A^\perp$, and we have the theorem.

It is important for certain applications to observe that there is a modified convergence theorem available whenever there is an enveloped representing measure, a remark that is contained in the first few lines of the proof of Theorem 23.8.

23.14 ***Theorem***

*If $\mu$ is an enveloped representing measure for the point $\varphi \in \Sigma(A)$, and if $f$ is a bounded function which lies in the $L^2(\mu)$-closure of $A$, then there is a sequence $\{g_n\}$ in $A$, $\|g_n\|_X \le \|f\|_\infty$, which converges* a.e. $[\lambda]$ *to $f$ for every $\lambda \in \mathcal{M}_\varphi$.*

PROOF

Since $\mu$ is dominant, it is enough to construct a sequence $\{g_n\}$ which converges a.e. $[\mu]$ to $f$. We assume that $\|f\|_\infty = 1$, that $f$ is the $L^2(\mu)$-limit of the sequence $\{f_n\}$ from $A$, and that $f_n \to f$ a.e. $[\mu]$. Arguing as in the first few lines of the proof of Theorem 23.8, we find that $\int \log^+|f_n|\, d\mu \to 0$. Since $\mu$ is enveloped, we can pass to a subsequence of $\{f_n\}$, if necessary, and suppose there to exist functions $h_n$ in $A$ with

$$|h_n| \le \exp(-\log^+|f_n|)$$

and

$$\int h_n\, d\mu \to 1.$$

Then $\exp(-\log^+|f_n|) \le 1$, so $\|h_n\|_X \le 1$. Since $\int h_n\, d\mu \to 1$ and $\|h_n\|_X \le 1$, it follows that $h_n \to 1$ a.e. $[\mu]$, so $f_n h_n \to f$ a.e. $[\mu]$. Also, $|h_n f_n| \le |f_n| \exp(-\log^+|f_n|) \le 1$, so $\|f_n h_n\|_X \le 1$, and we have the result.

This observation has important ramifications in the study of certain $H^\infty$-algebras. If $\mu$ is enveloped, then $H^\infty(\mu)$ is the weak * closure of $A$ in $L^\infty(\mu)$ where, as usual, $H^\infty(\mu)$ denotes the intersection of $L^\infty(\mu)$ with $H^2(\mu)$, the $L^2(\mu)$-closure of $A$. By the remarks just before Lemma 17.5, we know that $H^\infty(\mu)$ is a weak * closed subalgebra of $L^\infty(\mu)$. And if $h \in H^\infty(\mu)$, then since $h$ is the almost everywhere limit of a bounded sequence $\{g_n\}$ in $A$, it follows that $f$ lies in the weak * closure of $A$ in $L^\infty(\mu)$; for by the dominated convergence theorem, $\{g_n\}$ converges weak * to $h$.

There is a condition similar to that of being enveloped which characterizes strongly dominant representing measures. It is due in part to Glicksberg [7] and in part to Gamelin and Lumer [1].

23.15 ***Theorem***

*The representing measure $\mu$ for $\varphi$ is strongly dominant if and only if, given a nonnegative sequence $\{u_n\}$ in $\mathscr{C}_{\mathrm{R}}(X)$ with $\int u_n\, d\mu \to 0$, there is a sequence $\{v_n\}$ in* Re $A$ *with $v_n \geq u_n$ and $\int v_n\, d\mu \to 0$.*

PROOF

Assume $\mu$ to be strongly dominant so that for some $C$ we have that $\nu \in \mathscr{M}_\varphi$ implies that $\nu \ll \mu$ and $\|d\nu/d\mu\|_\infty \leq C$. Then we have for each $n$,

$$\inf\{\operatorname{Re} \varphi(f) : f \in A, \operatorname{Re} f \geq u_n\}$$

$$= \sup\left\{\int u_n\, d\nu : \nu \in \mathscr{M}_\varphi\right\}$$

$$\leq C \int u_n\, d\mu$$

by Lemma 7.19, and as $\operatorname{Re} \varphi(f) = \int \operatorname{Re} f_n\, d\mu$, it is clear that the desired sequence $\{v_n\}$ exists.

Conversely, assume that given $\{u_n\}$, the desired $\{v_n\}$ exists. Then $v_n = \operatorname{Re} f_n$, $f_n \in A$, and the function $g_n = \exp(-f_n)$ lies in $A$. We have that $|g_n| = e^{-v_n} \leq e^{-u_n}$, and

$$\left|\int g_n\, d\mu\right| = \left|\exp \int -f_n\, d\mu\right|$$

$$= \exp \int -v_n\, d\mu \to 1.$$

It follows that $\mu$ is enveloped and thus dominant. If it is not strongly dominant, there is a sequence $\mu_n \in \mathscr{M}_\varphi$ such that if $h_n = d\mu_n/d\mu$, then $\|h_n\|_\infty \to \infty$. We can choose $u_n \in \mathscr{C}_{\mathrm{R}}(X)$, $u_n \geq 0$, with

$$\int u_n\, d\mu \to 0 \qquad \text{and} \qquad \int u_n h_n\, d\mu \to \infty.$$

If $v_n \in \operatorname{Re} A$ satisfies $v_n \geq u_n$, then

$$\begin{aligned}\int v_n \, d\mu &= \int v_n h_n \, d\mu \\ &\geq \int u_n h_n \, d\mu \\ &\to \infty,\end{aligned}$$

so $\int v_n \, d\mu \to 0$ is impossible.

We shall see some further applications of the results of this section in the next chapter. Yet other applications may be found in the papers of Glicksberg [5, 7, 8], Garnett and Glicksberg [1], and Gamelin and Lumer [1].

# chapter five/ALGEBRAS ON PLANAR SETS

## 24 INTRODUCTION

In this chapter we shall study various algebras naturally attached to a compact set in the plane. The algebras to be studied consist of holomorphic functions and limits of sequences of holomorphic functions. If $X$ is a compact set in the plane, the following algebras are naturally associated with $X$:

$\mathscr{P}(X) = \{f \in \mathscr{C}(X) : f$ can be approximated uniformly on $X$ by polynomials$\}$,

$\mathscr{R}(X) = \{f \in \mathscr{C}(X) : f$ can be approximated uniformly on $X$ by rational functions$\}$,

$\overline{\mathscr{O}(X)} = \{f \in \mathscr{C}(X) : f$ can be approximated uniformly on $X$ by functions holomorphic on a neighborhood of $X\}$,

and

$\mathscr{A}(X) = \{f \in \mathscr{C}(X) : f$ is holomorphic on the interior of $X\}$.

These algebras are all uniform algebras on $X$, and it is clear that

$$\mathscr{P}(X) \subset \mathscr{R}(X) \subset \overline{\mathscr{O}(X)} \subset \mathscr{A}(X).$$

One of the central problems to be considered in this chapter is the problem of determining when these various inclusions degenerate into equalities. Thus, e.g., if $X$ has no interior so that $\mathscr{A}(X) = \mathscr{C}(X)$, we are asking when $\mathscr{P}(X) = \mathscr{C}(X)$ and when $\mathscr{R}(X) = \mathscr{C}(X)$. These questions are problems in classical analysis, problems of approximation. There are also algebraic problems of interest. For example, what are the spectra of these algebras and what are their Gleason parts? Let us mention in passing that these questions also make sense in higher dimensions, but although the theory in the plane has reached a certain degree of completion, our knowledge of the higher dimensional analogue of this theory is still exceedingly meager.

Let us begin our considerations by showing that in an essential way our approximation problems are nontrivial. We shall give an example of a compact planar set $X$ without interior for which $\mathscr{R}(X) \neq \mathscr{C}(X)$. The example we give is due to McKissick [2] and is a modification of one given by Mergelyan [2]. Let $\Delta$ denote the open unit disc.

24.1 **_Lemma_**

*Let* $\{D_j\}_{j=1}^{\infty}$ *be a sequence of open discs in the plane. If*

$$X = \bar{\Delta} \setminus \bigcup\{D_j : j = 1, 2, 3, \ldots\}$$

*and if the sum of the radii of the* $D_j$ *is less than* 1, *then* $\mathscr{R}(X) \neq \mathscr{C}(X)$.

In particular, then, if the $D_j$ are chosen so that $X$ has no interior, we have an example of the kind mentioned above.

PROOF OF THE LEMMA

To prove the lemma, we shall construct a nonzero measure $\mu$ on $X$ which annihilates $\mathscr{R}(X)$.

For $n = 1, 2, \ldots,$ let $X_n = \bar{\Delta} \setminus \bigcup\{D_j : j = 1, 2, \ldots, n\}$. The boundary of $X_n$ consists of the union of a finite collection of circular arcs, and we can define a measure $\mu_n$ on $\partial X_n$ by requiring that

$$\int f \, d\mu_n = \int_{\partial X_n} f(z) \, dz.$$

The length of $\partial X_n$ is less than $4\pi$, so $\|\mu_n\| < 4\pi$. Thus, $\{\mu_n\}_{n=1}^{\infty}$ has a weak * cluster point, $\mu$. Since $X = \bigcap X_n$, the measure $\mu$ is supported on $X$, and if $f$ is a national function with no poles on $X$, then for large values of $n$, $f$ has no poles on $X_n$; and so by the Cauchy integral theorem,

$$0 = \int_{\partial X_n} f(z) \, dz = \int f \, d\mu_n.$$

Thus,

$$0 = \int f \, d\mu$$

for all such $f$, and consequently $\mu$ annihilates $\mathscr{R}(X)$.

We now show that the measure $\mu$ is not the zero measure. For this purpose, denote by $A_n$ the area of $X_n$, and by $A$ that of $X$, so that by hypothesis $A \neq 0$. We have, using Green's theorem,

$$\begin{aligned} \int \bar{z} \, d\mu_n &= \int_{\partial X_n} (x - iy) \, dx + (ix + y) \, dy \\ &= 2i \int_{Xn} dx \, dy \\ &= 2iA_n. \end{aligned}$$

It follows that $\int \bar{z} \, d\mu = 2iA \neq 0$, so $\mu$ is not the zero measure, and we have the result.

The next matter we consider is Runge's theorem, which shows that for all compact sets $X$ in $\mathbb{C}$, $\overline{\mathcal{O}(X)} = \mathcal{R}(X)$. Runge's theorem is a classical result in function theory, but the proof we shall give is a functional analytic one in keeping with the general spirit of this book. I do not know the origin of the proof; I learned it from L. A. Rubel.

In the proof, we shall use certain integrals of Cauchy–Stieltjes type.

24.2 ***Definition***

*If $\mu$ is a finite measure on the plane, then $\tilde{\mu}$, the **Cauchy transform** of $\mu$, is the function defined by*

$$\tilde{\mu}(z) = \int \frac{d\mu(\zeta)}{\zeta - z}.$$

The function $\tilde{\mu}$ is defined at least on the complement of the support of $\mu$. This Cauchy transform will play a central role at several points in the theory we shall develop in the remainder of the chapter.

24.3 ***Theorem***

*Let $X$ be a compact set in the plane and let $\{x_j\}$ be a sequence of points, one from each bounded component of $\mathbb{C} \setminus X$. Every function holomorphic on a neighborhood of $X$ can be approximated uniformly on $X$ by rational functions with poles only among the points of the sequence $\{x_j\}$ and at infinity.*

PROOF

If the theorem is false, there is a function $f$ holomorphic on a neighborhood of $X$ and a measure $\mu$ on $X$ with $\int f\,d\mu \neq 0$, but $\int g\,d\mu = 0$ if $g$ is a polynomial or a polynomial in $(z - x_j)^{-1}$, $j = 1, 2, \ldots$. Let $x_j$ lie in the component $V_j$ of $\mathbb{C} \setminus X$. For fixed $j$, we have that if $|\zeta - x_j|$ is less than the distance from $x_j$ to $X$, then

$$\begin{aligned}\tilde{\mu}(\zeta) &= \int \frac{d\mu(z)}{z - \zeta} \\ &= \sum_{n=0}^{\infty} (\zeta - x_j)^n \int \frac{d\mu(z)}{(z - x_j)^{n+1}} \\ &= 0,\end{aligned}$$

so $\tilde{\mu}$ vanishes on a neighborhood of $x_j$. As $\tilde{\mu}$ is holomorphic on $V_j$, it follows that $\tilde{\mu}$ vanishes on the whole of $V_j$. Similarly, $\tilde{\mu}$ vanishes on $V_0$, the unbounded component of $\mathbb{C} \setminus X$.

If $\Gamma$ is a suitable system of oriented curves in $\mathbb{C} \setminus X$, then by the Cauchy integral formula we have for $z \in X$,

$$f(z) = \frac{1}{2\pi i} \int_{\Gamma} \frac{f(\zeta)}{\zeta - z} \, d\zeta,$$

and when we integrate this equation with respect to $\mu$, we find

$$\int f \, d\mu = \frac{1}{2\pi i} \int_{\Gamma} f(\zeta) \left\{ \int \frac{d\mu(z)}{z - \zeta} \right\} d\zeta.$$

By the choice of $\mu$ the left side of this equation is not zero, but since $\tilde{\mu}$ vanishes on $\mathbb{C} \setminus X$, the right side is zero, and we have a contradiction.

As a consequence of this result, we have the following fact, which is attributed to Bishop and Hoffman by Rossi [2].

**24.4** ***Corollary***
*The algebra $\mathscr{R}(X)$ is a doubly generated Banach algebra.*

PROOF
Let $V_1, V_2, \ldots$ be the components of $\mathbb{C} \setminus X$, and for each $n$, let $z_n$ be a point of $V_n$. Define a sequence $\{\varepsilon_n\}$ of positive numbers by requiring that $\varepsilon_1 = 1$, and for $n = 2, 3, \ldots,$

$$\varepsilon_n \max \left\{ 1, \left\| \frac{1}{z - z_n} \right\|_X, \left\| \frac{z - z_1}{z - z_n} \right\|_X, \ldots, \left\| \frac{z - z_{n-1}}{z - z_n} \right\|_X \right\} < 2^{-(n+1)} \varepsilon_{n-1}.$$

Let $g_0(z) = z$ and

$$g_1(z) = \sum_{n=1}^{\infty} \frac{\varepsilon_n}{z - z_n}.$$

We shall prove that $g_0$ and $g_1$ generate $\mathscr{R}(X)$. [The choice of the numbers $\varepsilon_n$ implies that the series defining $g_1$ converges uniformly on $X$, so $g_1$ does lie in $\mathscr{R}(X)$.] To prove the corollary, it suffices to show that for each $n$, the function $h_n$ given by $h_n(z) = 1/(z - z_n)$ lies in $A$, the algebra generated by $g_0$ and $g_1$. Since $A$ contains the constants, the function $(g_0 - z_1)g_1 - \varepsilon_1$; i.e.,

$$\sum_{n=2}^{\infty} \frac{\varepsilon_n(z - z_1)}{z - z_n}$$

lies in $A$. If $h_1$ does not lie in $A$, then $g_0 - z_1$ is not invertible in $A$, so there is $\varphi \in \Sigma(A)$ such that $\varphi(g_0) = z_1$. This implies that

$$\varepsilon_1 = \left| \varphi\left( \sum_{n=2}^{\infty} \frac{\varepsilon_n(z - z_1)}{z - z_n} \right) \right|. \tag{1}$$

Since $\|\varphi\| = 1$ and

$$\varepsilon_n \left\| \frac{z - z_1}{z - z_n} \right\|_X < 2^{-(n+1)}\varepsilon_1,$$

the equality (1) is impossible, for the right side is less than

$$\varepsilon_1 \sum_{n=2}^{\infty} 2^{-n} = \tfrac{1}{2}\varepsilon_1.$$

Thus, $h_1 \in A$, and so the function $g_2$ given by

$$g_2(z) = \sum_{n=2}^{\infty} \frac{\varepsilon_n}{z - z_n}$$

lies in $A$. If $h_2$ is not in $A$, then, as above, there is $\varphi \in \Sigma(A)$ with $\varphi(g_0) = z_2$, and applying $\varphi$ to $(g_0 - z_2)g_2 - \varepsilon_2$, we find that

$$\varepsilon_2 = \left| \varphi\left( \sum_{n=3}^{\infty} \frac{\varepsilon_n(z - z_2)}{z - z_n} \right) \right|.$$

The choice of $\varepsilon_3, \varepsilon_4, \ldots$ again leads to a contradiction. Thus, $h_2 \in A$. Continuing in this way we find that $A$ contains $h_n$ for all $n$, and so $A = \mathscr{R}(X)$, as we wished to prove.

The fact that $\mathscr{R}(X)$ is a doubly generated uniform algebra is quite useful in constructing counterexamples. Many plausible conjectures about finitely generated algebras can be refuted by appeal to a suitable $\mathscr{R}(X)$.

We now turn to the investigation of the spectra of the algebras $\mathscr{A}(X)$, $\mathscr{R}(X)$, and $\mathscr{P}(X)$.

24.5 ***Theorem***

(a) *The spectrum of $\mathscr{R}(X)$ is $X$.*

(b) *The spectrum of $\mathscr{P}(X)$ is the complement in $\mathbb{C}$ of the unbounded component of $\mathbb{C} \setminus X$.*

PROOF

Let $\mathscr{P}$ denote the algebra of polynomials in one complex variable. Every complex homomorphism, other than the zero homomorphism, is of the form $p \mapsto p(\zeta)$ for some fixed $\zeta \in \mathbb{C}$. Thus, if $\varphi \in \Sigma(\mathscr{R}(X))$, there exists a unique $\zeta \in \mathbb{C}$ with $\varphi(p) = p(\zeta)$ for all polynomials $p \in \mathscr{R}(X)$. The point $\zeta$ must lie in $X$, for if not set

$$f(z) = \frac{1}{z - \zeta}.$$

This is an invertible element of $\mathscr{R}(X)$, but our hypothesis implies that $\varphi(f^{-1}) = 0$. As this is impossible, $\zeta \in X$. If $g$ is a rational function without poles on $X$, then $g = g_0 g_1^{-1}$, $g_0, g_1$ polynomials, $g_1$ zero free on $X$, so $\varphi(g) = \varphi(g_0)\varphi(g_1^{-1}) = \varphi(g_0)\varphi(g_1)^{-1} = g(\zeta)$. It follows that $h$ is of the form $h \mapsto h(\zeta)$, so $\Sigma(\mathscr{R}(X)) = X$, as asserted.

We now consider $\mathscr{P}(X)$. This is a commutative Banach algebra with a single generator, so its spectrum is a certain polynomially convex subset of $\mathbb{C}$. Since $\Sigma(\mathscr{P}(X)) \supset X$, it follows that $\Sigma(\mathscr{P}(X)) \supset \hat{X}$, $\hat{X}$ the polynomially convex hull in $\mathbb{C}$ of $X$. By Example 5.5, $\hat{X}$ is the complement of the unbounded component of $\mathbb{C} \setminus X$. By the definition of $\hat{X}$ and the fact that every $\varphi \in \Sigma(\mathscr{P}(X))$ is of norm 1, it follows that no point of $\mathbb{C} \setminus \hat{X}$ can lie in $\Sigma(\mathscr{P}(X))$, and we have part (b) of the theorem.

It remains to consider the spectrum of $\mathscr{A}(X)$. The expected result that $X = \Sigma(\mathscr{A}(X))$ holds, but it is a much more difficult fact that the corresponding fact for $\mathscr{R}(X)$. The equality $X = \Sigma(\mathscr{A}(X))$ is due to Arens [1], though special cases had been treated by Hoffman and Singer [1] and by Royden. The result Arens proved is somewhat more general than $X = \Sigma(\mathscr{A}(X))$, and to formulate it, it is convenient to introduce some notation. Let $X$ be a compact set in the Riemann sphere $\mathbb{C}^*$, $X = \mathbb{C}^*$ is permitted, and let $\Omega$ be an open subset of $\mathbb{C}^*$ contained in $X$. We define

$$\mathscr{A}(X, \Omega) = \{f \in \mathscr{C}(X) : f|\Omega \text{ is holomorphic}\}.$$

When $\Omega$ is the interior of $X$, $\mathscr{A}(X, \Omega)$ is the algebra $\mathscr{A}(X)$ which we have already considered. Arens proved the following result.

24.6 ***Theorem***

*The spectrum of $\mathscr{A}(X, \Omega)$ is $X$ provided $\mathscr{A}(X, \Omega)$ contains a nonconstant function.*

The proof of this result depends on an approximation lemma.

24.7 ***Lemma***

*If $f \in \mathscr{A}(X, \Omega)$ is nonconstant, and if $z_0 \in X$, $z_0 \neq \infty$, then there is a sequence $\{h_n\}$ in $\mathscr{A}(X, \Omega)$ such that if $\tilde{h}_n(z) = (z - z_0)h_n(z)$, then $\tilde{h}_n \in \mathscr{A}(X, \Omega)$, and*

$$\|f - f(z_0) - \tilde{h}_n\|_X \to 0$$

*as $n \to \infty$.*

Let us assume the validity of this lemma and see how it implies the theorem.

PROOF OF THEOREM 24.6

If $\Omega$ is the empty set, then $\mathscr{A}(X, \Omega) = \mathscr{C}(X)$, and the result is known. Assume, therefore, that $\Omega \neq \varnothing$, and that $\mathscr{A}(X, \Omega)$ contains a nonconstant function. By a preliminary conformal map of the sphere onto itself, we may suppose that the point at infinity lies in $\Omega$ and thus that $X \setminus \Omega$ is a compact set in the complex plane.

The hypothesis that $\mathscr{A}(X, \Omega)$ contains nonconstant elements implies that it separates points on $X$. If $X$ omits an open subset of $\mathbb{C}^*$, this is clear; the general case is covered by Example 7.9.

Let $\varphi$ be a nonzero complex homomorphism of $\mathscr{A}(X, \Omega)$, and assume that $\varphi$ is not the homomorphism $f \mapsto f(\infty)$. Thus, for some $f_0 \in \mathscr{A}(X, \Omega)$, we have $\varphi(f_0) = 1$, $f_0(\infty) = 0$. The function $g$ given by $g(z) = zf_0(z)$ lies in $\mathscr{A}(X, \Omega)$, and if $\zeta_0 = \varphi(g)$, then for all $h \in \mathscr{A}(X, \Omega)$ for which $(z - \zeta_0)h \in \mathscr{A}(X, \Omega)$, we have $\varphi((z - \zeta_0)h) = 0$ since

$$\begin{aligned}\varphi((z - \zeta_0)h) &= \varphi((z - \zeta_0)h)\varphi(f_0)\\ &= \varphi(h)\varphi((z - \zeta_0)f_0)\\ &= 0.\end{aligned}$$

It follows that $\zeta_0 \in X$, for if not, $(z - \zeta_0)^{-1}$ is a well-defined element of $\mathscr{A}(X, \Omega)$, and we are led to the contradiction that

$$1 = \varphi(1) = \varphi((z - \zeta_0)(z - \zeta_0)^{-1}),$$

whereas the last term is zero by what we have just said. If $f \in \mathscr{A}(X, \Omega)$, then by the lemma there is a sequence $h_n \in \mathscr{A}(X, \Omega)$ such that if $\tilde{h}_n(z) = (z - \zeta_0)h_n(z)$, then $\tilde{h}_n \in \mathscr{A}(X, \Omega)$, and

$$\|f - f(\zeta_0) - \tilde{h}_n\|_X \to 0.$$

We have, therefore, that

$$\varphi(f) = f(\zeta_0) + \lim_{n \to \infty} \varphi(\tilde{h}_n),$$

and this yields $\varphi(f) = f(\zeta_0)$; i.e., $\varphi$ is point evaluation at the point $\zeta_0$.

It remains to prove the lemma.

PROOF OF LEMMA 24.7

Again, we suppose that $\infty \in \Omega$. If $n$ is a positive integer, denote by $D_n(z)$ the disc

$$\left\{\zeta \in \mathbb{C} : |z - \zeta| < \frac{1}{n}\right\}.$$

Consider $f \in \mathscr{A}(X, \Omega)$, and let $z_0 \in X$, $z_0 \neq \infty$. Extend $f$ to the whole of $\mathbb{C}^*$ by setting $f = 0$ off $X$, and then define, for $n = 1, 2, 3, \ldots$, a function $h_n$ by

$$h_n(z) = \frac{n^2}{\pi} \iint_{D_n(z_0)} \frac{f(\zeta) - f(z)}{\zeta - z} \, d\xi \, d\eta \qquad \zeta = \xi + i\eta.$$

As in Example 7.9, the functions $h_n$ are all continuous on $\mathbb{C}^*$. We shall show them to be holomorphic in $\Omega$. To prove this, let $\gamma$ be the boundary of a rectangle which, together with its interior, lies in $\Omega$. Then

$$\int_\gamma h_n(z) \, dz = \frac{h^2}{\pi} \iint_{D_n(z_0)} \left\{ \int_\gamma \frac{f(\zeta) - f(z)}{\zeta - z} \, dz \right\} d\xi \, d\eta.$$

For all choices of $\zeta$, the inner integral on the right is zero, so the left-hand side is zero, and the desired conclusion follows from Morera's theorem. Since $h_n$ is bounded, it is holomorphic at $\infty$.

We shall now show that if $\tilde{h}_n(z) = (z - z_0)h_n(z)$, then

$$\| f - f(z_0) - \tilde{h}_n \|_X \to 0.$$

For notational convenience we suppose $z_0 = 0$, and we write $D_n$ rather than $D_n(0)$. It is sufficient to consider the case that $f(0) = 0$. We have

$$\begin{aligned} zh_n(z) - f(z) &= \frac{n^2}{\pi} \iint_{D_n} \left\{ z \frac{f(\zeta) - f(z)}{\zeta - z} - f(z) \right\} d\xi \, d\eta \\ &= \frac{n^2}{\pi} \iint_{D_n} \frac{zf(\zeta) - \zeta f(z)}{\zeta - z} \, d\xi \, d\eta. \end{aligned}$$

If $M(r) = \sup\{|f(z)| : z \in X, |z| \leq r\}$, we have the estimate

$$\text{(2)} \qquad |zh_n(z) - f(z)| \leq \left\{ |z| M\left(\frac{1}{n}\right) + \frac{1}{n} |f(z)| \right\} \frac{n^2}{\pi} \iint_{D_n} \frac{d\xi \, d\eta}{|\zeta - z|}.$$

For the integral in this inequality, two estimates are possible:

$$\text{(3)} \qquad \frac{n^2}{\pi} \iint_{D_n} \frac{d\xi \, d\eta}{|\zeta - z|} \leq \begin{cases} (|z| - n^{-1})^{-1} & \text{if } |z| > \dfrac{1}{n}, \\ 4nk & \text{if } |z| \leq \dfrac{k}{n}, \qquad k \geq 1. \end{cases}$$

The latter estimate is obtained by noting that if $|z| \leq k/n$, then

$$\begin{aligned} \frac{n^2}{\pi} \iint_{D_n} \frac{d\xi \, d\eta}{|\zeta - z|} &\leq \frac{n^2}{\pi} \iint_{|\zeta| < 2k/n} |\zeta|^{-1} \, d\xi \, d\eta \\ &= 4nk. \end{aligned}$$

Choose $\varepsilon > 0$, and select $k > 1$ so that

$$2\|f\| < (k-1)\varepsilon.$$

Let $z \in X$, and consider first the case that $|z| \geq k/n$. For this choice of $z$, we obtain from (2) and (3) that

$$\begin{aligned}|zh_n(z) - f(z)| &\leq \frac{|z|}{|z| - n^{-1}} M\left(\frac{1}{n}\right) + \frac{1}{n}\|f\| \frac{1}{|z| - n^{-1}} \\ &\leq \frac{k}{k-1} M\left(\frac{1}{n}\right) + \|f\| \frac{1}{k-1} \\ &\leq \frac{k}{k-1} M\left(\frac{1}{n}\right) + \frac{\varepsilon}{2}.\end{aligned}$$

As $n \to \infty$, $M(1/n) \to 0$ because $f(0) = 0$ and $f$ is continuous on $X$. Thus, for $|z| \geq k/n$, $|zh_n(z) - f(z)| < \varepsilon$ if $n$ is big enough. For $|z| < k/n$, we have

$$\begin{aligned}|zh_n(z) - f(z)| &\leq \{k/nM(1/n) + 1/nM(k/n)\}4nk \\ &= 4k\{kM(1/n) + M(k/n)\}.\end{aligned}$$

Again, $M(1/n)$ and $M(k/n) \to 0$, so this quantity can be made small by taking $n$ large enough, and we have the result.

In his paper [2], Arens has given a version of this theorem valid for the corresponding algebras on certain Riemann surfaces. An abstract version has been given by Rosay [1].

We have identified the spectra of the algebras $\mathscr{A}(X)$, $\mathscr{R}(X)$, and $\mathscr{P}(X)$. What of their Shilov and minimal boundaries? The question of the minimal boundary for $\mathscr{P}(X)$ can be settled by reasonably simple means, as we shall see in the next section, but the question for $\mathscr{A}(X)$ and $\mathscr{R}(X)$ is considerably more difficult and depends on the theory of analytic capacity. It is quite easy to identify the Shilov boundaries for these algebras.

**24.8** ***Theorem***

*The topological boundary of $X$ is the Shilov boundary for $\mathscr{R}(X)$ and for $\mathscr{A}(X)$, and the Shilov boundary for $\mathscr{P}(X)$ is the topological boundary of the unbounded component of $\mathbb{C} \setminus X$.*

PROOF

If $x_0 \in \partial X$ and if $U$ is a neighborhood of $x_0$, then the function given by $f(z) = (z - z_1)^{-1}$, $z_1 \in U \setminus X$, is an element of $\mathscr{R}(X)$ which peaks at a point of $U$. Consequently, $x_0$ lies in the Shilov boundary for $\mathscr{R}(X)$.

This implies that the Shilov boundary for $\mathscr{R}(X)$ is the boundary of $X$. It follows that the Shilov boundary for $\mathscr{A}(X)$ is $\partial X$. If $x_0$ lies in the topological boundary of the unbounded component, $K$, of $\mathbb{C} \setminus X$, then, given a neighborhood $U$ of $x_0$, we can find a rational function with all its poles in $K$ and peaking (relative to $X$) at a point of $U$. If we approximate this rational function uniformly on $X$ by a polynomial—such approximation is possible because of Runge's theorem—we obtain an element of $\mathscr{P}(X)$ which peaks in $U$. Thus, $x_0$ lies in the Shilov boundary for $\mathscr{P}(X)$, and our assertion follows.

# 25 THE ALGEBRA $\mathscr{P}(X)$

The main result of this section is a theorem of Mergelyan which shows that $\mathscr{A}(X) = \mathscr{P}(X)$ if and only if $\mathbb{C} \setminus X$ is connected. The necessity of the condition is completely elementary, for if $V$ is a bounded component of $\mathbb{C} \setminus X$, if $z_0 \in V$, and if $C > \|z - z_0\|_X$, then the function $C(z - z_0)^{-1}$ lies in $\mathscr{A}(X)$ but not in $\mathscr{P}(X)$. To see this, suppose $p$ is a polynomial such that

$$|p(z) - C(z - z_0)^{-1}| < 1$$

for all $z \in X$. Then

$$\begin{aligned} |(z - z_0)p(z) - C| &\leq |z - z_0| \\ &< C \end{aligned}$$

holds for all $z \in X$ and hence, by the maximum modulus theorem, this inequality persists throughout $V$ and in particular at $z_0$. At $z_0$, however, it is $C < C$, an impossibility. Thus, if $\mathscr{A}(X) = \mathscr{P}(X)$, the set $\mathbb{C} \setminus X$ must be connected. The sufficiency of this condition lies much deeper.

The first step in our proof of Mergelyan's theorem consists of showing that if $\mathbb{C} \setminus X$ is connected, then $\mathscr{P}(X)$ is a Dirichlet algebra. We shall need certain elementary formulas which are consequences of Green's formula. Denote by $\Delta$ the Laplacian $(\partial^2/\partial x^2) + (\partial^2/\partial y^2)$ and by $\partial/\partial \bar{z}$ the operator $\frac{1}{2}[(\partial/\partial x) + i(\partial/\partial y)]$. If $R$ is a relatively compact region in the plane with piecewise continuously differentiable boundary, and if $P$ and $Q$ are continuously differentiable on a neighborhood of $\bar{R}$, then Green's theorem shows that

$$\text{(1)} \qquad \int_{\partial R} P\, dx + Q\, dy = \iint_R \left(\frac{\partial Q}{\partial x} - \frac{\partial P}{\partial y}\right) dx\, dy.$$

From this it follows that if $U$ and $V$ are twice continuously differentiable on a neighborhood of $\bar{R}$, then

(2) $$\iint_R (U\Delta V - V\Delta U)\,dx\,dy = -\int_{\partial R}\left(U\frac{\partial V}{\partial n} - V\frac{\partial U}{\partial n}\right)ds$$

if $\partial/\partial n$ is the inner normal derivative and $ds$ is the differential of arc length. Equation (2) is obtained from (1) by applying (1) first with $P = U(\partial V/\partial y)$, $Q = V(\partial U/\partial x)$, and then with $P = V(\partial U/\partial y)$ and $Q = U(\partial V/\partial x)$. If the first of the resulting equations is subtracted from the second the result is

(3) $$\iint_R \{U\,\Delta V - V\,\Delta U\}\,dx\,dy$$
$$= \int_{\partial R}\left\{V\frac{\partial U}{\partial y} - U\frac{\partial V}{\partial y}\right\}dx + \left\{U\frac{\partial V}{\partial x} - V\frac{\partial U}{\partial x}\right\}dy,$$

and the right-hand side of (3) is the right-hand side of (2), so we have (2). Also, if we apply (1) with $Q = iP$, $z = x + iy$, we find that

(4) $$\int_{\partial R} P\,dz = 2i\iint_R \frac{\partial P}{\partial \bar{z}}\,dx\,dy.$$

With these preliminary remarks, we can establish the formulas we need.

25.1 ***Lemma***

(a) *If $g \in \mathscr{C}^\infty(\mathbb{C})$ has compact support, then for all $\zeta \in \mathbb{C}$,*

$$g(\zeta) = -\frac{1}{2\pi}\iint_{\mathbb{C}} \Delta g(z)\log|z - \zeta|^{-1}\,dx\,dy \qquad (z = x + iy).$$

(b) *If $R$ is a relatively compact domain in the plane with piecewise continuously differentiable boundary, and if $h$ is continuously differentiable on a neighborhood of $\bar{R}$, then for all $\zeta \in R$,*

$$h(\zeta) = \frac{1}{2\pi i}\int_{\partial R}\frac{h(z)}{z - \zeta}\,dz - \frac{1}{\pi}\iint_R \frac{\partial h}{\partial \bar{z}}(z)\frac{1}{z - \zeta}\,dx\,dy.$$

(c) *If $g \in \mathscr{C}^\infty(\mathbb{C})$ has compact support, then*

$$g(\zeta) = -\frac{1}{\pi}\iint_{\mathbb{C}} \frac{\partial g}{\partial \bar{z}}(z)\frac{1}{z - \zeta}\,dx\,dy.$$

PROOF

(a) Let $\zeta \in \mathbb{C}$ and let $D_\varepsilon$ be the domain bounded by the circles $|z| = \rho$ and $|z - \zeta| = \varepsilon$ for large $\rho$ and small $\varepsilon$. Apply (2) with $D_\varepsilon$ for $R$, with $\log|z - \zeta|$ for $U$ and $g$ for $V$. The result is

$$\iint_{D_\varepsilon} \{g(z)\Delta \log|z - \zeta| - \Delta g(z) \log|z - \zeta|\}\, dx\, dy$$

$$= -\int_{\partial D_\varepsilon} \left\{\log|z - \zeta| \frac{\partial g}{\partial n}(z) - g(z)\frac{\partial \log|z - \zeta|}{\partial n}\right\} ds.$$

The first integral is

$$-\iint_{D_\varepsilon} \Delta g(z) \log|z - \zeta|\, dx\, dy$$

since $\log|z - \zeta|$ is harmonic on $D_\varepsilon$. If $\rho$ is big enough, $g$ will vanish on a neighborhood of the circle $|z| = \rho$, so the second integral is, for large $\rho$,

$$\int_{\gamma_\varepsilon} \log|z - \zeta| \frac{\partial g}{\partial n}(z)\, ds - \int_{\gamma_\varepsilon} g(z)\frac{\partial \log|z - \zeta|}{\partial n}\, ds$$

if $\gamma_\varepsilon$ is the circle $|z - \zeta| = \varepsilon$. (The signs are correct since $D_\varepsilon$ lies *outside* $\gamma_\varepsilon$.) Of these two integrals, the first is no more than $C\varepsilon \log \varepsilon$ for some constant $C$, and so it tends to zero as $\varepsilon \to 0$, and the second tends to $2\pi g(\zeta)$ as $\varepsilon \to 0$. Since as $\varepsilon \to 0$ and $\rho \to \infty$,

$$\iint_{D_\varepsilon} \Delta g(z) \log|z - \zeta|\, dx\, dy \to \iint_{\mathbb{C}} \Delta g(z) \log|z - \zeta|\, dx\, dy,$$

we have the desired formula.

(b) We let $R_\varepsilon = \{z \in R : |z - \zeta| > \varepsilon\}$, where $\varepsilon > 0$ is less than the distance from $\zeta$ to $\partial R$. Apply (4) to the function $h(z)(z - \zeta)^{-1}$ to find

$$\int_{\partial R\varepsilon} h(z)(z - \zeta)^{-1}\, dz = 2i \iint_{R_\varepsilon} \frac{\partial h}{\partial \bar{z}}(z)\frac{1}{z - \zeta}\, dx\, dy$$

$$+ 2i \iint_{R_\varepsilon} h(z)\frac{\partial}{\partial \bar{z}}\left(\frac{1}{z - \zeta}\right) dx\, dy.$$

Since $(z - \zeta)^{-1}$ is holomorphic in $R_\varepsilon$, the Cauchy–Riemann equations imply that the second integral on the right is zero. Also,

$$\int_{\partial R_\varepsilon} h(z)(z - \zeta)^{-1}\, dz = \int_{\partial R} h(z)(z - \zeta)^{-1}\, dz - i\int_0^{2\pi} h(\zeta + \varepsilon e^{i\theta})\, d\theta.$$

Since $(z - \zeta)^{-1}$ is integrable over $R$, we obtain the result by letting $\varepsilon \to 0$.

(c) Part (c) is a consequence of (b). Apply (b) with $R$ a disc $\{z : |z| \le \rho\}$ and with $g$ replacing $h$. The result is

$$g(\zeta) = \frac{1}{2\pi i}\int_{|z|=\rho} \frac{g(z)}{z - \zeta} dz - \frac{1}{\pi}\iint_R \frac{\partial g}{\partial \bar{z}}(z)\frac{1}{\zeta - z}\, dx\, dy.$$

Since $g$ has compact support, we obtain the result by letting $\rho \to \infty$.

For the remainder of this section, let $X$ be a compact set in the plane, and let $\Omega = \mathbb{C} \setminus X$. Our next lemma provides certain auxiliary measures.

25.2 ***Lemma***

*If $\Omega$ is connected and if $z_0 \in \partial\Omega$, then for every $\delta > 0$ there is a positive measure $\sigma_\delta$ on $\Omega$ such that*

$$\sigma_\delta(\{z : |z - z_0| > \delta\}) = 0,$$

*and such that if $0 < r < s \le \delta$, then*

$$\sigma_\delta(\{z : r < |z - z_0| < s\}) = s - r.$$

PROOF

Without loss of generality, we can suppose that $z_0 = 0$. The set $\Omega$ is connected, so for each $t > 0$, there is $z \in \Omega$ with $|z| = t$. Given a point $z \in \Omega$, there is a closed interval $L_z$ containing $z$ contained in $\Omega$ and lying in the line through the origin and $z$. It follows that we can find a sequence $\{L_j\}$ of closed intervals with the following properties:

(a) Each $L_j$ lies on a line through the origin and is contained in $\Omega$.
(b) If $L_j^* = \{|z| : z \in L_j\}$, then the $L_j^*$'s constitute a closed cover for the ray $(0, \infty)$.
(c) The intervals $L_j^*$ are pairwise disjoint except for the fact that adjacent $L_j^*$'s have a common endpoint.

Having made this preliminary construction, we define the measure $\sigma_\delta$ by requiring that if $E$ is an open set in $\Omega$, then $\sigma_\delta(E)$ is the Lebesgue measure of the set

$$\{|z| : |z| < \delta, z \in E \cap (\bigcup L_j)\},$$

a certain measurable set in the real line. The measure $\sigma_\delta$ obtained in this way has the desired properties.

25.3 ***Lemma***

*If $\alpha$ is a real measure on $X$, then for almost all $z$ in the plane, the integral*

$$u(z) = \int_X \log|z - \zeta|^{-1}\, d\alpha(\zeta)$$

*exists. If $u$ vanishes identically on $\Omega$, then $u(z) = 0$ for every $z \in \partial\Omega$ for which the integral*

$$v(z) = \int_X \log|z - \zeta|^{-1}\, d|\alpha|(\zeta)$$

*exists.*

PROOF

If $R > 0$, then since $\log|z - \zeta|^{-1}$ is Lebesgue integrable on compact sets in the plane, we have

$$\int_X \iint_{|z|<R} |\log|z - \zeta|^{-1}|\, dx\, dy\, d|\alpha|(\zeta) < \infty,$$

and thus $v(z)$ is finite almost everywhere in the plane. Consequently, the integral defining $u$ exists almost everywhere.

Assume now that $u$ vanishes in $\Omega$, and let $z_0 \in \partial\Omega$ be a point at which $v$ exists. We may suppose that $z_0 = 0$. Let $\delta$ be a small positive number, and let $\sigma_\delta$ be a measure on $\Omega$ as constructed in Lemma 25.2. The function $u$ vanishes identically in $\Omega$, so

$$0 = \frac{1}{\delta}\int u(z)\, d\sigma_\delta(z).$$

If we put

$$X_\rho^+ = \{\zeta \in X : |\zeta| \geq \rho\}$$

and

$$X_\rho^- = \{\zeta \in X : |\zeta| < \rho\},$$

we have that

$$0 = \int_{X_\rho^+} \left\{1/\delta \int \log|z - \zeta|^{-1}\, d\sigma_\delta(z)\right\} d\alpha(\zeta)$$
$$+ \int_{X_\rho^-} \left\{1/\delta \int \log|z - \zeta|^{-1}\, d\sigma_\delta(z)\right\} d\alpha(\zeta).$$

We now fix a value of $\rho \in (0,\frac{1}{2})$ and let $\delta \to 0$. For fixed $\zeta \neq 0$, we have that

$$\lim_{\delta \to 0} 1/\delta \int \log|z - \zeta|^{-1}\, d\sigma_\delta(z) \to \log|\zeta|^{-1}$$

since $\sigma_\delta$ is of total mass $\delta$ and is supported in $|z| \leq \delta$, and since $\log|z - \zeta|^{-1}$ is, for fixed $\zeta \neq 0$, continuous at 0. The limit relation is uniform in $\zeta$ on the set $X_\rho^+$. For $|\zeta| \leq \rho$, we have

$$\begin{aligned} 1/\delta \int \log|z - \zeta|^{-1}\, d\sigma_\delta(z) &\leq 1/\delta \int \log\big||z| - |\zeta|\big|^{-1}\, d\sigma_\delta(z) \\ &= 1/\delta \int_0^\delta \log\big|r - |\zeta|\big|^{-1}\, dr \\ &= \log|\zeta|^{-1} + 1/\delta \int_0^\delta \log\big|1 - r|\zeta|^{-1}\big|^{-1}\, dr \\ &\leq \log|\zeta|^{-1} + C, \end{aligned}$$

where

$$C = \sup_{T>0} 1/T \int_0^T \log|1 - t|^{-1}\, dt.$$

Thus, if we let $\delta \to 0$ in the equality

$$\begin{aligned} -\int_{X_\rho^+} &\left\{1/\delta \int \log|z - \zeta|^{-1}\, d\sigma_\delta(z)\right\} d\alpha(\zeta) \\ &= \int_{X_\rho^-} \left\{1/\delta \int \log|z - \zeta|^{-1}\, d\sigma_\delta(z)\right\} d\alpha(\zeta), \end{aligned}$$

we find that

$$\int_{X_\rho^+} \log|\zeta|^{-1}\, d\alpha(\zeta) \leq \int_{X_\rho^-} (\log|\zeta|^{-1} + C)\, d|\alpha|(\zeta).$$

By hypothesis, $v(0) < \infty$, so the right side tends, as $\rho \to 0$, to 0, and we have the lemma.

25·4 ***Lemma***

*With $\alpha$ and $u$ as in Lemma 25.2, if $u$ vanishes almost everywhere in the plane, then $\alpha$ is the zero measure.*

PROOF

We understand the *almost everywhere* of this lemma in the sense of planar Lebesgue measure.

Let $g$ be a real $\mathscr{C}^\infty$ function on the plane which has compact support. By Lemma 25.1,

$$-\frac{1}{2\pi}\iint \Delta g(z)\log|z-\zeta|^{-1}\,dx\,dy = g(\zeta),$$

so if we integrate with respect to $\alpha$ and interchange the order of integration in the left-hand integral, we find that

$$-\frac{1}{2\pi}\iint \Delta g(z)u(z)\,dx\,dy = \int g(\zeta)\,d\alpha(\zeta).$$

By hypothesis, the left integral is zero, so

$$0 = \int g(\zeta)\,d\alpha(\zeta)$$

for all choices of $g$. It follows that $\alpha$ must be the zero measure.

We shall need to make use of the harmonic measures associated with interior points of $X$. Let $\mathscr{C}_H(\partial\Omega)$ denote the collection of continuous functions on $\partial\Omega$ which admit harmonic extensions throughout the interior of $\Omega$. If $\varphi \in \mathscr{C}_H(\partial\Omega)$, let $U_\varphi$ denote the continuous function on $X$ which is harmonic in the interior of $X$ and which agrees on $\partial\Omega$ with $\varphi$. The maximum principle implies that $U_\varphi$ is uniquely determined by $\varphi$ and that the space $\mathscr{C}_H(\partial\Omega)$ is uniformly closed. If $a$ is an interior point of $X$, the map $\varphi \mapsto U_\varphi(a)$ is a positive linear functional of norm 1 on $\mathscr{C}_H(\partial\Omega)$; so by the Hahn–Banach and Riesz representation theorems, there exists a probability measure $\lambda_a$ on $\partial\Omega$ such that for $\varphi \in \mathscr{C}_H(\partial\Omega)$,

$$U_\varphi(a) = \int_{\partial\Omega} \varphi(\zeta)\,d\lambda_a(\zeta). \tag{5}$$

We should point out that on the basis of what we have done so far, it is not clear that the measure $\lambda_a$ is uniquely determined; it will follow from some of our later work that it is. The measure $\lambda_a$ is called the *harmonic measure* for $a$. For fixed $z \in \Omega$, (5) applied to $\log|z-\zeta|^{-1}$ regarded as a function of $\zeta$ yields

$$\log|z-a|^{-1} = \int_{\partial\Omega} \log|z-\zeta|^{-1}\,d\lambda_a(\zeta). \tag{6}$$

If $z_0 \in \partial\Omega$ and if we let $z \to z_0$ through $\Omega$, the left side of (6) tends to

$\log|z_0 - a|^{-1}$, and Fatou's lemma applies to the integral on the right to yield

$$\int_{\partial\Omega} \log|z_0 - \zeta|^{-1}\, d\lambda_a(\zeta) = \int_{\partial\Omega} \liminf_{\substack{z \to z_0 \\ z \in \Omega}} \log|z - \zeta|^{-1}\, d\lambda_a(\zeta)$$

$$\leq \liminf_{\substack{z \to z_0 \\ z \in \Omega}} \int_{\partial\Omega} \log|z - \zeta|^{-1}\, d\lambda_a(\zeta),$$

whence

$$\log|z_0 - a|^{-1} \geq \int_{\partial\Omega} \log|z_0 - \zeta|^{-1}\, d\lambda_a(\zeta).$$

If we define the measure $\nu_a$ on $X$ to be $\lambda_a - \delta_a$, $\delta_a$ the unit point mass at $a$, what we have just done implies that the potential

$$v_a(z) = \int_X \log|z - \zeta|^{-1}\, d|\nu_a|(\zeta)$$

exists at every point of $\partial\Omega$. If

$$u(z) = \int_X \log|z - \zeta|^{-1}\, d\nu_a(\zeta),$$

then $u(z)$ vanishes in $\Omega$; so by Lemma 25.3, $u(z) = 0$ if $z \in \partial\Omega$. Thus, for all $z_0 \in \partial\Omega$ and all $a \in \operatorname{int} X$,

$$\int_X \log|z_0 - \zeta|^{-1}\, d\lambda_a = \log|z_0 - a|^{-1}. \tag{7}$$

We can prove now that if $X$ does not separate the plane, then $\mathscr{P}(x)$ is a Dirichlet algebra.

25·5 ***Theorem*** (Walsh [3])

*If the compact set $X$ does not separate the plane, and if $\varphi \in \mathscr{C}_{\mathbb{R}}(\partial\Omega)$, then there exists a sequence $\{p_n\}$ of polynomials such that $\{\operatorname{Re} p_n\}$ converges uniformly on $X$ and converges on $\partial\Omega$ to $\varphi$.*

Note that the limit function is harmonic in the interior of $X$, so this theorem implies that certain Dirichlet problems are solvable.

PROOF

If the theorem is false, there is a real measure $\alpha$ on $\partial\Omega$ which annihilates all polynomials:

$$\int_{\partial\Omega} \zeta^n\, d\alpha(\zeta) = 0, \qquad n = 0, 1, 2, \ldots.$$

Thus, if $|z|$ is large, we have

$$
\begin{aligned}
0 &= \sum_{n=1}^{\infty} \frac{z^{-n}}{n} \int \zeta^n \, d\alpha(\zeta) \\
&= \int \sum_{n=1}^{\infty} \frac{1}{n}\left(\frac{\zeta}{z}\right)^n d\alpha(\zeta) \\
&= \int \log\left(1 - \frac{\zeta}{z}\right) d\alpha(\zeta).
\end{aligned}
$$

Since $\int d\alpha = 0$, we find from this last equality that

$$
\begin{aligned}
0 &= \int \{\log|z - \zeta|^{-1} - \log|z|^{-1}\} \, d\alpha(\zeta) \\
&= \int \log|z - \zeta|^{-1} \, d\alpha(\zeta).
\end{aligned}
$$

Let $u(z)$ denote the last integral. It is harmonic off $\partial\Omega$, and it vanishes for large values of $|z|$. Thus, it vanishes identically in $\Omega$, and so, by Lemma 25.3, it vanishes at almost all points of $\partial\Omega$. If $X$ has no interior, then $X = \partial\Omega$, and we may conclude from Lemma 25.4 that $\alpha$ is zero, and the theorem is proved. In the case that $X$ has an interior, we complete the proof by showing that $u$ vanishes on this interior.

Let $a \in \operatorname{int} X$, and let $\lambda_a$ be an associated harmonic measure. Then

$$
\int \left\{ \int \log|z - \zeta|^{-1} \, d\lambda_a(z) \right\} d|\alpha|(\zeta) = \int \log|\zeta - a|^{-1} \, d|\alpha|(\zeta).
$$

The last integral is finite, so it follows that $\lambda_a$ assigns zero measure to the set in $\partial\Omega$ on which the potential

$$
\int \log|z - \zeta|^{-1} \, d|\alpha|(\zeta)
$$

diverges. Consequently, by Lemma 25.3, $u(z) = 0$ a.e. $[\lambda_a]$, so

$$
\begin{aligned}
0 = \int u(z) \, d\lambda_a(z) &= \int \left\{ \int \log|z - \zeta|^{-1} \, d\lambda_a(z) \right\} d\alpha(\zeta) \\
&= \int \log|\zeta - a|^{-1} \, d\alpha(\zeta) \\
&= u(a).
\end{aligned}
$$

Thus, we have that $u$ vanishes almost everywhere, and the theorem follows from Lemma 25.4.

This theorem implies that the harmonic measures $\lambda_a$ are uniquely determined.

The next step in the proof of Mergelyan's theorem is a lemma on Cauchy transforms.

**25.6** ***Lemma***

*If $\mu$ is a finite measure on $\mathbb{C}$ with compact support, then the integral*

$$\tilde{\mu}(z) = \int \frac{d\mu(\zeta)}{\zeta - z}$$

*exists for almost all $z \in \mathbb{C}$. If $\tilde{\mu}$ vanishes almost everywhere, then $\mu$ is the zero measure.*

PROOF

If $R > 0$, Fubini's theorem yields

$$\iint_{|z| \leq R} \left\{ \int \frac{d|\mu|(z)}{|\zeta - z|} \right\} dx\, dy = \int \left\{ \iint_{|z| \leq R} \frac{dx\, dy}{|z - \zeta|} \right\} d|\mu|(\zeta)$$

$$< \infty.$$

Thus, $\tilde{\mu}$ exists almost everywhere.

If $\tilde{\mu}$ vanishes almost everywhere in the plane, then given an infinitely differentiable function $g$ with compact support in the plane, we have, by Lemma 25.1,

$$\begin{aligned} \int g(\zeta)\, d\mu(\zeta) &= -\int \left\{ \frac{1}{\pi} \iint \frac{\partial g}{\partial \bar{z}}(z) \frac{1}{z - \zeta}\, dx\, dy \right\} d\mu(\zeta) \\ &= -\frac{1}{\pi} \iint \frac{\partial g}{\partial \bar{z}}(z) \left\{ \int \frac{d\mu(\zeta)}{z - \zeta} \right\} dx\, dy \\ &= 0. \end{aligned}$$

As this holds for all choices of $g$, $\mu$ must be the zero measure.

We are able now to prove Mergelyan's theorem.

**25.7** ***Theorem***

*If $X$ is a compact set which does not separate the plane, then every function continuous on $X$ and holomorphic on the interior of $X$ can be approximated uniformly by polynomials, i.e., $\mathscr{A}(X) = \mathscr{P}(X)$.*

PROOF

Let $\mu$ be a measure on $\partial\Omega$ which is orthogonal to the polynomials. Since $\mathscr{P}(X)$ is Dirichlet, every point in $X$ admits a unique representing measure on $\partial\Omega$. By the abstract F. and M. Riesz theorem, there exists a sequence $\{z_n\}$ in $X$ and for each $n$ a measure $\mu_n$ absolutely continuous with respect to the representing measure for $z_n$ such that, in the sense of norm convergence,

$$\mu = \sum_{n=1}^{\infty} \mu_n + \sigma,$$

where $\sigma$ is singular with respect to every representing measure for $\mathscr{P}(X)$, and where $\sigma$ and all the $\mu_n$ are orthogonal to $\mathscr{P}(X)$.

To begin with, we shall prove that $\sigma$ is the zero measure, i.e., that for $\mathscr{P}(X)$, there are no completely singular orthogonal measures. We shall prove that for almost all $z_0$ in the plane,

$$\tilde{\sigma}(z_0) = \int \frac{d\sigma(z)}{z - z_0} = 0,$$

whence $\sigma = 0$ by Lemma 25.6. We know that for almost all points $z_0$ in the plane, the integral

$$\int \frac{d|\sigma|(z)}{|z - z_0|}$$

is finite. Let $z_0 \in X$ be such a point. If $p$ is a polynomial which vanishes at $z_0$, then $p(z)/(z - z_0)$ is a polynomial, so

$$\int \frac{p(z)}{z - z_0} d\sigma(z) = 0.$$

Consequently, if $g \in \mathscr{P}(X)$ vanishes at $z_0$,

$$\int \frac{g(z)}{z - z_0} d\sigma(z) = 0.$$

Let $\lambda_{z_0}$ be the harmonic measure for $z_0$, set $\alpha = \tilde{\sigma}(z_0)$, and put

$$d\nu(z) = -\alpha\, d\lambda_{z_0}(z) + \frac{d\sigma(z)}{z - z_0}.$$

If $f \in \mathscr{P}(X)$, then

$$\begin{aligned} \int f\, d\nu &= -\alpha f(z_0) + \int \frac{f(z)\, d\sigma(z)}{z - z_0} \\ &= \int \frac{f(z) - f(z_0)}{z - z_0} d\sigma(z) = 0, \end{aligned}$$

so $\nu \in \mathscr{P}(X)^{\perp}$. By the abstract F. and M. Riesz theorem,

$$\nu = \nu_c + \nu_s,$$

where $\nu_c \ll \lambda_{z_0}$, $\nu_s \perp \lambda_{z_0}$, and where both $\nu_c$ and $\nu_s$ are orthogonal to $\mathscr{P}(X)$. Since

$$d\nu_s = \frac{d\sigma}{z - z_0},$$

we have that $d\sigma(z)/(z - z_0) \in \mathscr{P}(X)^\perp$, whence in particular $\tilde{\sigma}(z_0) = 0$.

Thus, $\tilde{\sigma}$ vanishes almost everywhere on $X$. It also vanishes off $X$, for since

$$\tilde{\sigma}(\zeta) = \int \frac{d\sigma(z)}{z - \zeta},$$

$\tilde{\sigma}$ is holomorphic in $\Omega$; and for large values of $|\zeta|$, we have

$$\begin{aligned}\tilde{\sigma}(\zeta) &= -\frac{1}{\zeta}\int \frac{d\sigma(z)}{1 - z\zeta^{-1}}\\ &= -\sum_{n=0}^{\infty}\left(\frac{1}{\zeta}\right)^{n+1}\int z^n\, d\sigma(z).\end{aligned}$$

This series is zero term by term. Thus, $\tilde{\sigma}$ vanishes identically in $\Omega$. Consequently, $\tilde{\sigma}$ vanishes almost everywhere; so, by Lemma 25.6, $\sigma$ is the zero measure.

Next we treat the measures $\mu_n$. For each $n$, $\mu_n$ is absolutely continuous with respect to the representing measure for the point $z_n$ and is orthogonal to $\mathscr{P}(X)$. Since $\mathscr{P}(X)$ is Dirichlet on $\partial\Omega$, each point of $\partial\Omega$ is a peak point for $\mathscr{P}(X)$, so each of the points $z_n$ must lie in the interior of $X$. It follows that if $\lambda_n$ denotes the harmonic measure for the point $z_n$, $\mu_n \ll \lambda_n$.

Consider a fixed value of $n$ and an $f \in \mathscr{A}(X)$, and put $C = 2\|f\|_X$. Let $g$ be the branch of $\log(f + C)$ which is real for real values of $f + C$. Then $g \in \mathscr{A}(X)$, and since $\mathscr{P}(X)$ is Dirichlet, there is a sequence $\{P_k\}$ of polynomials such that

$$\|\mathrm{Re}(g - P_k)\|_X < 2^{-k}.$$

We may suppose that for all $k$, $P_k(z_n) = g(z_n)$. The function $(g - P_k)^2$ lies in $\mathscr{A}(X)$ and it vanishes at $z_n$, so

$$0 = \int (g - P_k)^2\, d\lambda_n.$$

The real part of this integral is zero, so

$$\begin{aligned}\int (\mathrm{Im}\, g - \mathrm{Im}\, P_k)^2\, d\lambda_n &= \int (\mathrm{Re}\, g - \mathrm{Re}\, P_k)^2\, d\lambda_n\\ &\le 2^{-k}.\end{aligned}$$

We may conclude that the series

$$\sum_{k=1}^{\infty} \int |g - P_k|^2 \, d\lambda_n$$

converges, and it follows that $P_k \to g$ a.e. $[\lambda_n]$. Since $\mu_n \ll \lambda_n$ and since $\mu_n \perp \mathscr{P}(X)$, we have by the dominated convergence theorem that

$$\begin{aligned}\int f \, d\mu_n &= \int (f + C) \, d\mu_n \\ &= \int e^{\log g} \, d\mu_n \\ &= \lim_{k \to \infty} \int e^{P_k} \, d\mu_n = 0.\end{aligned}$$

Thus, $\mu_n \perp \mathscr{A}(X)$, and we have proved that every measure $\mu$ on $\partial\Omega$ which is orthogonal to $\mathscr{P}(X)$ is also orthogonal to $\mathscr{A}(X)$. Therefore, $\mathscr{A}(X) = \mathscr{P}(X)$, and we have established the theorem.

The proof we have given for Mergelyan's theorem follows, in the main, that given by Carleson [4], though our treatment of the singular measure follows Wermer [12]. In particular, the proof of Walsh's theorem that $\mathscr{P}(X)$ is Dirichlet is due to Carleson. The theorem was proved first by Mergelyan [1], whose proof is entirely by classical means. Mergelyan's memoir [2] contains a discussion of the early history of the theorem. In his paper [2], Bishop presented for the first time a proof that utilizes the methods of functional analysis, and later Glicksberg and Wermer [1] found a proof in which the only remaining nonfunctional analytic element was the fact that $\mathscr{P}(X)$ is Dirichlet; for this they had to appeal to Walsh's theorem. A proof of Mergelyan's theorem along classical lines, but somewhat simpler than the original, is contained in Rudin's book [2].

Note, in particular, that if $X$ is an arc, then $\mathscr{P}(X) = \mathscr{C}(X)$. This is an extension of the classical Weierstrass approximation theorem, but, for general arcs, it lies much deeper than Weierstrass's theorem. This special case of Mergelyan's theorem was proved by Walsh [2].

Aside from being a theorem of great intrinsic beauty, Mergelyan's theorem has become, in recent years, a powerful tool in the theory of functions. For some of these applications, see Mergelyan [2]. For a striking application to the study of the boundary behavior of functions holomorphic in the unit disc, see Collingwood and Lohwater [1, pp. 163–166].

Combining Mergelyan's theorem and the generalized Stone–Weierstrass theorem we obtain the following result of Rudin [5].

25.8 ***Corollary***

*Let $K \subset \mathbb{C} \times R^N$ be compact. If for all $t = (t_1, \ldots, t_N) \in R^N$, the set $K_t = \{z \in \mathbb{C} : (z, t) \in K\}$ does not separate $\mathbb{C}$, then every function $f \in \mathscr{C}(K)$ such that $f(\cdot, t)$ is, for each $t \in R^N$, holomorphic on the interior of $K_t$ can be approximated uniformly on $K$ by polynomials in $z$ and $t_1, \ldots, t_N$.*

PROOF

Let $A$ be the algebra of those functions in $\mathscr{C}(K)$ with the properties required of $f$. If $E$ is a set of antisymmetry for the algebra $B$ of uniform limits on $K$ of polynomials in $z$ and $t_1, \ldots, t_N$, then $E$ must be contained in one of the slices

$$\{(z, t) \in K : t = t_0\},$$

$t_0$ fixed, of $K$. By Mergelyan's theorem, $f|E$ belongs to $B|E$, so $f \in B$ by the generalized Stone–Weierstrass theorem.

We saw in the proof of Mergelyan's theorem that there exist no completely singular measures orthogonal to $\mathscr{P}(X)$. Thus, each $v \in \mathscr{P}(X)^{\perp}$ is of the form $v = \Sigma h_j \lambda_j$, where $\lambda_j$ is the harmonic measure for a point $z_j$ in the interior of $X$, and where

$$h_j \in H_0^1(\lambda_j) = \left\{ f \in H^1(\lambda_j) : \int f \, d\lambda_j = 0 \right\}.$$

In some contexts it is important to know more about the measures orthogonal to $\mathscr{P}(X)$. We shall establish the following fact.

25.9 ***Theorem***

*Let $X$ be a compact set which does not separate the plane, and let $\lambda$ be the harmonic measure for the point $z_0$ in the interior of $X$. If $f \in H^1(\lambda)$, then either $f$ is the zero function or else*

$$-\infty < \int \log|f| \, d\lambda.$$

This result implies that if $h \in H_0^1(\lambda)$, $h$ not the zero function, then not only is the measure $v = h\lambda$, an element of $\mathscr{P}(X)^{\perp}$, absolutely continuous with respect to $\lambda$, but in fact $v$ and $\lambda$ are mutually absolutely continuous; for, by the theorem, $h$ cannot vanish on a set of positive $\lambda$-measure.

PROOF OF THE THEOREM

We shall consider the Cauchy transform, $\tilde{v}$, of the measure $v = f\lambda$.

Denote by $C$ the component of the interior of $X$ which contains $z_0$, and set $Y = \bar{C}$. The algebra $\mathscr{P}(Y)$ is Dirichlet on $\partial Y$ since $\mathscr{P}(X)$ is Dirichlet on $\partial X$. The measure $\lambda$ is supported on $\partial Y$.

Let us consider, to begin with, the case that $\int f\, d\lambda = 0$.

We know that the integral

$$I(z) = \int \frac{d|v|(\zeta)}{|\zeta - z|} = \int \frac{|f|\, d\lambda(\zeta)}{|\zeta - z|}$$

exists for almost all $z$ in the plane. We shall show that if $z_1$ is a point in $\partial Y$ at which $I$ exists, then $\tilde{v}(z_1) = 0$. This is so, for since $\mathscr{P}(Y)$ is Dirichlet on $\partial Y$, the point $z_1$ is a peak point for $\mathscr{P}(Y)$; so there exists a sequence $\{q_n\}$ of polynomials such that $q_n(z_1) = 0$ for all $n$, such that $\|q_n\|_Y \leq 2$, and such that $q_n(z) \to 1$ for all $x \in Y \setminus \{z_1\}$. Since $\int f\, d\lambda = 0$, we have $\int pf\, d\lambda = 0$ for all polynomials $p$, and consequently if $\tilde{q}_n$ is the polynomial such that $q_n(z) = (z - z_1)\tilde{q}_n(z)$, we have

$$0 = \int \tilde{q}_n f\, d\lambda = \int \frac{q_n(\zeta) f(\zeta)\, d\lambda(\zeta)}{\zeta - z_1}.$$

The choice of $q_n$ and the existence of the integral $I(z_1)$ allow us to invoke the dominated convergence theorem to find

$$\begin{aligned} 0 &= \int \frac{f(\zeta)}{\zeta - z_1}\, d\lambda(\zeta) \\ &= \tilde{v}(z_1). \end{aligned}$$

Thus, $\tilde{v}$ vanishes at almost every point of $\partial Y$.

We shall show that it also vanishes on all the components of $\mathbb{C} \setminus \partial Y$ other than $C$. (*Note*: Contrary to one's initial suspicion, there can be bounded components of $\mathbb{C} \setminus \partial Y$ other than $C$. This fact complicates the proof.) For large values of $|z|$, we have

$$\begin{aligned} \tilde{v}(z) &= -\sum_{n=0}^{\infty} z^{-(n+1)} \int f(\zeta)\zeta^n\, d\lambda(\zeta) \\ &= 0, \end{aligned}$$

so $\tilde{v}$ vanishes on the unbounded component of $\mathbb{C} \setminus \partial Y$. It also vanishes on the bounded components, other than $C$, of $\mathbb{C} \setminus \partial Y$. If $G$ is a bounded component of $\mathbb{C} \setminus Y$, $G \neq C$, then $G$ is a component of the interior of $X$. If $\tilde{v}$ does not vanish identically on $G$, choose $z_1 \in G$ with $\tilde{v}(z_1) \neq 0$.

There is a sequence $\{p_n\}$ of polynomials converging to $f$ in $H^1(\lambda)$, $p_n(z_0) = 0$, so there is a polynomial $p$ with

$$\int \frac{p(\zeta)}{\zeta - z_1} d\lambda(\zeta) \neq 0$$

and $p(z_0) = 0$. Since

$$\int \frac{p(\zeta)}{\zeta - z_1} d\lambda(\zeta) = \int \frac{p(\zeta) - p(z_1)}{\zeta - z_1} d\lambda(\zeta) + \int \frac{p(z_1)}{\zeta - z_1} d\lambda(\zeta)$$
$$= \frac{p(z_0) - p(z_1)}{z_0 - z_1} + p(z_1)\tilde{\lambda}(z_1),$$

and $p(z_0) = 0$, we see that if

$$\beta = p(z_1)\left(\tilde{\lambda}(z_1) - \frac{1}{z_0 - z_1}\right),$$

then $\beta \neq 0$. If $g$ is a polynomial,

$$\int \frac{p(\zeta)g(\zeta)}{\zeta - z_1} d\lambda(\zeta) = \int \frac{p(\zeta)g(\zeta) - p(z_1)g(z_1)}{\zeta - z_1} d\lambda(\zeta) + \int \frac{p(z_1)g(z_1)}{\zeta - z_1} d\lambda(\zeta)$$
$$= \frac{p(z_0)g(z_0) - p(z_1)g(z_1)}{z_0 - z_1} + g(z_1)p(z_1)\tilde{\lambda}(z_1),$$

so $\beta^{-1}g(\zeta)\,d\lambda(\zeta)$ is a complex representing measure for $z_1$ on $\mathscr{P}(X)$. Theorem 11.2 implies the existence of a positive representing measure $\mu \ll \lambda$, and thus, by Theorem 16.6, $z_0$ and $z_1$ lie in the same Gleason part for $\mathscr{P}(X)$. By Theorem 17.1, this part is a disc which is contained in $X$, which meets $C$ and $G$, but which misses $X$. No such disc can exist since $C$ and $G$ are distinct components of the interior of $X$. Thus $\tilde{v}$ vanishes on $G$.

So far, we have shown that $\tilde{v}$ vanishes almost everywhere off $C$. If, in addition, $\tilde{v}$ vanishes on $C$, then $\tilde{v}$ vanishes at almost every point of the plane, and we find that $\tilde{v}$ is the zero measure. This implies that $f$ is the zero function.

We assume henceforth that $\tilde{v}$ does not vanish identically on $C$. Let $\{p_n\}$ be a sequence of polynomials which converges in $H^1(\lambda)$ to $f$. Since $\int f\,d\lambda = 0$, we may suppose that $p_n(z_0) = 0$ for all $n$. If $z_1$ is a point in $C$ at which $\hat{v}$ does not vanish, $z_1 \neq z_0$, then

$$0 \neq \hat{v}(z_1) = \int \frac{f(\zeta)\,d\lambda(\zeta)}{\zeta - z_1}$$
$$= \lim_{n\to\infty} \left\{\int \frac{p_n(\zeta) - p_n(z_1)}{\zeta - z_1} d\lambda + p_n(z_1)\int \frac{d\lambda}{\zeta - z_1}\right\}.$$

The quotient $[p_n(\zeta) - p_n(z_1)]/(\zeta - z_1)$ is, as a function of $\zeta$, a polynomial with value $-p_n(z_1)(z_0 - z_1)^{-1}$ at $z_0$. As $\lambda$ is a representing measure for the point $z_0$, it follows that

$$\hat{v}(z_1) = \lim \left\{ -p_n(z_1)(z_0 - z_1)^{-1} + p_n(z_1) \int \frac{d\lambda}{\zeta - z_1} \right\},$$

and from this it follows that since $\hat{v}(z_1) \neq 0$, the sequence $\{p_n(z_1)\}$ does not converge to zero.

If $z_1 \in C$, the representing measure for $z_1$ is boundly absolutely continuous with respect to $\lambda$, and since $\{p_n\}$ converges in $L^1(\lambda)$, it follows that the sequence $\{p_n(z_1)\}$ converges. Thus, the sequence $\{p_n\}$ is pointwise convergent on $C$, and the convergence is evidently bounded on compacta in $C$. Moreover, the limit function, call it $F$, is not the zero function. Let $\Delta$ be a small closed disc about $z_0$ which lies entirely within $C$ and which has the property that the only zero of $F$ on $\Delta$ is the point $z_0$. If this zero is of order $k$, then for large enough $n$, we shall have a factorization,

$$p_n(z) = \tilde{p}_n(z)(z - \zeta_1^{(n)}) \cdots (z - \zeta_k^{(n)}),$$

where $\tilde{p}_n$ is a polynomial without zeros in $\Delta$ and $\zeta_1^{(n)}, \ldots, \zeta_k^{(n)}$ are certain, possibly repeated, points of $\Delta$. If $r_n$ denotes the polynomial

$$r_n(z) = (z - \zeta_1^{(n)}) \cdots (z - \zeta_k^{(n)}),$$

we have that $\lim_{n\to\infty} r_n(z) = (z - z_1)^k$. The points $\zeta_j^{(n)}$ are at positive distance from $\partial Y$, so it follows that the sequence $\{\tilde{p}_n\}$ converges in $H^1(\lambda)$ to a function, say $\tilde{f}$. We have then the decomposition

$$f(z) = \tilde{f}(z)(z - z_1)^k,$$

where $\tilde{f} \in H^1(\lambda)$ has the property that $\int \tilde{f}\, d\lambda \neq 0$.

It is clear that $\int \log|z - z_1|^k \, d\lambda$ exists, so for the remainder of the proof, we can restrict attention to the case that the function $f$ in question has the property that $\int f\, d\lambda \neq 0$.

Given $f \in H^1(\lambda)$ with $\int f\, d\lambda \neq 0$, let $\{p_n\}$ be a sequence of polynomials converging to $f$ in $L^1(\lambda)$. As $\lambda$ is a Jensen measure, it follows that for some absolute constant $C$ and all sufficiently large $n$,

$$\begin{aligned} -\infty < C &\leq \log|p_n(z_0)| \\ &\leq \int \log|p_n|\, d\lambda. \end{aligned}$$

We assume, as we may, that $p_n \to f$ a.e. $[\lambda]$. Write for a positive function $h$,

$$\log^+ h = \max(\log h, 0)$$

and

$$\log^- h = \min(\log h, 0).$$

By Fatou's lemma

$$\begin{aligned}\int \log^+|f|\,d\lambda &\leq \liminf \int \log^+|p_n|\,d\lambda \leq \limsup \int \log^+|p_n|\,d\lambda \\ &\leq \limsup \int |p_n|\,d\lambda \\ &= \int |f|\,d\lambda.\end{aligned}$$

This yields the fact that $\int \log^+|f|\,d\lambda$ is finite. Also, the fact that $\limsup \int \log^+|p_n|\,d\lambda \leq \int |f|\,d\lambda$ implies that for all large enough $n$,

$$-\infty < C - 1 - \int |f|\,d\lambda \leq \int \log^-|p_n|\,d\lambda.$$

As $\log^-|p_n| \leq 0$ and $\log^-|p_n| \to \log^-|f|$ a.e. $[\lambda]$, Fatou's lemma applies, and we find that $\int \log^-|f|\,d\lambda$ is finite. Thus, $\log|f|$ is integrable with respect to $\lambda$, as was to be shown.

The argument used in this proof is due, in part, to Sidney.

Our proof of the last theorem is of a mixed function theoretic-algebraic nature, and on first consideration, it may seem surprising that there is no proof depending only on the theory of uniform algebras. Such a proof cannot be given, for there are examples of representing measures $\mu$ for Dirichlet algebras such that not every $f \in H^1(\mu)$ has the property that $\log|f|$ is integrable with respect to $\mu$. Examples arise naturally in abelian harmonic analysis and may be found, e.g., in Chapter 8 of Rudin [1].

If the set $X$ of the theorem is the closed unit disc, the result we have just proved can be established by much simpler means if we are willing to use certain of the theory of the boundary behavior of functions in the Hardy class $H^1$ in the unit disc.

As a consequence of the theorem we have just proved, we are able to identify very explicitly the sets that are peak interpolation sets for $\mathscr{P}(X)$.

25.10 **Corollary**

(a) *A closed subset $E$ of $\partial X$ is a peak interpolation set for $\mathscr{P}(X)$ if and only if $\lambda(E) = 0$ for the harmonic measure $\lambda$ for every point of the interior of $X$.*

(b) *If $X$ is uncountable, there exist uncountable peak interpolation sets for $\mathscr{P}(X)$.*

PROOF

(a) This follows from the last theorem in combination with the fact that if $\{z_n\}$ is a sequence of points in $X$, one from each component of the interior of $X$, then each $\mu \in \mathscr{M}(\partial X)$ which annihilates $\mathscr{P}(X)$ is absolutely continuous with respect to the measure $\Lambda = \Sigma 2^{-n}\lambda_{z_n}$, and from Theorem 20.10.

(b) If $E$ is any closed subset of $\partial X$ which is a null set for the measure $\Lambda$, then $E$ is a peak interpolation set for $\mathscr{P}(X)$. If $X$ is uncountable, there exist uncountable closed $\Lambda$-null sets in $\partial X$, so (b) is proved.

In this corollary we have the implicit hypothesis that the set $X$ does not separate the plane. However, we see that conclusion (b) is correct for an arbitrary compact set: If $Y$ is a compact set, (b) applies to the compact set $X$ obtained by adjoining to $Y$ all the bounded components of $\mathbb{C} \setminus Y$. Part (b) in this extended form fills a gap left in the proof of Theorem 20.17.

Our general results on analytic structure in spectra enable us to determine the Gleason parts for the algebra $\mathscr{P}(X)$.

25.11 ***Theorem*** (Wermer [12])

*Let $X$ be a compact set that does not separate the plane. If $\Pi$ is a Gleason part for $\mathscr{P}(X)$, then either $\Pi$ is a point of $\partial X$ or else $\Pi$ is a component of the interior of $X$.*

PROOF

Since $\mathscr{P}(X)$ is Dirichlet, it follows that each point on $\partial X$ is a peak point and so is a one-point part. Also, from our general theory of parts for algebras which admit unique representing measures, it follows that every nontrivial part for $\mathscr{P}(X)$ is connected. Let $\Pi$ be a nontrivial part for $\mathscr{P}(X)$. Then $\Pi$ is contained within a component, $C$, of the interior of $X$ since $\Pi$ is connected and disjoint from $\partial X$. But also, since each $f \in \mathscr{P}(X)$ is holomorphic on $C$, it follows that $C$ is contained within a single part for $\mathscr{P}(X)$. Since $C \cap \Pi \neq \varnothing$, we have $C = \Pi$.

The final topic we shall treat in this section is a maximality theorem and some related results. Our first result was proved by Bishop [2]; it is a generalization of a well-known result of Wermer.

25.12 ***Theorem***

*Let $X$ be a compact set in the plane with connected complement and connected interior. The algebra $\mathscr{P}(X)$ is a maximal subalgebra of $\mathscr{C}(\partial X)$ in the sense that if $B$ is a closed subalgebra of $\mathscr{C}(\partial X)$ with $\mathscr{P}(X)|\partial X \subseteq B \subseteq \mathscr{C}(\partial X)$, then either $B = \mathscr{P}(X)|\partial X$ or else $B = \mathscr{C}(\partial X)$.*

For the proof of this result, we shall need to make use of a result from the next section, a result that will not depend on the circle of ideas we are going to develop.

PROOF

Let $\Omega$ and $D$ denote, respectively, the sets $\mathbb{C} \setminus X$ and $\mathbb{C} \setminus \bar{\Omega}$ so that $D$ is the interior of $X$ and both $\Omega$ and $D$ are connected by hypothesis. We consider two cases. The first is that in which there exists a point $z_0$ in $D$ such that the function $(z - z_0)^{-1}$ lies in $B$. In this case the algebra $B$ contains $\mathscr{R}(\partial X)$. This is so, for if $h$ is a rational function with no poles on $X \setminus D$, then $h = h' + h''$, where $h'$ and $h''$ are rational functions, $h'$ with no poles in $D$ and $h''$ with no poles in $\Omega$. The function $h'$ has no poles in $X$ and so lies in $\mathscr{P}(X)$, which is contained in $B$, and the function $h''$ is holomorphic on a neighborhood of $\mathbb{C}^* \setminus D$, $\mathbb{C}^*$ the Riemann sphere, and so it can be approximated uniformly on $\mathbb{C}^* \setminus D$ and *a fortiori* on $\partial X$ by polynomials in $z$ and $(z - z_0)^{-1}$. As these polynomials lie in $B$ and as $B$ contains $\mathscr{R}(X)$, it follows that $B$ contains $\mathscr{R}(\partial X)$. However, the fact that $\mathbb{C} \setminus \partial X$ has only two components implies, by a generalization of Mergelyan's theorem, Theorem 26.5, that $\mathscr{R}(\partial X) = \mathscr{C}(\partial X)$, whence $B = \mathscr{C}(\partial X)$.

The second case we consider is that in which $(z - z_0)^{-1}$ is an element of $B$ for no choice of $z_0 \in D$. Let $z_0$ be a fixed point of $D$. Since $(z - z_0)^{-1} \notin B$, it follows that for some choice of $\varphi \in \Sigma(B)$, $\varphi(z) = z_0$. Let $\mu_\varphi$ be a probability measure on $\partial X$ which represents $\varphi$. It follows that if $p$ is a polynomial, then

$$\int p \, d\mu = p(z_0),$$

so $\mu_\varphi$ is the unique measure on $\partial X$ which represents the point $z_0$ for the algebra $\mathscr{P}(X)$. If $f$ lies in $B$, we have that

$$\int (z - z_0)^n f \, d\mu_\varphi = 0$$

for $n = 1, 2, 3, \ldots$. It follows that if $h \in H^1_\varphi(\mu_\varphi)$, the closure in $L^1(\mu_\varphi)$ of the ideal in $\mathscr{P}(X)$ consisting of those functions which lie in the kernel of $\varphi$, i.e., which vanish at $z_0$, then

$$\int hf \, d\mu_\varphi = 0. \tag{8}$$

If $\nu$ is a measure on $\partial X$ which annihilates $\mathscr{P}(X)$, we know from Theorem 23.9 and from the fact that there are no completely singular measures orthogonal to $\mathscr{P}(X)$ that $d\nu = h \, d\mu_\varphi$, where $h \in H^1_\varphi(d\mu_\varphi)$. Equation (8) implies, therefore, that any measure on $\partial X$ which annihilates $\mathscr{P}(X)$ also annihilates $B$, so $B = \mathscr{P}(X)$.

As a very special case, we have Wermer's maximality theorem for the disc algebra:

**25.13** ***Corollary*** (Wermer [1])
*If $T$ denotes the unit circle and if $B$ is a uniform algebra on $T$ which contains $\mathscr{A}|T$, then either $B = \mathscr{C}(T)$ or else $B = \mathscr{A}|T$.*

It seems worthwhile to mention explicitly the fact that in the case of the disc algebra $\mathscr{A}$, it is unnecessary to appeal to Mergelyan's theorem and its generalization which we cited: A cursory examination of the proof above shows that in this case much simpler approximation results suffice. Several other proofs of Wermer's maximality theorem are known and are to be found in Wermer's original paper, in Cohen [1], in Srinivasan and Wang [1], and in Lumer [3].

The following is Cohen's very elementary proof of the corollary. By the Stone–Weierstrass theorem, it suffices to show that if $B \neq \mathscr{A}$, then the function $z$ is invertible in $B$, for on $T$, $z^{-1} = \bar{z}$, so it will follow that $B$ is closed under complex conjugation. Let $f \in B$, $f \notin \mathscr{A}$. Thus, if $f$ has the Fourier series $\sum_{n=-\infty}^{\infty} c_n e^{in\theta}$, $c_{-n} \neq 0$ for some $n > 0$. We may suppose that $c_{-n} = 1$. Then the Fourier series for $z^n f$ has constant term 1. Since polynomials in $z$ and $\bar{z}$ are dense in $\mathscr{C}(T)$, it follows that for some choice of polynomials $g_1$ and $g_2$, we have

$$z^n f = 1 + zg_1 + \bar{z}\bar{g}_2 + h, \tag{9}$$

where $h \in \mathscr{C}(T)$, $\|h\|_T < \frac{1}{2}$. Let $M = \|zg_2 - \bar{z}\bar{g}_2\|_T$. Since $zg_2 - \bar{z}\bar{g}_2$ is purely imaginary, we have for any $\delta > 0$,

$$\|1 + \delta(zg_2 - \bar{z}\bar{g}_2)\|_T \leq 1 + \delta^2 M^2. \tag{10}$$

By solving (9) for $\bar{z}\bar{g}_2$ and using (10), we derive the inequality

$$\|1 + \delta + z(\delta g_2 + \delta g_1 - \delta z^{n-1} f) + \delta h\|_T \leq 1 + \delta^2 M^2,$$

and from this, using $\|h\|_T \le \frac{1}{2}$, we get

$$\|1 + \delta + zp\|_T \le 1 + \delta^2 M^2 + \delta/2,$$

where $p = \delta g_2 + \delta g_1 - \delta z^{n-1} f$. This last inequality yields

$$\|1 + z(1 + \delta)^{-1} p\|_T \quad \le \quad (1 + \delta^2 M^2 + \delta/2)(1 + \delta)^{-1}.$$

For $\delta$ small enough, the right side is less than 1, so $z(1 + \delta)^{-1} p$ is invertible in $B$, whence $z$ itself is invertible in $B$, and we have the result.

Various other generalizations of Wermer's maximality theorem are known, and it has given rise to a certain theory of maximal algebras. We shall not enter into a general discussion of this theory. We refer the interested reader to the following relevant sources in addition to those already cited: Hoffman and Singer [2, 3], Bear [2], Gamelin and Rossi [1], Helson and Quigley [1], Glicksberg [3], and Wermer [2].

As a consequence of this maximality theorem, we are able to obtain a very short proof of a theorem of Rudin which has to do with *maximum modulus algebras* in the sense of the following definition.

25.14 ***Definition***

*If $X$ is a compact set in the plane, the subalgebra $A$ of $\mathscr{C}(X)$ is called a* ***maximum modulus algebra*** *if $\|f\|_X = \|f\|_{\partial X}$ for all $f \in A$.*

It is clear that if $A$ is a maximum modulus algebra on $X$, so is its closure in $\mathscr{C}(X)$, so we need deal only with uniformly closed maximum modulus algebras.

25.15 ***Lemma***

*If $A$ is a maximum modulus algebra on the planar set $X$ and if $A'$ is obtained from $A$ by adjoining the constants, then $A'$ is also a maximum modulus algebra on $X$.*

PROOF

We are to prove that if $f \in A$ and $\alpha \in \mathbb{C}$, then $|f + \alpha|$ achieves its maximum on $\partial X$. Denote by $H$ the convex hull of the set $f(\partial X)$. If $f(X) \subset H$, our assertion is clear for the range of $f + \alpha$ is merely a translate of that of $f$.

If $f(X)$ is not a subset of $H$, there is a point $w_0 \in f(X) \setminus H$ which we may take to be different from zero. As $H$ is convex and $w_0 \notin H$, there exist $\beta \in \mathbb{C}$ and $r > 0$ such that the set $\{z \in \mathbb{C} : |z - \beta| \le r\}$ contains $H$ but not $w_0$. There is an integer $n$ such that

$$|w_0|\,|w_0 - \beta|^n > (|\beta| + r)r^n.$$

The polynomial $p(w) = w(w - \beta)^n$ is larger at $w_0$ than at any point of $f(\partial X)$, and thus the function $f(f - \beta)^n$, an element of $A$, has larger modulus at an interior point of $X$ than at any point of $\partial X$. This contradiction implies that $f(X) \subset H$, and the lemma is proved.

25.16 ***Theorem*** (Rudin [4])

*If $A$ is a maximum modulus algebra on $\bar{\Delta}$, the closed unit disc, and if $A$ contains a function $\psi$ which is one to one on $\bar{\Delta}$, then each $f \in A$ is of the form $f^* \circ \psi$, where $f^*$ is continuous on the range of $\psi$ and holomorphic on the interior of this range. If, in addition, $A$ contains a single nonconstant function which is holomorphic on the interior of $\Delta$, then every $f$ in $A$ is holomorphic there.*

PROOF

By the lemma, we may suppose that $A$ contains the constants, and we may also suppose that $A$ is uniformly closed.

We consider first the case that $A$ contains the identity map of $\bar{\Delta}$ onto itself. Rudin treated this case by an elementary argument from first principles involving harmonic functions, but we shall obtain the result as an immediate consequence of the maximality theorem just proved. Since $A$ contains the identity map, it also contains the polynomials. It follows that $A$ contains $\mathscr{A}$, the disc algebra. The maximum modulus property of $A$ implies that it is isomorphic to $A|\partial\Delta$. We have that $\mathscr{A}|\partial\Delta \subseteq A|\partial\Delta \subseteq \mathscr{C}(\partial\Delta)$, and, by the maximality theorem, one of these inclusions must be an equality. Since the map $f \mapsto f(0)$ is a complex homomorphism of $A$ which is not of the form $f \mapsto f(z_0)$ for any $z_0 \in \partial(\Delta)$, and since $f \mapsto f|\partial\Delta$ is an isomorphism, it follows that $A|\partial\Delta \neq \mathscr{C}(\partial\Delta)$. Consequently, $A|\partial\Delta = \mathscr{A}|\partial\Delta$. If $f \in A$, there is $g \in \mathscr{A}$ with $f - g = 0$ on $\partial\Delta$. From this, from $\mathscr{A} \subset A$, and from the maximum modulus property of $A$, it follows that $f = g$, so $A \subset \mathscr{A}$ and we have the result in the special case we are considering.

For the general case, let $D = \psi(\bar{\Delta})$ so that $D$ is the union of a simple closed curve and its interior $D_0$, a simply connected domain. Let $g$ be a conformal mapping of $D_0$ onto the open unit disc. The function $g$ exists by the Riemann mapping theorem, and, moreover, $g$ extends to a homeomorphism of $D$ onto $\bar{\Delta}$. By Mergelyan's theorem $g$ can be approximated uniformly by polynomials, so $g \circ \psi \in A$.

The algebra $A_1 = \{f \circ \psi^{-1} \circ g^{-1} : f \in A\}$ is a maximum modulus algebra on $\bar{\Delta}$, and it contains $g \circ \psi \circ \psi^{-1} \circ g^{-1}$, the identity map. Thus, by the special case of the theorem already proved, $A_1 = \mathscr{A}$, and it follows that every $f \in A$ has the form $f = f^* \circ g \circ \psi, f^* \in \mathscr{A}$.

Consider finally the case that $A$ already contains a nonconstant holomorphic function $F$. By what we have done so far, we know that

$F = f \circ \psi$, $f$ holomorphic on the interior of the range of $\psi$. The function $F$ is nonconstant, and the equation $F = f \circ \psi$ implies that $\psi$ is holomorphic in $\Delta$ off $E$, the inverse image under $\psi$ of the zero set in $D_0$ of $f'$. The set $E$ is a discrete subset of $\Delta$, so the Riemann removable singularities theorem shows that $\psi$ is holomorphic in $\Delta$.

To obtain a local version of this result, we make the following definition.

**25.17** ***Definition***

*If $\Omega$ is a domain in $\mathbb{C}$, the subalgebra $A$ of $\mathscr{C}(\Omega)$ is a* ***local maximum modulus algebra*** *on $\Omega$ provided that if $f \in A$ and $K \subset \Omega$ is compact,* $\|f\|_K = \|f\|_{\partial K}$.

**25.18** ***Corollary*** (Rudin [4])

*If $A$ is a local maximum modulus algebra on the planar domain $\Omega$, and if $A$ contains a nonconstant holomorphic function, then $A$ consists entirely of holomorphic functions.*

For the purposes of this corollary, we understand by *domain* a connected open set.

PROOF

Let $f \in A$ be holomorphic and nonconstant, and let $E$ be the zero set of $f'$, a certain discrete set in $\Omega$. If $z_0 \in \Omega \setminus E$, then for sufficiently small $r$, the disc

$$D_r = \{z \in \Omega : |z - z_0| \leq r\}$$

is contained in $\Omega$, and $f$ is one to one there. By hypothesis, $A|D_r$ is a maximum modulus algebra, so each element of $A$ is holomorphic in the interior of $D_r$. It follows that every function in $A$ is holomorphic on $\Omega \setminus E$, and from this it follows by the Riemann removable singularities theorem that each element of $A$ is actually holomorphic in the whole of $\Omega$.

Hurd [2] has considered maximum modulus algebras from a somewhat different point of view. Among other things, he has observed that if we assume that $\Delta$ is contained as an open subset of $\Sigma(A)$, $A$ a maximum modulus algebra on $\bar{\Delta}$, then $A$ consists entirely of holomorphic functions under the sole hypothesis that it contains a single nonconstant holomorphic function. To draw this conclusion, let $f \in A$ be holomorphic and nonconstant, and let $E$ be the zero set of $f'$. If $z_0 \in \Delta \setminus E$, let $D$ be a small closed disc contained in $\Delta \setminus E$ on which $f$ is one to one. The local maximum modulus theorem, Theorem 9.8, implies that $A|D$ is a maximum modulus algebra on $D$, so every function in $A$ is holomorphic on the interior of $D$. Therefore, the algebra $A$ consists of holomorphic functions.

It is tempting to conjecture that if $A$ is a maximum modulus algebra on $\bar{\Delta}$ which contains nonconstant holomorphic functions, then $A$ must consist exclusively of holomorphic functions. Such a strengthened version of the result is false, as our next example shows.

25.19 ***Example*** (Rudin [6])

Let $E$ be a totally disconnected, perfect subset of the unit circle which has Lebesgue measure zero so that $E$ is a peak interpolation set for the disc algebra $\mathscr{A}$. The bicylinder

$$\bar{\Delta}^2 = \{(z, w) \in \mathbb{C}^2 : |z|, |w| \leq 1\}$$

is a compact, separable metric space and so is the image of $E$ under a continuous map, say $\varphi$, where $\varphi(z) = (\varphi_1(z), \varphi_2(z))$, $z \in E$. There is a function $F_1 \in \mathscr{A}$ with $F_1|E = \varphi_1$, $\|F_1\|_{\bar{\Delta}} = 1$. Choose $F_2 \in \mathscr{C}(\bar{\Delta})$, $F_2 \notin \mathscr{A}$, such that $F_2|E = \varphi_2$, $\|F_2\|_{\bar{\Delta}} = 1$. If $A$ is the subalgebra of $\mathscr{C}(\bar{\Delta})$ generated by $F_1$ and $F_2$, then $A$ contains both holomorphic and nonholomorphic functions. The algebra $A$ is a maximum modulus algebra on $\bar{\Delta}$ since for all $f \in A$, $\|f\|_{\Delta} = \|f\|_E$.

The algebra $A$ just constructed does not separate points on $\bar{\Delta}$, but in his paper [8], Rudin has constructed a maximum modulus algebra on $\Delta$ which separates points and which contains both holomorphic and nonholomorphic functions. The details of this example are considerably more involved than those of the example just given, so we shall refer to the original note for them.

Wermer's maximality theorem provides a very short proof of a classical theorem of Radó [1], which in turn implies a further interesting algebraic property of the algebras $\mathscr{A}(X)$.

25.20 ***Theorem***

*If $f$ is continuous on a domain $\Omega$ in $\mathbb{C}$ and if $f$ is holomorphic on $\Omega \setminus f^{-1}(0)$, then $f$ is holomorphic on $\Omega$.*

PROOF

The theorem is a local one, so it suffices to show that if $f$ is continuous on $\bar{\Delta}$, the closed unit disc, and holomorphic on $\bar{\Delta} \setminus f^{-1}(0)$, then $f$ belongs to the disc algebra. To do this let $B \subset \mathscr{C}(\bar{\Delta})$ be the algebra generated by $\mathscr{A}$ and $f$. Thus, each function in $B$ is holomorphic on $\Delta \setminus \partial E$, $E = f^{-1}(0)$, and it follows from the maximum modulus theorem that if $\Gamma$ is the Shilov boundary for $B$, then $\Gamma \subset \partial\Delta \cup \partial E$. Let $z_0 \in \Delta \setminus E$, and let $\mu$ be a Jensen measure for $z_0$. Since $f(z_0) \neq 0$, we have

$$-\infty < \log|f(z_0)| \leq \int \log|f|\, d\mu,$$

so $\mu(\partial E) = 0$, and it follows that $\mu$ is a measure on $\partial\Delta$. Since $\mu$ is a representing measure for $z_0$, it follows that if $g \in B$, then $|g(z_0)| \leq \|g\|_{\partial\Delta}$. This holds for all $z_0 \in \Delta \setminus E$, and therefore for all $z_0 \in \Delta \setminus E$. Since $\Gamma \subset \partial\Delta \cup \partial E$, it follows that $\|g\| \leq \|g\|_{\partial\Delta}$ for all $g \in B$. By Wermer's maximality theorem, $B|\partial\Delta = \mathscr{C}(\partial\Delta)$ or $B|\partial\Delta = \mathscr{A}|\partial\Delta$. The former case is impossible since $\Sigma(B) \supset \bar{\Delta}$ and $\Sigma(\mathscr{C}(\partial\Delta)) = \partial\Delta$, so $B|\partial\Delta = \mathscr{A}|\partial\Delta$, whence $f \in \mathscr{A}$, as desired.

This argument is Bishop's version (Bishop [6]) of a proof given by Glicksberg [3], which was suggested in turn by Wermer. Various abstract versions of Radó's theorem are contained in the cited paper of Glicksberg, and a stronger function theoretic version is due to Stout [2]. Kaufman [1] has given a short direct proof of Radó's theorem.

It is clear that Radó's theorem is local in nature so that it retains its validity if $f^{-1}(0)$ is replaced by $f^{-1}(E)$ for any finite or discrete set in $\mathbb{C}$.

As a consequence of Radó's theorem we have the fact that $\mathscr{O}(\Omega)$ is integrally closed in $\mathscr{C}(\Omega)$.

25.21 ***Corollary***

*If $f$ is a continuous function on the domain $\Omega$ which satisfies a monic polynomial equation*

$$f^q + g_1 f^{q-1} + \cdots + g_{q-1} f + g_q = 0, \tag{11}$$

*$g_1, \ldots, g_q$ holomorphic in $\Omega$, then $f$ itself is holomorphic in $\Omega$.*

PROOF

Let $Q(z, w) = w^m + g_1(z)w^{m-1} + \cdots + g_{m-1}(z)w + g_m(z)$ be a monic polynomial of minimal degree such that $Q(z, f(z)) = 0$. If $m = 1$ we are done, so suppose that $m > 1$. Define $Q_1$ by

$$Q_1(z, w) = mw^{m-1} + (m-1)g_1(z)w^{m-2} + \cdots + g_{m-1}(z).$$

The function $F$ given by $F(z) = Q_1(z, f(z))$ is continuous, and it is not identically zero because of the minimality of the degree of $Q$. If $E$ denotes the set on which $F$ vanishes, then given $z_0 \notin E$, the fact that $Q_1(z_0, f(z_0)) \neq 0$, taken together with the implicit-function theorem, implies that there is a neighborhood $W$ of $z_0$ in which there exists a uniquely defined continuous function $\varphi$ such that $\varphi(z_0) = f(z_0)$ and

$Q(z, \varphi(z)) = 0$, $z \in W$. Moreover, the function $\varphi$ is holomorphic. As $Q(z, f(z))$ vanishes identically, we find from the uniqueness assertion that $f = \varphi$, and it follows therefore that $f$ is holomorphic off $E$. Consequently, the same is true of the function $F$. Radó's theorem implies that $F$ is holomorphic throughout the domain $\Omega$, i.e., there exists $h$, a function holomorphic in $\Omega$, such that $Q_1(z, f(z)) = h(z)$. This leads to a contradiction, though, for the equation $Q_1(z, f(z)) - h(z) = 0$ yields a monic polynomial relation of degree $m - 1$ which $f$ satisfies, an impossibility in light of the choice of $Q$.

**25.22** ***Corollary***

*If $X$ is a compact set in the plane, the algebra $\mathscr{A}(X)$ is integrally closed in $\mathscr{C}(X)$: If $f \in \mathscr{C}(X)$ satisfies a monic polynomial equation with coefficients from $\mathscr{A}(X)$, then $f \in \mathscr{A}(X)$.*

Although it may be true, it does not seem obvious that for general compact sets $X$, $\mathscr{R}(X)$ is integrally closed in $\mathscr{C}(X)$.

Other theorems on the integral closure of uniform algebras are due to Glicksberg [3] and Helson and Quigley [2].

# 26 *THE ALGEBRA $\mathscr{R}(X)$*

We come now to our discussion of the algebra $\mathscr{R}(X)$. It is necessary for us to preface this discussion with a disclaimer: Our goal is not a comprehensive presentation of the whole theory of $\mathscr{R}(X)$, for such an exhaustive treatment would require its own, rather voluminous, treatise. Instead, we shall examine certain of the aspects of the theory of $\mathscr{R}(X)$ which are particularly closely related to the theory of uniform algebras and which, moreover, are accessible by more or less abstract methods. From the point of view of the theory of uniform algebras, $\mathscr{R}(X)$ is a very interesting object of study, for, on the one hand, abstract methods can provide results on $\mathscr{R}(X)$, and, on the other, the study of $\mathscr{R}(X)$ has helped delineate some of the boundaries of the more general theory. Certain very important results on $\mathscr{R}(X)$, most notably, perhaps, those of Vitushkin and Melnikov, are not included in our treatment, for they depend on the theory of analytic capacity, a theory whose systematic treatment would take us far afield.

One of the basic problems in the theory of $\mathscr{R}(X)$ is the analogue of the problem treated in the last section: When is $\mathscr{R}(X) = \mathscr{A}(X)$? Our first result shows that the problem can be localized, a fact that will lead to a very short proof of an extension of Mergelyan's theorem.

Let us fix for the duration of this section a compact set $X$ in $\mathbb{C}$.

**26.1** ***Theorem***

*If $f \in \mathscr{C}(X)$ and if every $x \in X$ has a closed neighborhood $N_x$ with the property that $f|(N_x \cap X) \in \mathscr{R}(N_x \cap X)$, then $f \in \mathscr{R}(X)$.*

This result is due to Bishop, though the proof we give follows Garnett [3]. It depends on some properties of the Cauchy transform of a measure.

**26.2** ***Lemma***

*If $\mu \in \mathscr{M}(X)$, then $\tilde{\mu}$ vanishes on $\mathbb{C} \setminus X$ if and only if $\mu \in \mathscr{R}(X)^{\perp}$.*

Observe that since $\tilde{\mu}$ is plainly continuous off the support of $\mu$, we can prove that $\tilde{\mu}$ vanishes off $X$ by showing that it vanishes almost everywhere (with respect to Lebesgue measure) off $X$.

PROOF

In one direction, this is obvious, for if $\mu \in \mathscr{R}(X)^{\perp}$, then since for fixed $z \in \mathbb{C} \setminus X, (\zeta - z)^{-1}$, a function of $\zeta$, belongs to $\mathscr{R}(X)$, we have $\tilde{\mu}(z) = 0$. Conversely, if $\tilde{\mu}(z) = 0$ for all $z \in \mathbb{C} \setminus X$, then given $f$, a rational function without poles on $X$, and given $\zeta \in X$, we can write

$$f(\zeta) = -\frac{1}{2\pi i}\int_\Gamma \frac{f(z)}{\zeta - z}\,dz,$$

where $\Gamma$ is an appropriate system of curves in $\mathbb{C} \setminus X$. This equation leads to

$$\int f(\zeta)\,d\mu(\zeta) = -\frac{1}{2\pi i}\int_\Gamma f(z)\tilde{\mu}(z)\,dz.$$

As $\Gamma \subset \mathbb{C} \setminus X$ and $\tilde{\mu}$ vanishes on $\mathbb{C} \setminus X$, the integral on the right is zero, so $\mu \in \mathscr{R}(X)^{\perp}$.

Note that his lemma makes explicit the idea used in the proof we gave for Runge's theorem in Section 24.

**26.3** ***Lemma***

*If $\varphi$ is a $\mathscr{C}^\infty$ function on $\mathbb{C}$ with compact support and if $\mu \in \mathscr{M}(X)$, then at almost every (with respect to Lebesgue measure) point of the plane*

$$\varphi\tilde{\mu} = (\varphi\mu)^{\sim} + \tilde{\sigma},$$

*where*

$$d\sigma = -\frac{1}{\pi}\tilde{\mu}\frac{\partial\varphi}{\partial\bar{z}}\,dx\,dy.$$

PROOF (Zalcman [1])

This lemma follows from Green's theorem. We have for almost every $\zeta$,

$$\begin{aligned}\tilde{\sigma}(\zeta) &= -\frac{1}{\pi}\iint\frac{1}{z-\zeta}\tilde{\mu}(z)\frac{\partial\varphi}{\partial\bar{z}}(z)\,dx\,dy \qquad z = x+iy\\ &= -\frac{1}{\pi}\iint\frac{1}{z-\zeta}\left\{\int\frac{d\mu(t)}{t-z}\right\}\frac{\partial\varphi}{\partial\bar{z}}(z)\,dx\,dy\\ &= -\frac{1}{\pi}\int\left\{\iint\frac{1}{z-\zeta}\,\frac{1}{t-z}\,\frac{\partial\varphi}{\partial\bar{z}}(z)\,dx\,dy\right\}d\mu(t).\end{aligned}$$

Since

$$\frac{1}{z-\zeta}\,\frac{1}{t-z} = \frac{1}{t-\zeta}\left(\frac{1}{z-\zeta}+\frac{1}{t-z}\right),$$

it follows that

$$\tilde{\sigma}(\zeta) = -\frac{1}{\pi}\int\frac{1}{t-\zeta}\left\{\iint\frac{\partial\varphi}{\partial\bar{z}}(z)\left(\frac{1}{z-\zeta}+\frac{1}{t-z}\right)dx\,dy\right\}d\mu(t).$$

By Lemma 25.1(c),

$$-\frac{1}{\pi}\iint\frac{\partial\varphi}{\partial\bar{z}}(z)\frac{1}{z-\zeta}\,dx\,dy = \varphi(\zeta),$$

so we find that

$$\tilde{\sigma}(\zeta) = \varphi(\zeta)\tilde{\mu}(\zeta) - (\varphi\mu)^{\sim},$$

as was to be shown.

PROOF OF THE THEOREM

Let $\mathscr{U} = \{U_1, \ldots, U_q\}$ be a finite open cover for $X$ such that $f \in \mathscr{R}(\overline{U}_j)$ for each $j$, and let $\{\varphi_1, \ldots, \varphi_q\}$ be a $\mathscr{C}^\infty$ partition of unity subordinate to $\mathscr{U}$. This may require a slight explanation: For each $j$, let $V_j$ be some relatively compact open set in $\mathbb{C}$ with $U_j = V_j \cap X$. We ask that $\varphi_j$ be infinitely differentiable on $\mathbb{C}$, that $\varphi_j$ vanish off $V_j$, and that $\Sigma\varphi_j(x) = 1$ if $x \in X$.

If $\mu \in \mathscr{R}(X)^\perp$, determine $\mu_j$ by

$$d\mu_j = \varphi_j\,d\mu - \frac{1}{\pi}\tilde{\mu}\frac{\partial\varphi_j}{\partial\bar{z}}\,dx\,dy.$$

The measure $\mu_j$ is concentrated on $U_j$ since $\varphi_j$ vanishes off $V_j$ and $\tilde{\mu}$ vanishes off $X$. The measure $\mu_j$ is in $\mathscr{R}(\overline{U}_j)^\perp$, because, by Lemma 26.3, $\tilde{\mu}_j = \varphi_j\tilde{\mu}$ almost everywhere so that $\tilde{\mu}_j$ vanishes off $\overline{U}_j$. As $\mu = \Sigma\mu_j$, it follows that

$$\int f\,d\mu = \sum \int f\,d\mu_j = 0$$

since $f \in \mathscr{R}(\overline{U}_j)$, so $f \in \mathscr{R}(X)$.

As a first application of the theorem on localization, we have the following fact.

**26.4** ***Theorem*** (Alexander [2])
*Let $\{X_n\}$ be a sequence of compact sets with compact union $X$. If $\mathscr{R}(X_n) = \mathscr{C}(X_n)$ for all $n$, then $\mathscr{R}(X) = \mathscr{C}(X)$.*

PROOF
Let $\mu \in \mathscr{M}(X)$ annihilate $\mathscr{R}(X)$, $\mu$ not the zero measure, and let $S$ be the closed support of $\mu$ so that $S$ is the minimal closed set with the property that $|\mu|(X \setminus S) = 0$. We shall show that $\mu \in \mathscr{R}(S)^\perp$. The sets $X_n$ have no interior, so by the Baire category theorem, $X$ has no interior. To prove that $\mu \in \mathscr{R}(S)^\perp$, it suffices to prove that $\tilde{\mu}$ vanishes indentically on $\mathbb{C} \setminus S$. This follows from the fact that $\tilde{\mu}$ vanishes on $\mathbb{C} \setminus X$ and the fact that each point of $\mathbb{C} \setminus S$ is the limit of a sequence from $\mathbb{C} \setminus X$. We have $S = \bigcup(S \cap X_n)$, so by category there is an open disc $D$ which meets $S$ such that $S \cap D = (S \cap X_n) \cap D$ for some $n$. If $D' \subset \overline{D}' \subset D$, $D'$ an open disc which meets $S$, then there is $f \in \mathscr{C}(\mathbb{C})$ with $f|D'$ identically 1 and $f$ identically zero on a neighborhood of $\mathbb{C} \setminus D$. The function $f|S$ belongs to $\mathscr{R}(S)$ locally since $\mathscr{R}(S \cap X_n) = \mathscr{C}(S \cap X_n)$, so $f \in \mathscr{R}(S)$. But then $f\mu \in \mathscr{R}(S)^\perp$, and this measure is supported in $X_n \cap S$. As $\mathscr{R}(X_n \cap S) = \mathscr{C}(X_n \cap S)$, the measure $f\mu$ must be the zero measure. This implies that $|\mu|(D' \cap X_n \cap S) = 0$, a contradiction to the minimality of $S$. This completes the proof.

The theorem on localization enables us to obtain with ease an extension of Mergelyan's theorem.

**26.5** ***Theorem***
*If every component of $\mathbb{C} \setminus X$ has diameter at least $\delta$, $\delta$ a fixed positive constant, then $\mathscr{A}(X) = \mathscr{R}(X)$.*

PROOF

If $x \in X$ and if $N(x)$ is a closed disc around $x$ of diameter less than $\delta$, then $\mathbb{C} \setminus (X \cap N(x))$ is connected, so by Theorem 25.7 we know that $\mathscr{P}(X \cap N(x)) = \mathscr{A}(X \cap N(x))$. Thus, every $f \in \mathscr{A}(X)$ belongs to $\mathscr{R}(X)$ locally and so belongs to $\mathscr{R}(X)$ by Theorem 26.1.

This theorem was obtained by Mergelyan [2]; for the proof we have followed Garnett [3]. The hypotheses of the theorem are plainly satisfied if the set $\mathbb{C} \setminus X$ has but finitely many components, but it also applies to many sets with infinitely many complementary domains.

As an easy corollary of the localization theorem, we have the following fact.

**26.6** ***Corollary***

*If $K$ is a component of $X$, then $K$ is a peak set for $\mathscr{R}(X)$ and $\mathscr{R}(X)|K = \mathscr{R}(K)$.*

PROOF

The hypothesis that $K$ is a component of $X$ implies that

$$K = \bigcap \{E_n : n = 1, 2, \ldots\},$$

where each $E_n$ is an open and closed subset of $K$, and $E_1 \supset E_2 \supset \cdots$, for $X$ is separable. Since $E_n$ is open and closed in $X$, the function $f$ which is identically zero on $X \setminus E_n$ and identically 1 on $E_n$ lies in $\mathscr{R}(X)$ since it is locally in $\mathscr{R}(X)$. Thus, $E_n$ is a peak set for $\mathscr{R}(X)$, whence $K$ is also a peak set for $\mathscr{R}(X)$. [That $E_n$ is a peak set for $\mathscr{R}(X)$ follows also from the fact that $X = \Sigma(\mathscr{R}(X))$ and the fact that holomorphic functions operate on Gelfand transforms, i.e., from a very weak version of the Shilov idempotent theorem.]

It remains to prove that $\mathscr{R}(K) = \mathscr{R}(X)|K$. Since $\mathscr{R}(X)|K$ is closed in $\mathscr{R}(K)$ as follows from the last paragraph, we see that it suffices to prove that if $h$ is a rational function with no poles on $K$, then for some choice of $n$ the function $\tilde{h}$, which is zero on $X \setminus E_n$ and which agrees with $h$ on $E_n$, lies in $\mathscr{R}(X)$. However, this follows from the choice of the sequence $\{E_n\}$ and the localization theorem, for since $\{E_n\}$ is a decreasing sequence, there is an $n$ such that the poles of $h$ lie off $E_n$. For this choice of $n$, the function $\tilde{h}$ just described belongs to $\mathscr{R}(X)$ locally and so belongs to $\mathscr{R}(X)$.

This observation and the following remarks are contained, essentially, in the paper of Glicksberg [7].

**26.7** ***Remarks***

As a consequence of this corollary, we see that if $\mu$ is a representing measure for the point $x \in K$, then $\mu$ is supported by $K$. If $f \in \mathscr{R}(X)$ peaks on $K$, then

$$1 = f^n(x) = \int_X f^n \, d\mu,$$

and, as $n \to \infty$, this sequence of integrals converges to $\mu(K)$.

Another observation along these lines is the fact that since $X = \Sigma(\mathscr{R}(X))$, every maximal antisymmetric set for $\mathscr{R}(X)$ is connected—recall the Remarks 12.7—so it follows if $f \in \mathscr{C}(X)$ and $f|K \in \mathscr{R}(K)$ for every component $K$ of $X$, then $f \in \mathscr{R}(X)$.

In his paper [1], Bishop established a necessary and sufficient condition for the equality of $\mathscr{R}(X)$ and $\mathscr{C}(X)$, a condition in terms of peak points. It is this result that we consider next.

**26.8** ***Theorem***

*The algebra $\mathscr{R}(X)$ is $\mathscr{C}(X)$ if and only if almost every point of $X$ is a peak point for $\mathscr{R}(X)$.*

Here *almost every* is understood in the sense of Lebesgue measure on the plane.

PROOF

Let $B$ be the set of peak points for $\mathscr{R}(X)$, i.e., the minimal boundary for $\mathscr{R}(X)$, and suppose that $\mathscr{R}(X) \neq \mathscr{C}(X)$. We shall show that $X \setminus B$ has positive Lebesgue measure.

The hypothesis implies that there is a measure $\mu$ on $X$ which annihilates $\mathscr{R}(X)$ but not $\mathscr{C}(X)$. The Cauchy transform $\tilde{\mu}$ of $\mu$ vanishes on $\mathbb{C} \setminus X$ since $\mu \in \mathscr{R}(X)^\perp$, but it does not vanish almost everywhere, for, if so, then, by Lemma 25.6, $\mu$ would have to be the zero measure. Let $E$ be a measurable set of positive Lebesgue measure on which $\tilde{\mu}$ exists and is nonzero. Since $\mu$ is a finite measure, it assigns nonzero mass to at most countably many points, so by excluding from $E$ a suitable countable set, we may suppose that $\mu(\{x\}) = 0$ for every $x \in E$.

We shall prove that $E \subset X \setminus B$. It is clear that $E \subset X$, for $\tilde{\mu}$ vanishes on $\mathbb{C} \setminus X$ but at no point of $E$. Let $z_0 \in E$ and set $c = \tilde{\mu}(z_0)$ so that $c$ is a nonzero complex number. We need to know that if $\delta$ denotes the unit point mass at $z_0$, and if $\nu \in \mathscr{M}(X)$ is defined by

$$d\nu = (z - z_0)^{-1} \, d\mu - c \, d\delta,$$

then $\nu \in \mathscr{R}(X)^{\perp}$. To see this, it suffices to prove that if $f \in \mathscr{R}(X)$, then

$$\int \frac{f(z)}{z - z_0}\, d\mu(z) = cf(z_0), \tag{1}$$

and, by continuity, it is enough to establish this equality for functions $f$ which are rational and pole free on $X$. If $f$ is such a function, so is the function $g$ given by

$$g(z) = \frac{f(z) - f(z_0)}{z - z_0}.$$

We have

$$\int \frac{f(z)}{z - z_0}\, d\mu(z) = \int g(z)\, d\mu(z) + f(z_0) \int \frac{d\mu(z)}{z - z_0},$$

so, since $\mu \in \mathscr{R}(X)^{\perp}$, this last equality yields (1).

If $z_0 \in B$, there is $f \in \mathscr{R}(X)$ which peaks at $z_0$. We have then

$$0 = \int f^n\, d\nu.$$

As $n \to \infty$, the right side tends to $\nu(\{z_0\})$ and since $\mu$ assigns zero mass to $\{z_0\}$, $\nu(\{z_0\}) = -c \neq 0$, and we have a contradiction. It follows that $z_0 \in X \setminus B$, and thus $X \setminus B$ has positive measure, as we wished to prove.

Bishop's theorem contains as a very special case a theorem of Hartogs and Rosenthal [1].

**26.9** ***Corollary***

*If $X$ has zero Lebesgue measure, then $\mathscr{R}(X) = \mathscr{C}(X)$.*

As Helson and Quigley [2] pointed out, the theorem of Hartogs and Rosenthal implies another characterization of $\mathscr{C}(X)$:

**26.10** ***Corollary***

*If $(A, Y)$ is a uniform algebra, if $Y = \Sigma(A)$, and if $A$ contains a subset $\mathscr{F}$ which separates points on $Y$ and which consists of functions with range of zero Lebesgue measure, then $A = \mathscr{C}(Y)$.*

PROOF

If $y, y' \in Y$, there is $f \in \mathscr{F}$ with $f(y) \neq f(y')$. If $E = f(Y)$, then $\mathscr{R}(E) = \mathscr{C}(E)$, so there is a real-valued $g \in \mathscr{R}(E)$ with $g(f(y)) \neq g(f(y'))$. The uniform closure of $A$ implies that $g \circ f \in A$, so $A$ contains enough real-valued functions to separate points. Thus, $A = \mathscr{C}(Y)$ by the Stone–Weierstrass theorem.

As a corollary to part of the proof of Bishop's theorem, we have the following result, an extension to $\mathscr{R}(X)$ of a fact already established in the proof of Theorem 25.7 for $\mathscr{P}(X)$ if $\mathbb{C} \setminus X$ is connected.

**26.11** ***Theorem*** (Wilken [1, 4])
*If $\mu \in \mathscr{R}(X)^{\perp}$ is singular with respect to every representing measure for $\mathscr{R}(X)$, then $\mu = 0$; i.e., there are no completely singular measures in $\mathscr{R}(X)^{\perp}$.*

PROOF
If $\mu \in \mathscr{R}(X)^{\perp}$ is not the zero measure, let $z_0$ be a point of the set

$$E = \{z \in \mathbb{C} : \tilde{\mu}(z) \text{ exists and is nonzero}\}.$$

As in the proof of Theorem 26.8, $E$ has positive Lebesgue measure, so it is nonempty. If $z_0 \in E$ and $c = \tilde{\mu}(z_0)$, then as in (1),

$$c^{-1} \int \frac{f(z)}{z - z_0} \, d\mu(z) = f(z_0),$$

so if $d\nu = c^{-1} \, d\mu/(z - z_0)$, then $\nu$ is a complex representing measure for $z_0$. By Theorem 11.2, there is a (positive) representing measure $\sigma$ for $z_0$ absolutely continuous with respect to $\nu$. The measures $\mu$ and $\sigma$ are not mutually singular.

We next take up the study of the Gleason parts for $\mathscr{R}(X)$. The parts for $\mathscr{R}(X)$ exhibit much more pathology than do the parts for $\mathscr{P}(X)$. We shall illustrate some of this pathology by examples after having established certain affirmative results. The first result in this direction that we shall consider is a theorem of Browder [4] which is an improvement of an earlier theorem of Wilken [1]. If $z, z' \in X$, let us understand by $\|z - z'\|$ the norm of the functional $f \mapsto f(z) - f(z')$ on $\mathscr{R}(X)$.

**26.12** ***Theorem***
*If $x \in X$, if $\varepsilon > 0$, if*

$$\Pi_\varepsilon(x) = \{z \in X : \|z - x\| < \varepsilon\},$$

*and if $x$ is not a peak point for $\mathscr{R}(X)$, then $x$ is a point of metric density for $\Pi_\varepsilon(x)$.*

Explicitly, the conclusion of the theorem is that if $m$ denotes Lebesgue measure on the plane and if $D(x, \delta) = \{z \in \mathbb{C} : |z - x| < \delta\}$, then

$$\lim_{\delta \to 0} \frac{m(\Pi_\varepsilon(x) \cap D(x, \delta))}{m(D(x, \delta))} = 1. \tag{2}$$

Since $\Pi_2(x)$ is the Gleason part for $\mathscr{R}(X)$ which contains $x$, we have the following fact.

**26.13 *Corollary***

*If $\Pi$ is a nontrivial Gleason part for $\mathscr{R}(X)$, every point of $\Pi$ is a point of metric density for $\Pi$. In particular, every nontrivial Gleason part for $\mathscr{R}(X)$ has positive Lebesgue measure.*

In his paper [1], Wilken proved that nontrivial parts for $\mathscr{R}(X)$ have positive Lebesgue measure and he observed the following consequences of this fact.

**26.14 *Corollary***

*A point of $X$ is a peak point for $\mathscr{R}(X)$ if and only if it is a Gleason part for $\mathscr{R}(X)$.*

PROOF

Clearly peak points are point parts, and the theorem implies that point parts are peak points.

Let us recall that the phenomenon of this last corollary is not universal: There are uniform algebras which have one-point parts $\{x\}$ for which $x$ is not a peak point. The algebra $\mathfrak{A}_\alpha$ of Theorem 18.1 provides an instance.

Another corollary of the theorem is the fact that a version of the peak point conjecture is true for $\mathscr{R}(X)$:

**26.15 *Corollary***

*If every point of $X$ is a Gleason part for $\mathscr{R}(X)$, then $\mathscr{R}(X) = \mathscr{C}(X)$.*

Again this is not a universal phenomenon, as the examples of Cole, Section 19, show.

We need to establish two lemmas before giving the proof of Theorem 26.12.

26.16 ***Lemma***

*Let $\mu$ be a finite measure on $X$, and let*

(3) $$M(z) = \int_X \frac{d|\mu|(w)}{|w - z|}.$$

*If $D(x, \delta)$ is as in (2), then*

$$\lim_{\delta\to 0} \frac{1}{m(D(x, \delta))} \iint_{D(x,\delta)} |z - x| M(z)\, dm(z) = |\mu(\{x\})|.$$

PROOF

If we set

$$F_\delta(w) = \frac{1}{m(D(x, \delta))} \iint_{D(x,\delta)} |(z - x)(z - w)^{-1}|\, dm(z),$$

we have the estimate

$$|F_\delta(w)| < \frac{1}{\pi\delta} \iint_{D(x,\delta)} |z - w|^{-1}\, dm(z)$$
$$\leq 2.$$

If $w \neq x$, then $\lim_{\delta\to 0} F_\delta(w) = 0$, for if $|x - w| > \delta$, then

$$|F_\delta(w)| \leq \frac{\delta}{|w - x| - \delta}$$

and the term on the right goes to zero as $\delta \to 0$. Since $F_\delta(x) = 1$, the functions $F_\delta$ converge boundedly and pointwise to the characteristic function of the set $\{x\}$. By Fubini's theorem

$$\frac{1}{m(D(x, \delta))} \iint_{D(x,\delta)} |z - x| M(z)\, dm(z) = \int_X F_\delta(w)\, d|\mu|(w),$$

so the desired result follows by letting $\delta \to 0$ and invoking the dominated convergence theorem.

26.17 ***Lemma***

*Let $x \in X$, let $\mu$ be a complex representing measure for $x$ [for the algebra $\mathscr{R}(X)$], and define $M$ as in the last lemma. If $\varepsilon > 0$, then $\|z - x\| < \varepsilon$ whenever $z \in X$ satisfies $|z - x| M(z) < \varepsilon(\|\mu\| + 1 + \varepsilon)^{-1}$.*

PROOF

Let $\eta = \varepsilon(\|\mu\| + 1 + \varepsilon)^{-1}$, and let

$$c = \int (w - x)(w - z)^{-1}\, d\mu(w)$$

$$= \int \{(w - z)(w - z)^{-1} + (z - x)(w - z)^{-1}\}\, d\mu(w)$$

$$= 1 + (z - x)\tilde{\mu}(z),$$

$\tilde{\mu}$ the Cauchy transform of $\mu$. Since $|\tilde{\mu}(z)| \leq M(z)$, it follows that if $|z - x|M(z) < \eta$, then

$$|c| > 1 - \eta.$$

Define a measure $\nu$ by

$$d\nu(w) = c^{-1}(w - x)(w - z)^{-1}\, d\mu(w).$$

We have that if $f$ is a rational function without poles on $X$, then

$$\int f\, d\nu = c^{-1} \int \frac{f(w) - f(z)}{w - z}(w - x)\, d\mu(w)$$

$$+ c^{-1}f(z) \int (w - z)^{-1}(w - x)\, d\mu(w)$$

$$= f(z)$$

since $\mu$ is a representing measure for $x$. Thus, $\nu$ is a complex representing measure for $z$. The measure $\nu - \mu$ represents the functional

$$f \mapsto f(z) - f(x)$$

on $\mathscr{R}(X)$, so

$$\|z - x\| \leq \|\nu - \mu\|.$$

We can estimate $\|\nu - \mu\|$, under the assumption that $|z - x|M(z) < \eta$, as follows:

$$\|\nu - \mu\| = \int |c^{-1}(w - x)(w - z)^{-1} - 1|\, d|\mu|(w)$$

$$= |c|^{-1} \int \left| \frac{(w - x) - (w - z)[1 + (z - x)\tilde{\mu}(z)]}{w - z} \right| d|\mu|(w)$$

$$= |c|^{-1}|z - x| \int \left| \frac{1}{w - z} - \tilde{\mu}(z) \right| d|\mu|(w)$$

$$\leq |c|^{-1}|z - x|(M(z) + |\tilde{\mu}(z)|\,\|\mu\|)$$

$$\leq \frac{\eta}{1 - \eta}(1 + \|\mu\|).$$

The definition of $\eta$ implies that the last quantity is $\varepsilon$. This proves the lemma.

26.18 ***Proof of Theorem 26.12***

If $x \in X$ is not a peak point for $\mathscr{R}(X)$, there exists a representing measure $\mu$ for $x$ with $\mu(\{x\}) = 0$. Let $M$ be the function associated with $\mu$ as in (3), and set $\eta = \varepsilon(\|\mu\| + 1 + \varepsilon)^{-1}$. We have

$$\begin{aligned} 1 &\geq \frac{m(\Pi_\varepsilon(x) \cap D(x, \delta))}{m(D(x, \delta))} \\ &= 1 - \frac{m(D(x, \delta) \setminus \Pi_\varepsilon(x))}{m(D(x, \delta))} \\ &= 1 - \frac{1}{m(D(x, \delta))} \iint_{D(x,\delta)} \chi(w)\, dm(w), \end{aligned} \tag{4}$$

where $\chi$ denotes the characteristic function of $\mathbb{C} \setminus \Pi_\varepsilon(x)$. By the last lemma, the inequality

$$\chi(z) \leq \frac{|z - x| M(z)}{\eta}$$

holds for $z \in X$. It also holds off $X$, for if $z \notin X$, then the function $f$ given by $f(w) = 1/(w - z)$ is in $\mathscr{R}(X)$, and since $\mu$ is a representing measure for $x$, we have the estimate

$$\begin{aligned} M(z) &\geq \left| \int \frac{1}{w - z}\, d\mu \right| \\ &= \left| \frac{1}{x - z} \right|. \end{aligned}$$

This inequality, together with the definition of $\eta$, shows that for $z \notin X$

$$\frac{|z - x| M(z)}{\eta} > 1.$$

It follows that

$$\frac{1}{m(D(x, \delta))} \iint_{D(x,\delta)} \chi(w)\, dm(w) \leq \frac{1}{\eta m(D(x, \delta))} \iint_{D(x,\delta)} |z - x| M(z)\, dm(z).$$

By Lemma 26.16, the quantity on the right approaches $(1/\eta)\mu(\{x\}) = 0$ as $\delta \to 0$, and the theorem follows from (4).

It is worth observing that Theorem 26.12 implies a metric criterion for a point $x \in X$ to be a peak point for $\mathscr{R}(X)$: If $x$ is not a point of metric density for $X$, then $x$ is a peak point for $\mathscr{R}(X)$. This is a rather weak condition since, e.g., it does not suffice to show that if $\gamma$ is a simple closed curve contained in $\partial X$ which constitutes the boundary of a component of $\mathbb{C} \setminus X$, then every point of $\gamma$ is a peak point, though this fact can be established easily by other means. The theorem of Melnikov is a much more refined metric result of this kind, for it characterizes in terms of analytic capacity the peak points for $\mathscr{R}(X)$.

Our next result and its corollaries are also due to Wilken [2].

**26.19** ***Theorem***

*If $\mu$ is a representing measure for the point $x \in X$ for the algebra $\mathscr{R}(X)$, if $S$ is the closed support of $\mu$, and if $\Pi$ is the Gleason part for $\mathscr{R}(X)$ which contains $x$, then $S \subset \overline{\Pi}$ and $\mu$ represents $x$ for the algebra $\mathscr{R}(\overline{\Pi})$.*

PROOF

Define a measure $\nu$ by $d\nu(z) = (z - x)\, d\mu(z)$. The measure annihilates $\mathscr{R}(X)$, and if

$$E = \{z \in \mathbb{C} : \tilde{\nu}(z) \text{ exists and is nonzero}\},$$

then $x \in E$. Since $\nu \in \mathscr{R}(X)^{\perp}$, $E \subset X$ by Lemma 26.2, and as in the proof of Theorem 26.11, each point $z_0 \in E$ has a representing measure which is not singular with respect to $\mu$. This implies that $E \subset \Pi$. The argument used in the proof of Lemma 25.6 shows that the support of a measure is contained within the complement of any open set on which the Cauchy transform of the measure vanishes almost everywhere, so it follows that the measure $\nu$ is supported in $\bar{E}$. Thus, $S \subset \overline{\Pi}$, as asserted. As $\tilde{\nu}$ vanishes off, $\nu \in \mathscr{R}(\overline{\Pi})^{\perp}$, whence the measure $\mu$ is a representing measure for $x$ for the algebra $\mathscr{R}(\overline{\Pi})$: If $f$ is a rational function without poles on $\overline{\Pi}$, then

$$\begin{aligned} \int f\, d\mu &= \int (f(z) - f(x))\, d\mu(z) + \int f(x)\, d\mu(z) \\ &= f(x) + \int \frac{f(z) - f(x)}{z - x}\, d\nu(x) \\ &= f(x). \end{aligned}$$

For our first corollary of this theorem, we need a simple observation which enables us to recognize that certain points are peak points for $\mathscr{R}(X)$.

**26.20** ***Lemma***

*If $\Omega$ is a component of $\mathbb{C} \setminus X$, every point of $\overline{\Omega} \setminus \Omega$ is a peak point for $\mathscr{R}(X)$.*

PROOF

Let $z_0$ be a point of $\Omega$, and let $Y$ be the complement in the Riemann sphere of $\Omega$. It follows from Theorem 25.5 that the uniform closure $B$ on $Y$ of polynomials in $z$ (if $\Omega$ is unbounded), or $1/(z - z_0)$ (if $\Omega$ is bounded), is a Dirichlet algebra, so every point of $\partial Y$ is a peak point for $B$. This implies that each point of $\partial\Omega = \partial Y$ is a peak point for $\mathscr{R}(X)$.

**26.21** ***Corollary***

*If $W$ is a connected relatively open subset of $X$ which lies in a part $\Pi$ for $\mathscr{R}(X)$, then each point in the topological boundary of $W$ lies in $\Pi$ or else is a peak point for $\mathscr{R}(X)$.*

PROOF

Let $x \in \partial W$ and let $\Pi_0$ be the part that contains $x$. If $\Pi_0 \neq \Pi$ and if $\mu$ represents $x$ for $\mathscr{R}(X)$, then $\mu$ is supported on $\bar{\Pi}_0$ and this is contained in $X \setminus W$. As $x$ lies in the boundary of a component of $\mathbf{C} \setminus \bar{\Pi}_0$, the preceding lemma implies that $x$ is a peak point for $\mathscr{R}(\bar{\Pi}_0)$. By the theorem, $\mu$ represents $x$ for $\mathscr{R}(\bar{\Pi}_0)$, so $\mu$ is a point mass. This statement is valid for every representing measure $\mu$ for $x$ for the algebra $\mathscr{R}(X)$, so $x$ is a peak point for $\mathscr{R}(X)$, as we wished to show.

**26.22** ***Corollary***

*If $\Pi$ is a Gleason part for $\mathscr{R}(X)$, $\bar{\Pi}$ is connected.*

PROOF

Assume that $\bar{\Pi} = E_0 \cup E_1$, this a disjoint union of closed, relatively open sets. Let $x_0 \in E_0 \cap \Pi$, $x_1 \in E_1 \cap \Pi$, and let $\mu_0$ and $\mu_1$ be mutually absolutely continuous representing measures for $x_0$ and $x_1$, respectively, for the algebra $\mathscr{R}(X)$. By the theorem, $\mu_0$ and $\mu_1$ represent $x_0$ and $x_1$ for $\mathscr{R}(\bar{\Pi})$. The choice of $E_0$ and $E_1$ implies that no Gleason part for $\mathscr{R}(\bar{\Pi})$ can meet both $E_0$ and $E_1$. However, since $\mu_0$ and $\mu_1$ are not mutually singular, $x_0$ and $x_1$ must lie in the same part for $\mathscr{R}(\bar{\Pi})$. This contradiction implies the corollary.

**26.23** ***Remark***

Needless to say, the preceding corollary is not the result we would like to prove. We would prefer to show that $\Pi$ itself is connected. However, examples are known which demonstrate the impossibility of such a result: There are Gleason parts for $\mathscr{R}(X)$, $X$ suitably chosen, which are not connected. Such an example has been given by Davie [1].

Our next two examples illustrate the pathology which can occur in the theory of the Gleason parts for $\mathscr{R}(X)$.

**26.24** ***Example***

For $n = 2, 3, \ldots$, and, $n$ being fixed, for $j = -2^n + 1, \ldots, 2^n - 1$, let $\gamma_{j,n}$ be the circle of radius $8^{-n}$ centered at the point $j2^{-n} + 2^{-(n+1)}i$, and let $D_{j,n}$ be the interior of $\gamma_{j,n}$. The discs $D_{j,n}$ all have disjoint closure, and these closures all lie in the interior $\Delta$ of the unit circle. We let

$$X = \bar{\Delta} \setminus \bigcup D_{j,n},$$

and we shall show that for the algebra $\mathscr{R}(X)$ there is one nontrivial Gleason part, $\Delta \setminus \bigcup \bar{D}_{j,n}$. If we denote by $S_+$ and $S_-$ the portions of $\Delta \setminus \bigcup \bar{D}_{j,n}$ lying, respectively, above and below the $x$-axis, then it is clear that $S_+$ is contained in a single Gleason part for $\mathscr{R}(X)$ as is $S_-$; but it is not clear at the outset that they are contained in the same Gleason part. These two sets are separated by the real diameter of the unit circle each point of which is a limit point of $\bigcup \gamma_{j,n}$. We shall show that every point of $(-1, 1)$ lies in the same Gleason part as $S_+$ and as $S_-$, a fact which implies that $S_+$ and $S_-$ both lie in the same Gleason part.

If $-1 < x < 1$, define a measure $m_x$ on $X$ by

$$\int f\, dm_x = \frac{1}{2\pi i} \int_{|\zeta| = 1} \frac{f(\zeta)}{\zeta - x}\, d\zeta - \sum_{j,n} \frac{1}{2\pi i} \int_{\gamma_{j,n}} \frac{f(\zeta)}{\zeta - x}\, d\zeta.$$

The measure $m_x$ defined in this way is a finite measure, for given $n$ and $j$,

$$\frac{1}{2\pi} \int_{\gamma_{j,n}} \frac{|d\zeta|}{|\zeta - x|} < 8^{-n}(2^{-(n+1)} - 8^{-n})^{-1}.$$

For fixed $n$, there are $2^{n+1} - 1$ corresponding terms, so it follows that

$$\sum_{j,n} \frac{1}{2\pi} \int_{\gamma_{j,n}} \frac{|d\zeta|}{|\zeta - x|} < \infty,$$

and therefore $m_x$ is a finite measure. By the Cauchy integral theorems, we see that if $f$ is a rational function with no poles on $X$, then

$$\int f\, dm_x = f(x).$$

Since $m_x$ is a complex representing measure for the point $x$ on the algebra $\mathscr{R}(X)$, it follows that $x$ is not a peak point for $\mathscr{R}(X)$.

There are several ways to conclude our discussion of this example. Perhaps the quickest, if not the most elementary, is to note that if

$\Pi_x$ is the Gleason part containing $x$, then since $x$ is a point of metric density for $\Pi_x$, $\Pi_x$ must meet both $S_+$ and $S_-$. Thus, it follows that $S_+ \cup S_- \cup (-1, 1)$ is the only nontrivial part for $\mathscr{R}(X)$.

26.25 ***Example***

Let $C_1$ and $C_2$ be rectifiable simple closed curves in $C$ which bound domains $\Omega_1$ and $\Omega_2$, respectively. Assume that $\Omega_1 \cap \Omega_2 = \varnothing$, and put $E = C_1 \cap C_2$. We are interested only in the case that $E \neq \varnothing$. Set $X = \bar{\Omega}_1 \cup \bar{\Omega}_2$, and consider the algebra $\mathscr{R}(X)$. It is clear that $\Omega_1$ lies in a single part for $\mathscr{R}(X)$, as does $\Omega_2$. We shall show that $\Omega_1$ and $\Omega_2$ lie in the same Gleason part for $\mathscr{R}(X)$ if the set $E$ has positive length.

For this, we must begin by explicating the notion of length with which we work. There are several equivalent possibilities. A very direct approach is possible through the notion of arc length, for since both $C_1$ and $C_2$ are rectifiable, there are measures $s_1$ and $s_2$ induced on $C_1$ and $C_2$, respectively, by arc length. We agree that $E$ has zero length if $s_1(E) = 0 = s_2(E)$, and that otherwise it has positive length.

Let $g_j(\cdot, z)$ be the Green's function for the domain $\Omega_j$ with singularity at $z$. The rectifiability of $C_j$ implies that at almost every (with respect to the measure $s_j$) point of $C_j$ the inner normal exists, and it is known that the inner normal derivative of $g_j(\cdot, z)$ exists and is strictly positive almost everywhere. Moreover, if $u$ is continuous on $\bar{\Omega}_j$ and harmonic throughout $\Omega_j$, then

$$u(z) = \frac{1}{2\pi} \int_{C_j} u(\zeta) \frac{\partial g_j(\zeta, z)}{\partial n} \, ds_j(\zeta).$$

With more stringent regularity conditions on $C_j$, this representation is classical. For the formula in the present degree of generality see Appendix D and the literature cited there.

This representation applies in particular to the case that $f \in \mathscr{R}(X)$, so if $z_1 \in \Omega_1$, $z_2 \in \Omega_2$, and if $E$ has positive length, then we have representing measures for $z_1$ and $z_2$ which are not mutually singular. It follows that $\Omega_1$ and $\Omega_2$ lie in the same Gleason part for $\mathscr{R}(X)$.

As a special case of this example, we consider a construction in the unit disc. Let $E \subset [-1, 1]$ be a closed set, and let $X_E$ be the set obtained from the closed unit disc by removing all the open discs with diameter one of the intervals constituting the set $[-1, 1] \setminus E$. If $E$ is nowhere dense, then the interior of $X_E$ consists of the union of two disjoint topological discs $\Omega_+$ and $\Omega_-$, the former lying above the real axis, the latter below.

It follows from what we have just said that if $E$ has positive length, then $\Omega_+$ and $\Omega_-$ lie in the same Gleason part for $\mathscr{R}(X)$. It can be shown that in case $E$ has zero length they lie in different parts and that each point of $E$ is a peak point.

In case $E$ has positive length, the problem of describing the part containing the points of $E$ remains. It follows from Lemma 26.20 that every point of $E$ which is an end point of one of the intervals constituting $[-1, 1] \setminus E$ is a peak point for $\mathscr{R}(X_E)$, but in general not every point of $E$ is a peak point, as is shown in Wilken's paper [3]. The example of Davie alluded to above is an example of this kind in which $E$ has positive length but every point of $E$ is a peak point for $\mathscr{R}(X_E)$. It follows then that $\Omega_+ \cup \Omega_-$ is a disconnected Gleason part for $\mathscr{R}(X_E)$.

A somewhat different treatment of this general example was given by Valskii [1, 2], and some related remarks are in the paper of Hoffman [4].

It is not clear what sort of situation obtains if we consider the case of general $\Omega_1$ and $\Omega_2$ bounded by not necessarily rectifiable curves.

We now turn from the consideration of general compact sets $X$ to the case that $X$ divides the plane into only finitely many disjoint pieces. We know from Theorem 26.5 that in this case $\mathscr{R}(X) = \mathscr{A}(X)$. Among the results we shall obtain are a characterization of the Gleason parts for $\mathscr{R}(X)$, a maximality theorem, and a theorem on the pointwise approximation of functions bounded and holomorphic on the interior of $X$ by bounded sequences of rational functions. The theory is due primarily to Ahern and Sarason [1, 2]. The first step in this development is the solution of certain Dirichlet problems.

**26.26** ***Theorem***

*If the bounded components of* $\mathbb{C} \setminus X$ *are* $\Omega_1, \ldots, \Omega_M$, *and if* $\zeta_j$ *is a point of* $\Omega_j$, *then the real linear span of* $\operatorname{Re} \mathscr{R}(X)$ *and the functions* $\log|z - \zeta_j|$, $1 \le j \le M$, *is dense in* $\mathscr{C}_{\mathbb{R}}(\partial X)$.

PROOF

Let $\alpha$ be a real measure on $\partial X$ which is orthogonal to $\operatorname{Re} \mathscr{R}(X)$ and to $\log|z - \zeta_j|$, $j = 1, \ldots, M$. Define $u$ on $\mathbb{C} \setminus X$ by

$$u(z) = \int \log|\zeta - z| \, d\alpha(\zeta).$$

If we use $\Omega_0$ to denote the unbounded component of $\mathbb{C} \setminus X$, then, as in the proof of Theorem 25.5, $u$ vanishes identically in $\Omega_0$. Also, if $|z - \zeta_j|$ is small enough, then

$$0 = \sum_{n=1}^{\infty} \frac{(z - \zeta_j)^n}{n} \int (\zeta - \zeta_j)^{-n} \, d\alpha(\zeta),$$

and computing as in the proof of Theorem 25.5, this leads to the fact that the function $u$ vanishes on a neighborhood of $\zeta_j$. The harmonicity of $u$ on $\Omega_j$ implies that $u$ vanishes identically on $\Omega_j$, and thus we have that $u$ vanishes on $\mathbb{C} \setminus X$. If we use $\Omega$ to denote one of the components of $\mathbb{C} \setminus X$, then we can apply Lemma 25.3 (extended in an obvious way) to the set $Y = \mathbb{C} \setminus \Omega$ to discover that the function $u$ exists at almost every boundary point of $\Omega$ and that it vanishes wherever it exists. As $\mathbb{C} \setminus X$ has only finitely many components, it is now clear that $u$ vanishes at almost every point of $\partial X$. The proof of our theorem can be completed now, as was the proof of Theorem 25.5.

**26.27** ***Corollary***

*If $\mathbb{C} \setminus X$ has finitely many components, then for every $\varphi \in \mathscr{C}(\partial X)$, there is a function $f \in \mathscr{C}(X)$ which agrees with $\varphi$ on $\partial X$ and which is harmonic in the interior of $X$.*

If $z$ is a point of $X$, we have the harmonic measure $\lambda_z$ on $\partial X$ given by

$$\int \varphi \, d\lambda_z = f(z),$$

where $f$ and $\varphi$ are related as in the corollary. These harmonic measures play an important role in the theory of $\mathscr{R}(X)$, in part because $\lambda_z$ is the unique Arens–Singer measure for the point $z$. To see this, note first that if $f \in \mathscr{R}(X)$ is invertible, then $\log|f|$ is in $\mathscr{C}(X)$ and is harmonic in the interior of $X$, so

$$\log|f(z)| = \int \log|f| \, d\lambda_z,$$

and this shows that $\lambda_z$ is an Arens–Singer measure. On the other hand, if $\mu \in \mathscr{M}(\partial X)$ satisfies

$$\log|f(z)| = \int \log|f| \, d\mu$$

for all invertible $f \in \mathscr{R}(X)$, then it follows that if $f \in \mathscr{R}(X)$,

$$\operatorname{Re} f(z) = \int \operatorname{Re} f \, d\mu$$

and also that

$$\log|z - \zeta_j| = \int \log|\zeta - \zeta_j| \, d\mu(\zeta), \qquad j = 1, 2, \ldots, M.$$

By the theorem proved above, it follows that $\mu$ is the harmonic measure $\lambda_z$.

It is important to note at this point that if $\mathbb{C} \setminus X$ has only finitely many components, then $\partial X$ is the minimal boundary for $\mathscr{R}(X)$: Under the hypothesis that $\mathbb{C} \setminus X$ has finitely many components, every point of $\partial X$ lies in the boundary for some component of $\mathbb{C} \setminus X$, so Lemma 26.20 applies. This observation implies that if $z$ is an interior point of $X$, then the harmonic measure $\lambda_z$ has no atoms.

Another important property of the harmonic measures $\lambda_z$ is that if $z$ and $w$ lie in distinct components of the interior of $X$, then $\lambda_z$ and $\lambda_w$ are mutually singular. The proof of this fact is based on an elementary topological observation and a function theoretic lemma.

**26.28** ***Definition***

*If $G$ is a domain in the plane, a point $z_0 \in \partial G$ is* ***arcwise accessible from*** $G$ *if there exists in $G \cup \{z_0\}$ an arc with $z_0$ as one of its end points.*

**26.29** ***Lemma***

*If $\mathbb{C} \setminus X$ has only finitely many components and if $\Omega_1$ and $\Omega_2$ are components of the interior of $X$, then there are at most finitely many points which are arcwise accessible from both $\Omega_1$ and $\Omega_2$.*

PROOF

Suppose that there exist $N$ points, $p_1, \ldots, p_N$, which are arcwise accessible from both $\Omega_1$ and $\Omega_2$. Let $q_1 \in \Omega_1$, $q_2 \in \Omega_2$, and let $\lambda_{1,j}$ be an arc in $\Omega_1 \cup \{p_j\}$ which has $q_1$ and $p_j$ as its end points, and let $\lambda_{2,j}$ be one in $\Omega_2 \cup \{p_j\}$ with end points $q_2$ and $p_j$. We may choose the $\lambda_{1,j}$ to be pairwise disjoint except for their common end point and we can choose $\lambda_{2,j}$ in a similar way. The set

$$\mathbb{C} \setminus (\lambda_{1,1} \cup \cdots \cup \lambda_{1,N} \cup \lambda_{2,1} \cup \cdots \cup \lambda_{2,N})$$

has $N$ components, say $S_1, \ldots, S_N$, and each of these components meets both $\Omega_1$ and $\Omega_2$. Thus, each $S_j$ must contain a point of $\partial X$, and as $\partial S_j \subset X$, it follows that each $S_j$ must contain a component of $\mathbb{C} \setminus X$. By hypothesis, $\mathbb{C} \setminus X$ has but finitely many components, so the lemma is proved.

We probably should mention explicitly the fact that the lemma does not imply that $\bar{\Omega}_1 \cap \bar{\Omega}_2$ is finite; simple examples show that $\bar{\Omega}_1 \cap \bar{\Omega}_2$ may be an arc.

**26.30** ***Lemma***

*If $\mathbb{C} \setminus X$ has only finitely many components and if $G$ is a component of the interior of $X$, then there is a measurable set $E \subset \partial G$ which consists exclusively of points which are arcwise accessible from $G$ and which has the further property that $\lambda_z(E) = 1$ for all $z \in G$.*

PROOF

To begin with, we have to resolve certain topological difficulties. We need the facts that $\mathbb{C} \setminus G$ has only finitely many components and that each component of $\mathbb{C}^* \setminus G$, $\mathbb{C}^*$ the Riemann sphere, is simply connected. These facts seem clear, but a proper proof of them requires care. If we denote by $U_1, \ldots, U_s$ the components of $\mathbb{C}^* \setminus X$, then $\partial G \subset \partial X \subset \partial U_1 \cup \cdots \cup \partial U_s$. Let $K$ be a component of $\mathbb{C}^* \setminus G$. If $x \in \partial K$, then $x \in \partial G$, and so $x \in \partial U_j$ for some $j$. Hence, the connected set $\bar{U}_j \cup K$ lies in $\mathbb{C}^* \setminus G$, and as $K$ is a component of $\mathbb{C}^* \setminus G$, $K \supset \bar{U}_j$. As distinct components of $\mathbb{C}^* \setminus G$ are disjoint and each contains one of the finitely many $\bar{U}_j$, we see that $\mathbb{C}^* \setminus G$ has only finitely many components, say $K_1, \ldots, K_N$, and moreover, $N \leq s$. Next, we show that the sets $K_j$ do not separate $\mathbb{C}$. If, e.g., $\mathbb{C}^* \setminus K_1 = C_1 \cup C_2$, $C_1$ and $C_2$ disjoint open, relatively closed sets, then one of them, say $C_2$, contains $G$. If $C_1'$ is a component of $C_1$, then $\partial C_1' \subset \partial C_1 \subset K_1$, so $K_1 \cup \bar{C}_1'$ is a connected set contained in $\mathbb{C}^* \setminus G$ and properly containing $K_1$. This contradicts the fact that $K_1$ is a component of $\mathbb{C}^* \setminus G$, so $\mathbb{C}^* \setminus K_1$ is connected. It now follows that the sets $\mathbb{C}^* \setminus K_j$ are all simply connected. (For a detailed, elementary discussion of simple connectivity which suffices in the present situation, see Rudin [2, Theorem 13.18].) Next we note that $\partial G = \partial K_1 \cup \cdots \cup \partial K_N$; this a disjoint union. Moreover, each of the sets $\partial K_j$ is connected. This final point requires substantially more effort than the preceding arguments, and we therefore content ourselves with a reference to Newman [1, Theorem V.14.4] for the necessary argument. (This reference and the preceding topological details were suggested by Sidney.)

We now find a domain $W$ bounded by $N$ analytic simple closed curves which is conformally equivalent to $G$. The existence of such a domain $W$ follows from the Riemann mapping theorem. Since $\mathbb{C}^* \setminus K_1$ is simply connected, there is a conformal homeomorphism $\Phi_1$ from $\mathbb{C}^* \setminus K_1$ onto the unit disc. Under $\Phi_1$, $G$ goes onto a set $G^{(1)}$ with boundary consisting of the unit circle, $T$, and the sets $\Phi_1(\partial K_2), \ldots, \Phi_1(\partial K_N)$. There is a conformal map $\Phi_2$ of $\mathbb{C}^* \setminus \Phi_1(K_2)$ onto the unit disc, and under $\Phi_2 \circ \Phi_1$, $G$ goes onto a domain $G^{(2)}$ whose boundary consists of the unit circle, the analytic simple closed curve $\Phi_2(T)$, and the continua $\Phi_2(\Phi_1(\partial K_3)), \ldots, \Phi_2(\Phi_1(\partial K_N))$. Iterating this process, we arrive in $N$ steps at a domain $W = G^{(N)}$ with the desired properties. (It is a classical fact in the theory of conformal mapping that we could find a suitable domain $W$ bounded by a collection of circles, but for our purposes this refined result is unnecessary. It may be found, e.g., in Hurwitz and Courant [1].)

Let $\Phi: W \to G$ be a conformal homeomorphism. The function $\Phi$ is bounded and holomorphic, so it follows that for almost all points $z_0$

in $\partial W$, in the sense of the measure induced by arc length, the limit $\lim \Phi(z) = \Phi(z_0)$ exists if $z$ approaches $z_0$ nontangentially through $W$. We regard $\Phi$ as being extended to almost all of $\partial W$ by means of these limiting values and obtain in this way a measurable function. The limiting values $\Phi(z_0)$ all lie in $\partial G$. To see this, suppose that $z_0$ is some point in $\partial W$ and that $\{z_n\}_{n=1}^{\infty}$ is a sequence in $W$ which converges to $z_0$ and which has the property that $\alpha = \lim \Phi(z_n)$ exists. Plainly, $\alpha \in \bar{G}$; if $\alpha \in G$, then since $\Phi$ is a homeomorphism of $W$ onto $G$, there is a neighborhood $Q$ of $\alpha$ whose closure in $G$ is compact and which is the image of a relative compact neighborhood $Q'$ of some point $z_1 \in G$. But then for large $n$, $z_n \notin Q'$, although $\Phi(z_n) \in Q$, contradicting the fact that $\Phi$ is one to one.

By Lusin's theorem, there is an increasing sequence $\{F_n\}$ of compact sets in $\partial W$ such that $\Phi$ has nontangential limit at every point of $F_n$, such that $\Phi$ is continuous on $F_n$, and such that $\partial W \setminus \bigcup F_n$ is a null set. The set $E = \bigcup \Phi(F_n)$ is a measurable set in $\partial G$ which consists entirely of points arcwise accessible from $G$, and we shall prove that if $z \in G$, then $\lambda_z(E) = 1$.

Let $K$ be a compact set in $\partial G \setminus E$. By Corollary 26.27, we can solve the Dirichlet problem on $X$; so it follows that for every $n$, there is a nonnegative function $u_n \in \mathscr{C}(X)$ which is harmonic on the interior of $X$, which is 1 on $K$ and zero on $\Phi(F_n)$, and which is bounded by 1. If $z \in G$, then $\lambda_z(K) \leq \limsup u_n(z)$. Let $z$ be the image of the point $w$ under $\Phi$, and define $U_n$ to be the function $u_n \circ \Phi$. As $U_n$ is a bounded harmonic function on $W$, it has nontangential limits at almost every point of $\partial W$, and (Appendix D) it admits the representation

$$u_n(z) = U_n(w) = \frac{1}{2\pi}\int_{\partial W} \frac{\partial g(\zeta, w)}{\partial n} U_n(\zeta)\, ds,$$

where $g(\cdot, w)$ is the Green's function for the domain $G$ with pole at $w$, where $\partial/\partial n$ denotes differentiation along the inner normal, and where $ds$ denotes the differential of arc length. As the functions $U_n$ are all bounded by 1 and tend to zero at almost every point of $\partial W$, it follows that $\lim U_n(w) = 0$ so that $\lim u_n(z) = 0$, whence $\lambda_z(K) = 0$. Regularity now implies that $\lambda_z(E) = 1$, as was to be proved.

We now have at our disposal enough information to prove the following fact.

26.31 ***Theorem***

*If $\mathbb{C} \setminus X$ has finitely many components, then the harmonic measures for points in distinct components of the interior of $X$ are mutually singular.*

PROOF

Let $G$ and $G'$ be distinct components of the interior of $X$, let $z \in G$, $z' \in G'$, and let $\lambda_z$ and $\lambda_{z'}$ be the associated harmonic measures. The last lemma applies to $G$ and $G'$ to yield Borel sets $E$ and $E'$ in $\partial G$ and $\partial G'$, respectively, such that $\lambda_z(E) = 1 = \lambda_{z'}(E')$, and such that $E$ (respectively $E'$) consists exclusively of points arcwise accessible from $G$ (respectively $G'$), and the next-to-last lemma now implies that $E \cap E'$ is finite. The measures $\lambda_z$ and $\lambda_{z'}$ have no atoms, so it follows that $\lambda_z(E \cap E') = \lambda_{z'}(E \cap E') = 0$, and the theorem is proved.

The next step in our study of the harmonic measures $\lambda_z$ is the proof that they are enveloped (recall Definition 23.12) and hence dominant representing measures. More is true: They are actually strongly dominant, but we shall not establish this point until later.

26.32 ***Theorem*** (Ahern and Sarason [1])

*If $\mathbb{C} \setminus X$ has only finitely many components and if $z$ is an interior point of $X$, then the harmonic measure $\lambda_z$ is an enveloped representing measure for $z$.*

PROOF

We shall write $\mathscr{M}_z$ for the set of representing measures for $\mathscr{R}(X)$ for the point $z$ which are supported on $\partial X$, and we let $S$ be the real linear span of the measures $\mu - \nu$, $\mu, \nu \in \mathscr{M}_z$. Since $\operatorname{Re} \mathscr{R}(X)$ has finite codimension in $\mathscr{C}(\partial X)$, $S$ is finite dimensional. Let $\dim S = s$. If $\sigma \in S$, and $f \in \mathscr{R}(X)$, then

$$\int \operatorname{Re} f \, d\sigma = 0,$$

so it follows from Theorem 26.26 that if $\sigma \in S$, then $\sigma = 0$ if

$$\int \log|z - \zeta_j| \, d\sigma(z) = 0$$

for $j = 1, \ldots, M$. (We understand by $\zeta_j$ the points considered in that earlier theorem.) Consequently, for some choice of $s$ of the points $\zeta_j$, say $\zeta_1, \ldots, \zeta_s$, it follows that $\sigma \in S$ and

$$\int \log|z - \zeta_j| \, d\sigma(z) = 0,$$

for $j = 1, \ldots, s$ imply that $\sigma = 0$. For the duration of the proof, we shall use $Z_j$ to denote the function $z - \zeta_j$, $1 \leq j \leq s$. These functions are invertible in $\mathscr{R}(X)$. There exist $s$ elements of $S$, say $\sigma_1, \ldots, \sigma_s$, such that

$$\int \log|Z_j| \, d\sigma_k = \delta_{jk},$$

$\delta_{j,k}$ the Kronecker delta.

Denote by $\mathbb{R}^s$ the space of $s$-tuples of real numbers, and if $(\beta_1, \ldots, \beta_s) = \beta \in \mathbb{R}^s$, let $|\beta|$ denote $\max(|\beta_1|, \ldots, |\beta_s|)$. By $\mathbb{Z}^s$ we denote the collection of elements of $\mathbb{R}^s$ with integral entries.

Recalling the definition of enveloped measure, we are to prove that if $\{u_n\}$ is a nonnegative sequence in $\mathscr{C}_{\mathbb{R}}(X)$ such that $\int u_n \, d\lambda_z \to 0$, there is a subsequence $\{u_{n_k}\}$ and a sequence $\{f_k\}$ in $\mathscr{R}(X)$ with

$$|f_k| \leq \exp(-u_{n_k}) \qquad \text{and} \qquad \int f_k \, d\lambda_z \to 1.$$

For the sake of notational simplicity, we replace $\{u_n\}$ by a subsequence, again to be denoted by $\{u_n\}$, with the property that

$$\left| \int u_n \, d\lambda_z \right| < n^{-s-1},$$

and such that each of the sequences

$$\left\{ \int u_n \, d\sigma_j \right\}$$

converges, modulo the integers, to a point $\alpha_j$, the convergence so rapid that

$$\alpha_j - \int u_n \, d\sigma_j < n^{-s-1} \quad (\text{modulo } \mathbb{Z})$$

for all $n$ and each $j$.

By a theorem of Dirichlet in diophantine approximation (see Appendix E), there exists a sequence of positive integers $\{q_n\}$ with $q_n \leq n^s$ such that each $q_n \alpha_j$ lies within $1/n$ of an integer.

Define $w_n$ by $w_n = q_n u_n$, and let $\beta_n \in \mathbb{R}^s$ have entries $\beta_{n,j} = \int w_n \, d\sigma_j$, $j = 1, \ldots, s$. We have that $\int w_n \, d\lambda_z \to 0$ and that there is a sequence $\{\gamma_n\}$ in $\mathbb{Z}^s$ such that $|\beta_n - \gamma_n| \to 0$. Put $\varepsilon_n = \beta_n - \gamma_n$ and

$$c = \|\log|Z_1|\,\| + \cdots + \|\log|Z_s|\,\|.$$

By Lemma 7.19 there exist functions $g_n \in \mathscr{R}(X)$ such that

$$\operatorname{Re} g_n \geq w_n - \sum_{j=1}^{s} \beta_{n,j} \log|Z_j| + |\varepsilon_n| c$$

and

$$\operatorname{Re} g_n(z) \le \frac{1}{n} + \int w_n \, d\lambda_z - \sum_{j=1}^{s} \beta_{n,j} \log|Z_j(z)| + |\varepsilon_n| c$$

since

$$\int \log|Z_j| \, d\lambda_z = \log|Z_j(z)|.$$

If $\gamma_n = (\gamma_{n,1}, \ldots, \gamma_{n,s})$, let

$$f_n = e^{-g_n} Z_1^{-\gamma_{n,1}} \cdots Z_s^{-\gamma_{n,s}},$$

a certain invertible element of $\mathscr{R}(X)$. We have that

$$\begin{aligned}
\log|f_n| &= -\operatorname{Re} g_n - \sum_{j=1}^{s} \gamma_{n,j} \log|Z_j| \\
&\le -w_n + \sum_{j=1}^{s} (\beta_{n,j} - \gamma_{n,j}) \log|Z_j| - |\varepsilon_n| c \\
&\le -w_n \le -u_n,
\end{aligned}$$

and this yields $|f_n| \le e^{-u_n}$. Also,

$$\begin{aligned}
\int \log|f_n| \, d\lambda_z &= -\operatorname{Re} g_n(z) - \sum_{j=1}^{s} \gamma_{n,j} \log|Z_j(z)| \\
&\ge -\frac{1}{n} - \int w_n \, d\lambda_z + \sum_{j=1}^{n} (\beta_{n,j} - \gamma_{n,j}) \log|Z_j(z)| - |\varepsilon_n| c \\
&\ge -\frac{1}{n} - \int w_n \, d\lambda_z - 2|\varepsilon_n| c \\
&\to 0.
\end{aligned}$$

We know that $\lambda_z$ is an Arens–Singer measure, so since $f_n$ is invertible in $A$, it follows that $\log|\int f_n \, d\lambda_z| \to 0$, i.e., that $|\int f_n \, d\lambda_z| \to 1$. Replacing $f_n$ by $\alpha_n f_n$, $\alpha_n \in \mathbb{C}$, $|\alpha_n| = 1$, we see that we can assume that $\int f_n \, d\lambda_z \to 1$. This finishes the proof.

We can prove, as an immediate corollary of this theorem, the promised fact about the Gleason parts for $\mathscr{R}(X)$.

26.33 ***Theorem***

*If $\mathbb{C} \setminus X$ has finitely many components, the nontrivial Gleason parts for the algebra $\mathscr{R}(X)$ are precisely the components of the interior of $X$.*

PROOF
It is clear that each component of the interior of $\mathscr{R}(X)$ is contained in a Gleason part. On the other hand, if $C_1$ and $C_2$ are distinct components of the interior of $X$, if $z_j \in C_j$ and $\lambda_{z_j}$ is the harmonic measure for $z_j$, then $\lambda_{z_1}$ and $\lambda_{z_2}$ are mutually singular. If $\mu_1$ and $\mu_2$ are any representing measures for $z_1$ and $z_2$, then the preceding theorem implies, by way of Theorem 23.14, that $\mu_1 \ll \lambda_{z_1}$, $\mu_2 \ll \lambda_{z_2}$, and hence that $\mu_1$ and $\mu_2$ are mutually singular. That $z_1$ and $z_2$ lie in different Gleason parts is now a consequence of Theorem 16.6.

Our next theorem is a maximality result. We let $\Omega$ be a component of the interior of $X$ where, as usual, $\mathbb{C} \setminus X$ has finitely many components, we choose a point $z_0 \in \Omega$, and we denote by $\lambda_0$ the harmonic measure for $z_0$. (Were we to be absolutely consistent in our notation, we would be obliged to write $\lambda_{z_0}$ rather than $\lambda_0$. As the simpler notation causes no confusion, we prefer to use it.) In conformity with notation we have used earlier, we denote by $H^2(\lambda_0)$ the closure of $\mathscr{R}(X)$ in $L^2(\lambda_0)$ and by $H^\infty(\lambda_0)$ the intersection $H^2(\lambda_0) \cap L^\infty(\lambda_0)$. Since $\lambda_0$ is an enveloped measure, the remarks following Theorem 23.15 show that $H^\infty(\lambda_0)$ is the weak * closure of $\mathscr{R}(X)$ in $L^\infty(\lambda_0)$, that it is an algebra, and that $\lambda_0$ is multiplicative on it.

26.34 ***Theorem*** (Ahern and Sarason [1])
*The algebra $H^\infty(\lambda_0)$ is maximal among the subalgebras of $L^\infty(\lambda_0)$ on which $\lambda_0$ is multiplicative.*

PROOF (Garnett [3])
We consider a subalgebra $B$ of $L^\infty(\lambda_0)$ containing $H^\infty(\lambda_0)$ on which $\lambda_0$ is multiplicative. Referring to the remarks just before Lemma 17.5, we see that we may suppose $B$ to be weak * closed. We denote by $H^2(B)$ the closure in $L^2(\lambda_0)$ of $B$, and for $p = 2, \infty$, we define $H_0^p(\lambda_0)$ to be

$$\left\{ f \in H^p(\lambda_0) : \int f \, d\lambda_0 = 0 \right\}.$$

The spaces $H^2(\lambda_0)$ and $\overline{H_0^2(\lambda_0)}$ are orthogonal in $L^2(\lambda_0)$, so there is a direct sum decomposition

$$L^2(\lambda_0) = H^2(\lambda_0) \oplus \overline{H_0^2(\lambda_0)} \oplus E$$

for some closed subspace $E$ of $L^2(\lambda_0)$. The space $E$ may be characterized as consisting of those $f \in L^2(\lambda_0)$ with the property that if $g \in \mathscr{R}(X)$, then

$$0 = \int fg \, d\lambda_0 = \int f\bar{g} \, d\lambda_0,$$

so if $f \in E$, the measure $f\lambda_0$ annihilates Re $\mathscr{R}(X)$. As Re $\mathscr{R}(X)$ has finite codimension in $\mathscr{C}(\partial X)$, we see that $E$ is finite dimensional.

We prove next that $H^2(B) \subseteq H^2(\lambda_0) \oplus E$. This amounts to proving that $H^2(B)$ is orthogonal to $\overline{H_0^2(\lambda_0)}$. If $g \in \overline{H_0^2(\lambda_0)}$, then $\bar{g} \in H_0^2(\lambda_0)$, so there is a sequence $\{g_n\}$ in $\mathscr{R}(X)$ which converges in $L^2(\lambda_0)$ to $\bar{g}$, and we may even choose $g_n$ so that

$$\int g_n \, d\lambda_0 = 0 \tag{5}$$

since $\int \bar{g} \, d\lambda_0 = 0$. Similarly, if $f \in H^2(B)$, $f$ is the $L^2(\lambda_0)$-limit of a sequence $\{f_n\}$ in $B$. Since $\lambda_0$ is assumed to be multiplicative on $B$, (5) implies that

$$\begin{aligned} 0 &= \int f_n \, d\lambda_0 \int g_n \, d\lambda_0 \\ &= \int f_n g_n \, d\lambda_0 \to \int f\bar{g} \, d\lambda_0. \end{aligned}$$

Therefore $H^2(B) \subseteq H^2(\lambda_0) \oplus E$, as asserted.

We may write

$$H^2(B) = H^2(\lambda_0) \oplus N,$$

$\dim N < \infty$. If $\{G_1, \ldots, G_q\}$ is an orthonormal basis for the space $N$, if $g_1, \ldots, g_q \in B$ are very close to $G_1, \ldots, G_q$, respectively, in the sense of the $L^2(\lambda_0)$ norm, and if $J$ is the linear span of $(g_1, \ldots, g_q)$, then

$$H^2(B) = H^2(\lambda_0) + J,$$

where the sum is direct though not necessarily orthogonal. Let $P_J$ be the projection of $H^2(B)$ onto $J$ which annihilates $H^2(\lambda_0)$.

For each $f \in \mathscr{R}(X)$ define $T_f \in \mathscr{L}(J)$ by

$$T_f(g) = P_J(fg).$$

The map $f \mapsto T_f$ is a homomorphism of $\mathscr{R}(X)$ into $\mathscr{L}(J)$: It is plainly linear, and if $f, f' \in \mathscr{R}(X)$ and $g \in J$, then

$$\begin{aligned} (T_{ff'} - T_f T_{f'})(g) &= P_J(ff'g) - P_J(fP_J(f'g)) \\ &= P_J(f(f'g - P_J(f'g))). \end{aligned}$$

The last term is zero because $P_J$ is the zero operator on $H^2(\lambda_0)$, $H^2(\lambda_0)$ is stable under multiplication by elements of $\mathscr{R}(X)$, and $f'g - P_J(f'g)$ is in $H^2(\lambda_0)$. Since $\mathscr{R}(X)$ is a commutative algebra and since $J$ is finite dimensional, there exists in $J$ a nonzero eigenvector common to all the operators $T_f, f \in \mathscr{R}(X)$, call it $g_0$. Thus, if $f \in \mathscr{R}(X)$, $T_f(g_0) = \lambda_f g_0$ for some $\lambda_f \in \mathbb{C}$. The map $f \mapsto \lambda_f$ is a linear functional which is, in fact, multiplicative, for if $f, f' \in \mathscr{R}(X)$, then

$$T_{ff'}(g_0) = T_f(T_{f'}g_0) = \lambda_{f'}\lambda_f g_0.$$

Denote this homomorphism by $\varphi$. We know that $\Sigma(\mathscr{R}(X)) = X$, so there is a $\zeta_0 \in X$ such that $\varphi(f) = f(\zeta_0)$ for all $f \in \mathscr{R}(X)$. Thus, $T_f(g_0) = f(\zeta_0)g_0$. If $f \in \mathscr{R}(X)$, then

$$(6) \qquad fg_0 = P_J(fg_0) + (I - P_J)(fg_0) = f(\zeta_0)g_0 + r,$$

$r \in H^2(\lambda_0)$. The function $g_0$ is uniformly bounded and $I - P_J$ is continuous and linear, so the function $r$ in (6) depends linearly and continuously, in the sense of the $L^2(\lambda_0)$ norm, on $f$.

It follows that the functional $\varphi$ extends to a bounded linear functional on $H^2(\lambda_0)$. To see this, consider an $f \in \mathscr{R}(X)$. The decomposition (6) of $fg_0$ implies that

$$\int fg_0\bar{g}_0\, d\lambda_0 = \varphi(f)\int g_0\bar{g}_0\, d\lambda_0 + \int r\bar{g}_0\, d\lambda_0.$$

Since $\int g_0\bar{g}_0\, d\lambda_0 \neq 0$ and since both $f \mapsto \int fg_0\bar{g}_0\, d\lambda_0$ and $f \mapsto \int r\bar{g}_0\, d\lambda_0$ are bounded in the sense of the $L^2(\lambda_0)$ norm, it follows that $\varphi(f)$ also is bounded in the sense of this norm, so $\varphi$ does extend to a bounded linear functional on $H^2(\lambda_0)$. Consequently, there is $F \in H^2(\lambda_0)$ with

$$f(\zeta_0) = \int f\bar{F}\, d\lambda_0$$

for all $f \in \mathscr{R}(X)$. Thus, the measure $\bar{F}\lambda_0$ is a complex representing measure for the point $\zeta_0$ which is absolutely continuous with respect to $\lambda_0$. From this it follows that $\zeta_0 \notin \partial X$, for $\partial X$ consists of peak points for $\mathscr{R}(X)$ and $\lambda_0$ is nonatomic.

If we apply the decomposition (6) to the function $f(z) = z$, we find that $(z - \zeta_0)g_0 \in H^2(\lambda_0)$; call this function $h$. Let $\{h_n\}$ be a sequence in $\mathscr{R}(X)$ which converges in $L^2(\lambda_0)$ to $h$. The point $\zeta_0$ lies in the interior of $X$, so if

$$r_n(z) = \frac{h_n(z) - h_n(\zeta_0)}{z - \zeta_0},$$

then $r_n \in \mathscr{R}(X)$, and $\{r_n\}$ converges in $L^2(\lambda_0)$ to $g_0 - c(z - \zeta_0)^{-1}$ if $c = \varphi(h)$.

If $c = 0$, this gives $g_0 \in H^2(\lambda_0)$, which is impossible since $g_0$ is not the zero function and lies in $J$. On the other hand, if $c \neq 0$, then $(z - \zeta_0)^{-1} \in H^2(B)$, and since $\zeta_0 \notin \partial X$, $(z - \zeta_0)^{-1}$ is in $L^\infty(\lambda_0)$. It follows that $(z - \zeta_0)^{-n} \in H^\infty(B)$ for all integers $n$. The codimension of $H^2(\lambda_0)$ in $H^2(B)$ is finite, so for some choice of $m$ and some $\alpha_m \neq 0$, we have $\sum_{j=1}^m \alpha_j(z - \zeta_0)^{-j} \in H^2(\lambda_0)$. Multiplying by $(z - \zeta_0)^{m-1}$ yields the fact that $(z - \zeta_0)^{-1} \in H^2(\lambda_0)$, and from this and $g_0 - c(z - \zeta_0)^{-1} \in H^2(\lambda_0)$, we again arrive at the contradiction $g_0 \in \mathrm{H}^2(\lambda_0)$.

It follows that $H^2(\lambda_0) = H^2(B)$, so $H^\infty(\lambda_0) \subset B \subset H^2(\lambda_0)$. By hypothesis $B \subset L^\infty(\lambda_0)$, so it follows that $B \subset H^\infty(\lambda_0)$, and we have the theorem.

The following corollary is the form in which we shall use this theorem. We denote by $i : \mathscr{R}(X) \to L^\infty(\lambda_0)$ the map which assigns to $f \in \mathscr{R}(X)$ the equivalence class $[f]$ of $\lambda_0$-essentially bounded functions agreeing a.e. $[\lambda_0]$ with $f$. The map $i$ is a homomorphism but it need not be one to one.

**26.35** ***Corollary***

*If $B$ is a subalgebra of $L^\infty(\lambda_0)$ on which $\lambda_0$ is multiplicative and which contains $i\mathscr{R}(X)$, then $B \subset H^\infty(\lambda_0)$.*

PROOF

Since $B$ contains $i\mathscr{R}(X)$, the intersection of its closure in $L^2(\lambda_0)$ with $L^\infty(\lambda_0)$ is a subalgebra of $L^\infty(\lambda_0)$ which contains $H^\infty(\lambda_0)$ and on which $\lambda_0$ is multiplicative, so the result follows from the theorem just proved.

We shall use this maximality theorem to establish a connection between the algebra $H^\infty(\lambda_0)$ and the algebra of bounded holomorphic functions on the component $\Omega$ of the interior of $X$ containing $z_0$. For this we need a topological preliminary.

As we have observed already in the proof of Lemma 26.30, there is a domain $W$ in $\mathbb{C}$ with $\partial W$ consisting of a finite set of analytic simple closed curves which is conformally equivalent to $\Omega$, say under the map $\Phi : W \to \Omega$, and we know that at almost every point of $\partial W$ the nontangential limit of $\Phi$ exists. We shall say that a point $p \in \partial\Omega$ is *ambiguous* if there exist at least two points in $\partial W$ at which $\Phi$ assumes the value $p$ as a nontangential limit.

**26.36** ***Lemma***

*There exist only finitely many ambiguous points in $\partial\Omega$, and each has only finitely many preimages under $\Phi$.*

PROOF

We can suppose that $\partial W$ has more than one component, for, if not, let $D$ be a small disc contained in $W$ with analytic boundary, and replace $W$ by $W' = W \setminus \bar{D}$ and $\Omega$ by $\Omega' = \Omega \setminus \Phi(\bar{D})$. A point in $\partial\Omega$ is ambiguous if and only if it is ambiguous in $\partial\Omega'$ (with respect to the domain $W'$ and the map $\Phi|W'$). For the rest of the proof, suppose that $\partial W$ is not connected.

Assume that $p \in \partial\Omega$ is an ambiguous point so that there exist two points $q', q'' \in \partial W$, at which the map $\Phi$ assumes the value $p$ as a nontangential limit. We need the fact that $q'$ and $q''$ lie in the same component of $\partial W$. If not, let $\lambda$ be an arc in $W \cup \{q', q''\}$ which approaches $q'$ and $q''$ nontangentially. Since $q'$ and $q''$ lie in different components of $\partial W$, the arc $\lambda$ does not disconnect $W$. However, $\Phi(\lambda)$ is a simple closed curve in $\Omega \cup \{p\}$ which divides $\Omega$ into two domains. This is impossible, though, for $\Phi$ is a homeomorphism of $W$ onto $\Omega$.

If there exist infinitely many ambiguous points, then given $m$, there exist points $p_1, \ldots, p_m$ with the property that the value $p_j$ is assumed as a nontangential limit of $\Phi$ at the distinct points $q'_j$ and $q''_j$; and we can choose $p_1, \ldots, p_m$ so that all the points $q'_j$ (and therefore all the points $q''_j$) lie in a single component, $C$, of $\partial W$. Let $\{q_1, \ldots, q_{2m}\}$ be an indexing of $\{q'_1, q''_1, \ldots, q'_m, q''_m\}$ with the property that the ordering induced by the indices is the same as the cyclic order induced from $C$. The points $q_j$ determine arcs $C_1, \ldots, C_{2m}$ of $C$, where $C_j$ has end points $q_{j-1}, q_j$ and contains no other $q_k$. (We understand by $q_0$ the point $q_{2m}$.) Let $\lambda_j$ be an arc with end points $q_{j-1}, q_j$ which are approached nontangentially and which lies, except for its end points, in $W$. The arcs $\lambda_j$ may be chosen to be disjoint except for their end points. The set $I = \lambda_1 \cup \cdots \cup \lambda_{2m}$ is a simple closed curve in $W \cup \{q_1, \ldots, q_{2m}\}$ which, by proper choice of the $\lambda$'s, may be supposed to separate $C$ from all other components of $\partial W$. The image $J = \Phi(I)$ is a closed curve in $\Omega \cup \{p_1, \ldots, p_m\}$ which divides the plane into $m + 2$ regions, $G_1, \ldots, G_{m+2}$. We shall show that each of the regions $G_j$ contains a component of $\mathbb{C} \setminus X$, a fact that will establish the first half of the lemma, for $\mathbb{C} \setminus X$ has only finitely many components.

In order to prove that each $G_j$ contains a component of $\mathbb{C} \setminus X$, we start by showing that each $G_j$ meets $\partial X$. For this purpose, define arcs, $L_j$, $j = 1, \ldots, 2m$, with interior contained in $W$. Require that $L_j$ meet $I$ in a single point, $a_j$, and that it approach nontangentially a point $x_j \in C_j$ and also a point $y_j$ in some other component of $\partial W$. We require, as we may, that $\Phi$ have nontangential limit at $x_j$ and at $y_j$ and that these limits not lie in the set $p_1, \ldots, p_m$. (This latter requirement can be met since the boundary values of $\Phi$ can be constant on no set of positive measure in $\partial W$.) Using these arcs $L_j$, we can show that no $G_j$ is contained in $\Omega$. If $\partial G_j$ contains, say, the arc $\Phi(\lambda_k)$, and if $L_k^+$ and $L_k^-$ are the two components of $L_k \setminus \{a_k\}$, then one of $\Phi(L_k^+)$, $\Phi(L_k^-)$ is contained entirely within $\bar{G}_j$. Thus, one of $\Phi(x_k)$, $\Phi(y_k)$ lies in $\bar{G}_j$. The boundary of $\bar{G}_j$ is contained in $J$, and the construction of the arc $L_k$ shows that neither of $\Phi(x_k)$, $\Phi(y_k)$ lies in the set $\{p_1, \ldots, p_m\}$. Also, as we saw in the proof of Lemma 26.30, the values of $\Phi$ at $x_k$ and $y_k$ lie in $\partial\Omega$, so we finally see that one of $\Phi(x_k)$, $\Phi(y_k)$ must lie in $G_j \cap \partial\Omega$. Since $\partial\Omega \subset \partial X$, there is a component, $U$, of $\mathbb{C} \setminus X$ which meets $G_j$. Now $U \cap \Omega = \varnothing$, so since $\partial G_j \setminus \{p_k\} \subset J \setminus \{p_k\} \subset \Omega$, $U \cap \partial G_j \subset \{p_k\}$. But also the openness of $U$ implies that none of the points $p_k$ lie in $U$. Thus, $U \cap \partial G_j = 0$, so $U \subset G_j$ by connectedness.

A minor modification of the argument just given shows that every ambiguous point has only finitely many preimages under $\Phi$.

We can now prove the following important fact.

26.37 ***Theorem***

*If $h$ is bounded and holomorphic in $\Omega$, there exists a function $h^* \in H^\infty(\lambda_0)$ such that $\|h^*\|_\infty = \|h\|_\Omega$ and*

$$\int h^* \, d\lambda_\zeta = h(\zeta)$$

*for every $\zeta \in \Omega$.*

PROOF

We shall effect the proof by using the conformal map $\Phi: W \to \Omega$ to reduce the question to one on the simpler domain $W$. Given $w \in W$, let $\mu_w$ denote the harmonic measure for $w$ with respect to $W$ so that (Appendix D)

$$d\mu_w = \frac{1}{2\pi} \frac{\partial g(\cdot, w)}{\partial n} \, ds,$$

where $g(\cdot, w)$ is the Green's function for $W$ with pole at $w$, $\partial/\partial n$ is differentiation along the inner normal, and $ds$ is the differential of arc length on $\partial W$. The function $\Phi$ is defined at almost every $[\mu_w]$ point of $\partial W$. Fix a $w \in W$, and let $z = \Phi(w)$.

Our first objective is to prove the inequality

$$\mu_w(\Phi^{-1}(F)) \leq \lambda_z(F)$$

for every closed subset $F$ of $\partial\Omega$. By Corollary 26.27, there is a sequence $\{u_k\}$ of functions continuous on $X$, harmonic on the interior of $X$, such that $0 \leq u_k \leq 1$, $u_k = 1$ on $F$, and $u_k \to 0$ pointwise on $\partial X \setminus F$. If $U_k = u_k \circ \Phi$, then $U_k$ is harmonic on $W$ and its values lie between

0 and 1. Consequently (Appendix D), $U_k$ has nontangential boundary values $U_k^*$ almost everywhere on $\partial W$, and

$$\int u_k \, d\lambda_z = u_k(z)$$

$$= U_k(w) = \int U_k^* \, d\mu_w.$$

The function $U_k^*$ is nonnegative and takes the value 1 on $\Phi^{-1}(F)$, so

$$\int u_k \, d\lambda_z \geq \mu_w(\Phi^{-1}(F)).$$

We obtain the desired inequality by letting $k$ go to infinity.

If $P$ denotes the set of ambiguous points in $\partial\Omega$, then, by Lemma 26.36, $P$ is finite, so $\lambda_z(P) = 0$ and consequently $\mu_w(\Phi^{-1}(P)) = 0$. As in the proof of Lemma 26.30, there is a sequence $\{F_n\}$ of compact sets in $\partial W$ such that $\Phi$ is defined and continuous on $F_n$ and such that if $E = \bigcup F_n$, then $\partial W \setminus E$ is a null set $[ds]$. By continuity, each set $F_n \cap \Phi^{-1}(P)$ is closed, and hence $F_n \setminus \Phi^{-1}(P)$ is the union of a sequence of compact sets. The set $E' = E \setminus \Phi^{-1}(P)$ is therefore a union of a sequence of compact sets on each of which $\Phi$ is defined and continuous, and $\partial W \setminus E'$ is a null set $[ds]$. Replacing $E$ by $E'$ if necessary, we may suppose that $\Phi(E) \cap P = \varnothing$. The map $\Phi$ is one to one on $E$.

Since $\Phi$ is a homeomorphism of $F_n$ onto the compact set $\Phi(F_n)$, $\Phi^{-1}:\Phi(F_n) \to F_n$ is continuous. The set $\Phi(E) = \bigcup \Phi(F_n)$ is measurable, and $\Phi^{-1}:\Phi(E) \to E$ is measurable.

If $K \subset E$ is compact, then $F_n \cap K$ is compact and therefore so is $\Phi(F_n \cap K)$, so by our inequality

$$\begin{aligned}\mu_w(F_n \cap K) &\leq \mu_w(\Phi^{-1}(\Phi(F_n \cap K))) \\ &\leq \lambda_z(\Phi(F_n \cap K)) \\ &\leq \lambda_z(\Phi(K)).\end{aligned}$$

If we let $n \to \infty$, we obtain

$$\mu_w(K) \leq \lambda_z(\Phi(K)).$$

It follows that if $S$ is any measurable subset of $E$ and $K$ is any compact subset of $S$, then

$$\mu_w(K) \leq \lambda_z(\Phi(K)) \leq \lambda_z(\Phi(S)),$$

so by regularity we get

$$\mu_w(S) \leq \lambda_z(\Phi(S)).$$

Applying this to both $S$ and $E \setminus S$, we find that

$$\begin{aligned} 1 = \mu_w(E) &= \mu_w(S) + \mu_w(E \setminus S) \\ &\leq \lambda_z(\Phi(S)) + \lambda_z(\Phi(E \setminus S)) \\ &= \lambda_z(\Phi(E)) \\ &\leq \lambda_z(\partial\Omega) = 1, \end{aligned}$$

so $\lambda_z(\Phi(E)) = 1$; and, moreover,

$$\mu_w(S) = \lambda_z(\Phi(S)) \tag{7}$$

for every measurable set $S \subset E$.

If $h$ is a bounded measurable function on $\partial\Omega$, then $h \circ \Phi$ is a bounded measurable function on $E$, and by what we have just done, it follows that

$$\begin{aligned} \int h\, d\lambda_z &= \int_{\Phi(E)} h\, d\lambda_z \\ &= \int_E (h \circ \Phi)\, d\mu_w = \int (h \circ \Phi)\, d\mu_w. \end{aligned} \tag{8}$$

If $f$ is a bounded holomorphic function on $\Omega$, then $f \circ \Phi$ is a bounded holomorphic function on $W$, so the nontangential limits $(f \circ \Phi)^*$ exist almost everywhere on $\partial W$ and satisfy

$$(f \circ \Phi)(w) = \int (f \circ \Phi)^*\, d\mu_w.$$

The function $f^* = (f \circ \Phi)^* \circ \Phi^{-1}$ is measurable on $\Phi(E)$, and (8) yields

$$\int f^*\, d\lambda_z = \int (f \circ \Phi)^*\, d\mu_w = f(z).$$

By (7),

$$\|f^*\|_\infty = \|(f \circ \Phi)^*\|_\infty \leq \|f \circ \Phi\|_W = \|f\|_\Omega,$$

while $z \in \Omega$ implies

$$|f(z)| \leq \int |f^*|\, d\lambda_z \leq \|f^*\|_\infty,$$

so

$$\|f^*\|_\infty = \|f\|_\Omega.$$

Observe that the map $f \mapsto f^*$ is a homomorphism.

We have to show that $f^* \in H^\infty(\lambda_0)$. To do this, let $B$ be the set of all $f^*$, a subalgebra of $L^\infty(\lambda_0)$. The measure $\lambda_0$ is multiplicative on $B$, for

$$\begin{aligned}\int f^*g^*\,d\lambda_0 &= \int (fg)^*\,d\lambda_0 = (fg)(z_0)\\ &= f(z_0)g(z_0) = \left(\int f^*\,d\lambda_0\right)\left(\int g^*\,d\lambda_0\right).\end{aligned}$$

The algebra $B$ contains $i\mathscr{R}(X)$, notation as in Corollary 26.35, so by that corollary, $B \subset H^\infty(\lambda_0)$, and we have the theorem.

**26.38** ***Corollary***

*$H^\infty(\lambda_0)$ is isomorphic to the algebra of bounded holomorphic functions on $\Omega$ under the map $f \mapsto f^*$.*

PROOF

In order to prove the corollary, it is necessary to show that every $f \in H^\infty(\lambda_0)$ is an $h^*$. Given $f \in H^\infty(\lambda_0)$, define $\hat{f}$ on $\Omega$ by

$$\hat{f}(z) = \int f\,d\lambda_z.$$

If $f \in i\mathscr{R}(X)$, then $\hat{f}$ is holomorphic on $\Omega$, and we can show that the same is true if $f$ is merely in $H^\infty(\lambda_0)$. Such an $f$ is the $L^2$-limit of a sequence $f_n$ from $i\mathscr{R}(X)$. If $K \subset \Omega$ is compact, Harnack's inequality implies the existence of a constant $C$ such that if $D_z$ denotes the Radon–Nikodym derivative of $\lambda_z$ with respect to $\lambda_0$, then $D_z \leq C$ if $z \in K$. Then

$$\begin{aligned}|f_n(z) - f_m(z)| &= \left|\int (f_n - f_m)\,d\lambda_z\right|\\ &= \left|\int (f_n - f_m)D_z\,d\lambda_0\right|\\ &\leq C\|f_n - f_m\|_2\end{aligned}$$

if we use $\|\cdot\|_2$ for the norm on $L^2(\lambda_0)$. Thus, the convergence of $\{f_n\}$ in $H^2(\lambda_0)$ implies uniform convergence on compacta in $\Omega$, so the function $\hat{f}$ is holomorphic.

To finish the proof, we have to prove that $\hat{f}^* = f$. To do this, note that $\hat{f}^*$ and $f$ are bounded measurable functions with the property that for every point $z \in \Omega$,

$$\int \hat{f}^*\,d\lambda_z = \hat{f}(z) = \int f\,d\lambda_z.$$

Invoking (8), it follows that $\hat{f}^* \circ \Phi$ and $f \circ \Phi$ are bounded measurable functions on $\partial W$ with

$$\frac{1}{2\pi}\int \hat{f}^*(\Phi(\zeta))\frac{\partial g(\zeta, w)}{\partial n}\,ds(\zeta) = \frac{1}{2\pi}\int f(\Phi(\zeta))\frac{\partial g(\zeta, w)}{\partial n}\,ds(\zeta)$$

for every $w \in W$. But this implies that $f \circ \Phi = \hat{f}^* \circ \Phi$ a.e. $[ds]$, for we know that if $u$ is a bounded measurable function on $\partial W$, then the harmonic function

$$U(w) = \frac{1}{2\pi}\int u(\zeta)\frac{\partial g(\zeta, w)}{\partial n}\,ds(\zeta)$$

has boundary values equal almost everywhere to $u$. (See Appendix D.) Thus, $f = \hat{f}^*$ a.e. $[\lambda_0]$, and the corollary is proved.

We prove next that the harmonic measures are strongly dominant, a result obtained by Ahern and Sarason [2] and Glicksberg [7]. We follow the former authors in our proof.

**26.39** ***Theorem***

*If $\mathbb{C} \setminus X$ has only finitely many components, and if $z$ lies in the interior of $X$, then the harmonic measure $\lambda_z$ is a strongly dominant representing measure for the point $z$.*

PROOF

Let us consider first the case of the set $\overline{W}$ utilized in the last theorem. The result we want can be obtained from this by way of the map $\Phi$, as we shall see. Let $\partial W$ be the union of the analytic simple closed curves $\Gamma_0, \ldots, \Gamma_q$, $\Gamma_0$ the boundary of the component of $\mathbb{C} \setminus W$ which contains the point at infinity. For $j = 1, \ldots, q$, let $\gamma_j$ be an analytic simple closed curve in $W$ which can be deformed through $W$ onto $\Gamma_j$ so that the set $\{\gamma_1, \ldots, \gamma_q\}$ constitutes a homology basis for $W$. Define measures $\nu_j$ on $\partial W$ by requiring that if $u$ is harmonic in $W$ and continuous on $\overline{W}$, then

$$\int u\,d\nu_j = \int_{\gamma_j}\frac{\partial u}{\partial n}\,ds,$$

where $\gamma_j$ is given the positive orientation induced from the plane and $\partial/\partial n$ is differentiation along the normal to $\gamma_j$ which when rotated through an angle of $\pi/2$ lies in the positive direction along $\gamma_j$. Thus, $1/2\pi \int u\,d\nu_j$ is the increment of the harmonic conjugate to $u$ along $\gamma_j$, and it follows that $\nu_j \in (\mathrm{Re}\,\mathscr{R}(\overline{W}))^\perp$. If $\zeta_j$ is a point of the component of $\mathbb{C} \setminus W$ bounded by $\Gamma_j$, then

$$\int \log|z - \zeta_j|\, d\nu_k(z) = 2\pi\delta_{kj},$$

which shows the measures $\nu_k$ to be linearly independent as functionals on $\mathrm{Re}\,\mathscr{R}(X)$. From the fact that $\mathrm{Re}\,\mathscr{R}(\overline{W})$ and $\log|z - \zeta_j|$, $j = 1, \ldots, q$, span $\mathscr{C}_{\mathbb{R}}(\partial W)$, it follows that the measures $\nu_k$ constitute a basis for $(\mathrm{Re}\,\mathscr{R}(\overline{W}))^\perp$ and, incidentally, that this latter space is $q$-dimensional.

We shall prove that each $\nu_j$ boundedly absolutely continuous with respect to $\mu_0$, the harmonic measure for the point $w_0 \in W$. To do this, let $D$ be an open set in $W$ which contains $\gamma_1 \cup \cdots \cup \gamma_q$ and whose closure is contained in $W$. As $\overline{D}$ is compact, Harnack's inequality implies the existence of a constant $C$ such that if $u$ is a positive harmonic function on $W$, then

$$\|u\|_D \leq Cu(w_0). \tag{9}$$

Moreover, there is a constant $C'$ such that if $u$ is harmonic on $D$ and bounded there, then

$$\left\|\frac{\partial u}{\partial x}\right\|_{\gamma_j} + \left\|\frac{\partial u}{\partial y}\right\|_{\gamma_j} \leq C'\|u\|_D, \tag{10}$$

as follows from the fact that $\gamma_j$ is contained in the interior of $D$, e.g., by applying the Poisson integral formula. Since the curves $\gamma_j$ are smooth, we may conclude from (9) and (10) that there is a constant, say $K$, such that if $u \in \mathscr{C}(\overline{W})$ is harmonic in $W$ and nonnegative, then

$$\left|\int u\, d\nu_j\right| \leq Ku(w_0)$$

$$= K\int u\, d\mu_0.$$

This inequality shows that the measures $\upsilon_j$ are each boundedly absolutely continuous with respect to $\mu_0$. As the measures $\upsilon_j$ constitute a basis for $(\mathrm{Re}\,\mathscr{R}(X))^\perp$, it follows that for some absolute constant $K_1$, $\sigma \in (\mathrm{Re}\,\mathscr{R}(X))^\perp$ implies that

$$\left\|\frac{d\sigma}{d\mu_0}\right\|_\infty \leq K_1\|\sigma\|.$$

Applying this to $\mu_0 - \nu$, where $\nu$ is a representing measure for $z_0$, yields the result that $\mu_0$ is strongly dominant.

To complete our proof we show that this special case of the theorem implies the general case. Let $\Omega$ be a component of the interior of $X$. We let $\Phi: W \to \Omega$ be the conformal mapping considered in the last proof. Let $z_0 \in \Omega$, $z_0 = \Phi(w_0)$. If $\nu \in \mathscr{M}(\partial X)$ is a representing measure for $z_0$, then $\nu \ll \lambda_0$, $\lambda_0$ harmonic measure for $z_0$, so $\nu = h\lambda_0$, $h \in L^1(\lambda_0)$. If $f \in \mathscr{R}(\overline{W})$, then $f \circ \Phi^{-1}$ is a bounded holomorphic function on $\Omega$, and if $(f \circ \Phi^{-1})^*$ is the element of $H^\infty(\lambda_0)$ associated with $f$ as in Theorem 26.37, then, by equation (8), we have

$$\int (f \circ \Phi^{-1})^* h \, d\lambda_0 = \int f(h \circ \Phi) \, d\mu_0 .$$

For every $g \in \mathscr{R}(X)$,

$$\begin{aligned} \int gh \, d\lambda_0 &= g(z_0) \\ &= \int g \, d\lambda_0 , \end{aligned}$$

so it follows that the equality

$$\int gh \, d\lambda_0 = \int g \, d\lambda_0$$

persists for all $g \in H^\infty(\lambda_0)$, for, as we know, $H^\infty(\lambda_0)$ is the weak $*$ closure of $\mathscr{R}(X)$. Consequently,

$$\int f(h \circ \Phi) \, d\mu_0 = f(w_0),$$

so the measure $(h \circ \Phi)\mu_0$ is a representing measure for $w_0$. The strong dominance of $\mu_0$, taken together with equation (8), now implies that $\lambda_0$ is strongly dominant, and the theorem is proved.

In the course of the preceding proof, we established a correspondence between representing measures for $\mathscr{R}(X)$ and representing measures for $\mathscr{R}(\overline{W})$. It is clear that a similar correspondence can be established between real annihilating measures. If $W$ is simply connected, then since $\partial W$ is, in this case, a single analytic simple closed curve, $\overline{W}$ is also simply connected, and hence $\mathscr{R}(\overline{W})$ is a Dirichlet algebra, as we saw in Section 25. In this case there are no real annihilating measures. Making use of the general F. and M. Riesz theorem and the non-existence of completely singular orthogonal measures, we see that the following theorem is true.

**26.40** ***Theorem***

*If $\mathbb{C} \setminus X$ has finitely many components and if every component of the interior of $X$ is simply connected, then $\mathscr{R}(X)$ is a Dirichlet algebra.*

There is a maximality theorem in this setting which is analogous to Theorem 25.12, the maximality theorem for $\mathscr{P}(X)$.

**26.41** ***Theorem***

*Let $X$ be a compact set in $\mathbb{C}$ which divides the plane into only finitely many components. If the interior of $X$ is connected, then $\mathscr{R}(X)$ is a maximal closed subalgebra of $\mathscr{C}(\partial X)$ in the sense that if $B$ is a closed subalgebra of $\mathscr{C}(\partial X)$ such that $\mathscr{R}(X)|\partial X \subset B$, then either $B = \mathscr{R}(X)|\partial X$ or else $B = \mathscr{C}(\partial X)$.*

PROOF

If $z \in \mathbb{C} \setminus \partial X$, let $g_z$ be defined by $g_z(\zeta) = (\zeta - z)^{-1}$. If $z \in \mathbb{C} \setminus X$, then $g_z \in \mathscr{R}(X)|\partial X$. If $g_z \in B$ for some $z \in \Omega$, $\Omega$ the interior of $X$, then by Runge's theorem (Theorem 24.3) $B \supset \mathscr{R}(\partial X)$, and since, by Theorem 26.5, $\mathscr{R}(\partial X) = \mathscr{C}(\partial X)$, we get $B = \mathscr{C}(\partial X)$.

We therefore assume henceforth that for no $z \in \Omega$ is $g_z \in B$. Choose $z_0 \in \Omega$, and let $\lambda_0$ be the harmonic measure (on $\partial X$) for $z_0$. If $h(\zeta) = \zeta - z_0$, then $h \in B$, but it is not invertible in $B$, so there is $\varphi \in \Sigma(B)$ with $\varphi(h) = 0$. If $\mu \in \mathscr{M}(\partial X)$ is an Arens–Singer measure for $\varphi$, then $\mu$ is, *a fortiori*, an Arens–Singer measure for $\varphi|\mathscr{R}(X)$, and since $\int h \, d\mu = 0$, we find that $\mu$ is a representing measure for the point $z_0$. The remark after Corollary 26.27 shows that $\mu = \lambda_0$, so $\lambda_0$ is multiplicative on $B$.

Denote by $B^*$ the intersection of $L^\infty(\lambda_0)$ with the $L^2(\lambda_0)$-closure of $B$. This is a subalgebra of $L^\infty(\lambda_0)$, as follows from the argument given just before Lemma 17.5, and $\lambda_0$ is multiplicative on $B^*$. Since $B^* \supset \mathscr{R}(X)|\partial X$, Theorem 26.34 shows that $B^* = H^\infty(\lambda_0)$.

If $\nu \in \mathscr{M}(\partial X)$ annihilates $\mathscr{R}(X)$, then by the general F. and M. Riesz theorem, $d\nu = p \, d\lambda_0$, $p \in L^1(\lambda_0)$. If $f \in B$, then $f \in H^\infty(\lambda_0)$, so, by Corollary 23.11 or Theorem 23.14, there is a bounded sequence $\{f_n\}$ in $\mathscr{R}(X)$ which converges pointwise a.e. $[\lambda_0]$ to $f$, so we have

$$0 = \lim_{n\to\infty} \int f_n p \, d\lambda_0$$

$$= \int f \, d\nu.$$

This shows that $\nu \in B^\perp$, so $B^\perp = \mathscr{R}(X)^\perp$, and it follows that $B = \mathscr{R}(X)|\partial X$, as we wished to show.

This proof follows rather closely one given by Merrill [1] in the case of simply connected $X$.

We now turn our attention to the promised bounded approximation theorem.

26.42 ***Theorem***

*If $X$ is a compact set in the plane with the property that $\mathbb{C} \setminus X$ has only finitely many components, and if $f$ is a function holomorphic and bounded in the interior $\Omega$ of $X$, then there exists a sequence $\{f_n\}$ of rational functions with no poles on $X$ such that $\|f_n\|_\Omega \leq \|f\|_\Omega$ for all $n$ and such that $f_n \to f$ pointwise on $\Omega$.*

The theorem in this degree of generality is due to Ahern and Sarason [2]; it generalizes the following result of Rubel and Shields [1], which in its turn is a generalized version of a theorem due to Farrell.

26.43 ***Corollary***

*If $G$ is a bounded open set in the plane which is the complement of the closure of the unbounded component of $\mathbb{C} \setminus \bar{G}$, then for every bounded holomorphic function $f$ on $G$, there is a sequence $\{p_n\}$ of polynomials which converges pointwise on $G$ to $f$ and which satisfies $\|p_n\|_G \leq \|f\|_G$ for all $n$.*

To obtain the corollary, we need merely apply the theorem to the set $\bar{G}$.

For some additional results in the general direction of the theorem, we refer to Gamelin and Garnett [1] and Fisher [1].

The proof of the bounded approximation theorem depends on a theorem of Banach which we state and whose proof may be found in Appendix C.

26.44 ***Theorem***

*If $E$ is a separable Banach space with dual space $E^*$, if $M$ is a subspace of $E^*$, and if $M_1$ is the set of all weak $*$ limits of sequences in $M$, then $M_1 = E^*$ if and only if there is a constant $c$ such that for each $x \in E$, there is $\varphi \in M$ with $|\varphi(x)| = \|x\|$ and $\|\varphi\| \leq c$. When this condition is satisfied, each $\psi \in E^*$ is the weak $*$ limit of a sequence in $M$ which is uniformly bounded by $c\|\psi\|$.*

PROOF OF THEOREM 26.42

Let $\{\Omega_j\}$ be the set of components of $\Omega$, and for each $j$ let $\lambda_j$ be the harmonic measure for a point $z_j$ of $\Omega_j$. The measure $\Lambda = \Sigma 2^{-n}\lambda_j$ is a finite regular Borel measure on $\partial X$. Each of the spaces $L^\infty(\lambda_j)$ is in a natural way a subspace of $L^\infty(\Lambda)$, for we know that the measures $\lambda_j$ are mutually singular. Let $P_j: L^\infty(\Lambda) \to L^\infty(\lambda_j)$ be the natural projection, and let

$$A = \{f \in L^\infty(\Lambda) : P_j f \in H^\infty(\lambda_j) \text{ for all } j\}.$$

It is clear that $A$ is a subalgebra of $L^\infty(\Lambda)$; we shall show that it is the weak * closure of $\mathscr{R}(X)$ in $L^\infty(\Lambda)$. To prove this assertion, let $\psi$ be a weak * continuous linear functional on $L^\infty(\Lambda)$ which annihilates $\mathscr{R}(X)$ so that $\psi$ is of the form

$$\psi(f) = \int fg \, d\Lambda,$$

$g \in L^1(\Lambda)$, and

$$\int fg \, d\Lambda = 0$$

if $f \in \mathscr{R}(X)$. Thus, the measure $g\Lambda$ is in $\mathscr{R}(X)^\perp$ and so is of the form $\Sigma g_j \lambda_j$, where each $g_j \lambda_j$ annihilates $\mathscr{R}(X)$. Since $g_j \lambda_j$ annihilates $\mathscr{R}(X)$, it also annihilates $H^\infty(\lambda_j)$, for every element of the latter algebra is the pointwise limit of a bounded sequence in $\mathscr{R}(X)$, by Theorem 23.8 or Theorem 23.14. Consequently, if $f \in A$, then

$$\begin{aligned}\psi(f) &= \sum \int g_j f \, d\lambda_j = \sum \int g_j P_j f \, d\lambda_j \\ &= 0.\end{aligned}$$

It follows that $A$ is contained in the weak * closure of $\mathscr{R}(X)$. $A$ is weak * closed because all the algebras $H^\infty(\lambda_j)$ are weak * closed in $L^\infty(\lambda_j)$, as we noted before Theorem 26.34, so we have the result.

Let $f$ be holomorphic and bounded in $\Omega$. By Theorem 26.37, there exists $h \in A$ with $\|h\|_\infty = \|f\|_\Omega$ and with the property that if $z \in \Omega$, then $f(z) = \int h \, d\lambda_z$.

We shall prove that if $c > 1$, there is a sequence $\{h_n\}$ in $\mathscr{R}(X)$ such that $\|h_n\|_\Omega \le c\|h\|_\infty$ and $h_n \to h$ weak * in $L^\infty(\Lambda)$. This follows from the theorem of Banach applied with $E^* = A$ so that $E$ is a certain quotient space of $L^1(\Lambda)$ and $M = i\mathscr{R}(X)$. We are to show that for $c$ greater than but arbitrarily near 1, the condition of Banach's theorem is satisfied, and this amounts to showing that if $f \in L^1(\Lambda)$, then the norm of the functional $f$ induces on $A$ is the same as the norm of the functional it induces on $\mathscr{R}(X)$ when $\mathscr{R}(X)$ is given the $L^\infty(\Lambda)$ norm. The $L^\infty(\Lambda)$ norm on $\mathscr{R}(X)$ is identical with the supremum norm on $\Omega$ and hence with the supremum norm on $\partial\Omega$. Let $N$ be the norm of the functional $f$ induces on $\mathscr{R}(X)$. By the Hahn–Banach and Riesz representation theorems, there is a measure $\mu$ on $\partial\Omega$ with $\|\mu\| = N$ and

$$\int gf\,d\Lambda = \int g\,d\mu$$

for all $g \in \mathscr{R}(X)$. Then $\mu - f\Lambda \in \mathscr{R}(X)^{\perp}$, so it follows that $\mu \ll \Lambda$. As $A$ is the weak * closure of $\mathscr{R}(X)$ in $L^{\infty}(\Lambda)$, the functional induced on $A$ by the function $d\mu/d\Lambda$ is the same as that induced by $f$, so this latter functional has norm no more than

$$\int \left|\frac{d\mu}{d\Lambda}\right| d\Lambda = \|\mu\| = N.$$

To complete the proof of the theorem, we choose, for $k = 2, 3, \ldots,$ a sequence $\{h_n^{(k)}\}$ in $\mathscr{R}(X)$ which converges weak * to $h$ and which satisfies $\|h_n^{(k)}\|_{\Omega} < (1 + k^{-1})\|h\|_{\infty}$. Let $\{D_1, D_2, \ldots\}$ be an increasing sequence of compact sets in $\Omega$ with $\bigcup D_j = \Omega$. The weak * convergence of $\{h_n^{(k)}\}$ implies that $h_n^{(k)}(z) \to f(z)$ for all $z \in \Omega$, and thus the convergence is uniform on each of the compact sets $D_j$. It follows that for each $k$, there is $N = N(k)$ such that if $n \geq N$, then

$$\|h_n^{(k)} - f\|_{D_k} < k^{-2}.$$

Let

$$g_k = k(k + 1)^{-1}h_{N(k)}^{(k)}.$$

We have $g_k \in \mathscr{R}(X)$, $\|g_k\|_{\Omega} < \|h\|_{\infty} = \|f\|_{\Omega}$, and

$$\begin{aligned}\|g_k - f\|_{D_k} &= \|h_{N(k)}^{(k)} - f - (k + 1)^{-1}h_{N(k)}^{(k)}\|_{D_k}\\ &\leq k^{-2} + (k + 1)^{-1}(1 + k^{-1}).\end{aligned}$$

If we now choose a rational function $f_k$ without poles on $X$ such that $\|g_k - f_k\|_X < k^{-1}$ and such that, in addition, $\|f_k\|_{\Omega} \leq \|h\|_{\infty}$, the sequence $\{f_k\}$ has the properties we seek.

This theorem is an interesting function theoretic result, but it is also interesting because it is an instance of a result in which the best possible qualitative result is obtained by the abstract methods of functional analysis.

It is important to remark that there exist domains for which the analogous bounded approximation fails.

26.45 ***Example*** (Fisher [1])

Let $E \subset [-\frac{1}{2}, \frac{1}{2}]$ be a closed set with the property that there exist nonconstant bounded holomorphic functions on $\mathbb{C}^* \setminus E$, $\mathbb{C}^*$ the Riemann sphere. By the Riemann mapping theorem, the interval

$[-\frac{1}{2}, \frac{1}{2}]$ itself has the property, and it is known that a closed set has the desired property if and only if it has positive (linear) Lebesgue measure. (The details of this point are in the book of Ahlfors and Sario [1, p. 254].)

Let $\{S_i\}_{i=1}^{\infty}$ be an infinite sequence of pairwise disjoint closed discs in the open unit disc $\Delta$ with the property that the sum of the radii of the $S_i$ is finite, and the additional property that

$$E = \left(\bigcup_i S_i\right)^{-} \setminus \left(\bigcup_i S_i\right).$$

Let $\Omega$ be the domain $\Delta \setminus (\bigcup_i S_i)^{-}$, and let $\Gamma = \partial\Omega \setminus E$. Thus, $\Gamma$ is a system of curves of finite total length, and if $f \in \mathscr{R}(\bar{\Omega})$, then $\int_\Gamma f(z)\,dz = 0$. Let $h$ be a function bounded and holomorphic on $\mathbb{C}^* \setminus E$ such that if $h(z) = \sum_{n=0}^{\infty} a_{-n} z^{-n}$ for large $|z|$, then $a_{-1} \neq 0$. We have that if $\gamma$ is the unit circle, then $a_{-1} = (1/2\pi i)\int_\gamma h(z)\,dz$. Assume that the function $h$ is the pointwise limit on $\Omega$ of a bounded sequence $\{g_n\}$ in $\mathscr{R}(\bar{\Omega})$. Then $\lim_n \int_\Gamma g_n(z)\,dz = \int_\Gamma h(z)\,dz$. To see this, let $\gamma'$ be a circle in $\Delta$ near the unit circle which surrounds $(\bigcup S_i)^{-}$, and for each $i$, let $\gamma_i$ be a circle in $\Omega$ which is near $\partial S_i$ and which separates $\partial S_i$ from the rest of $\partial\Omega$. We can choose the $\gamma_i$ to be pairwise disjoint and so that if $\Gamma' = \gamma' \cup (\bigcup_i \gamma_i)$, then $\Gamma'$ has finite length. By the Cauchy integral theorem, $\int_\Gamma g_n\,dz = \int_{\Gamma'} g_n\,dz$ and $\int_{\Gamma'} h\,dz = \int_{\Gamma'} h\,dz$, and the dominated convergence theorem yields $\lim_n \int_{\Gamma'} g_n\,dz = \int_{\Gamma'} h\,dz$. By the Cauchy integral theorem, $\int_\Gamma g_n\,dz = 0$, and since $h$ is holomorphic on $\Delta \setminus E$, it follows that $\int_{\partial S_i} h\,dz = 0$ so $\int_\Gamma h\,dz = \int_\gamma h\,dz$. As we have assumed that $a_{-1} \neq 0$, we have a contradiction.

We conclude this section with some bibliographical remarks. The theory of $\mathscr{R}(X)$ is vast and its literature formidable. An introduction to certain classical aspects of the theory is contained in the text of Markushevich [1, vol. 3], and the work of Walsh [1] covers a great deal of material related to this theory. Mergelyan's memoir [2] is also a valuable source. For some of the more recent developments, particularly the theory of analytic capacity, there are the book of Smirnov and Lebedev [1], the notes of Zalcman [1], and the survey article by Vitushkin [1]. Gamelin's book [1] contains a great deal of material concerning $\mathscr{R}(X)$ and various related topics, and Browder's book [1] also deals with some of this theory.

## 27 *A NORMAL UNIFORM ALGEBRA*

There is a class of Banach algebras which plays a rather important role in many fields of analysis, the class of *regular* Banach algebras.

It was a conjecture of some years standing that the algebras $\mathscr{C}(X)$ were the only regular uniform algebras. In this section we shall construct, following McKissick [1, 2] an example which shows this conjecture to be false. The existence of such an example is quite important for the theory of uniform algebras, as, for example, it completes the considerations of Section 19.

27.1 ***Definition***

*A function algebra $A$ on a locally compact space $X$ is said to be* **regular** *if given $x \in X$ and a closed subset $E$ of $X$ which does not contain $x$, there is $f \in A$ with $f|E = 0, f(x) = 1$. The algebra $A$ is said to be* **normal** *on $X$ if given a compact set $E_1$ and a closed set $E_0$ disjoint from $E_1$, there exists $f \in A$ with $f|E_0 = 0, f|E_1 = 1$. A commutative Banach algebra $A$ is said to be* **regular** *(or* **normal***) if the algebra $\hat{A}$ of Gelfand transforms is regular (or normal) on $\Sigma(A)$.*

27.2 ***Theorem***

*A regular commutative Banach algebra is normal.*

PROOF

We begin by proving that if $\varphi_0 \in \Sigma(A)$, then there is a neighborhood $V$ of $\varphi_0$ and an $a \in A$ with $\hat{a}|V = 1$. (This is trivial if $A$ has an identity.) Choose $b_0 \in A$ with $\hat{b}_0(\varphi_0) = 1$, and set $V = \{\varphi \in \Sigma(A) : |\hat{b}_0(\varphi)| \geq \frac{1}{2}\}$. This is a certain closed neighborhood of $\varphi_0$ in $\Sigma(A)$, and, by regularity,

$$V = \bigcap\{Z(\hat{a}) : a \in A, \hat{a}|V = 0\}$$

if $Z(\hat{a})$ denotes the zero set of $\hat{a}$. Let $I$ be the ideal in $A$ consisting of those $a$ whose Gelfand transforms vanish on $V$. The ideal $I$ is a closed ideal in $A$, and $\Sigma(A/I) = V$. The spectrum of the element $b_0 + I$ of $A|I$ can be identified with the range of $\hat{b}_0$ on $V$, and so it does not contain 0. Consequently, $A/I$ has an identity, say $c + I$, $c \in A$. We have then that $\hat{c}(\varphi) = 1$ if $\varphi \in V$, and thus we have completed the first step in the proof.

Next let $E_1$ be a compact set in $\Sigma(A)$. If $\varphi \in E_1$, let $V_\varphi$ be a neighborhood of $\varphi$ on which $\hat{a}_\varphi$ is identically 1 for some choice of $a_\varphi$ in $A$. Finitely many of the neighborhoods $V_\varphi$ cover $E_1$, say $V_1, \ldots, V_N$; let the associated $a_\varphi$ be $a_1, \ldots, a_N$. If $c_1 = a_1 + a_2 - a_1 a_2$, $\hat{c}_1$ takes the value 1 on $V_1 \cup V_2$. If $c_2 = a_3 + c_1 - a_3 c_1$, then $\hat{c}_2$ takes the value 1 on $V_1 \cup V_2 \cup V_3$. Iterating this process, we finally obtain an element $c$ which assumes the value 1 on $V_1 \cup \cdots \cup V_N$, a set that contains $E_1$.

To complete the proof, let $E_0$ be a closed subset of $\Sigma(A)$ which is disjoint from the compact set $E_1$, and put $I = \{a \in A : \hat{a}|E_0 = 0\}$. The ideal $I$ is a closed subalgebra of $A$ which, because of regularity, has $\Sigma(A) \setminus E_0$ as its spectrum. Moreover, $I$ is a regular Banach algebra, so the last paragraph implies the existence of an element $b \in I$ with $\hat{b}|E_1 = 1$. We have $b \in A$, $\hat{b}|E_0 = 0$, and $\hat{b}|E_1 = 1$.

The following observation about normal algebras is due to Bishop. (See Hoffman [1, pp. 190–191].)

27.3 ***Theorem***

*If $(A, X)$ is a uniform algebra, and if $A$ is normal on $X$, then $X = \Sigma(A)$.*

PROOF

If $\mu$ is a Jensen measure for the point $\varphi \in \Sigma(X) \setminus X$, then since $\varphi \notin X$, $\mu$ cannot be a point mass. Consequently, there exist disjoint closed sets $E_0$ and $E_1$ with $\mu(E_0) \neq 0$, $\mu(E_1) \neq 0$. By normality, there exists $f \in A$ with $f|E_0 = 0$, $f|E_1 = 1$. Then

$$\log|\varphi(f)| \leq \int \log|f| \, d\mu = -\infty,$$

so $\varphi(f) = 0$, but also

$$\log|\varphi(1 - f)| \leq \int \log|1 - f| \, d\mu = -\infty,$$

and this implies that $\varphi(f) = 1$. It follows that $\varphi$ cannot exist, so we have the theorem.

We now come to the example of McKissick.

27.4 ***Theorem***

*There exists a compact set $X$ in the plane such that $\mathscr{R}(X)$ is normal but is not $\mathscr{C}(X)$.*

Recall that $X$ is the spectrum of $\mathscr{R}(X)$.

For the construction of McKissick's example, we need to use a certain function which was investigated by Beurling [1]. Set[1]

$$\varphi(z) = \prod_{2}^{\infty} (1 - z^n \exp(n \log^{-1} n)).$$

The product defining $\varphi$ converges uniformly on compacta in the unit disc, so $\varphi$ is holomorphic there. Let $\alpha_1, \alpha_2, \ldots$ be the zeros of $\varphi$ arranged in order of nondecreasing magnitude. The following lemma gives the two properties of $\varphi$ which are crucial for our purposes.

[1] Here, and hereafter, $\log^{-1}$ refers to division, not to exponentiation.

27.5 ***Lemma***
*There exists an integer* $N_1$ *such that if* $n \geq N_1$, *then*

(a) $|\alpha_k| = \exp(-\log^{-1} n)$ *implies that* $|\varphi'(\alpha_k)| \geq \exp(n \log^{-3} n)$, *and*
(b) $|z| = \frac{1}{2}\{\exp(-\log^{-1} n) + \exp[-\log^{-1}(n+1)]\}$ *implies that* $|\varphi(z)| > \exp(n \log^{-3} n)$.

The verification of these two properties is elementary though not particularly simple. We shall defer it until the end of this section; in the meantime, we shall show how the desired example can be constructed using these properties of $\varphi$.

We shall apply this lemma to derive some of the properties of the series

$$\sum_{k=1}^{\infty} \frac{1}{\varphi'(\alpha_k)(z - \alpha_k)}. \tag{1}$$

We shall prove that if $|z| < 1$ and if $z$ is distinct from all the $\alpha_k$, then this series converges to $1/\varphi(z)$, and if $|z| > 1$, then it converges to zero. These observations follow immediately from the residue theorem. For notational convenience, set

$$R_n = \tfrac{1}{2}\{\exp(-\log^{-1} n) + \exp[-\log^{-1}(n+1)]\},$$

and put

$$I_n(z) = \frac{1}{2\pi i} \int_{|\zeta| = R_n} \frac{d\zeta}{\varphi(\zeta)(\zeta - z)}.$$

By the residue theorem, we have that if $|z| < R_n$ and if $z$ is distinct from all the $\alpha_k$, then

$$I_n(z) = \frac{1}{\varphi(z)} + \sum_{k=1}^{N(n)} \frac{1}{\varphi'(\alpha_k)(\alpha_k - z)},$$

where $N(n)$ is the integer such that $|\alpha_k| > R_n$ if $k > N(n)$, $|\alpha_k| < R_n$ if $k \leq N(n)$. Part (b) of the lemma implies that for fixed $z$, $I_n(z) \to 0$ as $n \to \infty$, so we have our first assertion. If $|z| > 1$, then

$$I_n(z) = \sum_{k=1}^{N(n)} \frac{1}{\varphi'(\alpha_k)(\alpha_k - z)}$$

and as $n \to \infty$, this tends to zero.

27.6 ***Lemma***
*Let* $D$ *be an open disc in the plane. If* $\varepsilon > 0$, *there is a sequence* $\{\Delta_k\}_{k=1}^{\infty}$ *of open discs in the plane and a sequence* $\{f_n\}$ *of rational functions with the following properties:*

(a) *If* $r_k$ *is the radius of* $\Delta_k$, *then* $\sum_{k=1}^{\infty} r_k < \varepsilon$.
(b) *The poles of* $f_n$ *lie in* $\cup\{\Delta_k : k = 1, 2, \ldots n\}$.
(c) *The sequence* $\{f_n\}$ *converges uniformly on the complement of* $\cup\{\Delta_k : k = 1, 2, \ldots\}$ *to a function that is identically zero outside D and zero free in* $D \setminus \cup\{\Delta_k : k = 1, 2, \ldots\}$.

PROOF

We may suppose that $D$ is the unit disc. Choose $N \geq 4$ so large that $\sum_{k=N}^{\infty} k^{-2} < \varepsilon/2$, and so that if $x \geq \log N$, then $\exp(-1/x) + \exp(-2x) < 1$. If $k \geq N$, let $\Delta_k$ be the open disc of radius $k^{-2}$ centered at the point $\alpha_k$. Since $|\alpha_k| < \exp(-\log^{-1} k)$ for $k \geq 4$, and since

$$\exp(-\log^{-1} k) + 1/k^2 = \exp(-\log^{-1} k) + \exp(-2 \log k),$$

the choice of $N$ shows that the disc $\Delta_k$ is contained in $D$. For $k = 1, 2, \ldots, N-1$, let $\Delta_k$ be an open disc contained in $D$ centered at $\alpha_k$ and so small that if its radius is $r_k$, then $\sum_{k=1}^{N-1} r_k < \varepsilon/2$.

If $z \in S = \mathbb{C} \setminus \cup\{\Delta_k : k = 1, 2, \ldots\}$, then, by Lemma 27.5(a),

$$\sum_{k=N}^{\infty} \left| \frac{1}{\varphi'(\alpha_k)(z - \alpha_k)} \right| \leq \sum_{k=N}^{\infty} k^2 \exp[-n(k) \log^{-3} n(k)]$$

where $n(k)$ satisfies $\exp[-\log^{-1} n(k)] = |\alpha_k|$. The estimate $n(k) \geq \frac{1}{2}(-1 + \sqrt{4k+9}) > \frac{1}{2}\sqrt{k}$ can be established easily, so since the series $\sum k^2 \exp[-\frac{1}{2}\sqrt{k} \log^{-3}(\frac{1}{2}\sqrt{k})]$ converges, it follows that series (1) converges uniformly on the set $S$. Thus, by our earlier remarks on the sum of series (1), it follows that for the sequence $\{f_n\}$ we can take the sequence of partial sums of that series.

## 27.7 *Proof of Theorem 27.4*

Let $\{D_m\}_{m=1}^{\infty}$ be a sequential arrangement of the family of all open discs in the plane with center $x_m + iy_m$, $x_m$ and $y_m$ rational and with rational radii. For fixed $m$, let $\{\Delta_k^m\}_{k=1}^{\infty}$ be a sequence of open discs contained in $D_m$ such that if $r_{m,k}$ is the radius of $\Delta_k^m$, then $\sum_k r_{m,k} < 2^{-m}$, $\{\Delta_k^m\}$ chosen as in Lemma 27.6 with respect to the disc $D_m$. We have that

$$\sum_{m,k} r_{m,k} < 1,$$

so if

$$X = \{z : |z| \leq 1\} \cup \{\Delta_k^m : k, m = 1, 2, \ldots\},$$

it follows from Lemma 24.1 that $\mathscr{R}(X) \neq \mathscr{C}(X)$. However, $\mathscr{R}(X)$ is regular, for if $K \subset X$ is compact, and if $x \in X \setminus K$, there exists a $D_m$ which contains $x$ but is disjoint from $K$. Thus, by the choice of $\{\Delta_k^m\}$, it follows that there is an $f \in \mathscr{R}(X)$ which vanishes on $K$ but not on $x$.

Thus, $\mathscr{R}(X)$ is regular and so, by Theorem 27.2 and the fact that $X = \Sigma(\mathscr{R}(X))$, $\mathscr{R}(X)$ is normal.

In order to complete the proof of the theorem, it is only necessary to obtain estimates (a) and (b) of Lemma 27.5.

## 27.8 *Proof of Lemma 27.5*

We shall verify the lemma only for even $n$; the case of odd $n$ may be handled by a parallel argument. The first step in the proof is to prove that

$$|\varphi'(\alpha_k)| \geq |\varphi'[\exp(-\log^{-1} n)]| \tag{2}$$

if $|\alpha_k| = \exp(-\log^{-1} n)$. For this purpose, write

$$\varphi(z) = (1 - z^n \exp n \log^{-1} n)[\varphi(z)(1 - z^n \exp n \log^{-1} n)^{-1}],$$

so, by the product rule for the derivative,

$$\varphi'(z) = (1 - z^n \exp n \log^{-1} n)\frac{d}{dz}[\varphi(z)(1 - z^n \exp n \log^{-1} n)^{-1}]$$
$$-(nz^{n-1} \exp n \log^{-1} n)[\varphi(z)(1 - z^n \exp n \log^{-1} n)^{-1}].$$

Thus, if $|\alpha_k| = \exp(-\log^{-1} n)$, we have that

$$|\varphi'(\alpha_k)| = n \exp \log^{-1} n \prod_{\substack{j=2 \\ j \neq n}}^{\infty} |1 - \alpha_k^j \exp j \log^{-1} j|$$
$$\geq n \exp \log^{-1} n \prod_{\substack{j=2 \\ j \neq n}}^{\infty} |1 - \exp j(\log^{-1} j - \log^{-1} n)|$$
$$= |\varphi'[\exp(-\log^{-1} n)]|,$$

and we have estimate (2). Next we must estimate $\varphi'[\exp(-\log^{-1} n)]$, i.e.,

$$\lim \varphi(z)[z - \exp(-\log^{-1} n)]^{-1}, \tag{3}$$

where the limit is taken as $z \to \exp(-\log^{-1} n)$.

We suppose that $n$ is even, say $n \pm 2m (m > 3)$, and we observe that the limit (3) can be written as a product of five factors, $G_1, \ldots, G_5$, where

$$G_1 = \lim(1 - z^n \exp n \log^{-1} n)[z - \exp(-\log^{-1} n)]^{-1}, \tag{4}$$

$$G_2 = \prod_{j=2}^{m} [\exp j(\log^{-1} j - \log^{-1} n) - 1], \tag{5}$$

(6) $$G_3 = \{\exp(m+1)[\log^{-1}(m+1) - \log^{-1} n]\} - 1,$$

(7) $$G_4 = \prod_{m+2}^{n-1} [\exp j(\log^{-1} j - \log^{-1} n) - 1]$$
$$\times \prod_{n+1}^{2n} [1 - \exp j(\log^{-1} j - \log^{-1} n)],$$

(8) $$G_5 = \prod_{2n+1}^{\infty} [1 - \exp j(\log^{-1} j - \log^{-1} n)].$$

We shall estimate these factors one at a time. The principal tools we use come from elementary calculus. Estimating $G_1$ is easy:

(9) $$G_1 = -n \exp \log^{-1} n.$$

For $G_2$, we shall establish the estimate, valid for large $n$,

(10) $$G_2 > \exp n^2 \log^{-3} n.$$

For this estimate, set

(11) $$h_n(x) = x(\log^{-1} x - \log^{-1} n).$$

We have that

$$h_n'(x) = -\log^{-2} x + \log^{-1} x - \log^{-1} n,$$

and thus $h_n'$ vanishes at points $\alpha(n)$ and $\beta(n)$ which satisfy

$$\log \alpha(n) = \tfrac{1}{2} \log n(1 + \sqrt{1 - 4\log^{-1} n})$$

and

$$\log \beta(n) = \tfrac{1}{2} \log n(1 - \sqrt{1 - 4\log^{-1} n}).$$

Now as $n$ becomes large, $\log \alpha(n) \to \infty$, so $\alpha(n) \to \infty$, and also $\log \beta(n) \to 1$, so $\beta(n) \to e$. For large values of $n$, $h_n(2)$ and $h_n(n/2)$ are greater than 1, and moreover

$$\lim_{n \to \infty} h_n(\beta(n)) = e;$$

so for large $n$, $h_n(\beta(n)) > 1$. It follows that $h_n > 1$ on the interval $[2, n/2]$ for sufficiently large $n$. Combining this with the fact that if $y > 1$, then $e^y - 1 > \exp \frac{1}{2}y$, we find that

$$G_2 > \prod_{j=2}^{m} \exp \tfrac{1}{2} j(\log^{-1} j - \log^{-1} n).$$

Since

$$\log^{-1} j - \log^{-1} n = (\log n - \log j)(\log n \log j)^{-1}$$
$$\geq (n - j)n^{-1} \log^{-2} n,$$

we have

(12) $$G_2 \geq \prod_{j=2}^{m} \exp \tfrac{1}{2} j(n - j)n^{-1} \log^{-2} n.$$

For $j \leq n/2$, we have $j^2 < (n - j)j$, so if we use the formula

$$\sum_{j=2}^{m} j^2 = \tfrac{1}{6} m(m + 1)(2m + 1) - 1,$$

estimate (12) yields

$$\begin{aligned} G_2 &\geq \exp \tfrac{1}{12}[(2m^3 + 3m^2 + m - 6)n^{-1} \log^{-2} n] \\ &= \exp \tfrac{1}{48}[(n^3 + 3n^2 + 2n - 24)n^{-1} \log^{-2} n], \end{aligned}$$

and for large values of $n$ this implies (10).

In order to estimate $G_3$, apply the mean-value theorem to obtain

$$\begin{aligned} (m + 1)(\log^{-1}(m + 1) - \log^{-1} n) &\geq (m + 1)(n - m - 1)n^{-1} \log^{-2} n \\ &= (n^2/4 - 1)n^{-1} \log^{-2} n. \end{aligned}$$

Taken together with $e^y - 1 \geq \exp y/2$ for $y \geq 1$, this yields the inequality

(13) $$G_3 \geq \exp \frac{n}{9} \log^{-2} n$$

for large values of $n$.

For the purposes of estimating $G_4$, let $H_1$ be its first factor, $H_2$ its second. Our earlier considerations of the function $h_n$ show that

$$\exp x(\log^{-1} x - \log^{-1} n)$$

decreases monotonically on the interval

$$(\tfrac{1}{2} \log n(1 + \sqrt{1 - 4 \log^{-1} n}), \infty).$$

For large enough values of $n$,

$$\tfrac{1}{2} \log n(1 + \sqrt{1 - 4 \log^{-1} n}) < n/2,$$

so we find that

(14) $$H_1 \geq \{\exp(n - 1)[\log^{-1}(n - 1) - \log^{-1} n] - 1\}^{m-2}.$$

Similarly,

(15) $$H_2 \geq \{1 - \exp(n + 1)[\log^{-1}(n + 1) - \log^{-1} n]\}^n.$$

We shall show that, for large $n$,

(16) $$\exp(n - 1)[\log^{-1}(n - 1) - \log^{-1} n) - 1 \geq 1/n.$$

This is equivalent to

(17) $\quad (n-1)[\log^{-1}(n-1) - \log^{-1} n] \geq \log(n+1) - \log n.$

By the mean-value theorem,

$$(n-1)[\log^{-1}(n-1) - \log^{-1} n] \geq (n-1)n^{-1}\log^{-2} n$$

and

$$\log(n+1) - \log n < 1/n.$$

For large values of $n$, these combine to yield (17) and so (16). From (16) and (14) we find that

(18) $$H_1 \geq (1/n)^{m-2}.$$

As above, we have that

$$1 - \exp(n+1)[\log^{-1}(n+1) - \log^{-1} n] > 1/n,$$

so

(19) $$H_2 \geq (1/n)^n,$$

and we find, finally, that

(20) $$G_4 \geq (1/n)^{(3/2)n-2}$$

for large values of $n$.

It remains to obtain an estimate for $G_5$. The estimate in this case is that if $\delta \in (0, 1)$, then for sufficiently large values of $n$,

(21) $$1 > G_5 > \delta.$$

To establish this, use the well-known fact from the theory of infinite products that

$$G_5 = \prod_{2n+1}^{\infty} [1 - \exp j(\log^{-1} j - \log^{-1} n)]$$

$$\geq 1 - \sum_{2n+1}^{\infty} \exp j(\log^{-1} j - \log^{-1} n).$$

We know that on an interval including $(2n, \infty)$ the function $\exp x(\log^{-1} x - \log^{-1} n)$ decreases monotonically, so we have

$$0 < \sum_{2n+1}^{\infty} \exp j(\log^{-1} j - \log^{-1} n)$$

$$< \int_{2n}^{\infty} \exp x(\log^{-1} x - \log^{-1} n)\, dx$$

$$\leq \int_{2n}^{\infty} \exp x(\log^{-1} 2n - \log^{-1} n)\, dx$$

$$= -\gamma(n)e^{2n\gamma(n)}, \qquad \gamma(n) = (\log 2n \log n)^{-1} \log \tfrac{1}{2},$$

and as $n \to \infty$, this quantity tends to 0. Thus, we obtain estimate (21) for $G_5$.

The estimates we have obtained for $G_1, \ldots, G_5$ now combine to provide estimate (a) of Lemma 27.5.

To obtain part (b) of Lemma 27.5, we again denote by $R_n$ the term

$$\tfrac{1}{2}\{\exp(-\log^{-1} n) + \exp[-\log^{-1}(n+1)]\}.$$

Since $|z| = R_n$ implies that $|\varphi(z)| \geq |\varphi(R_n)|$, it suffices to estimate $\varphi(R_n)$. We have

$$|\varphi(R_n)| = \varphi_n \psi_n,$$

where

$$\varphi_n = \prod_2^n [(R_n^j \exp j \log^{-1} j) - 1]$$

and

$$\psi_n = \prod_{n+1}^{\infty} [1 - (R_n^j \exp j \log^{-1} j)].$$

From

$$\exp(-\log^{-1} n) \leq R_n < \exp[-\log^{-1}(n+1)],$$

it follows that

$$\varphi_n \geq [(R_n^n \exp n \log^{-1} n) - 1] \times \prod_2^{n-1} \{[\exp j(\log^{-1} j - \log n)] - 1\},$$

and

$$\psi_n \geq \{1 - [R_n^{n+1} \exp(n+1) \log^{-1}(n+1)]\} \times \prod_{n+2}^{\infty} \{1 - (\exp j[\log^{-1} j - \log^{-1}(n+1)])\}.$$

In the notation of the proof of the first half of the lemma, we have

$$\varphi_n \geq \{[R_n^n \exp(n \log^{-1} n)] - 1\} G_2 G_3 H_1,$$

so we find from our earlier estimates that

$$\varphi_n \geq [R_n^n \exp(n \log^{-1} n) - 1] \times (1/n)^{(n/2)-2} \exp(n^2 \log^{-3} n) \exp\left(\frac{n}{9} \log^{-2} n\right).$$

In addition, our estimates on $H_2$ and $G_2$ yield

$$\psi_n \geq \delta\{1 - R_n^{n+1} \exp[(n+1)\log^{-1}(n+1)]\}\left(\frac{1}{n+1}\right)^{1/(n+1)},$$

where $0 < \delta < 1$ and $\delta$ can be made as near 1 as we wish by taking $n$ large enough. We finish the proof by showing that for large $n$,

(22) $$R_n^n \exp(n \log^{-1} n) - 1 > 1/n$$

and

(23) $$1 - R_n^{n+1} \exp[(n+1)\log^{-1}(n+1)] > 1/n.$$

Inequality (22) is equivalent to

$$R_n \exp(\log^{-1} n) > (1 + 1/n)^{1/n},$$

and recalling the definition of $R_n$, this inequality is

(24) $$1 + \exp[\log^{-1} n - \log^{-1}(n+1)] > 2(1 + 1/n)^{1/n}.$$

Since $\log^{-1} n - \log^{-1}(n+1) = \log(1 + 1/n)\log^{-1} n \log^{-1}(n+1)$, it follows by raising both sides to the $n$th power that (24) is equivalent to

(25) $$[1 + (1 + 1/n)^{(n+1)\beta_n}]^n > 2^n(1 + 1/n)$$

if

$$\beta_n = [(n+1)\log n \log(n+1)]^{-1}.$$

For large values of $n$, $(1 + 1/n)^n > \sqrt{e}$, so eventually the left side of (25) is greater than

$$(1 + e^{(1/2)\beta_n})^n > (1 + 1 + \tfrac{1}{2}\beta_n)^n,$$

and, by the binomial theorem,

$$(2 + \tfrac{1}{2}\beta_n)^n > 2^n(1 + \tfrac{1}{4}n\beta_n).$$

Since $\frac{1}{4}n\beta_n > 1/n$ for large $n$, inequality (25) is valid for large $n$, and so (22) holds if $n$ is large enough.

To prove (23), we use the fact that it is equivalent to

$$2(1 - 1/n)^{(n+1)^{-1}} > 1 + \exp[\log^{-1}(n+1) - \log^{-1} n].$$

For large values of $n$, the right-hand side is

$$1 + \exp\left[\log\left(\frac{n}{n+1}\right)\log^{-1}(n+1)\log^{-1} n\right]$$
$$= 2 + b_n + o(b_n),$$

if we write $b_n$ for the number

$$\log\left(\frac{n}{n+1}\right)\log^{-1}(n+1)\log^{-1}n,$$

and the left-hand side is

$$2\exp\left[(n+1)^{-1}\log\left(\frac{n-1}{n}\right)\right] = 2 + O\left[(n+1)^{-1}\log\left(\frac{n-1}{n}\right)\right]$$
$$= 2 + o(b_n).$$

Since $b_n < 0$, we have the result.

# *chapter six*/POLYNOMIAL CONVEXITY AND APPROXIMATION ON CURVES IN $\mathbb{C}^N$

## 28 *ABSTRACT FUNCTION THEORY*

At many points in our development of the preceding theory, we have used the theory of functions of several complex variables. It is not surprising that function theory should play an important role in our theory, for, as we know, the theory of uniform algebras coincides with the theory of the uniform closure of the polynomials on polynomially convex sets in $\mathbb{C}^\Lambda$, $\Lambda$ finite or infinite. Hitherto in our treatment we have not emphasized this point of view, but in this final chapter we shall bring it to the forefront. In particular, we shall study explicitly the theory of polynomial convexity.

In the first section, we shall develop a small portion of what may be termed *abstract function theory*, i.e., the development, in abstract function algebra settings of analogues of results from the theory of holomorphic functions of one or several variables. Not only do we obtain by this approach generalizations of the classical theory, but also the approach leads to significant new insights into the older theory. (In many cases the proofs of the general theorems, when specialized to the classical setting, are essentially different from the original proofs.) Although we shall consider only results relating to convexity, the reader should be aware of the existence of a considerably larger body of theory in this general direction. For this theory, we refer to the literature, particularly to Rickart [3, 4], Glicksberg [3, 4], Quigley [1, 2], Shauck [1], and Birtel [1].

In the second section of this chapter, we shall treat polynomial convexity in $\mathbb{C}^N$ and also the related notion of rational convexity. The final section is devoted to some topics in the theory of uniform approximation on curves in $\mathbb{C}^N$.

Our first objective is the study of a class of algebras of functions which are not necessarily uniform algebras but to which the methods of the theory of uniform algebras can be applied. This theory is due to Rickart [3].

**28.1** ***Definition***

*A **natural algebra** is a pair $(A, \Sigma)$ where $\Sigma$ is a Hausdorff space and* A *is a complex algebra of complex-valued functions on $\Sigma$ which contains the constants and which determines the topology on $\Sigma$ in the sense that the given topology is the weakest with respect to which all the elements of* A *are continuous. Finally, we require that every nonzero complex homomorphism of* A *which is continuous with respect to the topology of uniform convergence on compact sets be of the form $f \mapsto f(x)$ for some $x \in \Sigma$.*

Many examples of such natural algebras are familiar.

**28.2** ***Examples***

(a) If $A$ is a commutative Banach algebra with identity, then $(\hat{A}, \Sigma(A))$ is a natural algebra.
(b) If $\Lambda$ is an index set, then $(\mathscr{P}_\Lambda, \mathbb{C}^\Lambda)$ is a natural algebra. (Recall that $\mathscr{P}_\Lambda$ denotes the polynomials in $\Lambda$ variables. See Definition 5.3.)
(c) If $\mathscr{O}(\mathbb{C}^N)$ is the algebra of entire functions in $N$ complex variables, then $(\mathscr{O}(\mathbb{C}^N), \mathbb{C}^N)$ is a natural algebra.

The complex homomorphisms of $A$ considered in the definition of natural algebras are required to be continuous with respect to the topology of uniform convergence on compacta in $X$. A complex homomorphism $\varphi$ of $A$ enjoys this property if and only if there is a compact set $K \subset \Sigma$ such that for some constant $C$ and all $f \in A$,

$$(1) \qquad |\varphi(f)| \leq C\|f\|_K.$$

If (1) holds for all $f \in A$, then we also obtain, for $n = 1, 2, \ldots, |\varphi(f)| \leq C^{1/n}\|f\|_K$ by applying (1) to $f^n$. If we let $n$ become large, it follows that $|\varphi(f)| \leq \|f\|_K$ for all $f \in A$.

If $(A, \Sigma)$ is a natural algebra, the algebra $A$ with the topology of uniform convergence on compacta in $\Sigma$ will, in general, be neither metrizable nor complete. Example (b) is an instance in which the algebra is not complete.

In addition to natural algebras, we shall be interested in certain associated algebras. Generalizing Definition 14.8, we introduce locally $A$-approximable functions.

**28.3** ***Definition***

*If $(A, \Sigma)$ is a natural algebra and if $E$ is a subset of $\Sigma$, $f \in \mathscr{C}(E)$ is **locally $A$-approximable on** $E$ provided that for every point $x \in E$, there is an open set $V$ in $\Sigma$, $x \in V$, such that on $V \cap E$, $f$ is uniformly approximable by elements of $A$. The set of all such functions is denoted by $\mathscr{H}_A(E)$.*

Thus, if we take $(A, \Sigma)$ to be the natural algebra of example (b), $\Lambda$ a finite set, and if we take $E$ to be an open set in $\Sigma$, then $\mathscr{H}_A(E)$ is simply the algebra $\mathcal{O}(E)$ of holomorphic functions on $E$. The same result is obtained if we start with example (c).

The functions in $\mathscr{H}_A(E)$ were called *locally A-holomorphic* by Rickart.

We shall need a simple algebraic fact about the algebras $\mathscr{H}_A(E)$.

**28.4** ***Lemma***

*If $(A, \Sigma)$ is a natural algebra, if $E \subset \Sigma$, and if $f \in \mathscr{H}_A(E)$ is zero free, then $f^{-1} \in \mathscr{H}_A(E)$.*

PROOF

Let $x_0 \in E$, and choose a neighborhood $V$ of $x_0$ contained in $\{x : |f(x) - f(x_0)| < \frac{1}{4}|f(x_0)|\}$ on which $f$ is the uniform limit of the sequence $\{f_n|(V \cap E)\}$, $f_n \in A$. For large $n$, we have $f_n(x_0) \neq 0$, and $|f_n(x) - f_n(x_0)| < \frac{1}{2}|f_n(x_0)|$, $x \in V \cap E$. The series

$$f_n^{-1} = f_n^{-1}(x_0) \sum_{k=0}^{\infty} \left( \frac{f_n(x_0) - f_n}{f_n(x_0)} \right)^k$$

converges uniformly on $V \cap E$, so $f_n^{-1} \in \mathscr{H}_A(E)$. The sequence $\{f_n^{-1}\}$ converges uniformly to $f^{-1}$ on $V \cap E$, so if $\varepsilon > 0$, there is an $n$ such that $\|f_n^{-1} - f^{-1}\|_{V \cap E} < \varepsilon/2$, and then there is a partial sum $S$ of the series given above for $f_n^{-1}$ which is uniformly within $\varepsilon/2$ of $f_n^{-1}$ on $V \cap E$. Thus, $S$ is uniformly within $\varepsilon$ of $f^{-1}$ on $V \cap E$, so since $S$ lies in $A$, we find finally that $f^{-1} \in \mathscr{H}_A(E)$.

We are particularly interested in $\mathscr{H}_A(E)$ where $E$ is $A$-convex.

**28.5** ***Definition***

*If $(A, \Sigma)$ is a natural algebra and if $E \subset \Sigma$, the A-**convex hull** of $E$ is the set*

$$\hat{E} = \{x \in \Sigma : |f(x)| \leq \|f\|_E \text{ for all } f \in A\}.$$

*If $E = \hat{E}$, then $E$ is A-**convex**.*

When it is necessary in order to avoid confusion, we shall write $A$-hull $E$ instead of $\hat{E}$.

If $(A, \Sigma)$ is the natural algebra of polynomials on $\mathbb{C}$, then the compact, $A$-convex sets are the polynomially convex subsets of $\mathbb{C}$.

As a further extension of previously used notation, given a natural algebra $(A, \Sigma)$ and a compact set $E$ in $\Sigma$, we shall denote by $A_E$ the uniform closure in $\mathscr{C}(E)$ of the algebra of restrictions $\{f|E : f \in A\}$. Thus, $(A_E, E)$ is a uniform algebra. Of course, the algebras $A_E$ and $\mathscr{H}_A(E)$ will in general be distinct. The question of the equality of $A_E$ and $\mathscr{H}_A(E)$ leads immediately to many interesting, concrete approximation problems.

28.6 **Lemma**

*If $(A, \Sigma)$ is a natural algebra and if $E \subset \Sigma$ is compact, then $\hat{E}$ is compact, and $(A_{\hat{E}}, \hat{E})$ is a natural algebra.*

PROOF

The algebra $A_E$ is a uniform algebra on $E$. We shall identify $\hat{E}$ with the spectrum of $A_E$, and $A_{\hat{E}}$ with the algebra of Gelfand transforms of $A_E$. This will imply that $\hat{E}$ is compact and that $(A_{\hat{E}}, \hat{E})$ is a natural algebra.

If $\{f_n\}$ is a sequence in $A$ which converges uniformly on $E$, the definition of $\hat{E}$ implies that $\{f_n\}$ converges uniformly on the whole of $\hat{E}$, so each $f \in A_E$ admits a unique extension to an element, again denoted by $f$, of $A_{\hat{E}}$. In this way, we may identify $A_E$ and $A_{\hat{E}}$.

There is a natural map $i: \hat{E} \to \Sigma(A_E)$ given by $(ix)f = \hat{f}(x)$. This map is one to one and continuous. The topology on $\Sigma(A_E)$ is the weak topology determined by the algebra of Gelfand transforms of elements of $A_E$, and the topology on $\hat{E}$ is the weak topology determined by the algebra $A|\hat{E}$, which is dense in $A_{\hat{E}}$. It follows that the map $i$ is, in fact, a homeomorphism of $\hat{E}$ into $\Sigma(A_E)$.

It is onto. This is so, for a $\varphi \in \Sigma(A_E)$ is completely determined by its action on the algebra $A|E$, and we have

$$|\varphi(f)| \leq \|f\|_E \tag{2}$$

for all $f \in A_E$. Inequality (2) implies that $\varphi$ acts on $A$ as a nonzero complex homomorphism which is continuous with respect to the topology on $A$ of uniform convergence on compacta in $\Sigma$; so the assumption that $(A, \Sigma)$ is a natural algebra implies the existence of a unique point $x_0 \in \Sigma$ such that $\varphi(f) = f(x_0)$ for all $f \in A$, and (2) implies that $x_0 \in \hat{E}$. Thus, $\varphi$ is of the form $f \mapsto f(x_0)$, and $i$ is onto.

The last lemma implies that a version of the local maximum modulus theorem is applicable in the setting of natural algebras.

28.7 **Lemma**

*Let $(A, \Sigma)$ be a natural algebra, let $E$ be a compact, $A$-convex subset of $\Sigma$, and let $\Gamma$ be the Shilov boundary for $A$. If $U$ is open in $E$ and contained in $E \setminus \Gamma$, then for every $f \in A$,*

$$\sup\{|f(x)| : x \in U\} = \sup\{|f(x)| : x \in \partial_E U\}.$$

*Here $\partial_E U$ denotes the topological boundary of $U$ with respect to $E$.*

PROOF

This follows from the local maximum modulus theorem applied to the Banach algebra $A_E$.

At this point, the theory of holomorphic functions of several complex variables has entered our discussion in an implicit but nontrivial way.

The local maximum modulus theorem persists for functions in $\mathscr{H}_A(E)$.

**28.8 *Theorem***

*Let $(A, \Sigma)$ be a natural algebra, let $E \subset \Sigma$ be a compact A-convex set, and let $\Gamma$ be the Shilov boundary for $A_E$. If $U$ is an open subset of $E$, $U \subset E \setminus \Gamma$, and if $f \in \mathscr{H}_A(E)$, then*

$$\sup\{|f(x)| : x \in U\} = \sup\{|f(x)| : x \in \partial_E U\}.$$

PROOF

If the theorem is false, there is $x_0 \in U$ and $f \in \mathscr{H}_A(E)$ such that $|f(x_0)| > \|f\|_{\partial_E U}$. This will lead to a contradiction, for we shall show that if $\Gamma'$ is the Shilov boundary for the closed subalgebra $B$ of $\mathscr{C}(\overline{U})$ generated by $A_{\overline{U}}$ and $f|\overline{U}$, then $\Gamma' \cap U = \varnothing$.

If this latter assertion is false, then from the openness of $U$, we conclude the existence of a point $x_1 \in U$ which is a strong boundary point for $B$. There is a neighborhood $V$ of $x_1$, $V \subset U$, on which $f$ is the uniform limit of a sequence from $A$, so $B_{\overline{V}} = A_{\overline{V}}$. The hypothesis that $x_1$ is a strong boundary point for $B$ implies the existence of $g \in B_{\overline{V}} = A_{\overline{V}}$ such that $|g(x_1)| > \|g\|_{\partial_E V}$. If we approximate $g$ on $\overline{V}$ by an $h \in A$, we obtain a contradiction to the preceding lemma.

Note that this theorem is true also of functions in the uniform closure of $\mathscr{H}_A(E)$.

**28.9 *Corollary***

*If $E \subset \Sigma$ is compact and A-convex, then the Shilov boundary for $\mathscr{H}_A(E)$ is the same as the Shilov boundary for $A_E$.*

We come now to the notion of *variety* in this abstract setting.

**28.10 *Definition***

*If $(A, \Sigma)$ is a natural algebra and if $X$ is a subset of $\Sigma$, then the (relatively) closed subset $E$ of $X$ is an A-**variety in** $X$ if for every $x \in V$, there is a neighborhood $N$ of $x$ and a family $\mathscr{F} \subset \mathscr{H}_A(N \cap X)$ such that*

$$E \cap N = \{x \in N \cap X : f(x) = 0 \text{ for all } f \in \mathscr{F}\}.$$

This definition is plainly motivated by the usual definition of analytic varieties in complex analysis. It is to be observed that the set $\mathscr{F}$ in the definition need not be finite. In the complex analytic case, it follows from the Weierstrass preparation theorem that by reducing the neighborhood $N$ if necessary, it is possible to assume $\mathscr{F}$ finite. In the general case no such simplifying assumption is possible.

The main results of the present section are results relating properties of varieties to the theory of convexity. First we shall prove that, roughly put, varieties are convex, and then we shall characterize the convex hull of a set in terms of certain families of varieties. Both of these results are due to Rickart [3]; they are generalizations to an abstract setting of function theoretic results of Oka.

**28.11** ***Theorem***

*If $(A, \Sigma)$ is a natural algebra, if $X \subset \Sigma$ is compact and A-convex, and if $E$ is an A-variety in $X$, then $E$ is A-convex.*

Before giving the details of the proof of this theorem, let us comment on its content in the special case that the natural algebra in question is the algebra $\mathscr{P}_N$ of polynomials on $\mathbb{C}^N$. Let $\bar{\Delta}^N$ be the closed unit polydisc in $\mathbb{C}^N$. The theorem implies, e.g., that if $V$ is an analytic variety in a neighborhood of $\bar{\Delta}^N$, then $V \cap \bar{\Delta}^N$ is polynomially convex, a result that can be established by other means. Another case of the theorem is obtained if we consider a function $f \in \mathscr{A}(\Delta^N)$: The theorem implies that the zero set of $f$ is polynomially convex. This, too, can be proved by other means, for we can identify the zero set of $f$ with the spectrum of the quotient algebra $\mathscr{A}(\Delta_N)/(f)$, and, as spectra are polynomially convex (recall Theorem 5.8), the result follows. An instance in which our theorem applies but which does not seem accessible by other more direct means is the following. Let $V \subset \bar{\Delta}^N$ be a set such that if $x \in V$, then there is a closed neighborhood $N_x$ of $x$ in $\bar{\Delta}^N$ and a family $F_x$ of functions continuous on $N_x$ and holomorphic in its interior such that

$$V \cap N_x = \{y \in N_x : f(y) = 0 \text{ for all } f \in F_x\}.$$

By the theorem, $V$ is polynomially convex.

We turn now to the proof of the theorem. The proof depends on a preliminary lemma.

**28.12** ***Definition***

*If $X \subset Y \subset \Sigma$ and if $y_0 \in Y$, then the set $X$ is said to be* ***locally determining in X at*** *$y_0$ if for every neighborhood $U$ of $y_0$, there is a neighborhood $V$ of $y_0$ such that if $f \in \mathscr{H}_A(U \cap Y)$ and $f|(U \cap X) = 0$, then $f|(V \cap Y) = 0$.*

28.13 ***Lemma***

*If $X \subset \Sigma$ is compact but not A-convex, there exists a point in $X \cap (\hat{X} \setminus X)^-$ at which X is locally determining in $\hat{X}$.*

PROOF

By the local maximum modulus principle for $A$, the Shilov boundary for $A_{(\hat{X} \setminus X)^-}$ is contained in $X \cap (\hat{X} \setminus X)^-$. We shall show that if $x_0$ is a strong boundary point for this algebra, then $X$ is locally determining in $\hat{X}$ at $x_0$.

Let $U$ be a neighborhood of $x_0$. Since the topology for $\Sigma$ is that given by the functions in $A$, there is an open set $U^*$ containing $x_0$ with $\bar{U}^* \subset U$. The algebra $A|(\hat{X} \setminus X)^-$ is dense in $A_{(\hat{X} \setminus X)^-}$, so for some $f \in A$, $f(x_0) > 1$ and $|f(x)| < \frac{1}{3}$ if $x \in (\hat{X} \setminus X)^- \setminus U^*$. Let

$$V = \{x \in U^* : |f(x)| > 1\},$$

$$W = \{x \in U^* : |f(x)| > \tfrac{2}{3}\},$$

and

$$E = \bar{W} \cap \hat{X}.$$

We have then that $V \subset W \subset U$ and $E \subset U \cap \hat{X}$, and we shall show that if $h \in \mathscr{H}_A(U \cap \hat{X})$ vanishes on $U \cap X$, then $h$ vanishes on $V \cap \hat{X}$.

Suppose, on the contrary, that $h \in \mathscr{H}_A(U \cap \hat{X})$ vanishes on $U \cap X$, but $h(y_0) \neq 0$ for some $y_0 \in V \cap \hat{X}$. The function $h$ is continuous on $U \cap \hat{X}$ and so is bounded on the compact set $E$. Thus, for some positive integer $m$, we have

$$(\tfrac{2}{3})^m \|h\|_E \leq |h(y_0)|.$$

Define $g$ by

$$g(x) = f^m(x)h(x)$$

so that $g \in \mathscr{H}_A(U \cap \hat{X})$ and $g|(U \cap X) = 0$. Since $|f(y_0)| > 1$ and $y_0 \in E$, we have

$$\begin{aligned} |h(y_0)| &< |f^m(y_0)h(y_0)| \\ &= |g(y_0)| \leq \|g\|_E. \end{aligned}$$

The function $g$ lies in $\mathscr{H}_A(U \cap \hat{X})$, so Theorem 28.8 applies to yield that $\|g\|_E$ is the supremum of $|g|$ over the union of $X \cap E$ with the boundary, with respect to $\hat{X}$, of $E$. Since $g$ vanishes identically on $X \cap E$, there is $y_1$ in the boundary of $E$ (with respect to $\hat{X}$) such that $\|g\|_E = |g(y_1)|$. We have then that

$$|h(y_0)| < |g(y_1)|. \tag{3}$$

We assert that $|f(y_1)| = \frac{2}{3}$. Since $y_1 \in \overline{W}$, $|f(y_1)| \geq \frac{2}{3}$. We have that $g|(U \cap X) = 0$ and $y_1 \in \hat{X} \setminus X$, so if $y_1 \notin U^*$, we would have $|f(y_1)| < \frac{1}{3}$, a contradiction. Thus, $y_1 \in U^*$. If $|f(y_1)| > \frac{2}{3}$, then $y_1 \in W$ so that $y_1$ lies in the interior (with respect to $\hat{X}$) of $E$, a contradiction. Thus, $|f(y_1)| = \frac{2}{3}$. Also $y_1 \in E$, so we have

$$
\begin{aligned}
|g(y_1)| &= |f^m(y_1)||h(y_1)| \\
&= (\tfrac{2}{3})^m |h(y_1)| \\
&\leq (\tfrac{2}{3})^m \|h\|_E \\
&\leq |h(y_0)|.
\end{aligned} \tag{4}
$$

Inequalities (3) and (4) are inconsistent, so the lemma is proved.

PROOF OF THEOREM 28.11

The theorem is a fairly direct consequence of this lemma. Since $X$ is $A$-convex, the inclusion $E \subset X$ implies also the inclusion $\hat{E} \subset X$. The set $E$ is compact since $X$ is; so if $E \neq \hat{E}$, there is a point $x_0 \in (\hat{E} \setminus E)^- \cap E$ at which $E$ is locally determining. Since $E$ is a variety, there is a neighborhood $U$ of $x_0$ such that $U \cap E$ is an intersection of the zero sets of certain elements of $\mathscr{H}_A(U \cap X)$. Choose a neighborhood $V$ of $x_0$ in accordance with the local determining property so that if $f \in \mathscr{H}_A(U \cap X)$ vanishes on $E$, then necessarily $f$ vanishes on $V \cap \hat{E}$. This yields the inclusion $V \cap \hat{E} \subset E$, which is impossible since $x_0$ is a boundary point of $\hat{E} \setminus E$.

28.14 ***Definition***

*If $(A, \Sigma)$ is a natural algebra, an $A$-**polyhedron** is a set of form $\{x \in \Sigma : |f_i(x)| < 1, i = 1, \ldots, N\}$ where $f_1, \ldots, f_n \in A$.*

This notion is a generalization of that of polynomial polyhedra in $\mathbb{C}^N$.

The preceding theorem has an immediate consequence a further convexity property of varieties.

28.15 ***Corollary***

*If $P \subset \Sigma$ is an $A$-polyhedron, if $E$ is an $A$-variety in $P$, and if $X \subset E$ is compact, then $\hat{X} \subset E$.*

PROOF

Let $P = \{x \in \Sigma : |f_1(x)|, \ldots, |f_n(x)| < 1\}$, $f_1, \ldots f_n \in A$. By Lemma 28.6, $A$-hull $X$ is compact, and since $\|f_i\|_X < 1$, it follows that $\|f_i\|_{\hat{X}} < 1$, so $\hat{X} \subset P$. The set $E$ is an $A$-variety and so is a closed subset of $P$, so $E \cap \hat{X}$ is compact, and it follows that $E \cap \hat{X}$ is an $A$-variety in $\hat{X}$. Consequently, $E \cap \hat{X}$ is $A$-convex. As $X \subset E \cap \hat{X}$, we have that

$$\hat{X} \subset (E \cap \hat{X})\hat{\ } = E \cap \hat{X}$$
$$\subset E,$$

as was to be shown.

We come now to the second main result of this section, a characterization of hulls in terms of the way they are intersected by certain families of varieties.

28.16 ***Definition***
*If $(A, \Sigma)$ is a natural system and if $G \subset \Sigma$ is an open set, an $A$-**hypersurface** in $G$ is a set of the form*

$$\{x \in G : h(x) = 0\}$$

*for some fixed $h \in \mathscr{H}_A(G)$. A family $\{V_t : t \in [0, 1]\}$ of $A$-hypersurfaces in $G$ is called a **curve of $A$-hypersurfaces** in $G$ if there is a continuous function $F : [0, 1] \times G \to \mathbb{C}$ such that for each $t \in [0, 1]$, $F(t, \cdot) \in \mathscr{H}_A(G)$, and for each $t$, $V_t = \{x \in G : F(t, x) = 0\}$.*

28.17 ***Definition***
*If $X \subset \Sigma$ and if $\{V_t\}$ is a curve of $A$-hypersurfaces in $G$, then $\{V_t\}$ is said to **intersect** $X$ if $V_t \cap X \neq \varnothing$ for some $t$, and it is said to **intersect** $X$ **nontrivially** if $V_t \cap X$ is closed in $X$ for each $t$, and if, in addition, the set*

$$T = \{t : V_t \cap X \neq \varnothing\}$$

*is a proper, nonempty, closed subset of* $[0, 1]$.

We can formulate now our characterization of hulls.

28.18 ***Theorem***
*If $(A, \Sigma)$ is a natural algebra, and if $X$ is a compact set in $\Sigma$, then every curve of $A$-hypersurfaces which intersects $\hat{X}$ nontrivially must intersect $X$.*

Explicitly, if $G$ is an open set in $X$ and if $\{V_t\}$ is a curve of $A$-hypersurfaces in $G$ which intersects $\hat{X}$ nontrivially, then $V_t$ has to meet $X$. It is to be emphasized that the set $G$ is not assumed to be a neighborhood of $\hat{X}$ or of $X$; it is completely arbitrary.

Let us consider some examples.

28.19 ***Examples***

(a) In $\mathbb{C}$ let $X$ be a compact set with polynomially convex hull $\hat{X}$. If $F:[0, 1] \times \mathbb{C} \to \mathbb{C}$ is a continuous function with the property that for every $t$, the function $F(t, \cdot)$ is an entire function, and if, say, $F(0, \cdot)$ has no zeros in $\hat{X}$ while $F(1, \cdot)$ has a zero in $\hat{X}$, then for some $t \in [0, 1]$, $F(t, \cdot)$ has a zero in $X$. This is a perfectly elementary fact which does not depend on the theorem under discussion, but it does seem to be the most easily visualized special case of the theorem.

(b) The case of the theorem that relates to polynomially convex sets in $\mathbb{C}^N$ goes back to the work of Oka [2], where, however, one of the hypotheses is omitted. The difficulty there is rooted in the fact that the theorem is false if we omit from the definition of *intersect nontrivially* the requirement that the set $T$ be closed. This difficulty escaped the attention of writers on the subject until it was brought to light in the paper [3] of Rickart, where the observation of the problem is attributed to Garnett. An elementary example is as follows.

We work with the natural algebra $(\mathscr{P}, \mathbb{C})$ of polynomials in one complex variable. Let $X = \{z \in \mathbb{C} : |z| = 2\}$ so that $\hat{X}$ is the closed disc of radius 2 centered at the origin. Let $G = \{z \in \mathbb{C} : \operatorname{Re} z < 0\}$, and let $F:[0, 1] \times G \to \mathbb{C}$ be given by

$$F(t, z) = z + t.$$

If $V_t$ denotes the zero set of $F(t, \cdot)$, then $\{V_t\}$ is a curve of $\mathscr{P}$-hypersurfaces in $G$ which intersects $\hat{X}$. We have that $V_1 \cap \hat{X} = \{-1\}$, $V_0 \cap \hat{X} = \varnothing$, and that $V_t \cap \hat{X}$ is either a single point or the empty set. In either case, $V_t \cap \hat{X}$ is closed. However, $\{V_t\}$ does not meet $X$. This is not a contradiction to the theorem, for $\{V_t\}$ does not intersect $\hat{X}$ nontrivially in the sense of our definition, because the set

$$\{t : V_t \cap \hat{X} \neq \varnothing\}$$

is $(0, 1]$, not a closed set in $[0, 1]$.

The proof of the theorem is based on a lemma proved by purely Banach algebra techniques.

28.20 ***Lemma***

*Let $(A, \Sigma)$ be a natural system, let $X$ be a compact, $A$-convex subset of $\Sigma$, and let $\Gamma$ be the Shilov boundary for $A_X$. Further, let $Y$ be an open subset of $X \setminus \Gamma$ and let $\mathscr{F}$ be a connected set in $\mathbb{C}(\overline{Y})$ such that for each $f \in \mathscr{F}$, $f | Y \in \mathscr{H}_A(Y)$. If there are $f, g \in \mathscr{F}$ such that $f$ vanishes at a point of $Y$ but $g$ is zero free on $Y$, then there is an $h \in \mathscr{F}$ which vanishes at some point of the boundary of $Y$.*

PROOF

Let $B \subset \mathbb{C}(\bar{Y})$ consist of those functions $F$ with $F|Y \in \mathscr{H}_A(Y)$, and let $\bar{B}$ be its uniform closure so that $(\bar{B}, \bar{Y})$ is a uniform algebra. We have $\mathscr{F} \subset B$. By Lemma 28.4, the element $g$ is invertible in $B$ or else $g$ has a zero in $\bar{Y} \setminus Y$. In the latter case we are finished, so we assume henceforth that $g$ is regular in $B$. Since $f$ vanishes at a point of $Y$, it is not invertible in $\bar{B}$, so the connected set $\mathscr{F}$ must contain an element $g_0$ in the topological boundary (with respect to $\bar{B}$) of the group of invertible elements of $\bar{B}$. We know by Theorem 2.7 that $g_0$ is a topological divisor of zero in $\bar{B}$.

Now a topological division of zero in a uniform algebra has to have a zero on the Shilov boundary, for if $|F| \geq \varepsilon$ on the Shilov boundary but $FG_n \to 0$, then we have that $G_n \to 0$, for on the Shilov boundary $|FG_n| \geq \varepsilon|G_n|$. Thus, $F$ is not a topological zero divisor.

Our function $g_0$ therefore has a zero on the Shilov boundary for $\bar{B}$ which is the same as the Shilov boundary for $B$. However, the local maximum modulus theorem for elements of $\mathscr{H}_A(Y)$ shows that the Shilov boundary for $B$ lies in $\bar{Y} \setminus Y$, and we have the lemma.

PROOF OF THE THEOREM

Let $G$ be an open set in $\Sigma$ and let $\{V_t\}$ be a curve of $A$-hypersurfaces in $G$ which intersects $\hat{X}$ nontrivially. Assume that no $V_t$ meets $X$. Let $F:[0, 1] \times G \to \mathbb{C}$ be the function that determines $\{V_t\}$. The hypothesis implies that the set

$$T = \{t : V_t \cap \hat{X} \neq \varnothing\}$$

is closed, so there is an interval $(a, b) \subset [0, 1] \setminus T$ with one end point in $T$. Consequently, by a change of parameters we can suppose that the family $\{V_t\}$ has the property that

$$V_0 \cap \hat{X} \neq \varnothing, \qquad \text{but } V_t \cap \hat{X} = \varnothing \text{ for } 0 < t \leq 1.$$

The set $V_0 \cap \hat{X}$ is compact and by hypothesis it is contained in $G \setminus X$. Thus, there is a relatively open set $Y$ in $\hat{X}$ such that $\bar{Y} \subset G \setminus X$ and $V_0 \cap \hat{X} \subset Y$. The family $\{F_t\}$ given by

$$F_t(x) = F(t, x)$$

is a connected set in $\mathscr{C}(\bar{Y})$ to which the last lemma may be applied: $V_t \cap \bar{Y}$ meets the boundary of $Y$ for some $t$. This, however, is impossible, as $V_t \cap \hat{X} \subset Y$ for all $t$, so the theorem is proved.

As a consequence of the $\mathbb{C}^N$-version of the theorem just proved, we have the following fact.

28.21 ***Corollary*** (Stolzenberg [3])
*If $X \subset \mathbb{C}^N$ is compact, if $S \subset \hat{X}$ is compact, and if $f$ is holomorphic on a neighborhood of $S$, then*

$$|f(z_0)| \leq \sup\{|f(z)| : z \in \partial S \cup (S \cap X)\}$$

*for every $z_0 \in S$. (Here $\partial$ denotes boundary with respect to $\hat{X}$.)*

PROOF
Assume the result false so that for some $z_0 \in S$ and for some $f$ holomorphic on a neighborhood of $S$,

$$\begin{aligned} 1 = f(z_0) &= \|f\|_S \\ &> \|f\|_{\partial S \cup (S \cap X)}. \end{aligned} \tag{5}$$

There is a compact set $S_1$ in $\mathbb{C}^N$ with interior $S_1^0$ such that $S_1 \cap \hat{X} = S$, $S_1^0 \cap \hat{X} = S \setminus \partial S$, and such that $f$ is holomorphic on a neighborhood of $S_1$. The function $f$ is holomorphic and nonconstant on the component of $S_1^0$ which contains $z_0$, for otherwise $f$ would assume the value 1 at a point of $\partial S \cup (S \cap X)$. Therefore, $f(S_1^0)$ contains a neighborhood of 1 in $\mathbb{C}$, and consequently, if $\delta > 0$ is small enough, $f(S_1^0)$ contains the interval $[1, 1 + \delta]$.

Define $F : [1, 1 + \delta] \times S_1^0 \to \mathbb{C}$ by

$$F(t, \mathfrak{z}) = f(\mathfrak{z}) - t,$$

and let $V_t = \{\mathfrak{z} \in S_1^0 : F(t, \mathfrak{z}) = 0\}$ so that $\{V_t\}$ is a curve of $\mathcal{O}(\mathbb{C}^N)$-hypersurfaces in $S_1^0$.

We have $|f| \leq 1$ on $S$ and $V_t \cap \hat{X} \subset S$, so $V_t \cap \hat{X} = \varnothing$ if $t > 1$. The point $z_0$ lies in $V_1 \cap \hat{X}$, so this set is not empty. Also, $V_1 \cap \hat{X}$ is closed in $\hat{X}$, for if $w_n \in V_1 \cap \hat{X}$ and $w_n \to w_0$, then $w_0 \in S$ since this is a compact set, and since $f(w_n) = 1$, it follows that

$$f(w_0) = 1 > \|f\|_{\partial S \cup (S \cap X)},$$

whence $w_0 \in S \setminus \partial S = S_1^0 \cap \hat{X}$, so $w_0 \in S_1^0$. Thus, $w_0 \in V_1$, and we have that $V_1 \cap \hat{X}$ is closed. Consequently, $\{V_t\}$ intersects $\hat{X}$ nontrivially, and from the theorem we get that $\{V_t\}$ must meet $X$: For some $t$, $X \cap V_t \neq \varnothing$, and this implies, since $\hat{X} \cap V_t = \varnothing$ if $t > 1$, the existence of a point $w \in S \cap X$ at which $f$ assumes the value 1, a contradiction to (5).

28.22 ***Corollary***
*If $f$ and $S$ are as in the preceding corollary, then*

$$\partial_{\mathbb{C}} f(S) \subset f(\partial S \cup (S \cap X)).$$

Here $\partial_{\mathbb{C}}$ denotes boundary with respect to the complex plane and $\partial$ boundary with respect to $\hat{X}$.

PROOF

If $z_0 \in \partial_{\mathbb{C}} f(S) \setminus f(\partial S \cup (S \cap X))$, there is $\zeta_0 \in \mathbb{C} \setminus f(S)$ such that

$$|(z_0 - \zeta_0)^{-1}| > \sup\{|(\zeta - \zeta_0)^{-1}| : \zeta \in f(\partial S \cup (S \cap X))\}.$$

But then $(f - \zeta_0)^{-1}$ is holomorphic about $S$, and it contradicts the last corollary.

# 29 *POLYNOMIAL CONVEXITY*

In this section we turn our attention to slightly more concrete matters, the study of polynomially convex sets in $\mathbb{C}^N$. Whereas the study of polynomially convex sets in the plane is quite simple because a compact set in $\mathbb{C}$ is polynomially convex if and only if it does not separate $\mathbb{C}$, there is no correspondingly simple statement in $\mathbb{C}^N$. Of course, the maximum modulus principle does imply that a polynomially convex set in $\mathbb{C}^N$ does not separate $\mathbb{C}^N$, but easy examples show the converse to be false. We have mentioned only one other general geometric property of polynomially convex compact sets in $\mathbb{C}^N$, the theorem of Browder (Theorem 10.6), according to which $H^k(X, \mathbb{C}) = 0$ for $k \geq N$ if $X \subset \mathbb{C}^N$ is polynomially convex.

The theory of polynomial convexity in $\mathbb{C}^N$ is quite complicated, and as it stands the construction of counterexamples plays a central role in it. As we shall see, another major part of the theory is devoted to the relation between rational convexity and polynomial convexity.

One of the important tools in our study will be the *Oka approximation theorem.*

29.1 ***Theorem*** (Oka [1])

*If $X \subset \mathbb{C}^N$ is a compact polynomially convex set, and if $F$ is holomorphic on a neighborhood of $X$, then on $X$, $F$ is the uniform limit of a sequence of polynomials.*

PROOF

We are to prove that $F|X$ is an element of the algebra $\mathscr{P}(X)$. Since $X$ is the spectrum of the algebra $\mathscr{P}(X)$, $X$ being polynomially convex by hypothesis, and since $F$ is holomorphic on a neighborhood of this spectrum, the theorem is an immediate consequence of the Shilov–Arens–Calderon theorem (Theorem 8.3).

In the other direction, it is interesting to remark that a proof of the Shilov–Arens–Calderon theorem, at least in the case of uniform algebras, can be based on the Oka approximation theorem. This development may be found in the book of Gunning and Rossi [1].

29.2 ***Corollary***
*If $X \subset \mathbb{C}^N$ is compact, then $\partial_{\mathbf{C}} f(\hat{X}) \subset f(X)$ for every function $f$ holomorphic on a neighborhood of $\hat{X}$.*

PROOF
If the corollary is false, there are points $\zeta \in \partial_{\mathbf{C}} f(\hat{X}) \setminus f(X)$ and

$$\zeta_0 \in \mathbf{C} \setminus f(\hat{X})$$

with

$$|\zeta_0 - \zeta|^{-1} > \sup\{|z - \zeta|^{-1} : z \in f(X)\}.$$

The function $g = (f - \zeta_0)^{-1}$ is holomorphic on a neighborhood of $\hat{X}$, but it assumes at each point of the nonempty set $f^{-1}(\zeta)$ a value of modulus greater than $\|g\|_X$. By the approximation theorem, $g \in \mathscr{P}(X)$, so we have a contradiction.

The notion of rational convexity will be important in much of the sequel, and accordingly we introduce the following terminology.

29.3 ***Definition***
*If $X \subset \mathbb{C}^N$ is compact, its **rationally convex hull**, $\mathscr{R}$-hull $X$, is the set*

$$\mathscr{R}\text{-hull } X = \{\mathfrak{z} \in \mathbb{C}^N : \textit{if } p \textit{ and } q \textit{ are polynomials, } q \textit{ zero free on } X, \textit{ then } |p(\mathfrak{z})| \leq |q(\mathfrak{z})| \, \|pq^{-1}\|_X\}.$$

*The set $X$ is **rationally convex** if $X = \mathscr{R}$-hull $X$.*

There are certain alternative descriptions of the rationally convex hull of a set $X$ which will prove useful. First of all, there is another description in terms of polynomials.

29.4 ***Lemma***
*If $X$ is a compact set in $\mathbb{C}^N$, then $\mathscr{R}$-hull $X = \{\mathfrak{z} \in C^N : p(\mathfrak{z}) \in p(X)$ for all polynomials $p\}$.*

PROOF
Let $\mathfrak{z} \in \mathbb{C}^N$ and suppose that there is a polynomial $p$ such that $p(\mathfrak{z}) \notin p(X)$. Then $[p - p(\mathfrak{z})]^{-1}$ is a rational function which shows, in accordance with the definition, that $\mathfrak{z} \notin \mathscr{R}$-hull $X$. Conversely, if $\mathfrak{z} \notin \mathscr{R}$-hull $X$,

then there are polynomials $p$ and $q$, $q$ zero free on $X$, such that

$$|p(\mathfrak{z})| > |q(\mathfrak{z})| \, \|pq^{-1}\|_X.$$

The inequality is strict, so $p(\mathfrak{z}) \neq 0$. If $q(\mathfrak{z}) = 0$, we are done since $q$ is zero free on $X$. Otherwise, replacing $p$ by $\alpha p$ for some $\alpha \in \mathbb{C}$, we can suppose that $p(\mathfrak{z}) = q(\mathfrak{z})$, and therefore that $\|pq^{-1}\|_X < 1$. If $h = p - q$, then $h$ is a polynomial, $h(\mathfrak{z}) = 0$, but $h$ does not assume on $X$ the value zero; for if $\mathfrak{z}_1 \in X$ and $h(\mathfrak{z}_1) = 0$, then $p(\mathfrak{z}_1) = q(\mathfrak{z}_1)$, in contradiction to $\|pq^{-1}\|_X < 1$.

29.5 ***Corollary***

*If $X \subset \mathbb{C}^N$, $N \geq 2$, is rationally convex, then $X$ does not separate $\mathbb{C}^N$.*

PROOF

A proof may be based on the following simple fact: If $p$ is a polynomial in $N \geq 2$ variables, the zero set of $p$ has no compact, relatively open subsets. To see this, note first that by a linear change of variables, we can bring $p$ into the form

$$p(z_1, \ldots, z_N) = z_1^s + a_1(z_2, \ldots, z_N) z_1^{s-1} + \cdots + a_s(z_2, \ldots, z_N),$$

where the coefficients $a_j$ are polynomials and $s \geq 1$. Suppose that $K$ is a compact, relatively open subset of the zero set of $p$, and choose $\mathfrak{z}_0 = (z_1^0, \ldots, z_N^0) \in K$ so as to maximize $|z_2^0|$. Given an arbitrary $z_2 \in \mathbb{C}$, there exists a $z_1$ such that

$$p(z_1, z_2, z_3^0, \ldots, z_N^0) = 0,$$

and by the theorem regarding continuity of roots in one variable, $z_1$ can be chosen so that as $z_2 \to z_2^0$, $z_1 \to z_1^0$. Let $\Omega$ be an open subset in $\mathbb{C}^N$ whose intersection with the zero set of $p$ is $K$. We can choose $z_2$ near $z_2^0$ with $|z_2| > |z_2^0|$ such that for an appropriate $z_1$, the point $(z_1, z_2, z_3^0, \ldots, z_N^0)$ lies in the zero set of $p$ and also in $\Omega$. This contradicts the choice of $\mathfrak{z}_0$.

To derive the corollary from this fact, suppose $\Omega$ to be a bounded component of $\mathbb{C}^N \setminus X$. If $\mathfrak{z}_0 \in \Omega$, and if $p$ is a polynomial whose zero set, $V$, contains $\mathfrak{z}_0$, then unless $V$ intersects $X$, $V \cap \Omega$ is a relatively open, compact subset of $V$, contradicting the observation of the last paragraph. Consequently, $V$ meets $X$, and the preceding lemma therefore implies that $\mathfrak{z}_0 \in X$ since $X$ is rationally convex.

Another description of $\mathscr{R}$-hull $X$ is that it is the spectrum of the algebra $\mathscr{R}(X)$. This is a special case of a general fact set forth in the next lemma.

**29.6** ***Lemma***

*Let $(A, \Sigma)$ be a natural algebra, let $X \subset \Sigma$ be compact, and let $S$ be a subset of $A$ which is closed under multiplication and which consists entirely of functions with no zeros on $X$. Let $B$ be the uniform closure in $\mathscr{C}(X)$ of the algebra $\tilde{B}$ of functions of the form $F(x) = g(x)h^{-1}(x)$, $h \in S$, $g \in A$. Then $\Sigma(B)$ may be identified with the set*

$$Y = \{x \in \Sigma : |g(x)| \leq |h(x)| \, \|gh^{-1}\|_X \text{ for all } g \in A, h \in S\}.$$

PROOF

Let $x \in Y$. If $h \in S$, then by definition of $Y$,

$$1 \leq |h(x)| \, \|h^{-1}\|_X,$$

so $h(x) \neq 0$. Consequently, if $g \in A$ and $h \in S$, $g(x)h^{-1}(x)$ is a well-defined complex number, and the map $gh^{-1} \mapsto g(x)h^{-1}(x)$ is a well-defined multiplicative linear functional on $\tilde{B}$, call it $\varphi$. We have from the definition of $Y$ that $|\varphi(gh^{-1})| \leq \|gh^{-1}\|_X$, so since $\tilde{B}$ is dense in $B$, $\varphi$ extends to a unique element of $\Sigma(B)$. Thus, in a natural way, $\Sigma(B) \supset Y$.

Conversely, if $\varphi \in \Sigma(B)$, then $f \mapsto \varphi(f)$ is a nonzero complex homomorphism of $A$, and, as $\|\varphi(f)\| \leq \|f\|_X$, the fact that $(A, \Sigma)$ is a natural algebra implies the existence of an $x \in \Sigma$ such that $\varphi(f) = f(x)$ for all $f \in A$. We shall show that $x \in Y$ and that if $g \in A$, $h \in S$, then $\varphi(gh^{-1}) = g(x)h^{-1}(x)$. If $h \in S$, then $1 = hh^{-1}$, and $h \in A$, so

$$\begin{aligned} 1 &= \varphi(h)\varphi(h^{-1}) \\ &= h(x)\varphi(h^{-1}). \end{aligned}$$

Thus, for every $h \in S$, $\varphi(h^{-1}) = h(x)^{-1}$, and it follows that if $g \in A$, $h \in S$, $\varphi(gh^{-1}) = g(x)h^{-1}(x)$. The functional $\varphi$ is of norm 1, so

$$|g(x)h^{-1}(x)| \leq \|gh^{-1}\|_X.$$

This proves the lemma.

That $\mathscr{R}$-hull $X$ is the spectrum of $\mathscr{R}(X)$ follows from the lemma by taking the natural algebra $(A, \Sigma)$ to be the algebra of polynomials on $\mathbb{C}^N$, and the set $S$ to be the multiplicatively closed subset of the polynomials consisting of the polynomials which are zero free on $X$.

Motivated by the last characterization of $\mathscr{R}$-hull $X$, we are led to introduce yet another notion of convexity.

**29.7 *Definition***

*If $X \subset \mathbb{C}^N$ is compact, the $\mathcal{O}$-**hull** of $X$, $\mathcal{O}$-hull $X$, is the spectrum of the uniform algebra consisting of the uniform closure in $\mathscr{C}(X)$ of the set*

$$\{f|X : f \text{ is holomorphic on a neighborhood of } X\},$$

*and $X$ is $\mathcal{O}$-**convex** if $X = \mathcal{O}$-hull $X$.*

Let us note that although for a given compact set $X \subset \mathbb{C}^N$, the sets $X$ and $\mathscr{R}$-hull $X$ are, by their very definition, subsets of $\mathbb{C}^N$, it is not true that $\mathcal{O}$-hull $X$ lies in $\mathbb{C}^N$ for all $X$. Examples are known which show that if we denote by $z_1, \ldots, z_N$ the coordinate functions on $\mathbb{C}^N$, then the joint spectrum map $\varphi \mapsto (\hat{z}_1(\varphi), \ldots, \hat{z}_N(\varphi))$ from $\mathcal{O}$-hull $X$ to $\mathbb{C}^N$ is not necessarily one to one.

We shall consider the notion of $\mathcal{O}$-convexity from time to time in the sequel, but we shall not develop the subject systematically.

We turn now to an example which hinges on an application of the criterion of Lemma 29.4.

**29.8 *Example***

Let $E \subset \mathbb{C}$ be a Cantor set of locally positive measure, and choose $f_1$ and $f_2$ as in the last part of Example 7.9 so that the map $\Psi$ given by

$$\Psi(z) = (f_1(z), f_2(z))$$

carries $\mathbb{C}^*$ onto a set $S$ in $\mathbb{C}^2$, and $E_0$, the image of $E$ under $\Psi$, is a certain Cantor set. Thus, if $B$ denotes the closed subalgebra of $\mathscr{A}_E$ generated by $f_1$ and $f_2$, then $B$ may be identified with $\mathscr{P}(E_0)$, and $\Sigma(B)$ with $\hat{E}_0$. As the functions $f_1$ and $f_2$ are not locally constant, $\hat{E}_0$ contains a certain two-dimensional set, although $E_0$ is zero dimensional. [Note that we do not claim that $\hat{E}_0 = \Psi(\mathbb{C}^*)$. This may be true, but it is not obvious.]

For future reference, let us observe that by considering the tensor product of $k$ copies of $B$, we can obtain in $\mathbb{C}^{2k}$ zero-dimensional sets whose polynomially convex hulls have dimension at least $2k$.

Lemma 29.4 implies that the set $E_0$ is not even rationally convex: $\mathscr{R}$-hull $E_0$ contains the set $\Psi(\mathbb{C}^*)$. To see this, let $p$ be a polynomial in two variables. The function $h = p \circ \Psi$ lies in $\mathscr{A}_E$, so, by Example 7.9, $h(z) \in h(E)$ for every $z \in \mathbb{C}^*$. This implies that $p(\mathfrak{z}) \in p(E_0)$ for every $\mathfrak{z} \in S$. By the lemma, $\mathscr{R}$-hull $E_0 \supset S$, and $E_0$ is not rationally convex.

It is easy to see that $E_0$ and indeed every totally disconnected compact set $F$ of $\mathbb{C}^N$ is $\mathcal{O}$-convex. This follows from the fact that if $F = F' \cup F''$ is a decomposition of $F$ into the union of disjoint, open, closed subsets, then there is a function $h$ holomorphic on a neighborhood of $F$ which vanishes on $F_0$ and which is identically 1 on $F_1$.

As the characteristic functions of open, closed subsets of $F$ generate $\mathscr{C}(F)$ as a Banach algebra, it follows that $F$ is $\mathcal{O}$-convex.

If, in the preceding discussion, we start with an arc $E$ of locally positive measure, we obtain in $\mathbb{C}^3$ a nonrationally convex arc, $E_0$, but this time it is not clear that $E_0$ is $\mathcal{O}$-convex.

The fact that $\mathscr{R}$-hull $X$ is the spectrum of $\mathscr{R}(X)$ yields another approximation theorem.

**29.9** ***Theorem***

*If $X \subset \mathbb{C}^N$ is rationally convex and if $F$ is holomorphic on a neighborhood of $X$, then $F$ is the uniform limit on $X$ of a sequence $\{f_n\}$, $f_n = p_n q_n^{-1}$, where $p_n$ and $q_n$ are polynomials, $q_n$ zero free on $X$.*

PROOF

This follows from the Shilov–Arens–Calderon theorem, as in the proof of Theorem 29.1, for $X = \Sigma(\mathscr{R}(X))$.

This theorem, too, is due to Oka [1]. Observe that since every compact set in $\mathbb{C}$ is rationally convex, a weak form of Runge's theorem is a special case of the theorem.

Again since $\mathscr{R}$-hull $X = \Sigma(\mathscr{R}(X))$, it follows from Theorem 5.8 that a rationally convex set in $\mathbb{C}^N$ can be realized as a polynomially convex subset of $\mathbb{C}^\Lambda$, $\Lambda$ possibly infinite. There is a much better result due to Rossi [2], a result that extends to $\mathbb{C}^N$ the result of Corollary 24.4.

**29.10** ***Theorem***

*If $X$ is a compact set in $\mathbb{C}^N$, $\mathscr{R}(X)$ admits a system of $N + 1$ generators.*

PROOF

The proof of this theorem is essentially the same as that for Corollary 24.4. The set of generators we shall produce will be the coordinate functions $z_1, \ldots, z_N$ and a function $g_0$ to be constructed. The algebra $\mathscr{R}(X)$ is separable; let $\{F_n\}_{n=1}^{\infty}$ be a countable dense subset. We may suppose that each $F_n$ is of the form $p_n q_n^{-1}$, where $p_n$ and $q_n$ are polynomials, $q_n$ zero free on $X$. Define $\{\varepsilon_n\}_{n=1}^{\infty}$ by requiring that $\varepsilon_1 = 1$ and that for $n = 2, 3, \ldots,$

$$\varepsilon_n \max\{\|q_n^{-1}\|_X, \|q_1 q_n^{-1}\|_X, \ldots, \|q_{n-1} q_n^{-1}\|_X, 1\} < 2^{-(n+1)}\varepsilon_{n-1},$$

and set $g_0 = \sum_{n=1}^{\infty} \varepsilon_n q_n^{-1}$. The function $g_0$ does lie in $\mathscr{R}(X)$, and if $A$ denotes the algebra generated by $g_0$ and the coordinate functions, we shall prove that $A = \mathscr{R}(X)$. Since $p_n \in A$ for all $n$, it suffices to show that $q_n^{-1} \in A$ for all $n$.

The function $q_1 g_0 - \varepsilon_1 = \sum_{n=2}^{\infty} \varepsilon_n q_1 q_n^{-1}$ lies in $A$. If $q_1^{-1}$ does not lie in $A$, there is a nonzero complex homomorphism of $A$ which annihilates $q_1$. We have then that

$$-\varepsilon_1 = \varphi\left(\sum_{n=2}^{\infty} \varepsilon_n q_1 q_n^{-1}\right).$$

This is impossible, as $\|\varphi\| = 1$, but $\varepsilon_n \|q_1 q_n^{-1}\| < 2^{-(n+1)}\varepsilon_1$. Thus, $q_1^{-1} \in A$. Iterating this process as in Corollary 24.4, we find that all the $q_n^{-1}$ lie in $A$, so $A = \mathscr{R}(X)$, as we wished to prove.

It is worth noting that we have not proved that $\mathscr{R}(X)$ admits a system of $N + 1$ generators, every one of which is a rational function, and indeed this strengthened version is false, as is shown already in the plane: If $X \subset \mathbb{C}$ is compact and $\mathbb{C} \setminus X$ has infinitely many components, then $\mathscr{R}(X)$ cannot be generated by any finite family of rational functions even though it is doubly generated as a Banach algebra.

If $X \subset \mathbb{C}^N$ is rationally convex and $\{f_1, \ldots, f_{N+1}\}$ is a set of generators for $\mathscr{R}(X)$, then the set

$$\tilde{X} = \{(f_1(x), \ldots, f_{N+1}(x)) : x \in X\}$$

in $\mathbb{C}^{N+1}$ is polynomially convex. Thus, at least from one point of view, the study of rationally convex sets is subsumed under the study of polynomially convex sets. This geometric remark has another consequence for the topology of rationally convex sets.

**29.11** ***Corollary***

*If $X \subset \mathbb{C}^N$ is rationally convex, then $H^k(X, \mathbb{C}) = 0$ for $k \geq N + 1$.*

PROOF

The groups $H^k(X, \mathbb{C})$ depend only on the topology of $X$, and, as we just saw, $X$ is homeomorphic to a polynomially convex set in $\mathbb{C}^{N+1}$, whence the corollary by way of Browder's theorem (Theorem 10.6).

The next matter to which we turn our attention is a local description of $\mathscr{R}$-hull $X$ analogous to the description of $\hat{X}$ provided by Theorem 28.18. On the basis of that result, it would be reasonable to expect that if $V$ were a variety in an open set $\Omega$, $\Omega \supset \mathscr{R}$-hull $X$, and if $V \cap \mathscr{R}$-hull $X \neq \varnothing$, then $V \cap X \neq \varnothing$. A result of this kind was proved by Stolzenberg [3], but only under certain additional topological hypotheses, hypotheses that are necessitated by the geometry, not merely convenient for the proof. The proof involves the solution to a certain Cousin II problem. We shall need the notion of rational polyhedra.

29.12 ***Definition***
*A* ***rational polyhedron*** *in* $\mathbb{C}^N$ *is a relatively compact set of the form*

$$X = \{\mathfrak{z} \in \mathbb{C}^N : |p_j(\mathfrak{z})| < |q_j(\mathfrak{z})|, j = 1, 2, \ldots, m\},$$

*where*

$$p_1, \ldots, p_m \qquad \text{and} \qquad q_1, \ldots, q_m$$

*are polynomials.*

In analogy with a familiar result on polynomial convexity, we can approximate every rationally convex set by rational polyhedra:

29.13 ***Lemma***
*If* $X \subset \mathbb{C}^N$ *is rationally convex, there is a sequence* $\{Q_j\}$ *of rational polyhedra with* $X = \bigcap Q_j$.

PROOF
If $\mathfrak{z} \in \mathbb{C}^N \setminus X$, then since $X$ is rationally convex, there is a polynomial $p$ with $p(\mathfrak{z}) = 0$, $0 \notin p(X)$. Compactness implies that for some $\varepsilon > 0$, $X$ does not meet the neighborhood

$$W(\mathfrak{z}, p, \varepsilon) = \{\mathfrak{z}' \in \mathbb{C}^N : |p(\mathfrak{z}')| < \varepsilon\},$$

so for some $M$, $X$ is contained in the rational polyhedron

$$Q(p, \mathfrak{z}, \varepsilon) = \{\mathfrak{z}' \in \mathbb{C}^N : |z_1|, \ldots |z_N| < M, \varepsilon < |p(\mathfrak{z}')|\}.$$

A countable family of the neighborhoods $W(\mathfrak{z}, p, \varepsilon)$ covers $\mathbb{C}^N \setminus X$, so the result follows.

29.14 ***Theorem***
*Let* $X$ *be a compact set in* $\mathbb{C}^N$, *let* $\Omega$ *be an open set in* $\mathbb{C}^N$ *which contains* $\mathscr{R}$-hull $X$, *and assume that* $V \subset \Omega$ *is an analytic variety which is locally a hypersurface. If* $V \cap \mathscr{R}$-hull $X \neq \varnothing$, *and if* $H^2(\mathscr{R}\text{-hull } X, \mathbb{Z}) = 0$, *then* $V$ *meets* $X$.

As usual, we work with Čech cohomology.

The hypothesis that $V$ is locally a hypersurface means that if $\mathfrak{z}_0 \in V$, then there is a neighborhood $U$ of $\mathfrak{z}_0$ such that for some function $f$ holomorphic in $U$,

$$V \cap U = \{\mathfrak{z} \in U : f(\mathfrak{z}) = 0\}.$$

PROOF

Assume, to begin with, that $V$ is a hypersurface at least in *some* open set $\Omega'$, $\Omega' \supset \mathscr{R}$-hull $X$, say

$$V = \{\mathfrak{z} \in \Omega' : F(\mathfrak{z}) = 0\}$$

for some function $F$ holomorphic in $\Omega'$. If $V \cap X = \varnothing$, then $|F(\mathfrak{z})| \geq \varepsilon > 0$ for some $\varepsilon$ and all $\mathfrak{z} \in X$. Let $\mathfrak{z}_0 \in V \cap \mathscr{R}$-hull $X$ so that $F(\mathfrak{z}_0) = 0$. By the approximation result, Theorem 29.9, there is a rational function $F_1 = p/q$, $q$ zero free on $\mathscr{R}$-hull $X$, such that $\|F - F_1\|_{\mathscr{R}\text{-hull}X} < \varepsilon/2$ and $F_1(\mathfrak{z}_0) = 0$. Then it follows that $p(\mathfrak{z}_0) = 0$, but $|p(\mathfrak{z})| \neq 0$ if $\mathfrak{z} \in X$, a contradiction to Lemma 29.4.

This part of the argument has not used the toplogical hypothesis on $\mathscr{R}$-hull $X$. Its use comes about in showing that $V$ is a (globally defined) hypersurface in some neighborhood of $\mathscr{R}$-hull $X$. The argument that follows is simply an account of the usual cohomological discussion of the second Cousin problem.

Since $V$ is locally a hypersurface, there is a covering $\{\Omega_\alpha\}_{\alpha\in I}$ of $\Omega$ consisting of open polydiscs such that $V \cap \Omega_\alpha$ is the zero set of the function $f_\alpha \in \mathcal{O}(\Omega_\alpha)$ in the strong sense that if $g$ is holomorphic in a neighborhood $W \subset \Omega_\alpha$ of the point $z$ and vanishes on $W \cap V$, then in a possibly smaller neighborhood, $g = hf_\alpha$ for some holomorphic $h$. Since we are willing to permit arbitrarily small $\Omega_\alpha$, this assertion follows from the local theory of varieties. (See Gunning and Rossi [1, III.C.11].) On the convex set $\Omega_\alpha \cap \Omega_\beta$, there is a holomorphic branch of $\log f_\alpha f_\beta^{-1}$. For each pair $\alpha, \beta \in I$, let $h_{\alpha\beta}$ be a determination of $1/2\pi i \log f_\alpha f_\beta^{-1}$, the determinations subject only to the conditions that $h_{\alpha\beta} = -h_{\beta\alpha}$ and $h_{\alpha\beta} = 0$ if $\Omega_\alpha$ and $\Omega_\beta$ are disjoint.

If we put $\Omega'_\alpha = (\mathscr{R}\text{-hull } X) \cap \Omega_\alpha$, then $\{\Omega'_\alpha\}_{\alpha\in I} = \mathscr{U}$ is an open cover for $\mathscr{R}$-hull $X$; and if we define $c$ by

$$c(\Omega'_\alpha, \Omega'_\beta, \Omega'_\gamma) = c_{\alpha\beta\gamma} = h_{\alpha\beta} + h_{\beta\gamma} + h_{\gamma\alpha},$$

then since $c_{\alpha\beta\gamma}$ is, except for a factor of $1/2\pi i$, a determination of $\log 1$, it is an integer, and we see that $c$ is an integral 2-cochain for the nerve of the covering $\mathscr{U}$ of $\mathscr{R}$-hull $X$. The restrictions we have imposed on $h_{\alpha\beta}$ and the definition of $c$ show that $c$ is a cocycle: If $\alpha, \beta, \gamma, \delta \in I$ are given, then

$$c_{\beta\gamma\delta} - c_{\alpha\gamma\delta} + c_{\alpha\beta\delta} - c_{\alpha\beta\gamma} = 0.$$

By hypothesis, $H^2(\mathscr{R}\text{-hull } X, \mathbb{Z}) = 0$, so there are open sets $W_j$ in $\Omega$ such that if $W'_j = W_j \cap (\mathscr{R}\text{-hull } X)$, then $\mathscr{V} = \{W'_j\}_{j\in J}$ is an open cover for $\mathscr{R}$-hull $X$ which refines $\mathscr{U}$ and which has the property that the cochain which $c$ induces for the covering $\mathscr{V}$ is a coboundary. Explicitly, there is a map $\sigma : J \to I$ such that for each $j$, $W_j \subset \Omega_{\sigma(j)}$, and such that for each pair $i, j \in J$, there are integers $c_{ij}$ with the property that on $W'_i \cap W'_j \cap W'_k$,

$$c_{\sigma(i)\sigma(j)\sigma(k)} = c_{ij} + c_{jk} + c_{ki}.$$

We define holomorphic functions $\varphi_{ij}$ on $W_i \cap W_j$ by

$$\varphi_{ij} = h_{\sigma(i)\sigma(j)} - c_{ij}.$$

We have that

$$\varphi_{ij} = -\varphi_{ji}$$

and that on $W'_i \cap W'_j \cap W'_k$,

$$\varphi_{ij} + \varphi_{jk} + \varphi_{ki} = 0.$$

The last two equations mean that if we consider the sheaf $\mathcal{O}_{\mathscr{R}\text{-hull } X}$ of germs of holomorphic functions on $\mathscr{R}$-hull $X$, then $\varphi_{ij}$ is a 1-cocycle for the cover $\{W'_j\}_{j\in J}$ with values in $\mathcal{O}_{\mathscr{R}\text{-hull } X}$.

The rational convexity of $\mathscr{R}$-hull $X$ implies that this set is the intersection of a decreasing sequence of rational polyhedra, say $\{Q_m\}$, and from the fact that $H^1(Q_m, \mathcal{O}_{Q_m}) = 0$ for each $m$ (Appendix A), it follows that $H^1(\mathscr{R}\text{-hull } X, \mathcal{O}_{\mathscr{R}\text{-hull } X}) = 0$. (Two routes to this conclusion are possible. It can be established by a direct computation, or it can be proved by appeal to general properties of cohomology. We shall not enter into the details.) Thus, if we pass to yet a further refinement of the cover $\{W'_j\}_{j\in J}$, which we shall continue to denote by $\{W_j\}_{j\in J}$, there exist holomorphic functions $\varphi_i \in \mathcal{O}(W_i)$ with $\varphi_{ij} = \varphi_i - \varphi_j$ on $W_i \cap W_j$.

We define $F \in \mathcal{O}(\mathscr{R}\text{-hull } X)$ by $F = f_{\sigma(j)} \exp(-2\pi i \varphi_j)$ on $W'_j$. On $W'_j \cap W'_k$ we have

$$\begin{aligned} f_{\sigma(j)} f^{-1}_{\sigma(k)} \exp[-2\pi i(\varphi_j - \varphi_k)] &= f_{\sigma(j)} f^{-1}_{\sigma(k)} \exp(-2\pi i \varphi_{jk}) \\ &= f_{\sigma(j)} f^{-1}_{\sigma(k)} \exp[-2\pi i (h_{\sigma(j)\sigma(k)} - c_{jk})] \\ &= \exp 2\pi i c_{jk} \\ &= 1. \end{aligned}$$

This computation shows that the function $F$ is a well-defined element of $\mathcal{O}(\mathscr{R}\text{-hull } X)$. As it differs on $W'_j$ from $f_{\sigma(j)}$ only by an exponential, it follows that $F$ defines $V$ in some neighborhood of $X$, and the theorem is proved.

**29.15** ***Corollary***

*If $H^2(\mathscr{R}\text{-hull } X, \mathbb{Z}) = 0$ and $E \subset \mathscr{R}\text{-hull } X$ is compact, then $E \subset \mathscr{R}\text{-hull}((E \cap X) \cup \partial E)$. (Here $\partial E$ denotes the topological boundary of $E$ in $\mathscr{R}$-hull $X$.)*

This corollary may be regarded as a rational analogue of the local maximum modulus theorem. If $F$ is a closed subset of $\hat{X}$, the local maximum modulus theorem implies that $F \subset ((F \cap X) \cup \partial F)\hat{}$, where, this time, $\partial F$ is the boundary of $F$ with respect to $\hat{X}$. As Stolzenberg showed, this result, which may be regarded as a weak version of the local maximum modulus theorem, can be obtained from the local description of the polynomially convex hull given in Theorem 28.18.

PROOF OF THE COROLLARY

If $\mathfrak{z}_0 \in E \setminus \mathscr{R}$-hull$((E \cap X) \cup \partial E)$, then by Lemma 29.4 there is a polynomial $p$ such that $p(\mathfrak{z}_0) = 0$, $0 \notin p(\mathscr{R}$-hull$((E \cap X) \cup \partial E))$. Write $\mathscr{R}$-hull $X = \bigcap_{j=1}^{\infty} \Omega_j$, where $\{\Omega_j\}$ is a sequence of open sets with $\bar{\Omega}_{j+1} \subset \Omega_j$. Let $V_j$ be the component of the zero set, $V$, of $p$ in $\Omega_j$ which contains $\mathfrak{z}_0$. Then $\bigcap V_j$ is the component of $V \cap \mathscr{R}$-hull $X$ which contains $\mathfrak{z}_0$, and as $\bigcap V_j$ is disjoint from $(E \cap X) \cup \partial E$, it follows by connectedness that $\bigcap V_j \subset E \setminus X$. Consequently, there is an integer $j_0$ such that $V_{j_0} \cap X = \varnothing$. But then $V_{j_0}$ is a hypersurface in the open neighborhood $\Omega_{j_0}$ of $X$ which misses $X$ but which satisfies the hypotheses of the last theorem. This contradiction implies the corollary.

Before proceeding to further positive results in the theory of polynomial and rational convexity, we shall illustrate by various examples some of the pathology that occurs in the theory.

Given a class of sets, one of the first questions that arises is that of determining the transformations which leave the class invariant. Our first example is in this direction. It is clear that nonsingular affine transformations of $\mathbb{C}^N$ leave invariant the classes of polynomially convex and rationally convex sets. Can more be said? An example given by Wermer [7] shows that there are some surprising difficulties to be encountered.

**29.16** ***Example***

We shall construct, following Wermer, a polynomial map $\Phi$ of $\mathbb{C}^3$ into itself such that if $E$ is the bicylinder

$$E = \{(z, w, 0) \in \mathbb{C}^3 : |z|, |w| \leq 1\},$$

then $\Phi$ is a homeomorphism on a neighborhood of $E$, but yet $\Phi(E)$ is not polynomially convex.

Define $\Phi : \mathbb{C}^3 \to \mathbb{C}^3$ by

$$\Phi(z, w, t) = (z, zw + t, zw^2 - w + 2wt).$$

The Jacobian of this map is the function

$$J_\Phi(z, w, t) = 1 - 2t,$$

which is zero free off the hyperplane $t = \frac{1}{2}$ so that (Appendix $D$) $\Phi$ is locally one to one off this plane. We shall show that if $K$ is a compact set in the hyperplane $\{(z, w, t): t = 0\}$, then $\Phi$ is one to one on the set

$$K_\varepsilon = \{(z, w, t): (z, w) \in K, |t| < \varepsilon\}$$

provided $\varepsilon > 0$ is sufficiently small.

If not, then for $n = 1, 2, 3, \ldots$, there are points $\mathfrak{z}_n$ and $\mathfrak{z}'_n \in K_{1/n}$, $\mathfrak{z}_n \neq \mathfrak{z}'_n$, such that $\Phi(\mathfrak{z}_n) = \Phi(\mathfrak{z}'_n)$. If $\mathfrak{z}_n = (z_n, w_n, t_n)$, $\mathfrak{z}'_n = (z'_n, w'_n, t'_n)$, then $t_n$ and $t'_n$ tend to zero, and, by compactness of $K$, we may suppose that $\{\mathfrak{z}_n\}$ and $\{\mathfrak{z}'_n\}$ converge, say to $\mathfrak{z}_0 = (z_0, w_0, 0)$ and $\mathfrak{z}'_0 = (z'_0, w'_0, 0)$, respectively. Then $\Phi(\mathfrak{z}_0) = \Phi(\mathfrak{z}'_0)$, whence $z_0 = z'_0$ and $w_0 = w'_0$, and so $\mathfrak{z}_0 = \mathfrak{z}'_0$. Since $J_\Phi(\mathfrak{z}_0) \neq 0$, there is a neighborhood $V$ of $\mathfrak{z}_0$ on which $\Phi$ is one to one. This leads to a contradiction, though, because for large enough $n$, both $\mathfrak{z}_n$ and $\mathfrak{z}'_n$ lie in $V$, $\mathfrak{z}_n \neq \mathfrak{z}'_n$, but $\Phi(\mathfrak{z}_n) = \Phi(\mathfrak{z}'_n)$. Thus, for small enough $\varepsilon$, $\Phi$ is one to one on $K_\epsilon$.

We have proved that the map $\Phi$ carries a neighborhood of the polydisc $E$ homeomorphically into $\mathbb{C}^3$, but, nevertheless, the set $E' = \Phi(E)$ is not polynomially convex. To see this, consider in $\mathbb{C}^3$ the complex line

$$\pi = \{(z, w, t): w = 1, t = 0\}.$$

The intersection of $\pi$ and $E'$ is the circle $\gamma = \{(z, w, t) \in \pi: |z| = 1\}$. It is clear that $\gamma \subset E'$, for $\gamma = \{\Phi(e^{i\theta}, e^{-i\theta}, 0): \theta \text{ real}\}$. And if $\mathfrak{z} = (z, w, t) \in E' \cap \pi$, say $\mathfrak{z} = \Phi(\mathfrak{z}')$, $\mathfrak{z}' = (z', w', 0)$, then the definition of $\Phi$ shows that $z = z'$, $w = z'w'$. As $\mathfrak{z} \in \pi$ and $|z'|, |w'| \leq 1$, this implies that $|z'| = 1$ and $w' = 1/z'$, i.e., that $\mathfrak{z}' = (e^{i\theta}, e^{-i\theta}, 0)$ for some $\theta$. Thus, we have proved that $\pi \cap E' = \gamma$. It is now clear that $E'$ is not polynomially convex, for

$$\hat{E}' \supset \hat{\gamma} = \{(z, w, t): |z| \leq 1, w = 1, t = 0\},$$

and this latter set is not contained in $E$.

The next question which we treat is the following: If $X \subset \mathbb{C}^N$ is a small compact set, how large can $\hat{X}$ be? The question is admittedly vague, and as yet there are no definitive results, though, as we have seen in Example 29.8, there are zero-dimensional sets in $\mathbb{C}^{2k}$ such that $\dim \hat{X} \geq 2k$. It would be of interest to know whether there are zero-dimensional sets in $\mathbb{C}^N$ such that $\hat{X}$ is the closure of an open set or at least contains an open set. To the best of my knowledge no one has yet constructed such a set. If we are willing to settle for one-dimensional sets, there is the following example implicitly contained in the paper of Meyers [1].

29.17 ***Example***

In $\mathbb{C}^{N+1}$, $N \geq 1$, there exists a one-dimensional set $X$ such that $\hat{X}$ contains an open set. To construct such a set, let $E \subset [\frac{1}{2}, 1]$ be a Cantor set, and let $\varphi$ be a continuous map from $E$ onto the closed $N$-dimensional polydisc $\bar{\Delta}^N$. By Tietze's extension theorem, $\varphi$ extends to a continuous map, again denoted by $\varphi$, of $[0, 1]$ to $\bar{\Delta}^N$. Define $\Phi : \bar{\Delta} \to \bar{\Delta}^{N+1}$ by

$$\Phi(z) = (\varphi(|z|), z).$$

The map $\Phi$ is a homeomorphism, and if we set

$$E_c = \{z : |z| \in E\},$$

then $E_c$ is a compact, one-dimensional set, for topologically it is the product of a circle and a Cantor set. The set $X = \Phi(E_c)$ is also one dimensional.

Let $\mathfrak{z} = (z_1, \ldots, z_{N+1})$ be a point with $|z_j| < \frac{1}{2}$ for every $j$. The definition of $\varphi$ implies the existence of $t \in E$ such that $\varphi(t) = (z_1, \ldots, z_N)$. The set $X$ therefore contains the circle

$$\Lambda = \{(z_1, \ldots, z_N, te^{i\theta}) : \theta \text{ real}\},$$

and thus $\hat{\Lambda} \subset \hat{X}$. We have that

$$\hat{\Lambda} = \{(z_1, \ldots, z_N, \zeta) : |\zeta| \leq t\},$$

and as $t > \frac{1}{2}$, the point $\mathfrak{z}$ lies in $\hat{\Lambda}$. Thus, $\hat{X}$ contains the polydisc

$$\{(z_1, \ldots, z_{N+1}) : |z_1|, \ldots, |z_{N+1}| \leq \tfrac{1}{2}\},$$

the closure of an open set in $\mathbb{C}^N$.

It would be interesting to determine precisely what the set $\hat{X}$ is.

Following Meyers [1], constructions of the preceding type can be used to show the existence of closed subalgebras of $\mathscr{C}(\mathbb{C})$ which are algebraically and topologically isomorphic to the ring of entire functions in $N$ complex variables. We understand by $\mathscr{C}(\mathbb{C})$ the algebra of all continuous complex-valued functions on the plane endowed with the topology of uniform convergence on compacta. Thus, $\mathscr{C}(\mathbb{C})$ is a commutative Fréchet algebra, i.e., a locally convex, metrizable topological vector space in which multiplication is jointly continuous. (Banach algebras are Fréchet algebras.)

29.18 ***Theorem***

*For each* $N = 1, 2, \ldots, \mathscr{C}(\mathbb{C})$ *contains a closed subalgebra* $A$ *which contains* $\mathcal{O}(\mathbb{C})$ *and which is topologically and algebraically isomorphic to* $\mathcal{O}(\mathbb{C}^N)$.

Here, of course, $\mathcal{O}(\mathbb{C}^N)$ denotes the Fréchet algebra of entire functions in $N$ complex variables provided with the topology of uniform convergence on compacta.

PROOF

For $N = 1$, the theorem is obvious, so we assume henceforth that $N > 1$. For $k = 1, 2, \ldots$, let $E_k$ be a Cantor set contained in the interval $[k + \frac{1}{4}, k + \frac{3}{4}]$. Let $\varphi:\mathbb{R} \to \mathbb{C}^{N-1}$ be a continuous function such that $\varphi(E_k)$ contains the polycylinder

$$\bar{\Delta}_k^{N-1} = \{\mathfrak{z} \in \mathbb{C}^{N-1}:|z_1|, \ldots, |z_{N-1}| \leq k\},$$

and let $\Phi:\mathbb{C} \to \mathbb{C}^{N-1} \times \mathbb{C} = \mathbb{C}^N$ be given by

$$\Phi(z) = (\varphi(|z|), z).$$

Define $A \subset \mathscr{C}(\mathbb{C})$ by

$$A = \{f \circ \Phi : f \in \mathcal{O}(\mathbb{C}^N)\}.$$

The map $f \to f \circ \Phi$, call it $\Phi^*$, from $\mathcal{O}(\mathbb{C}^N)$ to $A$ is plainly continuous and an algebra homomorphism. It is also one to one and a topological isomorphism. This is so, for by the argument used in the preceding example, each $\bar{\Delta}_k^N$ is contained in the polynomially convex hull of $\Phi(S)$ for some compact set $S \subset \mathbb{C}$. From this it follows, first, that if $f \in \mathcal{O}(\mathbb{C}^N)$ vanishes on $\Phi(\mathbb{C})$, then it vanishes on an open set in $\mathbb{C}^N$ and so is the zero function, i.e., $\Phi^*$ is one to one; and, second, if $\{f_n\}$ is a sequence in $\mathcal{O}(\mathbb{C}^N)$ such that $\{\Phi^* f_n\} = \{f_n \circ \Phi\}$ converges in $\mathcal{O}(\mathbb{C})$, then $\{f_n\}$ converges uniformly on compacta in $\mathbb{C}^N$. Thus, $\Phi^*$ is a topological isomorphism, and $A$ is a closed subalgebra. As $A$ obviously contains $\mathcal{O}(\mathbb{C})$, the theorem is proved.

It is very easy to see that there are subalgebras of $\mathcal{O}(\mathbb{C})$ which are algebraically isomorphic to $\mathcal{O}(\mathbb{C}^N)$. To construct such an algebra, let $\{\mathfrak{z}_n\}_{n=1}^{\infty}$ be a countable dense set in $\mathbb{C}^N$, $\mathfrak{z}_n = (z_1^{(n)}, \ldots, z_N^{(n)})$, and for $j = 1, 2, \ldots, N$, let $\psi_j$ be an entire function of one complex variable such that $\psi_j(n) = z_j^{(n)}$. Define $\Psi:\mathbb{C} \to \mathbb{C}^N$ by

$$\Psi(\zeta) = (\psi_1(\zeta), \ldots, \psi_N(\zeta)).$$

The range of $\Psi$ is dense in $\mathbb{C}^N$, so the map $f \mapsto f \circ \Psi$ is an algebraic isomorphism of $\mathcal{O}(\mathbb{C}^N)$ onto a subalgebra of $\mathcal{O}(\mathbb{C})$.

The question now arises as to whether it is possible to combine the features of these two examples: Do there exist closed subalgebras of $\mathcal{O}(\mathbb{C})$ which are topologically isomorphic to $\mathcal{O}(\mathbb{C}^N)$? No such algebras exist, but the proof does not seem to be so easy. We shall obtain the result as a consequence of a deep theorem of Stolzenberg in the next section. (See Theorem 30.19.)

The next example was found by Stolzenberg [2] and may be regarded as arising from the search for analytic structure in spectra of Banach algebras. If $(A, X)$ is a uniform algebra with Shilov boundary $\Gamma$, how can we account for the presence of points in $\Sigma(A) \setminus \Gamma$? The only obvious reason for there to be points in $\Sigma(A) \setminus \Gamma$ is for this set to consist of certain analytic structures on which the Gelfand transforms of the elements of $A$ are all holomorphic. Translated into the language of polynomial convexity, this leads us to expect that if $Y$ is a compact set in $\mathbb{C}^N$, then $\hat{Y} \setminus Y$ should consist of analytic varieties (of positive dimension). Stolzenberg's example shows that this expectation is ill founded.

29.19 ***Theorem***

*There exists a compact set $X$ in $\mathbb{C}^2$ such that $\hat{X} \setminus X$ is nonempty but contains no analytic variety of positive dimension.*

The construction of the desired example will depend on the *open mapping theorem* for functions holomorphic on varieties: *If $V$ is an analytic variety without isolated points in an open set in $\mathbb{C}^N$, and if $f$ is holomorphic on $V$, then given a connected (relatively) open subset, $S$, of $V$, either $f(S)$ is open in $\mathbb{C}$, or else $f|S$ is constant.* This is a standard result in several complex variables and may be found, e.g., in Gunning and Rossi [1].

PROOF OF THE THEOREM

Let $\{\zeta_i\}_{i=1}^{\infty}$ be a countable dense set in the punctured disc $\{\zeta \in \mathbb{C} : 0 < |\zeta| < 1\}$, and, for each $i$, let

$$K_i = \{(z, w) \in \bar{\Delta}^2 : z = \zeta_i \text{ or } w = \zeta_i\}$$

so that each $K_i$ is the union of a pair of intersecting discs. Define entire functions $F_i$ inductively as follows. Let

$$F_1(\mathfrak{z}) = F_1(z, w) = \zeta_1^{-2}(z - \zeta_1)(w - \zeta_1).$$

We have that $F_1(0, 0) = 1$ and $F_1|K_1 = 0$. Let

$$L_1 = \{\mathfrak{z} \in \bar{\Delta}^2 : |\operatorname{Re} F_1(\mathfrak{z})| \le \tfrac{1}{2}\},$$

a certain closed neighborhood of $K_1$ in $\bar{\Delta}^2$. Assume that $F_1, \ldots, F_q$ have been defined so as to satisfy $F_j|K_j = 0$, $F_j(0, 0) = 1$, and $F_j(z, w) = F_j(w, z)$. We let

$$L_q = \{\mathfrak{z} \in \bar{\Delta}^2 : |\text{Re}\, F_q(\mathfrak{z})| \leq \tfrac{1}{2}\},$$

$$G_{q+1}(\mathfrak{z}) = \zeta_{q+1}^{-2}(z - \zeta_{q+1})(w - \zeta_{q+1}),$$

and

$$H_{q+1}(\mathfrak{z}) = \exp[F_q(\mathfrak{z}) - 1].$$

The function $H_{q+1}$ satisfies

$$\|H_{q+1}\|_{L_q} = \|\exp(\text{Re}\, F_q - 1)\|_{L_q} = e^{-1/2} < 1,$$

and the function $G_{q+1}$ is bounded on $L_q$; so if $N$ is a sufficiently large, positive integer,

$$\|H_{q+1}^N G_{q+1}\|_{L_q \cup K_{q+1}} < \tfrac{1}{4}.$$

We define $F_{q+1}$ and $L_{q+1}$ by

$$F_{q+1} = H_{q+1}^N G_{q+1},$$

$$L_{q+1} = \{\mathfrak{z} \in \bar{\Delta}^2 : |\text{Re}\, F_{q+1}(\mathfrak{z})| \leq \tfrac{1}{2}\}.$$

The function $F_{q+1}$ satisfies $F_{q+1}(0, 0) = 1$, $F_{q+1}(z, w) = F_{q+1}(w, z)$, and $F_{q+1}K_{q+1} = 0$. In addition, $L_q \cup K_{q+1}$ is contained in the interior (relative to $\bar{\Delta}^2$) of $L_{q+1}$.

For each $q$, let $W_q$ be the hypersurface

$$\{\mathfrak{z} \in \mathbb{C}^2 : F_q(\mathfrak{z}) = 1\}$$

so that $(0, 0) \in W_q$ and $W_q \cap L_q = \varnothing$. We let $\tilde{V}_q$ denote the component of $W_q$ which passes through the origin, and we let $V_q$ be the component of $\tilde{V}_q \cap \bar{\Delta}^2$ which contains the origin.

We shall need the following relations among the objects constructed so far:

(a) interior $L_1 \subset$ interior $L_2 \subset \cdots$ and $\bigcup_i$ (interior $L_i$) $\supset \bigcup_i K_i$.
(b) $V_p \cap L_q = 0$ if $p \geq q$.
(c) If $\mathfrak{z}_0 \in \bar{\Delta}^2$ has either component equal to $\zeta_i$, then there is a polynomial $h$ such that

$$|h(\mathfrak{z}_0)| > \sup\{|h(\mathfrak{z})| : \mathfrak{z} \in \bar{\Delta}^2, \tfrac{1}{2} \leq \text{Re}\, F_i(\mathfrak{z})\}.$$

Relations (a) and (b) follow immediately from the definitions. To prove (c), let $M_i = \{\mathfrak{z} \in \bar{\Delta}^2 : \tfrac{1}{2} \leq \text{Re}\, F_i(\mathfrak{z})\}$. We have then that

$$\begin{aligned}\|\exp(-F_i)\|_{M_i} &= \|\exp(-\text{Re}\, F_i)\|_{M_i} \\ &\leq \exp(-\tfrac{1}{2}) < 1 \\ &= \exp(-F_i(\mathfrak{z})), \qquad \mathfrak{z} \in K_i.\end{aligned}$$

As $F_i$ is entire, the desired polynomial $h$ can be obtained as a partial sum of the power series for $e^{-F_i}$.

Let $\delta$ denote the Hausdorff metric on the space of closed subsets of $\bar{\Delta}^2$ so that if $A, B \subset \bar{\Delta}^2$ are closed,

$$\delta(A, B) = \max_{a \in A} \min_{b \in B} \|a - b\| + \max_{b \in B} \min_{a \in A} \|a - b\|,$$

where $\|\cdot\|$ is the standard norm on $\mathbb{C}^2$. The space of closed subsets of $\bar{\Delta}^2$ is compact with respect to the topology induced by the metric $\delta$. (See Appendix A.) By passing to a subsequence if necessary, we may suppose that the sequence $\{V_q\}_{q=1}^{\infty}$ converges in the sense of the $\delta$-metric to a compact set $Y \subset \bar{\Delta}^2$.

For all $q$, $(0, 0) \in V_q$, so $(0, 0) \in Y$. Also, each $\tilde{V}_q$ is a connected variety, not a point, so it must meet $\partial\bar{\Delta}^2$, whence the same is true of the limit set $Y$. Finally, since each set $V_q$ is connected, $Y$ is connected. We shall prove that for the set $X$ of the theorem we can take $Y \cap \partial\bar{\Delta}^2$.

Denote by $\pi_1$ and $\pi_2$ the coordinate functions on $\mathbb{C}^2 : \pi_1(z, w) = z$, $\pi_2(z, w) = w$. We shall prove that neither $\pi_1(\hat{Y})$ nor $\pi_2(\hat{Y})$ contains an open set in $\mathbb{C}$. Granted these facts, the open mapping theorem implies $\hat{Y}$ contains no analytic variety of positive dimension. We have $\pi_1(\hat{Y}) \subset \bar{\Delta}$, so if $\pi_1(\hat{Y})$ contains an open set in $\mathbb{C}$, there is a $\zeta_q \in \pi_1(Y)$, and so a $w \in \bar{\Delta}$ with $(\zeta_q, w) \in \hat{Y}$. By property (b) and the fact that $\{V_q\}_{q=1}^{\infty}$ converges to $Y$, it follows that $Y$ is disjoint from the interior of every $L_n$, and therefore

$$Y \subset \{\mathfrak{z} \in \bar{\Delta}^2 : |\text{Re } F_q(\mathfrak{z})| > \tfrac{1}{2}\}.$$

We have $F_q(0, 0) = 1$, $(0, 0) \in Y$, so the connectedness of $Y$ implies that

$$Y \subset \{\mathfrak{z} \in \bar{\Delta}^2 : \text{Re } F_q(\mathfrak{z}) \geq \tfrac{1}{2}\} = M_q.$$

Relation (c) provides a polynomial $h$ with

$$\|h\|_Y \leq \|h\|_{M_q} < |h(\zeta_q, w)|,$$

a contradiction to $(\zeta_q, w) \in \hat{Y}$. Thus, $\pi_1(\hat{Y})$ contains no open subset of $\mathbb{C}$; the same assertion is true of $\pi_2(\hat{Y})$.

Finally, we prove that if $h$ is a polynomial, then $\|h\|_Y = \|h\|_X$ so that $\hat{X} = \hat{Y}$. To do this, suppose that $\mathfrak{z}_0 \in Y \setminus \partial\bar{\Delta}^2$ is such that for the polynomial $h$,

$$|h(\mathfrak{z}_0)| > \|h\|_X.$$

Then there exist disjoint open sets $U$ and $\Omega$ in $\mathbb{C}^2$, $\mathfrak{z}_0 \in U$, $X \subset \Omega$, with $|h(\mathfrak{z})| > \|h\|_\Omega$ for all $\mathfrak{z} \in U$. As $Y$ is the limit of $\{V_q\}$, there is a $q$ so large that $V_q \cap U \neq \varnothing$ and $V_q \cap \partial\bar{\Delta}^2 \subset \Omega$. This conclusion entails a contradiction to the open mapping theorem, so our contention is proved.

We have that $\hat{X} \setminus X$ is nonempty since it contains the origin, but $\hat{X} = \hat{Y}$, and $\hat{Y}$ contains no variety of positive dimension. Thus, our theorem is proved.

In the proof just given, we have constructed a set that contains no nontrivial variety as a limit of a sequence of varieties. Metric conditions are known which guarantee that the limit of a sequence of varieties is again a variety. For this theory, one may consult the notes of Stolzenberg [6] and the literature cited there.

**29.20** ***Remarks***

(a) Although $\pi_1(\hat{X})$ and $\pi_2(\hat{X})$ have no interior, they do have positive planar area. It is enough to prove that one of them does, for by construction, the set $X$, and hence also $\hat{X}$, is invariant under the automorphism $(z, w) \mapsto (w, z)$ of $\bar{\Delta}^2$. That at least one of these sets has positive measure follows from Corollary 26.10 and the fact that $\mathscr{P}(\hat{X}) \neq \mathscr{C}(\hat{X})$ since, by our construction, the Shilov boundary for $\mathscr{P}(\hat{X})$ is a proper subset of $\hat{X}$.

(b) Although $\hat{X}$ contains no variety of positive dimension and, *a fortiori*, no open subset of $\mathbb{C}^2$, $(\pi_1(X) \times \pi_2(X))\hat{}$ does contain an open subset of $\mathbb{C}^2$. To prove this fact, start with the simple observation that $(\pi_1(X) \times \pi_2(X))\hat{} = \pi_1(X)\hat{} \times \pi_2(X)\hat{}$ so that $(\pi_1(X) \times \pi_2(X))\hat{}$ contains an open set if and only if $\pi_1(X)\hat{}$ contains an open subset of $\mathbb{C}$. We know that $\mathbb{C} \setminus \pi_1(X)\hat{}$ is connected and that $(0, 0) \in \hat{X}_0 \subset \pi_1(X)\hat{} \times \mathbb{C}$. Let $\tilde{f} \in \mathscr{C}(\pi_1(X)\hat{})$ satisfy $\tilde{f}(0) = 1, |\tilde{f}(\zeta)| < 1$ if $\zeta \in \pi_1(X)\hat{} \setminus \{0\}$, and put $f = \tilde{f} \circ \pi_1$. If $\pi_1(X)\hat{}$ has no interior, then by Mergelyan's theorem $\tilde{f} \in \mathscr{P}(\pi_1(X)\hat{})$, so $f \in \mathscr{P}(\hat{X})$. The function $f$ attains its maximum over $\hat{X}$ on the slice $\hat{X}_0$ where, for a set $S \subset \mathbb{C}^2$ and a $\zeta \in \mathbb{C}$, we write

$$S_\zeta = S \cap \{\mathfrak{z} \in \mathbb{C}^2 : \pi_1(\mathfrak{z}) = \zeta\}.$$

We claim that $(0, 0) \in (X_0)\hat{}$. If not, there is a polynomial $h$ with $h(0, 0) = 1, |h(0, w)| < 1$ if $(0, w) \in X$. Put $E = \{\mathfrak{z} \in X : 1 \leq |h(\mathfrak{z})|\}$, a compact set disjoint from $X$. If $k$ is a large positive integer, then $|f^k h| < 1$ on $X$ since $|f| \leq 1$ on $X$, and on $E$, $|f|$ is strictly less than 1. However, $f^k h \in \mathscr{P}(\hat{X})$, $(0, 0) \in \hat{X}$, and $(f^k h)(0, 0) = 1$, which is inconsistent with $\|f^k h\|_X < 1$. Thus, $(0, 0) \in (X_0)\hat{}$, as desired. Since $(0, 0) \notin X_0$ and $X_0$ lies in the plane $z = 0$ it follows that $X_0$ bounds an open set $\Omega$ containing $(0, 0)$ in the plane $z = 0$. The set $\Omega$ is an analytic variety contained in $\hat{X}$, so we have a contradiction. Thus, $\pi_1(X)\hat{}$ does have an interior, and our result follows.

We come now to some phenomena discovered by Kallin [2].

Trivial examples, even in $\mathbb{C}$, show that the union of a finite family of polynomially convex sets need not be polynomially convex. On the other hand, in the complex plane, it is easy to see that a finite union of pairwise disjoint polynomially convex sets is polynomially convex. In higher dimensions, the analogous result fails: A finite disjoint union of polynomially convex sets in $\mathbb{C}^2$ need not be polynomially convex. A simple example is provided by the two circles

$$\Lambda_1 = \{(z, w) \in \mathbb{C}^2 : zw = 1, |z| = 1\},$$
$$\Lambda_2 = \{(z, w) \in \mathbb{C}^2 : zw = 1, |z| = \tfrac{1}{2}\}.$$

Each of these circles is polynomially convex, but the polynomially convex hull of $\Lambda_1 \cup \Lambda_2$ is an annulus:

$$(\Lambda_1 \cup \Lambda_2)\hat{} = \{(z, w) \in \mathbb{C}^2 : 1 \leq |w| \leq 2, zw = 1\}.$$

It seems then that if we are to obtain any result whatsoever on the polynomial convexity of unions, we must restrict in some way the class of sets to be considered. A restriction that presents itself naturally and sufficies to exclude the example just considered is that of (linear) convexity. It is clear that a compact, convex set in $\mathbb{C}^N$ is polynomially convex, for if $\mathfrak{z}_0 \in \mathbb{C}^N \setminus X$, there is a linear functional $\psi$ on $\mathbb{C}^N$ with $|\psi(\mathfrak{z}_0)| > \|\psi\|_X$. That the union of two disjoint compact, convex sets is polynomially convex is a direct consequence of the following lemma, for disjoint, compact, convex sets can be separated by a linear functional, i.e., a polynomial of degree 1.

29.21 ***Lemma***

(a) *Let $X_1$ and $X_2$ be compact, polynomially convex sets in $\mathbb{C}^N$, and let $p$ be a polynomial such that $p(X_1)\hat{} \cap p(X_2)\hat{} = \partial_{\mathbb{C}}(p(X_1)\hat{}) \cap \times\, \partial_{\mathbb{C}}(p(X_2)\hat{}) \subset \{0\}$. If $p^{-1}(0) \cap (X_1 \cup X_2)$ is polynomially convex, then $X_1 \cup X_2$ is polynomially convex.*

(b) *Let $X_1$ and $X_2$ be compact sets in $\mathbb{C}^N$ such that there exists a polynomial $p$ with $p(X_1)\hat{} \cap p(X_2)\hat{} = \varnothing$. Then $X_1 \cup X_2)\hat{} = \hat{X}_1 \cup \hat{X}_2$.*

PROOF

(b) is a simple consequence of (a), because $\hat{X}_i \subset p^{-1}(p(X_i)\hat{})$, so $p(\hat{X}_1)\hat{} \cap p(\hat{X}_2)\hat{} = \varnothing$. Thus, by (a), $(\hat{X}_1 \cup \hat{X}^2)\hat{} = \hat{X}_1 \cup \hat{X}_2$ and therefore $(X_1 \cup X_2)\hat{} \subset \hat{X}_1 \cup \hat{X}_2$. The opposite inclusion always holds, so we have part (b).

To prove (a), note first of all that if $A$ and $B$ are compact polynomially convex sets in $\mathbb{C}$ which meet only in one point, then $A \cup B$ is also polynomially convex: If neither $A$ nor $B$ separates the plane, and if $A \cap B$ is a point, then $A \cup B$ does not separate the plane.

Set $X = X_1 \cup X_2$, and suppose that $X$ is not polynomially convex. We may assume that $\|p\|_X \leq 1$. If $x_0 \in \hat{X} \setminus X$, then $p(x_0) \in p(X_1)\hat{} \cup p(X_2)\hat{}$, for otherwise, since this union is polynomially convex in $\mathbb{C}$, there is a polynomial in one complex variable, say $q$, such that $q(p(x_0)) = 1$, but $|q(\zeta)| < \frac{1}{2}$ if $\zeta \in p(X_1)\hat{} \cup p(X_2)\hat{}$. Then $q \circ p$ is a polynomial which shows that $x_0 \notin \hat{X}$ since $X \subset p^{-1}(p(X_1)\hat{} \cup (p(X_2)\hat{}))$.

Consider first the case that $p(x_0) \neq 0$. The polynomial convexity of $X_1$ yields a polynomial $q$ with $\|q\|_{X_1} < \frac{1}{3}$ and $q(x_0) = p(x_0)^{-1}$, $q$ not identically zero on $X_2$. Since $p(X_1)\hat{} \cap p(X_2)\hat{} \subset \{0\}$ and $p(X_1)\hat{} \cup p(X_2)\hat{}$ is polynomially convex, the function identically zero on $p(X_2)$ and the identity map on $p(X_1)\hat{}$ can be approximated uniformly by polynomials because of Mergelyan's theorem; so there is a polynomial $h$ is one complex variable such that

$$|h(\zeta)| < (2\|q\|_{X_2})^{-1}, \qquad \zeta \in p(X_2)$$

$$|h(\zeta) - \zeta| < \tfrac{1}{2}|p(x_0)|, \qquad \zeta \in p(X_1).$$

The polynomial $f$ given by

$$f(\mathfrak{z}) = h(p(\mathfrak{z}))q(\mathfrak{z})$$

is bounded by $\frac{1}{2}$ on $X$, and since $q(x_0) = 1/p(x_0)$ and $|h(p(x_0))| > \frac{1}{2}|p(x_0)|$, it follows that $|f(x_0)| > \frac{1}{2}$, and $f$ therefore contradicts the hypothesis that $x_0 \in \hat{X}$.

It follows that if $x_0 \in \hat{X} \setminus X$, then $p(x_0) = 0$, and in particular $\hat{X} \cap p^{-1}(0) \neq \varnothing$. In this case, choose a polynomial $q$ with $q(x_0) = 1$, $|q| < \frac{1}{3}$ on $p^{-1}(0) \cap X$, a choice that can be made because we have assumed $p^{-1}(0) \cap X$ to be polynomially convex. There is a $\delta > 0$ such that $|q(y)| < \frac{1}{2}$ if $y \in X$ and $|p(y)| < \delta$. Our hypotheses imply that 0 is a peak point for the algebra $\mathscr{P}(p(X))$, so there exists a polynomial in one complex variable, $h$, such that $h(0) = 1$, $\|h\|_{p(X)} \leq \frac{3}{2}$, and $|h(\zeta)| < (2\|q\|_X)^{-1}$ if $\zeta \in p(X)$ and $|\zeta| \geq \delta$. The polynomial $f$, given by

$$f(\mathfrak{z}) = q(\mathfrak{z})h(p(\mathfrak{z})),$$

takes the value 1 at $x_0$ but has norm no more than $\frac{3}{4}$ on $X$. Again we have a contradiction to $x_0 \in \hat{X}$, and the lemma is proved.

We know now that the union of two disjoint, compact, convex sets is polynomially convex. What about the union of three? The situation here is complicated. By a *sphere* in $\mathbb{C}^N$ we mean a set of the form $\{\mathfrak{z} : \|\mathfrak{z} - \mathfrak{z}_0\| = r\}$, where $\|\cdot\|$ is the usual norm on $\mathbb{C}^N$ and where $r > 0$. By a *ball* we mean a set of the form $\{\mathfrak{z} : \|\mathfrak{z} - \mathfrak{z}_0\| \leq r\}$. It is easy to see that if $S$ is a sphere in $\mathbb{C}^N$, then $\hat{S}$ is the ball that $S$ bounds.

29.22 ***Theorem*** (Kallin [2])
*If $S_1, S_2$, and $S_3$ are disjoint spheres in $\mathbb{C}^N$, then $(\bigcup S_i)\hat{} = \bigcup \hat{S}_i$.*

Thus, the union of three disjoint balls is polynomially convex. The corresponding problem for a union of more than three balls is apparently unresolved, and, even more surprisingly, the corresponding statement for the union of three disjoint polydiscs is false!

PROOF OF THE THEOREM

Each $\hat{S}_i$ is a ball, so we may as well assume that the $\hat{S}_i$ are disjoint.

Our proof will depend on part (b) of the lemma. We assume that $S_1$ is one of the largest of the spheres and that coordinates have been chosen so that $S_1$ has radius 1 and is centered at the origin. Thus, if $S_2$ and $S_3$ have, respectively, radii $\rho_2$ and $\rho_3$, neither of $\rho_2, \rho_3$ exceeds 1. Moreover, by proper choice of coordinates, we can suppose that the subspace spanned by the centers of $S_2$ and $S_3$ lies in the space

$$\{(z_1, \ldots, z_N) \in \mathbb{C}^N : z_3 = \cdots = z_N = 0\}$$

and that the center of $S_2$ has $(z_1, z_2)$ coordinates $(\zeta, 0)$ for some $\zeta$. We can now rotate the $z_1$-plane and the $z_2$-plane so that the $(z_1, z_2)$ coordinates of the center of $S_2$ are $(\alpha, \beta)$ with $\alpha$ and $\beta$ real. These rotations leave the center of $S_2$ at a point with $(z_1, z_2)$ coordinates $(\gamma, 0)$, $\gamma \in \mathbb{C}$. We shall show that with these normalizations the polynomial $p$ given by $p(\mathfrak{z}) = z_1^2 + z_2^2$ has the property that $(p(S_1))\hat{} \cap (p(S_2 \cup S_3))\hat{} = \varnothing$, so our theorem follows from the lemma just proved.

Under the polynomial $p$, $S_1$ goes onto the closed unit disc, and we shall prove that under $p$, $S_3$ goes to a set lying to the right of the line $\operatorname{Re} w = 1$, and that $S_2$ goes onto a set which can be separated from the unit disc by a line tangent to the unit circle. Granted these facts, it is clear that $(p(S_1))\hat{}$ and $(p(S_2 \cup S_3))\hat{}$ are disjoint, as we need to prove.

Consider $S_3$. We shall prove that

$$\operatorname{Re}(z_1^2 + z_2^2) > 1 \tag{1}$$

on the set $\{\mathfrak{z} \in \mathbb{C}^2 : |z_1 - \alpha|^2 + |z_2 - \beta|^2 \leq \rho_3^2\}$. We have that $\alpha$ and $\beta$ are real, and since $S_1$ and $S_3$ are disjoint and $\rho_3 \leq 1$, $\alpha^2 + \beta^2 > (1 + \rho_3)^2$. Write $z_j = x_j + iy_j$, $j = 1, 2$, so that

$$|z_1 - \alpha|^2 + |z_2 - \beta|^2 \leq \rho_3^2$$

is equivalent to

$$(x_1 - \alpha)^2 + y_1^2 + (x_2 - \beta)^2 + y_2^2 \leq \rho_3^2.$$

If $\eta = (\alpha^2 + \beta^2)^{1/2} - 1 - \rho_3$ and $\varepsilon = \rho_3 - [(x_1 - \alpha)^2 + (x_2 - \beta)^2]^{1/2}$, then

$$\begin{aligned}\operatorname{Re}(z_1^2 + z_2^2) &= x_1^2 + x_2^2 - y_1^2 - y_2^2 \\ &\geq x_1^2 + x_2^2 + (x_1 - \alpha)^2 + (x_2 - \beta)^2 - \rho_3^2 \\ &\geq (1 + \eta + \varepsilon)^2 + (\rho_3 - \varepsilon)^2 - \rho_3^2 \\ &= 1 + 2(1 - \rho_3)\varepsilon + 2\eta + (\eta + \varepsilon)^2 + \varepsilon^2 > 1.\end{aligned}$$

(The next-to-last step follows from the triangle inequality.) Thus, we have the desired fact about $p(S_3)$.

To treat $p(S_2)$, we show that $\operatorname{Re}(\theta(x_1^2 + z_2^2)) > 1$ on the set

$$\{\mathfrak{z} \in \mathbb{C}^2 : |z_1 - \gamma|^2 + |z_2|^2 \leq \rho_2^2\},$$

where $\theta = |\gamma|^2\gamma^{-2}$.
We have $|\gamma| > 1 + \rho_2$ and $\rho_2 \leq 1$, so our conclusion follows from (1) if we replace $z_1, z_2, \alpha, \beta$, and $\rho_3$ by $z_1|\gamma|\gamma^{-1}$, $z_2|\gamma|\gamma^{-1}$, $|\gamma|$, 0, and $\rho_2$, respectively.

We now prove, as mentioned above, that this theorem becomes false if we replace balls by polydiscs.

29.23 ***Example*** (Kallin [2])
There exist three disjoint polydiscs in $\mathbb{C}^3$ whose union is not polynomially convex.

To construct the desired polydiscs, we begin by considering in $\mathbb{C}^3$ the variety $V$ defined by

$$z_1 z_2 = 1$$

and

$$z_3(1 - z_1) = 1.$$

Let $M > 2$, and let $\gamma_1, \gamma_2$, and $\gamma_3$ be the curves in $V$ defined by intersecting $V$ with the tubes $T_1 = \{z \in \mathbb{C}^3 : |z_1| = M^{-1}\}$, $T_2 = \{z \in \mathbb{C}^3 : |z_1 - 1| = M^{-1}\}$, and $T_3 = \{\mathfrak{z} \in \mathbb{C}^3 : |z_1| = M\}$, respectively, so that the projections of the $\gamma_j$ into the $z_1$-plane are three disjoint circles; $\gamma_3$ projects onto a large circle which surrounds the projections of $\gamma_1$ and $\gamma_2$. If we project into the $z_2$-plane, we find that $\gamma_1, \gamma_2$, and $\gamma_3$ again go onto circles since the transformation

$$\zeta \mapsto \frac{1}{\zeta}$$

of the Riemann sphere to itself carries circles to circles. This time, however, the circle corresponding to $\gamma_1$ contains the others in its interior. Again, if we project into the $z_3$-plane, we get three circles, and in this case $\gamma_2$ projects to the exterior circle. It follows that the polynomially convex hulls of the curves $\gamma_i$ are pairwise disjoint since, e.g., that of $\gamma_1$ lies in the tube $T_1$ and that of $\gamma_2$ in the tube $T_2$.

The curves $\gamma_1, \gamma_2$, and $\gamma_3$ constitute the boundary of a domain $\Delta$ in the variety $V$. To understand this, it helps to consider the planar domain $D = \mathbb{C} \setminus \{0, 1\}$ and to define $\Phi : D \to \mathbb{C}^3$ by

$$\Phi(\zeta) = \left(\zeta, \frac{1}{\zeta}, \frac{1}{1-\zeta}\right).$$

The map $\Phi$ carries $D$ onto $V$, and if $\tilde{\Delta}$ is the set

$$\{\zeta \in D : |\zeta| < M, |\zeta| > M^{-1}, |\zeta - 1| < M^{-1}\},$$

then $\Phi$ carries $\tilde{\Delta}$ onto a relatively compact piece, $\Delta$ of $V$, with boundary $\gamma_1 \cup \gamma_2 \cup \gamma_3$. The maximum modulus theorem implies that

$$\|p\|_\Delta = \|p\|_{\gamma_1 \cup \gamma_2 \cup \gamma_3}$$

for every polynomial $p$, so $\bar{\Delta} \subset (\gamma_1 \cup \gamma_2 \cup \gamma_3)\hat{}$.

To complete our construction, it is enough to find disjoint closed polydiscs each of which contains one of the $\gamma_i$, for then if $S$ is the union of the polydiscs, $\hat{S}$ must contain $\Delta$, whence $S \neq \hat{S}$. We shall specify polydiscs $S_1$, $S_2$, and $S_3$ each with radius $M$ in each direction. Let

$S_1$ have center $(-M + M^{-1}, 0, M + M(M+1)^{-1})$,

$S_2$ have center $(M + 1 - M^{-1}, M + M(M+1)^{-1}, 0)$,

$S_3$ have center $(0, -M + M^{-1}, -M + (M+1)^{-1})$.

It is easily verified that the $S_i$ defined in this way are disjoint and that $S_i \supset \gamma_i$.

The last two examples have important consequences for the theory of polynomial approximation in $\mathbb{C}^N$. Recall that a domain $\Omega$ in $\mathbb{C}^N$ is a *Runge domain* if every function $F$ holomorphic on $\Omega$ can be approximated uniformly on compacta in $\Omega$ by polynomials. Open balls and polydiscs are simple examples, and Theorem 29.22, taken together with the Oka approximation theorem, shows that the disjoint union of three open balls is again a Runge domain. If we replace the closed polydiscs $S_i$ of the last example by slightly larger but still disjoint open polydiscs, we obtain a domain that is not a Runge domain.

Having dwelt at some length on the pathology in the theory of polynomial convexity, we now turn to some positive aspects of the theory. We shall be concerned especially with results which enable us to conclude that if a compact set is rationally convex and satisfies some further condition, then it is actually polynomially convex. Many of these results are from Stolzenberg's paper [3].

A topological condition which will play a role in our discussion is that of simple coconnectivity.

**29.24** ***Definition***

*A compact space is* ***simply coconnected*** *if* $H^1(X, \mathbb{Z}) = 0$.

A sufficient condition for simple coconnectivity is contractability. The condition is not equivalent to the banishing of the fundamental group, even for reasonable spaces. By the $\mathscr{C}(X)$ case of the Arens–Royden theorem (Theorem 10.1) we know that $X$ is simply coconnected if and only if every zero-free complex-valued continuous function on $X$ has a continuous logarithm.

**29.25** ***Definition***

*A compact set* $X \subset \mathbb{C}^N$ *satisfies the* ***generalized argument principle*** *if* $0 \notin p(\hat{X})$ *for every polynomial, p, which has a continuous logarithm on* $X$.

The relevance of this notion for the theory is established as follows.

**29.26** ***Theorem*** (Stolzenberg [3])

*If* $X \subset Y$ *are compact sets in* $\mathbb{C}^N$, *if* $X$ *satisfies the generalized argument principle, and if* $Y$ *is simply coconnected, then*

$$\hat{X} \subset \mathscr{R}\text{-hull } Y.$$

PROOF

If $\mathfrak{z} \notin \mathscr{R}$-hull $Y$, then by Lemma 29.4 there is a polynomial $p$ with $0 = p(\mathfrak{z}) \notin p(Y)$. Since $Y$ is simply coconnected, there is a continuous branch of $\log p$ on $Y$. The set $X$ satisfies the generalized argument principle and $\log p$ is defined on $X$, so $0 \notin p(\hat{X})$, which implies that $\mathfrak{z} \notin \hat{X}$.

**29.27** ***Corollary***

*If* $X$ *is simply coconnected and satisfies the generalized argument principle, then* $\hat{X} = \mathscr{R}$*-hull* $X$.

**29.28** ***Corollary***

*If* $X$ *is simply coconnected and rationally convex, and if it satisfies the generalized argument principle, then it is polynomially convex.*

It is clear that the second corollary has various defects, not the least of which is the fact that to apply it one needs a priori knowledge of $\hat{X}$. It is tempting to conjecture that this corollary can be strengthened on either of two grounds: (1) Perhaps every compact set satisfies the generalized argument principle; and (2) perhaps every simply coconnected, rationally convex set is polynomially convex. Neither of these conjectures is correct. Examples of compact sets that do not satisfy the generalized argument principle were constructed by Hoffman and by Bishop; Bishop's example is contained in Stolzenberg's paper [3]. Much more decisive, however, is the following fact, discovered by Stolzenberg [5].

**29.29** ***Theorem***

*There is a set $Y$ in $\mathbb{C}^2$ which consists of the disjoint union of two polynomially convex discs and which satisfies $\mathscr{R}(Y) = \mathscr{C}(Y)$ but which is not polynomially convex.*

Since $\mathscr{R}(Y) = \mathscr{C}(Y)$, $Y$ is rationally convex, and it is evident that $H^1(Y, \mathbb{Z}) = 0$.

**29.30** ***Corollary***

*The set $Y$ does not satisfy the generalized argument principle.*

We need a technical lemma in the proof of the theorem.

**29.31** ***Lemma***

*Let $X_1$ and $X_2$ be compact sets in $\mathbb{C}^N$, and let $S_1$ and $S_2$ be relatively open subsets of $X_1$ and $X_2$, respectively. Let $S_1 \cap S_2 = \varnothing$, and suppose that $S_2 \subset \hat{X}_1$, $S_1 \subset \hat{X}_2$. Then $S_1 \cup S_2 \subset ((X_1 \setminus S_1) \cup (X_2 \setminus S_2))\hat{}$.*

PROOF

Let $q$ be a polynomial with $\|q\|_{X_1 \cup X_2} = 1$, and suppose that $q$ assumes the value 1 on $X_1 \cup X_2$, say on the set $P_q$. We shall prove that

$$P_q \cap ((X_1 \setminus S_1) \cup (X_2 \setminus S_2)) \neq \varnothing, \tag{2}$$

which implies that

$$(X_1 \cup X_2) \subset ((X_1 \setminus S_1) \cup (X_2 \setminus S_2))\hat{}.$$

whence the desired result.

If (2) is false, then $P_q \subset S_1 \cup S_2$. By hypothesis, $S_1 \subset \hat{X}_2$, so

$$S_1 \cap P_q \subset \hat{X}_2 \cap P_q \subset (X_2 \cap P_q)\hat{}.$$

The last inclusion may be proved by the following simple argument. If it is false, there is $x \in \hat{X}_2 \cap P_q$ and a polynomial $h$ with $h(x) = 2$, $\|h\|_{X_2 \cap P_q} \leq 1$. Let $\mu$ be a probability measure on $X_2$ which represents $x$ for the algebra $\mathscr{P}(\hat{X}_2)$. If $\tilde{q} = \frac{1}{2}(1 + q)$, then $\tilde{q}$ takes on $P_q \cap X_2$ the value 1 and is strictly less than 1 in modulus on the remainder of $X_2$. Since $x \in P_q$, $\tilde{q}(x) = 1$, so for all $N$,

$$2 = h(x)\tilde{q}^N(x) = \int_{X_2} h\tilde{q}^N \, d\mu.$$

As $N \to \infty$, the integral approaches $\int_{X_2 \cap P_q} h \, d\mu$, and since $\|h\|_{X_2 \cap P_q} \leq 1$, we have a contradiction.

Now $(X_2 \setminus S_2) \cap P_q = \varnothing$ because we are assuming (2) false, so

$$S_1 \cap P_q \subset (S_2 \cap P_q)\hat{\ },$$

and by symmetry

$$S_2 \cap P_q \subset (S_1 \cap P_q)\hat{\ },$$

whence the equality $(S_1 \cap P_q)\hat{\ } = (S_2 \cap P_q)\hat{\ }$.

However, if $\Gamma$ is the Shilov boundary for the algebra $\mathscr{P}(\Sigma)$, $\Sigma = (S_1 \cap P_q)\hat{\ } = (S_2 \cap P_q)\hat{\ }$, then $\Gamma \subset S_1 \cap P_q$ and $\Gamma \subset S_2 \cap P_q$. As $\Gamma$ is nonempty but $S_1 \cap S_2 = \varnothing$, we have a contradiction, so the lemma is proved.

PROOF OF THE THEOREM

We begin by introducing several sets. Let $\Omega_1$ and $\Omega_2$ be domains in $\mathbb{C}$ bounded by simple closed curves such that if $I_+ = [1, \sqrt{3}]$, $I_- = [-\sqrt{3}, -1]$, then $\partial\Omega_1 \supset I_+$, $\partial\Omega_2 \supset I_-$. Require that $\partial\Omega_1$ and $\partial\Omega_2$ meet only twice, that $\Omega_1 \supset I_-$, $\Omega_2 \supset I_+$, and, finally, that $\partial\Omega_1 \cup \partial\Omega_2$ be the union of the boundary of the unbounded component of $\mathbb{C} \setminus (\partial\Omega_1 \cup \partial\Omega_2)$, together with the boundary of the component of this set which contains the origin.

We define

$$V_1 = \{(z, w) \in \mathbb{C}^2 : z^2 - w \text{ is real and lies in } [0, 1]\},$$

$$V_2 = \{(z, w) \in \mathbb{C}^2 : w \text{ is real and lies in } [1, 2]\},$$

$$X_1 = \{(z, w) \in V_1 : z \in \partial\Omega_2\},$$

$$X_2 = \{(z, w) \in V_2 : z \in \partial\Omega_1\},$$

$$Y_1 = (X_1 \setminus V_2)^-,$$

and

$$Y_2 = (X_2 \setminus V_1)^-.$$

We let $Y = Y_1 \cup Y_2$, and we shall show that the set $Y$ has the properties announced in the theorem.

The proof that $Y$ is not polynomially convex depends on the lemma we just proved. The sets $X_1 \setminus Y_1$ and $X_2 \setminus Y_2$ are relatively open in

$X_1$ and $X_2$, respectively, and we shall show that $X_1 \setminus Y_1 \subset \hat{X}_2$, $X_2 \setminus Y_2 \subset \hat{X}_1$. In doing this, it is useful to denote by $\pi: \mathbb{C}^2 \to \mathbb{C}$ the projection onto the first coordinate, $\pi(z, w) = z$. We have that

$$\begin{aligned} X_1 \setminus Y_1 &= X_1 \setminus (X_1 \setminus V_2)^- \\ &\subset X_1 \cap V_2 \\ &= \{(z, w): z \in \partial\Omega_2, 0 \leq z^2 - w \leq 1, 1 \leq w \leq 2\}. \end{aligned}$$

The two inequalities imply that $1 \leq z^2 \leq 3$, i.e., that $z \in I_+ \cup I_-$; and since $z \in \partial\Omega_2$, it follows that $z \in I_-$. Thus, $\pi(X_1 \cap V_2) \subset \Omega_1$, so $X_1 \setminus Y_1 \subset \pi^{-1}(\Omega_1)$. Also, $X_2 \setminus Y_2 \subset \pi^{-1}(\Omega_2)$, as a parallel argument shows. Thus, we have

$$X_1 \setminus Y_1 \subset \pi^{-1}(\Omega_1) \cap V_2$$

and

$$X_2 \setminus Y_2 \subset \pi^{-1}(\Omega_2) \cap V_1.$$

The maximum modulus theorem implies that

$$\pi^{-1}(\Omega_1) \cap V_2 \subset \hat{X}_2$$

and

$$\pi^{-1}(\Omega_2) \cap V_1 \subset \hat{X}_1,$$

for if, e.g., $(z, w) \in \pi^{-1}(\Omega_2) \cap V_1$ so that $z \in \Omega_2$ and $z^2 - w = t$, then the variety $W_t = \{(z, w): z^2 - w = t\}$ lies in $V_1$ and intersects $\pi^{-1}(\partial\Omega_2)$ in a certain curve $\Lambda_t$. We have $\Lambda_t \subset X_1$ and $(z, w) \subset \hat{\Lambda}_t$ by the maximum modulus theorem. Thus, $(z, w) \in \hat{X}_1$, so $\pi^{-1}(\Omega_2) \cap V_1 \subset \hat{X}_1$. In a similar way, $\pi^{-1}(\Omega_1) \cap V_2 \subset \hat{X}_2$.

We saw that $\pi(X_1 \setminus Y_1) \subset I_-$, and we used implicitly the fact that $\pi(X_2 \setminus Y_2) \subset I_+$; so since $I_-$ and $I_+$ are disjoint, so are the sets $X_1 \setminus Y_1$ and $X_2 \setminus Y_2$. Consequently, by the lemma proved above,

$$\begin{aligned} (X_1 \setminus Y_1) \cup (X_2 \setminus Y_2) &\subset \{(X_1 \setminus (X_1 \setminus Y_1)) \cup (X_2 \setminus (X_2 \setminus Y_2))\}\hat{} \\ &= (Y_1 \cup Y_2)\hat{}. \end{aligned}$$

This implies that $X_1 \cup X_2 \subset (Y_1 \cup Y_2)\hat{}$, so $(X_1 \cup X_2)\hat{} \subset (Y_1 \cup Y_2)\hat{}$.

The origin lies in $(X_1 \cup X_2)\hat{}$, for if $W_0 = \{(z, w): z^2 - w = 0\}$, then $W_0 \subset V_1$, $(0, 0) \in W_0$, and $W_0 \cap \pi^{-1}(\Omega_2) \subset \hat{X}_1$ by the maximum modulus theorem. Thus, the origin lies in $(Y_1 \cup Y_2)\hat{}$, and since $0 \notin \pi(Y_1 \cup Y_2)$, we find that $Y_1 \cup Y_2$ is not polynomially convex.

To prove that $Y = Y_1 \cup Y_2$ is rationally convex, note first that, by Theorem 26.5, $\mathscr{R}(\partial\Omega_1 \cup \partial\Omega_2) = \mathscr{C}(\partial\Omega_1 \cup \partial\Omega_2)$, so every maximal set of antisymmetry for $\mathscr{R}(Y)$ is contained in a fiber $\pi^{-1}(\zeta) \cap Y$, $\zeta \in (\partial\Omega_1 \cup \partial\Omega_2)$. However, given such a $\zeta$, $\pi^{-1}(\zeta) \cap Y$ is a subset of

$$E_\zeta = \{(\zeta, r): r \in [1, 2]\} \cup \{(\zeta, \zeta^2 - r): r \in [0, 1]\},$$

and it is clear that $\mathscr{P}(E_\zeta) = \mathscr{C}(E_\zeta)$. Thus, the maximal sets of antisymmetry for $\mathscr{R}(Y)$ are points, so $\mathscr{R}(Y) = \mathscr{C}(Y)$, and, a fortiori, $Y$ is rationally convex.

It follows immediately from the definition that the sets $Y_1$ and $Y_2$ are disjoint. We must show that they are polynomially convex topological discs.

In this direction, consider first the set $X_1$. The map $\varphi: \partial\Omega_2 \times [0, 1] \to X_1$ given by $\varphi(\zeta, t) = (\zeta, \zeta^2 - t)$ is continuous, one to one, and onto, so $X_1$ is, topologically, a planar annulus. The set $Y_1$ is mapped by the polynomial $p$ given by $p(z, w) = z^2 - w$ into the interval $[0, 1]$, so every maximal set of antisymmetry for $\mathscr{R}(Y_1)$ is contained in

$$Y_1 \cap p^{-1}(t_0) = \{(z, w): z^2 - w = t_0, z \in \Omega_2\}$$

for some $t_0 \in [0, 1]$. We shall show that $\pi(Y_1 \cap p^{-1}(t_0))$ is a proper subset of $\partial\Omega_2$. This will yield $\mathscr{P}(\pi(Y_1) \cap p^{-1}(t_0)) = \mathscr{C}(\pi(Y_1 \cap p^{-1}(t_0))$, so every maximal set of antisymmetry for $\mathscr{P}(Y_1)$ lies in a set $\pi^{-1}(z) \cap p^{-1}(t_0)$, a finite set. Thus, $\mathscr{P}(Y_1) = \mathscr{C}(Y_1)$, and $Y_1$ is polynomially convex. To prove our assertion about $\pi(Y_1 \cap p^{-1}(t_0))$, note that the definition of $Y_1$ implies that if $(z, w) \in Y_1$, then $w \notin (1, 2)$. If we assume further that $z_2 - w = t_0$, i.e., $p(z, w) = t_0$ for some $t_0 \in [0, 1]$, we find $w = z^2 - t_0$. As $z$ runs from $-\sqrt{3}$ to $-1$, $z^2 - t_0$ runs from $3 - t_0$ to $1 - t_0$ and, no matter what $t_0$ is, $t_0 \in [0, 1]$, the interval $[1 - t_0, 3 - t_0]$ contains $(1, 2)$. We may conclude that $\pi(Y_1 \cap p^{-1}(t_0))$ does not contain the interval $(-\sqrt{3}, -1)$, so it is a proper subset of $\partial\Omega_2$.

If $(\zeta, t) \in \partial\Omega_2 \times [0, 1]$, then $\varphi(\zeta, t) \in V_2$ if and only if $\zeta^2 - t \in [1, 2]$, that is, if and only if

$$\zeta \in [-\sqrt{2 + t}, -\sqrt{1 + t}] \cup [\sqrt{1 + t}, \sqrt{2 + t}].$$

This set is in $I_+ \cup I_-$, and as this latter set meets $\partial\Omega_2$ only in $I_-$, we find that $V_2 \cap X_1$ is the image under $\varphi$ of the set

$$\{(\zeta, t): 0 \leq t \leq 1, -\sqrt{2 + t} \leq \zeta \leq -\sqrt{1 + t}\},$$

and from this we find that $Y_1$ is a topological disc.

To treat $Y_2$, note first that $X_2 = \partial\Omega_1 \times [1, 2]$ and that under the map $(z, w) \mapsto w$, $X_2$ goes onto the interval $[1, 2]$, so every maximal set of antisymmetry for $\mathscr{P}(Y_2)$ is contained in a slice $Y_{2,t_0} = \{(\zeta, w) \in Y_2 : w = t_0\}$ for some $t_0 \in [1, 2]$. However, if $t_0 \in [1, 2]$ and $(z, t_0) \in Y_2$, then $z \in \partial\Omega_1$, and $z^2 - t_0 \notin (0, 1)$, whence $z^2 \notin (t_0, t_0 + 1)$ so that $z \notin (\sqrt{t_0},$

$\sqrt{t_0 + 1}$). Now $t_0 \in [1, 2]$, so $1 \leq \sqrt{t_0} \leq \sqrt{2} \leq \sqrt{3}$. It follows that under the projection $\pi$, $Y_2$ goes onto a subset of $\partial\Omega_1$ which omits a certain subinterval of the interval $[1, \sqrt{3}]$. Thus, $\mathscr{P}(\pi Y_2)) = \mathscr{C}(\pi(Y_2))$, and we find that each maximal set of antisymmetry for $\mathscr{P}(Y_2)$ is contained in $Y_{2,t_0} \cap \pi^{-1}(z)$ for some $z$. Such sets are points, so $\mathscr{P}(Y_2)$ $\mathscr{C}(Y_2)$, as we wished to show. Since $X_2 \cap V_1 = \{(z, w) : 1 \leq w \leq 2, \sqrt{w} \leq z \leq \sqrt{w + 1}\}$, $Y_2$ is evidently a disc.

Thus, not every compact set satisfies the generalized argument principle. In the positive direction, Stolzenberg [3] has shown that every compact analytic polyhedron in $\mathbb{C}^N$ does satisfy the generalized argument principle. Another strong result obtains for subsets of one-dimensional varieties, and it is to this situation that we now turn our attention. For these results, also due to Soltzenberg [3], we need to use certain facts from the theory of one-dimensional varieties, facts that are well known but are not conveniently derived at this point. If $\Omega$ is an open set in $\mathbb{C}^N$ and $V \subset \Omega$ is a one-dimensional variety, then, of course, $V$ is a closed subset of $\Omega$. There exists a discrete set, $\Sigma$, possibly empty, such that $V \setminus \Sigma$ is a complex submanifold of $\Omega \setminus \Sigma$. If $\Sigma_0$ is the smallest set with the indicated property, then $\Sigma_0$ is called the set of singularities of $V$, and the points of $V \setminus \Sigma_0$ are the regular points of $V$. One other fact that we shall need is that if $V \subset \Omega$ is a one-dimensional variety, then there exists a one-dimensional complex manifold $N$ and a proper holomorphic map $\eta$ from $N$ onto $V$ which is one to one on $\eta^{-1}(V \setminus \Sigma)$, $\Sigma$ the set of singularities of $V$. (Recall that a map is *proper* if the preimage of a compact set is compact.) For references to the literature containing these results, see Appendix D.

**29.32** ***Lemma***

*Let $\Omega \subset \mathbb{C}^N$ be an open set, let $V \subset \Omega$ be a one-dimensional variety, let $X \subset V$ be compact, and suppose that $\hat{X} \subset V$. Then $\partial_V \hat{X} \subset X$.*

PROOF

This can be proved by invoking a generalized version of Runge's theorem known to hold on one-dimensional varieties, but it is also a consequence of the local description of $\hat{X}$.

We know, e.g., by the Oka approximation theorem, that the isolated points of $\hat{X}$ belong to $X$. The set, $\Sigma$, of singularities of $V$ is a discrete set, so we need prove only that every regular point of $V$ contained in $\partial_V \hat{X}$ lies in $X$. Accordingly, let $\mathfrak{z}_0$ be a regular point of $V$ in $\partial_V \hat{X}$. There is a neighborhood $S$ of $\mathfrak{z}_0$ (in $V$) and a function $h$ holomorphic on a neighborhood, $S_0$, of $\bar{S}$ which maps $S_0$ homeomorphically onto an open set in $\mathbb{C}$. We have $\mathfrak{z}_0 \in \partial_V \hat{X}$, so $h(\mathfrak{z}_0) \in \partial_{\mathbb{C}} h(\bar{S} \cap \hat{X})$. By Corollary 28.22, we find that $h(\mathfrak{z}_0) \in h(\partial \bar{S})$ or else $h(\mathfrak{z}_0) \in h(X)$.

As $h$ is one to one on $S_0$, this yields $\mathfrak{z}_0 \in \partial \bar{S}$ or $\mathfrak{z}_0 \in X$. The former is impossible, so $\mathfrak{z}_0 \in X$.

As consequence of this result and the open mapping theorem on varieties, we have a very explicit description of $\hat{X}$, in the case just considered. This description should be compared with the situation in the plane.

29.33 ***Corollary***

*With $\Omega$, $V$, and $X$ as in the preceding lemma, $\hat{X} \setminus X$ consists of the union of $X$ and the relatively compact components of $V \setminus X$.*

We need to emphasize that here we understand by *relatively compact* the hypothesis that the closure *in* $V$ is compact.

PROOF

Suppose $C$ to be a component of $V \setminus X$ with $\bar{C}$ compact, $^-$ denoting closure with respect to $V$. Thus, $\bar{C} \setminus C \subset X$, and therefore, by the open mapping theorem, if $p$ is a polynomial, then $\|p\|_{\bar{C}} \leq \|p\|_X$, so $C \subset \hat{X}$. On the other hand, if $\mathfrak{z}_0 \in \hat{X} \setminus \hat{X}$, let $C_0$ be the component of $V \setminus X$ which contains it. If $\bar{C}_0$ is not compact, then there is a map

$$\varphi : [0, \infty] \to C_0$$

with $\varphi(0) = \mathfrak{z}_0$ and with closed, noncompact range. If $\mathfrak{z}_1 = \varphi(t_1)$, where $t_1$ is the last point $t$ such that $\varphi(t)$ lies in $\hat{X}$—such a point $t_1$ exists since $\hat{X} \cap \varphi([0, \infty])$ is compact but not empty—then $\mathfrak{z}_1 \in \partial \hat{X}$ and $\mathfrak{z}_1 \in C_0$. These conditions are incompatible as $\partial \hat{X} \subset X$ and $X \cap C_0 = \varnothing$. Thus, $C_0$ is compact, and the corollary is proved.

29.34 ***Definition***

*If $V$ is an analytic variety in the open set $\Omega \subset \mathbb{C}^N$, then $V$ is a* ***Runge variety*** *if $\hat{X} \subset V$ for every compact set $X \subset V$.*

Of course, not every one-dimensional variety is a Runge variety, but it is true that every subvariety of a polynomial polyhedron is a Runge variety as follows, e.g., from Corollary 28.15. One reason for our present interest in one-dimensional Runge varieties is that their compact subsets satisfy the generalized argument principle.

29.35 ***Theorem***

*If $V$ is a one-dimensional Runge variety in the open set $\Omega \subset \mathbb{C}^N$, and if $X \subset V$ is compact, then $X$ satisfies the generalized argument principle.*

The theorem follows easily from the following lemma.

29.36 ***Lemma***

*If $X$ is a compact subset of a one-dimensional variety $V$ in the open set $\Omega \subset \mathbb{C}^N$, if $f$ is holomorphic on a neighborhood of $X$, and if $\log f$ is defined on $\partial_V X$, then $0 \notin f(X)$.*

PROOF

We deal first with the case that $X$ has no singular points. As $f$ is holomorphic on a neighborhood of $X$ and $\log f$ is defined on $\partial X$, it follows that we can find a domain $Y$ containing $X$ with $\partial Y$ consisting of a finite family of smooth curves and with the property that $\partial Y$ lies so near $\partial X$ that $\log f$ exists on $\partial Y$. If we orient $\partial Y$ properly, then we have that if $N$ is the number of zeros of $f$ in $Y$, then

$$N = \frac{1}{2\pi i}\int_{\partial Y} d\log f = 0.$$

Thus, $f$ is zero free in $Y$, and so in $X$.

If $V$ has singularities, let $\tilde{V}$ be a one-dimensional manifold and let $\eta : \tilde{V} \to V$ be a proper holomorphic map. Since $\eta$ is proper, $\tilde{X} = \eta^{-1}(X)$ is compact. We have that $\tilde{f} = f \circ \eta$ is holomorphic on a neighborhood of $\tilde{X}$, and $\log \tilde{f}$ is defined on $\partial\tilde{X}$. Thus, $\tilde{f}$ is zero free on $\tilde{X}$ by what we have just done, so $f$ is zero free on $X$.

PROOF OF THE THEOREM

To derive the theorem from the lemma, note that since $V$ is a Runge variety by hypothesis, $\hat{X} \subset V$. Lemma 29.32 yields $\partial_V \hat{X} \subset X$, so the theorem follows from the last lemma.

29.37 ***Corollary***

*If $X$ is a simply coconnected set in $\mathbb{C}^N$, and if $V$ is a one-dimensional Runge variety in the open set $\Omega \subset \mathbb{C}^N$ such that $V \cap X$ is compact, then $(X \cap V)\hat{} \subset \mathscr{R}$-hull $X$.*

These last remarks imply a general property of certain rationally convex sets.

29.38 ***Definition***

*A compact set $X$ is* **polynomially convex in dimension one** *if for every one-dimensional Runge variety $V$ in an open set $\Omega \subset \mathbb{C}^N$ such that $X \cap V$ is compact, $X \cap V$ is polynomially convex.*

29.39 ***Corollary***

*Every simply coconnected, rationally convex set is polynomially convex in dimension 1.*

PROOF

By the theorem, if $V$ is a one-dimensional Runge variety such that $V \cap X$ is compact, then $(V \cap X)\hat{} \subset \mathscr{R}$-hull $X = X$, and since $V$ is a Runge variety, $(V \cap X)\hat{} \subset V$, so we get $(V \cap X)\hat{} \subset V \cap X$, whence the result.

In addition to conditions involving the generalized argument principle, there is another, rather geometric, result which guarantees the polynomial convexity of a set.

29.40 ***Theorem*** (Stolzenberg [3])

*Let $X$ be a compact set in $\mathbb{C}^N$, and let $f$ be holomorphic on a neighborhood of $\hat{X}$. If a continuous branch of* $\log f$ *exists on $X$, and if $X$ is open in $\hat{X} \cap f^{-1}(f(X))$, then $0 \notin f(\hat{X})$. In particular, if $f(X) \cap f(\hat{X} \setminus X) = \varnothing$ and if there is a continuous branch of* $\log f$ *on $X$, then, $0 \notin f(\hat{X})$.*

PROOF

The proof of this theorem depends on the local description of $\hat{X}$ given in the last section.

There is a neighborhood of $\hat{X}$ on which $f$ is holomorphic, and, as $\hat{X}$ is compact, we can suppose that the neighborhood in question is the disjoint union of a finite collection $\Omega_1, \ldots, \Omega_q$ of connected, open sets each of which meets $\hat{X}$. The Oka approximation theorem implies that each $\Omega_j$ meets $X$ and also that each of the sets $\hat{X} \cap \Omega_j$ is polynomially convex. If $f$ were constant on $\Omega_j$, then $f|(\Omega_j \cap X)$ would also be constant, and, as $f$ is zero free on $X$, it would follow that $0 \notin f(\Omega_j \cap \hat{X})$. Accordingly, we shall assume $f$ to be defined and nowhere locally constant on the neighborhood $\Omega'$ of $\hat{X}$. Since $\log f$ exists as a continuous function on $X$, there is a neighborhood $\Omega$ of $X$, $\Omega \subset \Omega'$, on which $\log f$ is defined and holomorphic.

The hypothesis that $X$ is open in $f^{-1}(f(X)) \cap \hat{X}$ implies that by shrinking $\Omega$ if necessary, we may suppose that

$$\Omega \cap f^{-1}(f(X)) \cap \hat{X} = X.$$

The function $f$ is holomorphic and nowhere locally constant on $\Omega$, so the set $f(\Omega)$ contains a neighborhood of $f(X)$ in $\mathbb{C}$, and we can show that there is a neighborhood $V$ of $f(X)$ with $\bar{V} \subset f(\Omega)$ and with the property that $f^{-1}(\bar{V}) \cap \hat{X} \cap \Omega$ is compact. To find such a $V$, argue by contradiction. If no such $V$ exists, there is a sequence $\{V_j\}$ of neighborhoods of $f(X)$ with $f(X) \subset \bar{V}_{j+1} \subset V_j \subset \bar{V}_j \subset f(\Omega)$ and with $\bigcap_j V_j = f(X)$ such that for each $j$, there exists a point

$$\mathfrak{z}_j \in f^{-1}(\bar{V}_j) \cap \hat{X} \cap \partial_{\mathbb{C}^N}\Omega.$$

By compactness we can suppose that $\{\mathfrak{z}_j\}$ converges to a point, $\mathfrak{z}_0$, in

$$\bigcap_j f^{-1}(\bar{V}_j) \cap \hat{X} \cap \partial_{\mathbb{C}^N}\Omega.$$

However, $f^{-1}(f(X)) = \bigcap_j f^{-1}(V_j)$, so $\mathfrak{z}_0 \in f^{-1}(f(X)) \cap \hat{X} \cap \bar{\Omega} = X$, an impossibility, for $\mathfrak{z}_0$ lies in $\partial_{\mathbb{C}^N}\Omega$, and $\Omega$ is a neighborhood of $X$. Thus, the desired neighborhood $V$ exists.

Set $S = f^{-1}(\bar{V}) \cap \hat{X} \cap \Omega$. This is a compact subset of $\Omega$, so $\partial_{\hat{X}}S$ is contained in $\partial_\Omega f^{-1}(\bar{V})$. From

$$\partial_\Omega f^{-1}(\bar{V}) \subset f^{-1}(\partial_{\mathbb{C}}\bar{V})$$

we find that

$$\partial_{\hat{X}}S \subset f^{-1}(\partial_{\mathbb{C}}\bar{V}).$$

Also, $S \cap X = X$ because $X \subset f^{-1}(\bar{V})$. It follows from Corollary 28.22, and this is where we use the local description of hulls, that

$$\partial_{\mathbb{C}} \log f(S) \subset \log f(f^{-1}(\partial_{\mathbb{C}}\bar{V})) \cup \log f(X).$$

Let $\mathscr{L}$ be the set-valued function which assigns to each $\zeta \in \mathbb{C} \setminus \{0\}$ the set of solutions of $e^x = \zeta$ so that $\mathscr{L}(\zeta)$ is the set of all logarithms of $\zeta$. For a set $E$, let $\mathscr{L}(E)$ be the set $\cup\{\mathscr{L}(\zeta) : \zeta \in E\}$. Then

$$\log f(f^{-1}(\partial_{\mathbb{C}}V)) \subset \mathscr{L}(\partial_{\mathbb{C}}\bar{V})$$

and

$$\log f(X) \subset \mathscr{L}(f(X)).$$

Thus,

$$\partial_{\mathbb{C}} \log f(S) \subset \mathscr{L}(\partial_{\mathbb{C}}\bar{V} \cup f(X)). \tag{3}$$

Let us suppose that the theorem is false so that $0 \in f(\hat{X})$. Let $V_0$ be a second neighborhood of $f(X)$ in $\mathbb{C}$ with $\bar{V}_0 \subset V$. By Corollary 29.2,

$$\partial_{\mathbb{C}} f(\hat{X}) \subset f(X),$$

so if

$$\Delta = f(\hat{X}) \setminus (\bar{V}_0 \cap f(\hat{X})),$$

then $\Delta$ is an open set in $\mathbb{C}$, and

$$\partial_{\mathbb{C}}\Delta \subset \partial_{\mathbb{C}}\bar{V}_0 \cap f(\hat{X}).$$

It follows that $\partial_{\mathbb{C}}\Delta$ does not meet $\partial_{\mathbb{C}}V \cup f(X)$. Moreover, $\bar{V}_0 \subset f(\Omega)$, and $\log f$ is defined on $\Omega$, so $0 \notin V_0$. From this it follows that $0 \in \Delta$.

We can invoke the argument principle from classical function theory to find a component, $\Gamma$, of $\partial\Delta$ on which $\log z$ is not defined as a continuous function. To do this explicitly, let $\Gamma$ denote the component of $\partial\Delta$

which meets the closure of the unbounded component of $\mathbb{C} \setminus \Delta$, and let $\Delta_0$ be the component of $\mathbb{C} \setminus \Gamma$ which contains 0. The set $\Delta_0$ is simply connected. If there exists a continuous branch of $\log z$ on $\Gamma$, then on some neighborhood, $W$, of $\Gamma$, $\log z$ is defined and holomorphic. Plainly, $0 \notin W$. Let $\psi : U \to \Delta_0$ be the mapping provided by the Riemann mapping theorem, $U$ the open unit disc. If $C_r = \{z : |z| = r\}$, $0 < r < 1$, then $\psi(C_r)$ is an analytic simple closed curve in $\Delta_0$, and if $r$ is close enough to 1, $C_r$ contains the compact set $\psi^{-1}(\Delta_0 \setminus W)$ in its interior. Of course, this implies that $\psi(C_r) \subset W$. But then, $\log \psi$ is a continuous function on $C_r$. This is impossible, though, for $\psi$ vanishes once inside $C_r$, so the argument principle implies that

$$\frac{1}{2\pi i}\Delta_{C_r} \log \psi = 1.$$

Since $\log z$ is not defined as a continuous function on $\Gamma$, the set $\mathscr{L}(\Gamma)$ is a connected, unbounded set in $\mathbb{C}$. Since $\partial\Delta \cap (\partial_{\mathbb{C}}\bar{V} \cup f(X)) = \varnothing$, we have

$$\mathscr{L}(\Gamma) \cap (\partial_{\mathbb{C}}\bar{V} \cup f(X)) = \varnothing. \tag{4}$$

Also, $\Gamma \subset \bar{V} \cap f(\hat{X}) = f(S)$, so $\mathscr{L}(\Gamma)$ meets $\log f(S)$.

However, $\mathscr{L}(\Gamma)$ is connected and unbounded, so it has to meet $\partial_{\mathbb{C}} \log f(S)$. By (3) it follows that $\mathscr{L}(\Gamma)$ meets $\mathscr{L}(\partial_{\mathbb{C}}\bar{V} \cup f(X))$, contradicting (4), so the theorem is proved.

29.41 ***Corollary***

*Let $X$ be a compact set in $\mathbb{C}^N$. If there is a simply coconnected set $Y$, $Y \supset X$, and a function $g$ holomorphic on a neighborhood of $\hat{X}$ and continuous on $Y \cup \hat{X}$ for which $X$ is open in $g^{-1}(g(Y)) \cap \hat{X}$, then $X$ is polynomially convex.*

As a special case, we see that if $X$ is simply coconnected and if there is a function g holomorphic on a neighborhood of $\hat{X}$ with the property that $g(X) \cap g(\hat{X} \setminus X) = \varnothing$, then $X$ is polynomially convex.

PROOF

If $g(\hat{X}) \setminus g(Y)$ is nonempty, choose $\mathfrak{z}_0 \in \hat{X}$ with $g(\mathfrak{z}_0) \notin \hat{X}$ with $g(\mathfrak{z}_0) \notin g(Y)$, and put $f = g - g(\mathfrak{z}_0)$. Then $0 = f(\mathfrak{z}_0)$, $0 \notin f(\hat{X})$, and as $Y$ is simply coconnected, there is a continuous branch of $\log f$ defined on $Y$ and hence on $X$. The preceding theorem implies that since $0 \in f(\hat{X})$, $X$ is not open in $f^{-1}(f(X)) \cap \hat{X}$, and as $f$ and g differ only by a constant, $X$ is not open in $g^{-1}(g(X)) \cap \hat{X}$, contrary to hypothesis. Thus, $g(\hat{X}) \subset g(Y)$, and therefore $g^{-1}(g(Y)) \cap \hat{X} = \hat{X}$. It follows that $X$ is open in $\hat{X}$, and this implies, by way of the Oka approximation theorem, that $X$ is polynomially convex, as we wished to show.

As an application of some of these ideas, we are able to prove that certain rather specific sets are polynomially convex.

29.42 ***Corollary***

*If $X \subset \mathbb{C}^2$ is an arc or a compact, simply coconnected, rationally convex set, and if there is a nonconstant polynomial $p$ such that $|p| = 1$ on $X$, then $X$ is polynomially convex.*

PROOF

If $X$ is rationally convex and simply coconnected, it is polynomially convex in dimension 1.

Suppose that $\mathfrak{z}_0 \in \hat{X} \setminus X$ is such that $p(\mathfrak{z}_0) \in p(X)$. Then $|p(\mathfrak{z}_0)| = 1$, and as $|p| \leq 1$ on $\hat{X}$, it follows that the set $V \cap \hat{X}$ is a peak set for $\mathscr{P}(\hat{X})$ if $V$ is the variety

$$V = \{\mathfrak{z} \in \mathbb{C}^2 : p(\mathfrak{z}) = p(\mathfrak{z}_0)\}.$$

Consequently,

$$\mathfrak{z}_0 \in V \cap \hat{X} \subset (V \cap X)\hat{}. \tag{5}$$

The variety $V$ is a Runge variety in $\mathbb{C}^2$, so the fact that $X$ is polynomially convex in dimension 1 implies that $V \cap X$ is polynomially convex. Thus, $\mathfrak{z}_0 \in X$, contrary to hypothesis. It follows that $p(X) \cap p(\hat{X} \setminus X) = \varnothing$, and the polynomial convexity of $X$ is implied by the last corollary. In the event that $X$ is an arc, we argue just as above until we reach (5), and, in this case, $V \cap X$ is polynomially convex because of the description of the polynomially convex hull of a compact subset of a one-dimensional Runge variety given in Corollary 29.33.

# 30 *APPROXIMATION ON CURVES* $\mathbb{C}^N$

In this concluding section we shall consider the problem of polynomial approximation on curves in $\mathbb{C}^N$ and some related questions. Although the problems of polynomial and rational approximation on planar sets have been treated in a reasonably definitive way, the corresponding theory in $\mathbb{C}^N$ is still in its infancy. This fact accounts, in part, for our restricting attention in the present section to problems on curves. In this restricted setting, some very impressive results have been obtained, results that invoke in an essential way some of the uniform algebra techniques and results developed in the preceding chapters.

Our results will have bearing on the following problem: If $A$ is a uniform algebra on the interval $I = [0, 1]$ with $I = \Sigma(A)$, is $A$ necessarily $\mathscr{C}(I)$? This, of course, admits an equivalent formulation: If $\lambda$ is a polynomially convex arc in $\mathbb{C}^\Lambda$, is $\mathscr{P}(\lambda) = \mathscr{C}(\lambda)$? Although this problem is still open in general, we shall see that if various conditions of regularity are imposed on $\lambda$, or, equivalently, on a set of generators for $A$, then $A = \mathscr{C}(I)$.

The first theorem we present was proved by Stolzenberg [4]. Before stating it, a few preliminary remarks are in order. If $\Lambda$ is a set, finite or infinite, we shall understand by a *smooth curve in* $\mathbb{C}^\Lambda$ a continuous map $F: I \to \mathbb{C}^\Lambda$, $I = [0, 1]$, with the property that each coordinate $f_j$ of $F$, $j \in \Lambda$, has a continuous derivative on $(0, 1)$.

Second, we need the notion of a one-dimensional analytic subset of an open set in $\mathbb{C}^\Lambda$ which is meaningful in case $\Lambda$ is infinite. If $\Omega$ is an open set in $\mathbb{C}^\Lambda$ and $E$ is a closed subset of $\Omega$, we shall say that $E$ is a *one-dimensional analytic subset of* $\Omega$ if given $\mathfrak{z}_0 \in E$, there exist coordinate projections $\pi_1, \ldots, \pi_q$ on $\mathbb{C}^\Lambda$ and positive numbers $a_1, \ldots, a_q$ such that if

$$N = \{\mathfrak{z} \in \mathbb{C}^\Lambda : |\pi_j(\mathfrak{z}) - \pi_j(\mathfrak{z}_0)| < \alpha_j,\ j = 1, \ldots, q\},$$

then $N \cap E$ is a finite union of sets of the form $\varphi(\Delta)$, $\varphi(\Delta)$ closed in $N$, where $\varphi$ is a nonconstant, $\mathbb{C}^\Lambda$-valued map homomorphic on a neighborhood of $\bar{\Delta}$, the closed unit disc, in the sense that there is a neighborhood $U$ of $\bar{\Delta}$ with the property that for every coordinate projection $\pi$ on $\mathbb{C}^\Lambda$, $\pi \circ \varphi$ is holomorphic on $U$. We are confronted immediately with a difficulty: If we are working in a finite-dimensional setting, i.e., if $\Lambda$ is a finite set, then we have the usual notion of one-dimensional analytic variety, and it is natural to ask if sets that are one-dimensional analytic sets in accordance with the definition just given are one-dimensional analytic varieties. There is also the problem of understanding the infinite-dimensional situation, which is complicated by the fact in an infinite-dimensional $\mathbb{C}^\Lambda$, there is no usual notion of one-dimensional analytic variety. It is important to explore this matter thoroughly not only for the sake of its own interest but also because we shall need to use some of the related facts both in the proof of the theorem and in certain subsequent applications. We shall unravel these questions in an appendix at the end of this section. Our reason for relegating this problem to an appendix is that its resolution would constitute a major digression if inserted at this point. For the moment, we shall content ourselves by stating that things work out as we would hope: If $\Lambda$ is finite and $E$ is a one-dimensional analytic subset of the open set $\Omega$ in $\mathbb{C}^\Lambda$, then $E$ is a one-dimensional variety in $\Omega$ in the usual function theoretic sense. In the case of an infinite $\Lambda$, we shall see that a one-dimensional analytic set (in our sense of the term) is, in a natural way, a one-dimensional analytic space (see Appendix D), and that the polynomials in $\Lambda$ variables are holomorphic on this analytic space.

A final preliminary we need has to do with $\mathscr{R}(X)$ when $X$ is a compact subset of an infinite-dimensional $\mathbb{C}^\Lambda$. We have dealt with the finite-dimensional case in some detail, but so far we have not stated the definition in the infinite-dimensional case. The definition is clear: $\mathscr{R}(X)$ is the closure in $\mathscr{C}(X)$ of the algebra of functions of the form $(p|X)(q|X)^{-1}$, $p, q \in \mathscr{P}_\Lambda$, $q$ zero free on $X$. The set $\mathscr{R}$-hull $X$ is defined as in the finite-dimensional case, and there persists in the general case the characterization that $\mathscr{R}\text{-hull } X = \{\mathfrak{z} \in \mathbb{C}^\Lambda : p(\mathfrak{z}) \in p(X) \text{ for all } p \in \mathscr{P}_\Lambda\}$. We also need a weak but immediate extension of Oka's approximation theorem to this setting: *If $X \subset \mathbb{C}^\Lambda$ is compact and polynomially convex then $\mathscr{R}(X) = \mathscr{P}(X)$.* To see this, it suffices to show that if $q \in \mathscr{P}_\Lambda$ has no zeros on $X$, then $(q|X)^{-1} \in \mathscr{P}(X)$. Since $q$ is a polynomial whose zero set, $Z(q)$, is disjoint from the polynomially convex set $X$, it follows that there is a finite set $\Lambda_0 \subset \Lambda$ such that if $\pi : \mathbb{C}^\Lambda \to \mathbb{C}^{\Lambda_0}$ is the canonical projection, then $q$ is constant on the fibers $\pi^{-1}(\mathfrak{z}), \mathfrak{z} \in \mathbb{C}^{\Lambda_0}$, and $\pi(Z(q)) \cap (\pi(X))\hat{} = \varnothing$. If we define $f$ on $\mathbb{C}^{\Lambda_0} \setminus \pi(Z(q))$ by $f(z) = (q(w))^{-1}$, $w \in \pi^{-1}(\mathfrak{z})$, then $f$ is well defined and holomorphic on a neighborhood of $(\pi(X))\hat{}$, and so, by the Oka approximation theorem, $f \in \mathscr{P}((\pi(X))\hat{})$. This implies that $(q|X)^{-1} \in \mathscr{P}(X)$, and we have the required generalization. (A more general infinite-dimensional version of this approximation theorem can be found in Rickart's paper [5].)

With these notions in mind, we can state Stolzenberg's theorem.

**30.1** ***Theorem***

*Let $X \subset \mathbb{C}^\Lambda$ be a compact polynomially convex set, and let $K = K_1 \cup \cdots \cup K_q$, each $K_j$ the locus of a smooth curve in $\mathbb{C}^\Lambda$.*

(A) *$(K \cup X)\hat{} \setminus (K \cup X)$ is a (possibly empty) one-dimensional analytic subset of $\mathbb{C}^\Lambda \setminus (K \cup X)$.*

(B) *$K \cup X$ is rationally convex.*

(C) *Every continuous function on $K \cup X$ which can be approximated uniformly by polynomials on $X$ lies in $\mathscr{R}(K \cup X)$.*

(D) *If the map $H^1(K \cup X, \mathbb{Z}) \to H^1(X, \mathbb{Z})$ induced by the inclusion $X \subset K \cup X$ is one to one (in particular if $K$ is simply coconnected and disjoint from $X$), then $K \cup X$ is polynomially convex.*

In this statement $\Lambda$ may be either finite or infinite, but $q$ is taken to be finite.

This theorem of Stolzenberg's is a continuation of a line of development which begins with work of Wermer [4, 5, 6] in which the case of empty $X$, and $K$ a single analytic arc or simple closed curve, is treated. The case of a general, real analytic $K$ (and empty $X$) was treated, in essentially different ways, by Royden [3] and Bishop [5], and Bishop [6] subsequently gave another treatment, from yet a different point of view, of the real analytic case.

The proof of this theorem is long and complicated, so we shall break it into several steps. First we prove (B) and (C)—their proofs are relatively simple—and then we show that (D) follows fairly easily from (A). The last step is the proof of (A). The proof of (A) will come in three, progressively more involved, parts. The first part will be almost purely algebraic, the second part will be general, though rather technical, lemmas on polynomial convexity, and the third part will be analysis special to our particular setting.

### 30.2 *Derivation of* (B), (C), *and* (D)

Part (C) will follow if we can show that for every $\mathfrak{z} \in K \setminus X$ and every $\mathfrak{z}' \in K \cup X$, $\mathfrak{z}' \neq \mathfrak{z}$, there is a real-valued $f \in \mathscr{R}(K \cup X)$ with $f(\mathfrak{z}') \neq f(\mathfrak{z})$; for then every maximal set of antisymmetry for $\mathscr{R}(K \cup X)$ is a point (in case it meets $K \setminus X$) or else is contained in $X$. Since $X$ is polynomially convex, it follows from the Oka approximation theorem that $\mathscr{R}(K \cup X)|X = \mathscr{P}(X)$, whence

$$\mathscr{R}(K \cup X) = \{f \in \mathscr{C}(K \cup X): f|X \in \mathscr{P}(X)\},$$

and we have (C). Suppose, therefore, that $\mathfrak{z} \in K \setminus X$, $\mathfrak{z}' \in K \cup X$, $\mathfrak{z} \neq \mathfrak{z}'$. Since $X$ is polynomially convex, so is the set $X \cup \{\mathfrak{z}'\}$, so there exists a polynomial $p$ with $\operatorname{Re} p \leq -1$ on $X \cup \{\mathfrak{z}'\}$, $p(\mathfrak{z}) = 1$. Now $p(K)$ is a certain network of loci of smooth curves in $\mathbb{C}$, so it follows, e.g., from Theorem 26.8, that $\mathscr{R}(p(K)) = \mathscr{C}(p(K))$. Consequently, if $h$ is a real-valued continuous function on $p(K \cup X)$ with $h = 0$ on $p(K \cup X) \cap \{\zeta: \operatorname{Re} \zeta \leq 0\}$, $h(1) = 1$, then $h$ belongs to $\mathscr{R}(p(K \cup X))$ locally and as $\mathscr{R}(E)$, $E \subset \mathbb{C}$, is always a local algebra, $h \in \mathscr{R}(p(K \cup X))$. Thus, $h \circ p \in \mathscr{R}(K \cup X)$, $h \circ p$ is real valued, and $h \circ p$ separates $\mathfrak{z}$ from $\mathfrak{z}'$. This establishes (C).

In light of the characterization of $\mathscr{R}$-hull $X$ given in Lemma 29.4, the first half of the next lemma yields the rational convexity of $K \cup X$.

### 30.3 *Lemma*

*If* $\mathfrak{z} \notin K \cup X$, *there is a polynomial* $p$ *with* $p(\mathfrak{z}) = 0$, $0 \notin p(K \cup X)$, *and* $\operatorname{Re} f \leq -1$ *on* $X$.

PROOF

By (C) with $X$ empty, $\mathscr{R}(K) = \mathscr{C}(K)$, so $K$ is rationally convex. Therefore, there is a polynomial $q$ with $q(\mathfrak{z}) = 0$, $0 \notin q(K)$. Since $X$ is polynomially convex, there is a polynomial $r$ with $r(\mathfrak{z}) = 0$, $\operatorname{Re} r < -1$ on $X$. It follows that for some small $\varepsilon > 0$, $\operatorname{Re}(r - \lambda q) < -1$ on $X$ for all $|\lambda| < \varepsilon$. The function $rq^{-1}$ is smooth on $K$, so its range omits some value $\lambda_0, |\lambda_0| < \varepsilon$. If $p = r - \lambda_0 q$, then $p(\mathfrak{z}) = 0$, but $p$ is zero free on $K \cup X$.

To prove (D), let $\mathfrak{z} \notin K \cup X$, and let $p$ be a polynomial as in the lemma. Then $p$ is zero free on $K \cup X$, and it has a continuous logarithm on $X$. Since the induced map $H^1(K \cup X, \mathbb{Z}) \to H^1(X, \mathbb{Z})$ is one to one, and since $H^1(X, \mathbb{Z})$ can be identified with the group of units of $\mathscr{C}(X)$ modulo the group of exponentials (recall Theorem 10.1), it follows that there is a neighborhood $\Omega$ of $K \cup X$ on whose closure $p$ has a logarithm. For notational convenience, let $E$ be $(K \cup X)\hat{} \setminus (K \cup X)$. The set $E \setminus \Omega$ is compact, and by the maximum modulus theorem, $E \setminus \Omega \subset (E \cap \partial\Omega)\hat{}$. On the set $E \cap \partial\Omega$, $p$ has a logarithm, so, if we are in the finite-dimensional case where we know $E$ to be a variety, we can invoke Lemma 29.36 to conclude that $0 \notin p((E \cap \partial\Omega)\hat{})$, whence $0 \notin p(E \setminus \Omega)$. As $0 \notin p(\Omega)$, we conclude that $0 \notin p((K \cup X)\hat{})$, and therefore $\mathfrak{z} \notin (K \cup X)\hat{}$. In the not necessarily finite case, we know that $E$ is a one-dimensional analytic space. As such, it is the image of a one-dimensional complex manifold under a proper holomorphic map (see Appendix D), so the *proof* of Lemma 29.36 is applicable, and we find once more that $\mathfrak{z} \notin (K \cup X)\hat{}$. Thus, $K \cup X$ is polynomially convex, and we have proved (D).

We now turn to the proof of (A).

## 30.4 *Proof of (A): Algebra*

We consider a complex commutative algebra with identity, $A$, upon which we impose no topology. If $D$ is a planar domain, a function $L: A \times D \to \mathbb{C}$ which is linear on $A$ and holomorphic on $D$ in the sense that for each $\zeta \in D$, $L(\cdot, \zeta)$ is a linear functional on $A$, and, for each $x \in A$, $L(x, \cdot)$ is holomorphic on $D$, is called an *analytic linear functional*. An analytic linear functional $L$ is an *analytic character of order* $d$ provided there is a discrete subset, $E$, of $D$ such that for every $\zeta \in D \setminus E$ there are $d$ distinct algebra homomorphisms $\varphi_1(\zeta), \ldots, \varphi_d(\zeta)$, none of which is the zero homomorphism, and $d$ nonzero complex numbers $c_1(\zeta), \ldots, c_d(\zeta)$ *such that*

$$L(x, \zeta) = \sum_{j=1}^{d} c_j(\zeta)\varphi_j(\zeta)(x). \tag{1}$$

It is important, for the sake of the definition, to observe that the representation (1) is unique if it exists at all. This amounts to the remark that if $\Sigma\alpha_j\psi_j$ is a finite linear combination of distinct $\mathbb{C}$-homomorphisms of $A$ which is the zero functional, then each $\alpha_j = 0$.

The facts we shall need concerning analytic characters may be summarized as follows.

**30.5 *Lemma***

*Let $D$ be a connected open set in $\mathbb{C}$, and let $L: A \times D \to \mathbb{C}$ be an analytic linear functional.*

(a) *$L$ is an analytic character of order $d$ if and only if*
  (1) *For all $e > d$ and all pairs of e-tuples $(x_1, \ldots, x_e)$ and $(y_1, \ldots, y_e)$, $\det[L(x_i y_j, \zeta)] = 0$ for every $\zeta \in D$.*
  (2) *There exist $x_1, \ldots, x_d, y_1, \ldots, y_d$, and $z$ in $A$ such that the polynomial $P_\zeta$ given by*

$$P_\zeta(\lambda) = \det[L(x_i(z - \lambda)y_j, \zeta)]$$

  *has, for some choice of $\zeta \in D$, $d$ distinct roots.*

(b) *If, for some nonempty open subset $D_0 \subset D$, $L|A \times D_0$ is an analytic character of order $d$, then $L$ is an analytic character of order $d$ on the whole of $A \times D$.*

(c) *If $L$ is an analytic character of order $d$ on $A \times D$, then there is a discrete set $E \subset D$ such that about every point of $D \setminus E$, there is a disc $\Omega$ in which there exist zero-free holomorphic functions $c_j$ and functions $\varphi_j$ taking values in the space of nonzero complex homomorphisms of $A$ which satisfy $L(x, \zeta) = \sum_{j=1}^d c_j(\zeta)\varphi_j(\zeta)(x)$. Moreover, for each $\zeta$, the homomorphisms $\varphi_j(\zeta)$ are all distinct, and for each $x \in A$, the map $\zeta \mapsto \varphi_j(\zeta)(x)$ is holomorphic.*

These facts about analytic characters were given by Royden [3].

PROOF

Part (b) follows from (a), for $\det[L(x_{ij}, \zeta)]$ is a holomorphic function of $\zeta$ on the connected set $D$ for every choice of $x_{ij} \in A$, and so it vanishes identically or else has only a discrete set of zeros.

To prove (a), we observe that one implication is fairly direct. If $L$ is an analytic character of order $d$, then $\det[L(x_i y_j, \zeta)] = 0$ for all pairs of $e$-tuples $(x_1, \ldots, x_e)$ and $(y_1, \ldots, y_e)$; for if

$$L(x, \zeta) = \sum_{k=1}^{d} c_k(\zeta)\varphi_k(\zeta)(x),$$

then

$$L(x_iy_j, \zeta) = \sum_{k=1}^{d} c_k(\zeta)\varphi_k(\zeta)(x_i)\varphi_k(\zeta)(y_j),$$

and thus the $e \times e$ matrix $[L(x_iy_j, \zeta)]$ is the product of the $d \times e$ matrix $[c_k(\zeta)\varphi_k(\zeta)(x_i)]$ $(k = 1, \ldots, d, i = 1, \ldots, e)$ and the $e \times d$ matrix $[\varphi_k(\zeta)(y_i)]$ $(i = 1, \ldots, e, k = 1, \ldots, d)$. Such an $e \times e$ matrix is singular since $e > d$, so (1) holds. Also, (2) can be satisfied if we choose $\zeta_0 \in D \setminus E$, if we choose $x_1, \ldots, x_d \in A$ with $\varphi_j(x_i) = \delta_{ij}$, if we take $y_i = x_i$, and if we set $z = \sum_{i=1}^{d} ix_i$.

We now suppose that $L$ is an analytic linear functional on $A \times D$ which satisfies (1) and (2). Let $d(\zeta)$ be the discriminant of the polynomial $P_\zeta(\lambda)$ so that $d(\zeta_0) \neq 0$ for some $\zeta_0$, and consequently the set

$$E_1 = \{\zeta \in D : d(\zeta) = 0\}$$

is discrete. The polynomial $P_\zeta$ has $d$ distinct roots for every $\zeta \in D \setminus E_1$, and it is of degree $d$, so the coefficient of $\lambda^d$ in $P_\zeta$ is not zero. This coefficient is, except possibly for sign, $\det[L(x_iy_j, \zeta)]$, so the matrix $[L(x_iy_j, \zeta)]$ has, for every $\zeta \in D \setminus E_1$, an inverse, $[A_{ij}(\zeta)]$; and as the entries of $[L(x_iy_j, \zeta)]$ are holomorphic in $\zeta$, the entries of $[A_{ij}(\zeta)]$ are holomorphic. Let

$$w_i(\zeta) = \sum_{j=1}^{d} A_{ij}(\zeta)x_j, \qquad i = 1, \ldots, d.$$

We have then that

$$L(w_i(\zeta)y_k, \zeta) = \sum_j A_{ij}(\zeta)(L(x_jy_k, \zeta)) = \delta_{ik},$$

for $[A_{ij}(\zeta)]$ is inverse to $[L(x_iy_j, \zeta)]$. It follows that

$$\begin{aligned} &\det[L(w_i(\zeta)(z - \lambda)y_j, \zeta)] \\ &\quad = \det[L(w_i(\zeta)zy_j, \zeta) - \lambda L(w_i(\zeta)y_j, \zeta)] \\ &\quad = \det[L(w_i(\zeta)zy_j, \zeta) - \lambda\delta_{ij}], \end{aligned}$$

so for $\zeta \in D \setminus E_1$, the eigenvalues of the matrix $[L(w_i(\zeta)zy_j, \zeta)]$ are the $d$ roots of $P_\zeta(\lambda)$. Since these roots are distinct, the matrix $[L(w_i(\zeta)zy_j, \zeta)]$ is diagonalizable. Let $[B_{ij}(\zeta)]$ and $[C_{ij}(\zeta)]$ be matrices inverse to each other such that

$$[B_{ij}(\zeta)][L(w_i(\zeta)zy_j, \zeta)][C_{ij}(\zeta)]$$

is a diagonal matrix. If we set

$$X_i(\zeta) = \sum_{j=1}^{d} B_{ij}(\zeta)w_j(\zeta)$$

and

$$Y_i(\zeta) = \sum_{j=1}^{d} C_{ji}(\zeta)y_j,$$

then

$$L(X_i(\zeta)Y_j(\zeta), \zeta) = \delta_{ij}.$$

We shall prove that the map $\rho$ from $A$ into the ring of $d \times d$ matrices given by

$$\rho(x) = [L(X_i(\zeta)xY_j(\zeta), \zeta)]$$

is an algebra homomorphism. That it is linear is clear, and that it is multiplicative may be seen as follows: Apply the hypothesis (1) to the determinant of the matrix

$$\begin{bmatrix} L(xy, \zeta) & L(X_1 y, \zeta) & \cdots & L(X_d y, \zeta) \\ L(xY_1, \zeta) & L(X_1 Y_1, \zeta) & \cdots & L(X_d Y_1, \zeta) \\ \vdots & \vdots & \ddots & \vdots \\ L(xY_d, \zeta) & L(X_1 Y_d, \zeta) & \cdots & L(X_d Y_d, \zeta) \end{bmatrix}$$

and use the fact that $L(X_i(\zeta)Y_j(\zeta), (\zeta) = \delta_{ij}$ to find that

$$L(xy, \zeta) = \sum_{k=1}^{d} L(xY_k, \zeta)L(X_k y, \zeta).$$

If we replace $x$ by $X_i x$ and $y$ by $yY_j$, we find that

$$L(X_i xyY_j, \zeta) = \sum_{k=1}^{d} L(X_i xY_k, \zeta)L(X_k yY_j, \zeta),$$

and this implies that $\rho(xy) = \rho(x)\rho(y)$. Thus, $\rho$ is an algebra homomorphism. We have that $\rho(z)$ is a diagonal matrix with distinct eigenvalues, so since each element of $\rho(A)$ commutes with $\rho(z)$, it follows that $\rho(A)$ is an algebra of diagonal matrices. If we define $\varphi_i(\zeta)(x)$ to be the $i$th diagonal element of $\rho(x)$, i.e., $\varphi_i(\zeta)(x) = L(X_i(\zeta)xY_i(\zeta), \zeta)$, then $\varphi_i(\zeta)$ is a complex homomorphism of $A$; and since $\rho(z)$ has distinct eigenvalues, the homomorphisms $\varphi_i$ are all distinct. If we define $c_j$ by $c_j(\zeta) = L(X_j(\zeta), \zeta)L(Y_j(\zeta), \zeta)$, then applying the expression given above for $L(xy, \zeta)$ to $L(x, \zeta) = L(x \cdot 1, \zeta)$, and using the fact that

$L(X_j(\zeta)xY_k(\zeta),\zeta) = 0$ if $k \neq j$, we find by a short calculation that if $x \in A$, then

$$L(x,\zeta) = \Sigma\, c_j(\zeta)\varphi_j(\zeta)(x),$$

and we have shown that $L$ is an analytic character of order $d$.

The final step is the proof of (c). This will be achieved if we can prove that the functions $c_j$ are holomorphic, a fact which can be proved by showing that the coefficients $B_{ij}(\zeta)$ can be chosen to depend, at least locally, in a holomorphic way on $\zeta$. If $\zeta_0 \in D \setminus E_1$, then for all $\zeta$ in a disc $\Delta$ in $D \setminus E_1$ about $\zeta_0$, the $d$ distinct roots of $P_\zeta$ can be parameterized as holomorphic functions, $\lambda_1(\zeta), \ldots, \lambda_d(\zeta)$. Let $\mathfrak{z}_k(\zeta_0)$ be the eigenvector of $[L(w_i(\zeta_0)zy_i, \zeta_0]$ corresponding to $\lambda_k(\zeta_0)$ whose first nonzero entry, say the $s_k$th, is 1. For $\zeta$ close to $\zeta_0$, let $\mathfrak{z}_k(\zeta)$ be the eigenvector whose $s_k$th entry is 1. Cramer's rule applied in a sufficiently small disc about $\zeta_0$ shows that the vector $\mathfrak{z}_k(\zeta)$ depends holomorphically on $\zeta$, i.e., that each of its coordinates is a holomorphic function of $\zeta$. It follows that the matrix $B_{ij}(\zeta)$ does have entries that are holomorphic functions of $\zeta$.

### 30.6 *Proof of (A): Polynomial Convexity*

This portion of the proof consists of five lemmas on polynomial convexity. The final one depends on the theory of analytic characters.

### 30.7 *Lemma*

*If $X \subset \mathbb{C}^N$ is a compact set and $p$ is a polynomial, and if $\zeta$ lies in the boundary of $\Omega$, the unbounded component of $\mathbb{C} \setminus p(X)$, then the set $M_\zeta = \{\mathfrak{z} \in \hat{X} : p(\mathfrak{z}) = \zeta\}$ satisfies*

$$M_\zeta = (M_\zeta \cap X)\hat{}.$$

PROOF

Let $E = \mathbb{C} \setminus \Omega$ so that $E$ is a compact set in $\mathbb{C}$ with connected complement. The algebra $\mathscr{P}(E)$ is Dirichlet on $\partial E = \partial\Omega$, recall Theorem 25.5, so the point $\zeta$ is a peak point for $\mathscr{P}(E)$. This implies that $M_\zeta$ is a peak set for $\mathscr{P}(X)$, so if $\mu \in \mathscr{M}(X)$ is a representing measure for a point $\mathfrak{z} \in M_\zeta$, we must have $\mu$ supported on $X \cap M_\delta$. Thus, $\mathfrak{z} \in (X \cap M_\delta)\hat{}$, whence $M_\zeta \subset (X \cap M_\zeta)\hat{}$. The opposite inclusion is clear, so we have the lemma.

### 30.8 *Lemma*

*Let $E \subset \mathbb{C}^N$ be compact, let $\mathfrak{z} \in \mathbb{C}^N$, and let $\mu, \sigma \in \mathscr{M}(E)$. If $\int p\, d\mu = p(\mathfrak{z})$ and $\int p\, d\sigma = 0$ for all polynomials $p$, then for any open subset $U$ of $E$ such that $\mathfrak{z} \notin (E \setminus U)\hat{}$, $\sigma_U \neq \mu_U$.*

Recall that $\sigma_U$ denotes the restriction of the measure $\sigma$ to the set $U: \sigma_U(S) = \sigma(U \cap S)$.

PROOF

As $\mathfrak{z} \notin (E \setminus U)\hat{}$, there is a polynomial $p$ with $1 = p(\mathfrak{z}) > \|p\|_{E \setminus U}$. This yields

$$1 = \int p^m \, d\mu = \int_U p^m \, d\mu + \int_{E \setminus U} p^m \, d\mu$$

and

$$0 = \int p^m \, d\sigma = \int_U p^m \, d\sigma + \int_{E \setminus U} p^m \, d\sigma.$$

As $\|p\|_{E \setminus U} < 1$, the dominated convergence theorem yields the fact that the two integrals over $E \setminus U$ are small if $m$ is large enough, so for large $m$, $\int_U p^m \, d\mu$ is nearly 1 and $\int_U p^m \, d\sigma$ is nearly zero. This implies the lemma.

30.9 ***Lemma***

*Let $E \subset \mathbb{C}^N$ be compact, and let $p$ be a polynomial which maps $E$ into a simple closed curve $\lambda$ in $\mathbb{C}$. If there is a nonempty, open arc $\lambda_0 \subset \lambda$ over which $p|E$ is one to one and onto, then $p$ is one to one on $\hat{E} \setminus E$.*

PROOF

Suppose that $\mathfrak{z}_1$ and $\mathfrak{z}_2$ are points of $\hat{E} \setminus p^{-1}(\lambda \setminus \lambda_0)$ identified by $p$. For $i = 1, 2$, let $\nu_i$ be a positive representing measure for $\mathfrak{z}_i$ supported by $E$. Define measures $\nu_i^*$ on $\lambda$ by

$$\nu_i^*(S) = \nu_i(p^{-1}(S))$$

for every Borel set $S \subset \lambda$. If $h$ is a polynomial in one complex variable, then

$$\int_\lambda h \, d\nu_i^* = \int_E h \circ p \, d\nu_i = h(p(\mathfrak{z}_i)),$$

and thus $\nu_i^*$ is a representing measure for the point $p(\mathfrak{z}_1) = p(\mathfrak{z}_2)$ for the algebra $\mathscr{P}(\lambda)$. But $\mathscr{P}(\lambda)$ is Dirichlet on $\lambda$, so representing measures are unique, and we get $\nu_1^* = \nu_2^*$. There is a polynomial $q$ with $q(\mathfrak{z}_1) = 0$, $q(\mathfrak{z}_2) = 1$, so the measures $\mu$ and $\sigma$ given by $\mu = q\nu_2$, $\sigma = q\nu_1$ are, respectively, a representing measure for $\mathfrak{z}_2$ and a measure orthogonal to $\mathscr{P}(E)$. This contradicts the preceding lemma since on $p^{-1}(\lambda_0)$, $\mu$ and $\sigma$ agree.

30.10 **Lemma**

*Let E, $\lambda$, and p be as in the preceding lemma, and let $\Omega$ be the interior of $\lambda$. Either $p(\hat{E}) \supset \lambda$ or else p carries $(\hat{E} \setminus E) \cap p^{-1}(\Omega)$ one to one onto $\Omega$ in such a way that for every polynomial q on $\mathbb{C}^{\Lambda}$, $q \circ p^{-1}$ is holomorphic on $\Omega$.*

PROOF

We have $p(\hat{E}) \subset \lambda \cup \Omega$. Assume that $p(\hat{E}) \cap \Omega \neq \varnothing$ but that $p(\hat{E})$ does not contain $\Omega$. Since $p(\hat{E})$ is the spectrum of the element $p$ of the Banach algebra $\mathscr{P}(E)$, we know that if $f$ is holomorphic on a neighborhood of $p(\hat{E})$, then $f \circ p \in \mathscr{P}(E)$. By choosing for $f$ a function of the form

$$f(\zeta) = \frac{1}{\zeta - \zeta_0}$$

for a suitable $\zeta_0 \in \Omega$, we obtain a function holomorphic on $p(E)$ which has the property that

$$\|f\|_{p(\hat{E})} > \|f\|_{\lambda}.$$

As $\lambda \supset p(E)$, this leads to $\|f \circ p\|_{\hat{E}} > \|f \circ p\|_E$. Thus, either $p(\hat{E}) \cap \Omega = \varnothing$ or $p(\hat{E}) \supset \Omega$.

In the latter case we know from the last lemma that $p$ is one to one on $\hat{E} \cap p^{-1}(\Omega)$. Consider a small disc $D \subset \Omega$ with $\partial D$ the circle $\gamma$. The set $\hat{E} \cap p^{-1}(\Omega)$ is open in $\hat{E}$, so by the local maximum modulus theorem we find that if $q$ is any polynomial on $\mathbb{C}^{\Lambda}$, then

$$\|q\|_{p^{-1}(\bar{D}) \cap \hat{E}} = \|q\|_{p^{-1}(\gamma) \cap \hat{E}}.$$

Consequently, the closure in $\mathscr{C}(\bar{D})$ of the algebra

$$B = \{q \circ p^{-1}|\bar{D} : q \text{ a polynomial}\}$$

contains the identity map, $p \circ p^{-1}$, of $\bar{D}$ onto itself and is a maximum modulus algebra on $\bar{D}$; so Theorem 25.16 shows that $B$ consists exclusively of functions holomorphic in $D$, and we have the lemma.

30.11 **Lemma**

*Let $Y \subset \mathbb{C}^{\Lambda}$ be compact, and let p be a polynomial such that $p(\hat{Y}) \setminus p(Y)$ is nonempty and connected. Let $Y_0 = p^{-1}(p(Y)) \cap \hat{Y}$. If there is an open disc $\Delta_*$ contained in $p(\hat{Y}) \setminus p(Y)$ such that $p^{-1}(\Delta_*) \cap \hat{Y}$ is a one-dimensional analytic subset of $p^{-1}(\Delta_*)$, then $\hat{Y} \setminus Y_0$ is a one-dimensional analytic subset of $\mathbb{C}^{\Lambda} \setminus Y_0$.*

PROOF

First, note that $p$ is a proper mapping on $p^{-1}(\Delta_*) \cap \hat{Y}$, for if $E \subset \Delta_*$ is compact, then $p^{-1}(K)$ is closed in $\mathbb{C}^N$, so $p^{-1}(K) \cap \hat{Y}$ is compact. By hypothesis, $p^{-1}(\Delta_*) \cap \hat{Y}$ is a one-dimensional analytic set, and by our remarks on such sets just before the statement of Theorem 30.1, such sets are analytic varieties (or spaces). Let $\Sigma$ be the set of singularities of the analytic space $p^{-1}(\Delta_*) \cap \hat{Y}$ so that $\Sigma$ is discrete. Since $p$ is a proper map from $p^{-1}(\Delta_*) \cap \hat{Y}$ to $\Delta_*$, $p(\Sigma)$ is a discrete subset of $\Delta_*$. Let $z_1$ be a point of $\Delta_* \setminus p(\Sigma)$, and let $\Delta'$ be a disc about $z_1$ which is contained in $\Delta_*$ and which misses $p(\Sigma)$. Then $p^{-1}(\Delta') \cap \hat{Y}$ is an open subset of $p^{-1}(\Delta_*) \cap \hat{Y}$ which misses $\Sigma$, so it is a one-dimensional complex manifold, $M$.

We need to remark that locally $M$ has only finitely many components. Specifically, if $K \subset \Delta'$ is compact, then $p^{-1}(K)$ can meet only finitely many components of $M$. To see this, suppose that $M_n, n = 1, 2, \ldots$, is a sequence of distinct components of $M$, suppose that for each $n$, $z_n$ is a point of $M_n$, and suppose that the sequence $\{p(z_n)\}$ converges to a point of $\Delta'$. Since $p$ is proper on $M$ and since the set $\{p(z_n)\}$ is relatively compact in $\Delta'$, the sequence $\{z_n\}$ must have a convergent subsequence. This, however, is impossible since distinct $z_n$'s lie in distinct components.

Next we note that if $M_0$ is a component on which $p$ is nonconstant, then $p$ carries $M_0$ onto $\Delta'$. If not, there is a point $w \in \Delta' \setminus p(M_0)$ which is the limit point of a sequence $\{p(z_n)\}$, $z_n \in M_0$. Again, the properness of $p$ yields a limit point, $z_0$, of the set $\{z_n\}$, and as $M_0$ is both open and closed in $M$, $z_0$ lies in $M_0$. Thus, $w = p(z_0)$ and we reach a contradiction.

It follows that the components of $M$ fall into two classes: those on which $p$ is constant and those which are mapped by $p$ onto $\Delta'$. The second class cannot be empty, for, by hypothesis, $\Delta_* \subset p(\hat{Y}) \setminus p(Y)$, and by the properness of $p$ on $\Delta'$, the range of $p$ on the union of those components of $M$ on which $p$ is constant is a discrete set.

We can find a disc $\Delta \subset \Delta'$ which has the property that $p$ is constant on no component of $M \cap p^{-1}(\Delta)$. Repeating the argument given above, we see that every component of $M \cap p^{-1}(\Delta)$ maps onto $\Delta$ and that there can be only finitely many of these components, say $D_1, \ldots, D_d$. Since $p$ is nonconstant on each of the sets $D_j$, the differential of $p$ on each $D_j$ has only a discrete set of zeros, and this discrete set is mapped by $p$ onto a discrete set in $\Delta$. Consequently, by choosing $\Delta$ properly, we can suppose that the differential of $p$ has no zeros on any of the sets $D_j$. But then each of the $D_j$ is a covering space of the disc (by way of the map $p$) and so must be a disc, and $p$ must map it homeomorphically onto $\Delta$. Let $p_j$, $j = 1, 2, \ldots, d$, be the restriction of $p$ to $D_j$. Choose distinct closed discs $\Delta_2 \subset \Delta_1 \subset \Delta$ centered at the center, $\zeta_0$, of $\Delta$ and let $\Delta_j^0$ be the interior of $\Delta^j$.

We shall prove that if $\zeta_1 \in p(\hat{Y}) \setminus (p(Y) \cup \Delta_1)$, then there is an open disc $\Delta(\zeta_1)$ about $\zeta_1$ with the property that $p^{-1}(\Delta(\zeta_1)) \cap \hat{Y}$ is a one-dimensional analytic set in $p^{-1}(\Delta(\zeta_1))$.

Given a $\zeta_1$, the connectedness of $p(\hat{Y}) \setminus p(Y)$ implies the existence of a simple closed curve $\Gamma$ contained, with its interior, in $p(\hat{Y}) \setminus p(Y)$ with the property that $\zeta_1$ lies in the interior of $\Gamma$, and $\Gamma$ intersects $\Delta_1$ in a diameter. Let $\gamma$ be a closed interval in $\Gamma \cap \Delta_2^0$, and let $\Delta_\gamma = \Delta_1^0 \setminus \Gamma_*$ where $\Gamma_* = \Gamma \setminus (\text{interior}\, \gamma)$ so that $\Delta_\gamma$ is a dense, open, connected subset of $\Delta_1^0$. Let $B = p^{-1}(\Gamma) \cap \hat{Y}$, let $B_* = p^{-1}(\Gamma_*) \cap \hat{Y}$, and let $\beta_j = p_j^{-1}(\gamma)$, $j = 1, \ldots, d$.

The first major step in our proof is the demonstration of the fact that if $\xi \in \hat{\Gamma} \setminus \Gamma$, i.e., if $\xi$ is a point of the interior of $\Gamma$, then $\hat{Y} \cap p^{-1}(\xi)$ contains no more than $d$ points. Let $\mathfrak{z}_1, \ldots, \mathfrak{z}_e$ be distinct points of $\hat{Y} \cap p^{-1}(\xi)$. We are to prove that $e \leq d$. [Note that we do not assert a priori that $p^{-1}(\xi) \cap \hat{Y}$ is finite.] The local maximum modulus theorem implies that $\mathfrak{z}_j \in \hat{B} \setminus B$, so there are distinct representing measures $\mu_j$ on $B$ for the points $\mathfrak{z}_j$: $\int f\, d\mu_j = f(\mathfrak{z}_j)$ for every $f \in \mathscr{P}(\hat{B})$. Set $\mu = \sum_{j=1}^{e} \mu_j$, $\mu_* = \mu_{B*}$, and $\nu_j = \mu_{\beta_j}$. Define measures $\nu_j^+$ on $\gamma$ by

$$\nu_j^+(E) = \nu_j(p_j^{-1}(E))$$

for every Borel set $E \subset \gamma$. Finally, we let $\psi_j$ be the Cauchy transform of $\nu_j^+$:

$$\psi_j(\zeta) = \int \frac{1}{z - \zeta}\, d\nu_j^+(z).$$

We shall need to consider certain analytic linear functionals on $\mathscr{P}(\hat{Y})$:

$N: \mathscr{P}(\hat{Y}) \times (\Delta \setminus \gamma) \to \mathbb{C}$ is given by

$$N(g, \zeta) = \frac{1}{2\pi i}(\xi - \zeta) \sum_{j=1}^{d} \psi_j(\zeta) g(p_j^{-1}(\zeta)).$$

$O_i: \mathscr{P}(\hat{Y}) \times (\mathbb{C} \setminus \partial\Delta_i) \to \mathbb{C}$ is given by

$$O_i(g, \zeta) = \int_{\partial\Delta_i} \frac{N(g, z)}{z - \zeta}\, dz, \qquad i = 1, 2.$$

In the definition of $O_i$ we take the positive orientation on $\partial\Delta_i$.

$P: \mathscr{P}(\hat{Y}) \times (\mathbb{C} \setminus \Gamma_*) \to \mathbb{C}$ is given by

$$P(g, \zeta) = \int_{B*} \frac{p - \xi}{p - \zeta}\, g\, d\mu.$$

$Q:\mathscr{P}(\hat{Y}) \times (\mathbb{C} \setminus \Gamma) \to \mathbb{C}$ is given by

$$Q(g, \zeta) = \int_B \frac{p - \xi}{p - \zeta} g \, d\mu.$$

We shall exploit certain relations among these functionals. The first of these is perfectly straightforward. $O_1 - O_2 = 2\pi i N$ on $\mathscr{P}(\hat{Y}) \times A$ if $A$ is the annulus $\Delta_1^0 \setminus \Delta_2$. This fact follows from the Cauchy integral formula.

The Cauchy integral formula also implies that on $\mathscr{P}(\hat{Y}) \times (A \setminus \Gamma)$,

$$Q = P + O_2, \tag{2}$$

a formula whose proof requires a bit of computation. If $\zeta \in A \setminus \Gamma$ and $g \in \mathscr{P}(\hat{Y})$, then

$$\begin{aligned} Q(g, \zeta) - P(g, \zeta) &= \int_{p^{-1}(\gamma)} \frac{p - \xi}{p - \zeta} g \, d\mu \\ &= \sum_{j=1}^{d} \int_{p^{-1}(\gamma)} \frac{p - \xi}{p - \zeta} g \, d\nu_j \\ &= \sum_{j=1}^{d} \int_{\gamma} \frac{z - \xi}{z - \zeta} g(p_j^{-1}(z)) \, d\nu_j^+(z). \end{aligned}$$

Also,

$$\begin{aligned} O_2(g, \zeta) &= \sum_{j=1}^{d} \frac{1}{2\pi i} \int_{\partial\Delta_2} (\xi - z)\psi_j(z) g(p_j^{-1}(z)) \frac{1}{z - \zeta} dz \\ &= \sum_{j=1}^{d} \int \left\{ \frac{1}{2\pi i} \int_{\partial\Delta_2} \frac{z - \xi}{z - \tau} g(p_j^{-1}(z)) \frac{1}{z - \zeta} dz \right\} d\nu_j^+(\tau). \end{aligned}$$

The function $(z - \xi)(z - \zeta)^{-1} g(p_j^{-1}(z))$ is holomorphic in $z$ on a neighborhood of $\bar{\Delta}_2$ since $\zeta \in A$, so the Cauchy integral formula shows that

$$O_2(g, \zeta) = \sum_{j=1}^{d} \int \frac{\tau - \xi}{\tau - \zeta} g(p_j^{-1}(\tau)) \, d\nu_j^+(\tau),$$

and this gives $Q - P = O_2$ on $\mathscr{P}(\hat{Y}) \times A$.

A final fact we need is that $Q = 0$ on $\mathscr{P}(\hat{Y}) \times (\mathbb{C} \setminus \hat{\Gamma})$. This is so, for if $\zeta$ is outside $\Gamma$, then $(p - \zeta)^{-1}$ is uniformly approximable by polynomials on $B$, so if $g \in \mathscr{P}(\hat{Y})$

$$\begin{aligned} Q(g, \zeta) &= \int \frac{p - \xi}{p - \zeta} g \, d\mu \\ &= \sum_{j=1}^{e} \frac{p(\mathfrak{z}_j) - \xi}{p(\mathfrak{z}_j) - \zeta} g(\mathfrak{z}_j) \\ &= 0. \end{aligned}$$

[Recall that $p(\mathfrak{z}_j) = \xi$.]

It follows from the relations just established that the analytic linear functional $P + O_1$ on $\mathscr{P}(\hat{Y}) \times \Delta_\gamma$ agrees on $\mathscr{P}(\hat{Y}) \times (A \setminus \hat{\Gamma})$ with $2\pi i N$, which is an analytic character of order $d_1 \leq d$. By the second part of Lemma 30.5, it follows that $P + O_1$ is an analytic character of order $d_1$ on the whole of $\mathscr{P}(\hat{Y}) \times \Delta_\gamma$. Moreover, on $\mathscr{P}(\hat{Y}) \times (A \cap (\hat{\Gamma} \setminus \Gamma))$, $Q = P + O_1 - 2\pi i N$, so $Q$ is an analytic character of some order $d_2 \leq d$ on $\mathscr{P}(\hat{Y}) \times (\hat{\Gamma} \setminus \Gamma)$.

We can prove now that $e \leq d$. By the first part of Lemma 30.5 applied to the analytic character $L$ given by

$$L(g, \zeta) = Q(g, \xi) = \sum_{j=1}^{e} g(\mathfrak{z}_j),$$

there exist $f_1, \ldots, f_e, g_1, \ldots, g_e, h \in \mathscr{P}(\hat{Y})$ and $\lambda \in \mathbb{C}$, such that

(3) $$\det[Q(f_i(h - \lambda)g_j, \xi)] \neq 0.$$

However, if $e > d$, then using the fact that $Q$ is an analytic character of order no more than $d$, Lemma 30.5 implies that the determinant in (3) is zero. Thus, $e \leq d$.

Thus, we have shown that every point $\xi$ of $\hat{\Gamma} \setminus \Gamma$ has at most $d$ preimages in $\hat{Y}$ under the map $p$.

If the point $\xi$ considered above lies in $\Delta \cap (\hat{\Gamma} \setminus \Gamma)$, then $p^{-1}(\xi) \cap \hat{Y}$ consists of precisely $d$ distinct points, so it follows that $Q$ is, in fact, an analytic character of order exactly $d$ on $\mathscr{P}(\hat{Y}) \times (\hat{\Gamma} \setminus \Gamma)$.

Let $E \subset \hat{\Gamma} \setminus \Gamma$ be the discrete set associated with $Q$ by the third part of Lemma 30.5 so that if $\xi_0$ lies in $\hat{\Gamma} \setminus (\Gamma \cup E)$, then in a small disc $D$ about $\xi_0$, $Q$ has a representation

$$Q(f, \zeta) = \sum_{j=1}^{d} c_j(\zeta)\varphi_j(\zeta)(g),$$

where $\varphi_j(\zeta) \in \Sigma(\mathscr{P}(\hat{Y}))$, and for each $g \in \mathscr{P}(\hat{Y})$, $\varphi_j(\zeta)(g)$ is holomorphic in $\zeta$. The $\varphi_j(\zeta)$ are distinct, and the functions $c_j$ have no zeros in $D$. Since $\Sigma(\mathscr{P}(\hat{Y})) = \hat{Y}$, $\zeta \mapsto \varphi_j(\zeta)$ is a map from $D$ to $\hat{Y}$ which is holomorphic. The definition of $Q$ shows that if $g \in \mathscr{P}(\hat{Y})$, then

$$Q(pg, \zeta) = \zeta Q(g, \zeta).$$

We apply this for a $g \in \mathscr{P}(\hat{Y})$ such that $\varphi_k(\zeta)(g) = \delta_{kj}(c_j(\zeta))^{-1}$ to find that $p(\varphi_j(\zeta)) = \zeta$. Now $p^{-1}(\zeta) \cap \hat{Y}$ consists of at most $d$ points, and for each $\zeta \in D$, $\varphi_1(\zeta), \ldots, \varphi_d(\zeta)$ are distinct points in $p^{-1}(\zeta) \cap \hat{Y}$, so it follows that $p^{-1}(\zeta) \cap \hat{Y} = \{\varphi_1(\zeta), \ldots, \varphi_d(\zeta)\}$. Let $\Omega_1, \ldots, \Omega_d$ be disjoint neighborhoods of $\varphi_1(\zeta), \ldots, \varphi_d(\zeta)$. Note that none of the maps $\varphi_j$ can be constant. Consider, e.g., $\varphi_1$, and let $\pi_1$ be one of the coordinate functions on $\mathbb{C}^\Lambda$ such that $\pi_1 \circ \varphi_1$ is not constant.

The function $\pi_1 \circ \varphi_1$ maps a neighborhood $W$ of $\xi_0$ onto a neighborhood of $\pi_1(\varphi_1(\xi_0))$, and consequently if we choose certain further coordinate functions $\pi_2, \ldots, \pi_m$ on $\mathbb{C}^\Lambda$ and $\varepsilon_1, \ldots, \varepsilon_m$ sufficiently small but positive, then $\Omega_1$ contains the neighborhood

$$N_1 = \{\mathfrak{z} \in \mathbb{C}^\Lambda : |\pi_k(\mathfrak{z}) - \pi_k(\varphi_1(\xi_0))| < \varepsilon_k, k = 1, \ldots, m\},$$

and for some domain $G_1$ with an analytic simple closed curve as boundary, $\xi_0 \subset G_1 \subset \bar{G}_1 \subset D$, $N_1 \cap \hat{Y} = \{\varphi_1(\zeta) : \zeta \in G_1\}$. It follows that $p^{-1}(D) \cap \hat{Y}$ is a one-dimensional analytic set, and, indeed, a one-dimensional manifold.

We consider finally a point $\xi_0$ in $E$. Let $D$ be a disc about $\xi_0$ contained in $\hat{\Gamma} \setminus \Gamma$ and containing no other point of $E$. Since $\hat{\Gamma} \setminus \Gamma \subset p(\hat{Y}) \setminus p(Y)$, $D \cap p(Y)$ is empty. Let $V$ be a component of $p^{-1}(D \setminus \{\xi_0\}) \cap \hat{Y}$ so that $V$ is a connected open subset of a one-dimensional complex manifold. Since $p(\varphi_j(\zeta)) = \zeta$, it follows that $p$ is a covering map from $p^{-1}(D \setminus \{\zeta_0\}) \cap \hat{Y}$ onto $D \setminus \{\xi_0\}$, so for some integer $d_1 \leq d$, $p$ is a $d_1$-to-one map from $V$ onto $D \setminus \{\xi_0\}$, and the $d_1$ locally defined inverses to $p$ are all analytic continuations of each other. It follows that if $D'$ is the disc about $\xi_0$ whose radius is the $d_1$th root of that of $D$, then there is a holomorphic map $\Phi : D' \setminus \{\xi_0\} \to V$ such that $p(\Phi(z - \xi_0)) = (z - \xi_0)^{d_1}$. If $\pi$ is a coordinate function on $\mathbb{C}^\Lambda$, then $\pi \circ \Phi$ is a bounded holomorphic function on $D' \setminus \{\xi_0\}$ and so can be continued to a map from $D'$ into $\bar{V}$, closure taken with respect to $p^{-1}(D)$. This gives rise to an extension of $\Phi$ to a map, still denoted by $\Phi$, from $D'$ to $\bar{V}$. Note that $\bar{V}$ differs from $V$ by a single point.

There are only a finite number of the components $V$, say $V_1, \ldots, V_s$, $s \leq d$. Let $\bar{V}_j = V_j \cup \{\mathfrak{z}_j\}$. Then as before $\bigcup \bar{V}_j$ is a one-dimensional analytic set in $p^{-1}(D)$ which contains $p^{-1}(D \setminus \{\xi_0\}) \cap \hat{Y}$. We have that $p(\mathfrak{z}_j) = \xi_0$, so we shall be finished if we can show that $p^{-1}(\xi_0) \cap \hat{Y} = \{\mathfrak{z}_1, \ldots, \mathfrak{z}_s\}$. If not, let $\mathfrak{z}_0 \in p^{-1}(\xi_0) \cap \hat{Y}$, $\mathfrak{z}_0 = \mathfrak{z}_j$ for no $j = 1, \ldots, s$. Then as $p^{-1}(\xi_0) \cap \hat{Y}$ is finite and $\bigcup V_j$ is open in $\hat{Y}$, $\{\mathfrak{z}_0\}$ is open and closed in $\hat{Y}$. Consequently, $\mathfrak{z}_0$ is a peak point for $\mathscr{P}(\hat{Y})$ and so must lie in $Y$. However, $\mathfrak{z}_0 \in p^{-1}(D)$, and $D \cap p(Y) = \varnothing$. This completes the proof of the lemma.

## 30.12 *Proof of (A): Conclusion*

In this final portion of the proof we shall apply the machinery developed in the last few paragraphs to the derivation of (*A*). For this purpose, we need a very special case of Sard's lemma. The general case may be found, e.g., in Milnor [1]. Recall that if $f$ is a real-valued differentiable function on an interval in $\mathbb{R}$, then the *critical set* of $f$ is the set on which $f'$ vanishes, and the *critical values* of $f$ are the values $f(x)$, $x$ in the critical set.

**30.13** ***Lemma***

*If $f$ is a continuously differentiable, real-valued function on* $(0, 1)$, *the set of critical values of $f$ has zero Lebesgue measure.*

PROOF

Let $C = \{x \in (0, 1): f'(x) = 0\}$. If $\varepsilon > 0$, let $\{I_{\varepsilon,n}\}_{n=1}^{\infty}$ be a sequence of closed intervals in $(0, 1)$ which are pairwise disjoint except possibly for end points and which have the property that

$$C \subset \bigcup I_{\varepsilon,n} \subset \{x : |f'(x)| < \varepsilon\}.$$

Then if $I_{\varepsilon,n} = [x_n, x_n']$ and if $y \in I_{\varepsilon,n}$, we have

$$\begin{aligned} |f(y) - f(x_n)| &= \left| \int_{x_n}^{y} f'(t)\, dt \right| \\ &\leq (y - x_n)\varepsilon. \end{aligned}$$

It follows that the set $f(C)$ is contained in a set of measure no more than $2\varepsilon\Sigma(x_n' - x_n) \leq 2\varepsilon$, and as this is true of all $\varepsilon > 0$, $f(C)$ is a Lebesgue null set.

The following simple corollary is the form in which we shall use this result.

**30.14** ***Corollary***

*If $E$ is a closed, totally disconnected subset of* $[0, 1]$ *and if* $f:[0, 1] \to \mathbb{R}$ *is continuous and continuously differentiable in* $(0, 1)$, *then $f(E)$ is totally disconnected.*

PROOF

Let $C$ be the critical set of $f$. Then $f(E) = f(E \setminus C) \cup f(E \cap C)$, and since $f(C)$ has zero measure by the lemma just proved, the same is true of $f(E \cap C)$. Also, $(0, 1) \setminus C$ is a union $\bigcup J_n$ where each $J_n$ is an interval on which $f'$ is zero free. Thus, $f$ carries $E \cap J_n$ homeomorphically into $\mathbb{R}$, and the set $(f(E \cap J_n))^-$ is totally disconnected. It follows that $f(E)$, a compact set which is the union of countably many totally disconnected, compact sets, is itself totally disconnected.

In the notation of Theorem 30.1, we are given that each of the sets $K_j$, $j = 1, \ldots, q$, is the locus of a smooth curve. Let $I = [0, 1]$, and let $\varphi_i : I \to K_i$ be a smooth map with $\varphi_i(I) = K_i$.

**30.15 *Lemma***

*Let $\mathfrak{z}_0 \notin K \cup X$, and let p be a polynomial such that*

$$p(\mathfrak{z}_0) = 0 \notin p(K \cup X),$$

$$\text{Re}\, p \leq -1 \text{ on } X,$$

$$\{\mathfrak{z} \in K : \text{Re}\, p(\mathfrak{z}) > 0\} \neq \emptyset,$$

*and*

$$\text{Arg}\, p \text{ is nonconstant on each } K_j.$$

*There exist real numbers r and s and a complex number ε with $-1 < \text{Re}\,\varepsilon < 0$, $-\pi/2 < r < s < \pi/2$ such that if*

$$S = \{\zeta \in \mathbb{C} : r < \text{Arg}(\zeta - \varepsilon) < s\}$$

*and $J = p^{-1}(S) \cap K$, then $0 \in S$, and $J = J_1 \cup \cdots \cup J_k \neq \emptyset$, where the $J_j$ are open arcs whose closures $\bar{J}_j$ are disjoint, each $\bar{J}_j$ is mapped one to one onto $[r, s]$ by $\text{Arg}(p - \varepsilon)$, any two of the $p(\bar{J}_j)$ are disjoint or else coincident, each $J_j$ is of the form $\varphi_m(I_0)$ for one of the $\varphi_m$ and some open interval $I_0$ in I, and p is nonsingular on each $J_j$ in the sense that if $J_j = \varphi_m(I_0)$, then $p \circ \varphi_m$ has nonvanishing derivative on $I_0$.*

Lemma 30.3 enables us to construct such polynomials $p$.

PROOF

Let $V_i$ be the set of numbers in $[-\pi, \pi]$ equal, modulo $2\pi$ to a critical value of some (and hence any) branch of $\arg(p \circ \varphi_i)$ or to the value of this function at 0 or 1. The set $V_i$ is compact and totally disconnected, so the union $V = \bigcup V_i$ is, too.

Let $I_i = \{t \in I : \text{Re}\, p(\varphi_i(t)) > 0\}$, and define $A_i : I_i \to [-\pi/2, \pi/2]$ by $A_i(t) = \text{Arg}\, p(\varphi_i(t))$. There is an interval $[a, b] \subset (-\pi/2, \pi/2) \setminus V$ such that at least one set $A_i^{-1}([a, b])$ is nonempty. For each $i$ we can write

$$A_i^{-1}([a, b]) = \bigcup_{j=1}^{k(j)} I_{ij},$$

where each of the $I_{ij}$ is a closed interval carried by $A_i$ one to one onto $[a, b]$, where, for fixed $i$, the $I_{ij}$ are disjoint for distinct $j$, where $k(i) > 0$ for at least one $i$, and where $k(i) < \infty$ by a compactness argument. Evidently, $\varphi_i$ and $p \circ \varphi_i$ are one to one on $I_{ij}$ since $A_i$ is. Similarly, since $A_i'$ has no zeros on $I_{ij}$, the same is true for $\varphi_j'$ and $(p \circ \varphi_i)'$ as follows from the chain rule. For notational ease, we relabel the pairs $(I_{ij}, \varphi_i)$ as $(I_1', \psi_1), \ldots, (I_m', \psi_m)$ so that each $I_k'$ is one of the $I_{ij}$ and $\psi_k$ is the corresponding $\varphi_i$. Let $K_i' = \psi_i(I'_i)$, $K' = K_1' \cup \cdots \cup K_m'$, and $F_i = p(K_i')$. Thus, $K' = \{\mathfrak{z} \in K : \operatorname{Arg} p(\mathfrak{z}) \in [a, b]\}$. The sets $K_i'$ are all arcs in $\mathbb{C}^\wedge$, and $F_i$ is an arc in $\mathbb{C}$.

Let $\delta_{ij}$ be the boundary of $K_i' \cap K_j'$ in $K_i'$ so that if $\delta = \bigcup_{i,j}\delta_{ij}$, the set $K' \setminus \delta$ is a disjoint union of relatively (with respect to $K'$) open arcs. In addition, let $\beta_{ij}$ be the boundary of $F_i \cap F_j$ in $F_i$, and let $b_{ij}$ be the boundary of $(p \circ \psi_i)^{-1}(\beta_{ij})$ in $I_i'$ so that $p \circ \psi_i(b_{ij}) = \beta_{ij}$. Let $\beta = \bigcup_{i,j}\beta_{ij}$. Then $(F_1 \cup \cdots \cup F_m) \setminus \beta$ is a disjoint union of open (relative to $F_1 \cup \cdots \cup F_m$) arcs. The set $\operatorname{Arg}(p(\delta)) \cup \operatorname{Arg}(\beta)$ is a compact, totally disconnected set in $[a, b]$, so there is an interval $[c, d]$ in $[a, b] \setminus (\operatorname{Arg}(p(\delta)) \cup \operatorname{Arg}(\beta))$.

If $K'' = \{\mathfrak{z} \in K : \operatorname{Arg} p(\mathfrak{z}) \in [c, d]\}$, then $K''$ is a finite union of arcs each of which is mapped by $\operatorname{Arg} p$ one to one onto $[c, d]$, and any two images of these arcs under $p$ are either disjoint or identical.

The set $p(K)$ is compact and does not contain 0, and the set of critical values of $\operatorname{Arg} p \circ \varphi_j$ is disjoint from $[c, d]$, so if we choose $r < s$ with $c < r < s < d$, then for $\varepsilon = \varepsilon' + i\varepsilon''$, $|\varepsilon|$ sufficiently small and $\varepsilon' < 0$, the critical values of $\operatorname{Arg}((p - \varepsilon) \circ \varphi_j)$ will miss $[r, s]$, and the set $\{\mathfrak{z} \in K : r < \operatorname{Arg}(p - \varepsilon) < s\}$ will be nonempty and contained in $K''$. The numbers $r$ and $s$ have the properties required in the lemma.

**30.16** ***Lemma***

*Let $\mathfrak{z}_0$, $p$, $\varepsilon$, $r$, and $s$ be as in the last lemma. There exist $t$ and $u$, $r < t < u < s$, such that if for each $j \le k$, we set $J_j^* = \{\mathfrak{z} \in J_j : t < \operatorname{Arg}(p(\mathfrak{z}) - \varepsilon) < u\}$, then each $J_j^*$ which is not contained in $((K \cup X) \setminus J_j^*)^\wedge$ admits the following description:*

*There exists a polynomial $p_j$ and an interval $[r_j, s_j]$ in $(-\pi/2, \pi/2)$ with $\operatorname{Re} p_j \le -1$ on $X$, $0 \notin p_j(K \cup X)$, $\operatorname{Arg} p_j$ mapping $J_j^*$ one to one onto $(r_j, s_j)$, $p_j$ nonsingular on $J_j^*$, $p_j(J_j^*)$ a relatively open subset of $p_j(K \cup X)$, and $p_j(J_j^*)$ contained in the closure of the unbounded component of $\mathbb{C} \setminus p_j(K \cup X)$.*

We remark that the principal new ingredient of this lemma is the statement about $p_j(J_j^*)$.

PROOF

We shall choose $r = t_0 \le t_1 \le \cdots \le t_k < u_k \le \cdots \le u_1 \le u_0 = s$ so that if $J_j^i = \{\mathfrak{z} \in J_j : t_i < \operatorname{Arg}(p(\mathfrak{z}) - \varepsilon) < u_i$, then whenever $j \le i$ and $J_j^i$ is not contained in $((K \cup X) \setminus J_j^i)^\wedge$, there is a polynomial $p_j^i$, as in the statement of the lemma, with $J_j^i$ in place of $J_j^*$. Suppose that $1 \le i_0 \le k$ and that we have already treated those $i$ with $0 \le i < i_0$.

If $J_{i_0}^{i_0-1} \subset ((K \cup X) \setminus J_{i_0}^{i_0-1})^\wedge$, let $t_{i_0} = t_{i_0-1}$, $u_{i_0} = u_{i_0-1}$, and $p_j^{i_0} = p_j^{i_0-1}$ for all those $j < i_0$ for which $p_j^{i_0-1}$ exists.

So assume that $J_{i_0}^{i_0-1}$ is not contained in $((K \cup X) \setminus J_{i_0}^{i_0-1})\hat{} = X_{i_0}$. Fix $\mathfrak{z}_1 \in J_{i_0}^{i_0-1} \setminus X_{i_0}$. There is a polynomial $\tilde{p}$ such that $\tilde{p}(\mathfrak{z}_1) = 0$ and $\operatorname{Re} \tilde{p} < -1$ on $X_{i_0}$. If $K^{i_0} = (J_{i_0}^{i_0-1})^-$, $\tilde{p}(K^{i_0})$ is a smooth curve, so we can find a complex number $\alpha$ as near 0 as we like with $\operatorname{Re} \alpha < 0$ and $\alpha$ not in this curve. By taking $\alpha$ small enough, we can arrange that still $\operatorname{Re}(\tilde{p} - \alpha) < -1$ on $X_{i_0}$; of course, now $\operatorname{Re}(\tilde{p} - \alpha)(\mathfrak{z}_1) > 0$ and $\tilde{p} - \alpha$ has no zeros on $X_{i_0} \cup K^{i_0}$. If we modify $\tilde{p} - \alpha$ by adding a polynomial that is small on $X_{i_0} \cup K^{i_0}$, we obtain a polynomial $p_{i_0}^{i_0}$ whose real part is greater than 0 at $\mathfrak{z}_1$ and is less than $-1$ on $X_{i_0}$, which has no zeros on $X_{i_0} \cup K^{i_0}$, and whose argument is nonconstant on $K^{i_0}$.

Now we can apply Lemma 30.15 to get numbers $r_{i_0}$ and $s_{i_0}$ with $-\pi/2 < r_{i_0} < s_{i_0} < \pi/2$, and a complex number $\varepsilon_{i_0}$ with $\operatorname{Re} \varepsilon_{i_0} < 0$ such that if

$$S_{i_0} = \{\zeta \in \mathbb{C} : r_{i_0} < \operatorname{Arg}(\zeta - \varepsilon_{i_0}) < s_{i_0}\},$$

then $J^{i_0} = p_{i_0}^{i_0-1}(S_{i_0}) \cap K^{i_0} \neq \varnothing$, and $J^{i_0} = L_1 \cup \cdots \cup L_m$, a disjoint union of open arcs each of which is mapped one to one onto $(r_{i_0}, s_{i_0})$ by $\operatorname{Arg}(p_{i_0}^{i_0} - \varepsilon_{i_0})$, and any two of the images under $p_{i_0}^{i_0}$ are distinct or identical. One of these image arcs, say that corresponding to $L_p$, is farthest from the origin. Since $J^{i_0} \subset J_{i_0}^{i_0-1}$, if we set $J_{i_0}^{i_0} = L_p$, then $J_{i_0}^{i_0} \subset J_{i_0}^{i_0-1}$.

There are numbers $t_{i_0}$, $u_{i_0}$, $t_{i_0-1} \leq t_{i_0} < u_{i_0} \leq u_{i_0-1}$ with $J_{i_0}^{i_0} = \{\mathfrak{z} \in J_{i_0}^{i_0-1} : t_{i_0} < \operatorname{Arg}(p(\mathfrak{z}) - \varepsilon_{i_0}) < u_{i_0}\}$. Define the $J_{j_0}^{i}$ as in the first sentence of the proof, and for those $j < i_0$ with $J_j^{i_0}$ not contained in $((K \cup X) \setminus J_j^{i_0})\hat{}$, automatically $J_j^{i_0-1} \not\subset ((K \cup X) \setminus J_j^{i_0-1})\hat{}$, and we may set $p_j^{i_0} = p_j^{i_0-1}$. We have only to verify that $p_j^{i_0}(J_j^{i_0})$ lies in the closure of the unbounded component of $\mathbb{C} \setminus p_j^{i_0}(K \cup X)$, and this is true because $p_j^{i_0}(J_j^{i_0}) \subset p_j^{i_0-1}(J_j^{i_0-1})$ and $p_j^{i_0}(K \cup X) = p_j^{i_0-1}(K \cup X)$.

Finally, set $t = t_k$, $u = u_k$, $J_j^* = J_j^k$, and, for those $j$ such that $J_j^* \not\subset ((K \cup X) \setminus J_j^*)$, set $p_j = p_j^k$. The proof is complete.

30.17 ***Lemma***

*Let $p$, $\varepsilon$, $t$, and $u$ be as in the preceding lemma. Let $K_0$ be the union of all the $J_j^*$ for which*

$$J_j^* \subset ((K \cup X) \setminus J_j^*)\hat{},$$

*and let $L = K \setminus K_0$. Then $(L \cup X)\hat{} = (K \cup X)\hat{}$.*

PROOF

$L \cup X = \bigcap\{(K \cup X) \setminus J_j^*\}$, the intersection over all those $j$ for which $J_j^* \subset K_0$. However, if $J_j^* \subset K_0$, then $(K \cup X) \setminus J_j^*$ contains the Shilov boundary for $\mathscr{P}(K \cup X)$, and therefore so does the intersection $L \cup X$.

The next step will be to produce *some* analytic structure in $(K \cup X)\hat{}$.

**30.18 Lemma**

*Let $p$, $\varepsilon$, $t$, $u$, and $L$ be as the last lemma. If $\zeta \in \mathbb{C}$ satisfies $t < \operatorname{Arg}(\zeta - \varepsilon) < u$, there is a closed disc $\Delta$ centered at $\zeta$ with boundary $\gamma$ such that if*

$$D = p^{-1}(\Delta) \cap (L \cup X)\hat{},$$

$$D_1 = \text{union of the components of } D \text{ which meet } L \cup X,$$

$$D_2 = D \setminus D_1,$$

$$\delta = p^{-1}(\gamma) \cap (L \cup X)\hat{},$$

and

$$\delta_i = \delta \cap D_i,\ i = 1, 2,$$

*then $D_1$ and $D_2$ are open and closed in $D$, $D_2 = \hat{\delta}_2$, and $D_1 \setminus (\delta_1 \cup L)$ is a one-dimensional analytic subset of $p^{-1}(\Delta) \setminus p^{-1}(\gamma \cup L)$.*

PROOF

If $\zeta \notin p((L \cup X)\hat{})$, we are done, as by taking $\Delta$ small enough, we can arrange $D = \emptyset$. We assume, therefore, that $\zeta \in p((L \cup X)\hat{})$.

Only finitely many points of $L \cup X$ are carried to $\zeta$ by $p$, and all these points lie in $L$, so each of them is contained in one of the $J_j^*$. If $\mathfrak{z} \in L$ goes onto $\zeta$ under $p$, let $\lambda(\mathfrak{z})$ be the $J_j^*$ containing it. Let $p_j$ be the polynomial associated with $\lambda(\mathfrak{z})$ as in Lemma 30.16. The set $p_j(\lambda(\mathfrak{z}))$ is an arc which lies in the closure of the unbounded component of $\mathbb{C} \setminus p_j(L \cup X)$, so Lemma 30.7 applies to yield the fact that for each $z \in p_j(\lambda(\mathfrak{z}))$, $p_j^{-1}(z) \cap (L \cup X)\hat{} = (p_j^{-1}(z) \cap (L \cup X))\hat{}$. The last set is the finite set $p_j^{-1}(z) \cap (L \cup X)$ which is contained in $L$. We have therefore that $p_j^{-1}(p_j(\lambda(\mathfrak{z}))) \cap (L \cup X)\hat{} \subset L$. The description of the $J_j^*$ given in Lemma 30.16 shows that $\lambda(\mathfrak{z})$ is open in $L \cup X$.

Let $p_j^{-1}(p_j(\mathfrak{z})) \cap (L \cup X)\hat{} = \{\mathfrak{z}, \mathfrak{z}_1, \ldots, \mathfrak{z}_s\}$, and choose $B, B_1, \ldots, B_s$ open, pairwise disjoint balls about these points. Choose $B$ so small that $B \cap (L \cup X) \subset \lambda(\mathfrak{z})$, a choice that is possible since $\lambda(\mathfrak{z})$ is open in $L \cup X$.

Let $\Delta'$ be a small domain about $\zeta = p_j(\mathfrak{z})$ with $\partial\Delta'$ a simple closed curve meeting $p_j(\lambda(\mathfrak{z}))$ twice and such that $\Delta' \cap p_j(\lambda(\mathfrak{z}))$ is an arc. We can choose $\Delta'$ so small that $p_j^{-1}(\Delta') \cap (L \cup X) \subset B \cup B_1 \cup \cdots \cup B_s$. The portion, $D'$, of $p_j^{-1}(\Delta') \cap (L \cup X)$ in $B$ is open in $(L \cup X)\hat{}$, and it meets $L \cup X$ in a subarc $\lambda'(\mathfrak{z})$ of $\lambda(\mathfrak{z})$.

Denote by $\delta'$ the boundary of $D'$ in $(L \cap X)\hat{}$. According to the local maximum modulus theorem, $D' \subset (\delta' \cup \lambda'(\mathfrak{z}))\hat{}$. Moreover, if $\gamma'$ is that part of the boundary of $\Delta'$ which does not lie in the unbounded component of $\mathbb{C} \setminus p_j(L \cup X)$ so that it is a certain arc, then $p_j(\delta')$ is contained in $\gamma'$, and $\gamma' \cup p_j(\lambda'(\mathfrak{z}))$ is a simple closed curve, $C$.

There are two cases to consider now. If $p_j((\delta' \cup \lambda'(\mathfrak{z}))\hat{\ }) \subset C$, then $p_j(D') \subset C \cap \Delta' = p_j(\lambda'(\mathfrak{z}))$, and by Lemma 30.9 we conclude that $D' = \lambda'(\mathfrak{z})$. If, on the other hand, $p_j((\delta' \cup \lambda'(\mathfrak{z}))\hat{\ })$ is not contained in $C$, then Lemma 30.10 says that $D' \setminus \lambda'(\mathfrak{z})$ is mapped by $p_j$ one to one onto the interior of $C$ in such a way that $q \circ p_j^{-1}$ is holomorphic on this interior for every polynomial $q$.

Thus, either $D' = \lambda'(\mathfrak{z})$ or else $D' \setminus \lambda'(\mathfrak{z})$ is an analytic disc.

It follows that $\mathfrak{z}$ is an isolated point of the zero set of $(p - p(\mathfrak{z}))|(L \cup X)\hat{\ }$. Thus, if $\Delta_\mathfrak{z}$ is a sufficiently small open disc about $p(\mathfrak{z}) = \zeta$, the component $D_\mathfrak{z}$ of $p^{-1}(\Delta_\mathfrak{z}) \cap (L \cup X)$ that contains $\mathfrak{z}$ is open in $D'$, and therefore $D_\mathfrak{z}$ is open in $(L \cup X)\hat{\ }$. If $D' = \lambda'(\mathfrak{z})$, then $D_\mathfrak{z} \subset L$, and, if not, $D_\mathfrak{z} \setminus L$ is open in $D' \setminus \lambda'(\mathfrak{z})$, and so is a one-dimensional analytic set.

As the points $\mathfrak{z}$ under consideration are finite in number, there is a closed disc $\Delta$ centered at $\zeta$ and contained in the intersection of the discs $\Delta_\mathfrak{z}$. Then the sets $D_1$ and $D_2$ defined as in the lemma are open and closed in $D = p^{-1}(\Delta) \cap (L \cup X)\hat{\ }$, and $D_1$ is the finite union of those components of $D$ that contain a $\mathfrak{z}$ with $p(\mathfrak{z}) = \zeta$. Therefore, $D_1 \setminus (\delta_1 \cup L)$ is open in $\cup(D_\mathfrak{z} \setminus L)$, and each $D_\mathfrak{z} \setminus L$ is a one-dimensional analytic set in $p^{-1}(\Delta_\mathfrak{z}) \setminus L$. Thus $D_1 \setminus (\delta_1 \cup L)$ is a perhaps empty one-dimensional analytic set in $p^{-1}(\Delta) \setminus p^{-1}(\gamma \cup L)$. Finally, as $D$ is polynomially convex, the local maximum modulus theorem yields $D_2 = \hat{\delta}_2$.

Having discovered some analytic structure in $(X \cup K)\hat{\ } \setminus (X \cup K)$, we can complete the proof by invoking Lemma 30.11 in the following way. Let the point $\mathfrak{z}_0$ of Lemma 30.15 lie in $(K \cup X)\hat{\ } \setminus (K \cup X)$, choose $p$, $\varepsilon$, $t$, $u$, and $L$ as in Lemma 30.16, and set

$$\Sigma = \{\zeta \in \mathbb{C} : t < \operatorname{Arg}(\zeta - \varepsilon) < u\}.$$

Let $\Sigma'$ be the set of points $\zeta \in \Sigma$ for which there is an open disc $\Delta$ about $\zeta$ with $p^{-1}(\Delta) \cap (L \cup X)\hat{\ } \setminus (L \cup X)$ a possibly empty one-dimensional analytic subset of $p^{-1}(\Delta) \cap (L \cup X)$. The set $\Sigma'$ is open in $\Sigma$, and $\Sigma'$ is not empty since if $\zeta \in \Sigma$ has sufficiently large modulus, $p^{-1}(\zeta) \cap (K \cup X)\hat{\ }$ is empty. Also, $\Sigma'$ is closed in $\Sigma$. To see this, consider a point $\zeta$ in the closure of $\Sigma'$ in $\Sigma$. Let $\Delta$ be a disc about $\zeta$ as in the last lemma.

In the notation of that lemma,

$$p^{-1}(\Delta) \cap D_2$$

is open in $p^{-1}(\Delta) \cap ((L \cup X)\hat{\ } \setminus (L \cup X))$, and so is a one-dimensional analytic set in $p^{-1}(\Delta)$. Also, $D_2 = \hat{\delta}_2$, so by Lemma 30.11, $D_2 \setminus \delta_2$ is a one-dimensional analytic subset of $\mathbb{C}^\Lambda \setminus \delta$. This yields $\Sigma' = \Sigma$.

Since $0 = p(\mathfrak{z}_0) \in \Sigma$, we have the desired conclusion that $(X \cup K)\hat{\ }$ is a one-dimensional analytic set at $\mathfrak{z}_0$, and the theorem is proved.

The theorem of Stolzenberg is deep and impressive, but a natural question arises in connection with it: Can the smoothness hypotheses be relaxed? In particular, a perfectly natural extension would be to replace the condition of continuous differentiability by that of rectifiability. A result of this kind has been proved by Alexander [3], who has shown that *if X is a rectifiable arc in* $\mathbb{C}^N$, *then X is polynomially convex and* $\mathscr{P}(X) = \mathscr{C}(X)$. Somewhat more generally he proved that *if X is a compact subset of* $\mathbb{C}^N$ *which is contained in a connected compact set of finite linear measure, then* $\hat{X} \setminus X$ *is a possibly empty pure one-dimensional analytic subset of* $\mathbb{C}^N \setminus X$. *If, in addition,* $H^1(X, \mathbb{Z}) = 0$, *then X is polynomially convex, and* $\mathscr{P}(X) = \mathscr{C}(X)$. These results depend on the theorem of Stolzenberg. For their proofs, we refer to the original papers.

In the direction of conditions of a metric nature for approximation in $\mathbb{C}^N$, it may be worth noting that the obvious $N$-dimensional formulation of the Hartogs–Rosenthal theorem (Corollary 26.9) is correct. Denote by $H_2^N$ the two-dimensional Hausdorff measure on $\mathbb{C}^N$ (with respect to the usual metric).

30.19 ***Theorem***

*If E is a compact set in* $\mathbb{C}^N$ *with* $H_2^N(E) = 0$, *then E is rationally convex, and* $\mathscr{R}(E) = \mathscr{C}(E)$.

PROOF

We proceed by induction on $N$. If $N = 1$, then since $H_2^1$ is, to within a constant multiple, planar Lebesgue measure, $E$ has zero Lebesgue measure and so, by the Hartogs–Rosenthal theorem, $\mathscr{R}(E) = \mathscr{C}(E)$.

Assume that the theorem is true for all compact sets of zero $H_2^N$-measure in $\mathbb{C}^N$, and let $E \subset \mathbb{C}^{N+1}$ be a compact set of zero $H_2^{N+1}$-measure. Denote by $\pi : \mathbb{C}^{N+1} \to \mathbb{C}^N$ the projection given by

$$\pi(z_1, \ldots, z_{N+1}) = (z_1, \ldots, z_N).$$

The map $\pi$ does not increase distances, so $H_2^N(\pi(E)) = 0$. The induction hypothesis implies that $\pi(E)$ is rationally convex and that $\mathscr{R}(\pi(E)) = \mathscr{C}(\pi(E))$. It follows that each maximal set of antisymmetry for $\mathscr{R}(E)$ is contained in a fiber $E_\mathfrak{z} = E \cap \pi^{-1}(\mathfrak{z})$ for some $\mathfrak{z} \in \mathbb{C}^N$. However, $\mathscr{R}(E)|E_\mathfrak{z} = \mathscr{C}(E_\mathfrak{z})$ for each $\mathfrak{z}$.

To see this, let $\xi : \mathbb{C}^{N+1} \to \mathbb{C}$ be the projection which takes $(z_1, \ldots, z_{N+1})$ to $z_{N+1}$. Since $\pi$ is constant on $E_\mathfrak{z}$, $\xi$ carries $E_\mathfrak{z}$ homeomorphically into $\mathbb{C}$. The set $\xi(E)$ has zero $H_2^1$-measure, so $\mathscr{R}(\xi(E)) = \mathscr{C}(\xi(E))$, by Corollary 26.9. If $f \in \mathscr{C}(E_\mathfrak{z})$, there is $F \in \mathscr{C}(\xi(E))$ with $F \circ \xi = f$ on $E_\mathfrak{z}$. But then $F \in \mathscr{R}(\xi(E))$, so $F \circ \xi \in \mathscr{R}(E)$. Thus, $f \in \mathscr{R}(E)|E_\mathfrak{z}$, whence $\mathscr{R}(E)|E_\mathfrak{z} = \mathscr{C}(E_\mathfrak{z})$, and the result is proved.

Stolzenberg's theorem has important applications in the study of spectra of algebras of holomorphic functions.

Suppose that $M$ is a one-dimensional complex manifold, suppose that $\gamma \subset M$ is a compact set which consists of the union of a finite family of continuously differentiable curves, and suppose that $A$ is an algebra of holomorphic functions on $M$. The set $\gamma$ meets only finitely many of the components of $M$. Let $\|\cdot\|_\gamma$ be the norm

$$\|f\|_\gamma = \sup\{|f(\zeta)| : \zeta \in \gamma\},$$

and let $B$ be the completion of $A$ with respect to $\|\cdot\|_\gamma$ so that $B$ is the closure of $A|\gamma$ in $\mathscr{C}(\gamma)$. We do not suppose that $A$ separates points on $\gamma$, so $B$ need not be a uniform algebra on $\gamma$. However, $B$ is a uniformly closed subalgebra of $\mathscr{C}(\gamma)$, and we ask what can be said of $\Sigma(B)$. Stolzenberg's theorem provides an immediate answer as follows.

Let $\Phi : \gamma \to \mathbb{C}^A$ be the map defined by requiring that for all $f \in A$, the $f$th coordinate of $\Phi(\zeta)$, $\zeta \in \gamma$, is $f(\zeta)$. Then $B$ can be identified with $\mathscr{P}(\Phi(\gamma))$, and $\Sigma(B)$ can be identified with the polynomially convex hull of the set $\Phi(\gamma)$. Since $\Phi(\gamma)$ is a finite union of continuously differentiable curves, we see that $\Sigma(B) \setminus \Phi(\gamma)$ is a possibly empty, purely one-dimensional analytic space.

More information on the spectra of algebras of holomorphic functions on one-dimensional complex manifolds, i.e., Riemann surfaces, can be found in the papers of Bishop [5, 6] and Royden [3]. The higher-dimensional analogue of this theory is much more complicated than the one-dimensional theory. In this connection, see Bishop [6], Grauert [1], and Rossi and Stolzenberg [1].

As a consequence of the description of the hull of a smooth curve, we can justify an assertion made in the remarks following Theorem 29.18.

30.20 ***Theorem***

*If $A$ is a closed subalgebra of $\mathscr{C}(\mathbb{C})$ which is algebraically isomorphic to $\mathscr{O}(\mathbb{C}^N)$, $N > 1$, then $A$ contains functions which are not holomorphic.*

PROOF

Let $\Phi$ be an algebraic isomorphism of $\mathscr{O}(\mathbb{C}^N)$ onto the closed subalgebra $A$ of $\mathscr{C}(\mathbb{C})$. We shall show that $\Phi$ is, in fact, continuous, and hence, by the open mapping theorem, bicontinuous. This depends on the fact that every nonzero complex homomorphism of $\mathscr{O}(\mathbb{C}^N)$ is of the form $f \mapsto f(\mathfrak{z}_0)$ for some fixed $\mathfrak{z}_0 \in \mathbb{C}^N$, a well-known and easily proved fact. Let $\varphi : \mathscr{O}(\mathbb{C}^N) \to \mathbb{C}$ be a nonzero complex homomorphism. There is

$\mathfrak{z}_0 \in \mathbb{C}^N$ such that $\varphi(p) = p(\mathfrak{z}_0)$ for every polynomial $p$. If $f \in \mathcal{O}(\mathbb{C}^N)$, we can write

$$f(\mathfrak{z}) = f(\mathfrak{z}_0) + (z_1 - z_1^0)g_1 + \cdots + (z_N - z_N^0)g_N,$$

$g_1, \ldots, g_N \in \mathcal{O}(\mathbb{C}^N)$. [Here we suppose that $\mathfrak{z}_0 = (z_1^0, \ldots, z_N^0)$.] Thus,

$$\varphi(f) = f(\mathfrak{z}_0) + \Sigma\varphi(z_j - z_j^0)\varphi(g_i) = f(\mathfrak{z}_0),$$

whence $\varphi$ is evaluation at $\mathfrak{z}_0$.

The algebra $A$ is plainly semisimple in the sense that the only element annihilated by all complex homomorphisms is zero. Since we have assumed $A$ closed, it follows from the closed graph theorem, just as in the proof of Theorem 4.12, that $\Phi$ is continuous and hence bicontinuous.

Now let $z_1, \ldots, z_N$ be the $N$ coordinate functions on $\mathbb{C}^N$, and define a dual map $\Phi' : \mathbb{C} \to \mathbb{C}^N$ by

$$\Phi'(\zeta) = ((\Phi z_1)(\zeta), \ldots, (\Phi z_N)(\zeta))$$

so that $\Phi'$ is a holomorphic map.

We shall prove that if $K \subset \mathbb{C}^N$ is compact, then $K \subset [\Phi'(E)]\hat{}$ for some compact set $E$ in $\mathbb{C}$. If not, then for $n = 1, 2, \ldots$, we let

$$D_n = \{\zeta \in \mathbb{C} : |\zeta| \le n\},$$

and for each $n$, $K \setminus [\Phi'(D_n)]\hat{}$ is nonempty. It follows that there are polynomials $\{p_n\}$ such that

$$\|p_n\|_K = 1 \qquad \text{and} \qquad \|p_n\|_{\Phi'(D_n)} < 1/n.$$

The sequence $\{\Phi p_n\}$ converges to zero in $\mathscr{C}(\mathbb{C})$, but $\{p_n\}$ does not converge uniformly on compact sets in $\mathbb{C}^N$ to the zero function. This contradiction to the continuity of $\Phi^{-1}$ proves our contention.

Consequently, for some $n$, $[\Phi'(D_n)]\hat{}$ contains an open set in $\mathbb{C}^N$. By the maximum modulus theorem,

$$[\Phi'(D_n)]\hat{} = [\Phi'(\partial D_n)]\hat{},$$

and since $\partial D_n$ is a circle and $\Phi'$ is holomorphic, $\Phi'(\partial D_n)$ is a certain closed analytic curve in $\mathbb{C}^N$. Thus,

$$[\Phi'(\partial D_n)]\hat{} \setminus \Phi'(\partial D_n)$$

is a purely one-dimensional variety and so cannot contain an open subset of $\mathbb{C}^N$. So our theorem is proved.

It may be of interest to contrast the preceding result with the situation for the disc algebra. For every $N = 2, 3, \ldots$, the disc algebra $\mathscr{A}$ contains closed subalgebras isomorphic to the polydisc algebra $\mathscr{A}_N$.

This is an easy consequence of the Rudin–Carleson theorem and elementary topology, as we have seen, in effect, in Example 25.19.

We turn next to some results established by Alexander [1, 2]. These results depend on Theorem 30.1 in its full generality and also on some of the techniques used in its proof. They hinge on a rather technical lemma that isolates the most difficult part of the theory.

30.21 ***Lemma***

*Let $X$ be a compact set in $\mathbb{C}^N$ and let $\pi:\mathbb{C}^N \to \mathbb{C}$ be a coordinate projection. Assume there to exist a compact, totally disconnected set $J \subset \pi(X)$ with the following properties:*

(a) *At every point of $\pi(X) \setminus J$, $\pi(X)$ has the structure of an open arc.*
(b) *If $E$ is a compact set in $\pi(X) \setminus J$, then there is an integer $e$ with the property that no point of $E$ has more than $e$ preimages under $\pi$ in $X$.*
(c) *Every point of $\mathbb{C} \setminus J$ can be joined to infinity by an arc in $\mathbb{C} \setminus J$ which meets $\pi(X)$ in a finite set.*
(d) *If $\zeta \in J$, the set $\pi^{-1}(\zeta) \cap X$ is polynomially convex.*
(e) *At every point of $X \setminus \pi^{-1}(J)$, $X$ has the structure of an open arc, and $\pi$ is locally one to one on $X \setminus \pi^{-1}(J)$.*

*Granted these hypotheses, $\hat{X} \setminus X$ consists of a possibly empty one-dimensional analytic subset of $\mathbb{C}^N \setminus X$.*

It is to be observed that the set $X \setminus \pi^{-1}(J)$ is required to satisfy no smoothness conditions whatsoever.

Also, we probably should comment on the exact content of hypothesis (c). There we speak of connecting a point to infinity by an arc with certain properties. This, of course, is a brief way of saying that the point in question can be connected to the unbounded component of the set $\mathbb{C} \setminus \pi(X)$ by an arc with the desired properties.

In the proof of Lemma 30.21, we shall have to make use of a fact about proper holomorphic maps from one-dimensional analytic varieties (or spaces) to domains in $\mathbb{C}$. The result we need is a special instance of a general result in the theory of *analytic covers*, a theory which is developed in detail in Gunning and Rossi [1]. The fact we shall use may be formulated as follows. *Let $D$ be a connected open set in $\mathbb{C}$, let $V$ be a one-dimensional analytic space on which the holomorphic functions separate points, and let $h: V \to D$ be a proper holomorphic map. There is a discrete set $E \subset D$ and a positive integer $\lambda$ such that $h$ is a $\lambda$-sheeted covering map from $V \setminus h^{-1}(E)$ onto $D \setminus E$. The set $V \setminus h^{-1}(E)$ is dense in $V$, and for every $z \in D$, $h^{-1}(z)$ is a discrete set.* We shall summarize this situation by saying that $V$ is a $\lambda$-sheeted branched covering of $D$.

PROOF OF LEMMA 30.21

If $\zeta \in \mathbb{C} \setminus J$, define $v(\zeta)$ to be the least number of points in $\lambda \cap \pi(X)$, where $\lambda$ ranges over the arcs in $\mathbb{C} \setminus J$ which connect $\zeta$ to infinity. Our proof will involve an induction on the integer $v(\zeta)$.

Consider a point $\mathfrak{z}_0 \in \hat{X} \setminus X$, and assume to begin with that $\pi(\mathfrak{z}_0) \notin J$ so that $v(\pi(\mathfrak{z}_0))$ is defined. For no choice of $\mathfrak{z}_0$ is $v(\pi(\mathfrak{z}_0)) = 0$. Let us assume that near any point $\mathfrak{z}$ of $\hat{X} \setminus X$ with $v(\pi(\mathfrak{z})) \leq k - 1$, $\hat{X}$ has the structure of a one-dimensional analytic set, and suppose that $v(\pi(\mathfrak{z}_0)) = k$. It seems necessary to consider three cases.

(A) The case that $\pi(\mathfrak{z}_0)$ lies in the common boundary of two different components, $\Omega_1$ and $\Omega_2$ of $\mathbb{C} \setminus \pi(X)$. We may suppose that $v$ is identically $k - 1$ in $\Omega_2$, a supposition that entails the inequality $v \geq k - 1$ in $\Omega_1$. It follows from the inductive hypothesis that $\hat{X} \cap \pi^{-1}(\Omega_2)$ is empty or a purely one-dimensional analytic set. If $X \cap \pi^{-1}(\Omega_2) \neq \varnothing$, the map $\pi$ carries this set properly onto $\Omega_2$, so it follows from the result quoted above that $\hat{X} \cap \pi^{-1}(\Omega_2)$ is a branched covering, say with $d$ sheets, of the domain $\Omega_2$. Set $\zeta_0 = \pi(\mathfrak{z}_0)$ and let $\Gamma$ be a simple closed curve which bounds a domain, $D$, whose closure is contained in $\Omega_1 \cup \Omega_2 \cup \pi(X)$. We require that (1) $\bar{D} \cap \pi(X)$ be an arc $\gamma$ with $\gamma \cap \Gamma$ consisting of the end points of $\gamma$, that (2) $\zeta_0$ lie in the interior of $\gamma$, that (3) $(D \cap \Omega_1)$ be a domain bounded by a simple closed curve, and that (4) $\Gamma \cap \Omega_2$ contain a closed straight-line segment $\alpha$.

The set $X_0 = \pi^{-1}(\bar{D}) \cap \hat{X}$ is polynomially convex since $\bar{D}$ is a polynomially convex subset of the plane, and it follows from the local maximum modulus theorem (explicitly, Corollary 9.13) that

$$\pi^{-1}(\bar{D}) \cap \hat{X} = [(\pi^{-1}(\Gamma) \cap \hat{X}) \cup (\pi^{-1}(D) \cap X)]\hat{}.$$

Let $\alpha^0$ be the interior of $\alpha$ and set

$$X_1 = (\pi^{-1}(\Gamma \setminus \alpha^0) \cap \hat{X}) \cup (\pi^{-1}(\gamma) \cap X).$$

The point $\zeta_0$ lies in the boundary of the unbounded component of $\mathbb{C} \setminus \pi(X_1)$, so, by Lemma 30.7,

$$\pi^{-1}(\zeta_0) \cap \hat{X}_1 = (\pi^{-1}(\zeta_0) \cap X_1)\hat{}.$$

However, by hypothesis $\zeta_0 \notin J$, so assumption (b) of the lemma implies that $\pi^{-1}(\zeta_0) \cap X_1$ is finite and hence polynomially convex. Thus, $\pi^{-1}(\zeta_0) \cap \hat{X}_1 = \pi^{-1}(\zeta_0) \cap X_1$. Since $\mathfrak{z}_0 \notin X_1$, we have that $\mathfrak{z}_0 \notin \hat{X}_1$.

The segment $\alpha$ is real analytic, and $\pi$ is holomorphic on $\hat{X} \cap \pi^{-1}(\Omega_2)$, so since $\hat{X} \cap \pi^{-1}(\Omega_2)$ is a branched cover of $\Omega_2$, it follows that $\pi^{-1}(\alpha) \cap \hat{X}$ is a finite union of real analytic curves.

The definition of $X_1$ and the choice of $D$ show that

$$(\pi^{-1}(\Gamma) \cap \hat{X}) \cup (\pi^{-1}(D) \cap X) \subset \hat{X}_1 \cup (\pi^{-1}(\alpha) \cap \hat{X}),$$

whence

$$[(\pi^{-1}(\Gamma) \cap \hat{X}) \cup (\pi^{-1}(D) \cap X)]\hat{} \subset [\hat{X}_1 \cup (\pi^{-1}(\alpha) \cap \hat{X})]\hat{}.$$

The opposite inclusion is also correct, as follows from the definition of $X_1$. Thus, the two sets are equal. Stolzenberg's theorem (Theorem 30.1) implies that

$$[\hat{X}_1 \cup (\pi^{-1}(\alpha) \cap \hat{X})]\hat{} \setminus (\hat{X}_1 \cup \pi^{-1}(\alpha) \cap \hat{X})$$

is a possibly empty purely one-dimensional analytic set. As $\mathfrak{z}_0$ is a point of this set, we are finished with case (A).

(B) The case that $\zeta_0 = \pi(\mathfrak{z}_0) \in \pi(X)$ but it does not lie in the common boundary of two components of $\mathbb{C} \setminus \pi(X)$. As we are assuming that $\pi(\mathfrak{z}_0) \notin J$, there exists a neighborhood, $U$, of $\pi(\mathfrak{z}_0)$ in $\mathbb{C}$ such that $U \cap J = \varnothing$, $\bar{U} \cap \pi(X)$ is an arc, and $U \setminus (U \cap \pi(X))$ is contained in a component $\Omega$ of $\mathbb{C} \setminus \pi(X)$. Since $v(\pi(\mathfrak{z}_0)) = k$, we have $v \equiv k - 1$ in $\Omega$, so, by induction, $\pi^{-1}(\Omega) \cap \hat{X}$ is a one-dimensional analytic set in $\pi^{-1}(\Omega)$. By hypothesis, $\pi^{-1}(\zeta_0) \cap X$ is finite and so polynomially convex. It follows that there exists a neighborhood $V$ of $\zeta_0$ in $\mathbb{C}$ such that $\mathfrak{z}_0 \notin (\pi^{-1}(\bar{V}) \cap X)\hat{}$, for if $\{V_n\}$ is a sequence of neighborhoods of $\zeta_0$ which shrink to $\zeta_0$, then the sets $(\pi^{-1}(\bar{V}_n) \cap X)\hat{}$ shrink to $(\pi^{-1}(\zeta_0) \cap X)\hat{}$, and the latter set does not contain $\mathfrak{z}_0$. For $V$ we can take a domain with boundary a simple closed curve $\Gamma$ which meets $\pi(X)$ at two points $\zeta_1$ and $\zeta_2$, which are the end points of the arc $\gamma = V \cap \pi(X)$. Let $\alpha$ consist of the union of two straight-line segments, one contained in each component of $\Gamma \setminus \{\zeta_1, \zeta_2\}$, and let $\alpha^0$ be obtained from $\alpha$ by removing the end points. The local maximum modulus theorem again yields

$$\pi^{-1}(\bar{V}) \cap \hat{X} = [(\pi^{-1}(\Gamma) \cap \hat{X}) \cup (\pi^{-1}(\gamma) \cap X)]\hat{}. \tag{4}$$

Let

$$X_1 = (\pi^{-1}(\Gamma \setminus \alpha^0) \cap \hat{X}) \cup (\pi^{-1}(\gamma) \cap X).$$

The set $\pi^{-1}(\Gamma \setminus \alpha^0) \cap \hat{X}$ is polynomially convex, and since if $\zeta \in \bar{\gamma}$, then $\pi^{-1}(\zeta) \cap X$ is finite by hypothesis, it follows that $\pi^{-1}(\bar{\gamma}) \cap X$ is polynomially convex. Also, two applications of Lemma 29.21 show that the set $X_1$ itself is polynomially convex. The choice of $V$ and the fact that $\pi^{-1}(\bar{\gamma}) \cap X = \pi^{-1}(\bar{V}) \cap X$ show that $\mathfrak{z}_0 \notin X_1$. Since $X_1$ differs from $(\pi^{-1}(\Gamma) \cap \hat{X}) \cup (\pi^{-1}(\gamma) \cap X)$ by the finitely many analytic curves which constitute $\pi^{-1}(\alpha)$,

Theorem 30.1, together with (4), implies that near $\mathfrak{z}_0$, $\hat{X}$ has the structure of a one-dimensional analytic set.

(C) The case that $\zeta_0 = \pi(\mathfrak{z}_0)$ lies in $\mathbb{C} \setminus \pi(X)$. In this case we consider an arc, $\lambda$, joining $\zeta_0$ to infinity which lies in $\mathbb{C} \setminus J$ and which meets $\pi(X)$ $k = \nu(\zeta_0)$ times. Let $\zeta_1$ be the first point (counting from $\zeta_0$) at which $\lambda$ meets $\pi(X)$. The point $\zeta_1$ lies in the boundary of two distinct components $\Omega_1$ and $\Omega_2$ of $\mathbb{C} \setminus \pi(X)$, for otherwise there would be an arc from $\zeta_0$ to infinity meeting $X$ in fewer than $k$ points. Let the indexing be such that $\nu = k$ on $\Omega_1$, $\nu = k - 1$ on $\Omega_2$. As before, $\hat{X} \cap \pi^{-1}(\Omega_2)$ is a branched cover of $\Omega_2$, say of the order $d$.

Let $\gamma$ be a short arc in $\pi(X) \setminus J$ which contains the point $\zeta_1$ in its interior. By case (A) of our proof, $\hat{X}$ is a one-dimensional analytic set near each point of $(\hat{X} \setminus X) \cap \pi^{-1}(\gamma)$. We shall show that if $\mathfrak{z}_1 \in \pi^{-1}(\zeta_1) \cap (\hat{X} \setminus X)$, then $\pi$ maps a small neighborhood of $\mathfrak{z}_1$ finite to one onto a neighborhood of $\zeta_1$ in $\mathbb{C}$. This is so, for $\hat{X}$ is a one-dimensional analytic set at $\mathfrak{z}_1$, so there is a neighborhood $N$ of $\mathfrak{z}_1$ such that

$$N \cap \hat{X} = \varphi_1(\Delta) \cup \cdots \cup \varphi_q(\Delta),$$

as in our definition of a one-dimensional analytic set. We can suppose that $\mathfrak{z}_1 \in \varphi_j(\Delta)$ for each $j$. If $\pi \circ \varphi_j$ is not constant, then under $\pi \circ \varphi_j$, $\Delta$ goes finitely many to one onto a neighborhood of $\zeta_1$. (Recall that $\pi \circ \varphi_1$ is supposed to be holomorphic on a neighborhood of $\bar{\Delta}$.) Thus, to prove our contention, it is enough to prove that for no $j$ is $\pi \circ \varphi_j$ constant. As $\hat{X} \setminus X$ is an analytic set near each point of $\pi^{-1}(\zeta_1)$, there is an open set $\Omega$ in $\mathbb{C}^N$ which is disjoint from $X$ and which has the property that $\Omega \cap (\hat{X} \setminus X)$ is an analytic set in $\Omega$ and $\Omega \supset \pi^{-1}(\zeta_1) \setminus X$. There is a one-dimensional manifold $M$ and a proper holomorphic map $\eta : M \to \Omega \cap (\hat{X} \setminus X)$. Assume that $\pi \circ \varphi_j$ is constant and that $M_0$ is a component of $M$ such that $\eta(M_0)$ contains an open subset of $\varphi_j(\Delta)$. Then necessarily $\eta(M_0) \subset \pi^{-1}(\zeta_1)$, and $\overline{\eta(M_0)}$ is compact. The maximum modulus theorem implies that $\eta(M_0) \subset [\overline{\eta(M_0)} \setminus \eta(M_0)]\hat{}$, and the fact that $\eta$ is a proper map shows that no point of $\overline{\eta(M_0)} \setminus \eta(M_0)$ can lie in $\eta(M)$. Thus, $\overline{\eta(M_0)} \setminus \eta(M_0) \subset X \cap \pi^{-1}(\zeta_0)$. By hypothesis this latter set is finite and so polynomially convex, and we have reached a contradiction. This establishes the desired fact that some neighborhood of $\mathfrak{z}_1$ is carried onto a neighborhood of $\zeta_1$, and, indeed, our argument shows that $\mathfrak{z}_1$ has arbitrarily small neighborhoods with this property.

Since $\hat{X} \cap \pi^{-1}(\Omega_2)$ is a $d$-sheeted branched cover of $\Omega_2$ and since each neighborhood of every $\zeta$ in the interior of $\gamma$ meets $\Omega_2$, we see that if $\zeta$ lies in the interior of $\gamma$, then $\pi^{-1}(\zeta) \cap (\hat{X} \setminus X)$

consists of no more than $d$ points, for otherwise some points of $\Omega_2$ would be covered more than $d$ times. Denote by $d_1$ the integer, $d_1 \leq d$, such that for all $\zeta$ in $\gamma$, $\pi^{-1}(\zeta) \cap (\hat{X} \setminus X)$ has at most $d_1$ points, but such that for some point $\zeta_1$, $\pi^{-1}(\zeta) \cap (\hat{X} \setminus X)$ has $d_1$ points.

By choosing $\zeta_1$ properly and shrinking $\gamma$ to a shorter arc about $\zeta_1$, if necessary, we can arrange that $\pi^{-1}(\gamma) \cap (\hat{X} \setminus X)$ is a union of $d_1$ disjoint arcs on each of which $\pi$ is one to one. Moreover, by hypotheses (b) and (e), we can shrink $\gamma$ further, if necessary, so that $\pi^{-1}(\gamma) \cap X$ consists of a finite number of arcs $\gamma_{d_1+1}, \ldots, \gamma_s$, each of which is carried homeomorphically onto $\gamma$ by $\pi$.

Let $\delta$ be an arc contained in $\Omega_1$ except for its end points, which we require to coincide with the end points of $\gamma$, and denote by $D$ the region bounded by the simple closed curve $\delta \cup \gamma$. We can choose $\delta$ close enough to $\gamma$ that no component of $\pi^{-1}(\bar{D}) \cap \hat{X}$ contains more than one of the arcs which constitute $\pi^{-1}(\gamma) \cap \hat{X}$. Denote by $\mathscr{C}$ the collection of those components $C$ of $\pi^{-1}(\bar{D}) \cap \hat{X}$ with the property that $\pi(C)$ is not contained in $\delta \setminus \gamma$. If $C \in \mathscr{C}$, then $C$ necessarily contains a component of $\pi^{-1}(\gamma) \cap \hat{X}$. In establishing this, it suffices to prove that $\pi(C)$ meets $\gamma$. But if $\pi(C)$, $C \in \mathscr{C}$, does not meet $\gamma$, then since $\pi(C)$ is not contained in $\delta \setminus \gamma$, it follows that there exists a point $z_0 \in D \cap \partial\pi(C)$ and a neighborhood $U$ of $z_0$, $U \subset D$, such that for some function $f$ holomorphic in $U$, $f(z_0) = 1$, $|f(z)| < 1$ if $z \in (\pi(C) \cap U) \setminus \{z_0\}$. But then $f \circ \pi$ is holomorphically approximable in $C \cap \pi^{-1}(U)$ by the algebra $\mathscr{P}(C)$, so $C \cap \pi^{-1}(z_0)$ is a peak set for $\mathscr{P}(C)$ by Theorem 9.3. The set $C$ is a component of the polynomially convex set $\pi^{-1}(\bar{D}) \cap \hat{X}$, so $C \cap \pi^{-1}(z_0)$ is a peak set for $\mathscr{P}(\pi^{-1}(\bar{D}) \cap \hat{X})$, and the local peak set theorem now implies that some element of $\mathscr{P}(\hat{X})$ peaks on a set near $C \cap \pi^{-1}(z_0)$. As this latter phenomenon cannot occur since $z_0 \notin \pi(X)$, we have a contradiction. Thus, if $C \in \mathscr{C}$, $\pi(C)$ meets $\gamma$, so each element of $\mathscr{C}$ contains one (and only one) of the arcs which constitute $\pi^{-1}(\gamma)$. Since $\pi$ is one to one on each arc lying above $\gamma$, Lemma 30.10 is applicable: For every component $C \in \mathscr{C}$, $\pi(C) \subset \gamma \cup \delta$, or else $\pi$ maps $\pi^{-1}(D) \cap C$ one to one onto $D$ and $\pi^{-1}(D) \cap C$ is an analytic set in $\mathbb{C}^N \setminus \pi^{-1}(\gamma \cup \delta)$.

Choose a straight-line segment $\alpha$ in $D$ and let $\Gamma$ be a simple closed curve contained in $\Omega_1$ which contains $\alpha$ as a subarc and which surrounds the point $\zeta_0 = \pi(\mathfrak{z}_0)$. Let $D_0$ be the interior of $\Gamma$, and let $\alpha^0$ be the interior of $\alpha$; i.e., $\alpha^0$ is the set obtained by removing the end points of $\alpha$. The local maximum modulus theorem shows that $\pi^{-1}(\bar{D}_0) \cap \hat{X} = (\pi^{-1}(\Gamma) \cap \hat{X})\hat{}$. The set $\pi^{-1}(\alpha) \cap \hat{X}$ consists of a finite family of analytic arcs, and the set $\pi^{-1}(\Gamma \setminus \alpha^0) \cap \hat{X}$ is polynomially convex. Theorem 30.1 is applicable and shows that $\pi^{-1}(D) \cap \hat{X}$ is a one-dimensional analytic set, so $\hat{X}$ has the structure of a one-dimensional set in a neighborhood of the point $\mathfrak{z}_0$.

There remains one step in the proof. We have to show that if $\mathfrak{z}_0 \in \hat{X} \setminus X$ lies in $\pi^{-1}(J)$, then $\hat{X}$ is a one-dimensional analytic set near $\mathfrak{z}_0$. By hypothesis, the set $\pi^{-1}(\zeta_0) \cap X$, $\zeta_0 = \pi(\mathfrak{z}_0)$, is polynomially convex, and, as $\mathfrak{z}_0 \notin X$, it follows that there is a neighborhood $U$ of $\zeta_0$ such that $\mathfrak{z}_0 \notin (\pi^{-1}(\bar{U}) \cap X)\hat{}$. Since $J$ is compact and totally disconnected, there is a simple closed curve $\Gamma$ contained in $U \setminus J$ which contains $\zeta_0$ in its interior. We can choose $\Gamma$ to be piecewise analytic and to have the further property that $\Gamma \cap \pi(X)$ is a finite set, though the construction of a $\Gamma$ with this latter property, while elementary, requires some care. We shall return to the geometric details of this latter construction after completing the analytic argument.

Let $D$ be the interior of $\Gamma$. The local maximum modulus theorem implies that

$$\pi^{-1}(\bar{D}) \cap \hat{X} = [(\pi^{-1}(\bar{D}) \cap X) \cup (\pi^{-1}(\Gamma) \cap \hat{X})]\hat{}.$$

The set $\pi^{-1}(\Gamma) \cap \hat{X}$ consists of the union of a finite number of smooth curves by the portion of the proof which we have completed, and we have that

$$\mathfrak{z}_0 \notin (\pi^{-1}(\bar{D}) \cap X)\hat{}$$

by choice of $U$. Writing

$$\pi^{-1}(\bar{D}) \cap \hat{X} = [(\pi^{-1}(\bar{D}) \cap X)\hat{} \cup (\pi^{-1}(\Gamma) \cap \hat{X})]\hat{}$$

and invoking Theorem 30.1, we see that near $\mathfrak{z}_0$ the set $\pi^{-1}(\bar{D}) \cap \hat{X}$ is a one-dimensional analytic set. The same is true, therefore, of $\hat{X}$ itself, and the argument is complete except for the geometric detail of the existence of $\Gamma$.

There is a simple closed curve $\Gamma'$ in $U$ which surrounds $\zeta_0$ and which is disjoint from $J$. Let $\Gamma''$ be a simple closed curve contained in the interior of $\Gamma'$ which has the same properties and which, together with $\Gamma$, bounds an annular domain $R$ containing no point of $J$. Of the open arcs and simple closed curves which constitute $\pi(X) \setminus J$, only finitely many, say $L_1, \ldots, L_t$, meet $\bar{R}$. Let us say that an arc $\lambda$ is a *cross cut* if it is contained in one of the $L_j$ and if, moreover, its interior lies in $R$ and one of its end points lies in $\Gamma'$, the other in $\Gamma''$. There are only finitely many cross cuts. To see this, let $\gamma$ be a simple closed curve which is contained in $R$ and which divides $R$ into two annular regions: $\gamma$ separates $\Gamma'$ from $\Gamma''$. The curve $\gamma$ intersects every cross cut, so if there are infinitely many, there exists a point $z \in \gamma$ which is a limit point of a sequence of distinct cross cuts. This, taken with the fact that, except possibly for end points, the cross cuts are disjoint and connect $\Gamma'$ to $\Gamma''$, contradicts the local connectedness of $\pi(X)$ at $z$.

Let the finitely many cross cuts be $C_1, \ldots, C_g$, where we take the indices to be integers mod $g$. Let $C_j$ have end points $x'_j$ and $x''_j$ in $\Gamma'$ and $\Gamma''$, respectively. We assume the indexing to be such that the order on $\{x'_1, \ldots, x'_g\}$ induced by the indices is the same as the cyclic order induced from $\Gamma'$. [Note that for some choices of $j$, $x'_j = x'_{j+1}$ is possible, but if $x'_j = x'_{j+1}$, then since $\pi(X) \setminus J$ is a disjoint union of arcs, $x'_{j-1} \neq x'_j$ and $x'_{j+2} \neq x'_j$, unless, of course, $j - 1 = j + 1$ and $j + 2 = j$, a case that arises if $g = 2$.] It follows that the order on $\{x''_1, \ldots, x''_g\}$ is that induced by the order on $\Gamma'$.

The cross cuts $C_j$ divide $R$ into $g$ open cells $\Delta_1, \ldots, \Delta_g$, where the labeling is such that $\partial\Delta_j \supset C_j \cup C_{j+1}$. Each $\Delta_j$ is a (possibly degenerate) curvilinear rectangle. Let $\beta'_j = \partial\Delta_j \cap \Gamma'$, $\beta''_j = \partial\Delta_j \cap \Gamma''$. By construction, no component of $\pi(X) \setminus J$ contains an arc $\mu$ with end points in $\beta'_j$ and $\beta''_j$ and interior contained in $\Delta_j$: Such a $\mu$ would be a cross cut, but the only cross cuts are the $C_j$, and they are disjoint from all the $\Delta_j$.

The set $\Delta_j$ is simply connected with boundary the simple closed curve $C_j \cup C_{j+1} \cup \beta'_j \cup \beta''_j$, and its intersection with $\pi(X)$ consists of a union of open arcs $\lambda$ with the property that the end points of each $\lambda$ both lie in $\beta'_j$ or else in $\beta''_j$.

It follows that there is an arc $v_1$ which connects a point $p_1$ in the interior of $C_1$ to a point $p_2$ in the interior of $C_2$, which is contained in $\Delta_1 \cup \{p_1, p_2\}$ and which, moreover, meets $\pi(X)$ only at $p_1$ and $p_2$. We can choose $v_1$ to be analytic at every point of its interior though not necessarily at its end points. Next we find an arc $v_2$ connecting $p_2$ to a point $p_3$ in the interior of $C_3$, $v_3 \subset (\Delta_2 \setminus \pi(X)) \cup \{p_2, p_3\}$. Iterating this process we find $g$ arcs $v_1, \ldots, v_g$ whose union is a simple closed curve $\Gamma$ with the properties we seek. This completes the proof of Lemma 30.21.

We now proceed to some consequences of the lemma just proved.

30.22 ***Corollary***

*If $X \subset \mathbb{C}^{\Lambda}$ is a finite union of arcs $\lambda_1, \ldots, \lambda_q$ such that the coordinate function $\pi$ is locally one to one on each $\lambda_j$, then $\hat{X} \setminus X$ consists of a possibly empty, purely one-dimensional analytic subset of $\mathbb{C}^{\Lambda} \setminus X$.*

PROOF

Since $\pi$ is locally one to one on each $\lambda_j$, we may suppose, by subdividing the $\lambda_j$'s as necessary, that $\pi$ is actually one to one on each $\lambda_j$. Thus, each of the sets $L_j = \pi(\lambda_j)$ is an arc. For $i \neq j$, let $J'_{ij}$ be the boundary of $L_i \cap L_j$ in $L_i$ and let $J'_{ii}$ be the set of end points of $L_i$. The sets $J'_{ij}$ are totally disconnected and so is their union, $J'$. Let $J$ be any compact, totally disconnected subset of $\mathbb{C}$ which contains $J'$. The set $\pi(X)$ has at every point of $\pi(X) \setminus J$ the structure of an arc. Let $E$ be the set of points in $\mathbb{C} \setminus J$ which can be joined to infinity by an arc which meets $\pi(X) \setminus J$ only finitely often. The set $E$ is plainly open since $\pi(X) \setminus J$ is locally an arc, and it is closed in $\mathbb{C} \setminus J$. If $z_0 \in \mathbb{C} \setminus J$ is a limit point $E$, then there is some disc $V$ about $z_0$ which is disjoint from $J$ and which meets $\pi(X)$ only in an arc, if at all. The set $V$ contains a point $z_1$ of $E$, and $z_1$ can be joined to $z_0$ by an arc which meets $\pi(X)$ at most twice. Thus, $z_0 \in E$, so $E$ is closed in $\mathbb{C} \setminus J$. As $J$ is totally disconnected, $\mathbb{C} \setminus J$ is connected, and so $E = \mathbb{C} \setminus J$: Every point of $\mathbb{C} \setminus J$ can be joined to infinity by an arc in $\mathbb{C} \setminus J$ which meets $\pi(X)$ in at most a finite set.

For $i \neq j$, let $K_{ij}$ be the boundary of $\lambda_i \cap \lambda_j$ in $\lambda_i$, and let $K_{ii}$ be the set of end points of $\lambda_i$. For fixed $i$, $\bigcup_j K_{ij}$ is a compact and totally disconnected subset of $\lambda_i$, and as $\pi$ is one to one, $\pi(\bigcup_i K_{ij})$ is compact and totally disconnected. Let $J'$ be as in the last paragraph, and let $J = J' \cup \bigcup_{i,j} \pi(K_{ij})$. With this $J$, we can apply Lemma 30.21 to find that $\hat{X} \setminus X$ is either empty or a one-dimensional analytic set.

30.23 ***Theorem***

*If $X \subset \mathbb{C}^N$ is an arc on which the coordinate projection $\pi$ is locally one to one, then $X$ is polynomially convex and $\mathscr{P}(X) = \mathscr{C}(X)$.*

This theorem admits a reformulation in terms of uniform algebras:

30.24 ***Corollary***

*If $A$ is a uniform algebra on the interval $[0, 1]$ which contains an element which is locally one to one, then $A = \mathscr{C}([0, 1])$.*

PROOF OF THE THEOREM

Suppose, to begin with, that $X$ is polynomially convex. Then $\pi(X)$ is the union of a finite family of arcs, so by Theorem 26.4, $\mathscr{R}(\pi(X)) = \mathscr{C}(\pi(X))$. Consequently, every maximal antisymmetric set for $\mathscr{P}(X)$ is contained in one of the fibers $\pi^{-1}(\pi(\mathfrak{z}))$, $\mathfrak{z} \in X$. As maximal sets of antisymmetry for algebras on their spectra—and here we use again the assumed polynomial convexity of $X$—are connected, and as $\pi^{-1}(\pi(\mathfrak{z}))$ is finite, we see that the maximal antisymmetric sets for $\mathscr{P}(X)$ are points. Thus, $\mathscr{P}(X) = \mathscr{C}(X)$.

To prove that $X$ is polynomially convex, suppose it is not so that Corollary 30.22 implies that $\hat{X} \setminus X$ is a one-dimensional analytic set, $V$. Let $\mathfrak{z}_0 \in \hat{X} \setminus X$ so that some neighborhood $N$ of $\mathfrak{z}_0$ is of the form $\varphi_1(\Delta) \cup \cdots \cup \varphi_n(\Delta)$, in accordance with the definition of one-dimensional analytic set. Then, as in the proof of Lemma 30.21, $\pi$ is constant on none of the sets $\varphi_j(\Delta)$, and so it maps $N$ onto a neighborhood of $\pi(\mathfrak{z}_0)$ in $\mathbb{C}$. The set $\pi(X)$ is nowhere dense since it is a finite union of arcs, so there is a point $\alpha \in \pi(\hat{X}) \setminus \pi(X)$. The function $\pi - \alpha$ is zero free on $X$ and so has a holomorphic logarithm on a neighborhood of $X$. However, this is impossible because $\pi - \alpha$ is holomorphic on $V$ and has a zero there, and $\partial V \subset X$. (Recall Lemma 29.36.) This concludes the proof.

There is a corresponding result for algebras on circles, though in this case we have to assume a priori that the circle is polynomially convex.

30.25 ***Theorem***

*If $X \subset \mathbb{C}^{\Lambda}$ is a simple closed curve on which one of the coordinate functions is locally one to one, and if $X$ is polynomially convex, then $\mathscr{P}(X) = \mathscr{C}(X)$.*

PROOF

This is a consequence of Corollary 30.22, as in the first half of the last proof.

As with Theorem 30.23, this result can be reformulated in terms of uniform algebras on circles.

The next result is a minor generalization of one proved by Alexander. It is most conveniently formulated in terms of uniform algebras on $I$, the closed unit interval.

30.26 ***Theorem***

*Let $(A, I)$ be a uniform algebra such that for some closed set $E \subset I$ and some $f \in A$, $f|E$ is constant, $A|E$ is dense in $\mathscr{C}(E)$, and $f$ separates each pair of points of $I$ not both of which lie in $E$. Then $A = \mathscr{C}(I)$.*

PROOF

We assume that $f$ vanishes identically on $E$. It follows then that $f(I)$ consists of a countable family of simple closed curves together with not more than two arcs which are pairwise disjoint except for the origin which is common to them all. Moreover, they shrink to the origin.

Let $\{g_\alpha\}_{\alpha\in\Lambda} \cup \{f\}$ be a set of generators for $A$ and define $\Phi: I \to \mathbb{C}^\Lambda \times \mathbb{C}$ by $\Phi(x) = (G(x), f(x))$, where $G(x)$ is the point of $\mathbb{C}^\Lambda$ with $\alpha$th coordinate $g_\alpha(x)$. Let $\pi: \mathbb{C}^\Lambda \times \mathbb{C} \to \mathbb{C}$ be the projection such that $\pi \circ \Phi(x) = f(x)$. If $X = \Phi(I)$, then $X$ is an arc, and $\pi(X) = f(I)$. The projection $\pi$ is one to one on $\pi^{-1}(\pi(X) \setminus \{0\})$, and since $A|E$ is dense in $\mathscr{C}(E)$, $\pi^{-1}(0) = X \cap (C^\Lambda \times \{0\})$ is a polynomially convex set. Lemma 30.21 may be invoked; it implies that either $X$ is polynomially convex or else that $\hat{X} \setminus X$ is a nonempty one-dimensional analytic set. The latter case may be excluded on the basis of the argument principle as in the proof of Corollary 30.24, so $X$ is polynomially convex. By Theorem 26.4, $\mathscr{R}(\pi(X)) = \mathscr{C}(\pi(X))$, so every maximal antisymmetric set for $\mathscr{P}(X)$ is contained in a set of the form $\pi^{-1}(\pi(\mathfrak{z}))$, $\mathfrak{z} \in X$. If $\pi(\mathfrak{z}) \neq 0$, the hypotheses on $f$ show that $\pi^{-1}(\pi(\mathfrak{z})) \cap X$ is a singleton. Also, since $\mathscr{R}(\pi(X)) = \mathscr{C}(\pi(X))$, the origin is a peak point for $\mathscr{R}(\pi(X))$, whence $\Phi(E) = \pi^{-1}(0) \cap X$ is one for $\mathscr{P}(X)$; and as $\mathscr{P}(X)|\Phi(E)$ is dense in $\mathscr{C}(\Phi(E))$, we find that $\mathscr{P}(X)|\Phi(E) = \mathscr{C}(\Phi(E))$. We may conclude that every maximal set of antisymmetry for $\mathscr{P}(X)$ is a point. Thus, $\mathscr{P}(X) = \mathscr{C}(X)$, whence $A = \mathscr{C}(I)$, as we were to prove.

**30.27** ***Corollary***

*If $A$ is a uniform algebra on $I$ generated by two functions $f$ and $g$, if $E \subset I$ is a closed set on which $f$ vanishes, and if $f$ separates every pair of points not both of which lie in $E$, then $A = \mathscr{C}(I)$.*

PROOF

We see that $g$ is one to one on $E$, so by Mergelyan's theorem, $\mathscr{P}(g(E)) = \mathscr{C}(g(E))$, so $A|E$ is dense in $\mathscr{C}(E)$. The corollary now follows from the theorem just proved.

Alexander also proved the following theorem.

**30.28** ***Theorem***

*Let $f_1, \ldots, f_n \in \mathscr{C}(I)$, $I = [0, 1]$, and let $f_1, \ldots, f_{n-1}$ be piecewise smooth. If $E$ is a closed set in $I$ on which $f_1, \ldots, f_n$ separate points, then the set $X = \{(f_1(x), \ldots, f_n(x)) : x \in E\}$ is polynomially convex and $\mathscr{P}(X) = \mathscr{C}(X)$.*

Of course, the case of primary interest is that in which $E = I$, but it is convenient, for the sake of induction, to prove the general result.

The proof of this theorem depends on Lemma 30.21, but to see that we can invoke that lemma, we need a preliminary geometric construction.

30.29 **Lemma**

*Let $f: I \to \mathbb{C}$ be piecewise smooth, let $E \subset I$ be a closed set, and set $F = f(E)$. There exists a compact, totally disconnected set $J$ in $F$ with the following properties:*

(a) *At each point of $F \setminus J$, $F$ has the structure of an open arc.*
(b) *If $C \subset F \setminus J$ is compact, then $f$ maps $f^{-1}(C) \cap E$ in a finite-to-one way onto $C$.*
(c) *Every point of $\mathbb{C} \setminus J$ can be connected to infinity by an arc in $\mathbb{C} \setminus J$ which meets $F$ only finitely often.*

PROOF

Note first that if $Q$ is the critical set of $f$, then $f(Q)$ is a totally disconnected, compact set in $\mathbb{C}$, for if $f(Q)$ is not totally disconnected, one of the sets $(\text{Re} f)(Q)$, $(\text{Im} f)(Q)$ contains an interval, and we have a contradiction to Lemma 30.13.

If $B$ is the boundary of $(I \setminus Q) \cap E$ in $I$, then $B$ is compact and totally disconnected; and it follows that $f(B)$ is totally disconnected since $B$ is the union of $B \cap Q$ and a countable number of compact, totally disconnected sets on which $f$ is one to one. The set $f(Q)$ is totally disconnected, so the same is true of $J_1 = f(B) \cup f(E \cap Q)$. The set $E \setminus f^{-1}(J_1)$ consists of a (possibly empty) countable union of pairwise disjoint open arcs $\lambda_n$ on each of which $f$ is locally one to one.

If we consider a compact set $C \subset F = f(E)$, which is disjoint from $J_1$, then $f^{-1}(C) \cap E$ is compact, and therefore it can meet only finitely many of the arcs $\lambda_n$. Since $f$ is locally one to one on each $\lambda_n$, it follows that $C$ is contained in a finite union of arcs and that no point of $C$ has more than $q$ preimages in $E$ under the map $f$ for some integer $q$. Further, if we choose for $C$ a compact neighborhood of a point $\zeta \in F \setminus J_1$, then since $C = L_1 \cup \cdots \cup L_p$, each $L_j$ an arc, it follows, as in our proof of Corollary 30.22, that we can write $C = J_C \cup L_C$, where $J_C$ is totally disconnected and $L_C$ is a set that locally has the structure of an open arc. It follows that if we denote by $J_2$ the set of points at which $F$ fails to have the structure of an arc, then the set $J = J_1 \cup J_2$ is totally disconnected and has the properties we seek.

That $J$ enjoys property (c) follows by a connectedness argument. Since $J$ is totally disconnected, $\mathbb{C} \setminus J$ is connected. Denote by $S$ the subset of $\mathbb{C} \setminus J$ consisting of those points which can be connected to infinity by an arc in $\mathbb{C} \setminus J$ which meets $F$ only finitely often. The set $S$ is open. To prove this, let $z \in S$, and let $D$ be an open disc about $z$ which misses $J$. If $z \notin F$, we can choose $D$ to miss $F$, and then it is contained in $S$. If $z \in F$, let $L$ be the component of $F \cap D$ which contains $z$ so that $L$ is an open arc. The component $W$ of $D \setminus (F \setminus L)$ which contains $L$ is a neighborhood of $z$ contained in $S$. Thus, $S$ is open. In the same way it is closed, for if $z \in \mathbb{C} \setminus J$ but $z \notin S$, there is a neighborhood of $z$ which misses $J \cup S$. As $S$ is not empty, $S = \mathbb{C} \setminus J$, as we desire.

We now establish the theorem.

PROOF OF THE THEOREM 30.28
By Mergelyan's theorem, the theorem is true if $n = 1$. For the purpose of induction, assume that the theorem is true with $n$ replaced by $n - 1$. Let $\pi: \mathbb{C}^n \to \mathbb{C}$ be the first coordinate projection: $\pi(z_1, \ldots, z_n) = z_1$.

The first step is to show that $\pi(X) = \pi(\hat{X})$. We have that $\pi(X) = f_1(E)$; let $J$ be the set associated with $f_1$ by the lemma just proved. If $\zeta_0 \in J$, the set $\pi^{-1}(\zeta_0) \cap X$ is polynomially convex, for if

$$E_0 = \{x \in E : f_1(x) = \zeta_0\},$$

then the map

$$x \mapsto (\zeta_0, f_2(x), \ldots, f_n(x)), \qquad x \in E_0$$

is a map from $E_1$ to $\mathbb{C}^{n-1} = \{\zeta_0\} \times \mathbb{C}^{n-1} \subset \mathbb{C}^n$ of the kind contemplated in the theorem; so by the induction hypothesis, this set, which is $\pi^{-1}(\zeta_0) \cap X$, is polynomially convex. Thus, Lemma 30.21 applies: $\hat{X} \setminus X$ is a one-dimensional analytic set. If $\pi(\hat{X}) \neq f_1(E)$, then for any $\alpha \in \pi(\hat{X}) \setminus f_1(E)$, $\pi - \alpha$ has a holomorphic logarithm on $X$ but $\pi - \alpha$ vanishes at a point in $\hat{X}$. We have reached a contradiction to the argument principle, so $\pi(\hat{X}) = f_1(E)$.

Since $f_1$ is piecewise smooth, $f_1(E)$ has zero area, so, by Corollary 26.9, $\mathscr{R}(f_1(E)) = \mathscr{C}(f_1(E))$, and it follows that every set of antisymmetry for $\mathscr{P}(X)$ is contained in a set $X_{\zeta_0} = X \cap \pi^{-1}(\zeta_0)$ for some $\zeta_0 \in f_1(E)$. However, the induction hypothesis is applicable to the sets

$$X_{\zeta_0} : \mathscr{P}(X_{\zeta_0}) = \mathscr{C}(X_{\zeta_0}),$$

and the theorem is proved.

Using Alexander's results, it is possible to prove a theorem which gives a geometric condition on a simple closed curve $\gamma$ in $\mathbb{C}^\Lambda$ sufficient for $\mathscr{P}(\gamma) = \mathscr{C}(\gamma)$.

30.30 ***Theorem***
*Let $\gamma \subset \mathbb{C}^\Lambda$ be a simple closed curve with the property that one of the coordinate projections, $\pi$, of $\mathbb{C}^\Lambda$ is locally one to one on $\gamma$. If, in addition,*

(a) *$\pi(\gamma)$ is not a simple closed curve, and*

(b) *$\pi(\gamma)$ is the boundary of the unbounded component of $\mathbb{C} \setminus \pi(\gamma)$, then $\gamma$ is polynomially convex, and $\mathscr{P}(\gamma) = \mathscr{C}(\gamma)$.*

A theorem of this general kind was established by L. A. Markushevich [1] under somewhat stronger hypotheses. He assumed that the set of pairs of points $(\mathfrak{z}, \mathfrak{z}')$, $\mathfrak{z}, \mathfrak{z}' \in \gamma$, such that $\pi(\mathfrak{z}) = \pi(\mathfrak{z}')$ is finite, and that the curve $\pi(\gamma)$ is rectifiable. The hypothesis of rectifiability was made so that the theory of the boundary behavior of integrals of Cauchy type could be used.

PROOF OF THE THEOREM

Consider first the case that $\pi(\gamma)$ does not separate the plane. Since $\pi$ is locally one to one, $\pi(\gamma)$ is a finite union of arcs and so has no interior. Mergelyan's theorem implies that $\mathscr{P}(\pi(\gamma)) = \mathscr{C}(\pi(\gamma))$, and therefore every maximal set of antisymmetry for the algebra $\mathscr{P}(\gamma)$ is contained in the finite set $\pi^{-1}(\zeta)$, $\zeta \in \pi(\gamma)$. It follows that $\mathscr{P}(\gamma) = \mathscr{C}(\gamma)$ and that $\gamma$ is polynomially convex.

We now consider the case that $\pi(\gamma)$ does separate the plane. The main difficulty in this case, granted the theory we have developed so far, is a topological one. We need to use the following fact:

30.31 ***Lemma***

*There is a point $\zeta_0 \in \pi(\gamma)$ such that $\pi(\gamma) \setminus \{\zeta_0\}$ is not connected.*

This seems to be a perfectly obvious point, but, as is the case with many points in plane topology, its proof requires a certain amount of care. We shall postpone its proof for the moment and establish the theorem.

Let $\pi(\gamma) \setminus \{\zeta_0\} = E_1' \cup E_2'$, a disjoint union of nonempty sets, and let $E_1 = E_1' \cup \{\zeta_0\}$, $E_2 = E_2' \cup \{\zeta_0\}$. We shall show that $\hat{E}_1 \cap \hat{E}_2 = \{\zeta_0\}$. As a first step in this direction, we show that $E_1 \cap \hat{E}_2 = \{\zeta_0\}$. If this is false, let $q \in E_1 \cap \hat{E}_2 \setminus E_2$. Since $q$ lies in $E_1$ and in a bounded component of $\mathbb{C} \setminus E_2$, say $\Omega$, we have a contradiction to hypothesis (b), for $\Omega$ is also a bounded component of $\mathbb{C} \setminus \pi(\gamma)$. Thus, $E_1 \cap \hat{E}_2 = \{\zeta_0\}$, and, similarly, $\hat{E}_1 \cap E_2 = \{\zeta_0\}$. Consequently,

$$(\hat{E}_1 \cap \hat{E}_2) \setminus \{\zeta_0\} = (\hat{E}_1 \setminus E_1) \cap (\hat{E}_2 \setminus E_2).$$

Let $p$ be a point of this latter set, and let $\Omega_1$ be the component of $\hat{E}_1 \setminus E_1$ which contains $p$, $\Omega_2$ the component of $\hat{E}_2 \setminus E_2$ which contains $p$ so that $\Omega_1 \cap \Omega_2 \neq \varnothing$. As $\partial\Omega_1 \subset E_1 \subset \pi(\gamma)$ and $\Omega_2$ is a bounded component of $\mathbb{C} \setminus \pi(\gamma)$, it follows that $\partial\Omega_1 \cap \Omega_2 = \varnothing$, so, by connectedness, $\Omega_2 \subset \Omega_1$. By symmetry, $\Omega_1 \subset \Omega_2$, so $\Omega_1 = \Omega_2$, and thus $\partial\Omega_1 = \partial\Omega_2$. Now $\partial\Omega_1 \subset E_1$, $\partial\Omega_2 \subset E_2$, and $E_1 \cap E_2 = \{\zeta_0\}$, so we have a contradiction, for the boundary of a bounded connected open set in $\mathbb{C}$ cannot consist of a single point. Thus, $\hat{E}_1 \cap \hat{E}_2 = \{\zeta_0\}$, as we were to prove.

As $\gamma$ is a simple closed curve, there is a short open arc $\lambda$ in $\gamma$ such that $\pi(\lambda) \cap E_2 = \varnothing$. The set $\gamma \setminus \lambda$ is an arc, $\Lambda$, to which Alexander's results apply: $\mathscr{P}(\Lambda) = \mathscr{C}(\Lambda)$. Consequently, e.g., by Tietze's extension theorem, $\mathscr{P}(\gamma \cap \pi^{-1}(E_2)) = \mathscr{C}(\gamma \cap \pi^{-1}(E_2))$, and the set $\gamma \cap \pi^{-1}(E_2)$ is polynomially convex. Similarly, the set $\gamma \cap \pi^{-1}(E_1)$ is polynomially convex. Thus, Lemma 29.21(a) implies that $\gamma$ itself is polynomially convex.

Since $\pi(\gamma)$ is a finite union of arcs, $\mathscr{R}(\pi(\gamma)) = \mathscr{C}(\pi(\gamma))$ by Theorem 26.4, so the maximal sets of antisymmetry for $\mathscr{R}(\gamma)$ lie in the finite sets $\pi^{-1}(\zeta) \cap \gamma$, $\zeta \in \pi(\gamma)$. It follows that $\mathscr{R}(\gamma) = \mathscr{C}(\gamma)$. Since $\gamma$ is polynomially convex, the Oka approximation theorem (Theorem 29.1) shows that $\mathscr{C}(\gamma) = \mathscr{P}(\gamma)$ and our theorem is proved.

PROOF OF THE LEMMA

We have to prove that the set $\pi(\gamma)$ has a cut point. If it does not, then by the cyclic connectivity theorem (Appendix A), every pair of points of $\pi(\gamma)$ lies in a simple closed curve contained in $\pi(\gamma)$. Let $\Omega_0$ be a bounded component of $\mathbb{C} \setminus \pi(\gamma)$ and let $\zeta_1, \zeta_2 \in \partial\Omega_0$. Denote by $\Lambda$ a simple closed curve in $\pi(\gamma)$ which contains $\zeta_1$ and $\zeta_2$. Then int $\Lambda$, the interior of $\Lambda$, is disjoint from $\pi(\gamma)$ since $\pi(\gamma)$ is the boundary of the unbounded component of $\mathbb{C} \setminus \pi(\gamma)$, and thus int $\Lambda$ is a bounded component of $\mathbb{C} \setminus \pi(\gamma)$.

Suppose that $\Lambda \neq \partial\Omega_0$ so that there is $\zeta_0 \in \partial\Omega_0 \setminus \Lambda$. There is an arc $\tilde{\lambda}_1$ in $\pi(\gamma)$ with one end point $\zeta_0$ and one end point in $\Lambda$. Let $\tilde{\lambda}_1$ be the first point of $\tilde{\lambda}_1$ (starting from $\zeta_0$) at which $\tilde{\lambda}_1$ meets $\Lambda$, and let $\lambda_1$ be the subarc of $\tilde{\lambda}_1$ connecting $\zeta_0$ and $\xi_1$. Thus, $\lambda_1 \cap \Lambda = \{\xi_1\}$. By hypothesis, $\pi(\gamma)$ has no cut point, so there is an arc $\tilde{\lambda}_2$ from $\zeta_0$ to $\Lambda$ which lies entirely in $\pi(\gamma) \setminus \{\xi_1\}$, and we can find an arc $\lambda_2 \subset \tilde{\lambda}_2$ which connects $\zeta_1$ to $\Lambda$ and which meets $\Lambda$ at only one point, $\{\xi_2\}$. Then, in $\lambda_1 \cup \lambda_2$ there is an arc, $\lambda$, from $\xi_1$ to $\xi_2$. The points $\xi_1$ and $\xi_2$ determine two arcs $\Lambda^+$ and $\Lambda^-$ in $\Lambda$ such that $\Lambda = \Lambda^+ \cup \Lambda^-$, $\Lambda^+ \cap \Lambda^- = \{\xi_1, \xi_2\}$. The sets $\Lambda^+ \cup \lambda$ and $\Lambda^- \cup \lambda$ are both simple closed curves, and if the labeling is chosen correctly, the interior of $\Lambda^+ \cup \lambda$ will contain $\Lambda^- \setminus \{\xi_1, \xi_2\}$. This is impossible, though, for no point of the interior of $\Lambda^+ \cup \lambda$ is in the boundary of the unbounded component of $\mathbb{C} \setminus \pi(\gamma)$. Thus, $\Lambda = \partial\Omega_0$.

We have assumed that $\pi(\gamma)$ is not a simple closed curve, so there is a point $p \in \pi(\gamma) \setminus \partial\Omega_0$, and an arc $L$ from $p$ to $\partial\Omega_0$ in $\pi(\gamma)$. Let $p_0$ be the first point of $L$ contained in $\partial\Omega_0$. The point $p_0$ is a cut point of $\pi(\gamma)$. This assertion follows by a repetition of the argument just given:

If $p_0$ is not a cut point, we can find a simple closed curve $C$ in $\pi(\gamma)$ which contains an open subarc of $\partial\Omega_0$ in its interior. This completes the proof of the lemma.

Our final result is an extension of a theorem of Cirka [1]. In its statement and proof we shall use concepts from dimension theory, but the reader who is unfamiliar with this theory will obtain an intelligible statement (and proof) if he assumes throughout that the set $E$ of the theorem is an arc in $\mathbb{C}^{N+1}$.

**30.32** ***Theorem***

*Let $E \subset \mathbb{C}^{N+1}$ be a compact, locally connected set of dimension one such that $H^1(E, \mathbb{Z}) = 0$, and such that if $\pi_j:\mathbb{C}^{N+1} \to \mathbb{C}$ is given by*

$$\pi_j(z_1, \ldots, z_{N+1}) = z_j,$$

*then for $j = 1, \ldots, N$, $\pi_j(E)$ is nowhere dense in $\mathbb{C}$. Then every continuous function on $E$ can be approximated uniformly by functions holomorphic on a neighborhood of $E$.*

The theorem given by Cirka is the case that $E$ is an arc and $N = 1$.

PROOF

The set $E$ is locally connected and compact, so it has only finitely many components. Thus, we may assume it connected so that it is a Peano continuum. (See Appendix A.)

We are to prove that if $A$ is the uniform closure in $\mathscr{C}(E)$ of the set $\{f|E : f \text{ is holomorphic in some neighborhood of } E\}$, then $A = \mathscr{C}(E)$.

Let $A_0$ be the subalgebra of $\mathscr{C}(E)$ generated algebraically by the coordinate function $\pi_1$. Generally, $A_0$ will not be closed in $A$; it will be closed if and only if $\pi_1(E)$ is a finite set, a case which can occur under the stated hypotheses. Let $A_1$ be the integral closure of $A_0$ in $A$ so that $f \in A$ lies in $A_1$ if and only if it satisfies an equation of the form

$$f^\mu + g_1 f^{\mu-1} + \cdots + g_{\mu-1} f + g_\mu = 0 \tag{5}$$

with coefficients $g_j \in A_0$. It is an elementary algebraic fact that $A_0$ is an algebra (Zariski and Samuel [1]).

**30.33** ***Lemma***

*If $f \in A_1$, then $f(E)$ is nowhere dense.*

PROOF

By hypothesis, $f$ satisfies a monic polynomial equation of the form (5). The coefficients $g_j$ lie in $A_0$ and so are polynomials, so we can define a polynomial, $P$, in two indeterminants by

$$P(X_1, X_2) = X_2^\mu + \sum_{j=1}^{\mu-1} g_j(X_j)X_2^j.$$

Let $V$ be the variety

$$\{(\zeta_1, \zeta_2) \in \mathbb{C}^2 : P(\zeta_1, \zeta_2) = 0\}.$$

The map $x \mapsto (\pi_1(x), f(x))$ carries $E$ onto a set $F \subset V$, and $F$ has the property that its projection into the $\zeta_1$-plane is $\pi_1(E)$, a nowhere dense set, and its projection into the $\zeta_2$-plane is $f(E)$.

There is a positive integer $\nu \leq \mu$ and a discrete set $S_0$ in the $\zeta_1$-plane with the property that $\mathbb{C} \setminus S_0$ can be written as a countable union of *closed* discs, $D_j$, such that for each $j$, there exist $\nu$ functions $\varphi_{j,1}, \ldots, \varphi_{j,\nu}$ holomorphic and one to one on a neighborhood of $D_j$ with the property that

$$V \setminus \pi_1^{-1}(S_0) = \bigcup_{j,k} \{(\zeta, \varphi_{j,k}(\zeta)) : \zeta \in D_j\}.$$

Consequently, the range of $f$, $f(E)$, with the exception of the countable set

$$\{f(x) : x \in \pi_1^{-1}(S_0)\}$$

is contained in

$$\bigcup_{j,k} \{\varphi_{j,k}(\pi_1(E) \cap D_j)\}.$$

The functions $\varphi_{j,k}$ are all one to one on $D_j$ and the set $\pi_1(E) \cap D_j$ is nowhere dense, so it follows that $\varphi_{j,k}(\pi_1(E) \cap D_j)$ is nowhere dense. Thus, the set $f(E)$ is of the first category in $\mathbb{C}$, and, as it is closed, it contains no open set; i.e., it is nowhere dense. This finishes the proof of the lemma.

It now follows that $\bar{A}_1$, the uniform closure of $A_1$, is closed under the extraction of square roots in the sense that every $f \in \bar{A}_1$ has, in $\bar{A}_1$, a square root. To prove this, consider first the case of an $f \in A_1$. The function $f$ has a square root in $A$, for since the range of $f$ is nowhere dense, $f$ is the limit of a sequence $\{f_n\}$ of zero-free elements of the algebra $A$, and we can even assume that each $f_n$ is the restriction to $E$ of a function holomorphic on some neighborhood of $E$. Since $H^1(E, \mathbb{Z}) = 0$, it follows that each $f_n$ has a holomorphic logarithm, and, *a fortiori*, a holomorphic square root on some neighborhood of $E$, so $f_n = h_n^2$ for some $h_n \in A$. It follows now, from Lemma 13.16, that $f$ has a square root, $h$, in $A$. Since $A_1$ is the integral closure in $A$ of $A_0$, $A_1$ is integrally closed in $A$, so $h \in A_1$. Thus, every element of $A_1$ has a square root in $A_1$, so, again by Lemma 13.16, every element of $\bar{A}_1$ has a square root in $\bar{A}_1$.

We would like to invoke Cirka's theorem (Theorem 13.15) to conclude that $\bar{A}_1 = \mathscr{C}(E)$, but we do not know that $A_1$ separates points on $E$. We circumvent this difficulty as follows: Let $\Theta: E \to \mathbb{C}^{A_1}$ be given by requiring that if $f \in A_1$, then the $f$th coordinate of $\Theta(x)$ be $f(x)$. We can identify $\bar{A}_1$ with $\mathscr{P}(\Theta(E))$, and since the continuous image of a Peano continuum is again a Peano continuum, $\Theta(E)$ is locally connected, so Cirka's theorem implies that $\mathscr{P}(\Theta(E)) = \mathscr{C}(\Theta(E))$. Consequently, every set of antisymmetry for $A$ is contained in one of the fibers $\Theta^{-1}(\mathfrak{z})$, $\mathfrak{z} \in \Theta(E)$. The algebra $A_1$ contains $\pi_1$, so it follows that every set of antisymmetry for $A$ is contained in a fiber $\pi_1^{-1}(\zeta)$ for some $\zeta \in \pi_1(E)$.

We can treat in the same way the projections $\pi_2, \ldots, \pi_n$, and we finally reach the conclusion that if $P$ is a set of antisymmetry for $A$, then $P$ is contained in a set of the form

$$E_{\mathfrak{z}} = \Pi_{\mathfrak{z}} \cap E,$$

where $\Pi_{\mathfrak{z}}$ is the plane

$$\{(z_1, \ldots, z_N, \zeta) : \zeta \in \mathbb{C}\}$$

for fixed $\mathfrak{z} = (z_1, \cdots, z_N) \in \mathbb{C}^N$. Thus, if $E$ has no interior relative to $\Pi_{\mathfrak{z}}$ and does not separate $\Pi_{\mathfrak{z}}$, Mergelyan's theorem shows that polynomials in $z_{N+1}$, certainly elements of $A$, are dense in $\mathscr{C}(E_{\mathfrak{z}})$, so the theorem follows.

If $E$ is an arc, then $E_{\mathfrak{z}}$ has the desired properties, for then it is a subset of an arc and so cannot separate $\Pi_{\mathfrak{z}}$ and has no interior relative to $\Pi_{\mathfrak{z}}$. In the general case, we have to use the hypotheses that $\dim E = 1$ and $H^1(E, \mathbb{Z}) = 0$ to prove that $E_{\mathfrak{z}}$ has the properties we seek. Since $\dim E = 1$ and $E_{\mathfrak{z}} \subset E$, it follows that $\dim E_{\mathfrak{z}} \leq 1$, so (Hurewicz–Wallman [1], IV.3 and III.1) $E_{\mathfrak{z}}$ has no interior in $\Pi_{\mathfrak{z}}$. Also, since $\dim E = 1$ and $H^1(E, \mathbb{Z}) = 0$, it follows that $H^1(E_{\mathfrak{z}}, \mathbb{Z}) = 0$, for by Hurewicz–Wallman VI.4, every zero-free continuous complex-valued $f$ on $E_{\mathfrak{z}}$ extends to a zero-free continuous $F$ on $E$. As $H^1(E_{\mathfrak{z}}, \mathbb{Z}) = 0$, $F$ is an exponential and hence $f$ is an exponential. Now use the $\mathscr{C}(X)$ case of the Arens–Royden theorem. Finally, since $H^1(E_{\mathfrak{z}}, \mathbb{Z}) = 0$, $E_{\mathfrak{z}}$ does not separate the plane $\Pi_{\mathfrak{z}}$ (Hurewicz–Wallman, VI.13), and our proof is accomplished.

## *APPENDIX*: ONE-DIMENSIONAL ANALYTIC SETS

In connection with Theorem 30.1, there arose the problem of showing that a set which is a one-dimensional analytic set, in the sense of our definition of this term, is a one-dimensional variety or, in the infinite-dimensional case, a one-dimensional analytic space. In the present appendix we shall explore this matter in detail.

The discussion of the infinite-dimensional case depends on some results about modules of holomorphic functions, and it is this theory that we develop in the first part of the appendix.

We continue to use $\Delta$ to denote the open unit disc, and we let $\bar{\Delta}$ be its closure. We denote by $\mathcal{O}(\bar{\Delta})$ the usual ring of germs of holomorphic functions on $\bar{\Delta}$. The ring $\mathcal{O}(\bar{\Delta})$ has rather simple algebraic structure: It is a principal ideal domain. This follows easily from the fact that if $F$ is any function holomorphic on $\bar{\Delta}$, there exist a polynomial $p$ and a function $f$ holomorphic and zero free on $\bar{\Delta}$ such that $F = fp$.

Let $\eta \in \mathcal{O}(\bar{\Delta})$ have the property that $|\eta|$ is constant on $\partial\Delta$, say $|\eta| = b$ on $\partial\Delta$, $\eta$ not the constant function. [Of course there is a certain abuse of language here since the elements of $\mathcal{O}(\bar{\Delta})$ are germs and not functions. As a practical matter, we shall not go astray if we speak as if $\mathcal{O}(\bar{\Delta})$ consisted of functions.] It is not difficult to see that $\eta'$ is zero free on $\partial\Delta$, and it follows from the argument principle that for some integer $m$, every value of modulus no more than $b$ is achieved as a value of $\eta$ at precisely $m$ points, due account being taken of multiplicities. If $f$ is a function holomorphic in $\Delta$, let $p_f(X;\zeta)$ be the polynomial

$$p_f(X;\zeta) = \prod_{j=1}^{m} \{X - f(z_j(\zeta))\}, \tag{1}$$

where for fixed $\zeta \in \Delta_b = \{z \in \mathbb{C} : |z| < b\}$, $z_1(\zeta), \ldots, z_m(\zeta)$ are the points, counted with multiplicities, in the set $\eta^{-1}(\zeta)$. If we expand the polynomial (1), we obtain an expression of the form $p_f(X;\zeta) = X^m + \sigma_1(\zeta)X^{m-1} + \cdots + \sigma_{m-1}(\zeta)X + \sigma_m(\zeta)$, where the coefficients $\sigma_j(\zeta)$ are symmetric functions of the numbers $f(z_1(\zeta)), \ldots, f(z_m(\zeta))$. Thus, they are well-defined functions of $\zeta$, and, indeed, they are holomorphic. It is easy to see that if $f \in \mathcal{O}(\bar{\Delta})$, then the coefficients $\sigma_j$ are holomorphic on a neighborhood of $\bar{\Delta}_b$, and if $f$ admits a continuous extension to $\bar{\Delta}$, then the coefficients extend continuously to $\bar{\Delta}_b$. By the very definition of $p_f(X;\zeta)$, we see that $p_f(f(z);\eta(z))$ vanishes identically. If we define a homomorphism $\eta^* : \mathcal{O}(\bar{\Delta}_b) \to \mathcal{O}(\bar{\Delta})$ by $\eta^* f = f \circ \eta$, our remarks show that $\mathcal{O}(\bar{\Delta})$ is integral over its subring $\eta^*\mathcal{O}(\bar{\Delta}_b)$: Each $f \in \mathcal{O}(\bar{\Delta})$ satisfies a monic polynomial equation of degree $m$ with coefficients from $\eta^*\mathcal{O}(\bar{\Delta}_b)$.

As usual we denote by $\mathscr{A}$ the disc algebra. By $\mathscr{A}^b$ we shall denote the corresponding algebra on the disc $\bar{\Delta}_b$. For our purposes it is useful to consider $\mathscr{A}$ as a module over $\eta^*\mathscr{A}^b = \{f \circ \eta : f \in \mathscr{A}^b\}$.

**30.A.1** ***Theorem***

*The algebra $\mathscr{A}$ is a free module of rank $m$ over $\eta^*\mathscr{A}^b$, and the functions $1, z, \ldots, z^{m-1}$ constitute a free basis for $\mathscr{A}$ as an $\eta^*\mathscr{A}^b$-module.*

PROOF

We are to show that if $f \in \mathscr{A}$, there exist uniquely determined $f_0, \ldots, f_{m-1} \in \mathscr{A}^b$ such that for all $z \in \bar{\Delta}$

(2) $$f(z) = f_0(\eta(z)) + zf_1(\eta(z)) + \cdots + z^{m-1}f_{m-1}(\eta(z)).$$

To prove the existence of such a decomposition for $f$, we begin by considering a $\zeta \in \bar{\Delta}_b$ such that $\eta^{-1}(\zeta)$ consists of $m$ distinct points, $z_0(\zeta), \ldots, z_{m-1}(\zeta)$, and we let $d(\zeta)$ be the Vandermonde determinant $\det(z_j(\zeta)^k)_{j,k=0,\ldots,m-1}$. We know that $d(\zeta) \neq 0$. The determinant $d(\zeta)$ is not a well-defined function of $\zeta$ since it depends not only on $\zeta$ but also on the ordering of the $z_j(\zeta)$, since an odd permutation of $z_0(\zeta), \ldots, z_{m-1}(\zeta)$ induces a change of sign in $d(\zeta)$. However, $d^2(\zeta)$ is a well-defined function of $\zeta$, and it extends to be holomorphic on a neighborhood of $\bar{\Delta}$. If $d(\zeta) \neq 0$, then the system of equations

(3) $$f(z_k(\zeta)) = f_0(\zeta) + z_k(\zeta)f_1(\zeta) + \cdots + z_k(\zeta)^{m-1}f_{m-1}(\zeta)$$

($k = 0, \ldots, m-1$) has a unique solution $f_0(\zeta), \ldots, f_{m-1}(\zeta)$ which can be computed by Cramer's rule, and Cramer's rule shows that the functions $f_0, \ldots, f_{m-1}$ determined in this way are holomorphic functions of $\zeta$ off the set where $d^2$ vanishes. On this set, they could conceivably have poles. However, the functions $d^2 f_j$ are seen to be holomorphic in $\Delta_b$. Our conclusion is that $B$, the set of all $f \in \mathscr{A}$ which admit an expansion of the form (2), contains the ideal generated by $d^2 \circ \eta$. Since $d^2$ has no zero on $\partial\Delta_b$, it follows that the ideal in $\mathscr{A}$ generated by $d^2 \circ \eta$ is closed and, moreover, it has finite codimension in $\mathscr{A}$. It follows that $B$ itself is closed.

It is easy to see that $B$ contains every power of $z$. It plainly contains $1, z, \ldots, z^{m-1}$. To see that it contains $z^m$, recall there is a monic polynomial equation

(4) $$z^m + g_1(\eta(z))z^{m-1} + \cdots + g_{m-1}(\eta(z))z + g_m(\eta(z)) = 0,$$

so $z^m \in B$. Multiplying (4) by $z$ and using the fact that $z^m \in B$, we find that $z^{m+1} \in B$. Continuing inductively, we get that $B$ contains every power of $z$ and so is dense in $\mathscr{A}$. As $B$ is closed, $B = A$. This implies the existence of decompositions of the form (2) for each $f \in \mathscr{A}$. The uniqueness follows by solving (3) with Cramer's rule for $f_0(\zeta), \ldots, f_{m-1}(\zeta)$.

The proof just given shows that if $f \in \mathcal{O}(\bar{\Delta})$, then the coefficients $f_0, \ldots, f_{m-1}$ in (2) are in $\mathcal{O}(\bar{\Delta}_b)$. This remark yields the following fact:

30.A.2 ***Corollary***

*$\mathcal{O}(\bar{\Delta})$ is a free module of rank $m$ over $\eta^*\mathcal{O}(\bar{\Delta}_b)$.*

30.A.3 ***Corollary***

*If $\mathscr{F} \subset \mathcal{O}(\bar{\Delta})$ is any set and $M_{\mathscr{F}}$ is the $\eta^*\mathcal{O}(\bar{\Delta}_b)$-module generated by $\mathscr{F}$, then $M_{\mathscr{F}}$ is a free module of rank no more than $m$ over $\eta^*\mathcal{O}(\bar{\Delta}_b)$.*

PROOF

This is a purely algebraic fact: Submodules of finitely generated free modules over principal ideal domains are finitely generated free modules of no higher rank. This is a standard result which can be found, e.g., in Zariski–Samuel [1, p. 247].

For a systematic exploitation of the algebraic ideas contained in Theorem 30.A.1 and its first corollary, we refer to two papers of Alling ([1, 2]) and the paper by Rudin and Stout [2].

We now turn to the geometric problem which is the main concern of this appendix. Recall that we are using the term *one-dimensional analytic set* in the sense introduced just before the statement of Theorem 30.1. The terms *one-dimensional variety* and *one-dimensional analytic space* are taken in their usual function theoretic sense. (See Appendix D.)

**30.A.4** ***Theorem***

*Let E be a one-dimensional analytic subset of the open set $\Omega$ in $\mathbb{C}^\Lambda$. If $\Lambda$ is finite, E is a one-dimensional analytic variety in $\Omega$. If $\Lambda$ is infinite, E has, in a natural way, the structure of a one-dimensional analytic space on which the polynomials in $\Lambda$ variables are holomorphic.*

PROOF

We consider a point $\mathfrak{z}_0 \in E$ which, without loss of generality, we assume to be the origin of $\mathbb{C}^\Lambda$. By definition, there exist positive numbers $\alpha_1, \ldots, \alpha_q$ and coordinate projections $\pi_{\lambda_1}, \ldots, \pi_{\lambda_q}$ on $\mathbb{C}^\Lambda$ such that if

$$N = \{\mathfrak{z} \in \mathbb{C}^\Lambda : |\pi_{\lambda_j}(\mathfrak{z})| < a_j, j = 1, \ldots, q\},$$

then $N \cap E$ is a finite union of closed subsets $S_1, \ldots, S_t$ of $N$ where each $S_k$ is of the form $\varphi_k(\Delta)$, $\Delta$ the open unit disc, $\varphi_k$ a nonconstant map from $\Delta$ to $\mathbb{C}^\Lambda$ which is holomorphic on a neighborhood of $\bar{\Delta}$. We assume, as we may, that the sets $\varphi_k(\Delta)$ are pairwise distinct in that $\varphi_k(\Delta) = \varphi_j(\Delta)$ implies that $k = j$.

We shall prove that for each $j = 1, \ldots, q$, there is a corresponding $k$, $1 \leq k \leq q$, such that $|\pi_{\lambda_k} \circ \varphi_j| \equiv \alpha_k$ on $\partial\Delta$. To establish such an identity, it is enough to prove that $|\pi_{\lambda_k} \circ \varphi_j| = \alpha_k$ on an infinite set, for then the analytic curves $\pi_{\lambda k}(\varphi_j(\partial\Delta))$ and $\{\zeta : |\zeta| = \alpha_k\}$ intersect infinitely often and so must coincide. Consider $j = 1$. The set $\varphi_1(\Delta)$ is closed in $N$, but it cannot be compact, for if $\pi$ is some coordinate function on $\mathbb{C}^\Lambda$ for which $\pi \circ \varphi_1$ is nonconstant, then compactness of $\varphi_1(\Delta)$ would imply the existence of a point $\mathfrak{z} = \varphi_1(\zeta) \in \Delta$ with the property that

$$|\pi(\mathfrak{z})| = \sup\{|\pi(\mathfrak{z}')| : \mathfrak{z}' \in \varphi_1(\Delta)\},$$

and this is impossible by the maximum modulus theorem applied to the nonconstant holomorphic function $\pi \circ \varphi_1$. Thus, $\varphi_1(\Delta)$ is not compact. A first consequence of this observation is the fact that not all the functions $\pi_{\lambda_j} \circ \varphi_1$, $j = 1, \ldots, q$, can be constant, for if they were, the form of $N$ together with the boundedness of all the compositions $\pi \circ \varphi_1$, $\pi$ a coordinate function, would imply compactness of $\varphi_1(\Delta)$. The same line of reasoning implies that for some $j = 1, \ldots, q$, there is $\zeta$ *a* $\partial\Delta$ with $|\pi_{\lambda_j}(\varphi_1(\zeta))| = \alpha_j$.

Suppose, e.g., that $|\pi_{\lambda_1} \circ \varphi_1|$ assumes the value $\alpha_1$ on $\partial\Delta$. Since $\varphi_1(\Delta) \subset N$, $\pi_{\lambda_1} \circ \varphi_1$ cannot be constant. Two cases arise: It may happen that $|\pi_{\lambda_1} \circ \varphi_1| \equiv \alpha_1$ on $\partial\Delta$. If so, the first step of our proof is complete. If not, there are infinitely many points $\zeta \in \partial\Delta$ at which $|\pi_{\lambda_1}(\varphi_1(\zeta))| < \alpha_1$. Since $\varphi_1(\Delta)$ is a closed subset of $N$, to each such $\zeta$ there corresponds an integer $j = 2, \ldots,$ or $q$ such that $|\pi_{\lambda_j}(\varphi_1(\zeta))| = \alpha_j$. This is so, for otherwise the set $\pi_{\lambda_1}(\varphi_1(\Delta))$ is a proper open subset of the disc $\Delta_1 = \{z \in \mathbb{C} : |z| < \alpha_1\}$ whose closure meets $\partial\Delta_1$ only finitely often and whose boundary, $C$, is a piecewise analytic curve. If $z_0 \in C$ but $|z_0| < \alpha_1$, then there is a sequence $\{\zeta_n\}$ in $\Delta$ with limit $\zeta_0$ such that $\pi_{\lambda_1}(\varphi_1(\zeta_0)) = z_0$. The point $\zeta_0$ lies in $\partial\Delta$ since $\pi_{\lambda_1} \circ \varphi_1$ is an open map. For some $j$, we must have $|\pi_{\lambda_j}(\varphi_1(\zeta_0))| = \alpha_j$, for, if not, $\varphi_1(\zeta_0)$ is a point of $N$ which lies in $\overline{\varphi_1(\Delta)} \setminus \varphi_1(\Delta)$ contrary to the assumed closure of $\varphi_1(\Delta)$. By assumption, we have infinitely many points $z_0$ at our disposal, so one of the functions $\pi_{\lambda_j} \circ \varphi_1$ takes values constantly of modulus $\alpha_j$ infinitely often on $\partial\Delta$.

Denote by $\Psi$ the map from $\mathbb{C}^\Lambda$ to $\mathbb{C}^q$ given by

$$\Psi(\mathfrak{z}) = (\pi_{\lambda_1}(\mathfrak{z}), \ldots, \pi_{\lambda_q}(\mathfrak{z})).$$

The image of $N$ under $\Psi$ is the polycylinder

$$\Delta_\alpha^q = \{(z_1, \ldots, z_q) : |z_1| < \alpha_1, \ldots, |z_q| < \alpha_q\}.$$

We shall prove that $\Psi(N \cap E)$ is a one-dimensional variety in $\Delta_\alpha^q$. For this purpose, it suffices to show that each of the sets $\Psi(\varphi_j(\Delta))$ is a variety. In establishing this, there are two possible routes we can follow. One is short but decidedly non-elementary. The map $\Psi \circ \varphi_1$ is a holomorphic map from $\Delta$ into $\Delta_\alpha^q$, and since one of the functions $\pi_{\lambda_j} \circ \varphi_1$ is of constant modulus on $\partial\Delta$, $\Psi \circ \varphi_1$ is proper in that if $C \subset \Delta_\alpha^q$ is compact, then $(\Psi \circ \varphi_1)^{-1}(C)$ is compact in $\Delta$. It follows from general function theoretic facts (the *proper mapping theorem*, Gunning and Rossi [1, p. 162]) that $\Psi(\varphi(\Delta))$ is a variety; it is clearly one dimensional. A longer, but elementary, proof of this fact may be obtained from our algebraic considerations at the beginning of this appendix. Let the coordinates of $\Psi \circ \varphi_1$ be $\psi_1, \ldots, \psi_q$, and suppose, to be specific, that $|\psi_1| = \alpha_1$ on $\partial\Delta$. Then for each $j = 2, \ldots, q$, there is an irreducible monic polynomial of the form

$$p_j(z_j; z_1) = z_j^{\mu_j} + f_1^{(j)}(z_1)z_j^{\mu_j - 1} + \cdots + f_{\mu_j - 1}^{(j)}(z_1)z_j + f_{\mu_j}^{(j)}(z)$$

with coefficients $f_i^{(j)}$ holomorphic in $\{z : |z| < \alpha_1\}$ such that for all $z \in \Delta$, $p_j(\psi_j(z); \psi_1(z)) = 0$. It follows that if $V = \{\mathfrak{z} \in \Delta_\alpha^q : p_2(z_2; z_1) = \cdots = p_q(z_q; z_1) = 0\}$, then $\Psi(\varphi_1(\Delta)) \subset V$. The variety $V$ is a one-dimensional variety, and it is not difficult to see that, in fact, $V = \Psi(\varphi_1(\Delta))$. (Since $\Psi \circ \varphi_1$, is proper, its range is closed in $V$. It is also open, whence equality.) Thus, we have that each of the sets $\Psi(\varphi_1(\Delta))$ is a variety in $\Delta_\alpha^q$. Slightly more is true, as follows from either of the proofs, because of the holomorphicity of each $\varphi_i$ on a neighborhood of $\Delta$: For each $j$, there is a variety $V_j$ in a neighborhood of $(\Delta_\alpha^q)^-$ such that $\Psi(\varphi_j(\Delta)) = V_j \cap \Delta_\alpha^q$. This observation is important, for from it follows the fact that any two of the sets $\Psi(\varphi_i(\Delta))$ coincide or else have at most a finite set of points in common.

We see now that if we restrict attention to the finite-dimensional case, we are essentially done. If, say, we are working in $\mathbb{C}^m$, and if the $q$ we have considered is $m$, our result is proved. If $m > q$, then suppose the additional coordinate projections on $\mathbb{C}^m$ are $\pi_{\lambda_{q+1}}, \ldots, \pi_{\lambda_m}$. In this case, we replace $N$ by the slightly smaller

$\tilde{N} = \{\mathfrak{z} \in N : |\pi_{\lambda_{q+1}}(\mathfrak{z})|, \ldots, |\pi_{\lambda_m}(\mathfrak{z})| < K\}$ for some large positive $K$ and we have $N \cap E = \tilde{N}(E)$. The argument just used (with $m$ in place of $q$) yields the result.

We continue now with the general case. Since the sets $\varphi_j(\Delta)$ are, by hypothesis, pairwise distinct, it follows that there exists a finite set $S = \{\pi_{\lambda_{q+1}}, \ldots, \pi_{\lambda_r}\}$ of projections of $\mathbb{C}^\Lambda$ such that if $\Psi' : \mathbb{C}^\Lambda \to \mathbb{C}^r$ is given by

$$\Psi'(\mathfrak{z}) = (\pi_{\lambda_1}(\mathfrak{z}), \ldots, \pi_{\lambda_v}(\mathfrak{z})),$$

then the sets $\Psi'(\varphi_j(\Delta))$ are pairwise distinct. Moreover, we can choose $S$ so large that if $k(j)$ is one of $\{1, \ldots, q\}$ for which $|\pi_{\lambda_{k(j)}} \circ \varphi_j| \equiv \alpha_{k(j)}$ on $\partial\Delta$, then given a coordinate projection $\pi$ on $\mathbb{C}^\Lambda$, the function $\pi \circ \varphi_j$ has an expansion in the form

$$(5) \qquad \pi(\varphi_j(z)) = \Sigma\{f_{\pi,m}(\pi_{\lambda_{k(j)}}(\varphi_j(z)))\pi_{\lambda_m}(\varphi_j(z)) : m = 1, \ldots, r, m \neq k(j)\}.$$

These expansions follow from Corollary 30.A.3 applied to the $(\pi_{\lambda_{k(j)}} \circ \varphi_j)^*\mathcal{O}(\bar{\Delta}_{\alpha_{k(j)}})$-submodule of $\mathcal{O}(\bar{\Delta})$ generated by the set $\{\pi \circ \varphi_j : \pi$ a coordinate projection on $\mathbb{C}^\Lambda\}$.

With this choice of $\pi_{\lambda_{q+1}}, \ldots, \pi_{\lambda_r}$, it is easily seen that the projection $\Psi'$ is one to one on each of the sets $\varphi_j(\Delta)$, for if $\Psi'(\varphi_j(z)) = \Psi'(\varphi_j(w))$, then (5) implies that $\pi(\varphi_j(z)) = \pi(\varphi_j(w))$ for every coordinate function $\pi$, whence $\varphi_j(z) = \varphi_j(w)$. We would like to assert that $\Psi'$ is one to one on $\bigcup \varphi_j(\Delta) = N \cap E$, but this need not be the case. However, what is true is that since the sets $\Psi'(\varphi_j(\Delta))$ are of the form $V'_j \cap \Psi'(N)$, $V'_j$ a variety in a neighborhood of $(\Psi'(N))^-$, a given pair $\Psi'(\varphi_j(\Delta))$ and $\Psi'(\varphi_k(\Delta))$ can have only a finite intersection. Thus, we can find finitely many additional projections $\pi_{\lambda_{r+1}}, \ldots, \pi_{\lambda_t}$ on $\mathbb{C}^\Lambda$ such that if $\Psi'' : \mathbb{C}^\Lambda \to \mathbb{C}^t$ is the projection defined by $\pi_{\lambda_1}, \ldots, \pi_{\lambda_t}$, then $\Psi''$ takes $N \cap E$ one to one onto $\Psi''(N \cap E)$. Our construction shows that $\Psi''(N \cap E)$ is a one-dimensional variety $V''$ in $\Psi''(N)$ and that $\Psi''$ is a homeomorphism from $N \cap E$ onto $V''$. Let $\xi : V'' \to N \cap E$ be the inverse of $\Psi''$.

To provide $E$ with the structure of a one-dimensional analytic space, we must specify a sheaf $\mathscr{S}$ of germs of continuous functions and show that each point of $E$ has a neighborhood $U$ such that for some one-dimensional analytic variety $W$ in an open subset of some $\mathbb{C}^m$, there is a homeomorphism $U \to W$ which induces an isomorphism between the sheaves $\mathscr{S}|U$ and $\mathcal{O}_W$, the sheaf of germs of holomorphic functions on $W$. The mechanism we have developed enables us to do exactly this.

We define the sheaf $\mathscr{S}$ on $E$ by requiring that if $U$ is an open set in $E$, then $\mathscr{S}(U)$ is the ring of all those $f \in \mathscr{C}(U)$ which are locally approximable by polynomials in $\Lambda$ variables: Given $\mathfrak{z} \in U$, there is a neighborhood $\tilde{U}$ of $\mathfrak{z}$ on which $f$ is the uniform limit of a sequence of polynomials in the coordinate functions on $\mathbb{C}^\Lambda$.

We have the neighborhood $N \cap E$ in $E$, the variety $V''$, the homeomorphism $\Psi'' : N \cap E \to V''$, and the inverse $\xi : V'' \to N \cap E$. If $U \subset V''$ is an open set and $f$ is holomorphic on $U$, then $f \circ \Psi'' \in \mathscr{S}(\Psi''^{-1}(U)) = \mathscr{S}(\xi(U))$, for in $U$, $f$ is locally approximable by polynomials in the $t$ coordinate functions on $\mathbb{C}^t$, so $f \circ \Psi''$ is locally approximable by polynomials in the coordinate functions $\pi_{\lambda_1}, \ldots, \pi_{\lambda_t}$ on $\mathbb{C}^\Lambda$. Conversely, let $U \subset N \cap E$ be an open set, and let $F \in \mathscr{S}(U)$ so that $F$ is locally approximable on $U$ by polynomials. Then $F \circ \xi$ is continuous on $\Psi''(U) = \xi^{-1}(U)$. Moreover, the representation (5) of coordinate functions $\pi$

on the sets $\varphi_j(\Delta)$ shows that every coordinate function is locally approximable on $\varphi_j(\Delta)$ by the coordinate functions $\pi_{\lambda_1}, \ldots, \pi_{\lambda_t}$, whence the same is true of $F$. It follows that $F \circ \xi$ is holomorphic on the intersection of $\xi^{-1}(U)$ with every branch of $V''$, and hence $F \circ \xi \in \mathcal{O}(\xi^{-1}(U))$. This concludes the proof of Theorem 30.A.4.

# APPENDIXES

We include five appendixes, which deal with topology, measure and integration, functional analysis, complex variables, and number theory. Our intent is not to give systematic treatments of these subjects but rather to collect in a form suitable for our purposes certain information which, although well known, may be unfamiliar to some readers. Proofs are included only rarely; those included are primarily proofs of results that do not seem to be available in standard English-language references. Each appendix concludes with a short list of relevant sources.

# *appendix a*/TOPOLOGY

We assume that the reader is reasonably well versed in elementary topology. We shall use standard terminology from this subject with the possible exception that, following Bourbaki, we shall understand that a space which is locally compact or compact is, by definition, Hausdorff.

It is convenient to use nets to deal with questions of convergence. A *directed set* is a set $S$ together with an order relation $\leq$ on it which has the properties that (1) if $a,b,c \in S$ and $a \leq b$, $b \leq c$, then $a \leq c$; (2) if $a \in S$, then $a \leq a$; and (3) if $a,b \in S$, then there exists $c \in S$ with $a \leq c$, $b \leq c$. If $X$ is a topological space, a *net in* $X$ is a function with domain a directed set, range in $X$. If $s$ is a net in $X$ with domain the directed set $S$, $s$ *converges* to the point $x_0 \in X$ if for every neighborhood $U$ of $x_0$, there is $a \in S$ such that $s(b) \in U$ for all $b \in S$ such that $a \leq b$. More generally, a net $s$ has the point $x_0$ as a *cluster point* (or *limit point*) provided that for every neighborhood $U$ of $x_0$ and for every $a \in S$, there is $b \in S$, $a \leq b$, with $s(b) \in U$.

Compact spaces can be characterized in terms of nets, as the following theorem shows.

***Theorem***

*A Hausdorff space is compact if and only if every net in it has a cluster point.*

If $X$ is a Hausdorff space, a *compactification* of $X$ is a space $X^*$ with the property that there exists a homeomorphism of $X$ onto a dense subspace of $X^*$. The simplest general example of a compactification is the *Alexandroff one-point compactification* of a locally compact space. Let $X$ be a locally compact space, denote by $\infty$ the set $\{X\}$, and let $X^*$ be defined, as a set, to be $X \cup \{\infty\}$. If we topologize $X^*$ by taking for the open sets the open sets of $X$ and the sets of the form $(X \setminus K) \cup \{\infty\}$, $K$ a compact set in $X$, then we obtain a compact space which contains $X$ as a dense subspace.

For completely regular spaces, there is the *Stone–Čech* (or *Čech*) *compactification*, which is conventionally denoted by $\beta(X)$. [Recall that a space is *completely regular* if, given a closed set $E$ and a point $x \in E$, there is a continuous, real-valued function $f$ with $f(x) = 0, f(y) = 1$ for all $y \in E$.]

***Theorem***

*If $X$ is a completely regular space, there exists a compactification $\beta(X)$ of $X$ with the property that if $Y$ is a compact space and $f : X \to Y$ is a* *continuous map, then $f$ extends to a continuous map $f^{\beta}: \beta(X) \to Y$. The compactification $\beta(X)$ is uniquely determined up to homeomorphism.*

In the statement of this result we are regarding $X$ as a subset of $\beta(X)$.

A topological space is said to be of the *first category* if it is the union of a countable family of sets no one of which is dense in an open set. Otherwise, it is of *second category*. According to the *Baire category theorem, a complete metric space is of second category, and so is a locally compact space.*

If we are given a compact metric space $X$, with metric $\rho$, it is possible to endow the collection $\mathscr{K}_X$ of closed subsets of $X$ with the structure of a metric space by defining a metric $d$ on $\mathscr{K}_X$ by

$$d(A, B) = \max(\sup_{a \in A} \rho(a, B), \sup_{b \in B} \rho(b, A)),$$

where $\rho(a, B) = \inf_{b \in B} \rho(a, b)$. With respect to the metric topology induced by the metric $d$, $\mathscr{K}_X$ is a compact metric space. The metric $d$ is called the *Hausdorff metric on the space of closed subsets of $X$*.

The *Cantor set* or the *Cantor ternary set* is the compact totally disconnected, perfect set $\bigcap_{n=1}^{\infty} E_n$, where the sets $E_n \subset [0, 1]$ are defined inductively as follows: $E_1$ is obtained from $[0, 1]$ by deleting the open middle third of $[0, 1]$, i.e., the open interval $(\frac{1}{3}, \frac{2}{3})$. The set $E_2$ is obtained from $E_1$ by removing the open middle thirds of the two intervals that constitute $E_1$. Generally, $E_n$ consists of $2^{n-1}$ closed subintervals, and $E_{n+1}$ is obtained by removing from $E_n$ the open middle thirds of each of these intervals. This set is homeomorphic to the product of countably many discrete two-point spaces. Occasionally, we speak of a set which "is a Cantor set." By this we mean that the set in question is homeomorphic to the Cantor ternary set. There is a simple topological characterization of such sets.

***Theorem***

*A topological space is homeomorphic to the Cantor set if and only if it is a totally disconnected, compact, perfect, metric space.*

The Cantor set enjoys a certain universal property among compact metric spaces.

***Theorem***

*Every compact metric space is the continuous image of the Cantor set.*

If $E$ denotes the Cantor set, it is easy to see that $\mathscr{C}(E)$ is separable: Finite linear combinations with algebraic coefficients of characteristic functions of sets of the form $E \cap [a, b]$, $a$ and $b$ rational, not in $E$, are dense in $\mathscr{C}(E)$. It follows rather quickly that *if $X$ is a compact space, then $\mathscr{C}(X)$ is separable if and only if $X$ is metric.* If $X$ is metric, let $\varphi: E \to X$ be a continuous map with range the whole of $X$. Then $\mathscr{C}(X)$ is isometrically isomorphic to the subspace $\{f \circ \varphi : f \in \mathscr{C}(X)\}$ of $\mathscr{C}(E)$, and thus it is separable. Conversely, if $\mathscr{C}(X)$ is separable, it admits a countable number of generators as a Banach algebra; so, by invoking the fact that $X$ is the spectrum of $\mathscr{C}(X)$ and Theorem 5.8, we see that $X$ is homeomorphic to a subset of the cartesian product of countably many copies of $\mathbb{C}$, and thus it must be metrizable.

We conclude our discussion of results from general topology with some matters from the theory of curves. A topological space is a *Peano continuum* or a *continuous curve* or a *curve* if it is Hausdorff and the continuous image of the closed unit interval. An *arc* is a topological space that is homeomorphic to the unit interval, and a *simple closed curve* is a homeomorph of a circle.

The Hahn–Mazurkiewicz theorem characterizes abstractly Peano continua:

**Theorem**

*A nonempty Hausdorff space is a Peano curve if and only if it is compact, connected, locally connected, and metrizable.*

Say that a topological space $X$ is *cyclic* provided $X \setminus \{x\}$ is connected for every choice of $x \in X$. The following theorem is the *cyclic connectivity theorem.*

**Theorem**

*A Peano continuum $X$ is cyclic if and only if every pair of points in $X$ lies in a simple closed curve contained in $X$.*

Finally, a result about pathological arcs.

**Theorem**

*There exist arcs in the plane all of whose subarcs have positive two-dimensional Lebesgue measure.*

Thus, locally, these arcs have positive area.

We shall devote the second half of this appendix to some facets of the theory of sheaves. There are two rather different ways to define sheaves, each of which is useful in its place. According to one definition, a sheaf is a functor, according to the other a topological space.

We consider sheaves first as functors. Thus, if $X$ is a topological space, a *presheaf of abelian groups on* $X$ is a contravariant functor from the category of open sets (and inclusion maps) in $X$ to the category of abelian groups. Explicitly, a presheaf $\mathscr{S}$ of abelian groups on $X$ is an object that assigns to each open set $U$ in $X$ an abelian group $\mathscr{S}(U)$ and to each pair of open sets $U$ and $V$ in $X$, with $U \supset V$, a homomorphism $r_{VU}:\mathscr{S}(U) \to \mathscr{S}(V)$. These homomorphisms satisfy the condition that if $W \subset V \subset U$, then $r_{WU} = r_{WV} \circ r_{VU}$. In a similar way, we can define a presheaf of rings on $X$. If $\mathscr{R}$ is a presheaf of rings on $X$, a *presheaf* $\mathscr{M}$ *of* $\mathscr{R}$*-modules* is a presheaf of abelian groups such that for each open set $U$, $\mathscr{M}(U)$ is an $\mathscr{R}(U)$-module, and such that if $V \subset U$ are open sets and $r_{VU}$ is the homomorphism associated with $\mathscr{M}$ and $p_{VU}$ that associated with $\mathscr{R}$, then for each $a \in \mathscr{R}(U)$ and each $m \in \mathscr{M}(U)$, $r_{VU}(am) = p_{VU}(a)r_{VU}(m)$.

***Definition*** $S_f$

*A* ***sheaf of abelian groups*** *is a presheaf* $\mathscr{S}$ *of abelian groups which enjoys the following properties:*

(a) *If* $\{U_i\}_{i\in I}$ *is a family of open sets in* $X$ *with union* $U$, *and if* $s,s' \in \mathscr{S}(U)$ *satisfy* $r_{U_i,U}(s) = r_{U_i,U}(s')$ *for every* $i$, *then* $s = s'$.

(b) *If* $\{U_i\}_{i\in I}$ *is a family of open sets in* $X$ *with union* $U$, *and if for each* $i$, *the elements* $s_i \in \mathscr{S}(U_i)$ *satisfy* $r_{U_i\cap U_j,U_j}(s_j) = r_{U_i\cap U_j,U_i}(s_i)$, *then there exists an* $s \in \mathscr{S}(U)$ *such that for every* $i$, $r_{U_i,U}(s) = s_i$.

The first of these says, roughly, that globally defined objects which agree locally agree globally, and the second that locally defined objects which are defined consistently can be pieced together to yield a global object.

If $\mathscr{S}$ is a sheaf, $\mathscr{S}(U)$ is often denoted by $\Gamma(U, \mathscr{S})$, and its elements are called the *sections of* $\mathscr{S}$ *over* $U$.

Two examples come to mind immediately. Let $G$ be a fixed abelian group, and for each open set $U$ in $X$, let $\mathscr{S}(U) = G$. Define the restriction homomorphisms $r_{VU}$ to be the identity for every choice of open sets $V$ and $U$ in $X$. In this way we construct a sheaf, the *constant sheaf* with values in $G$.

For the second example, let $\mathscr{C}(U)$ be the ring of all continuous complex-valued functions on $U$, and define the homomorphism $r_{VU}:\mathscr{C}(U) \to \mathscr{C}(V)$ to be the restriction mapping so that if $f \in \mathscr{C}(U)$, $r_{VU}(f) = f|V$. In this way we construct on $X$ a sheaf, the sheaf of *germs of functions continuous on* $X$.

If $\mathscr{S}$ is a sheaf of abelian groups, we define the *stalk* $\mathscr{S}_x$, for $x \in X$, to be the direct limit of the directed system

$$\{\mathscr{S}(U), r_{VU}, U \text{ and } V \text{ neighborhoods of } x\}.$$

Thus, for each neighborhood $U$ of $x$, there is a well-defined homomorphism $r_{x,U}$ from $\mathscr{S}(U)$ to the stalk $\mathscr{S}_x$.

The alternative definition of sheaf in terms of topological spaces may be formulated as follows.

***Definition*** $S_s$

*If $X$ is a topological space, a sheaf of abelian groups on $X$ is a topological space $\mathscr{S}$ together with a continuous map $\pi:\mathscr{S} \to X$ such that*

(a) *$\pi$ is a local homeomorphism.*
(b) *For each $x \in X$, $\pi^{-1}(x)$ is an abelian group.*
(c) *The group operations are continuous in the following sense: If $\mathscr{S} \vee \mathscr{S} = \{(s, t) \in \mathscr{S} \times \mathscr{S} : \pi(s) = \pi(t)\}$, then the map $\mathscr{S} \vee \mathscr{S} \to \mathscr{S}$ given by $(s, t) \mapsto s - t$ is continuous.*

If $E$ is a subset of $X$, a *section* of $\mathscr{S}$ over $E$ is a continuous map $\varphi:E \to \mathscr{S}$ such that $\pi \circ \varphi$ is the identity map on $E$. The set $\Gamma(E, \mathscr{S})$ of all sections of $\mathscr{S}$ over $E$ is, in a natural way, a group.

If we are given a sheaf $\mathscr{S}$ on $X$, a sheaf in the sense of our second definition, the map $U \mapsto \Gamma(U, \mathscr{S})$ gives rise to a functor from open sets to groups if we take for the homomorphisms $r_{VU}$ the restriction maps. In this way we are led from a sheaf in the topological space sense to a sheaf in the functorial sense.

In the opposite direction, if $\mathscr{S}$ is a (functorially defined) sheaf over $X$, there is a canonical construction for a topological space $S$ (the *espace étale* associated with $\mathscr{S}$) and a projection $\pi:S \to X$ such that the pair $(S, \pi)$ is a sheaf in the topological space sense, and such that $\mathscr{S}$ is, in a natural sense, the sheaf of sections of $S$. The space $S$ is defined, as a set, to be the union $\bigcup_{x \in X} \mathscr{S}_x$ of all the stalks of $\mathscr{S}$ over points of $X$, and the projection $\pi:S \to X$ takes all the points of $\mathscr{S}_x$ to the point $x$. If $U$ is an open set in $X$ and $f \in \mathscr{S}(U)$, there is a map $U \to S$ sending each point $x \in U$ to the germ, $f_x$, of $f$ at $x$, i.e., to $r_{x,U}(f)$. The space $S$ is given the finest topology which renders all these maps continuous. In this way, $S$ is made into a topological space, and it can be verified that the pair $(S, \pi)$ is a sheaf in the topological space sense.

If $\mathscr{S}$ and $\mathscr{T}$ are sheaves of abelian groups on $X$, a *homomorphism* $\varphi:\mathscr{S} \to \mathscr{T}$ is a collection $\{\varphi_U\}_{U \text{ open in } X}$ of homomorphisms, $\varphi_U:\mathscr{S}(U) \to \mathscr{T}(U)$, which are compatible with the restriction maps in that $\varphi_V(r_{VU}(s)) = r_{VU}(\varphi_U(s))$ for every $s \in \mathscr{S}(U)$. (Here we have abused the notation slightly in that we have written $r_{V,U}$ for the restriction homomorphism associated with the sheaf $\mathscr{S}$, and also for the corresponding homomorphism associated with $\mathscr{T}$.) The compatibility requirement implies that a sheaf homomorphism $\varphi:\mathscr{S} \to \mathscr{T}$ induces a homomorphism from $\mathscr{S}_x$ to $\mathscr{T}_x$. A sequence

$$\mathscr{S}^{(1)} \to \mathscr{S}^{(2)} \to \mathscr{S}^{(3)}$$

of sheaves of abelian groups is said to be *exact* if the corresponding sequence of homomorphisms of stalks is exact at every $x \in X$.

Given a locally finite open cover $\mathscr{U} = \{U_i\}$ of $X$ and a sheaf $\mathscr{S}$ of abelian groups on $X$, a *partition of unity of subordinate to* $\mathscr{U}$ is a family $\{\eta_i\}$ of homomorphisms of $\mathscr{S}$ to itself with the following properties:

(a) For each $i$, there is a neighborhood $V_i$ of $X \setminus U_i$ such that for every open set $W \subset V$, the homomorphism $\eta_i:\mathscr{S}(W) \to \mathscr{S}(W)$ is the zero homomorphism, and
(b) $\Sigma \eta_i$ is the identity map of $\mathscr{S}$ to itself.

The canonical example of such a partition of unity arises in connection with the sheaf $\mathscr{C}_X$ of germs of continuous functions on a paracompact space. Let $\mathscr{U} = \{U_i\}$ be a locally finite open cover of $X$, and let $\{\psi_i\}$ be a collection of continuous functions on $X$ each of which takes values in the closed interval $[0, 1]$ and which satisfy the conditions that $\psi_i$ vanishes on a neighborhood of $X \setminus U_i$, and $\Sigma\psi_i(x) = 1$ for every $x \in X$. Thus, $\{\psi_i\}$ is a partition of unity in the usual topological sense. For each $i$, define a homomorphism $\eta_i:\mathscr{C}_X \to \mathscr{C}_X$ by requiring that if $f \in \mathscr{C}(U)$, then $\eta_i(f) = \psi_i f$. The family $\{\eta_i\}$ is a partition of unity in the sheaf theoretic sense.

A sheaf is said to be *fine* if it admits partitions of unity subordinate to every locally finite cover of the base space.

Given a paracompact space $X$, it is possible to construct a Čech-type cohomology theory for sheaves. We shall not carry out this construction in detail, but, instead, we shall only recall certain of the features of such a theory and indicate very briefly how the theory is constructed. By a *sheaf cohomology theory* on $X$ is meant a correspondence that assigns to each sheaf $\mathscr{S}$ of abelian groups on $X$, and each nonnegative integer, an abelian group $H^p(X, \mathscr{S})$, *the* $q$th *cohomology group of* $X$ *with coefficients in* $\mathscr{S}$. Among others, these cohomology groups are required to have the following properties:

(a) $H^0(X, \mathscr{S}) = \Gamma(X, \mathscr{S}) = \mathscr{S}(X)$.
(b) $H^q(X, \mathscr{S}) = 0$ for all $q > 0$ if $\mathscr{S}$ is a fine sheaf.
(c) If $\varphi:\mathscr{S} \to \mathscr{T}$ is a sheaf homomorphism, there are induced homomorphisms $\varphi_q:H^q(X, \mathscr{S}) \to H^q(X, \mathscr{T})$.

(d) If $0 \to \mathscr{S} \xrightarrow{\varphi} \mathscr{S}' \xrightarrow{\psi} \mathscr{S}'' \to 0$ is an exact sequence of sheaves, there exist homomorphisms $\delta_q : H^q(X, \mathscr{S}'') \to H^{q+1}(X, \mathscr{S})$ such that the sequence

$$0 \to H^0(X, \mathscr{S}) \xrightarrow{\varphi_0} H^0(X, \mathscr{S}') \xrightarrow{\psi_0} H^0(X, \mathscr{S}'') \xrightarrow{\delta_0} H^1(X, \mathscr{S})$$

$$\cdots \xrightarrow{\delta_{q-1}} H^q(X, \mathscr{S}) \xrightarrow{\varphi_q} H^q(X, \mathscr{S}') \xrightarrow{\psi_q} H^q(X, \mathscr{S}'') \xrightarrow{\delta_q} H^{q+1}(X, \mathscr{S}) \cdots$$

is exact.

If certain additional conditions of naturality are imposed, conditions that various natural diagrams commute, it can be shown that there exists one, and, up to isomorphism, only one sheaf cohomology theory on a paracompact space.

One way of constructing the cohomology groups $H^q(X, \mathscr{S})$ for a paracompact space $X$ is as follows. Start with a locally finite cover $\mathfrak{A}$ of $X$. Denote by $C^p(X, \mathscr{U}, \mathscr{S})$ the group of $p$-cochains associated with the covering $\mathscr{U}$ and the sheaf $\mathscr{S}$ so that $f \in C^p(X, \mathscr{U}, \mathscr{S})$ is a function that assigns to each ordered intersection $U_{i_0} \cap U_{i_1} \cap \cdots \cap U_{i_p}$, $U_{i_j} \in \mathscr{U}$, an element $f_{i_0 i_1 \cdots i_p} \in \Gamma(U_{i_0} \cap \cdots \cap U_{i_p}, \mathscr{S})$ in such a way that if $\sigma$ is a permutation of the set $\{i_0, \ldots, i_p\}$, then $f_{i_0 \cdots i_p} = (\operatorname{sign} \sigma) f_{\sigma(i_0) \cdots \sigma(i_p)}$. A coboundary map $\delta : C^p(X, \mathscr{U}, \mathscr{S}) \to C^{p+1}(X, \mathscr{U}, \mathscr{S})$ is defined by

$$(\delta f)_{i_0 \cdots i_{p+1}} = \sum_{j=0}^{p+1} (-1)^j f_{i_0 \cdots \hat{\imath}_j \cdots i_{p+1}},$$

where the symbol $\hat{\imath}_j$ indicates that the term $i_j$ is omitted. It can be verified that $\delta \circ \delta : C^p(X, \mathscr{U}, \mathscr{S}) \to C^{p+2}(X, \mathscr{U}, \mathscr{S})$ is the zero map. By definition a $p$-cochain is a *coclycle* if $\delta f = 0$, and it is a *coboundary* if $f = \delta g$ for some $(p-1)$-cochain $g$. The groups of $p$-cocyles and coboundaries are denoted by $B^p(X, \mathscr{U}, \mathscr{S})$ and $Z^p(X, \mathscr{U}, \mathscr{S})$, respectively, and the cohomology group $H^p(X, \mathscr{U}, \mathscr{S})$ is defined to be the quotient $B^p(X, \mathscr{U}, \mathscr{S})/Z^p(X, \mathscr{U}, \mathscr{S})$. Now if $\mathscr{V}$ is a refinement of $\mathscr{U}$, it can be shown that there is a well-defined natural homomorphism $H^p(X, \mathscr{U}, \mathscr{S}) \to H^p(X, \mathscr{U}, \mathscr{S})$ for each $p$. By definition then, $H^p(X, \mathscr{S})$ is the direct limit of the groups $H^p(X, \mathscr{U}, \mathscr{S})$. It can be shown that the assignment to $\mathscr{S}$ of the cohomology groups $H^p(X, \mathscr{S})$ constructed in this way constitutes a sheaf cohomology theory on $X$.

*References*

Bredon, G., *Sheaf Theory*, McGraw-Hill, New York, 1967.

Gelbaum, B. R., and J. H. M. Ohmsted, *Counterexamples in Analysis*, Holden-Day, San Francisco, 1964.

Godement, R., *Topologie algébrique et théorie des faisceaux* (Actualités Scientifiques et Industrielles No. 1257), Hermann, Paris, 1964.

Hall, D. W., and G. L. Spencer, II, *Elementary Topology*, Wiley, New York, 1955.

Kelley, J. L., *General Topology*, Van Nostrand Reinhold, New York, 1955.

Osgood, W. F., A Jordan curve of positive area, *Trans. Amer. Math. Soc.* **4** (1903), 102–112; correction, p. 370.

For the set theoretic topology generally useful to functional analyists, including the theory of nets, the best single source is Kelley's book. Plane topology can be found in the book by Hall and Spencer. Arcs of positive measure were constructed by Osgood in his 1903 *Transactions* paper. Another source is the book by Gelbaum and Ohmsted. The theory of sheaves is expounded exhaustively by Godement and by Bredon. A short treatment is given by Gunning and Rossi [1].

# *appendix b*/MEASURE AND INTEGRATION

The setting for abstract measure and integration theory is a set $X$ together with a $\sigma$-algebra $\Sigma$ of subsets of $X$. Thus, $\Sigma$ is a collection of subsets of $X$ which contains $X$ and which is closed under the formation of countable unions and under complementation. The elements of $\Sigma$ are referred to as the *measurable sets*. The pair $(X, \Sigma)$ is called a measurable space. A particularly important example of this notion arises when for $X$ we take a locally compact topological space and for $\Sigma$ we take the $\sigma$-algebra of *Borel sets*, i.e., the smallest $\sigma$-algebra which contains the open sets.

If $(X, \Sigma)$ is a measurable space, a complex-valued function $f$ on $X$ is measurable if for every open set $V \subset \mathbb{C}$, the set $f^{-1}(V)$ is measurable.

Given a measurable space $(X, \Sigma)$, a *positive measure* on this space is a function $m$ on $\Sigma$ which takes values in $[0, \infty]$ and which satisfies $m(\bigcup_n E_n) = \sum_n m(E_n)$ for every sequence $\{E_n\}$ of pairwise disjoint measurable sets. A *real measure* is a countably additive function on $\Sigma$ which takes real values, and a *measure* is a countably additive complex-valued set function.

If we are given a measure space $(X, \Sigma, m)$ so that $(X, \Sigma)$ is a measurable space and $m$ is a positive measure, we have the notion of integral with respect to $m$. We shall denote by $\mathscr{L}(m)$ the collection of functions integrable with respect to $m$, and, if $f \in \mathscr{L}(m)$, its integral is denoted by $\int_X f\, dm$ or $\int f\, dm$. The map $f \mapsto \int f\, dm$ is a linear functional on the linear space $\mathscr{L}(m)$.

Abstract integration theory handles many limiting operations with great ease. We have the following convergence theorems.

***Theorem***

*Let $(X, \Sigma, m)$ be a measure space, and let $f$ and $f_1, f_2, \ldots$ be measurable functions.*

(a) *Lebesgue's Monotone Convergence Theorem: If $0 \le f_1(x) \le f_2(x) \le \cdots \le \infty$ for almost every $x$, and if $f_n(x) \to f(x)$ for almost every $x$, then*

$$\lim_{n\to\infty} \int_X f_n\, dm = \int_X f\, dm.$$

(b) *Lebesgue's Dominated Convergence Theorem: If $f(x) = \lim_{n\to\infty} f_n(x)$ for almost every $x$, and if $|f_n| \le g$ for some $g \in \mathscr{L}(m)$*

*and all n, then* $f \in \mathscr{L}(m)$, *and*

$$\lim_{n\to\infty} \int_X |f_n - f|\, dm = 0,$$

*and*

$$\lim_{n\to\infty} \int_X f_n\, dm = \int_X f\, dm.$$

(c) *Fatou's Lemma: If each* $f_n$ *is nonnegative, then*

$$\int_X (\liminf f_n)\, dm \leq \liminf \int_X f_n\, dm.$$

Given $(X, \Sigma)$ and a positive measure $m$, we can define an equivalence relation on the collection of measurable functions by saying that $f \sim g$ if $f(x) = g(x)$ for all $x$ outside a set $E \in \Sigma$ with $m(E) = 0$. The space $L^p(m)$, $0 < p < \infty$, is defined to be the space of equivalence classes which contain an element $f$ such that $|f|^p \in \mathscr{L}(m)$. It is meaningful, in an obvious way, to write $\int f\, dm$ if $f \in L^1(m)$, even though, strictly construed, $f$ is not a function. For $f \in L^p(m)$, we write

$$\|f\|_p = \left( \int |f|^p\, dm \right)^{1/p}.$$

Certain important inequalities hold for the norms $\|\cdot\|_p$.

**Theorem**

(a) *Minkowski's Inequality: If* $f, g \in L^p(m)$, *then* $f + g \in L^p(m)$, *and if* $1 \leq p$, *then* $\|f\|_p + \|g\|_p \geq \|f + g\|_p$.

(b) *Holder's Inequality: If* $(1/p) + (1/q) = 1$, *and if* $f \in L^p(m)$ *and* $g \in L^q(m)$, *then* $fg \in L^1(m)$, *and* $\|fg\|_1 \leq \|f\|_p \|g\|_q$.

(c) *Jensen's Inequality: If m is of total mass* 1, *and if g is a real valued element of* $L^1(m)$, *then* $\int e^g\, dm \leq \exp \int g\, dm$.

The following very simple proof of (c) is from Hoffman [2], where it is attributed to S. Orszag. We can suppose that $\int g\, dm = 0$. But then since $e^g \geq 1 + g$, $\int e^g\, dm \geq \int (1 + g)\, dm = 1 = \exp \int g\, dm$.

A measurable function is *m-essentially bounded* if there exists a constant $C$ such that

$$m(\{x \in X : |f(x)| > C\}) = 0.$$

The *essential supremum* of $|f|$ is the greatest lower bound of all the permissible constants $C$. The space $L^\infty(m)$ is the space of equivalence classes of measurable functions which contain an essentially bounded function. It is plainly closed under addition and multiplication. For $f \in L^{\infty}(m)$, we write $\| f \|_{\infty}$ = essential supremum $|f|$.

***Theorem***
*If $\{f_n\}$ is a sequence in $L^p(m)$ such that $\| f_n - f \|_p \to 0$, there is a subsequence of $\{f_n\}$ which converges almost everywhere to $f$.*

***Theorem***
*The space $L^p(m)$ is complete: If $\{f_n\}$ is a sequence in $L^p(m)$ such that $\lim_{i,j} \| f_i - f_j \|_p \to 0$, then there is $f \in L^p(m)$ such that*

$$\lim_{n \to \infty} \| f_n - f \|_p = 0.$$

If $(X, \Sigma, m)$ is a measure space, $m$ is said to be *finite* if $m(X) < \infty$ and *$\sigma$-finite* if $X = \bigcup_n X_n$, where $m(X_n) < \infty$ for every $n$.

It is possible to identify the dual space of $L^p(m)$:

***Theorem***
*Let $\varphi$ be a linear functional on $L^p(m)$ with*

$$\|\varphi\| = \sup\{|\varphi(f)| : \| f \|_p \le 1\}$$

*finite. If $1 < p < \infty$ and $(1/p) + (1/q) = 1$, there is a unique $g \in L^q(m)$ such that for all $f \in L^p(m)$,*

$$\varphi(f) = \int fg \, dm \tag{1}$$

*and such that $\|g\|_q = \|\varphi\|$. If $p = 1$ and $m$ is $\sigma$-finite, there is a unique $g \in L^{\infty}(m)$ such that the representation (1) holds for all $f \in L^1(m)$ and such that $\|g\|_{\infty} = \|\varphi\|$.*

The dual space of $L^{\infty}(m)$ is much more complicated.

If $(X, \Sigma, m)$ is a measure space, the measure $\mu$ on $X$ is said to be *absolutely continuous with respect to $m$*, $\mu \ll m$, if $\mu(E) = 0$ for every measurable set $E$ with $m(E) = 0$, and $\mu$ is *singular* with respect to $m$ if there are disjoint measurable sets $E$ and $F$ such that $m$ is concentrated on $F$ in the sense that $m(S) = m(S \cap F)$ for all measurable sets $S$ and $\mu$ is concentrated on $E$. The measure $\mu$ is said to be *continuous* if it vanishes on every singleton. A point $x$ of $X$ is an *atom* for $\mu$ if $\mu(\{x\}) \neq 0$, and $\mu$ is *purely atomic* if it is concentrated on a set of atoms. Every measure can be written as the sum of a continuous measure and a purely atomic measure.

***Theorem*** (Radon–Nikodym Theorem)
*If m is a σ-finite positive measure and if μ is a measure absolutely continuous with respect to m, then there is a unique* $f \in L^1(m)$ *such that*

$$\mu(E) = \int_E f\, dm$$

*for every measurable set E.*

The function $f$ is called the *Radon–Nikodym derivative* of $\mu$ with respect to $m$ and is denoted by $d\mu/dm$.

If we have a measure $\mu$, there is an associated positive measure, $|\mu|$, defined by

$$|\mu|(E) = \sup \sum_i |\mu(E_i)|,$$

where the supremum is taken over all sequences $\{E_i\}$ of pairwise disjoint measurable subsets of $E$ whose union is $E$. The measure $\mu$ is said to be $\sigma$-finite if its total variation, $|\mu|$, is $\sigma$-finite.

The following is a simple consequence of the Radon–Nikodym theorem.

***Corollary***
*If μ is a σ-finite measure, there exists a measurable function h such that* $|h| = 1$ *a.e.* $[|\mu|]$ *and*

$$\mu(E) = \int_E h\, d|\mu|$$

*for all measurable sets E.*

In other words, the Radon–Nikodym derivative $d\mu/d|\mu|$ is of modulus 1 almost everywhere.

There are certain decompositions for measures which are used frequently.

***Theorem*** (Lebesgue Decomposition)
*If m is a positive measure and μ a measure on* $(X, \Sigma)$, *both σ-finite, there is a decomposition* $\mu = \mu_0 + \mu_1$, *where* $\mu_0 \ll m$ *and* $\mu_1 \perp m$.

A real measure $\mu$ is said to be *positive with respect to the measurable set E* if $\mu(E') \geq 0$ for every measurable set $E' \subset E$, and it is *negative with respect to E* if $-\mu$ is positive with respect to $E$.

***Theorem*** (Hahn Decomposition)
*If $\mu$ is a real measure, there exist two complementary measurable sets $E_+$ and $E_-$ such that $\mu$ is positive with respect to $E_+$, negative with respect to $E_-$.*

The most important single fact about integration on locally compact spaces is the Riesz representation theorem. The formulation of this result requires the notion of regularity for Borel measures.

***Definition***
*If $X$ is a locally compact space and $m$ a positive Borel measure on $X$, a Borel set $E$ is m-**inner regular** if it has the property that $m(E) = \sup \{m(K): K$ a compact subset of $E\}$, and it is m-**outer regular** if $m(E) = \inf\{m(V): V$ is an open set containing $E\}$. If $E$ is both m-inner and m-outer regular, it is termed m-**regular**. The measure m is said to be **regular** if every Borel set is m-regular. A complex measure is said to be **regular** if its total variation is a regular measure.*

If a locally compact space is not too large, certain Borel measures are automatically regular.

***Theorem***
*If $X$ is a locally compact space in which every open set is $\sigma$-compact, then a positive Borel measure m on $X$ is regular provided $m(K) < \infty$ for every compact set $K$.*

It is probably worth explicit mention that not every $\sigma$-compact (or compact, for that matter) space $X$ satisfies the hypotheses of the last theorem.

(Riesz Representation Theorem)
*If $X$ is a locally compact space and $\varphi$ is a linear functional on $\mathscr{C}_0(X)$ such that*

$$\|\varphi\| = \sup\{|\varphi(f)| : f \in \mathscr{C}_0(X), \|f\|_X \leq 1\}$$

*is finite, there is a unique finite regular Borel measure m on $X$ such that for all $f \in \mathscr{C}_0(X)$*

$$\varphi(f) = \int_X f\, dm.$$

*The measure m satisfies $\|m\| = \|\varphi\|$.*

It is necessary sometimes to deal with positive linear functionals $\varphi$ on $\mathscr{C}_c(X)$, the space of continuous functions with compact support on the locally compact space $X$. We understand by *positive* the qualification that $\varphi(f) \geq 0$ if $f \in \mathscr{C}_c(X)$ is nonnegative.

***Theorem***

*If $X$ is a locally compact space and $\varphi$ is a positive linear functional on $\mathscr{C}_c(X)$, there exists a $\sigma$-algebra $\Sigma$ of subsets of $X$ which contains all the Borel sets and a unique positive measure $m$ on $\Sigma$ such that $\varphi(f) = \int f\,dm$ for all $f \in \mathscr{C}_c(X)$. Every set in $\Sigma$ is outer regular and every open set and every set of finite measure is inner regular. Moreover, if $E \subset F$, $F \in \Sigma$, and $m(F) = 0$, then $E \in \Sigma$.*

In general, for locally compact spaces that are too large, it is not possible to find a regular measure that represents a given positive linear functional on $\mathscr{C}_c(X)$.

In the theory of Borel functions on locally compact spaces, Lusin's theorem, an approximation theorem, plays an important role.

***Theorem***

*Let $X$ be a locally compact space, let $f$ be a Borel measurable function on $X$, and let $m$ be a measure with the properties of the last theorem. If $f$ vanishes off a set $E$ with $m(E) < \infty$, and if $\varepsilon > 0$, there is $g \in \mathscr{C}_c(X)$ with $m(\{x : f(x) \neq g(x)\}) < \varepsilon$ and $\sup_{x \in X} |g(x)| \leq \sup_{x \in X} |f(x)|$.*

A *topological group* is a group on which there is defined a topology with respect to which the group operations are continuous. If $G$ is an abelian topological group, its *dual group* is the group of all *characters* of $G$, i.e., the multiplicative group of all homomorphisms of $G$ into $T$, the group of complex numbers of modulus 1, the group operation being multiplication.

A theorem basic for most of analysis on groups affirms the existence of invariant integrals. Let $G$ be a locally compact group with the operation written multiplicatively, and let $f$ be a complex-valued function on $G$. If $x \in G$, define the *left* (*right*) *translate*, $f_x(f^x)$, of $f$ by $f_x(y) = f(xy)$ $(f^x(y) = f(yx))$ for all $y \in Y$. A linear functional $\varphi$ on $\mathscr{C}_c(G)$ is called *left* (*right*) *invariant* if $\varphi(f_x) = \varphi(f)$ $(\varphi(f^x) = \varphi(f))$ for all $x \in G$, $f \in \mathscr{C}_c(G)$.

***Theorem***

*If $G$ is a locally compact group, there exists on $\mathscr{C}_c(G)$ an invariant positive linear functional and, up to a multiplicative constant factor, only one.*

If we combine this fact with the representation theorem for positive linear functionals, we obtain the Haar measure on $G$, and we see in particular that if $G$ is compact and abelian, there is a unique regular Borel probability measure $m$ on $G$ such that $\int_G f_x\,dm = \int_G f\,dm$ for all $f \in \mathscr{C}(X)$. The Haar measure has the property that if $V$ is open in $G$, then $m(V) > 0$. Since Haar measure is essentially unique, the notation $L_p(G)$ may be used unambiguously: It denotes the $L_p$-space associated with some (and therefore every) Haar measure on $G$.

The existence and uniqueness of Haar measure on compact groups can be used to produce some interesting examples of nonmeasurable sets. Let $G_\sigma$ be a compact abelian group with Haar measure $m_\sigma$ ($m_\sigma(G_\sigma) = 1$) for each $\sigma \in S$, $S$ an uncountable index set, and let $G = \Pi_{\sigma\in S} G_\sigma$ be the product endowed with the product topology and the product group structure so that $G$ is also a compact abelian group. Denote by $m$ the Haar measure on $G$ normalized so that $m(G) = 1$. If $\sigma \in S$, let $\pi_\sigma : G \to G_\sigma$ be the projection onto the $\sigma$th coordinate, and if $S_0$ is a finite set in $S$, say $S_0 = \{\sigma_1, \ldots, \sigma_k\}$, let $\pi_{S_0} : G \to G_{\sigma_1} \times \cdots \times G_{\sigma_k}$ be the projection $\pi_{S_0}(x) = (\pi_{\sigma_1}(x), \ldots, \pi_{\sigma_k}(x))$. For each choice of $S_0$, the map

$$f \mapsto \int (f \circ \pi_{S_0})\,dm$$

is a positive linear functional on $\mathscr{C}(G_{\sigma_1} \times \cdots \times G_{\sigma_k})$ which is translation invariant, so it follows that the Haar measure $m_{S_0}$ on $G_{\sigma_1} \times \cdots \times G_{\sigma_k}$ satisfies $m_{S_0}(E) = m(\pi_{S_0}^{-1}(E))$. For each $\sigma \in S$, let $E_\sigma \subset G$ be a Borel set with $m_\sigma(E_\sigma) = 1$ but $E_\sigma \neq G_\sigma$. (This imposes the implicit restriction that each $G_\sigma$ be infinite.) The set $E = \Pi_{\sigma\in S} E_\sigma$ is not a Borel set, as we can show by invoking the regularity of $m$. If $U \subset G$ is an open set which contains $E$, then, by the definition of the product topology, $U \supset \pi_{S_0}^{-1}(\pi_{S_0}(E))$ for some finite $S_0 \subset S$, and thus $m(U) \geq m_{S_0}(\pi_{S_0}(E)) = 1$. On the other hand, if $K \subset E$ is compact, then since for no $\sigma$ is $E_\sigma = G_\sigma$, $m_\sigma(\pi_\sigma(K)) < 1$ holds for all $\sigma$, so there is a sequence $\sigma_1, \sigma_2, \ldots$ in $S$ and an $\alpha < 1$ with $m_{\sigma_j}(\pi_{\sigma_j}(K)) < \alpha$ for $j = 1, 2, \ldots$. Then if $S_k = \{\sigma_1, \ldots, \sigma_k\}$, $K \subset \pi_{S_k}^{-1}(\pi_{S_k}(E))$, so it follows that

$$\begin{aligned} m(K) &\leq m_{S_k}(\pi_{S_k}(E)) \\ &= m_{\sigma_1}(E_{\sigma_1}) \cdots m_{\sigma_k}(E_{\sigma_k}) \\ &\leq \alpha^k. \end{aligned}$$

As this inequality holds for all $k$, we have $m(K) = 0$. The fact that $m$ is a regular Borel measure now implies that $E$ is not a Borel set.

If $G$ is a locally compact abelian group with Haar measure $m$ and dual group $\Gamma$, $\Gamma$ has a natural topology, that of pointwise convergence so that a net $\{\chi_\alpha\}$ of characters on $G$ converges to $\chi$ if and only if $\chi_\alpha(g) \to \chi(g)$ for every $g \in G$. With this topology, $\Gamma$ itself is a locally compact group. If $G$ is compact, $\Gamma$ is discrete, and if $G$ is discrete, $\Gamma$ is compact.

The *Pontryagin duality theorem* states that in a natural way, $G$ is the dual of $\Gamma$: *Every character of* $\Gamma$ is of the form $\gamma \mapsto \langle g, \gamma \rangle$ *for some fixed* $g \in G$. (Here we use the symbol $\langle g, \gamma \rangle$ to denote the value of the character $\gamma$ at the point $g$ of $G$.)

If $\mathcal{M}(G)$ denotes the space of finite regular Borel measures on $G$, then given $\sigma \in \mathcal{M}(G)$, $\hat{\sigma}$, the *Fourier* or *Fourier–Stieltjes transform* of $\sigma$ is the function on $\Gamma$ given by $\hat{\sigma}(\gamma) = \int_G \langle -g, \gamma \rangle \, dm(g)$. The *Fourier transform*, $\hat{f}$, of the integrable function $f$ on $G$ is defined to be the Fourier–Stieltjes transform of the measure $fm$ so that

$$\hat{f}(\gamma) = \int_G \langle -g, \gamma \rangle f(g) \, dm(g).$$

The Fourier transform is an extremely powerful analytic tool, of whose properties we shall recall but three. First, there is a uniqueness theorem: *If* $\sigma \in \mathcal{M}(G)$, *the Fourier–Stieltjes transform* $\hat{\sigma}$ *vanishes identically if and only if* $\sigma$ *is the zero measure.* [If $\mathcal{M}(G)$ is provided with convolution for multiplication, it becomes a commutative Banach algebra, and it is a consequence of the uniqueness theorem that it is semisimple.] Second, there is the *inversion theorem.* If $G$ is compact, it is natural to require that the Haar measure be a probability measure, and if $G$ is discrete the Haar measure is most naturally required to assign to each point mass 1. If, however, $G$ is neither compact nor discrete, there is no a priori obvious way of normalizing the Haar measure. The inversion theorem provides a natural way of normalizing the Haar measure on $\Gamma$ granted a choice of Haar measure on $G$. Denote by $B(G)$ the set of all functions on $G$ which are of the form $f(g) = \int_\Gamma \langle g, \gamma \rangle \, d\sigma(\gamma)$ for some $\sigma \in \mathcal{M}(\Gamma)$.

***Theorem***

*If* $f \in L_1(G) \cap B(G)$, *then* $\hat{f} \in L_1(\Gamma)$, *and if the Haar measure on* $G$ *is fixed, the Haar measure on* $\Gamma$, *call it* $\mu$, *can be normalized so that for all* $f \in L_1(G) \cap B(G)$,

$$f(g) = \int_\Gamma \hat{f}(\gamma) \langle g, \gamma \rangle \, d\mu(\gamma).$$

We can state now the Plancherel theorem.

***Theorem***

*If the Haar measures on* $G$ *and* $\Gamma$ *are normalized as in the inversion theorem, the map from* $L_1(G) \cap L_2(G)$ *given by* $f \mapsto \hat{f}$ *is an isometry (with respect to the* $L_2$ *norm) onto a dense subspace of* $L_2(\Gamma)$ *and so extends uniquely to an isometry of* $L_2(G)$ *onto* $L_2(\Gamma)$.

For certain purposes it is convenient to use a process given by Carathéodory for the construction of measures. Let $X$ be a set and let $m$ be an *outer measure on* $X$ so that $m$ is a function defined on all subsets of $X$ and satisfies (a) $m(\varnothing) = 0$, (b) $m(E) \leq m(F)$ if $E \subset F$, and (c) $m(\bigcup E_n) \leq \sum m(E_n)$ for every sequence $\{E_n\}$ of subsets of $X$. In this context a set $E$ is said to be $m$-measurable if

$$m(E \cap F) + m(E^c \cap F) = m(F)$$

for every set $F$. It can be shown that the set $\Sigma$ of $m$-measurable sets is a $\sigma$-algebra and that $m$ is countably additive on it. In this way we pass from an outer measure $m$ on $X$ to a measure space $(X, \Sigma, m)$.

An important case of this construction occurs in connection with Hausdorff measures on metric spaces. Let $X$ be a metric space with metric $\rho$. For every $p > 0$, we shall define an outer measure $\Lambda_p$ on $X$. To do this, start by defining for each $\varepsilon > 0$ and each $E \subset X$,

$$\Lambda_p^\varepsilon(E) = \inf \sum (\delta(E_j))^p,$$

where the infimum extends over all partitions $\{E_j\}_{j=1}^\infty$ of $E$ with the property that for each $j$, $\delta(E_j)$, the diameter of $E_j$, is no more than $\varepsilon$. If no partitions of this kind exist for $E$, set $\Lambda_p^\varepsilon(E) = \infty$. If $\varepsilon < \varepsilon'$, then $\Lambda_p^\varepsilon(E) \geq \Lambda_p^{\varepsilon'}(E)$, so $\lim_{\varepsilon \to 0^+} \Lambda_p^\varepsilon(E)$ exists either as a real number or as $+\infty$. Call this limit $\Lambda_p(E)$. It can be shown that the set function $\Lambda_p$ is a Carathéodory outer measure on $X$. It is called the *p-dimensional Hausdorff measure on X (computed with respect to the metric* $\rho$). If we take for $X$ the usual euclidean $n$-dimensional space $\mathbb{R}^n$ and for $\rho$ the usual euclidean metric, then the induced Hausdorff measure is, except for a normalizing factor, Lebesgue measure.

*References*

Hewitt, E., and K. Ross, *Abstract Harmonic Analysis*, Springer-Verlag, Berlin, two vols., 1963 and 1969.

Hewitt, E., and K. Stromberg, *Real and Abstract Analysis*, Springer-Verlag, Berlin, 1965.

Saks, S., Theory of the integral, *Monografie Matematyczne*, vol. 7, Warsaw, 2nd ed., 1937; reprinted by Hafner, New York, and by Dover, New York.

The theory of measure and integration can be found in many places. Our outline follows the treatment in Hewitt and Stromberg and in Rudin [2], and most of the quoted results can be found in either of these sources. For harmonic analysis on groups, see Rudin [1], and for an exhaustive treatment, Hewitt and Ross. The theory of Hausdorff measures can be found in the book of Saks. The point about nonmeasurable sets is from Hewitt and Ross.

# *appendix c*/FUNCTIONAL ANALYSIS

In this appendix we deal with real and complex topological vector spaces. A topological vector space over $k = \mathbb{C}$ or $\mathbb{R}$ is a vector space $E$ together with a topology such that the map $(x, y) \mapsto x - y$ from $E \times E$ to $E$ is continuous, and the map $(\alpha, x) \mapsto \alpha x$ from $k \times E$ to $E$ is also continuous. The space $E^*$ *dual* to $E$ is the space of $k$-valued, continuous linear functionals. In general, $E^* = \{0\}$, but if we deal with locally convex spaces, $E^*$ is large enough to be useful.

Define a *seminorm* on a vector space $E$ to be a positive-valued, positive homogeneous, sublinear functional $\|\cdot\|$:

$$0 \leq \|x\| \text{ for all } x \in E,$$

$$\|x + y\| \leq \|x\| + \|y\|,$$

and

$$\|\lambda x\| = |\lambda|\|x\|.$$

A seminorm $\|\cdot\|$ is a norm if $\|x\| = 0$ implies that $x = 0$. A topological vector space is *locally convex* if there exists a family $\mathcal{S}$ of seminorms such that the sets

$$N_{s,\alpha} = \{x : s(x) < \alpha\} \qquad s \in \mathcal{S}, \alpha > 0,$$

constitute a basis for the topology at the origin. Thus, a net $\{x_\gamma\}$ converges to $x$ if and only if $\lim_\gamma s(x_\gamma - x) = 0$ for every $s \in \mathcal{S}$.

Among the locally convex spaces are the Fréchet, normed, and Banach spaces. If $k = \mathbb{C}$ or $\mathbb{R}$, a *Fréchet space* over $k$ is a locally convex space $E$ in which the topology is metrizable, a condition which is equivalent to the condition that there exists a *sequence* $\{\|\cdot\|_n\}_{n=1}^{\infty}$ of seminorms which define the topology and which, moreover, is complete in the sense that every sequence which is Cauchy with respect to each of the seminorms $\|\cdot\|_n$ converges. A space $E$ is *normable* if there exists a single norm on $E$ which defines the topology of $E$, and a *normed space* is a space $E$ together with a norm which defines the topology. A *Banach space* is a normed space which is complete with respect to the norm. If $E$ is a Banach space with norm $\|\cdot\|$, the dual $E^*$ is a Banach space under the norm $\|\cdot\|^*$ given by

$$\|\varphi\|^* = \sup\{|\varphi(x)| : x \in E, \|x\| \leq 1\}.$$

According to the Hahn–Banach theorem, a locally convex topological vector space has a rich supply of continuous linear functionals.

***Theorem***

*If $E$ is a real or complex vector space and $s$ is a seminorm on $E$, then given a subspace $E_0$ of $E$ and a linear functional $\varphi$ on $E_0$ which satisfies $|\varphi(x)| \leq s(x)$ for $x \in E_0$, there is a linear functional $\psi$ on $E$ with $\psi|E_0 = \varphi$ and $|\psi(x)| \leq s(x)$ for all $x \in E$.*

***Corollary***

*If $s$ is a seminorm on $E$, then for every $x_0 \in E$, there is a linear functional $\varphi$ with $|\varphi(x)| \leq s(x)$ for all $x$ and $\varphi(x_0) = s(x_0)$.*

Closely related to the Hahn–Banach theorem are certain separation theorems. Call a subset $K$ of a real or complex vector space *convex* if given $x, y \in K$ and $\lambda \in [0, 1]$, the point $\lambda x + (1 - \lambda)y$ also lies in $K$. If $S$ and $T$ are sets in the vector space $E$ over $k = \mathbb{C}$ or $\mathbb{R}$, a linear functional $\varphi$ is said to *separate $S$ and $T$* if

(1) $$\sup\{\text{Re } \varphi(x) : x \in S\} \leq \inf\{\text{Re } \varphi(x) : x \in T\}.$$

If the inequality (1) is strict, $\varphi$ is said to separate $S$ and $T$ *strongly*. [Of course, if $k = \mathbb{R}$, the Re in (1) is redundant.]

***Theorem***

(a) *If $A$ and $B$ are nonvoid convex sets in the locally convex space $E$, $A$ with nonempty interior, $A^0$, and if $B$ is disjoint from $A^0$, there is a continuous linear functional which separates $A$ and $B$.*
(b) *If $A$ and $B$ are disjoint, $A$ compact and $B$ closed, then $A$ and $B$ can be separated strongly by a linear functional.*
(c) *If, moreover, $B$ is* ***circled*** *in that $\alpha \in k, |\alpha| \leq 1$, and $x \in B$ imply $\alpha x \in B$, then there is a continuous linear functional $\varphi$ such that*

$$\sup\{|\varphi(x)| : x \in B\} < \inf\{|\varphi(x)| : x \in A\}.$$

If $K$ is a convex set in a vector space $E$, a point $x \in K$ is said to be an *extreme point* if $x = \lambda y' + (1 - \lambda)y''$ with $\lambda \in [0, 1]$, $y', y'' \in K$ implies that $x = y' = y''$. Generally, a convex set has no extreme points, but the fundamental Krein–Milman theorem provides extreme points in certain important cases.

***Theorem***

*If $K$ is a compact, convex subset of a locally convex topological vector space $E$, $K$ has extreme points, and indeed, $K$ is the closed convex hull* *of its set of extreme points in the sense that finite convex combinations* $\lambda_1 x_1 + \cdots + \lambda_n x_n$, *where* $x_1, \ldots, x_n$ *are extreme points and* $\lambda_1, \ldots, \lambda_n$ *are nonnegative numbers with sum* 1, *are dense in* $K$.

An extension of the Krein–Milman theorem is the Choquet–Bishop–deLeeuw theorem. Let $K$ be a compact, convex subset of the locally convex space $E$. If $x_0 \in K$, say that the probability measure $m$ on $K$ *represents* $x_0$ if for every continuous linear functional $\varphi$ on $E$, $\varphi(x_0) = \int \varphi(x)\, dm(x)$. For a given point in $K$ there are always representing measures, e.g., the point mass at the given point. The Krein–Milman theorem implies that it is possible to choose $m$ supported on the closure of the set of extreme points. More is true.

***Theorem*** (Choquet–Bishop–deLeeuw)
*If $K$ is a compact convex set in a locally convex topological vector space and if $x_0 \in K$, there exists a probability measure, i.e., a regular positive Borel measure of total mass* 1, $m$, *on $K$ which represents $x_0$ and which vanishes on every Baire subset and on every $G_\delta$ subset of $K$ disjoint from the set of extreme points of $K$.*

Recall that the Baire sets of a compact space are the elements of the smallest $\sigma$-ring of sets containing the compact sets of type $G_\delta$.

In the Choquet–Bishop–deLeeuw theorem, it would be desirable to obtain a representing measure supported on the set of extreme points of $K$, but generally this is not possible because this set is not necessarily a Borel set. In case $E$ is a Fréchet space, it can be shown that the set of extreme points is a Borel set, so the desired improvement can be obtained.

For Fréchet spaces and, in particular, for Banach spaces there are certain important results which depend on the Baire category theorem.

***Theorem*** (Open Mapping Theorem)
*If $E$ and $F$ are Fréchet spaces and $T: F \to F$ is a continuous linear operator with $T(E) = F$, then $T$ is an open map in that if $U$ is open in $E$, $T(U)$ is open in $F$.*

***Theorem*** (Closed Graph Theorem)
*If $E$ and $F$ are Fréchet spaces, a linear map $T: E \to F$ is continuous if its graph $\{(x, T(x)) : x \in E\}$ is closed in $E \times F$.*

***Theorem*** (Banach–Steinhaus Theorem)
*If $\mathscr{F}$ is a family of continuous linear functionals on a topological vector space $E$, and if $\mathscr{F}$ is pointwise bounded on a set of second category in $E$, then $\mathscr{F}$ is equicontinuous.*

In Banach spaces, this result reads as follows:

***Corollary***
*If $\mathscr{F}$ is a family of continuous linear functionals on the Banach space E, and if $\mathscr{F}$ is pointwise bounded, then $\mathscr{F}$ is uniformly bounded: There is C such that if $\varphi \in \mathscr{F}$, then $\|\varphi\| \leq C$.*

Given a topological vector space $E$ with dual space $E^*$, there are several topologies on $E$ and $E^*$ which arise naturally. The *weak topology on E induced by E** is the topology induced by the seminorms $\|\cdot\|_\varphi$, $\varphi \in E^*$, given by $\|x\|_\varphi = |\varphi(x)|$, $x \in E$, and the *weak topology on E* induced by E* is the topology induced by the seminorms $\|\cdot\|_x$, $x \in E$, given by $\|\varphi\|_x = |\varphi(x)|$ for all $\varphi \in E^*$. In the case of Banach spaces $E$, the weak topology on $E^*$ induced by $E$ is frequently referred to as the *weak * topology.*

***Theorem***
*If E is a locally convex topological vector space with dual E*, a linear functional L on E* is continuous with respect to the weak topology on E* induced by E if and only if for some $x_0 \in E$, and all $\varphi \in E^*$, $L(\varphi) = (x_0)$.*

In the Banach-space setting there is the simple but important theorem of Alaoglu which follows easily from the Tychonoff theorem.

***Theorem***
*If E is a Banach space with dual E*, the closed unit ball of E* is compact in the weak * topology.*

If $E$ and $F$ are topological vector spaces and $T: E \to F$ is a continuous linear operator, there is a naturally defined dual map $T^*: F^* \to E^*$ given by $(T^*\varphi)(x) = \varphi(Tx)$ for $x \in E$, $\varphi \in F^*$. An important example of the interrelation between the properties of $T$ and $T^*$ is the following fact.

***Theorem***
*If E and F are Fréchet spaces and $T: E \to F$ is a continuous linear map, the range of T is closed if and only if the range of $T^*$ is.*

Another frequently used, though more elementary, criterion for the closure of the range of a linear operator $T: E \to F$, $E$ and $F$ Banach spaces, is the fact that $T(E)$ is closed if and only if for some constant $\alpha$, given $x \in E$, there is $y \in E$ with $T(y) = T(x)$ and $\|y\| \leq \alpha \|T(x)\|$.

The Krein–Smulian theorem also enables one to conclude that certain subspaces are closed.

***Theorem***
*Let $E$ be a Fréchet space over $k$, the reals or complexes, and let $E^*$ be the dual space. A convex set $K$ in $E^*$ is closed in the weak topology on $E^*$ induced by $E$ if and only if $M \cap K$ is closed in this topology for every closed, convex, equicontinuous subset $M$ which is circled.*

***Corollary*** (Banach)
*If $E$ is a Banach space with dual $E^*$, a linear subspace $K$ of $E$ is weak * closed if and only if its intersection with the unit ball of $E^*$ is weak * closed.*

We need another, somewhat more technical, theorem of Banach. Its proof will depend on the fact that if $E$ is a separable Banach space, then the unit ball of $E^*$, taken with the weak * topology, is metrizable. This follows quickly from the observation that if $\{x_n\}_{n=1}^{\infty}$ is a countable dense subset of the unit ball in $E$, and if we define seminorms $\|\cdot\|_n$ on $E^*$ by $\|\varphi\|_n = |\varphi(x_n)|$, then a net $\{\varphi_\gamma\}_{\gamma\in\Gamma}$ in the unit ball of $E^*$ converges to 0 if and only if $\lim_{\gamma\in\Gamma} \|\varphi_\gamma\|_n = 0$ for every $n$.

***Theorem*** (Banach)
*Let $E$ be a separable Banach space with dual space $E^*$, and let $M$ be a subspace of $E^*$. In order that every element of $E^*$ be the weak* limit of a sequence in $M$, it is necessary and sufficient that for some positive constant $c$ and for every $x \in E$, there be a $\varphi \in M$ with $\|\varphi(x)\| = \|x\|$ and $\|\varphi\| \le c$. When this condition is met, every $\varphi \in E^*$ is the weak * limit of a sequence in $M$ uniformly bounded in norm by $c\|\varphi\|$.*

PROOF
*Necessity.* For every $N$, let $D_N$ be the subset of $E^*$ consisting of those functionals which are weak * limits of sequences each element of which is bounded by $N$ in norm. Under the assumption that $M = E^*$, we have that $\bigcup D_N = E^*$, because, by the Banach–Steinhaus theorem, a weak * convergent net in $E^*$ is norm bounded and because the unit ball in $E^*$ is metrizable with respect to the weak * topology.

We want to invoke the Baire category theorem, and to do this, we note that all the sets $D_N$ are norm closed. To see this, let $\{x_k\}_{k=1}^{\infty}$ be a countable dense set in $E$. If $\{\varphi_m\}_{j=1}^{\infty}$ is a sequence in $D_N$ with $\|\varphi_m - \varphi\| \to 0$, and if, for every $m$, $\{\varphi_{m,j}\}_{j=1}^{\infty}$ is a sequence in $M$ such that $\|\varphi_{m,j}\| \le N$ and $\varphi_{m,j}(x) \to \varphi_m(x)$ for all $x \in E$, a diagonal process yields a sequence $\{\varphi_{m_s,j_s}\}_{s=1}^{\infty}$ which converges weak * to $\varphi$. Thus, $\varphi \in D_N$, and $D_N$ is norm closed. By category, there is an $N_0$ such that $D_0 = D_{N_0}$ contains a ball, say the ball $B(\psi, r) = \{\varphi \in E^* : \|\varphi - \psi\| \le r\}$ for some $r > 0$. If $x \in E$, the Hahn–Banach theorem provides $\varphi \in E^*$ such that $|\varphi(x)| = \|x\|$, $\|\varphi\| = 1$. If $\lambda = r(1 + \|\psi\|)^{-1}$ and $\eta = \lambda\varphi + (1 - \lambda)\psi$, then

$$\begin{aligned}\|\eta - \psi\| &= \|\lambda\varphi + (1 - \lambda)\psi - \psi\| \\ &= \lambda\|\varphi - \psi\| \\ &= r(1 + \|\psi\|)^{-1}\|\varphi - \psi\| \\ &\le r\end{aligned}$$

so that $\eta \in B(\psi, r) \subset D_0$. It follows that there exist sequences $\{\eta_k\}_{k=1}^{\infty}$ and $\{\psi_k\}_{k=1}^{\infty}$ in $M$ which converge weak * to $\eta$ and $\psi$, respectively, and which are bounded by $N_0$ in norm.

If $\varphi_k = \lambda^{-1}\eta_k + (1 - \lambda)\lambda^{-1}\psi_k$, then $\varphi_k \in M$ and $\{\varphi_k\}$ converges weak * to $\varphi$. Since $\|\varphi\| = 1$ and $|\varphi(x)| = \|x\|$, there is an index $k_0$ with

$$\lambda^{-1}\eta_{k_0} - (1 - \lambda)\lambda^{-1}\psi_{k_0} = \alpha\|x\|$$

for some $\alpha \in (\frac{1}{2}, 2)$. If $\tilde{\varphi} = \alpha^{-1}(\lambda^{-1}\eta_{k_0} - (1 - \lambda)\lambda^{-1}\psi_{k_0})$, then $\tilde{\varphi} \in M$, $|\tilde{\varphi}(x)| = \|x\|$, and $\|\tilde{\varphi}\| \le c$, where $c$ is the constant

$$2N_0(2^{-1} + 1) = 2N_0 r^{-1}(2 + 2\|\psi\| + r),$$

a constant that is independent of $x$.

*Sufficiency.* Assume that the constant $c$ has the properties mentioned in the statement of the theorem. Let $D$ be the unit ball of $M$, and let $\{\varphi_k\}_{k=1}^{\infty}$ be a countable dense subset of $D$. If we denote, as usual, the space of bounded sequences of complex numbers (with the supremum norm) by $l_\infty$, and if we define $T: E \to l_\infty$ by

$$T(x) = (\varphi_1(x), \varphi_2(x), \dots),$$

then $\|T\| \le 1$, and we shall show that $T$ has closed range. If $x$ is given, there is $\varphi \in M$ with $\|\varphi\| \le c$ and $|\varphi(x)| = \|x\|$. The functional $\tilde{\varphi} = c^{-1}\varphi$ lies in $D$, so there is a subsequence $\{\varphi_{k_j}\}_{j=1}^{\infty}$ of $\{\varphi_k\}$ which converges weak * to $\tilde{\varphi}$. We have $\lim_{j\to\infty}\varphi_{k_j}(x) = |\varphi(x)|$ so that $\limsup|\varphi_k(x)| \ge \|\tilde{\varphi}(x)\| \ge c^{-1}\|x\|$, which implies that $\|T(x)\| \ge c^{-1}\|x\|$. From this we conclude that $T$ is one to one and has closed range. It follows that $T^{-1}: T(E) \to E$ is a bounded linear operator. We have $\|T(x)\| \ge c^{-1}\|T^{-1}(T(x))\|$ so that $\|T^{-1}\| \le c$, and thus $\|(T^{-1})^*\| \le c$ where we use * to denote the dual or adjoint map.

We denote by $\beta(\mathbb{N})$ the Čech compactification of the positive integers, and if $X = (x_1, x_2, \dots)$ is in $l_\infty$, we let $\hat{X}$ be the canonical extension of

$X$ to an element of $\mathscr{C}(\beta(\mathbb{N}))$. For $j = 1, 2, \ldots$, let $m_j$ be the unit point mass at the positive integer $j$ regarded as a probability measure on $\beta(\mathbb{N})$. For each $j$ we have that

$$\varphi_j(x) = \int (Tx)\hat{}\, dm_j.$$

If $\varphi \in E^*$, $\|\varphi\| \leq 1$, then $\varphi = (T^{-1})^*\psi$ for some $\psi \in (T(E))^*$, $\|\psi\| \leq c$. The Hahn–Banach and Riesz representation theorems yield a measure $m$ on $\beta(\mathbb{N})$ of norm no more than $c$ such that

$$\varphi(x) = \int (T(x))\hat{}\, dm.$$

Since $\mathbb{N}$ is dense in $\beta(\mathbb{N})$, the measure $m/\|m\|$ is the weak $*$ limit of a net of measures of the form $\sum_{j=1}^{s} \alpha_j m_j$, where the coefficients $\alpha_j$ satisfy $\sum_{j=1}^{s} |\alpha_j| < 1$. From this it follows that the functional $\varphi$ is the weak $*$ limit of a sequence of functionals of the form $\sum_{j=1}^{s} \gamma_j \varphi_j$, each of which is of norm no more than $c$. As all the functionals $\varphi_j$ lie in $M$, we have the theorem.

A great deal of elementary calculus can be done in Banach spaces. From the calculus in this context, we need only one result, the implicit-function theorem. If $E$ and $F$ are Banach spaces and $f$ is an $F$-valued function on the open set $U$ of $E$, call $f$ *differentiable* at $x_0$ if there exists a continuous linear transformation $T_{x_0}: E \to F$ such that

$$\|f(x) - f(x_0) - T_{x_0}(x - x_0)\| \leq r(x - x_0)\|x - x_0\|,$$

where $r$ is a continuous real-valued function on $E$ which vanishes at the origin. If such an approximation is possible, the operator $T_{x_0}$ is called the *derivative*, $f'(x_0)$, of $f$ at $x_0$. The function $f$ is said to be *continuously differentiable in* $U$ if it is differentiable at every point of $U$ and if the map $x \mapsto f'(x)$ from $U$ to $\mathscr{L}(E, F)$ is continuous.

If we are given Banach spaces $E_1, E_2$, and $F$ and an open set $U \subset E_1 \times E_2$, then given $(x_1, x_2) \in U$, let $U_{x_2}$ be the open set $\{x \in E_1 : (x, x_2) \in U\}$. If $f$ is an $F$-valued function on $U$, the map $\tilde{f}: U_{x_2} \to F$ given by $\tilde{f}(x) = f(x, x_2)$ is well defined. If it has a derivative at $x_1$, we denote this derivative by $f_1(x_1, x_2)$ and call it the *partial derivative of* $f$ *with respect to* $x_1$. The function $f_1$ maps $U$ into $\mathscr{L}(E_1, F)$. In a similar way, the partial derivative of $f$ with respect to $x_2$ is defined.

As in finite-dimensional calculus the implicit-function theorem is valid.

***Theorem***

*Let $U \subset E_1 \times E_2$ be an open set, and let $f: U \to F$ be a continuously differentiable map, $E_1, E_2$, and $F$ Banach spaces. Let $f(x_1^0, x_2^0) = 0$,*

*$(x_1^0, x_2^0) \in U$, and suppose that the partial derivative $f_2(x_1^0, x_2^0)$ is invertible in $\mathscr{L}(E_2, F)$. In a neighborhood $W$ of $x_1^0 \in E_1$ there exists a unique, continuously differentiable, map $\varphi: W \to E_2$ with $\varphi(x_1^0) = x_2^0$ and* $f(x, \varphi(x)) = 0$.

Finally, we cite a result from elementary linear algebra.

### Theorem

*Let $V$ be a finite-dimensional complex vector space. Every commuting family, $\mathscr{F}$, of linear transformations of $V$ into itself has a common eigenvalue: There is $\lambda \in \mathscr{C}$ such that for each $F \in \mathscr{F}$, there is $x_F \in V$ with $F(x_F) = \lambda x_F$.*

In this result the complex field can be replaced by any algebraically closed field, and, moreover, for a suitable choice of basis in $V$, the matrices representing the elements of $\mathscr{F}$ are all upper triangular.

*References*


Banach, W., *Opérations linéaires*, Chelsea, New York, 1955.

Dieudonné, J., *Foundations of Modern Analysis*, Academic Press, New York, 1960.

Dunford, N., and J. T. Schwartz, *Linear Operators*, Wiley (Interscience), New York, 1958.

Horváth, J. *Topological Vector Spaces and Distributions*, vol. 1, Addison–Wesley, Reading, Mass., 1966.

Kelley, J. L., and I. Namioka, *Linear Topological Vector Spaces*, Van Nostrand Reinhold, New York, 1963.

Supruneko, D. A., and R. I. Tyskevich, *Commutative Matrices*, Academic Press, New York, 1968.


There are many sources for functional analysis. The books by Dunford and Schwartz, Horváth, and Kelley and Namioka contain most of what we have cited. The theorem of Banach which we proved is contained in his book, page 213. For calculus in Banach spaces, see Dieudonné. A proof of the theorem on commuting matrices may be found in Supruneko and Tyskevich, page 15.

# *appendix d*/COMPLEX FUNCTION THEORY

We collect in this appendix certain results related to the theory of holomorphic functions. The first portion is devoted to some results from the classical theory of functions of one complex variable, the second to some topics from several complex variables. Of course, we suppose the reader to be reasonably familiar with the elements of complex variables. One bit of notation will be used consistently: $\Delta$ will stand for the open unit disc in the complex plane.

We begin by recalling some results centering about Schwarz's lemma. *If $f$ is holomorphic and bounded by one in $\Delta$ and if $f(0) = 0$, then $|f(z)| \leq |z|$ for all $z$ in $\Delta$; strict inequality persists at all points unless $f(z) = cz$, $c$ a constant of modulus* 1. By using conformal maps of the disc onto itself, it follows easily that if we set $[z, w] = |(z - w)/(1 - z\bar{w})|$, and *if $f$ is a function holomorphic and bounded by* 1 *in $\Delta$ which vanishes at the point $w_0$, then $|f(z)| \leq [z, w_0]$ with strict inequality for all $z$ unless $f(z) = c[(z - w_0)/(1 - z\bar{w}_0)]$ for some unimodular constant, c.*

It can be verified that the mapping $(z, w) \mapsto [z, w]$ from $\Delta \times \Delta$ to $\mathbb{R}$ is a metric on $\Delta$ which gives rise to the usual topology. That the map is a metric can be established by direct calculation, though verification of the triangle inequality necessitates a certain amount of care. A more conceptual proof is obtained by observing that if $H^\infty(\Delta)$ denotes the Banach space of bounded holomorphic functions on $\Delta$ with the supremum norm, then the map $z \mapsto \psi_z$, evaluation at $z$, carries $\Delta$ into the unit ball of $H^\infty(\Delta)^*$, the dual space of $H^\infty(\Delta)$. It follows easily from the Schwarz lemma that $[z, w]$ is nothing but the norm of the functional $\psi_z - \psi_w$, so $[z, w]$ does satisfy the triangle inequality. The metric $[\ ,\ ]$ has the property of being *conformally invariant* in the sense that if $\varphi$ is a conformal map of $\Delta$ onto itself, then $[\varphi(z), \varphi(w)] = [z, w]$. The *hyperbolic metric* $\rho$ on $\Delta$ is the metric given by

$$\rho(z, w) = \tfrac{1}{2}\log\frac{1 + [z, w]}{1 - [z, w]}.$$

Again, $\rho$ is conformally invariant, and if $f \in H^\infty(\Delta)$, $\|f\|_\Delta \leq 1$, then $\rho(f(z), f(w)) \leq \rho(z, w)$. The metric $\rho$ has an important additivity property: *If* $z, z', z'' \in \Delta$ *are three points lying on a circle orthogonal to the unit circle (or on one of its diameters) with* $z'$ *between* $z$ *and* $z''$, *then* $\rho(z, z'') = \rho(z, z') + \rho(z', z'')$.

According to the *Riemann mapping theorem, if* $\Omega$ *is a simply connected domain in the plane,* $\Omega \neq \mathbb{C}$, *then there exists a holomorphic function* $\varphi$ *on* $\Omega$ *which is a homeomorphism of* $\Omega$ *onto* $\Delta$. *Moreover, if* $\partial\Omega$ *is a simple closed curve, then* $\varphi$ *admits an extension to a homeomorphism from* $\bar{\Omega}$ *onto* $\bar{\Delta}$.

The Phragmén–Lindelöf theorem may be regarded as a generalization of the maximum modulus theorem. We formulate it for half-planes, though many other versions may be stated.

***Theorem***

*If* $f$ *is a function holomorphic on the half-plane* $\Pi_+ = \{z \in \mathbb{C} : \operatorname{Re} z > 0\}$, *if* $\limsup_{x \to 0^+} |f(x + iy)| \leq M$ *for all* $y \in \mathbb{R}$, *and if* $f$ *is bounded, then* $|f(z)| \leq M$ *for all* $z \in \Pi_+$.

The same conclusion can be drawn if instead of boundedness, we suppose only that $|f(z)| = 0(e^{|z|^b})$ for some $b < 1$.

The theory of the boundary behavior of holomorphic and harmonic functions in $\Delta$ is a very interesting and delicate study from which we need certain results. If $p > 0$ and $\Omega$ is a planar domain, denote by $H^p(\Omega)$ the space of those holomorphic functions $f$ on $\Omega$ with the property that $|f|^p$ admits a harmonic majorant, and let $H^\infty(\Omega)$ be the space of bounded holomorphic functions. In the disc case the condition that $f \in H^p(\Omega)$ is equivalent to the boundedness of the means

$$M_p(f;r) = \frac{1}{2\pi} \int_{-\pi}^{\pi} |f(re^{i\theta})|^p \, d\theta, \qquad 0 \leq r < 1.$$

***Theorem***

*If* $f \in H^p(\Delta)$, $p > 0$, *then for almost all points* $e^{i\theta} \in T$, *the unit circle, the limit* $f^*(e^{i\theta}) = \lim_{z \to e^{i\theta}} f(z)$ *exists if* $z$ *is constrained to approach* $e^{i\theta}$ *nontangentially through* $\Delta$. *The boundary function* $f^*$ *belongs to* $L^p(T)$, *and* $f = 0$ *if* $f^* = 0$ *on a set of positive measure.*

The nontangential restriction is understood in the sense that $z$ approaches $e^{i\theta}$ through the angle between two rays emanating from $e^{i\theta}$ which are not tangent to the unit circle and which approach $e^{i\theta}$ through $\Delta$.

For functions of class $H^1(\Delta)$, the Cauchy integral formula holds.

***Theorem***

*If $f \in H^1(\Delta)$ and $z \in \Delta$,*

$$f(z) = \frac{1}{2\pi}\int_{-\pi}^{\pi} \frac{f^*(e^{i\theta})}{e^{i\theta} - z} e^{i\theta}\, d\theta.$$

There are analogous results for harmonic functions.

***Theorem***

*If $f$ is a harmonic function in $\Delta$ such that the means $M_p(f;r)$, $p \geq 1$, are bounded uniformly in $r$, $0 \leq r < 1$, then for almost all $\theta$, the nontangential limit $f^*(e^{i\theta}) = \lim_{z \to e^{i\theta}} f(z)$ exists and the boundary function lies in $L^p(T)$.*

The *Dirichlet problem* for a region $\Omega$ in $\mathbb{C}$ is the problem of finding a harmonic function on $\Omega$ which assumes given boundary values. For the disc there is a very explicit solution.

***Theorem***

*If $f$ is a harmonic function in $\Delta$ such that the means $M_p(f;r)$ are uniformly bounded in $r$, $p > 1$, then given $re^{i\varphi} \in \Delta$,*

$$f(re^{i\varphi}) = \frac{1}{2\pi}\int_{-\pi}^{\pi} \frac{1 - r^2}{1 - 2r\cos(\theta - \varphi) + r^2} f^*(e^{i\theta})\, d\theta.$$

*Conversely, if $g \in L^1(T)$ and if*

$$G(re^{i\varphi}) = \frac{1}{2\pi}\int_{-\pi}^{\pi} \frac{1 - r^2}{1 - 2r\cos(\theta - \varphi) + r^2} g(e^{i\theta})\, d\theta,$$

*then $G$ is harmonic in $\Delta$, and it assumes the boundary values $g$ a.e. nontangentially and uniformly on any closed interval in $T$ on which $g$ is continuous.*

The kernel $(1 - r^2)/[1 - 2r\cos(\theta - \varphi) + r]$, which figures in the theorem just stated, is the *Poisson kernel.*

For positive harmonic functions in $\Delta$ there is a very explicit representation, as provided by the *theorem of Herglotz.*

***Theorem***

*If $U$ is a positive harmonic function in $\Delta$, there exists a unique finite positive measure $\mu_2$ on $T$ such that*

$$U(re^{i\varphi}) = \int \frac{1 - r^2}{1 - 2r\cos(\theta - \varphi) + r^2} d\mu(\theta).$$

Most of the theory of harmonic and $H^p$ functions extends rather directly from the unit disc to reasonable planar regions. For certain of these considerations it is useful to introduce the notion of the Green's function of a domain. Let $\Omega$ be a planar domain, and let $z_0$ be a point of $\Omega$. The *Green's function for* $\Omega$ *with pole at* $z_0$, $G(\cdot, z_0)$, is that function, if it exists, which is harmonic in $\Omega \setminus \{z_0\}$, which tends to zero at every point of the boundary of $\Omega$, and which becomes positively infinite near $z_0$, but in such a way that the function $G(z, z_0) + \log|z - z_0|$ is harmonic throughout $\Omega$. Not every domain admits a Green's function, but, e.g., every domain bounded by finitely many simple closed curves does.

Suppose that $\Omega$ is a domain in the plane whose boundary $\Gamma$ consists of the union of a finite number of pairwise disjoint analytic simple closed curves, $\gamma_1, \ldots, \gamma_q$. For the domain $\Omega$ there are integral representations of harmonic functions in terms of the Green's function of the domain. There is a natural measure on $\Gamma$, that induced by the differential of arc length, $ds$. If $u$ is a harmonic function defined on $\Omega$ with the property that $|u|^p$, $p \geq 1$, has a harmonic majorant, then $u$ has nontangential limits $u^*(\zeta)$ at almost every (in terms of the natural measure on $\Gamma$) point $\zeta$ of $\Gamma$, and if $p > 1$, $u$ has the representation

$$u(z_0) = \frac{1}{2\pi}\int_\Gamma u^*(\zeta)\frac{2G(\zeta, z_0)}{\partial n}\,ds,$$

where $\partial/\partial n$ denotes differentiation along the inner normal to $\Gamma$. In the opposite direction, if $u^*$ is a given integrable function on $\Gamma$, this integral defines a harmonic function $u$ which assumes at almost every point the nontangential limit $u^*$. Since holomorphic functions are harmonic, these results hold also for functions of class $H^p$ on $\Omega$. For holomorphic functions of class $H^1$, the Cauchy integral formula holds in this context: $f \in H^1(\Omega)$ can be recaptured from its boundary values by the formula

$$f(z) = \frac{1}{2\pi i}\int_\Gamma \frac{f^*(\zeta)}{\zeta - z}\,d\zeta.$$

The analogous theory for domains with rectifiable rather than analytic boundary requires considerably more effort to develop. From this more general theory we shall cite only the following fact.

***Theorem***

*Let $\Gamma$ be a simple closed rectifiable curve, and let $z_0$ be a point inside $\Gamma$. If $G(\cdot, z_0)$ is the Green's function for the interior of $\Gamma$ with pole at $z_0$, then the inner normal derivative $\partial G( \;, z_0)/\partial n$ exists and is strictly positive* *at almost every point of* $\Gamma$. *If* $u$ *is a bounded harmonic function defined inside* $\Gamma$ *with nontangential boundary values* $u^*$, *then*

$$u(z_0) = \frac{1}{2\pi} \int_\Gamma u^*(\zeta) \frac{G(\zeta, z_0)}{\partial n} ds.$$

From several complex variables we shall quote certain nontrivial results. We begin by recalling the definition of holomorphicity for functions of several variables. A function $F$, defined and complex-valued on an open set $\Omega$ in $\mathbb{C}^m$, is said to be *holomorphic* in $\Omega$ provided that about each point $(z_1^0, \ldots, z_m^0)$ of $\Omega$, $F$ admits an absolutely convergent series representation of the form

$$f(z_1, \ldots, z_m) = \textstyle\sum_{k_1,\ldots,k_m \geq 0} C(k_1, \ldots, k_m)(z_1 - z_1^0)^{k_1} \cdots (z_m - z_m^0)^{k_m}.$$

A famous theorem of Hartogs shows that this notion of holomorphic function is equivalent to the requirements that $F$ be holomorphic in each variable separately. We denote by $\mathcal{O}(\Omega)$ the collection of all holomorphic functions defined on the open set in $\mathbb{C}^m$.

Let $\Omega_1$ and $\Omega_2$ be open sets in $\mathbb{C}^m$, and let $F:\Omega_1 \to \Omega_2$ be a mapping which is holomorphic in the sense that

$$F(\mathfrak{z}) = (f_1(\mathfrak{z}), \ldots, f_m(\mathfrak{z}))$$

for all $\mathfrak{z} \in \Omega_1$, the functions $f_1, \ldots, f_m$ holomorphic. If $\mathfrak{z}_0 \in \Omega_1$ and if the Jacobian determinant $\det(\partial f_i/\partial z_j)_{1 \leq i,j \leq m}$ is not zero at $\mathfrak{z}_0$, then in some neighborhood $W$ of $F(\mathfrak{z}_0) = w_0$, there is a holomorphic map $\Phi: W \to \Omega_1$ which takes $\omega_0$ to $\mathfrak{z}_0$, which takes $W$ onto a neighborhood of $\mathfrak{z}_0$, and which is inverse to $F: F(\Phi(\omega)) = \omega$ if $\omega \in W$, and $\Phi(F(\mathfrak{z})) = \mathfrak{z}$ if $\mathfrak{z}$ lies in $\Phi(W)$.

An $m$-dimensional complex analytic manifold $M$ is a Hausdorff space together with an open covering $\mathcal{U} = \{U_\alpha\}$ of $M$ such that for each $\alpha$ there is a homeomorphism $\varphi_\alpha$ of $U_\alpha$ onto an open subset of $\mathbb{C}^m$. We require, in addition, that the functions $\varphi_\alpha$ be holomorphically compatible in the sense that on the set $\varphi_\alpha(U_\alpha \cap U_\beta)$, $\varphi_\beta \circ \varphi_\alpha^{-1}$ is a holomorphic map no matter what the choice of $\alpha$ and $\beta$. A $\mathbb{C}$-valued function on $M$ is holomorphic if for each $\alpha$, $f \circ \varphi_\alpha^{-1}$ is holomorphic on $\varphi_\alpha(U_\alpha)$.

A one-dimensional complex manifold is frequently called a *Riemann surface.*

If $M$ and $N$ are complex manifolds defined by open coverings $\mathcal{U} = \{U_\alpha\}$ and $\mathcal{V} = \{V_\beta\}$ and maps $\varphi_\alpha: U_\alpha \to \mathbb{C}^m$ and $\psi_\beta: V_\beta \to \mathbb{C}^n$, respectively, a map $\Phi: M \to N$ is said to be *holomorphic* if for each choice of $\alpha$ and $\beta$, the map $\psi_\beta \circ \Phi \circ \varphi_\alpha^{-1}$ is holomorphic on $\varphi_\alpha(U_\alpha)$. (A map from an open set in $\mathbb{C}^m$ to $\mathbb{C}^n$ is holomorphic provided each of its coordinates is a holomorphic function.) If $\Phi$ is a holomorphic map of $M$ onto $N$ with holomorphic inverse, then $\Phi$ is said to be *biholomorphic*, and $M$ and $N$ are said to be *biholomorphically equivalent*.

If $M$ is an $n$-dimensional complex manifold whose structure is defined by means of an open cover $\{U_\alpha\}$ and corresponding maps $\varphi_\alpha : U_\alpha \to \mathbb{C}^n$, a subset $V$ of $M$ is an $m$-dimensional submanifold of $M$ if given $\mathfrak{z}_0 \in V$, say $\mathfrak{z}_0 \in U_\alpha$, there is a neighborhood $\Omega$ of $\mathfrak{z}_0$ and a holomorphic map $F = (f_1, \ldots, f_m)$ from $\Omega$ to $\mathbb{C}^m$ with the property that

$$V \cap \Omega = \{\mathfrak{z} \in \Omega : F(\mathfrak{z}) = 0\}$$

and the further property that the Jacobian determinant

$$\det\left(\frac{f_j \circ \varphi_\alpha^{-1}}{\partial z_k}\right)$$

is of rank $m$ at $\varphi_\alpha(\mathfrak{z}_0)$. Such an object is a complex manifold, as can be shown by way of the implicit-function theorem.

A natural class of complex manifolds on which to do function theory is the class of *Stein manifolds*. An $m$-dimensional Stein manifold is an $m$-dimensional complex manifold which is separable and which enjoys the following properties:

(a) If $\mathfrak{z}$ and $\mathfrak{z}'$ are distinct points of $M$, there is $f \in \mathcal{O}(M)$ such that $f(\mathfrak{z}) \neq f(\mathfrak{z}')$.
(b) If $K \subset M$ is a compact set, $\hat{K} = \{\mathfrak{z} \in M : |f(\mathfrak{z})| \leq \sup_{\mathfrak{z}' \in K} |f(\mathfrak{z}')|\}$ is also compact.
(c) If $\mathfrak{z}_0 \in M$, there is a neighborhood $U$ of $\mathfrak{z}_0$ and there are $m$ functions $f_1, \ldots, f_m \in \mathcal{O}(M)$ such that the map $\mathfrak{z} \mapsto (f_1(\mathfrak{z}), \ldots, f_m(\mathfrak{z}))$ carries $U$ biholomorphically onto an open set in $\mathbb{C}^m$.

By their very definition, Stein manifolds have a rich supply of holomorphic functions. Every complex submanifold of $\mathbb{C}^m$ is a Stein manifold, and one of the important results in the theory of Stein manifolds is that, conversely, every Stein manifold is biholomorphically equivalent to a complex submanifold of some $\mathbb{C}^m$.

We shall state two additional results on Stein manifolds, one an extension theorem, the other a cohomology vanishing theorem.

***Theorem***

*If $M$ is a Stein manifold and $N$ is a complex submanifold, then given $f \in \mathcal{O}(N)$, there is $F \in \mathcal{O}(M)$ with $f = F|N$.*

***Theorem***

*If $M$ is a Stein manifold and if $\{U_\alpha\}$ is an open cover for $M$, then given for each pair $\alpha$, $\beta$ functions $f_{\alpha\beta} \in \mathcal{O}(U_\alpha \cap U_\beta)$ such that $f_{\alpha\beta} + f_{\beta\alpha} = 0$*

*and* $f_{\alpha\beta} + f_{\beta\gamma} + f_{\gamma\alpha} = 0$ *on* $U_\alpha \cap U_\beta \cap U_\gamma$, *there exist functions* $\pi \in \mathcal{O}(U_\alpha)$ *such that on* $U_\alpha \cap U_\beta$, $f_{\alpha\beta} = g_\alpha - g_\beta$.

The second of these results is sometimes described by saying that on a Stein manifold, the *first problem of Cousin* is solvable. It shows also that if $\mathcal{O}_M$ denotes the sheaf of germs of functions holomorphic on $M$, then $H^1(M, \mathcal{O}_M) = 0$.

In our applications of these results, we are working on polynomial or rational polyhedra, objects that are surely Stein manifolds.

In addition to the notion of submanifold, we need also the idea of subvariety of a complex submanifold. A closed subset $E$ of a complex manifold $M$ is called an *analytic subvariety* or simply a *subvariety* of $M$ if for every $\mathfrak{z} \in E$, there is a neighborhood $W$ of $\mathfrak{z}$ and a family $\mathcal{F}$ of functions holomorphic on $W$ such that

$$E \cap W = \{w \in W : f(w) = 0 \text{ for every } f \in \mathcal{F}\}.$$

It can be shown that if $F$ is a variety in $M$, there is a closed, nowhere dense subset $S$ of $E$ such that $E \setminus S$ has at each point the structure of a complex manifold. (The dimension of this manifold may vary from component to component.) If $E$ is a variety without isolated points and if the manifold just mentioned is one dimensional at every point, $E$ is said to be *one dimensional.*

If $V$ is a variety in the open set $W \subset \mathbb{C}^n$, a continuous function $h$ on $V$ is said to be *holomorphic* provided that for all $\mathfrak{z} \in V$, there is a neighborhood $N$ of $\mathfrak{z}$ (in $W$) such that $f|(V \cap N) = F|(V \cap N)$ for some $F \in \mathcal{O}(N)$. The functions holomorphic on $V$ constitute a ring denoted by $\mathcal{O}(V)$.

A generalization of the notion of one-dimensional analytic variety is that of *one-dimensional analytic space.* Such a space is defined to be a pair $(X, \mathcal{O}_X)$ where $X$ is a separable Hausdorff space and where $\mathcal{O}_X$ is a subsheaf of the sheaf $\mathcal{C}_X$ of germs of continuous functions on $X$. We require, moreover, that for each $x \in X$ there exist a neighborhood $U$ of $x$, a one-dimensional variety $V$ in an open set $\Omega$ of some $\mathbb{C}^n$, and a homeomorphism $\psi$ of $U$ onto $V$ such that if $U_0$ is an open subset of $U$, then

$$\mathcal{O}_X(U_0) = \{f \circ \psi : f \in \mathcal{O}(\psi(U_0))\}.$$

For a given open set $U$ in $X$, the elements of the ring $\mathcal{O}_X(U)$ are called the *holomorphic functions* on $U$. If $(X, \mathcal{O}_X)$ and $(Y, \mathcal{O}_Y)$ are one-dimensional analytic spaces, a holomorphic map from $X$ to $Y$ is a continuous function $f: X \to Y$ such that for every open set $V \subset Y$, $\mathcal{O}_X(f^{-1}(V)) \supset \{g \circ f : g \in \mathcal{O}_Y(V)\}$. Riemann surfaces, provided with their sheaves of germs of holomorphic functions, are particular cases of one-dimensional analytic spaces.

Locally, a one-dimensional analytic space has the structure of a one-dimensional variety in some open subset of a $\mathbb{C}^m$. From the known structure of one-dimensional varieties, it follows that there is a discrete subset $\Sigma$ of $X$ such that $X \setminus \Sigma$ has the structure of a Riemann surface. The smallest such set $\Sigma$ is called the set of singularities of the space.

Concerning the global structure of one-dimensional analytic spaces, there is the following result.

### *Theorem*

*If $(X, \mathcal{O}_X)$ is a one-dimensional analytic space with singular set $\Sigma$, there is a not necessarily connected Riemann surface $R$ and a proper holomorphic map $\eta: R \to X$ which maps $R \setminus \eta^{-1}(\Sigma)$ biholomorphically onto $X \setminus \Sigma$ and which has the property that if $s \in \Sigma$, $\eta^{-1}(s)$ is a finite set.*

For $R$ one can take the normalization of $(X, \mathcal{O}_X)$.

*References*

Hille, E., *Analytic Function Theory*, Ginn, Boston, vol. I and II, 1959 and 1962.
Narasimhan, R., *Introduction to the Theory of Analytic Spaces* (Springer Lecture Notes, vol. 25), Springer-Verlag, Berlin, 1966.
Rudin, W., Analytic functions of class $H_p$, *Trans. Amer. Math. Soc.* **78** (1955).
Verblunski, S., On a fundamental theorem of potential theory, *J. London Math. Soc.* **26** (1951), 25–30.

A good source for much of classical function theory is the two-volume work of Hille. The boundary theory of harmonic and holomorphic functions in the unit disc may be found in Hoffman [1]. For $H^p$ theory on finitely connected planar domains, see the paper of Rudin. The theorem on Green's function for rectifiably bounded domains does not seem to appear in the standard texts, but it can be found in the paper of Verblunski. For several complex variables, see Gunning and Rossi [1] or Hörmander [1]. These sources do not deal with normalizations, but the theory of normalization can be found in the notes of Narasimhan.

# *appendix e*/NUMBER THEORY

This short appendix contains two results from number theory.

***Theorem***

*If $v_1, \ldots, v_n$ are real numbers linearly independent over the rationals, then the set $\{(e^{iv_1t}, \ldots, e^{iv_nt}) : t \in \mathbb{R}, t > 0\}$ is dense in the n-dimensional torus $T^n = \{(z_1, \ldots, z_n) \in \mathbb{C}^n : |z_1| = |z_2| = \cdots = |z_n| = 1\}$.*

PROOF

The assumed rational independence of the $v_j$ implies that no $v_j$ is zero. The numbers $\mu_j = 2\pi(v_j/v_1)$ are rationally independent since the $v_j$ are, and we will prove that the set

$$E = \{(e^{ik\mu_1}, \ldots, e^{ik\mu_n}) : k = 0, 1, \ldots\}$$

is dense in $T^n$. This will prove the theorem. To prove $E$ dense in $T^n$, it is enough to prove that for every $f \in \mathscr{C}(T^n)$,

$$(1) \qquad \lim_{M \to \infty} \frac{1}{M+1} \sum_{j=0}^{M} f(e^{ij\mu_1}, \ldots, e^{ij\mu_n}) = \int_{T^n} f\, dm_n,$$

$m_n$ the normalized Haar measure on $T^n$; for if $E$ is not dense, there is an $f \in \mathscr{C}(T^n)$ which vanishes identically on $E$ but which has strictly positive integral, in contradiction to (1). In proving (1), it is enough to consider only functions $f$ which are trigonometric polynomials, and thus, by linearity, we need only prove (1) for exponentials. Suppose, therefore, that

$$f(e^{it_1}, \ldots, e^{it_n}) = e^{i(t_1 s_1 + \cdots + t_n s_n)}$$

for some integers $s_1, \ldots, s_n$. If all the integers $s_j$ are zero, both sides of (1) are 1, and if not, the left side is

$$\frac{1}{M+1} \sum_{j=0}^{M} e^{ij(s_1\mu_1 + \cdots + s_n\mu_n)} = \frac{1}{M+1} \frac{1 - e^{i(M+1)(s_1\mu_1 + \cdots + s_n\mu_n)}}{1 - e^{i(s_1\mu_1 + \cdots + s_n\mu_n)}}.$$

We are assured that the denominator is not zero because of the rational independence of the $\mu$'s. As $M \to \infty$, the last term tends to zero, and the theorem is proved.

The theorem just proved is also an easy corollary of a theorem of Kronecker:

***Theorem***

*If* $v_1, \ldots v_n$ *are real numbers linearly independent over the rationals, then given* $N$ *and* $\varepsilon > 0$, *there exists* $n > N$ *such that for some integers,* $p_1, \ldots, p_n$,

$$|nv_j - p_j - a_j| < \varepsilon, \qquad j = 1, 2, \ldots, n.$$

***Theorem*** (Dirichlet's Theorem)

*If* $a_1, \ldots, a_n$ *are real numbers, then for* $k = 1, \ldots,$ *there exists a positive integer* $q_k$, *and for each* $j$, *a positive integer* $p_{j,k}$ *such that*

$$|p_{j,k} - q_k| < 1/k$$

*and* $q_k \leq k^n$.

PROOF

Let $Q$ be the unit cube in $R^n$. Taking coordinates reduced modulo 1, each of the $k^n + 1$ vectors $(sa_1, \ldots, sa_n)$, $s = 1, \ldots, k^n + 1$, determines a point $v_s$ of $Q$. If $Q$ is divided into $k^n$ congruent subcubes with sides parallel the coordinate planes, then one of the resulting subcubes must contain two of the points $v_s$, say $v_s$ and $v_s{}'$, with say, $s - s' > 0$. Thus, $0 < s - s' \leq k^n$, and modulo integers $|(s - s')a_j| < 1/k$ for $j = 1, \ldots, n$.

This is Dirichlet's famous *box principle.*

*Reference*

Hardy, G. H., and E. M. Wright, *An Introduction to the Theory of Numbers,* Oxford University Press, London, 4th ed., 1960.

# REFERENCES


AHERN, P. R.

[1] On the generalized F. and M. Riesz theorem, *Pacific J. Math.* **15** (1965), 373–376.

AHERN, P. R., AND D. SARASON

[1] The $H^p$ spaces of a class of function algebras, *Acta Math.* **117** (1967), 123–163; MR 36 #689.

[2] On some hypo-Dirichlet algebras of analytic functions, *Amer. J. Math.* **89** (1967), 932–941; MR 36 #4338.

AHLBERG, J.

[1] Algebraic properties of topological significance, Ph.D. dissertation, Yale University, 1956.

AHLFORS, L. V., AND L. SARIO

[1] *Riemann Surfaces*, Princeton University Press, Princeton, N.J., 1960, MR 22 #5729.

ALEXANDER, H.

[1] Uniform algebras on curves, *Bull. Amer. Math. Soc.*, **75** (1969), 1269–1272.

[2] Polynomial approximation and analytic structure, *Duke Math. J.* **38** (1971), 123–135.

[3] Polynomial approximation and hulls of sets of finite linear measure in $\mathbb{C}^n$, *Amer. J. Math.* **93** (1971), 65–75.

ALLAN, G. R.

[1] A form of local characterization of Gelfand transforms, *J. London Math. Soc.* **43** (1968), 623–625; MR 37 #6756.

[2] An extension of the Silov–Arens–Calderón theorem, *J. London Math. Soc.* **44** (1969), 595–601; MR 39 #4678.

[3] A note on the holomorphic functional calculus in a Banach algebra, *Proc. Amer. Math. Soc.*, **22** (1969), 77–81; MR40 #405.

[4] On lifting analytic relations in commutative Banach algebras, *J. Functional Analysis* **5** (1970), 37–43.

[5] An extension of Rossi's local maximum modulus principle, in press.

ALLING, N. L.

[1] Extensions of meromorphic function rings over non-compact Riemann surfaces, I, *Math Z.* **89** (1965), 273–299; MR 32 #6252.

[2] Extensions of meromorphic function rings over non-compact Riemann surfaces. II, *Math. Z.* **93** (1966), 345–394; MR 33 #7572.

ARENS, R.

[1] The maximal ideals of certain function algebras, *Pacific J. Math.* **8** (1958), 641–648; MR 22 #8315.

[2] The closed maximal ideals of algebras of functions holomorphic on a Riemann surface, *Rend. Cir. Mat. di Palermo* (2)**7** (1958), 245–260; MR 21 #4242.

[3] The analytic functional calculus in commutative topological algebras, *Pacific J. Math.* **11** (1961), 405–429; MR 25 #4373.

[4] The problem of locally $A$ functions in a commutative Banach algebra $A$, *Trans. Amer. Math. Soc.* **104** (1962), 24–36; MR 26 #4197.

[5] The group of invertible elements of a commutative Banach algebra, *Studia Math. Zeszyt* (Ser. Specjalna) **1** (1963), 21–23; MR 26 #4198.

[6] To what extent does the space of maximal ideals determine the algebra? *Function Algebras* (Proc. Intern. Symp. Function Algebras, Tulane Univ., 1965), Scott, Foresman, Chicago, 1966, pp. 164–168; MR 33 #3126.

ARENS, R., AND A. P. CALDERÓN

[1] Analytic functions of several Banach algebra elements, *Ann. Math.* (2)**62** (1955), 204–216. MR 17, p. 177.

ARENS, R., AND I. M. SINGER

[1] Function values as boundary integrals, *Proc. Amer. Math. Soc.* **5** (1954), 735–745; MR 16, p. 373.

ARENSON, E. L.

[1] Certain properties of algebras of continuous functions, *Soviet Math.* **7** (1966), 1522–1524; MR 34 #6560.

BADÉ, W. G., AND P. C. CURTIS

[1] Banach algebras on *F*-spaces, *Function Algebras* (Proc. Intern. Symp. Function Algebras, Tulane Univ., 1965), Scott, Foresman, Chicago, 1966, pp. 90–92; MR 33 #3127.

BANASCHEWSKI, B.

[1] Analytic discs in the maximal ideal space of a Banach algebra, *Bull. Acad. Polonaise Sci. Ser. Sci. Math. Ast. Phy.* **14** (1966), 137–144.

BARTLE, R. G. AND L. M. GRAVES

[1] Mappings between function spaces, *Trans. Amer. Math. Soc.* **72** (1952), 400–413; MR 13, p. 951.

BASENER, R. F.

[1] An example concerning peak points, *Notices Amer. Math. Soc.* **18** (1971), 415–416.

BAUER, H.

[1] Silovscher Rand und Dirichletsches Problem, *Ann. Inst. Fourier* (*Grenoble*) **11** (1961), 89–136; MR 25 #443.

BEAR, H. S.

[1] Complex function algebras, *Trans. Amer. Math. Soc.* 90 (1959), 383–393; MR 21 #5889.

[2] A strong maximum modulus theorem for maximal function algebras, *Trans. Amer. Math. Soc.* **92** (1959), 464–469; MR 22 #A5908.

[3] The Silov boundary for a linear space of continuous functions, *Amer. Math. Monthly* **68** (1961), 483–485; MR 23 #A4001.

[4] *Lectures on Gleason Parts* (Lecture Notes in Mathematics, Vol. 121), Springer-Verlag, Berlin, 1970.

BERNARD, A.

[1] Caractérisations de certaines parties d'un espace compact muni d'un espace vectoriel ou d'une algèbra de fonctions continues, *Ann. Inst. Fourier* (*Grenoble*) **17** (1967), 359–382; MR 36 #6912.

[2] Une caractérisation de $\mathscr{C}(X)$ parmi les algèbres de Banach, *Compt. Rend.* **A267** (1968), 634–635; MR 38 #2601.

BEURLING, A.

[1] Sur les fonctions limites quasi analytiques des fonctions rationelles, VIII Congrès des Mathématiciens Scandinaves, Stockholm, 1934.

[2] Sur les intégrales de Fourier absolument convergentes et leur application à une transformation fonctionelle, IX Congrès des Mathématiciens Scandinaves, Helsingfors, 1938.

BIRTEL, F. T.

[1] Uniform algebras with unbounded functions, *Complex Analysis* (Proc. Conf. Complex Analysis, Rice Univ., 1967), *Rice Univ. Studies* **54** (1968), no. 4; MR 38 #6101.

BIRTEL, F. T., AND E. DUBINSKI

[1] Bounded analytic functions of two complex variables, *Math. Z.* **93** (1966), 299–310; MR 34 #1873.

BISHOP, E.

[1] A minimal boundary for function algebras, *Pacific J. Math.* **9** (1959), 629–642; MR 22 #A191.

[2] Boundary measures of analytic differentials, *Duke Math. J.* **27** (1960), 331–340; MR 22 #A9621.

[3] A generalization of the Stone–Weierstrass theorem, *Pacific J. Math.* **11** (1961), 777–783; MR 24 #A3502.

[4] A general Rudin–Carleson theorem, *Proc. Amer. Math. Soc.* **13** (1962), 140–143; MR 24 #A3293.

[5] Analyticity in certain Banach algebras, *Trans. Amer. Math. Soc.* 102 (1962), 507–544; MR 25 #5410.

[6] Holomorphic completions, analytic continuation, and the interpolation of seminorms, *Ann. Math.* (2)**78** (1963), 468–500; MR 27 #4958.
[7] Representing measures for points in a uniform algebra, *Bull. Amer. Math. Soc.* **70** (1964), 121–122; MR 28 #1510.
[8] Abstract dual extremal problems, *Notices Amer. Math. Soc.* **12** (1965), 123.

BISHOP, E., AND K. DELEEUW
[1] The representation of linear functionals on sets of extreme points, *Ann. Inst. Fourier* (*Grenoble*) **9** (1959), 305–331; MR 22 #4945.

BLUMENTHAL, R. G.
[1] The geometric structure of the spectrum of a function algebra, Ph.D. dissertation, Yale University, 1968.

BONNARD, M.
[1] Sur le calcul fonctionnel holomorphe multiforme dans les algèbres topologiques, *Ann. Scient. d'École Nor. Sup.* (4)**2** (1969), 397–422.

BOURBAKI, N.
[1] *Éléments de mathématiques, fasc. XXXII, Théories spectrales*: Chapitre I, Algèbres normées; Chapitre II, Groupes localement compacts commutatifs (Actualités Scientifiques et Indistrielles No. 1332), Hermann, Paris, 1967; MR 35 #4725.

DEBRANGES, L.
[1] The Stone–Weierstrass theorem, *Proc. Amer. Math. Soc.* **10** (1959), 822–824; MR 22 #3970.

BROWDER, A.
[1] *Introduction to Function Algebras*, W. A. Benjamin, New York, 1969.
[2] Cohomology of maximal ideal spaces, *Bull. Amer. Math. Soc.* **67** (1961), 515–516; MR 24 #A440.
[3] On a theorem of Hoffman and Wermer, *Function Algebras* (Proc. Intern. Symp. Function Algebras, Tulane Univ., 1965), Scott, Foresman, Chicago, 1966, pp. 88–89; MR 33 #4707.
[4] Point derivations on function algebras, *J. Functional Analysis* **1** (1967), 22–27; MR 35 #2144.

CARLESON, L.
[1] On bounded analytic functions and closure problems, *Ark. Mat.* **2** (1952), 283–291; MR 14, p. 630.
[2] Representations of continuous functions, *Math. Z.* **66** (1957), 447–451; MR 18, p. 798.
[3] Interpolation by bounded analytic functions and the corona problem, *Ann. Math.* (2)**76** (1962), 547–560; MR 25 #5186.
[4] Mergelyan's theorem on uniform polynomial approximation, *Math. Scand.* **15** (1964), 167–175; MR 33 #6368.

CIRKA, E. M.
[1] Approximation of continuous functions by functions holomorphic on Jordan arcs in $\mathbb{C}^n$, *Soviet Math.* **7** (1966), 336–338; MR 34 #1563.

COHEN, P. J.
[1] A note on constructive methods in Banach algebras, *Proc. Amer. Math. Soc.* **12** (1961), 159–163; MR 23 #A1827.

COLE, B.
[1] One point parts and the peak point conjecture, Ph.D. dissertation, Yale University, 1968.

COLLINGWOOD, E. F., AND A. J. LOHWATER
[1] *The Theory of Cluster Sets* (Cambridge Tract No. 56), Cambridge University Press, New York, 1966; MR 38 #325.

DAVIE, A. M.
[1] Real annihilating measures for $\mathscr{R}(K)$, *J. Funct. Anal.* **6** (1970), 357–386.

DIEUDONNÉ, J.
[1] Sur les homomorphismes d'espaces normés, *Bull. Sci. Math.* (2)**67** (1943), 72–84; MR 7, p. 124.

DOUADY, A.
[1] Le problème des modules pour les sous-espaces analytiques compacts d'un espace analytique donné, *Ann. Inst. Fourier (Grenoble)* **16** (1966), 1–95; MR 34 #2940.

EILENBERG, S., AND N. STEENROD
[1] *Foundations of Algebraic Topology*, Princeton University Press, Princeton, N.J., 1952; MR 14, p. 398.

EPE, R.
[1] Characterizierung des Schilowrandes von Holomorphiegebieten, *Schr. Math. Inst. Münster* **25** (1963); MR 28 #2251.

FATOU, P.
[1] Séries trigonométriques et séries de Taylor, *Acta Math.* **30** (1906), 335–400.

FISHER, S.
[1] Bounded approximation by rational functions, *Pacific J. Math.* **28** (1969), 319–326; MR 39 #1663.

FORELLI, F.
[1] Analytic measures, *Pacific J. Math.* **13** (1963), 571–578; MR 28 #429.
[2] Measures orthogonal to polydisc algebras, *J. Math. Mech.* **17** (1968), 1073–1086; MR 36, #6940.
[3] Lectures about Analytic Measures, *Summer School in Harmonic Analysis*, Mathematics Institute, University of Warwick, 1968.

FORSTER, O.

[1] Banach-Algebren stetiger Funktionen auf lompakten Räumen, *Math. Z.* **81** (1963), 1–34; MR 27 #2857.

FUKS, B. A.

[1] *Special Chapters in the Theory of Analytic Functions of Several Complex Variables* (Translations of Mathematical Monographs, vol. 14), American Mathematical Society, Providence, 1965.

GAMELIN, T. W.

[1] *Uniform Algebras*, Prentice-Hall, Englewood Cliffs, N.J., 1969.

[2] Restrictions of subspaces of $\mathscr{C}(X)$, *Trans. Amer. Math. Soc.* **112** (1964), 278–286; MR 28 #5331.

[3] Embedding Riemann surfaces in maximal ideal spaces, *J. Functional Analysis* **2** (1968), 123–146; MR 36 #6941.

GAMELIN, T. W., AND J. GARNETT

[1] Constructive techniques in rational approximation, *Trans. Amer. Math. Soc.* **143** (1969), 187–200.

GAMELIN, T. W., AND G. LUMER

[1] Theory of abstract Hardy spaces and the universal Hardy class, *Advances in Math.* **2** (1968), 118–174; MR 37 #1982.

GAMELIN, T. W., AND H. ROSSI

[1] Jensen measures and algebras of analytic functions, *Function Algebras* (Proc. Intern. Symp. Function Algebras, Tulane Univ., 1965), Scott, Foresman, Chicago, 1966, pp. 15–35; MR 33 #7879.

GAMELIN, T. W., AND D. R. WILKEN

[1] Closed partitions of maximal ideal spaces, *Ill. J. Math.* **13** (1969), 789–795; MR 40 #4767.

GARNETT, J.

[1] Disconnected Gleason parts, *Bull. Amer. Math. Soc.* **72** (1966), 490–492; MR 32 #6258.

[2] A topological characterization of Gleason parts, *Pacific J. Math.* **20** (1967), 59–64; MR 34 #4942.

[3] On a theorem of Mergelyan, *Pacific J. Math.* **26** (1968), 461–467; MR 38 #1532.

GARNETT, J., AND I. GLICKSBERG

[1] Algebras with the same multiplicative measures, *J. Functional Analysis* **1** (1967), 331–341; MR 36 #691.

GELBAUM, B. R.

[1] Tensor products and related questions, *Trans. Amer. Math. Soc.* **103** (1962), 525–548; MR 25 #2406.

GELFAND, I. M.
[1] Normierte Ringe, *Mat. Sb.* **9** (1941), 3–24; MR 3, p. 51.

GELFAND, I. M., D. A. RAIKOV, AND G. E. SHILOV
[1] Commutative normed rings. *Uspekhi Mat. Nauk* (N.S.) **1** (1946), 48–146 (in Russian); English translation, *Amer. Math. Soc. Translation*, (II) **5** (1957), 115–220; MR 10, p. 258.
[2] *Commutative Normed Rings* (in Russian), Gosudarstv. Izdat. Fiz.-Mat. Lit., Moscow, 1960; English translation, Chelsea, New York, 1964; MR 23 #A1242.

GLEASON, A. M.
[1] Function algebras, *Seminar on Analytic Functions*, vol. II, Institute for Advanced Study, Princeton, N.J., 1957.
[2] Finitely generated ideals in Banach algebras, *J. Math. Mech.* **13** (1964), 125–132; MR 28 #2458.

GLICKSBERG, I.
[1] Measures orothogonal to algebras and sets of antisymmetry, *Trans. Amer. Math. Soc.* **105** (1962), 415–435; MR 30 #4164.
[2] Function algebras with closed restrictions, *Proc. Amer. Math. Soc.* **14** (1963), 158–161; MR 26 #616.
[3] Maximal algebras and a theorem of Radó, *Pacific J. Math.* **14** (1964), 919–941; MR 29 #6337. Correction, *Pacific J. Math.* **19** (1966), 587; MR 34 #1874.
[4] A Phragmén–Lindelöf theorem for function algebras, *Pacific J. Math.* **22** (1967), 401–406; MR 35 #5943.
[5] The abstract F. and M. Riesz theorem, *J. Functional Analysis* **1** (1967), 109–122; MR 35 #2146.
[6] On two consequences of a theorem of Hoffman and Wermer, *Math. Scand.* **23** (1968), 188–192.
[7] Dominant representing measures and rational approximation, *Trans. Amer. Math. Soc.* **130** (1968), 425–462; MR 37 #744.
[8] Extensions of the F. and M. Riesz theorem, *J. Functional Analysis* **5** (1970), 125–136.

GLICKSBERG, I., AND J. WERMER
[1] Measures orothogonal to Dirichlet algebras, *Duke Math. J.* **30** (1963), 661–666; MR 27 #6150. Errata, *Duke Math. J.* **31** (1964), 717; MR 29 #5124.

GONCHAR, A. A.
[1] On the minimal boundary of *A*(*E*), *Izv. Akad. Nauk. SSSR Ser. Mat.* **27** (1963), 949–955 (in Russian); MR 27 #3812.

GORIN, E. A.
[1] Moduli of invertible elements in a normed algebra, *Vestnik Moskov. Univ. Ser. I. Mat. Mek.* **1965**, no. 5, 35–39 (in Russian, English summary); MR 32 #8206.

GRAUERT, H.
[1] Bemerkenswerte pseudokonvexe Mannigfaltigkeiten, *Math. Z.* **81** (1963), 377–391; MR 29 #6054.

GULICK, S. L.
[1] The minimal boundary of $\mathscr{C}(X)$, *Trans. Amer. Math. Soc.* **131** (1968), 303–314; MR 36 #4341.

GUNNING, R. C., AND H. ROSSI
[1] *Analytic Functions of Several Complex Variables*, Prentice-Hall, Englewood Cliffs, N.J., 1965; MR 31 #4927.

HALMOS, P. R.
[1] *Measure Theory*, Van Nostrand Reinhold, New York, 1950; MR 11, p. 504.

HARTOGS, F., AND A. ROSENTHAL
[1] Über Folgen analytischen Funktionen, *Math. Ann.* **104** (1931), 606–611.

HEARD, E. A., AND J. H. WELLS
[1] An interpolation problem for subalgebras of $H^\infty$, *Pacific J. Math.* **28** (1969), 543–553; MR 39 #4681.

HELSON, H.
[1] On a theorem of F. and M. Riesz, *Colloq. Math.* **3** (1955), 113–117; MR 16, p. 1016.

HELSON, H., AND D. LOWDENSLAGER
[1] Prediction theory and Fourier series in several variables, *Acta Math.* **99** (1958), 165–201; MR 20 #4155.

HELSON, H., and F. QUIGLEY
[1] Maximal algebras of continuous functions, *Proc. Amer. Math. Soc.* **8** (1957), 111–114; MR 18, p. 911.
[2] Existence of maximal ideals in algebras of continuous functions, *Proc. Amer. Math. Soc.* **8** (1957), 115–119; MR 18, p. 911.

HOFFMAN, K.
[1] *Banach Spaces of Analytic Functions*, Prentice-Hall, Englewood Cliffs, N.J., 1962; MR 24 #A2844.
[2] Analytic functions and logmodular Banach algebras, *Acta Math.* **108** (1962), 271–317; MR 26 #6820.
[3] Lectures on sup norm algebras, *Summer School on Topological Algebra Theory*, Bruges, 1966.
[4] Parts and analyticity, *Function Algebras* (Proc. Intern. Symp. Function Algebras, Tulane Univ., 1965) Scott, Foresman, Chicago, 1966, pp. 1–5; MR 33 #3129.

[5] Bounded analytic functions and Gleason parts, *Ann. Math.* (2)**86** (1967), 74–111; MR 35 #5945.

HOFFMAN, K., AND A. RAMSEY
[1] Algebras of bounded sequences, *Pacific J. Math.* **5** (1965), 1239–1248; MR 33 #6442.

HOFFMAN, K., AND H. ROSSI
[1] Extensions of positive weak* continuous functionals, *Duke Math. J.* **34** (1967), 453–466; MR 37 #763.

HOFFMAN, K., AND I. M. SINGER
[1] On some problems of Gelfand, *Uspekhi Mat. Nauk* (N.S.)**14** (1957), no. 3, 99–114 (in Russian); English translation, *Amer. Math. Soc. Translations* (II) **27**, 143–157; MR 22 #8316.
[2] Maximal subalgebras of $\mathscr{C}(\Gamma)$, *Amer. J. Math.* **79** (1957), 295–305; MR 19, p. 46.
[3] Maximal algebras of continuous functions, *Acta Math.* **103** (1960), 217–241; MR 22 #8318.

HOFFMAN, K., AND J. WERMER
[1] A characterization of $\mathscr{C}(X)$, *Pacific J. Math.* **12** (1962), 941–944; MR 27 #325.

HOLLADAY, J. C.
[1] Boundary conditions for algebras of continuous functions, Ph.D. dissertation, Yale University, 1953.

HÖRMANDER, L.
[1] *An Introduction to Complex Analysis in Several Variables*, Van Nostrand Reinhold, New York, 1966; MR 34 #2933.

HUA, L. K.
[1] *Harmonic Analysis of Functions of Several Complex Variables in the Classical Domains*, American Mathematical Society, Providence, 1963; MR 33 #1483.

HURD, A. E.
[1] Applications of sheaf theory to function algebras, Ph.D. dissertation, Stanford University, 1962.
[2] Maximum modulus algebras and local approximation in $\mathbb{C}^n$, *Pacific J. Math.* **13** (1963), 597–602; MR 27 #4067.

HUREWICZ, H., AND H. WALLMAN
[1] *Dimension Theory*, Princeton University Press, Princeton, N.J., 1948; MR 3, p. 312.

HURWITZ, A., AND R. COURANT

[1] *Funktionentheorie*, vierte Auflage, Springer-Verlag, Berlin, 1964; MR 30 #3959.

KALLIN, E.

[1] A nonlocal function algebra, *Proc. Nat. Acad. Sci. U.S.A.* **49** (1963), 821–824; MR 27 #2878.

[2] Polynomial convexity: The three spheres problem, *Proceedings of the Conference on Complex Analysis held in Minneapolis*, 1964, Springer-Verlag, Berlin, 1965, pp. 301–304; MR 31 #3631.

[3] Fat polynomially convex sets, *Function Algebras* (Proc. Intern. Symp. Function Algebras, Tulane Univ., 1965), Scott, Foresman, Chicago, 1966, pp. 149–152; MR 33 #2828.

KATZNELSON, Y.

[1] A characterization of the algebra of all continuous functions on a compact Hausdorff space, *Bull. Amer. Math. Soc.* **66** (1960), 313–315; MR 22 #12404.

KAUFMAN, R.

[1] A theorem of Radó, *Math. Ann.* **169** (1967), 282; MR 34 #4468.

KERZMAN, N., AND A. NAGEL

[1] Finitely generated ideals in certain function algebras, *J. Funct. Anal.* **7** (1971).

KÖNIG, H.

[1] On certain applications of the Hahn-Banach and minimax theorems, in press.

KÖNIG, H., AND G. SEEVER

[1] The abstract F. and M. Riesz theorem, *Duke Math. J.* **36** (1969), 791–797.

KOPPELMAN, W.

[1] The Cauchy integral for functions of several complex variables, *Bull. Amer. Math. Soc.* **73** (1967), 373–377; MR 35 #416.

KRA, I.

[1] On the ring of holomorphic functions on an open Riemann surface, *Trans. Amer. Math. Soc.* **132** (1968), 231–244; MR 37 #1633.

LEWITTES, J.

[1] A note on parts and hyperbolic geometry, *Proc. Amer. Math. Soc.* **17** (1966), 1087–1090; MR 33 #7880.

LORCH, E. R.

[1] The theory of analytic functions in normed abelian vector rings, *Trans. Amer. Math. Soc.* **54** (1943), 414–425; MR 5, p. 100.

LUMER, G.

[1] Analytic functions and Dirichlet problem, *Bull. Amer. Math. Soc.* **70** (1964), 98–104; MR 28 #1509.

[2] *Algèbres de fonctions et espaces de Hardy* (Lecture Notes in Mathematics, vol. 75), Springer-Verlag, Berlin, 1968.

[3] On Wermer's maximality theorem, *Inv. Math.* **8** (1969), 236–237; MR 40 #4769.

MARKUSHEVICH, A. I.

[1] *Theory of Functions of a Complex Variable*, vol. 3, translated by R. A. Silverman, Prentice-Hall, Englewood Cliffs, N.J., 1967; MR 35 #6798.

MARKUSHEVICH, L. A.

[1] Approximation by polynomials of continuous functions on Jordan arcs in the space $K^n$ of $n$ complex variables, *Sb. Mat. Z.* **2** (1961), 237–260; English translation, *Amer. Math. Soc. Translation* (II) **57**, 144–170; MR 27 #4959. #4959.

McKISSICK, R.

[1] Existence of a nontrivial normal function algebra, Ph.D. dissertation, Massachusetts Institute of Technology, 1963.

[2] A nontrivial normal sup norm algebra, *Bull. Amer. Math. Soc.* **69** (1963), 391–395; MR 26 #4166.

MERGELYAN, S. N.

[1] On the representation of functions by series of polynomials on closed sets, *Dokl. Akad. Nauk SSSR* **78** (1951), 405–408 (in Russian); English translation, *Amer. Math. Soc. Translations* (I)**85**; MR 13, p. 85.

[2] Uniform approximations to functions of a complex variable, *Uspekhi Mat. Nauk* (N.S.)**7** (1952), 31–122 (in Russian); English translation, *Amer. Math. Soc. Translations* (I)**101**; MR 14, p. 547.

MERRILL, S.

[1] Maximality of certain algebras $H^\infty(dm)$, *Math. Z.* **106** (1968), 261–266; MR 38 #2606.

MEYERS, W. E.

[1] Montel algebras on the plane, *Can. J. Math.* **22** (1970), 116–122; MR 40 #7817.

MICHAEL, E. A.

[1] Continuous selections, I, *Ann. Math.* (2)**63** (1956), 361–382; MR 18, p. 325.

[2] Continuous selections. II, *Ann. Math.* (2)**64** (1956), 562–580; MR 18, p. 325.

[3] Continuous selections. III, *Ann. Math.* (2)**65** (1957), 375–390; MR 18, p. 750.

MICHAEL, E. A., AND A. PEŁCZYŃSKI

[1] Peaked partition subspaces of $\mathscr{C}(X)$, *Ill. J. Math.* **11** (1967), 555–562; MR 36 #670.

[2] A linear extension theorem, *Ill. J. Math.* **11** (1967), 563–579; MR 36 #671.

MILNOR, J.

[1] *Topology from the Differentiable Viewpoint*, University Press of Virginia, Charlottesville, 1965; MR 37 #2239.

MULLINS, R.

[1] The essential set of function algebras, *Proc. Amer. Math. Soc.* **18** (1967), 271–273; MR 34 #6564.

NATZITZ, B.

[1] A note on interpolation, *Proc. Amer. Math. Soc.* **25** (1970), 918.

NEGREPONTIS, S.

[1] On a theorem of Hoffman and Ramsey, *Pacific J. Math.* **20** (1967), 281–282; MR 35 #741.

NELSON, E.

[1] The distinguished boundary of the unit operator ball, *Proc. Amer. Math. Soc.* **12** (1961), 994; MR 24 #A2830.

NEWMAN, M. H. A.

[1] *Elements of the Topology of Plane Sets of Points*, Cambridge University Press, New York, 1951; MR 13, p. 483.

OKA, K.

[1] Sur les fonctions analytiques de plusieurs variables, I. Domaines convexes par rapport aux fonctions rationnelles, *J. Sci. Hiroshima Univ.* **6** (1936), 245–255. (This and eight other memoirs of Oka have been published together under the title *Sur les fonctions analytiques de plusieurs variables*, Iwanami Shoten, Tokyo, 1961.)

[2] Sur les fonctions analytiques de plusieurs variables, II. Domaines d'holomorphie, *J. Sci. Hiroshima Univ.* **7** (1937), 115–130.

O'NEILL, B. V.

[1] Parts and one dimensional analytic spaces, *Amer. J. Math.* **90** (1968), 84–87; MR 36 #5699.

O'NEILL, B. V., AND J. WERMER

[1] Parts as finite sheeted coverings of the disc, *Amer. J. Math.* **90** (1968), 98–107; MR 36 #5700.

PEŁCZYŃSKI, A.

[1] On simultaneous extensions of continuous functions: A generalization of the theorems of Rudin–Carleson and Bishop, *Studia Math.* **24** (1964), 285–304; MR 30 #5184a.

[2] Supplement to my paper "On simultaneous extensions of continuous functions," *Studia Math.* **25** (1964), 157–161; MR 30 #5184b.

[3] Some linear topological properties of separable function algebras, *Proc. Amer. Math. Soc.* **18** (1967), 652–661; MR 35 #4737.

PHELPS, R. R.
[1] *Lectures on Choquet Theory*, Van Nostrand Reinhold, New York, 1966; MR 33 #1690.

QUIGLEY, F.
[1] Approximation by algebras of functions, *Math. Ann.* **135** (1958), 81–92; MR 20 #4180.
[2] Generalized Phragmén–Lindelöf theorems, *Function Algebras* (Proc. Intern. Symp. Function Algebras, Tulane Univ., 1965), Scott, Foresman, Chicago, 1966, pp. 36–41; MR 34 #3374.

RADÓ, T.
[1] Über eine nicht fortsetzbare Riemannsche Mannigfaltigkeit, *Math. Z.* **20** (1924), 1–6.

RAINWATER, J.
[1] Note on the preceding paper, *Duke J. Math.* **36** (1969), 799–800.

READ, T. T.
[1] Analytic structure in the spectrum of a Banach algebra, Ph.D. dissertation, Yale University, 1969.
[2] The powers of a maximal ideal in a Banach algebra and analytic structure, in press.

REITER, H.
[1] Contributions to harmonic analysis. VI, *Ann. Math.* (2)**77** (1966), 552–562.

RICKART, C. E.
[1] *General Theory of Banach Algebras*, Van Nostrand Reinhold, New York, 1960; MR 22 #5903.
[2] The maximal ideal space of functions locally approximable in a function algebra, *Proc. Amer. Math. Soc.* **17** (1966), 1320–1326; MR 34 #1876.
[3] Analytic phenomena in general function algebras, *Pacific J. Math.* **18** (1966), 361–377; MR 33 #6438.
[4] Holomorphic convexity for general function algebras, *Canad. J. Math.* **20** (1968), 272–290; MR 37 #3362.
[5] Analytic functions of an infinite number of complex variables, *Duke Math. J.* **36** (1969), 581–597; MR 40 #7819.

ROSAY, J.-P.
[1] Spectre de certaines sous-algèbres de l'algébre du disque, *Compt. Rend.* **A268** (1969), 221–223; MR 39 #7440.

ROSSI, H.

[1] The local maximum modulus principle, *Ann. Math.* (2)**72** (1960), 1–11; MR 22 #8317.

[2] Holomorphically convex sets in several complex variables, *Ann. Math.* (2)**74** (1961), 470–493; MR 24 #A3310.

ROSSI, H., AND G. STOLZENBERG

[1] Analytic function algebras, *Function Algebras* (Proc. Intern. Symp. Function Algebras, Tulane Univ., 1965), Scott, Foresman, Chicago, 1966, pp. 138–148; MR 33 #6439.

ROYDEN, H. L.

[1] Function algebras, *Bull. Amer. Math. Soc.* **69** (1963), 281–298; MR 26 #6817.

[2] One-dimensional cohomology of domains of holomorphy, *Ann. Math.* (2)**78** (1963), 197–200; MR 27 #2650.

[3] Algebras of bounded analytic functions on Riemann surfaces, *Acta Math.* **114** (1965), 113–141; MR 30 #3972.

RUBEL, L. A., AND A. L. SHIELDS

[1] Bounded approximation by polynomials, *Acta Math.* **112** (1964), 145–162; MR 30 #5104.

RUDIN, W.

[1] *Fourier Analysis on Groups*, Wiley (Interscience), New York, 1962; MR 27 #2808.

[2] *Real and Complex Analysis*, McGraw-Hill, New York, 1966; MR 35 #1420.

[3] *Function Theory in Polydiscs*, W. A. Benjamin, New York, 1969; MR 41 #501.

[4] Analyticity and the maximum modulus principle, *Duke Math. J.* **20** (1953), 449–457; MR 15, p. 21.

[5] Subalgebras of spaces of continuous functions, *Proc. Amer. Math. Soc.* **7** (1956), 825–830; MR 18, p. 587.

[6] Boundary values of continuous analytic functions, *Proc. Amer. Math. Soc.* **7** (1956), 808–811; MR 18, p. 472.

[7] Continuous functions on spaces without perfect sets, *Proc. Amer. Math. Soc.* **8** (1957), 39–42; MR 19, p. 46.

[8] On the structure of maximum modulus algebras, *Proc. Amer. Math. Soc.* **9** (1958), 708–712; MR 20 #3449.

[9] Essential boundary points, *Bull. Amer. Math. Soc.* **70** (1964), 321–324; MR 28 #3167.

RUDIN, W., AND E. L. STOUT

[1] Boundary properties of functions of several complex variables, *J. Math. Mech.* **14** (1965), 991–1006; MR 32 #230.

[2] Modules over polydisc algebras, *Trans. Amer. Math. Soc.* **138** (1969) 327–342; MR 39 #3034.

SCHARK, I. J.
[1] Maximal ideals in an algebra of bounded analytic functions, *J. Math. Mech.* **10** (1961), 735–746; MR 23 #A2744.

SCHATTEN, R.
[1] A nonapproximation theorem, *Amer. Math. Monthly* **69** (1963), 745–750.

SERRE, J.-P.
[1] Une propriété topologique des domains de Runge, *Proc. Amer. Math. Soc.* **6** (1955), 133–134; MR 16, p. 736.

SHAUCK, M. E.
[1] Algebras of holomorphic functions in ringed spaces. I, *Canad. J. Math.* **21** (1969), 1281–1293; MR 40 #4474.

SHILOV, G. E.
[1] On the decomposition of a commutative normed ring into a direct sum of ideals, *Mat. Sb.* **32** (1954), 353–364 (in Russian); English translation, *Amer. Math. Soc. Translations* (II) **1** (1955), and Chapter 9 of Gelfand, Raikov, and Shilov [2]; MR 17, p. 512.
[2] Letter to the editor: On locally analytic functions, *Uspekhi Mat. Nauk* **21** (1966), 177–182 (in Russian); MR 36 #1696.

SIDNEY, S. J.
[1] Point derivations in certain sup-norm algebras, *Trans. Amer. Math. Soc.* **131** (1968), 119–127; MR 36 #4344.
[2] Properties of the sequence of closed powers of a maximal ideal in a sup-norm algebra, *Trans. Amer. Math. Soc.* **131** (1968), 128–148; MR 36 #5701.
[3] Fréchet function algebras: Examples and antisymmetry properties, in press.
[4] High-order non-local function algebras, *Proc. London Math. Soc.*, in press.

SIDNEY, S. J., AND E. L. STOUT
[1] A note on interpolation, *Proc. Amer. Math. Soc.* **19** (1968), 380–382; MR 36 #6944.

SMIRNOV, V. I., AND N. A. LEBEDEV
[1] *Functions of a Complex Variable: Constructive Theory*, M.I.T. Press, Cambridge, Mass., 1968; MR 37 #5369.

SRINIVASAN, T. P., AND J.-K. WANG
[1] On the maximality theorem of Wermer, *Proc. Amer. Math. Soc.* **14** (1963), 997–998; MR 28 #2459.

STEIN, K.
[1] Analytische Funktionen mehrere komplexer Veränderlichen zu vorgegeben Periodizitätsmoduln und das zweite Cousinsche Problem, *Math. Ann.* **123** (1951), 201–222; MR 13, p. 224.

STOLZENBERG, G.

[1] The maximal ideal space of the functions locally in a function algebra, *Proc. Amer. Math. Soc.* **14** (1963), 342–345; MR 26 #4196.

[2] A hull with no analytic structure, *J. Math. Mech.* **12** (1963), 103–112; MR 26 #627.

[3] Polynomially and rationally convex sets, *Acta Math.* **109** (1963), 259–289; MR 26 #3929.

[4] Uniform approximation on smooth curves, *Acta Math.* **115** (1966) 185–198; MR 33 #307.

[5] On the analytic part of the Runge hull, *Math. Ann.* **164** (1966), 286–290; MR 34 #2938.

[6] *Volumes, Limits, and Extensions of Analytic Sets* (Lecture Notes in Mathematics, Vol. 19), Springer-Verlag, Berlin, 1966; MR 34 #6156.

STONE, M.

[1] The generalized Weierstrass approximation theorem, *Math. Mag.* **21** (1948), 167–184, 237–254; MR 10, p. 255.

STOUT, E. L.

[1] On some restriction algebras, *Function Algebras* (Proc. Intern. Symp. Function Algebras, Tulane Univ., 1965), Scott, Foresman, Chicago, 1966, 6–11; MR 35 #3447.

[2] A generalization of a theorem of Radó, *Math. Ann.* **177** (1968), 339–340; MR 37 # 6447.

[3] The second Cousin problem with bounded data, *Pacific J. Math.* **26** (1968), 379–387; MR 38 #3467.

TITCHMARSH, E. C.

[1] *Theory of Functions*, Oxford University Press, New York, 2nd ed., 1958.

TOMIYAMA, J.

[1] Tensor products of commutative Banach algebras, *Tôhoku Math. J.* **12** (1960), 147–154; MR 22 #5910.

[2] Some remarks on antisymmetric decompositions of function algebras, *Tôhoku Math. J.* **16** (1964), 340–344; MR 30 #5157.

VALSKII, R. E.

[1] Gleason's parts for algebras of analytic functions and measures orthogonal to these algebras, *Soviet Math. Dokl.* **8** (1967), 300–303; MR 35 #3448.

[2] Parts of algebras of analytic functions and measures orthogonal to these algebras, *Siberian Math. J.* **8** (1967), 935–944; MR 36 #1986.

VITUSHKIN, A. G.

[1] Analytic capacity of sets in problems of approximation theory, *Russian Math. Surveys* **22** (1967), 139–201; MR 37 #5404.

WAELBROECK, L.

[1] Le calcul symbolique dans les algèbres commutatives, *J. Math. Pures Appl.* **33** (1954), 147–186; MR 17, p. 513.

[2] *Théorie des algèbres de Banach et des algèbres localement convexes*, Les Presses de l'Université Montréal, Montréal, 2nd ed., 1965.

WALSH, J. L.

[1] *Interpolation and Approximation by Rational Functions in the Complex Domain*, American Mathematical Society, Providence, 3rd ed., 1960.

[2] Über die Entwicklung einer analytischer Funktion nach Polynomen, *Math. Ann.* **96** (1926), 437–450.

[3] Über die Entwicklung einer harmonische Funktion nach harmonischen Polynomen, *J. Reine Angew. Math.* **159** (1928), 197–209.

WANG, J.-K.

[1] *Banach Algebras*, Yale lecture notes, New Haven, 1965.

WATTS, C.

[1] Alexander–Spanier cohomology and rings of continuous functions, *Proc. Nat. Acad. Sci. U.S.A.* **54** (1965), 1027–1028; MR 34 #3571.

WEIL, A.

[1] L'intégral de Cauchy et les fonctions de plusieurs variables, *Math. Ann.* 111 (1935), 178–182.

WERMER, J.

[1] On algebras of continuous functions, *Proc. Amer. Math. Soc.* **4** (1953), 866–869; MR 15, p. 440.

[2] Subalgebras of the algebra of all complex-valued continuous functions on the circle, *Amer. J. Math.* **78** (1956), 225–242; MR 18, p. 911.

[3] Polynomial approximation on an arc in $\mathbb{C}^3$, *Ann. Math.* (2)**62** (1955), 269–270; MR 17, p. 255.

[4] Rings of analytic functions, *Ann. Math.* (2)**67** (1958), 497–516; MR 20 #3299.

[5] The hull of a curve in $\mathbb{C}^n$, *Ann. Math.* (2)**68** (1958), 550–561; MR 20 #6536.

[6] Function rings and Riemann surfaces, *Ann. Math.* (2)**67** (1958), 45–71; MR 20 #109.

[7] An example concerning polynomial convexity, *Math. Ann.* **139** (1959), 147–150; MR 22 #768.

[8] Dirichlet algebras, *Duke Math. J.* **27** (1960), 373–381; MR 22 #12405.

[9] Banach algebras and analytic functions, *Advances in Math.* **1** (1961), 51–102; MR 26 #629.

[10] The space of real parts of a function algebra, *Pacific J. Math.* **13** (1963), 1423–1426; MR 26 #6152.

[11] Analytic discs in maximal ideal spaces, *Amer. J. Math.* **86** (1964), 161–170; MR 28 #5355.

[12] *Seminar über Funktionen-Algebren* (Lecture Notes in Mathematics, vol. 1), Springer-Verlag, Berlin, 1964; MR 35 #5947.

WIENER, N.

[1] Tauberian theorems, *Ann. Math.* (2)**33** (1932), 1–100.

WILKEN, D. R.

[1] Lebesgue measure of parts for $R(X)$, *Proc. Amer. Math. Soc.* **18** (1967), 508–512.

[2] The support of representing measures for $R(X)$, *Pacific J. Math.* **26** (1968), 621–626; MR 38 #5008.

[3] Remarks on the string of beads, *Proc. Amer. Math. Soc.* **23** (1969), 133–135; MR 40 # 3319.

[4] Representing measures for harmonic functions, *Duke Math. J.* **35** (1968), 283–290; MR 40 #2005.

ZALCMAN, L.

[1] *Analytic Capacity and Rational Approximation* (Lecture Notes in Mathematics, vol. 50), Springer-Verlag, Berlin, 1968; MR 37 #3018.

[2] Null sets for a class of analytic functions, *Amer. Math. Monthly* **75** (1968), 462–470; MR 37 #6448.

ZARISKI, O., AND P. SAMUEL

[1] *Commutative Algebra*, vol. 1, Van Nostrand Reinhold, New York, 1958.

# INDEX